Lecture Notes in Computer Science 12538

More information about this subseries at http://www.springer.com/series/7412

Adrien Bartoli · Andrea Fusiello (Eds.)

Computer Vision – ECCV 2020 Workshops

Glasgow, UK, August 23–28, 2020
Proceedings, Part IV

Editors
Adrien Bartoli
University of Clermont Auvergne
Clermont Ferrand, France

Andrea Fusiello
Università degli Studi di Udine
Udine, Italy

ISSN 0302-9743 ISSN 1611-3349 (electronic)
Lecture Notes in Computer Science
ISBN 978-3-030-66822-8 ISBN 978-3-030-66823-5 (eBook)
https://doi.org/10.1007/978-3-030-66823-5

LNCS Sublibrary: SL6 – Image Processing, Computer Vision, Pattern Recognition, and Graphics

This Springer imprint is published by the registered company Springer Nature Switzerland AG
The registered company address is: Gewerbestrasse 11, 6330 Cham, Switzerland

Foreword

Hosting the 2020 European Conference on Computer Vision was certainly an exciting journey. From the 2016 plan to hold it at the Edinburgh International Conference Centre (hosting 1,800 delegates) to the 2018 plan to hold it at Glasgow's Scottish Exhibition Centre (up to 6,000 delegates), we finally ended with moving online because of the COVID-19 outbreak. While possibly having fewer delegates than expected because of the online format, ECCV 2020 still had over 3,100 registered participants.

Although online, the conference delivered most of the activities expected at a face-to-face conference: peer-reviewed papers, industrial exhibitors, demonstrations, and messaging between delegates. As well as the main technical sessions, the conference included a strong program of satellite events, including 16 tutorials and 44 workshops.

On the other hand, the online conference format enabled new conference features. Every paper had an associated teaser video and a longer full presentation video. Along with the papers and slides from the videos, all these materials were available the week before the conference. This allowed delegates to become familiar with the paper content and be ready for the live interaction with the authors during the conference week. The 'live' event consisted of brief presentations by the 'oral' and 'spotlight' authors and industrial sponsors. Question and Answer sessions for all papers were timed to occur twice so delegates from around the world had convenient access to the authors.

As with the 2018 ECCV, authors' draft versions of the papers appeared online with open access, now on both the Computer Vision Foundation (CVF) and the European Computer Vision Association (ECVA) websites. An archival publication arrangement was put in place with the cooperation of Springer. SpringerLink hosts the final version of the papers with further improvements, such as activating reference links and supplementary materials. These two approaches benefit all potential readers: a version available freely for all researchers, and an authoritative and citable version with additional benefits for SpringerLink subscribers. We thank Alfred Hofmann and Aliaksandr Birukou from Springer for helping to negotiate this agreement, which we expect will continue for future versions of ECCV.

August 2020

Vittorio Ferrari
Bob Fisher
Cordelia Schmid
Emanuele Trucco

Preface

Welcome to the workshops proceedings of the 16th European Conference on Computer Vision (ECCV 2020), the first edition held online. We are delighted that the main ECCV 2020 was accompanied by 45 workshops, scheduled on August 23, 2020, and August 28, 2020.

We received 101 valid workshop proposals on diverse computer vision topics and had space for 32 full-day slots, so we had to decline many valuable proposals (the workshops were supposed to be either full-day or half-day long, but the distinction faded away when the full ECCV conference went online). We endeavored to balance among topics, established series, and newcomers. Not all the workshops published their proceedings, or had proceedings at all. These volumes collect the edited papers from 28 out of 45 workshops.

We sincerely thank the ECCV general chairs for trusting us with the responsibility for the workshops, the workshop organizers for their involvement in this event of primary importance in our field, and the workshop presenters and authors.

August 2020

Adrien Bartoli
Andrea Fusiello

Organization

General Chairs

Vittorio Ferrari	Google Research, Switzerland
Bob Fisher	The University of Edinburgh, UK
Cordelia Schmid	Google and Inria, France
Emanuele Trucco	The University of Dundee, UK

Program Chairs

Andrea Vedaldi	University of Oxford, UK
Horst Bischof	Graz University of Technology, Austria
Thomas Brox	University of Freiburg, Germany
Jan-Michael Frahm	The University of North Carolina at Chapel Hill, USA

Industrial Liaison Chairs

Jim Ashe	The University of Edinburgh, UK
Helmut Grabner	Zurich University of Applied Sciences, Switzerland
Diane Larlus	NAVER LABS Europe, France
Cristian Novotny	The University of Edinburgh, UK

Local Arrangement Chairs

Yvan Petillot	Heriot-Watt University, UK
Paul Siebert	The University of Glasgow, UK

Academic Demonstration Chair

Thomas Mensink	Google Research and University of Amsterdam, The Netherlands

Poster Chair

Stephen Mckenna	The University of Dundee, UK

Technology Chair

Gerardo Aragon Camarasa	The University of Glasgow, UK

Tutorial Chairs

Carlo Colombo	University of Florence, Italy
Sotirios Tsaftaris	The University of Edinburgh, UK

Publication Chairs

Albert Ali Salah	Utrecht University, The Netherlands
Hamdi Dibeklioglu	Bilkent University, Turkey
Metehan Doyran	Utrecht University, The Netherlands
Henry Howard-Jenkins	University of Oxford, UK
Victor Adrian Prisacariu	University of Oxford, UK
Siyu Tang	ETH Zurich, Switzerland
Gul Varol	University of Oxford, UK

Website Chair

Giovanni Maria Farinella	University of Catania, Italy

Workshops Chairs

Adrien Bartoli	University Clermont Auvergne, France
Andrea Fusiello	University of Udine, Italy

Workshops Organizers

W01 - Adversarial Robustness in the Real World

Adam Kortylewski	Johns Hopkins University, USA
Cihang Xie	Johns Hopkins University, USA
Song Bai	University of Oxford, UK
Zhaowei Cai	UC San Diego, USA
Yingwei Li	Johns Hopkins University, USA
Andrei Barbu	MIT, USA
Wieland Brendel	University of Tübingen, Germany
Nuno Vasconcelos	UC San Diego, USA
Andrea Vedaldi	University of Oxford, UK
Philip H. S. Torr	University of Oxford, UK
Rama Chellappa	University of Maryland, USA
Alan Yuille	Johns Hopkins University, USA

W02 - BioImage Computation

Jan Funke	HHMI Janelia Research Campus, Germany
Dagmar Kainmueller	BIH and MDC Berlin, Germany
Florian Jug	CSBD and MPI-CBG, Germany
Anna Kreshuk	EMBL Heidelberg, Germany

Peter Bajcsy	NIST, USA
Martin Weigert	EPFL, Switzerland
Patrick Bouthemy	Inria, France
Erik Meijering	University New South Wales, Australia

W03 - Egocentric Perception, Interaction and Computing

Michael Wray	University of Bristol, UK
Dima Damen	University of Bristol, UK
Hazel Doughty	University of Bristol, UK
Walterio Mayol-Cuevas	University of Bristol, UK
David Crandall	Indiana University, USA
Kristen Grauman	UT Austin, USA
Giovanni Maria Farinella	University of Catania, Italy
Antonino Furnari	University of Catania, Italy

W04 - Embodied Vision, Actions and Language

Yonatan Bisk	Carnegie Mellon University, USA
Jesse Thomason	University of Washington, USA
Mohit Shridhar	University of Washington, USA
Chris Paxton	NVIDIA, USA
Peter Anderson	Georgia Tech, USA
Roozbeh Mottaghi	Allen Institute for AI, USA
Eric Kolve	Allen Institute for AI, USA

W05 - Eye Gaze in VR, AR, and in the Wild

Hyung Jin Chang	University of Birmingham, UK
Seonwook Park	ETH Zurich, Switzerland
Xucong Zhang	ETH Zurich, Switzerland
Otmar Hilliges	ETH Zurich, Switzerland
Aleš Leonardis	University of Birmingham, UK
Robert Cavin	Facebook Reality Labs, USA
Cristina Palmero	University of Barcelona, Spain
Jixu Chen	Facebook, USA
Alexander Fix	Facebook Reality Labs, USA
Elias Guestrin	Facebook Reality Labs, USA
Oleg Komogortsev	Texas State University, USA
Kapil Krishnakumar	Facebook, USA
Abhishek Sharma	Facebook Reality Labs, USA
Yiru Shen	Facebook Reality Labs, USA
Tarek Hefny	Facebook Reality Labs, USA
Karsten Behrendt	Facebook, USA
Sachin S. Talathi	Facebook Reality Labs, USA

W06 - Holistic Scene Structures for 3D Vision

Zihan Zhou	Penn State University, USA
Yasutaka Furukawa	Simon Fraser University, Canada
Yi Ma	UC Berkeley, USA
Shenghua Gao	ShanghaiTech University, China
Chen Liu	Facebook Reality Labs, USA
Yichao Zhou	UC Berkeley, USA
Linjie Luo	Bytedance Inc., China
Jia Zheng	ShanghaiTech University, China
Junfei Zhang	Kujiale.com, China
Rui Tang	Kujiale.com, China

W07 - Joint COCO and LVIS Recognition Challenge

Alexander Kirillov	Facebook AI Research, USA
Tsung-Yi Lin	Google Research, USA
Yin Cui	Google Research, USA
Matteo Ruggero Ronchi	California Institute of Technology, USA
Agrim Gupta	Stanford University, USA
Ross Girshick	Facebook AI Research, USA
Piotr Dollar	Facebook AI Research, USA

W08 - Object Tracking and Its Many Guises

Achal D. Dave	Carnegie Mellon University, USA
Tarasha Khurana	Carnegie Mellon University, USA
Jonathon Luiten	RWTH Aachen University, Germany
Aljosa Osep	Technical University of Munich, Germany
Pavel Tokmakov	Carnegie Mellon University, USA

W09 - Perception for Autonomous Driving

Li Erran Li	Alexa AI, Amazon, USA
Adrien Gaidon	Toyota Research Institute, USA
Wei-Lun Chao	The Ohio State University, USA
Peter Ondruska	Lyft, UK
Rowan McAllister	UC Berkeley, USA
Larry Jackel	North-C Technologies, USA
Jose M. Alvarez	NVIDIA, USA

W10 - TASK-CV Workshop and VisDA Challenge

Tatiana Tommasi	Politecnico di Torino, Italy
Antonio M. Lopez	CVC and UAB, Spain
David Vazquez	Element AI, Canada
Gabriela Csurka	NAVER LABS Europe, France
Kate Saenko	Boston University, USA
Liang Zheng	The Australian National University, Australia

Xingchao Peng	Boston University, USA
Weijian Deng	The Australian National University, Australia

W11 - Bodily Expressed Emotion Understanding

James Z. Wang	Penn State University, USA
Reginald B. Adams, Jr.	Penn State University, USA
Yelin Kim	Amazon Lab126, USA

W12 - Commands 4 Autonomous Vehicles

Thierry Deruyttere	KU Leuven, Belgium
Simon Vandenhende	KU Leuven, Belgium
Luc Van Gool	KU Leuven, Belgium, and ETH Zurich, Switzerland
Matthew Blaschko	KU Leuven, Belgium
Tinne Tuytelaars	KU Leuven, Belgium
Marie-Francine Moens	KU Leuven, Belgium
Yu Liu	KU Leuven, Belgium
Dusan Grujicic	KU Leuven, Belgium

W13 - Computer VISion for ART Analysis

Alessio Del Bue	Istituto Italiano di Tecnologia, Italy
Sebastiano Vascon	Ca' Foscari University and European Centre for Living Technology, Italy
Peter Bell	Friedrich-Alexander University Erlangen-Nürnberg, Germany
Leonardo L. Impett	EPFL, Switzerland
Stuart James	Istituto Italiano di Tecnologia, Italy

W14 - International Challenge on Compositional and Multimodal Perception

Alec Hodgkinson	Panasonic Corporation, Japan
Yusuke Urakami	Panasonic Corporation, Japan
Kazuki Kozuka	Panasonic Corporation, Japan
Ranjay Krishna	Stanford University, USA
Olga Russakovsky	Princeton University, USA
Juan Carlos Niebles	Stanford University, USA
Jingwei Ji	Stanford University, USA
Li Fei-Fei	Stanford University, USA

W15 - Sign Language Recognition, Translation and Production

Necati Cihan Camgoz	University of Surrey, UK
Richard Bowden	University of Surrey, UK
Andrew Zisserman	University of Oxford, UK
Gul Varol	University of Oxford, UK
Samuel Albanie	University of Oxford, UK

Kearsy Cormier	University College London, UK
Neil Fox	University College London, UK

W16 - Visual Inductive Priors for Data-Efficient Deep Learning

Jan van Gemert	Delft University of Technology, The Netherlands
Robert-Jan Bruintjes	Delft University of Technology, The Netherlands
Attila Lengyel	Delft University of Technology, The Netherlands
Osman Semih Kayhan	Delft University of Technology, The Netherlands
Marcos Baptista-Ríos	Alcalá University, Spain
Anton van den Hengel	The University of Adelaide, Australia

W17 - Women in Computer Vision

Hilde Kuehne	IBM, USA
Amaia Salvador	Amazon, USA
Ananya Gupta	The University of Manchester, UK
Yana Hasson	Inria, France
Anna Kukleva	Max Planck Institute, Germany
Elizabeth Vargas	Heriot-Watt University, UK
Xin Wang	UC Berkeley, USA
Irene Amerini	Sapienza University of Rome, Italy

W18 - 3D Poses in the Wild Challenge

Gerard Pons-Moll	Max Planck Institute for Informatics, Germany
Angjoo Kanazawa	UC Berkeley, USA
Michael Black	Max Planck Institute for Intelligent Systems, Germany
Aymen Mir	Max Planck Institute for Informatics, Germany

W19 - 4D Vision

Anelia Angelova	Google, USA
Vincent Casser	Waymo, USA
Jürgen Sturm	X, USA
Noah Snavely	Google, USA
Rahul Sukthankar	Google, USA

W20 - Map-Based Localization for Autonomous Driving

Patrick Wenzel	Technical University of Munich, Germany
Niclas Zeller	Artisense, Germany
Nan Yang	Technical University of Munich, Germany
Rui Wang	Technical University of Munich, Germany
Daniel Cremers	Technical University of Munich, Germany

W21 - Multimodal Video Analysis Workshop and Moments in Time Challenge

Dhiraj Joshi	IBM Research AI, USA
Rameswar Panda	IBM Research, USA
Kandan Ramakrishnan	IBM, USA
Rogerio Feris	IBM Research AI, MIT-IBM Watson AI Lab, USA
Rami Ben-Ari	IBM-Research, USA
Danny Gutfreund	IBM, USA
Mathew Monfort	MIT, USA
Hang Zhao	MIT, USA
David Harwath	MIT, USA
Aude Oliva	MIT, USA
Zhicheng Yan	Facebook AI, USA

W22 - Recovering 6D Object Pose

Tomas Hodan	Czech Technical University in Prague, Czech Republic
Martin Sundermeyer	German Aerospace Center, Germany
Rigas Kouskouridas	Scape Technologies, UK
Tae-Kyun Kim	Imperial College London, UK
Jiri Matas	Czech Technical University in Prague, Czech Republic
Carsten Rother	Heidelberg University, Germany
Vincent Lepetit	ENPC ParisTech, France
Ales Leonardis	University of Birmingham, UK
Krzysztof Walas	Poznan University of Technology, Poland
Carsten Steger	Technical University of Munich and MVTec Software GmbH, Germany
Eric Brachmann	Heidelberg University, Germany
Bertram Drost	MVTec Software GmbH, Germany
Juil Sock	Imperial College London, UK

W23 - SHApe Recovery from Partial Textured 3D Scans

Djamila Aouada	University of Luxembourg, Luxembourg
Kseniya Cherenkova	Artec3D and University of Luxembourg, Luxembourg
Alexandre Saint	University of Luxembourg, Luxembourg
David Fofi	University Bourgogne Franche-Comté, France
Gleb Gusev	Artec3D, Luxembourg
Bjorn Ottersten	University of Luxembourg, Luxembourg

W24 - Advances in Image Manipulation Workshop and Challenges

Radu Timofte	ETH Zurich, Switzerland
Andrey Ignatov	ETH Zurich, Switzerland
Kai Zhang	ETH Zurich, Switzerland
Dario Fuoli	ETH Zurich, Switzerland
Martin Danelljan	ETH Zurich, Switzerland
Zhiwu Huang	ETH Zurich, Switzerland

Hannan Lu	Harbin Institute of Technology, China
Wangmeng Zuo	Harbin Institute of Technology, China
Shuhang Gu	The University of Sydney, Australia
Ming-Hsuan Yang	UC Merced and Google, USA
Majed El Helou	EPFL, Switzerland
Ruofan Zhou	EPFL, Switzerland
Sabine Süsstrunk	EPFL, Switzerland
Sanghyun Son	Seoul National University, South Korea
Jaerin Lee	Seoul National University, South Korea
Seungjun Nah	Seoul National University, South Korea
Kyoung Mu Lee	Seoul National University, South Korea
Eli Shechtman	Adobe, USA
Evangelos Ntavelis	ETH Zurich and CSEM, Switzerland
Andres Romero	ETH Zurich, Switzerland
Yawei Li	ETH Zurich, Switzerland
Siavash Bigdeli	CSEM, Switzerland
Pengxu Wei	Sun Yat-sen University, China
Liang Lin	Sun Yat-sen University, China
Ming-Yu Liu	NVIDIA, USA
Roey Mechrez	BeyondMinds and Technion, Israel
Luc Van Gool	KU Leuven, Belgium, and ETH Zurich, Switzerland

W25 - Assistive Computer Vision and Robotics

Marco Leo	National Research Council of Italy, Italy
Giovanni Maria Farinella	University of Catania, Italy
Antonino Furnari	University of Catania, Italy
Gerard Medioni	University of Southern California, USA
Trivedi Mohan	UC San Diego, USA

W26 - Computer Vision for UAVs Workshop and Challenge

Dawei Du	Kitware Inc., USA
Heng Fan	Stony Brook University, USA
Toon Goedemé	KU Leuven, Belgium
Qinghua Hu	Tianjin University, China
Haibin Ling	Stony Brook University, USA
Davide Scaramuzza	University of Zurich, Switzerland
Mubarak Shah	University of Central Florida, USA
Tinne Tuytelaars	KU Leuven, Belgium
Kristof Van Beeck	KU Leuven, Belgium
Longyin Wen	JD Digits, USA
Pengfei Zhu	Tianjin University, China

W27 - Embedded Vision

Tse-Wei Chen	Canon Inc., Japan
Nabil Belbachir	NORCE Norwegian Research Centre AS, Norway

Stephan Weiss	University of Klagenfurt, Austria
Marius Leordeanu	Politehnica University of Bucharest, Romania

W28 - Learning 3D Representations for Shape and Appearance

Leonidas Guibas	Stanford University, USA
Or Litany	Stanford University, USA
Tanner Schmidt	Facebook Reality Labs, USA
Vincent Sitzmann	Stanford University, USA
Srinath Sridhar	Stanford University, USA
Shubham Tulsiani	Facebook AI Research, USA
Gordon Wetzstein	Stanford University, USA

W29 - Real-World Computer Vision from inputs with Limited Quality and Tiny Object Detection Challenge

Yuqian Zhou	University of Illinois, USA
Zhenjun Han	University of the Chinese Academy of Sciences, China
Yifan Jiang	The University of Texas at Austin, USA
Yunchao Wei	University of Technology Sydney, Australia
Jian Zhao	Institute of North Electronic Equipment, Singapore
Zhangyang Wang	The University of Texas at Austin, USA
Qixiang Ye	University of the Chinese Academy of Sciences, China
Jiaying Liu	Peking University, China
Xuehui Yu	University of the Chinese Academy of Sciences, China
Ding Liu	Bytedance, China
Jie Chen	Peking University, China
Humphrey Shi	University of Oregon, USA

W30 - Robust Vision Challenge 2020

Oliver Zendel	Austrian Institute of Technology, Austria
Hassan Abu Alhaija	Interdisciplinary Center for Scientific Computing Heidelberg, Germany
Rodrigo Benenson	Google Research, Switzerland
Marius Cordts	Daimler AG, Germany
Angela Dai	Technical University of Munich, Germany
Andreas Geiger	Max Planck Institute for Intelligent Systems and University of Tübingen, Germany
Niklas Hanselmann	Daimler AG, Germany
Nicolas Jourdan	Daimler AG, Germany
Vladlen Koltun	Intel Labs, USA
Peter Kontschieder	Mapillary Research, Austria
Yubin Kuang	Mapillary AB, Sweden
Alina Kuznetsova	Google Research, Switzerland
Tsung-Yi Lin	Google Brain, USA
Claudio Michaelis	University of Tübingen, Germany
Gerhard Neuhold	Mapillary Research, Austria

Matthias Niessner	Technical University of Munich, Germany
Marc Pollefeys	ETH Zurich and Microsoft, Switzerland
Francesc X. Puig Fernandez	MIT, USA
Rene Ranftl	Intel Labs, USA
Stephan R. Richter	Intel Labs, USA
Carsten Rother	Heidelberg University, Germany
Torsten Sattler	Chalmers University of Technology, Sweden and Czech Technical University in Prague, Czech Republic
Daniel Scharstein	Middlebury College, USA
Hendrik Schilling	rabbitAI, Germany
Nick Schneider	Daimler AG, Germany
Jonas Uhrig	Daimler AG, Germany
Jonas Wulff	Max Planck Institute for Intelligent Systems, Germany
Bolei Zhou	The Chinese University of Hong Kong, China

W31 - The Bright and Dark Sides of Computer Vision: Challenges and Opportunities for Privacy and Security

Mario Fritz	CISPA Helmholtz Center for Information Security, Germany
Apu Kapadia	Indiana University, USA
Jan-Michael Frahm	The University of North Carolina at Chapel Hill, USA
David Crandall	Indiana University, USA
Vitaly Shmatikov	Cornell University, USA

W32 - The Visual Object Tracking Challenge

Matej Kristan	University of Ljubljana, Slovenia
Jiri Matas	Czech Technical University in Prague, Czech Republic
Ales Leonardis	University of Birmingham, UK
Michael Felsberg	Linköping University, Sweden
Roman Pflugfelder	Austrian Institute of Technology, Austria
Joni-Kristian Kamarainen	Tampere University, Finland
Martin Danelljan	ETH Zurich, Switzerland

W33 - Video Turing Test: Toward Human-Level Video Story Understanding

Yu-Jung Heo	Seoul National University, South Korea
Seongho Choi	Seoul National University, South Korea
Kyoung-Woon On	Seoul National University, South Korea
Minsu Lee	Seoul National University, South Korea
Vicente Ordonez	University of Virginia, USA
Leonid Sigal	University of British Columbia, Canada
Chang D. Yoo	KAIST, South Korea
Gunhee Kim	Seoul National University, South Korea
Marcello Pelillo	University of Venice, Italy
Byoung-Tak Zhang	Seoul National University, South Korea

W34 - "Deep Internal Learning": Training with no prior examples

Michal Irani	Weizmann Institute of Science, Israel
Tomer Michaeli	Technion, Israel
Tali Dekel	Google, Israel
Assaf Shocher	Weizmann Institute of Science, Israel
Tamar Rott Shaham	Technion, Israel

W35 - Benchmarking Trajectory Forecasting Models

Alexandre Alahi	EPFL, Switzerland
Lamberto Ballan	University of Padova, Italy
Luigi Palmieri	Bosch, Germany
Andrey Rudenko	Örebro University, Sweden
Pasquale Coscia	University of Padova, Italy

W36 - Beyond mAP: Reassessing the Evaluation of Object Detection

David Hall	Queensland University of Technology, Australia
Niko Suenderhauf	Queensland University of Technology, Australia
Feras Dayoub	Queensland University of Technology, Australia
Gustavo Carneiro	The University of Adelaide, Australia
Chunhua Shen	The University of Adelaide, Australia

W37 - Imbalance Problems in Computer Vision

Sinan Kalkan	Middle East Technical University, Turkey
Emre Akbas	Middle East Technical University, Turkey
Nuno Vasconcelos	UC San Diego, USA
Kemal Oksuz	Middle East Technical University, Turkey
Baris Can Cam	Middle East Technical University, Turkey

W38 - Long-Term Visual Localization under Changing Conditions

Torsten Sattler	Chalmers University of Technology, Sweden, and Czech Technical University in Prague, Czech Republic
Vassileios Balntas	Facebook Reality Labs, USA
Fredrik Kahl	Chalmers University of Technology, Sweden
Krystian Mikolajczyk	Imperial College London, UK
Tomas Pajdla	Czech Technical University in Prague, Czech Republic
Marc Pollefeys	ETH Zurich and Microsoft, Switzerland
Josef Sivic	Inria, France, and Czech Technical University in Prague, Czech Republic
Akihiko Torii	Tokyo Institute of Technology, Japan
Lars Hammarstrand	Chalmers University of Technology, Sweden
Huub Heijnen	Facebook, UK
Maddern Will	Nuro, USA
Johannes L. Schönberger	Microsoft, Switzerland

Pablo Speciale	ETH Zurich, Switzerland
Carl Toft	Chalmers University of Technology, Sweden

W39 - Sensing, Understanding, and Synthesizing Humans

Ziwei Liu	The Chinese University of Hong Kong, China
Sifei Liu	NVIDIA, USA
Xiaolong Wang	UC San Diego, USA
Hang Zhou	The Chinese University of Hong Kong, China
Wayne Wu	SenseTime, China
Chen Change Loy	Nanyang Technological University, Singapore

W40 - Computer Vision Problems in Plant Phenotyping

Hanno Scharr	Forschungszentrum Jülich, Germany
Tony Pridmore	University of Nottingham, UK
Sotirios Tsaftaris	The University of Edinburgh, UK

W41 - Fair Face Recognition and Analysis

Sergio Escalera	CVC and University of Barcelona, Spain
Rama Chellappa	University of Maryland, USA
Eduard Vazquez	Anyvision, UK
Neil Robertson	Queen's University Belfast, UK
Pau Buch-Cardona	CVC, Spain
Tomas Sixta	Anyvision, UK
Julio C. S. Jacques Junior	Universitat Oberta de Catalunya and CVC, Spain

W42 - GigaVision: When Gigapixel Videography Meets Computer Vision

Lu Fang	Tsinghua University, China
Shengjin Wang	Tsinghua University, China
David J. Brady	Duke University, USA
Feng Yang	Google Research, USA

W43 - Instance-Level Recognition

Andre Araujo	Google, USA
Bingyi Cao	Google, USA
Ondrej Chum	Czech Technical University in Prague, Czech Republic
Bohyung Han	Seoul National University, South Korea
Torsten Sattler	Chalmers University of Technology, Sweden and Czech Technical University in Prague, Czech Republic
Jack Sim	Google, USA
Giorgos Tolias	Czech Technical University in Prague, Czech Republic
Tobias Weyand	Google, USA

Xu Zhang	Columbia University, USA
Cam Askew	Google, USA
Guangxing Han	Columbia University, USA

W44 - Perception Through Structured Generative Models

Adam W. Harley	Carnegie Mellon University, USA
Katerina Fragkiadaki	Carnegie Mellon University, USA
Shubham Tulsiani	Facebook AI Research, USA

W45 - Self Supervised Learning – What is Next?

Christian Rupprecht	University of Oxford, UK
Yuki M. Asano	University of Oxford, UK
Armand Joulin	Facebook AI Research, USA
Andrea Vedaldi	University of Oxford, UK

Contents – Part IV

W25 - Assistive Computer Vision and Robotics

W26 - Computer Vision for UAVs Workshop and Challenge

W24 - Advances in Image Manipulation Workshop and Challenges

W24 - Advances in Image Manipulation Workshop and Challenges

The second edition of the AIM workshop was organized jointly with ECCV 2020. The success of AIM 2020 was contributed by 27 organizers, 67 PC members, 6 sponsors, 300+ authors with submitted papers, 4 invited speakers and thousands of participants in 8 associated challenges. AIM 2020 attracted 94 paper submissions to its two tracks meant for early and regular papers and late and challenge papers. We had 12 submissions of rejected papers from ECCV 2020 and BMVC 2020, out of which 7 were accepted. In total, 60 papers were accepted for publication. Each submission (except the challenge reports) was reviewed on average by three reviewers. The pool of reviewers comprised PC members, organizers, and volunteer reviewers. AIM 2020 had 8 associated challenges on: scene relighting and illumination estimation, image extreme inpainting, learned image signal processing pipeline, rendering realistic bokeh, real image super-resolution, efficient super-resolution, video temporal super-resolution, and video extreme super-resolution. The challenges were hosted by the CodaLab platform. From thousands of registered participants, hundreds entered in the final test phases and submitted results, factsheets, and codes/executables for reproducibility. 22 teams were awarded certificates and the top-ranking winners received prizes: money or an Nvidia Titan RTX GPU. We are grateful to our sponsors: Huawei, Qualcomm AI Research, MediaTek, Nvidia, Google, and CVL/ETH Zurich. AIM 2020 had 4 highly impactful invited talks provided by David Bau (MIT) on "Reflected Light and Doors in the Sky: Rewriting a GAN's Rules", Richard Zhang (Adobe Research) on "Style and Structure Disentanglement for Image Manipulation", Ravi Ramamoorthi (UCSD) on "Light Fields and View Synthesis from Sparse Images: Revisiting Image-Based Rendering", and Peyman Milanfar (Google Research) on "Modern Computational Photography". We would like to express our gratitude to all our colleagues in the community for submitting papers, to our PC members for their support and help with the reviewing process, to all the challenge participants, to CodaLab for hosting our challenges, to the invited speakers for sharing their research, to our sponsors, and last but not least, to the attendees for their active, positive attitude. This volume contains sixteen papers from the workshop. The remaining papers were published in the preceding ECCVW volume, LNCS 12537.

August 2020

Radu Timofte
Andrey Ignatov
Kai Zhang
Dario Fuoli
Martin Danelljan
Zhiwu Huang
Hannan Lu
Wangmeng Zuo
Shuhang Gu
Ming-Hsuan Yang
Majed El Helou
Ruofan Zhou
Sabine Süsstrunk
Sanghyun Son
Jaerin Lee
Seungjun Nah
Kyoung Mu Lee
Eli Shechtman
Evangelos Ntavelis
Andres Romero
Siavash Bigdeli
Pengxu Wei
Liang Lin
Ming-Yu Liu
Roey Mechrez
Luc Van Gool

DeepGIN: Deep Generative Inpainting Network for Extreme Image Inpainting

Chu-Tak Li[1], Wan-Chi Siu[1(✉)], Zhi-Song Liu[2], Li-Wen Wang[1], and Daniel Pak-Kong Lun[1]

[1] Centre for Multimedia Signal Processing, Department of Electronic and Information Engineering, The Hong Kong Polytechnic University, Kowloon, Hong Kong
enwcsiu@polyu.edu.hk

[2] LIX, Ecole Polytechnique, CNRS, IP Paris, Paris, France

Abstract. The degree of difficulty in image inpainting depends on the types and sizes of the missing parts. Existing image inpainting approaches usually encounter difficulties in completing the missing parts in the wild with pleasing visual and contextual results as they are trained for either dealing with one specific type of missing patterns (mask) or unilaterally assuming the shapes and/or sizes of the masked areas. We propose a deep generative inpainting network, named DeepGIN, to handle various types of masked images. We design a Spatial Pyramid Dilation (SPD) ResNet block to enable the use of distant features for reconstruction. We also employ Multi-Scale Self-Attention (MSSA) mechanism and Back Projection (BP) technique to enhance our inpainting results. Our DeepGIN outperforms the state-of-the-art approaches generally, including two publicly available datasets (FFHQ and Oxford Buildings), both quantitatively and qualitatively. We also demonstrate that our model is capable of completing masked images in the wild.

Keywords: Image inpainting · Attention · Back projection

1 Introduction

Image inpainting (also called image completion) is a task of predicting the values of missing pixels in a corrupted/masked image such that the completed image looks realistic and is semantically close to the reference ground truth even though it does not exist in real-world situations. This task would be useful for repairing corrupted photos or erasing unwanted parts from photos. It could also serve applications of restoration of photos and footage of films, scratch removal, automatic modifications to images and videos, and so forth. Because of the wide-ranging applications, image inpainting has been an overwhelming research topic in the computer vision and graphics communities for decades.

Inspired by the recent success of deep learning approaches at the tasks of image recognition [7,29], image super-resolution [5,18,30], visual place recognition and localization [1,15], image enlightening [27], image synthesis [10,28]

A. Bartoli and A. Fusiello (Eds.): ECCV 2020 Workshops, LNCS 12538, pp. 5–22, 2020.
https://doi.org/10.1007/978-3-030-66823-5_1

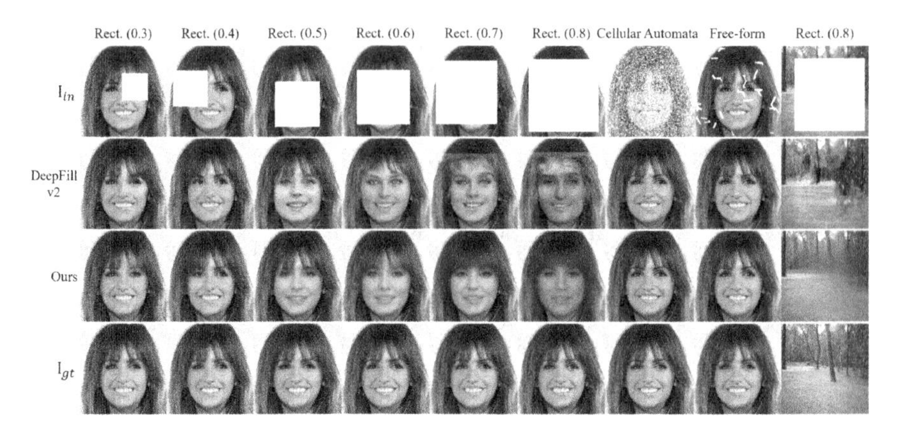

Fig. 1. Degree of difficulty in extreme image inpainting. From top to bottom: the first row shows the input masked images $\mathbf{I}_{in}$ with the corresponding mask described on top of them. Rect. (α) represents a random rectangular mask with the height and width rate of α of each dimension. The randomly generated mask based on cellular automata is introduced in the AIM 2020 Extreme Image Inpainting Challenge [21], and the free-form mask is proposed in DeepFillv2 [34]. The second and third rows are the completed images using DeepFillv2 and our proposed DeepGIN respectively. The last row displays the ground truth images. Please zoom in to see the examples at the 6th and the last column especially

and many others, a growing number of CNN based methods of image inpainting [9,16,20,23,32–34] have been proposed to fill images with holes in an end-to-end manner. For example, Iizuka et al. [9] employed dilated convolutions instead of standard convolutions to widen the receptive field at each layer for better conservation of the spatial structure of an image. Yu et al. [33] proposed a two-stage generative network with a contextual attention layer to intentionally consider correlated feature patches at distant spatial locations for coherent estimation of local missing pixels. Liu et al. [16] suggested a partial convolutional layer to identify the non-hole regions at each layer such that the convolutional results are derived only from the valid pixels. However, the effectiveness of these strategies depends highly on the scales and forms of the missing regions as well as the contents of both the valid and invalid pixels as shown in Fig. 1. Based on this observation, we aim for a generalized inpainting network which can complete masked images in the wild.

In this paper, we present a coarse-to-fine Deep Generative Inpainting Network (DeepGIN) which consists of two stages, namely coarse reconstruction stage and refinement stage. Similar to the network design of previous studies [20,33, 34], the coarse reconstruction stage is responsible for rough estimation of the missing pixels in an image while the refinement stage is responsible for detailed decoration on the coarse reconstructed image. In order to obtain a realistic and coherent completed image, Spatial Pyramid Dilation (SPD), Multi-Scale Self-Attention (MSSA) and Back Projection (BP) techniques are redesigned and embedded in our proposed network. The main function of SPD is to extensively allow for different receptive fields such that information gathered from both surrounding and distant spatial locations can contribute to the prediction of local missing regions. The concept of SPD is applied to both stages while MSSA and BP are integrated into the refinement stage. The core idea of MSSA is that it takes the self-similarity of the image itself at multiple levels into account for the coherence on the completed image while BP enforces the alignment of the predicted and given pixels in the completed image.

The contributions made in this work are summarized as follows:

- We propose a Spatial Pyramid Dilation (SPD) block to deal with different types of masks with various shapes and sizes.
- We stress the importance of self-similarity to image inpainting and we significantly improve our inpainting results by employing the strategy with Multi-Scale Self-Attention (MSSA).
- We design a Back Projection (BP) strategy for obtaining the inpainting results with better alignment of the generated patterns and the reference ground truth.

2 Related Work

Existing approaches to image inpainting can be classified into two categories, namely conventional and deep learning based methods. PatchMatch [2] is one of the representative conventional methods in which similar patches from a target image or other source images are copied and pasted into the missing regions of the target image. However, it is computationally expensive even a fast approximate nearest-neighbor search algorithm has been adopted to alleviate the problem of costly patch search iterations.

Regular Mask. Context Encoder [23] is the first deep learning based inpainting algorithm that employs the framework of Generative Adversarial Networks (GANs) [4] for more realistic image completion. For GAN based image inpainting, a generator is designed for filling the missing regions with semantic awareness and a discriminator is responsible for distinguishing the completed image and the reference ground truth. Based on this setting, the generator and discriminator are alternately optimized to compete against each other, as a result, the completed image given by the generator would be visually and semantically close to the reference ground truth. Specifically, Pathak et al. [23] resize images to 128 $\times$ 128 and assume a 64 $\times$ 64 rectangular center missing region for the task of inpainting. The encoded feature of the image with the center hole is then decoded to reconstruct a 64 $\times$ 64 image for the center hole.

Based on this early work, Yang et al. [32] attached the pre-trained VGG-19 [26] as their proposed texture network to perform the task in a coarse-to-fine manner. The output of the context encoders is then passed to the texture network for local texture refinement to enhance the accuracy of the reconstructed hole content. Iizuka et al. [9] suggested an approach in which two auxiliary discriminators are designed for ensuring both the global and local consistency of the completed image. The global discriminator takes the entire image as input for differentiation between real and completed images while the local discriminator examines only the local area around the filled region. To further alleviate the visual artifacts in the filled region, they also employed Poisson image blending as a simple post-processing. Expanding on this idea and the above-mentioned idea of PatchMatch [2], Yu et al. [33] also adopted a two-stage coarse-to-fine approach with global and local discriminators. A contextual attention layer is proposed and applied to the second refinement network which plays the similar role of the post-processing.

Irregular Mask. For the early stage of deep learning based methods of image inpainting, authors focused on the rectangular types of masks and this assumption limits the effectiveness of these methods in real-world situations. Liu et al. [16] addressed this problem by suggesting a partial convolutional layer, in which a binary mask for indicating the missing regions is automatically updated along with the convolutional operations inside their model for guiding the reconstruction. Nazeri et al. [20] forced an image completion network to generate images with fine-grained details by providing guidance for filling the missing regions with the use of their proposed edge generator. The edge generator is responsible for predicting a full edge map of the masked image. With the estimated edge map as additional information, their trained model can be extended to an interactive image editing tool in which users can sketch the outline of the missing regions to obtain tailor-made completed images. Combining the concept of partial convolution with optional user-guided image inpainting, Yu et al. [34] improved their previous model [33] by proposing gated convolution for free-form image inpainting. They modify the hard-assigned binary mask in partial convolution to a learnable soft-gated convolutional layer. The soft gating layer can be achieved by using convolutional filters with size 1×1 followed by a sigmoid function. However, additional soft gating layers introduce additional parameters and the effectiveness still depends on the scales of the masked areas.

Our work echoes the importance of information given by distant spatial locations and self-similarity of the image itself to image inpainting. We increase the number of receptive fields and apply multi-scale self-attention strategy to handle various types of masks in the wild. Our multi-scale self-attention strategy is derived from the non-local network [29], in which the correlation between features is emphasized and it has been used in image super-resolution [5,17]. For achieving better coherency of the completed images, we also adopt the back projection technique [6,17] to encourage better alignment of the generated and valid pixels. We weight the back projected residual instead of using the parametric back projection blocks as in [6,17] to avoid more additional parameters.

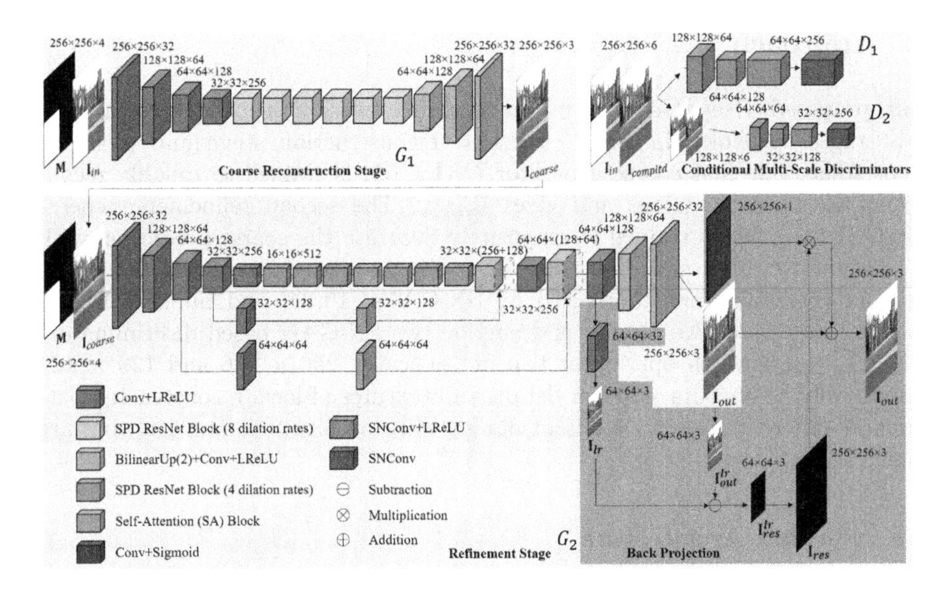

Fig. 2. Architecture of our proposed model for image inpainting. Our proposed model consists of two generators and two discriminators. The coarse generator G_1 at Coarse Reconstruction Stage and the second refinement generator G_2 at Refinement Stage constitute our DeepGIN which is used in both training and testing. The two discriminators D_1 and D_2 located within Conditional Multi-Scale Discriminators area are only used in training as an auxiliary network for generative adversarial training

3 Problem Formulation

Let us start to define an input RGB masked image and a binary mask image as $\mathbf{I}_{in} \in \mathbb{R}^{H \times W \times 3}$ and $\mathbf{M} \in \mathbb{R}^{H \times W}$ respectively. The pixel values input to our model are normalized between 0 and 1 and pixels with value 1 in $\mathbf{M}$ represent the masked regions. $\mathbf{I}_{coarse} \in \mathbb{R}^{H \times W \times 3}$ denotes the output of our coarse generator G_1 at the first coarse reconstruction stage. We also define the output of our refinement generator G_2 at the second refinement stage and the reference ground truth image as $\mathbf{I}_{out} \in \mathbb{R}^{H \times W \times 3}$ and $\mathbf{I}_{gt} \in \mathbb{R}^{H \times W \times 3}$ respectively. Note that H and W are the height and width of an input/output image and we fix the input to 256×256 for inpainting. Our objective is straightforward. We would like to complete $\mathbf{I}_{in}$ conditioned on $\mathbf{M}$ and produce a completed image $\mathbf{I}_{out}$ ($\mathbf{I}_{compltd}$) which should be both visually and semantically close to the reference ground truth $\mathbf{I}_{gt}$. $\mathbf{I}_{compltd}$ is the same as $\mathbf{I}_{out}$ except the valid pixels are directly replaced by the ground truth. We propose a coarse-to-fine network trained under the framework of generative adversarial learning with training data $\{\mathbf{I}_{in}, \mathbf{M}, \mathbf{I}_{gt}\}$ where $\mathbf{M}$ is randomly generated with arbitrary sizes and shapes. Generator G_1 takes $\mathbf{I}_{in}$ and $\mathbf{M}$ as input and generates $\mathbf{I}_{coarse}$ as output. Subsequently, we feed $\mathbf{I}_{coarse}$ and $\mathbf{M}$ to generator G_2 to obtain the completed image $\mathbf{I}_{out}$ ($\mathbf{I}_{compltd}$).

4 Approach

Our proposed Deep Generative Inpainting Network (DeepGIN) consists of two stages as shown in Fig. 2, a coarse reconstruction stage and a refinement stage. The first coarse generator $G_1(\mathbf{I}_{in}, \mathbf{M})$ is trained to roughly reconstruct the masked regions and gives $\mathbf{I}_{coarse}$. The second refinement generator $G_2(\mathbf{I}_{coarse}, \mathbf{M})$ is trained to exquisitely decorate the coarse prediction with details and textures, and eventually forms the completed image $\mathbf{I}_{out}$ ($\mathbf{I}_{compltd}$). For our discriminators, motivated by SN-GANs [19,34] and multi-scale discriminators [10,28], we modify and employ two SN-GAN based discriminators $D(\mathbf{I}_{in}, \mathbf{I}_{compltd})$ which operate at two image scales, 256×256 and 128×128 respectively, to encourage better details and textures of local reconstructed patterns at different scales. Details of our network architecture and learning are shown below.

4.1 Network Architecture

Coarse Reconstruction Stage. Recall that G_1 is our coarse generator and it is responsible for rough estimation of the missing pixels in a masked image. Referring to the previous section, we concatenate $(\mathbf{I}_{in}, \mathbf{M}) \in \mathbb{R}^{H \times W \times (3+1)}$ as the input to G_1 and then obtain the coarse image $\mathbf{I}_{coarse}$. G_1 follows an encoder-decoder structure. As the scales of the masked regions are randomly determined, we propose a Spatial Pyramid Dilation (SPD) ResNet block with various dilation rates to enlarge the receptive fields such that information given by distant spatial locations can be included for reconstruction. Our SPD ResNet block is an improved version of the original ResNet block [7] as shown in Fig. 3, and in total, 6 SPD ResNet blocks with 8 different dilation rates are used at this stage.

Refinement Stage. Generator G_2 is designed for refinement of $\mathbf{I}_{coarse}$ and it is similar to generator G_1. At this stage, we have 6 SPD ResNet blocks with 4 different dilation rates and a Self-Attention (SA) block in between at the middle layers. Apart from the SPD ResNet block, Multi-Scale Self-Attention (MSSA) blocks [17,29] are used for self-similarity consideration. The SA block used in this paper is exactly the same as the one proposed in [29]. One similarity between the SA block and the contextual attention layer [33,34] is that they both have the concept of self-similarity which is useful for amending the reconstructed patterns based on the remaining ground truth in a masked image. We apply MSSA instead of single scale SA to enhance the coherency of the completed image $\mathbf{I}_{out}$ by attending on the self-similarity of the image itself at three different scales, namely 16×16, 32×32 and 64×64 as shown in Fig. 2. To avoid an excessive increase in additional parameters, we simply use standard convolutional layers to reduce the channel size before connecting to the SA blocks. The idea of Back Projection (BP) [6,17] is also redesigned and it is used at the last decoding process of this stage (see the shaded Back Projection region in Fig. 2). At the layer with spatial size of 64×64, we output a low-resolution (LR) completed

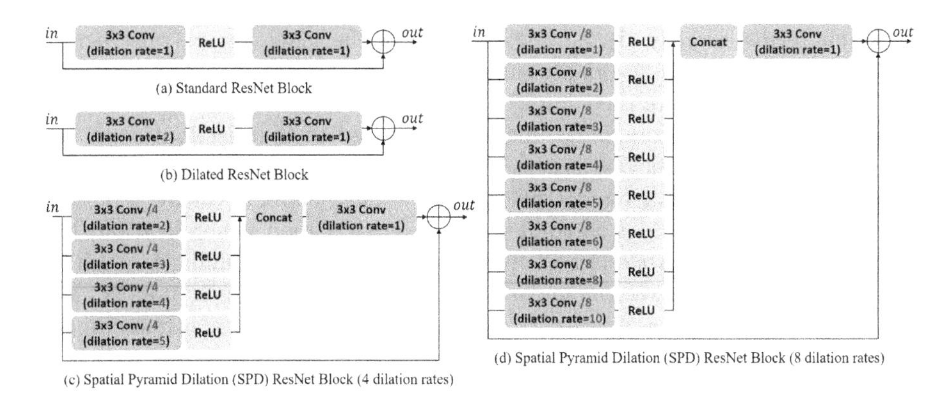

Fig. 3. Variations of ResNet Block. From top to bottom, left to right: (a) Standard ResNet block [7], (b) Dilated ResNet block used in [9,20,33] which adopts a dilation rate of 2 of the first convolutional layer, (c) The proposed SPD ResNet block with 4 dilation rates and (d) 8 dilation rates. To avoid additional parameters, we split the number of input feature channels into equal parts according to the number of dilation rates employed. As shown in (c), if 4 dilation rates are used, the output channel size of the first convolutions equals a quarter of the input channel size

image $\mathbf{I}_{lr}$ and perform BP with $\mathbf{I}_{out}$. By learning to weight the BP residual and adding it back to update $\mathbf{I}_{out}$, the generated patterns can have better alignments with the reference ground truth and hence $\mathbf{I}_{out}$ looks more coherent.

Conditional Multi-scale Discriminators. Two discriminators D_1 and D_2 at two input scales (i.e. 256 × 256 and 128 × 128) are trained together with the generators to stimulate details of the filled regions. Combining the idea of multi-scale discriminators [28] with SN-GANs [19] and PatchGAN [10,34], our $D_1(\mathbf{I}_{in}, \mathbf{I})$ and $D_2(\mathbf{I}_{in}, \mathbf{I})$ take the concatenation result of two RGB images as input ($\mathbf{I}$ is either $\mathbf{I}_{compltd}$ or $\mathbf{I}_{gt}$, recall that $\mathbf{I}_{compltd}$ is the same as $\mathbf{I}_{out}$ except the valid pixels are directly replaced by the ground truth) and output a set of feature maps with size of $H/2^2 \times W/2^2 \times c$ where c represents the number of feature maps. Note that each value on these output feature maps represents a local region in the input image at two different scales. By training D_1 and D_2 to discriminate between real and fake local regions, $\mathbf{I}_{out}$ would gradually be close to its reference ground truth $\mathbf{I}_{gt}$ in terms of both appearance and semantic similarity. For achieving stable generative adversarial learning, we employ the spectral normalization layer described in [19] after each convolutional layer in D_1 and D_2.

4.2 Network Learning

We design our loss function based on consideration to both quantitative accuracy and visual quality of the completed images. Our loss function consists of five major terms, namely (i) a *L1 loss* to ensure the pixel-wise reconstruction accuracy especially if using quantitative evaluation metrics such as PSNR and mean L1 error to evaluate the completed images; (ii) an *adversarial loss* to urge the distribution of the completed images to be close to the distribution of the real images; (iii) the *feature perceptual loss* [11] that encourages each completed image and its reference ground truth image to have similar feature representations as computed by a well-trained network with good generalization, like VGG-19 [26]; (iv) the *style loss* [3] to emphasize the style similarity such as textures and colors between completed images and real images; and (v) the *total variation loss* used as a regularizer [11] to guarantee the smoothness in the completed images by penalizing its visual artifacts or discontinuities.

L1 Loss. Our *L1 loss* is derived from three image pairs, namely $\mathbf{I}_{coarse}$ and $\mathbf{I}_{gt}$; $\mathbf{I}_{out}$ and $\mathbf{I}_{gt}$; and $\mathbf{I}_{lr}$ and $\mathbf{I}_{gt}^{lr}$. Note that $\mathbf{I}_{gt}^{lr}$ is obtained by down-sampling $\mathbf{I}_{gt}$ by 4 times. We sum the L1-norm distances of these three image pairs and define our *L1 loss*, $\mathcal{L}_{L1}$, as follows:

$$\mathcal{L}_{L1} = \lambda_{hole}\mathcal{L}_{hole} + \mathcal{L}_{valid} \tag{1}$$

where $\mathcal{L}_{hole}$ and $\mathcal{L}_{valid}$ are the sums of the distances which are calculated only from the missing pixels and the valid pixels respectively. λ_{hole} is a weight to the pixel-wise loss within the missing regions.

Adversarial Loss. For generative adversarial learning, our discriminators are trained to rightly distinguish $\mathbf{I}_{compltd}$ from $\mathbf{I}_{gt}$ while our generators strive to cheat the discriminators of incorrect classification. We employ the hinge loss to train our model, $\mathcal{L}_{adv,G}$ and $\mathcal{L}_{adv,D}$ are computed as:

$$\mathcal{L}_{adv,G} = -\mathbb{E}_{\mathbf{I}_{in}\sim\mathbb{P}_i}[D_1(\mathbf{I}_{in}, \mathbf{I}_{compltd})] - \mathbb{E}_{\mathbf{I}_{in}\sim\mathbb{P}_i}[D_2(\mathbf{I}_{in}, \mathbf{I}_{compltd})] \tag{2}$$

$$\mathcal{L}_{adv,D} = \mathbb{E}_{\mathbf{I}_{in}\sim\mathbb{P}_i}\left[\sum_{d=1}^{2}[\mathrm{ReLU}(1 - D_d(\mathbf{I}_{in}, \mathbf{I}_{gt})) + \mathrm{ReLU}(1 + D_d(\mathbf{I}_{in}, \mathbf{I}_{compltd}))]\right] \tag{3}$$

where $\mathbb{P}_i$ represents the data distribution of $\mathbf{I}_{in}$, ReLU is the rectified linear unit defined as $f(x) = \max(0, x)$.

Perceptual Loss. Let ϕ be the well-trained loss network, VGG-19 [26], and $\phi_l^{\mathbf{I}}$ be the activation maps of the lth layer of network ϕ given an image $\mathbf{I}$. We choose five layers of the pre-trained VGG-19, namely *conv*1_1, *conv*2_1, *conv*3_1, *conv*4_1, and *conv*5_1 for computing this loss. Our $\mathcal{L}_{perceptual}$ is calculated as:

$$\mathcal{L}_{perceptual} = \sum_{l=1}^{L}\frac{\|\phi_l^{\mathbf{I}_{out}} - \phi_l^{\mathbf{I}_{gt}}\|_1}{N_{\phi_l^{\mathbf{I}_{gt}}}} + \sum_{l=1}^{L}\frac{\|\phi_l^{\mathbf{I}_{compltd}} - \phi_l^{\mathbf{I}_{gt}}\|_1}{N_{\phi_l^{\mathbf{I}_{gt}}}} \tag{4}$$

where $N_{\phi_l^{\mathbf{I}_{gt}}}$ indicates the number of elements in $\phi_l^{\mathbf{I}_{gt}}$ and L equals 5 as five layers are used. Here, we compute the L1-norm distance between the high-level feature representations of $\mathbf{I}_{out}$, $\mathbf{I}_{compltd}$ and $\mathbf{I}_{gt}$ given by network ϕ.

Style Loss. Let $(\phi_l^{\mathbf{I}})^\top(\phi_l^{\mathbf{I}})$ be the Gram matrix [3] which computes the feature correlations between pixels of each activation map, say the lth layer of ϕ given $\mathbf{I}$, and this is also called the auto-correlation matrix. We then calculate the *style loss* ($\mathcal{L}_{style}$) using $\mathbf{I}_{out}$, $\mathbf{I}_{compltd}$ and $\mathbf{I}_{gt}$ as:

$$\mathcal{L}_{style} = \sum_{\mathbf{I}}^{\mathbf{I}_{out},\mathbf{I}_{compltd}} \sum_{l=1}^{L} \frac{1}{C_l C_l} \left\| \frac{1}{C_l H_l W_l}((\phi_l^{\mathbf{I}})^\top(\phi_l^{\mathbf{I}}) - (\phi_l^{\mathbf{I}_{gt}})^\top(\phi_l^{\mathbf{I}_{gt}})) \right\|_1 \quad (5)$$

where C_l denotes the number of activation maps of the lth layer of ϕ. H_l and W_l are the height and width of each activation map of the lth layer of ϕ. Note that we use the same five layers of the VGG-19 as mentioned for this loss as well.

Total Variation (TV) Loss. We also used the total variation regularization to ensure the smoothness in $\mathbf{I}_{compltd}$.

$$\mathcal{L}_{tv} = \sum_{x,y}^{H-1,W} \frac{\|\mathbf{I}_{compltd}^{x+1,y} - \mathbf{I}_{compltd}^{x,y}\|_1}{N_{\mathbf{I}_{compltd}}^{row}} + \sum_{x,y}^{H,W-1} \frac{\|\mathbf{I}_{compltd}^{x,y+1} - \mathbf{I}_{compltd}^{x,y}\|_1}{N_{\mathbf{I}_{compltd}}^{col}} \quad (6)$$

where H and W are the height and width of $\mathbf{I}_{compltd}$. $N_{\mathbf{I}_{compltd}}^{row}$ and $N_{\mathbf{I}_{compltd}}^{col}$ are the number of pixels in $\mathbf{I}_{compltd}$, not counting the pixels at the last row and the last column of $\mathbf{I}_{compltd}$ respectively.

Total Loss. Our total loss function for the generators is the weighted sum of the five major loss terms:

$$\mathcal{L}_{total} = \mathcal{L}_{L1} + \lambda_{adv}\mathcal{L}_{adv,G} + \lambda_{perceptual}\mathcal{L}_{perceptual} + \lambda_{style}\mathcal{L}_{style} + \lambda_{tv}\mathcal{L}_{tv} \quad (7)$$

where λ_{adv}, $\lambda_{perceptual}$, λ_{style}, and λ_{tv} are the hyper-parameters which indicate the significance of each term.

5 Experimental Work

We have participated in the AIM 2020 Extreme Image Inpainting Challenge [21] of the ECCV 2020 (please find in our github for details and qualitative results of the challenge). In designing our proposed model, we take reference to the networks in [10,11,28]. We have attached our improved SPD ResNet block to our DeepGIN. We have also modified and applied the ideas of MSSA and BP in our proposed model. Inspired by ESRGAN [30], we remove all batch normalization layers in the model to smooth out the related visual artifacts. We have used discriminators at two different scales which share the same architecture. Also, we have adjusted the number of layers of each discriminator and applied spectral normalization layers [19,34] after the convolutional layers for training stability.

5.1 Training Procedure

Random Mask Generation. Three different types of masks are used in our training. The first type is a rectangular mask with the height and width between 30–70% of each dimension [9,21,23,32,33]. The second type is the free-form mask proposed in [34]. The third type of masks is introduced in the AIM 2020 Image Inpainting Challenge [21], for which masks are randomly generated based on cellular automata. During training, each mask was randomly generated and we applied the three types of masks to each training image to get three different masked images. We observed that this can balance the three types of masks to achieve more stable training.

Training Batch Formation. As the size of training images could be very diverse, we resized all training images to the size of 512×512 and adopted a sub-sampling method [25] to randomly select a sub-image with size of 256×256. We then apply the random mask generation as stated above to obtain three masked images. Therefore, each training image becomes three training images. We set a batch size of 4 and this means that there are 12 training images in a batch.

Two-Stage Training. Our training process is divided into two stages, namely a warm-up stage and then the main stage. First, we trained only the generators by using the *L1 loss* for 10 epochs. We used the initialization method mentioned in [30], using a smaller initialization for ease of training a very deep network. The trained model at the warm-up stage was used as an initialization for the main stage. This *L1*-oriented pre-trained model provides a reasonable initial point for training GANs, for which a balance between quantitative accuracy of the reconstruction and visual quality of the output is required. For the main stage, we trained the generators alternately with the discriminators for 100 epochs. We used Adam [14] with momentum 0.5 for both stages. The initial learning rates for generators and discriminators were set to 0.0001 and 0.0004 respectively. We trained them for 10 epochs with the initial rates and linearly decayed the rates to zero over the last 90 epochs. The hyper-parameters of the loss terms in Eq. (1) and Eq. (7) were set to $\lambda_{hole} = 5.0$, $\lambda_{adv} = 0.001$, $\lambda_{perceptual} = 0.05$, $\lambda_{style} = 80.0$, and $\lambda_{tv} = 0.1$. We developed our model using Pytorch 1.5.0 [22] and trained it on two NVIDIA GeForce RTX 2080Ti GPUs.

5.2 Training Data

ADE20K Dataset. We trained our model on the subset of ADE20K dataset [36,37] for participating in the AIM challenge [21]. This dataset is collected for scene parsing and understanding, in which it contains images from various scene categories. The subset is provided by the organizers of the challenge and it consists of 10,330 training images with diverse resolutions roughly, from 256×256 to 3648×2736. We took around two and a half days for training on this dataset.

CelebA-HQ Dataset. Beyond the ADE20K dataset, we also trained our model on the CelebA-HQ dataset [12] that contains 30K high-quality face images with a standard size of 1024 × 1024. We randomly split this dataset into two groups, 27,000 images for training and 3,000 images for testing. This required approximately 6 days to train our model on this dataset.

6 Analysis of Experimental Results

We have thoroughly evaluated our proposed model. We first provide evidence in our model analysis to show the effectiveness of our suggested strategies for using Spatial Pyramid Dilation (SPD) ResNet block, Multi-Scale Self-Attention (MSSA), and Back Projection (BP). We then compare our model with state-of-the-art approaches, namely DeepFillv1 [33] and DeepFillv2 [34], which are known to have a good generalization. We demonstrate that our model is able to handle images in the wild by testing it on two publicly available datasets, namely Flickr-Faces-HQ (FFHQ) dataset [13] and The Oxford Buildings (Oxford) dataset [24]. Related materials are available at: https://github.com/rlct1/DeepGIN.

6.1 Model Analysis

We first evaluate the effectiveness of the three proposed strategies, namely SPD, MSSA, and BP. Refer to the proposed architecture as shown in Fig. 2, our baselines are denoted as StdResBlk (Coarse only, using only the Coarse Reconstruction Stage) and StdResBlk (a conventional ResNet for inpainting), for which all SA blocks and BP branch are eliminated and all SPD ResNet blocks are replaced by standard ResNet blocks (see Fig. 3(a)). DilatedResBlk or SPDResBlk represents StdResBlk with standard ResNet blocks replaced by Dilated or SPD ResNet blocks. Please refer to Fig. 3(b), (c), and (d). SA or MSSA indicates whether single SA block or MSSA is used and the use of BP is denoted as BP. We conducted the model analysis on CelebA-HQ dataset [12] using the 3K testing images. Note that the testing images were randomly masked by the three types of masks and the same set of masked images was used for each variation of our model. During testing, for images with size larger than 256 × 256, we divided the input into a number of 256 × 256 sub-images using the sub-sampling method [25] and obtained the completed sub-images using the proposed model. We then regrouped the sub-images to form the completed image by using the reverse sub-sampling method. We finally replaced the valid pixels by the ground truth.

Quantitative Comparisons. As the lack of good quantitative evaluation metric for inpainting [16,33,34], we report several numerical metrics which are commonly used in image manipulation, namely PSNR, SSIM [31], mean L1 error, Fréchet Inception Distance (FID) [8], and Learned Perceptual Image Patch Similarity (LPIPS) [35], for a comprehensive analysis of the performance. The results are listed in Table 1 and higher PSNR, SSIM and smaller L1 err. mean better

Table 1. Model analysis of our proposed model on CelebA-HQ dataset. The best results are in **bold** typeface

Variations of our model	Number of parameters	PSNR	SSIM	L1 err. (%)	FID	LPIPS
StdResBlk (Coarse only)	8.168M	31.55	0.925	4.690	23.824	0.182
StdResBlk	40.850M	31.34	0.923	4.710	19.436	0.191
StdResBlk-SA	41.376M	31.60	0.925	4.510	18.239	0.180
StdResBlk-MSSA	42.892M	32.66	0.933	4.067	12.843	0.148
DilatedResBlk-MSSA	42.892M	32.71	0.933	4.034	12.548	0.149
SPDResBlk-MSSA	42.892M	32.88	0.935	3.884	12.335	0.143
SPDResBlk-MSSA-BP	42.930M	**33.26**	**0.939**	**3.666**	**11.424**	**0.132**

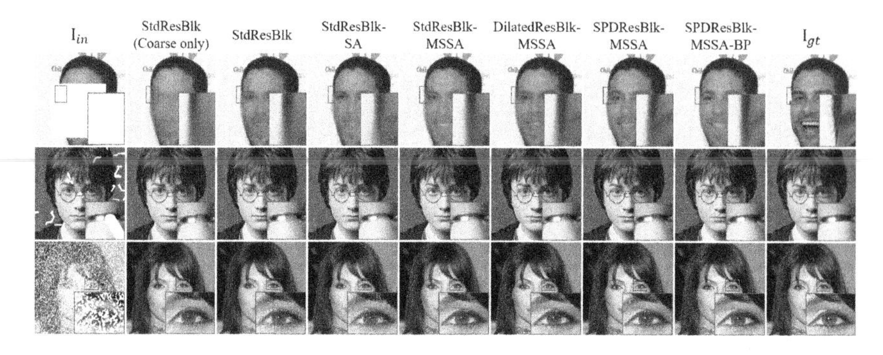

Fig. 4. Comparisons of test results of the variations of our model on CelebA-HQ dataset. Three different types of masked images are displayed from top to bottom. The first and the last columns show $\mathbf{I}_{in}$ and $\mathbf{I}_{gt}$ respectively. The variations of our model are indicated on top of Fig. 4. Our full model (the 8th column), SPDResBlk-MSSA-BP (GAN based), provides high quality results with both the best similarity and visual quality to the ground truth images. Please zoom in for a better view

pixel-wise reconstruction accuracy. FID and LPIPS are also used to estimate the visual quality of the output, the smaller the better. It is obvious that our full model, SPDResBlk-MSSA-BP, gives the best performance on these numerical metrics. The employment of MSSA brings an 1.06 dB increase in PSNR compared to StdResBlk-SA. This reflects the importance of multi-scale self-similarity to inpainting. Our SPD ResNet blocks and the adoption of BP also bring about 0.22 dB and 0.38 dB improvement in PSNR respectively.

Qualitative Comparisons. Figure 4 shows the comparisons of the variations of our model on CelebA-HQ dataset. Without the second refinement stage, the completed images lack of facial details like the first example of the 2nd column in Fig. 4. It can also be observed that the use of MSSA greatly enhances the visual quality as compared to the two which are without SA block and with only a single SA block (see the 3rd and 4th columns). Apart from this, with the

Table 2. Comparisons of DeepFillv1 [33] and DeepFillv2 [34] on both FFHQ and Oxford datasets with two sets of masked images. One set only contains the rectangular masks while another set includes all the three types of masks. Our DeepGIN is denoted as Ours (i.e. the full model, SPDResBlk-MSSA-BP in the previous section). (OS) and (256) mean that the testing images are with the original sizes and size of 256 × 256 respectively. The best results are in **bold** typeface

Method	PSNR	SSIM	L1 err. (%)	FID	LPIPS
Flickr-Faces-HQ Dataset (FFHQ), random rectangular masks					
DeepFillv1 (OS)	20.22	0.872	16.523	97.630	0.173
DeepFillv2 (OS)	20.95	0.903	14.607	92.070	0.170
Ours (OS)	**26.05**	**0.923**	**7.183**	**20.849**	**0.137**
DeepFillv1 (256)	21.55	0.836	13.631	26.276	0.144
DeepFillv2 (256)	22.52	0.845	12.029	**19.336**	**0.128**
Ours (256)	**24.36**	**0.867**	**9.797**	37.577	0.142
The Oxford Buildings Dataset (Oxford), random rectangular masks					
DeepFillv1 (OS)	19.20	0.767	20.322	67.193	0.187
DeepFillv2 (OS)	18.58	0.766	21.204	77.636	0.192
Ours (OS)	**21.92**	**0.861**	**12.067**	**63.744**	**0.170**
DeepFillv1 (256)	19.49	0.795	16.851	**58.588**	**0.169**
DeepFillv2 (256)	18.88	0.789	18.308	66.615	0.174
Ours (256)	**21.90**	**0.819**	**12.995**	74.866	0.185
Flickr-Faces-HQ Dataset (FFHQ), random three types of masks					
DeepFillv1 (OS)	25.12	0.839	11.363	64.534	0.232
DeepFillv2 (OS)	29.70	0.912	7.994	36.940	0.188
Ours (OS)	**32.36**	**0.929**	**4.071**	**14.327**	**0.156**
DeepFillv1 (256)	22.87	0.683	16.812	80.952	0.310
DeepFillv2 (256)	22.75	0.716	17.472	75.555	0.293
Ours (256)	**24.71**	**0.760**	**13.417**	**64.542**	**0.274**
The Oxford Buildings Dataset (Oxford), random three types of masks					
DeepFillv1 (OS)	21.48	0.741	23.460	61.958	0.237
DeepFillv2 (OS)	24.68	0.802	19.195	38.315	**0.179**
Ours (OS)	**27.57**	**0.871**	**7.268**	**38.016**	0.191
DeepFillv1 (256)	21.64	0.686	18.835	81.009	0.284
DeepFillv2 (256)	20.80	0.702	20.687	82.671	0.266
Ours (256)	**23.60**	**0.744**	**14.659**	**79.927**	**0.265**

SPD ResNet blocks and BP technique, the completed images are with better color coherency and alignment of the generated features. For example, see the spectacle frames in the 2nd row.

Fig. 5. Comparisons of test results on FFHQ and Oxford Buildings datasets. Each column shows an example of the test results. From top to bottom: the first row displays various masked input images from both datasets. The second to the fourth rows show the completed images by DeepFillv1, v2 and our DeepGIN. The reference ground truth images are also provided at the last row. Zoom in for a better view

6.2 Comparison with Previous Works

In order to test the generalization of our model, we compare our best model against some state-of-the-art approaches, DeepFillv1 [33] and DeepFillv2 [34], on the two publicly available datasets, FFHQ [13] and Oxford Buildings [24]. It is worth noting that both DeepFillv1 and v2 are known to have good generalization for dealing with images in the wild. We directly used their provided pre-trained models[1] for comparison. The FFHQ dataset is similar to the CelebA-HQ dataset and it contains 70K high-quality face images at 1024 × 1024 resolution. We randomly selected 1,000 images for the testing on this dataset. For the Oxford dataset, it consists of 5,062 images of Oxford landmarks with a wide variety of styles. The images include buildings, suburban areas, halls, people, etc. We also randomly selected 523 testing images on this dataset for comparison.

Similarly, testing images were randomly masked by the three types of masks. For DeepFillv1 and v2, the authors divided an image into a number of grids to perform inpainting and indicated that their models were trained with images of resolution 256 × 256. Note also that DeepFillv1 was trained only for the

[1] https://github.com/JiahuiYu/generative_inpainting.

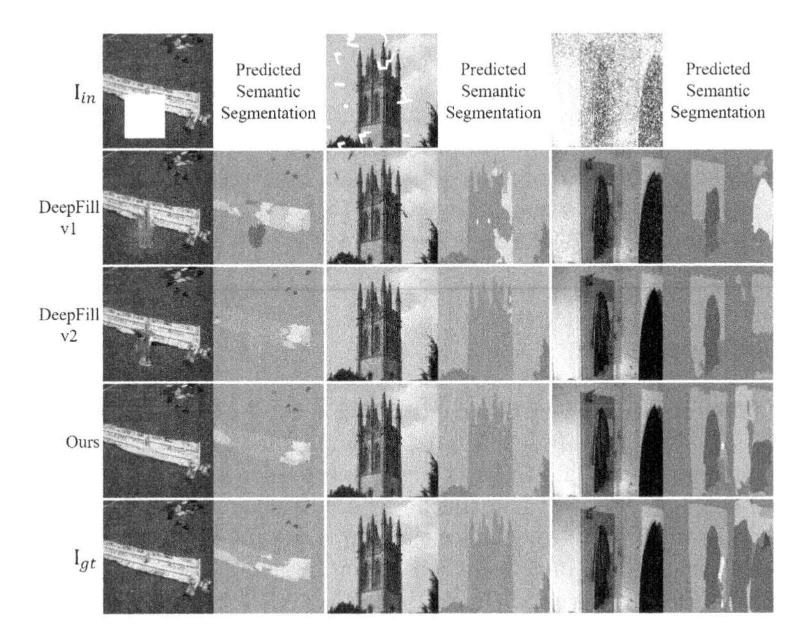

Fig. 6. Visualizations of predicted semantic segmentation test results on Oxford Buildings dataset. 2nd to 4th rows show the completed images by different methods and the corresponding predicted semantic segmentation obtained using the trained network [37]. The ground truth images are also attached to the last row for reference. Please zoom in for a better view

rectangular types of masks. For fair comparison, we also conducted experiments in which testing images were randomly masked only by the rectangular masks.

Quantitative Comparisons. Table 2 shows the comparisons with DeepFillv1 and v2 on the two datasets with two sets of masked images. It is clear that our model outperforms DeepFillv1 and v2 in all the experiments on the two datasets in terms of the pixel-wise reconstruction accuracy. Our model achieves better PSNR compared with DeepFillv1 and v2 in the range of 1.84–5.1 dB and offers better SSIM and L1 error. For FID and LPIPS, we attain better performance on the testing images with the original sizes. For the testing images with size of 256×256 and masked by random rectangular masks, we are also comparable to the other two approaches.

Qualitative Comparisons. Figure 5 displays the test results on both FFHQ and Oxford datasets. It can be seen that DeepFillvl and v2 fail to achieve satisfactory visual quality on the large rectangular masks as shown in the first and fourth columns in Fig. 5. For the other two types of masked images, our model also provides the completed images with better color and content coherency. Note that our model tends to produce blurry images and the reason is that our model was trained to be more PSNR-oriented than the DeepFillv1 and v2. We seek a balance between the pixel-wise accuracy and the visual quality to avoid some strange generated patterns like the completed image by DeepFillv2 of the last example in Fig. 5. To show that our model offers better pixel-wise accuracy, we provide the predicted semantic segmentation test results in Fig. 6. It is obvious that our results are semantically closer to $\mathbf{I}_{gt}$ than that of the other two methods, see for example, the intersection of the newspaper and the lawn in Fig. 6.

7 Conclusions

We have presented a deep generative inpainting network, called DeepGIN. Unlike the existing works, we propose a Spatial Pyramid Dilation (SPD) ResNet block to include more receptive fields for utilizing information given by distant spatial locations. This is important to inpainting especially when the masked regions are too large to be filled. We also enhance the significance of self-similarity consideration, hence we employ Multi-Scale Self-Attention (MSSA) strategy to enhance our performance. Furthermore, Back Projection (BP) is strategically used to improve the alignment of the generated and valid pixels. We have achieved performance better than the state-of-the-art image inpainting. This research work participated in the AIM 2020 Extreme Image Inpainting Challenge, which requires the right balance of pixel-wise reconstruction accuracy and visual quality. We believe that our DeepGIN is able to achieve the right balance and we encourage scholars in the field to give more attention in this direction.

Acknowledgments. Authors would like to thank the support fr. PolyU & PC (PTeC:P19-0319).

References

1. Anoosheh, A., Sattler, T., Timofte, R., Pollefeys, M., Gool, L.V.: Night-to-day image translation for retrieval-based localization. In: 2019 International Conference on Robotics and Automation (ICRA), May 2019. https://doi.org/10.1109/icra.2019.8794387
2. Barnes, C., Shechtman, E., Finkelstein, A., Goldman, D.B.: PatchMatch: a randomized correspondence algorithm for structural image editing. ACM Trans. Graph. **28**(3), 24 (2009)
3. Gatys, L.A., Ecker, A.S., Bethge, M.: A neural algorithm of artistic style. arXiv preprint arXiv:1508.06576 (2015)
4. Goodfellow, I., et al.: Generative adversarial nets. In: Advances in Neural Information Processing Systems, pp. 2672–2680 (2014)
5. Gu, S., et al.: Aim 2019 challenge on image extreme super-resolution: methods and results. In: 2019 IEEE/CVF International Conference on Computer Vision Workshop (ICCVW), pp. 3556–3564. IEEE (2019)
6. Haris, M., Shakhnarovich, G., Ukita, N.: Deep back-projection networks for super-resolution. In: Proceedings of the IEEE Conference on Computer Vision and Pattern Recognition, pp. 1664–1673 (2018)

7. He, K., Zhang, X., Ren, S., Sun, J.: Deep residual learning for image recognition. In: Proceedings of the IEEE Conference on Computer Vision and Pattern Recognition, pp. 770–778 (2016)
8. Heusel, M., Ramsauer, H., Unterthiner, T., Nessler, B., Hochreiter, S.: GANs trained by a two time-scale update rule converge to a local Nash equilibrium. In: Advances in Neural Information Processing Systems, pp. 6626–6637 (2017)
9. Iizuka, S., Simo-Serra, E., Ishikawa, H.: Globally and locally consistent image completion. ACM Trans. Graph. **36**(4) (2017). https://doi.org/10.1145/3072959.3073659
10. Isola, P., Zhu, J.Y., Zhou, T., Efros, A.A.: Image-to-image translation with conditional adversarial networks. In: Proceedings of the IEEE Conference on Computer Vision and Pattern Recognition, pp. 1125–1134 (2017)
11. Johnson, J., Alahi, A., Fei-Fei, L.: Perceptual losses for real-time style transfer and super-resolution. In: Leibe, B., Matas, J., Sebe, N., Welling, M. (eds.) ECCV 2016. LNCS, vol. 9906, pp. 694–711. Springer, Cham (2016). https://doi.org/10.1007/978-3-319-46475-6_43
12. Karras, T., Aila, T., Laine, S., Lehtinen, J.: Progressive growing of gans for improved quality, stability, and variation. arXiv preprint arXiv:1710.10196 (2017)
13. Karras, T., Laine, S., Aila, T.: A style-based generator architecture for generative adversarial networks. In: Proceedings of the IEEE Conference on Computer Vision and Pattern Recognition, pp. 4401–4410 (2019)
14. Kingma, D.P., Ba, J.: Adam: A method for stochastic optimization. arXiv preprint arXiv:1412.6980 (2014)
15. Li, C., Siu, W., Lun, D.P.K.: Vision-based place recognition using convnet features and temporal correlation between consecutive frames. In: 2019 IEEE Intelligent Transportation Systems Conference (ITSC), pp. 3062–3067 (2019)
16. Liu, G., Reda, F.A., Shih, K.J., Wang, T.-C., Tao, A., Catanzaro, B.: Image inpainting for irregular holes using partial convolutions. In: Ferrari, V., Hebert, M., Sminchisescu, C., Weiss, Y. (eds.) ECCV 2018. LNCS, vol. 11215, pp. 89–105. Springer, Cham (2018). https://doi.org/10.1007/978-3-030-01252-6_6
17. Liu, Z.S., Wang, L.W., Li, C.T., Siu, W.C., Chan, Y.L.: Image super-resolution via attention based back projection networks. In: 2019 IEEE/CVF International Conference on Computer Vision Workshop (ICCVW), pp. 3517–3525. IEEE (2019)
18. Lugmayr, A., Danelljan, M., Timofte, R.: NTIRE 2020 challenge on real-world image super-resolution: methods and results. In: Proceedings of the IEEE/CVF Conference on Computer Vision and Pattern Recognition Workshops, pp. 494–495 (2020)
19. Miyato, T., Kataoka, T., Koyama, M., Yoshida, Y.: Spectral normalization for generative adversarial networks. arXiv preprint arXiv:1802.05957 (2018)
20. Nazeri, K., Ng, E., Joseph, T., Qureshi, F.Z., Ebrahimi, M.: EdgeConnect: generative image inpainting with adversarial edge learning. arXiv preprint arXiv:1901.00212 (2019)
21. Ntavelis, E., Romero, A., Bigdeli, S., Timofte, R., et al.: AIM 2020 challenge on image extreme inpainting. In: European Conference on Computer Vision Workshops (2020)
22. Paszke, A., et al.: Pytorch: an imperative style, high-performance deep learning library. In: Advances in Neural Information Processing Systems, pp. 8026–8037 (2019)
23. Pathak, D., Krahenbuhl, P., Donahue, J., Darrell, T., Efros, A.A.: Context encoders: feature learning by inpainting. In: Proceedings of the IEEE Conference on Computer Vision and Pattern Recognition, pp. 2536–2544 (2016)

24. Philbin, J., Chum, O., Isard, M., Sivic, J., Zisserman, A.: Object retrieval with large vocabularies and fast spatial matching. In: 2007 IEEE Conference on Computer Vision and Pattern Recognition, pp. 1–8. IEEE (2007)
25. Shi, W., et al.: Real-time single image and video super-resolution using an efficient sub-pixel convolutional neural network. In: Proceedings of the IEEE Conference on Computer Vision and Pattern Recognition, pp. 1874–1883 (2016)
26. Simonyan, K., Zisserman, A.: Very deep convolutional networks for large-scale image recognition. arXiv preprint arXiv:1409.1556 (2014)
27. Wang, L.W., Liu, Z.S., Siu, W.C., Lun, D.P.: Lightening network for low-light image enhancement. IEEE Trans. Image Process. (2020). https://doi.org/10.1109/TIP.2020.3008396
28. Wang, T.C., Liu, M.Y., Zhu, J.Y., Tao, A., Kautz, J., Catanzaro, B.: High-resolution image synthesis and semantic manipulation with conditional GANs. In: Proceedings of the IEEE Conference on Computer Vision and Pattern Recognition, pp. 8798–8807 (2018)
29. Wang, X., Girshick, R., Gupta, A., He, K.: Non-local neural networks. In: Proceedings of the IEEE Conference on Computer Vision and Pattern Recognition, pp. 7794–7803 (2018)
30. Wang, X., et al.: ESRGAN: enhanced super-resolution generative adversarial networks. In: Leal-Taixé, L., Roth, S. (eds.) ECCV 2018. LNCS, vol. 11133, pp. 63–79. Springer, Cham (2019). https://doi.org/10.1007/978-3-030-11021-5_5
31. Wang, Z., Bovik, A.C., Sheikh, H.R., Simoncelli, E.P.: Image quality assessment: from error visibility to structural similarity. IEEE Trans. Image Process **13**(4), 600–612 (2004)
32. Yang, C., Lu, X., Lin, Z., Shechtman, E., Wang, O., Li, H.: High-resolution image inpainting using multi-scale neural patch synthesis. In: Proceedings of the IEEE Conference on Computer Vision and Pattern Recognition, pp. 6721–6729 (2017)
33. Yu, J., Lin, Z., Yang, J., Shen, X., Lu, X., Huang, T.S.: Generative image inpainting with contextual attention. In: Proceedings of the IEEE Conference on Computer Vision and Pattern Recognition, pp. 5505–5514 (2018)
34. Yu, J., Lin, Z., Yang, J., Shen, X., Lu, X., Huang, T.S.: Free-form image inpainting with gated convolution. In: Proceedings of the IEEE International Conference on Computer Vision, pp. 4471–4480 (2019)
35. Zhang, R., Isola, P., Efros, A.A., Shechtman, E., Wang, O.: The unreasonable effectiveness of deep features as a perceptual metric. In: Proceedings of the IEEE Conference on Computer Vision and Pattern Recognition, pp. 586–595 (2018)
36. Zhou, B., Zhao, H., Puig, X., Fidler, S., Barriuso, A., Torralba, A.: Scene parsing through ade20k dataset. In: Proceedings of the IEEE Conference on Computer Vision and Pattern Recognition (2017)
37. Zhou, B., Zhao, H., Puig, X., Xiao, T., Fidler, S., Barriuso, A., Torralba, A.: Semantic understanding of scenes through the ade20k dataset. Int. J. Comput. Vis. **127**(3), 302–321 (2019)

AIM 2020 Challenge on Video Temporal Super-Resolution

Sanghyun Son[1(✉)], Jaerin Lee[1], Seungjun Nah[1], Radu Timofte[2], Kyoung Mu Lee[1], Yihao Liu[3], Liangbin Xie[3], Li Siyao[3], Wenxiu Sun[3], Yu Qiao[3], Chao Dong[3], Woonsung Park[3], Wonyong Seo[3], Munchurl Kim[3], Wenhao Zhang[3], Pablo Navarrete Michelini[3], Kazutoshi Akita[3], and Norimichi Ukita[3]

[1] CVLab, Seoul National University, Seoul, Korea
sonsang35@gmail.com
[2] Computer Vision Lab, ETH Zürich, Zurich, Switzerland
[3] Challenge Participants, Glasgow, SEC, Scotland

Abstract. Videos in the real-world contain various dynamics and motions that may look unnaturally discontinuous in time when the recorded frame rate is low. This paper reports the second AIM challenge on Video Temporal Super-Resolution (VTSR), a.k.a. frame interpolation, with a focus on the proposed solutions, results, and analysis. From low-frame-rate (15 fps) videos, the challenge participants are required to submit higher-frame-rate (30 and 60 fps) sequences by estimating temporally intermediate frames. To simulate realistic and challenging dynamics in the real-world, we employ the REDS_VTSR dataset derived from diverse videos captured in a hand-held camera for training and evaluation purposes. There have been 68 registered participants in the competition, and 5 teams (one withdrawn) have competed in the final testing phase. The winning team proposes the enhanced quadratic video interpolation method and achieves state-of-the-art on the VTSR task.

Keywords: Video temporal super-resolution · Frame interpolation

1 Introduction

The growth of broadcasting and streaming services have brought needs for high-fidelity videos in terms of both spatial and temporal resolution. Despite the advances in modern video technologies capable of playing videos in high frame-rates, e.g., 60 or 120 fps, videos tend to be recorded in lower frame-rates due

---

S. Son (sonsang35@gmail.com, Seoul National University), J. Lee, S. Nah, R. Timofte, K. M. Lee are the challenge organizers, while the other authors participated in the challenge. Appendix A contains the authors' teams and affiliations.
AIM 2020 webpage: https://data.vision.ee.ethz.ch/cvl/aim20/.
Competition webpage: https://competitions.codalab.org/competitions/24584.

A. Bartoli and A. Fusiello (Eds.): ECCV 2020 Workshops, LNCS 12538, pp. 23–40, 2020.
https://doi.org/10.1007/978-3-030-66823-5_2

to the recording cost and quality adversaries [25]. First, fast hardware components are required to process and save large numbers of pixels in real-time. For example, memory and storage should be fast enough to handle at least 60 FHD (1920×1080) video frames per second, while the central processor has to encode those images into a compressed sequence. Meanwhile, relatively high compression ratios are required to save massive amounts of pixels into slow storage. Line-skipping may occur at the sensor readout time to compensate for the slow processor. Also, small camera sensors in mobile devices may suffer from high noise due to the short exposure time as the required frame rate goes higher. Such limitations can degrade the quality of recorded videos and make them difficult to be acquired in practice.

To overcome those constraints, video temporal super-resolution (VTSR), or video frame interpolation (VFI), algorithms aim to enhance the temporal smoothness of a given video by interpolating missing frames. For such purpose, several methods [17,24,29,30,38,39] have been proposed to reconstruct smooth and realistic sequences from videos of low-frame-rate, which looks discontinuous in the temporal domain. Nevertheless, there also exist several challenges in conventional VTSR methods. First, real-world motions are highly complex and nonlinear, not easy to estimate with simple dynamics. Also, since the algorithm has to process videos of high-quality and high-resolution, efficient approaches are required to handle thousands of frames swiftly in practice.

To facilitate the development of robust and efficient VTSR methods, we have organized AIM 2020 VTSR challenge. In this paper, we report the submitted methods and the benchmark results. Similarly to the last AIM 2019 VTSR challenge [26], participants are required to recover 30 and 60 fps video from low-frame-rate 15 fps input sequences from the REDS_VTSR [26] dataset. 5 teams have submitted the final solution, and the XPixel team won the first-place prize with the Enhanced Quadratic Video Interpolation method. The following sections describe the related works and introduce AIM 2020 Video Temporal Super-Resolution challenge (VTSR) in detail. We also present and discuss the challenge results with the proposed approaches.

2 Related Works

While some approaches synthesize intermediate frames directly [6,22], most methods embed motion estimation modules into neural network architectures. Roughly, the motion modeling could be categorized into phase, kernel, and flow-based approaches. Several methods try to further improve inference accuracy adaptively to the input.

Phase-Based, Kernel-Based Methods: In an early work of Meyer et al. [24], the temporal change of frames is modeled as phase shifts. The PhaseNet [23] model has utilized steerable pyramid filters for encoding larger dynamics with a phase shift. On the other hand, kernel-based approaches use convolutional kernels to shift the input frames to an intermediate position. The AdaConv [29] model estimates spatially adaptive kernels, and the SepConv [27] improves computational efficiency by factorizing the 2D spatial kernel into 1D kernels.

Flow-Based Approaches: Meanwhile, optical flow-based piece-wise linear models have grown popular. The DVF [21] and SuperSloMo [17] models estimate optical flow between two input frames and warp them to the target intermediate time. Later, the accuracy of SuperSloMo is improved by training with cycle consistency loss [34]. The TOF [39], CtxSyn [27], and BMBC [32] models additionally apply image synthesis modules on the warped frames to produce the intermediate frame. To interpolate high-resolution videos in fast speed, the IM-Net [33] uses block-level horizontal/vertical motion vectors in multi-scale.

It is essential to handle the occluded area when warping with the optical flow as multiple pixels could overlap in the warped location. Jiang *et al.* [17] and Yuan *et al.* [40] estimate occlusion maps to exclude invalid contributions. Additional information is explored, such as depth [1] and scene context [27] to determine proper local warping weights. Niklaus *et al.* [28] suggests forward warping with the softmax splatting method, while many other approaches use backward warping with sampling.

Several works tried mixed motion representation of kernel-based and flow-based models. The DAIN [1] and MEMC-Net [2] models use both the kernel and optical flow to apply adaptive warping. The AdaCof [18] unifies the combined representation via an adaptive collaboration of flows in a similar formulation as deformable convolution [7]. To enhance the degree of freedom in the representation, the local weight values differ by position.

Despite the success of motion estimation with optical flow, complex motions in the real world are hard to be modeled due to the simplicity of optical flow. Most of the methods using the optical flow between the two frames are limited to locally linear models. To improve the degree of freedom in the motion model, higher-order representations are proposed. Quadratic [19,38] and cubic [4] motion flow is estimated from multiple input frames. Such higher-order polynomial models could reflect accelerations and nonlinear trajectories. In AIM 2019 Challenge on Video Temporal Super-Resolution, the quadratic video interpolation method [19,38] achieved the best accuracy.

Refinement Methods: On top of the restored frames, some methods try to improve the initial estimation with various techniques. Gui *et al.* [10] propose a 2-stage estimation method. Initial interpolation is obtained with a structure guide in the first stage and the second stage refines the texture. Choi *et al.* [5] perform test-time adaptation of the model via meta-learning. Motivated by classical pyramid energy minimization, Zhang *et al.* [41] propose a recurrent residual pyramid network. Residual displacements are refined via recurrence.

3 AIM 2020 VTSR Challenge

This challenge is one of the AIM 2020 associated challenges on: scene relighting and illumination estimation [8], image extreme inpainting [31], learned image signal processing pipeline [15], rendering realistic bokeh [16], real image super-resolution [37], efficient super-resolution [42], and video extreme super-resolution [9]. Our development phase has started on May 1st, and the test

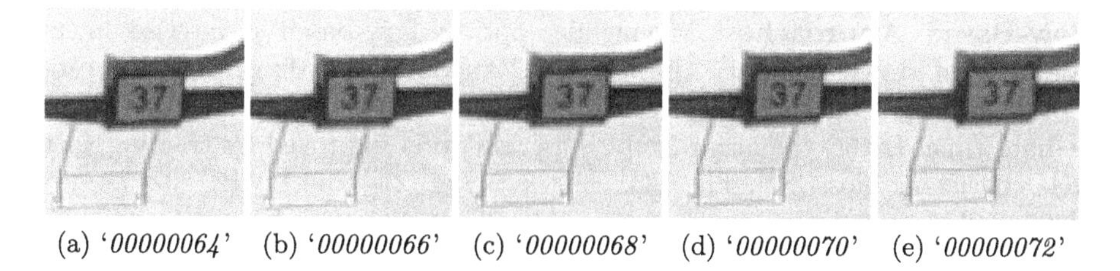

(a) '*00000064*' (b) '*00000066*' (c) '*00000068*' (d) '*00000070*' (e) '*00000072*'

Fig. 1. An example of the VTSR task. In the 15 fps → 30 fps task, i.e., ×2 VTSR, (c) should be estimated from (a) and (e). In the challenging 15 fps → 60 fps tasks, i.e., ×4 VTSR, the goal is to predict (b), (c), and (d) using (a) and (e) only. We provide a sample sequence from the REDS_VTSR [25] test split. Each label indicates the name of the file from a test sequence '*022*,' and patches are cropped from the original full-resolution images for better visualization. We note that the ground-truth frames (b), (c), and (d) are not provided to participants.

phase is opened on July 10th. Participants are required to prepare and submit their final interpolation results in one week. In AIM 2020 VTSR challenge, there have been a total of 68 participants registered in the CodaLab. 5 teams, including one withdrawn, have submitted their final solution after the test phase.

3.1 Challenge Goal

The purpose of the AIM 2020 VTSR challenge is to develop state-of-the-art VTSR algorithms and benchmark various solutions. Participants are required to reconstruct 30 and 60 fps video sequences from 15 fps inputs. We provide 240 training sequences and 30 validation frames from the REDS_VTSR [25] dataset with 60 fps ground-truth videos. The REDS_VTSR dataset is captured by GoPro 6 camera and contains HD-quality (1280 × 720) videos with 43,200 independent frames for training. At the end of the test phase, we evaluate the submitted methods on disjoint 30 test sequences. Figure 1 shows a sample test sequence.

3.2 Evaluation Metrics

In conventional image restoration problems, PSNR and SSIM between output and ground-truth images are considered standard evaluation metrics. In the AIM 2020 VTSR challenge, we use the PSNR as a primary metric, while SSIM values are also provided. We also require all participants to submit their codes and factsheets at the submission time for the fair competition. We then check consistency between submitted result images and reproduced outputs by challenge organizers to guarantee reproducible methods. Since VTSR models are required to process thousands of frames swiftly in practice, one of the essential factors in the algorithms is their efficiency. Therefore, we measure the runtime of each method in a unified framework to provide a fair comparison. Finally, this year, we have experimentally included a new perceptual metric, LPIPS [43], to assess the visual quality of result images. Similar to the PSNR and SSIM, the reference-based LPIPS metric is also calculated between interpolated and ground-truth frames. However, it is known that the LPIPS shows a better correlation with human perception [43] than the conventional PSNR and SSIM. We describe more details in Sect. 4.2.

Table 1. AIM 2020 Video Temporal Super-Resolution Challenge results on the REDS_VTSR test data. Teams are sorted by PSNR (dB). The runtime is measured using a single RTX 2080 Ti GPU based on the submitted codes. However, due to the tight memory constraint, we execute the TTI team's method (†) on a Quadro RTX 8000 GPU, which is ~10% faster than the RTX model. SenseSloMo is the winning team from the last AIM 2019 VTSR challenge [26]. SepConv [30] and Baseline overlay methods are also provided for reference. The best and second-best are marked with red and blue, respectively.

Team	15fps → 30fps		15fps → 60fps		Runtime
	PSNR↑	SSIM↑	PSNR↑	SSIM↑	(s/frame)
XPixel [20] **(Challenge Winner)**	24.78	0.7118	25.69	0.7425	12.40
KAIST-VICLAB	24.69	0.7142	25.61	0.7462	1.57
BOE-IOT-AIBD (Reproduced)	24.49	0.7034	25.27	0.7326	1.00
TTI	23.59	0.6720	24.36	0.6995	6.45†
BOE-IOT-AIBD (Submission)	24.40	0.6972	25.19	0.7269	-
Withdrawn team	24.29	0.6977	25.05	0.7267	2.46
SenseSloMo (AIM 2019 Winner)	24.56	0.7065	25.47	0.7383	-
SepConv [30]	22.48	0.6287	23.40	0.6631	-
Baseline (overlay)	19.68	0.6384	20.39	0.6625	-

4 Challenge Results

Table 1 shows the overall summary of the proposed methods from the AIM 2020 VTSR challenge. The XPixel team has won the first prize, while the KAIST-VICLAB team follows by small margins. We also provide visual comparisons between the submitted methods. Figure 2 illustrates how various algorithms perform differently on a given test sequence.

4.1 Result Analysis

In this section, we focus on some test examples where the proposed methods show interesting behaviors. First, we pick several cases where different methods show high performance, i.e., PSNR, variance. Such analysis can show an advantage of a specific method over the other approaches. In the first row of Fig. 3, the XPixel team outperforms all the other by a large margin by achieving 32.64 dB PSNR. Compared to the other results, it is clear that an accurate reconstruction of the

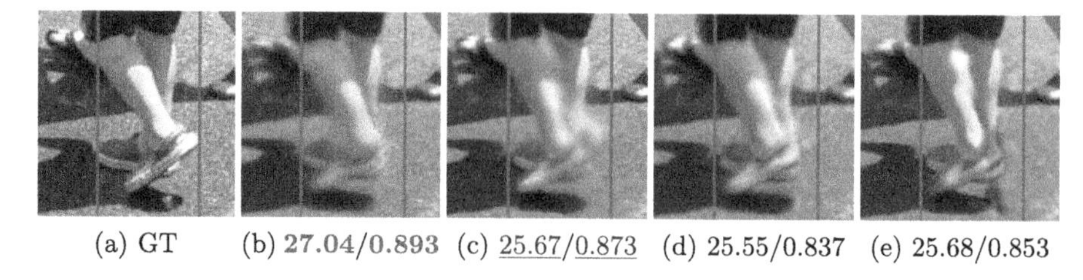

(a) GT (b) **27.04/0.893** (c) 25.67/0.873 (d) 25.55/0.837 (e) 25.68/0.853

Fig. 2. 15 → 60 fps VTSR results on the REDS_VTSR dataset. We provide PSNR/SSIM values of each sample for reference. (a) is a ground-truth frame, and (b) ∼ (e) show result images from the XPixel, KAIST-VICLAB, BOE-IOT-AIBD, and TTI team, respectively. Red lines are drawn to compare alignments with the ground-truth image. Each frame is cropped from the test example '*007/00000358*.' Best viewed in digital zoom.

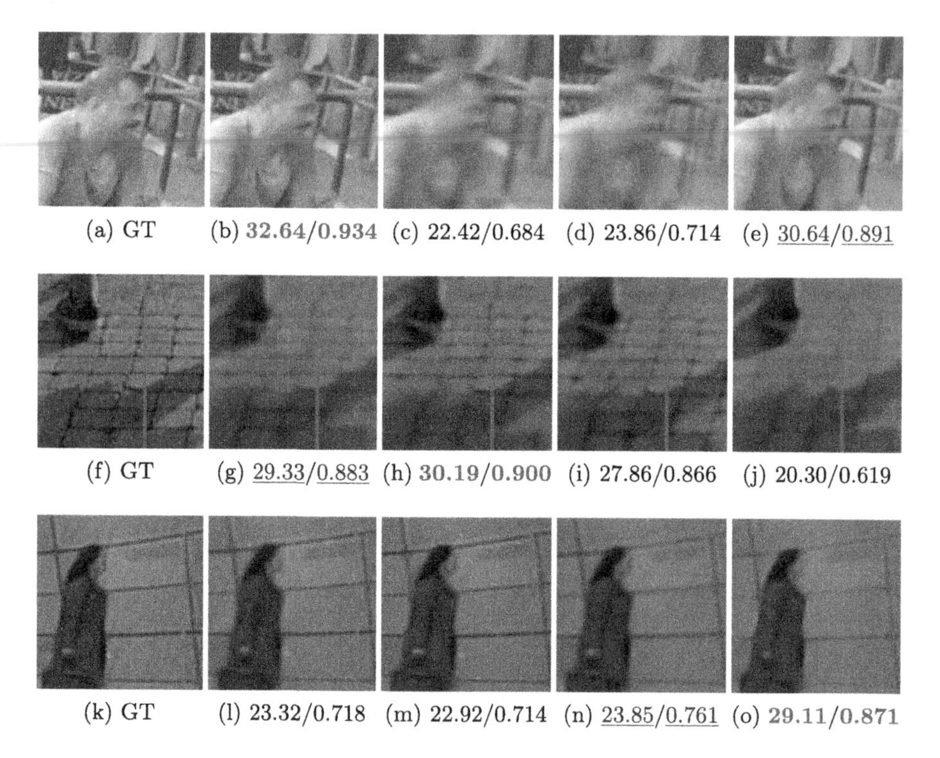

(a) GT (b) **32.64/0.934** (c) 22.42/0.684 (d) 23.86/0.714 (e) 30.64/0.891

(f) GT (g) 29.33/0.883 (h) **30.19/0.900** (i) 27.86/0.866 (j) 20.30/0.619

(k) GT (l) 23.32/0.718 (m) 22.92/0.714 (n) 23.85/0.761 (o) **29.11/0.871**

Fig. 3. 15 → 60 fps VTSR results with high performance variance on the REDS_VTSR dataset. (b) ∼ (e), (g) ∼ (j), and (l) ∼ (o) show result images from the XPixel, KAIST-VICLAB, BOE-IOT-AIBD, and TTI team on the test input '*016/00000358*,' '*019/00000182*,' and '*029/00000134*,' respectively.

intermediate frame has contributed to high performance. On the other hand, the KAIST-VICLAB and TTI team show noticeable performance in the second and third row of Fig. 3, respectively. In those cases, precise alignments play a critical

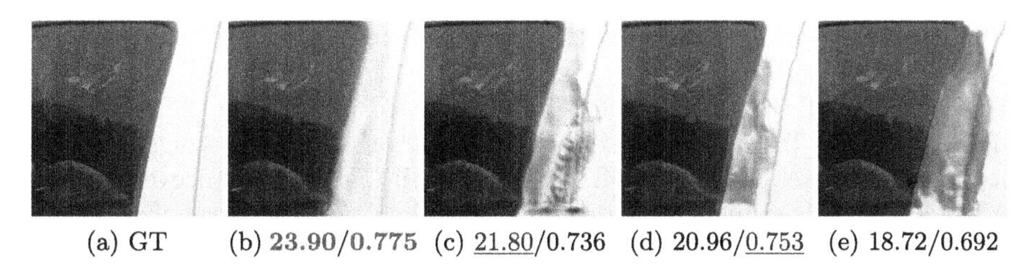

(a) GT (b) **23.90/0.775** (c) 21.80/0.736 (d) 20.96/0.753 (e) 18.72/0.692

Fig. 4. Failure cases of 15 → 60 fps VTSR on the REDS_VTSR dataset. (a) is a ground-truth frame, and (b) ∼ (e) show result images from the XPixel, KAIST-VICLAB, BOE-IOT-AIBD, and TTI team, respectively. Each frame is cropped from the test example '*021/00000218.*'

role in achieving better performance. Specifically, Fig. 3(i) looks perceptually more pleasing than Fig. 3(h) while there is a significant gap between PSNR and SSIM of those interpolated frames. Figure 3(o) shows over +5 dB gain compared to all the other approaches. Although the result frames all look very similar, an accurate alignment algorithm has brought such large gain to the TTI team.

In Fig. 4, we also introduce a case where all of the proposed methods are not able to generate clean output frames. While the submitted models can handle edges and local structures to some extent, they cannot deal with large plain regions and generate unpleasing artifacts. Such limitation shows the difficulty of the VTSR task on real-world videos and implies several rooms to be improved in the following research.

4.2 Perceptual Quality of Interpolated Frames

Recently, there have been rising needs for considering the visual quality of output frames in image restoration tasks. Therefore, we have experimentally included the LPIPS [43] metric to evaluate how the result images are perceptually similar to the ground-truth. Table 2 compares the LPIPS score of the submitted methods. The KAIST-VICLAB team has achieved the best (lowest) LPIPS on both 15 fps → 30 fps and 15 fps → 60 fps VTSR tasks even they show lower PSNR compared to the XPixel team. However, we have also observed that the perceptual metric has a weakness to be applied to video-related tasks directly. For example, the result from KAIST-VICLAB team in Fig. 3(m) has an LPIPS of 0.100, while the TTI team in Fig. 3(o) shows an LPIPS of 0.158. As the score shows, the LPIPS metric does not consider whether the output frame is accurately aligned with the ground-truth. In other words, an interpolated sequence with a better perceptual score may look less natural, regardless of how each frame looks realistic. In future challenges, it would be interesting to develop a video-specific perceptual metric to overcome the limitation of the LPIPS.

5 Challenge Methods and Teams

This section briefly describes each of the submitted methods in the AIM 2020 VTSR challenge. Teams are sorted by their final ranking. Interestingly, the top three teams (XPixel, KAIST-VICLAB, and BOE-IOT-AIBD) have leveraged the QVI model as a baseline and improved the method by their novel approaches. We briefly describe each method base on the submitted factsheets.

Table 2. LPIPS$_\downarrow$ of the submitted methods on the REDS_VTSR dataset. All values are calculated from reproduced results.

Task\Team	XPixel	KAIST-VICLAB	BOE-IOT-AIBD	TTI
15 fps → 30 fps	0.268	0.222	<u>0.249</u>	0.289
15 fps → 60 fps	<u>0.214</u>	0.181	0.230	0.253

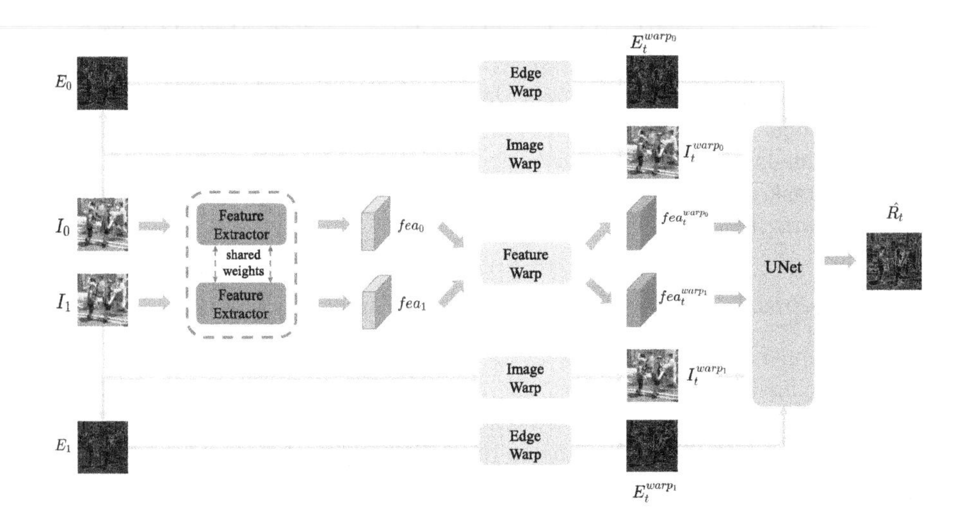

Fig. 5. XPixel: The residual contextual synthesis network.

5.1 XPixel

Method. The XPixel team has proposed the Enhanced Quadratic Video Interpolation (EQVI) [20] method. The algorithm is built upon the AIM 2019 VTSR Challenge winning approach, QVI [19,38], with three main components: rectified quadratic flow prediction (RQFP), residual contextual synthesis network (RCSN), and multi-scale fusion network (MS-Fusion). To ease the overall optimization procedure and verify the performance gain of each component, four-stage training, and fine-tuning strategies are adopted. First, the baseline QVI model is trained with a ScopeFlow [3] flow estimation method instead of the PWC-Net [36]. Then, the model is fine-tuned with an additional residual contextual synthesis network. In the third stage, the rectified quadratic flow prediction is adopted to improve the model with fine-tuning all the modules except the optical flow estimation network. Finally, all the modules are assembled into the multi-scale fusion network and fine-tuned.

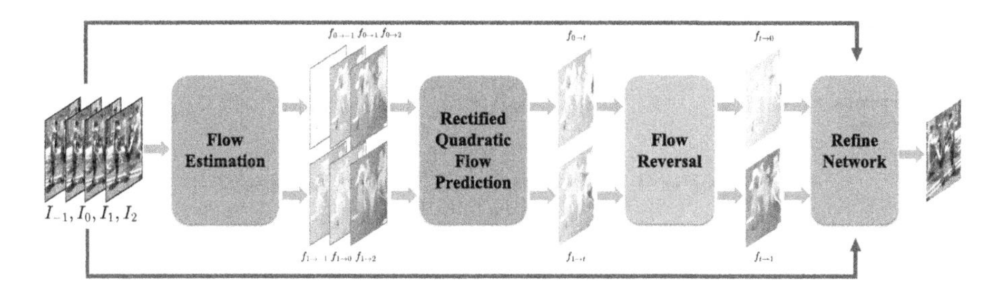

Fig. 6. XPixel: The rectified quadratic flow prediction.

Residual Contextual Synthesis Network. Inspired by [27], a residual contextual synthesis network (RCSN) is designed to warp and exploit the contextual information in high-dimension feature space. Specifically, the *conv*1 layer of the pre-trained ResNet18 [14] is adopted to capture the contextual information of input frame I_0 and I_1. Then we apply back warping on the features with $f_{0\to t}$ to attain pre-warped features. The edge information of I_0 and I_1 is also extracted and warped to preserve and leverage more structural information. For simplicity, we calculate the gradient of each channel of the input frame as edge information. Afterward, we feed the warped images, edges, and features into a small network to synthesize a residual map $\hat{R}_t$. The refined output is obtained by $\hat{I}_t^{refined} = \hat{I}_t + \hat{R}_t$. Figure 5 shows the overall organization of the network.

Rectified Quadratic Flow Prediction. Given four input consecutive frames I_{-1}, I_0, I_1 and I_2, the goal of video temporal super-resolution is to interpolate an intermediate frame I_t, where $t \in (0, 1)$. Following the quadratic model [38], a least square estimation method is utilized to improve the accuracy of quadratic frame interpolation further. Unlike the original QVI [38], all four frames (or three equations) are used for the quadratic model. If the motion of the input four frames basically conforms to the uniform acceleration model, the following equations should hold or approximately hold:

$$\begin{aligned} f_{0\to -1} &= -v_0 + 0.5a, \\ f_{0\to 1} &= v_0 + 0.5a, \\ f_{0\to 2} &= 2v_0 + 2a, \end{aligned} \tag{1}$$

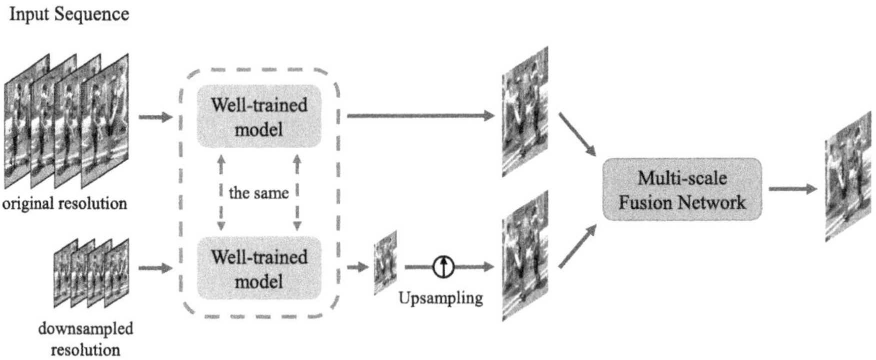

Fig. 7. XPixel: The multi-scale fusion network.

where $f_{0\to t}$ denotes the displacement of the pixel from frame 0 to frame t, v_0 represents the velocity at frame 0, and a is the acceleration of the quadratic motion model. The equations above can be transformed into a matrix form:

$$\underbrace{\begin{bmatrix} -1 & 0.5 \\ 1 & 0.5 \\ 2 & 2 \end{bmatrix}}_{A} \begin{bmatrix} v_0 \\ a \end{bmatrix} = \underbrace{\begin{bmatrix} f_{0\to -1} \\ f_{0\to 1} \\ f_{0\to 2} \end{bmatrix}}_{b}, \tag{2}$$

where the solution $x^* = [v_0^* \quad a^*]^T$ can be derived as $x^* = [A^T A]^{-1} A^T b$. Then, the intermediate flow could be formulated as $f_{0\to t} = v_0^* t + \frac{1}{2} a^* t^2$. When the motion of input four frames approximately fits the quadratic assumption, the LSE solution can make a better estimation of $f_{0\to t}$. As the real scenes are usually complex, several simple rules are adopted to discriminate whether the motion satisfies the quadratic assumption. For any pair picked from I_{-1}, I_1 and I_2, an acceleration a_i is calculated as:

$$\begin{aligned} a_1 &= f_{0\to -1} + f_{0\to 1} \\ a_2 &= \frac{2}{3} f_{0\to -1} + \frac{1}{3} f_{0\to 2} \\ a_3 &= f_{0\to 2} + 2 f_{0\to 2} \end{aligned} \tag{3}$$

Theoretically, for quadratic motion, a_1, a_2 and a_3 should be in the same direction and approximately equal to each other. If the orientation of those accelerations is not consistent, the model will be directly degenerated to the original QVI; otherwise, we adopt the following weight function to fuse the rectified flow and the original quadratic flow according to the proximity of quadratic model:

$$\alpha(z) = -\frac{1}{2} \left[\frac{e^{\omega(z-\gamma)} - e^{-\omega(z-\gamma)}}{e^{\omega(z-\gamma)} + e^{-\omega(z-\gamma)}} \right] + \frac{1}{2}, \tag{4}$$

where $z = |a_1 - a_2|$, ω is the axis of symmetry, and γ is the stretching factor. Empirically, ω and γ are set to 5 and 1, respectively. The final v_0 and a is obtained by:

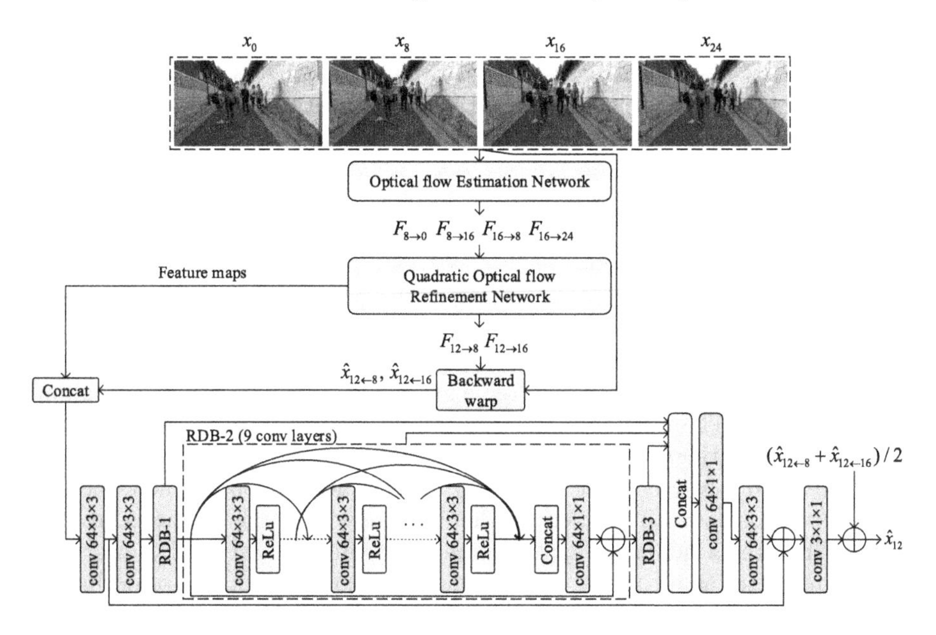

Fig. 8. KAIST-VICLAB: The architecture of the multi-frame synthesis network. The optical flow estimation network is composed of the PWC-net [36], and the quadratic optical flow refinement network is composed of the U-Net model similar to the original QVI method [38].

$$\begin{aligned} v_0 &= \alpha * v_0^{lse} + (1-\alpha) * v_0^{ori}, \\ a &= \alpha * a^{lse} + (1-\alpha) * a^{ori}, \end{aligned} \tag{5}$$

where ^{lse} and ^{ori} represent the rectified LSE prediction and original prediction, respectively. Figure 6 demonstrates how the rectified quadratic flow is estimated.

Multi-scale Fusion Network. Finally, a novel multi-scale fusion network is proposed to capture various motions at a different level and further boost the performance. This fusion network can be regarded as a learnable augmentation process at different resolutions. Once we obtain a well-trained interpolation model from previous stages, we feed the model with one input sequence and its downsampled counterpart. Then a fusion network is trained to predict a pixel-wise weighted map M to fuse the results.

$$M = F(Q(I_{in}), Up(Q(Down(I_{in})))), \tag{6}$$

where Q represents the well-trained QVI model, $Down(.)$ and $Up(.)$ denote the downsampling and upsampling operations, respectively. Then, the final output interpolated frame is formulated as follows, as illustrated in Fig. 7:

$$\hat{I}_t^{final} = M * Q(I_{in}) + (1-M) * Up(Q(Down(I_{in}))). \tag{7}$$

The proposed method is implemented under Python 3.6 and PyTorch 1.2 environments. The training requires about 5 days using 4 RTX 2080 Ti GPUs.

5.2 KAIST-VICLAB

Method. KAIST-VICLAB team proposes the quadratic video frame interpolation with a multi-frame synthesis network. Inspired by the quadratic video interpolation [38] and the context-aware synthesis for video interpolation [27], the proposed approach leverages these types of flow-based method. The REDS_VTSR dataset [26] is composed of high-resolution frames with much more complex motion than typical frame interpolation dataset [39]. Since the optical flow is vulnerable to complex motion or occlusion, the proposed method attempts to handle the vulnerability by a nonlinear synthesis method with more information from neighbor frames. Therefore, the proposed algorithm first estimates the optical flows between the intermediate frame and the neighboring frames using four adjacent frames. The method for estimating the optical flows follows the same method for estimating the optical flows of the intermediate frame by a quadratic method based on the PWC-net [36] in [38]. Unlike the QVI [38] method, the proposed method utilizes all the four neighboring frames for the synthesis network and incorporates a nonlinear synthesis network into the last part of our total network to better adapt to the frames containing complex motion or occlusion.

Details. The proposed model is composed of the residual dense network [44], and the inputs of the synthesis network are not only the feature maps obtained by estimating the optical flows using four neighboring frames, but also the warped neighboring frames. Also, the output of the synthesis network is the corresponding intermediate frame. Motivated by TOFlow [39], the PWC-net for the optical flow estimation is fine-tuned on the REDS_VTSR dataset. More details about the proposed network are illustrated in Fig. 8. The intermediate frames are optimized using the Laplacian loss [27] on three different scales and VGG loss. The method is developed under Python 3.6 and PyTorch 1.4 environments, using one TITAN Xp GPU. The training takes about 3 days.

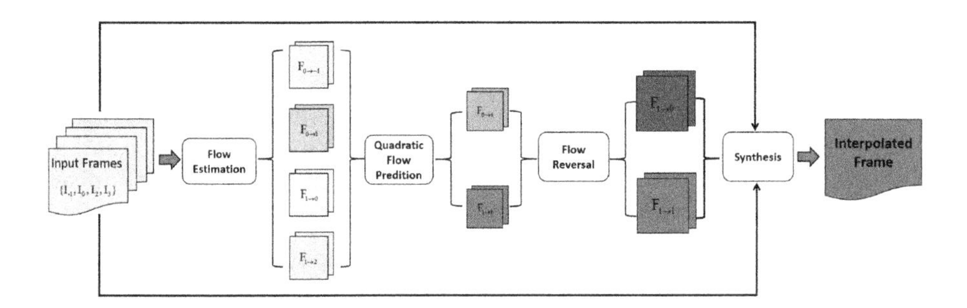

Fig. 9. BOE-IOT-AIBD: The architecture of the MSQI model.

5.3 BOE-IOT-AIBD

Method. BOE-IOT-AIBD team has proposed the Multi Scale Quadratic Interpolation (MSQI) approach for the VTSR task. The model is trained on the REDS_VTSR dataset [26] of three different scales. At each scale, the proposed MSQI employs PWC-Net [36] to extract optical flows and further refines the flow continuously through three modules: quadratic acceleration, flow reverse, and U-Net [35] refine module. Finally, a synthesis module interpolates the output frame by warping in-between inputs and refined flows. During the inference, a post-processing method is applied to grind the interpolated frame.

Details. Figure 9 illustrates an overview of the proposed MSQI method. Firstly, the MSQI model has trained on sequences from the REDS_VTSR dataset [26], using ×4 lower resolution (320 × 180) and frame rate (15 fps) for quick convergence. Then, the model is fine-tuned on sequences of higher resolution (640 × 360). In the last stage, the MSQI model is trained using full-size images (1280 × 720) with a 15fps frame rate. During the inference time, a smoothing function is adopted for post-processing. Also, quantization noise is injected during the training and inference phase. The proposed method also adds the quantization noise to input frames at the inference phase and employs a smooth function to grind the interpolated frame. The method adopts the pre-trained off-the-shelf optical flow algorithm, PWC-Net [36], from the SenseSlowMo [19,38] model. Compared to the baseline SenseSlowMo [19,38] architecture, the proposed MSQI converges better and focuses on multiple resolutions. Python 3.7 and PyTorch environments are used on a V100 GPU server with a total of 128 GB VRAM.

5.4 TTI

Method. TTI team has proposed a temporal super-resolution method that copes large motions by reducing an input sequence's spatial resolution. The approach is inspired by STARnet [11] model. With the idea that space and time are related, STARnet jointly optimizes three tasks, i.e., spatial, temporal, and spatio-temporal super-resolution.

Details. Figure 10 illustrates the concept of the STARnet. The model takes two LR frames I_t^{lr} and I_{t+1}^{lr}, with bidirectional dense motion flows of these frames $F_{t\to t+1}$ and $F_{t+1\to t}$ as input. The output consists of an in-between LR I_{t+n}^{lr} and three SR frames I_t^{sr}, I_{t+n}^{sr}, and I_{t+1}^{sr}, where $n \in [0,1]$ denotes the temporal interpolation rate. While the STARnet model can be fine-tuned only for temporal super-resolution as proposed in [11], the TTI team has employed the original spatio-temporal framework. This is because large motions observed in the REDS [25] dataset make it difficult to estimate accurate optical flows, which take an essential role for the VTSR task. Therefore, the optical flows on the LR domain might support VTSR of HR frames in the REDS sequences. In the proposed strategy, two given frames are resized to LR frames I_t^{lr} and I_{t+1}^{lr}, and then fed into the network shown in Fig. 10 in order to acquire I_{t+n}^{sr} in the

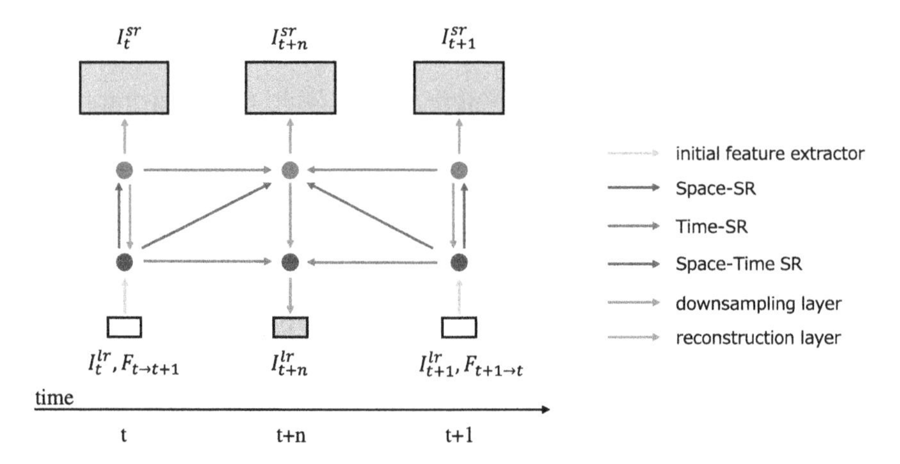

Fig. 10. TTI: The architecture of the STARnet model. White and gray rectangles indicate input and output frames, respectively. For temporal super-resolution, two given frames are regarded as HR frames and downscaled to I_t^{lr} and I_{t+1}^{lr}. From I_t^{lr} and I_{t+1}^{lr}, the proposed STARnet acquires I_{t+n}^{sr} as an interpolated frame in the original HR resolution.

original resolution. For $\times 4$ temporal super-resolution, n is set to $[0.25, 0.5, 0.75]$. To realize those three temporal interpolation rates with one network, input flow maps $F_{t\to t+1}$ and $F_{t+1\to t}$ are scaled according to n as follows:

$$\begin{aligned} \hat{F}_{t\to t+1} &= nF_{t\to t+1}, \\ \hat{F}_{t+1\to t} &= (1-n)F_{t+1\to t}, \end{aligned} \tag{8}$$

where $\hat{F}_{t\to t+1}$ and $\hat{F}_{t+1\to t}$ denote the re-scaled flow maps that are used as input, respectively. For higher accuracy, two changes are made to the original STARnet model. First, a deformable convolution [7] is added at the end of T- and ST-SR networks to explicitly deal with object motions. Second, the PWC-Net [36] is used to get flow maps, which show better performance than the baseline.

During training, each input frame is downscaled by half using bicubic interpolation. Those frames are super-resolved to its original resolution by the S-SR network. The $\times 2$ S-SR network, which is based on the DBPN [12] and RBPN [13] models, is pre-trained on the Vimeo-90k dataset [39]. Various data augmentations, such as rotation, flipping, and random cropping, are applied to input and target pairs. All experiments are conducted using Python 3.6 and PyTorch 1.4 on Tesla V100 GPUs.

6 Conclusion

In the AIM 2020 VTSR challenge, 5 teams competed to develop state-of-the-art VTSR methods with the REDS_VTSR dataset. Top 3 methods leverage the quadratic motion modeling [19,38], demonstrating the importance of accurate motion prediction. The winning team XPixel proposes Enhanced Quadratic Video Interpolation framework, which improves the QVI [19,38] method with three novel components. Compared to AIM 2019 VTSR challenge [26], there has been a significant PSNR improvement of 0.22 dB on the 15 → 60 fps task over the QVI method [19]. We also compare the submitted methods in a unified framework and provide a detailed analysis with specific example cases. In our future challenge, we will encourage participants to develop 1) More efficient algorithms, 2) Perceptual interpolation methods, 3) Robust models on challenging real-world inputs.

Acknowledgments. We thank all AIM 2020 sponsors: Huawei Technologies Co. Ltd., MediaTek Inc., NVIDIA Corp., Qualcomm Inc., Google, LLC and CVL, ETH Zürich.

A Teams and Affiliations

AIM 2020 Team

Title: AIM 2020 Challenge on Video Temporal Super-Resolution

Members: Sanghyun Son[1] (sonsang35@gmail.com), Jaerin Lee[1], Seungjun Nah[1], Radu Timofte[2], Kyoung Mu Lee[1]

Affiliations:

[1] Department of ECE, ASRI, Seoul National University (SNU), Korea

[2] Computer Vision Lab, ETH Zürich, Switzerland

XPixel

Title: Enhanced Quadratic Video Interpolation

Members: Yihao Liu[1,2] (liuyihao14@mails.ucas.ac.cn), Liangbin Xie[1,2], Li Siyao[3], Wenxiu Sun[3], Yu Qiao[1], Chao Dong[1]

Affiliations:

[1] Shenzhen Institutes of Advanced Technology, CAS

[2] University of Chinese Academy of Sciences

[3] SenseTime Research

KAIST-VICLAB

Title: Quadratic Video Frame Interpolation with Multi-frame Synthesis Network

Members: Woonsung Park[1] (pys5309@kaist.ac.kr), Wonyong Seo[1] (wyong0122@kaist.ac.kr), Munchurl Kim[1] (mkimee@kaist.ac.kr)

Affiliations:

[1] Video and Image Computing Lab, Korea Advanced Institute of Science and Technology (KAIST)

BOE-IOT-AIBD

Title: Multi Scale Quadratic Interpolation Method

Members: Wenhao Zhang[1] (zhangwenhao@boe.com.cn), Pablo Navarrete Michelini[1]
Affiliations:
[1] BOE, IOT AIBD, Multi Media Team

TTI
Title: STARnet for Video Frame Interpolation
Members: Kazutoshi Akita[1] (sd19401@toyota-ti.ac.jp), Norimichi Ukita[1]
Affiliations:
[1] Toyota Technological Institute (TTI)

References

1. Bao, W., Lai, W.S., Ma, C., Zhang, X., Gao, Z., Yang, M.H.: Depth-aware video frame interpolation. In: CVPR (2019)
2. Bao, W., Lai, W.S., Zhang, X., Gao, Z., Yang, M.H.: MEMC-Net: motion estimation and motion compensation driven neural network for video interpolation and enhancement. IEEE TPAMI (2019)
3. Bar-Haim, A., Wolf, L.: ScopeFlow: dynamic scene scoping for optical flow. In: CVPR (2020)
4. Chi, Z., Mohammadi Nasiri, R., Liu, Z., Lu, J., Tang, J., Plataniotis, K.N.: All at once: temporally adaptive multi-frame interpolation with advanced motion modeling. In: Vedaldi, A., Bischof, H., Brox, T., Frahm, J.M. (eds.) ECCV 2020. LNCS, vol. 12372, pp. 107–123. Springer, Cham (2020). https://doi.org/10.1007/978-3-030-58583-9_7
5. Choi, M., Choi, J., Baik, S., Kim, T.H., Lee, K.M.: Scene-adaptive video frame interpolation via meta-learning. In: CVPR (2020)
6. Choi, M., Kim, H., Han, B., Xu, N., Lee, K.M.: Channel attention is all you need for video frame interpolation. In: AAAI (2020)
7. Dai, J., et al.: Deformable convolutional networks. In: ICCV (2017)
8. El Helou, M., Zhou, R., Süsstrunk, S., Timofte, R., et al.: AIM 2020: scene relighting and illumination estimation challenge. In: ECCV Workshops (2020)
9. Fuoli, D., et al.: AIM 2020 challenge on video extreme super-resolution: methods and results. In: Bartoli, A., Fusiello, A. (eds.) ECCV 2020. LNCS, vol. 12538, pp. 57–81. Springer, Cham (2020)
10. Gui, S., Wang, C., Chen, Q., Tao, D.: FeatureFlow: robust video interpolation via structure-to-texture generation. In: CVPR (2020)
11. Haris, M., Shakhnarovich, G., Ukita, N.: Space-time-aware multi-resolution video enhancement. In: CVPR (2020)
12. Haris, M., Shakhnarovich, G., Ukita, N.: Deep back-projection networks for super-resolution. In: CVPR (2018)
13. Haris, M., Shakhnarovich, G., Ukita, N.: Recurrent back-projection network for video super-resolution. In: CVPR (2019)
14. He, K., Zhang, X., Ren, S., Sun, J.: Deep residual learning for image recognition. In: CVPR (2016)
15. Ignatov, A., Timofte, R., et al.: AIM 2020 challenge on learned image signal processing pipeline. In: ECCV Workshops (2020)

16. Ignatov, A., Timofte, R., et al.: AIM 2020 challenge on rendering realistic bokeh. In: ECCV Workshops (2020)
17. Jiang, H., Sun, D., Jampani, V., Yang, M.H., Learned-Miller, E., Kautz, J.: Super SloMo: high quality estimation of multiple intermediate frames for video interpolation. In: CVPR (2018)
18. Lee, H., Kim, T., Chung, T.y., Pak, D., Ban, Y., Lee, S.: AdaCoF: adaptive collaboration of flows for video frame interpolation. In: CVPR (2020)
19. Li, S., Xu, X., Pan, Z., Sun, W.: Quadratic video interpolation for VTSR challenge. In: ICCV Workshops (2019)
20. Liu, Y., Xie, L., Siyao, L., Sun, W., Qiao, Y., Dong, C.: Enhanced quadratic video interpolation. In: Bartoli, A., Fusiello, A. (eds.) ECCV 2020. LNCS, vol. 12538, pp. 41–56. Springer, Cham (2020)
21. Liu, Z., Yeh, R.A., Tang, X., Liu, Y., Agarwala, A.: Video frame synthesis using deep voxel flow. In: ICCV (2017)
22. Long, G., Kneip, L., Alvarez, J.M., Li, H., Zhang, X., Yu, Q.: Learning image matching by simply watching video. In: Leibe, B., Matas, J., Sebe, N., Welling, M. (eds.) ECCV 2016. LNCS, vol. 9910, pp. 434–450. Springer, Cham (2016). https://doi.org/10.1007/978-3-319-46466-4_26
23. Meyer, S., Djelouah, A., McWilliams, B., Sorkine-Hornung, A., Gross, M., Schroers, C.: PhaseNet for video frame interpolation. In: CVPR (2018)
24. Meyer, S., Wang, O., Zimmer, H., Grosse, M., Sorkine-Hornung, A.: Phase-based frame interpolation for video. In: CVPR (2015)
25. Nah, S., et al.: NTIRE 2019 challenges on video deblurring and super-resolution: dataset and study. In: CVPR Workshops (2019)
26. Nah, S., Son, S., Timofte, R., Lee, K.M., et al.: AIM 2019 challenge on video temporal super-resolution: methods and results. In: ICCV Workshops (2019)
27. Niklaus, S., Liu, F.: Context-aware synthesis for video frame interpolation. In: CVPR (2018)
28. Niklaus, S., Liu, F.: Softmax splatting for video frame interpolation. In: CVPR (2020)
29. Niklaus, S., Mai, L., Liu, F.: Video frame interpolation via adaptive convolution. In: CVPR (2017)
30. Niklaus, S., Mai, L., Liu, F.: Video frame interpolation via adaptive separable convolution. In: ICCV (2017)
31. Ntavelis, E., Romero, A., Bigdeli, S.A., Timofte, R., et al.: AIM 2020 challenge on image extreme inpainting. In: ECCV Workshops (2020)
32. Park, J., Ko, K., Lee, C., Kim, C.-S.: BMBC: bilateral motion estimation with bilateral cost volume for video interpolation. In: Vedaldi, A., Bischof, H., Brox, T., Frahm, J.-M. (eds.) ECCV 2020. LNCS, vol. 12359, pp. 109–125. Springer, Cham (2020). https://doi.org/10.1007/978-3-030-58568-6_7
33. Peleg, T., Szekely, P., Sabo, D., Sendik, O.: IM-Net for high resolution video frame interpolation. In: CVPR (2019)
34. Reda, F.A., et al.: Unsupervised video interpolation using cycle consistency. In: ICCV (2019)
35. Ronneberger, O., Fischer, P., Brox, T.: U-Net: convolutional networks for biomedical image segmentation. In: Navab, N., Hornegger, J., Wells, W.M., Frangi, A.F. (eds.) MICCAI 2015. LNCS, vol. 9351, pp. 234–241. Springer, Cham (2015). https://doi.org/10.1007/978-3-319-24574-4_28
36. Sun, D., Yang, X., Liu, M.Y., Kautz, J.: PWC-Net: CNNs for optical flow using pyramid, warping, and cost volume. In: CVPR (2018)

37. Wei, P., Lu, H., Timofte, R., Lin, L., Zuo, W., et al.: AIM 2020 challenge on real image super-resolution. In: ECCV Workshops (2020)
38. Xu, X., Siyao, L., Sun, W., Yin, Q., Yang, M.H.: Quadratic video interpolation. In: NeurIPS (2019)
39. Xue, T., Chen, B., Wu, J., Wei, D., Freeman, W.T.: Video enhancement with task-oriented flow. Int. J. Comput. Vision **127**(8), 1106–1125 (2019). https://doi.org/10.1007/s11263-018-01144-2
40. Yuan, L., Chen, Y., Liu, H., Kong, T., Shi, J.: Zoom-in-to-check: boosting video interpolation via instance-level discrimination. In: CVPR (2019)
41. Zhang, H., Zhao, Y., Wang, R.: A flexible recurrent residual pyramid network for video frame interpolation. In: ICCV (2019)
42. Zhang, K., Danelljan, M., Li, Y., Timofte, R., et al.: AIM 2020 challenge on efficient super-resolution: methods and results. In: ECCV Workshops (2020)
43. Zhang, R., Isola, P., Efros, A.A., Shechtman, E., Wang, O.: The unreasonable effectiveness of deep features as a perceptual metric. In: CVPR (2018)
44. Zhang, Y., Tian, Y., Kong, Y., Zhong, B., Fu, Y.: Residual dense network for image super-resolution. In: CVPR (2018)

Enhanced Quadratic Video Interpolation

Yihao Liu[1,2](✉), Liangbin Xie[1,2], Li Siyao[3], Wenxiu Sun[3], Yu Qiao[1], and Chao Dong[1]

[1] ShenZhen Key Lab of Computer Vision and Pattern Recognition, SIAT-SenseTime Joint Lab, Shenzhen Institutes of Advanced Technology, Chinese Academy of Sciences, Shenzhen, China
{lb.xie,yu.qiao,chao.dong}@siat.ac.cn

[2] University of Chinese Academy of Sciences, Beijing, China
liuyihao14@mails.ucas.ac.cn

[3] SenseTime Research, Beijing, China
{lisiyao1,sunwenxiu}@tetras.ai

Abstract. With the prosperity of digital video industry, video frame interpolation has arisen continuous attention in computer vision community and become a new upsurge in industry. Many learning-based methods have been proposed and achieved progressive results. Among them, a recent algorithm named quadratic video interpolation (QVI) achieves appealing performance. It exploits higher-order motion information (*e.g.* acceleration) and successfully models the estimation of interpolated flow. However, its produced intermediate frames still contain some unsatisfactory ghosting, artifacts and inaccurate motion, especially when large and complex motion occurs. In this work, we further improve the performance of QVI from three facets and propose an enhanced quadratic video interpolation (EQVI) model. In particular, we adopt a rectified quadratic flow prediction (RQFP) formulation with least squares method to estimate the motion more accurately. Complementary with image pixel-level blending, we introduce a residual contextual synthesis network (RCSN) to employ contextual information in high-dimensional feature space, which could help the model handle more complicated scenes and motion patterns. Moreover, to further boost the performance, we devise a novel multi-scale fusion network (MS-Fusion) which can be regarded as a learnable augmentation process. The proposed EQVI model won the first place in the AIM2020 Video Temporal Super-Resolution Challenge. Codes are available at https://github.com/lyh-18/EQVI.

Keywords: Video frame interpolation · Least squares method · Rectified quadratic flow prediction · Contextual information · Multi-scale fusion

1 Introduction

Video frame interpolation has attracted increasing attention in recent years, due to its urgent needs in video enhancement market. It is intrinsically an ill posed

Y. Liu and L. Xie—Co-first authors.

A. Bartoli and A. Fusiello (Eds.): ECCV 2020 Workshops, LNCS 12538, pp. 41–56, 2020.
https://doi.org/10.1007/978-3-030-66823-5_3

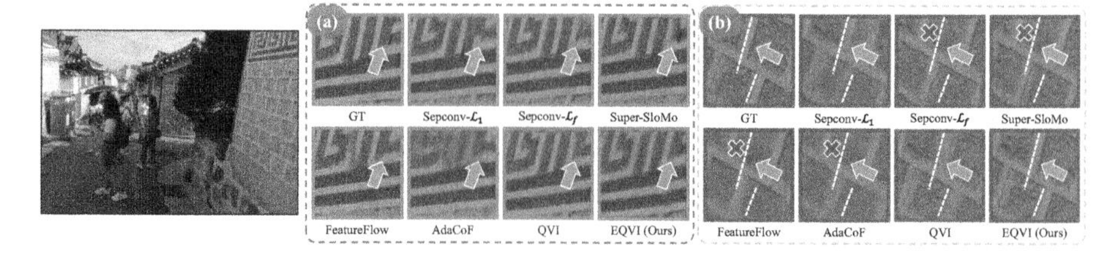

Fig. 1. Video frame interpolation examples. Compared with other state-of-the-art methods, EQVI can generate more visually pleasing intermediate frames with more accurate textures and fewer artifacts. In patch (a), the results of other methods are prone to be blurry or contain artifacts in regions with complex structures (indicated by the arrows); In patch (b), other methods may result in blur and misalignment, due to the inaccurate motion estimation (indicated by the dashed lines and red crosses). (Color figure online)

problem that predicts intermediate/missing frames from a low frame-rate video. With the help of deep learning techniques, recent video interpolation algorithms have achieved significant improvement in both evaluation metrics and visual quality. To further promote the progress, the AIM 2020 held the challenge on video interpolation [20]. Our team won the challenge with the proposed enhanced quadratic video interpolation (EQVI) method. We will describe our winning entry in this paper, and compare with other state-of-the-arts.

Existing deep-learning-based methods can be roughly divided into two groups. The first one, which is summarized as "flow-based" method, tries to estimate the optical flow of the intermediate frame. Then, the corresponding pixels from adjacent frames are warped according to the estimated flow. The representative methods are DVF [11], Super-SloMo [9], DAIN [2] and QVI [24]. While the second group, concluded as "kernel-based" method, includes AdaConv [16], Sepconv [17] and AdaCoF [10], which directly generate isolated sampling kernel maps for each pixel of the intermediate frame. As most of the above methods perform video interpolation with only two adjacent frames, they inevitably adopt the linear motion assumption, which does not always hold in real scenarios. Among them, QVI model [24] applies a quadratic interpolation formula for predicting more accurate intermediate frames.

Our method is built upon QVI, and improves it from three aspects. 1) From the aspect of formulation, we refine the original quadratic flow prediction module by rectifying it with least squares method. This rectified quadratic flow prediction formula (RQFP) could correct the estimation error caused by calculating with incomplete data samples, and improve the accuracy of the interpolated optical flow when the frame motion satisfies the quadratic assumption. 2) To employ more information, we introduce another residual contextual synthesis network (RCSN), which can incorporate contextual information from pre-extracted high-dimensional features (ResNet-18 [8]). This additional module could alleviate the problem of inaccurate motion estimation and object occlusion, and further push the performance by at least 0.18 dB. 3) As a post-processing step, we propose a novel learnable augmentation module, named multi-scale fusion network. It is based on the observation that the performance can be further boosted by fusing results of different resolution levels. This component could improve PSNR value, but at the expense of SSIM. These three strategies are complementary to each other, thus can be adopted independently or simultaneously.

To maximize the final performance, we propose an efficient training strategy to sequentially use the above components. First, we train a standard QVI as our baseline. Then the RCSN is added and finetuned to exploit the contextual information in high-dimensional feature space. After that, we adopt the RQFP formula to rectify the interpolated optical flow.[1] At last, we employ multi-scale fusion network to further push the PSNR results.

Our full model, named Enhanced Quadratic Video Interpolation (EQVI), won the first place in AIM2020 VTSR challenge, and is superior to the second and third place by 0.08 dB and 0.5 dB, respectively. The proposed EQVI improves the original QVI by 0.38 dB, and outperforms all recent state-of-the-art methods on REDS_VTSR dataset (see Table 2). We have also conducted extensive experiments to evaluate each proposed component.

2 Related Work

Video frame interpolation is a long-standing topic and has been extensively studied in the computer vision community. Recently, a variety of learning-based algorithms have been developed for video frame interpolation.

In 2016, Niklaus *et al.* [16] proposed a spatially-adaptive convolution to tackle video frame interpolation. They postulated that the per-pixel interpolation can be modeled with a convolutional operation over the corresponding patches from the two input frames. Subsequently, Sepconv [17] was introduced to separately estimate the convolutional kernels in different orientations. Afterwards, Liu *et al.* [11] constructed an encoder-decoder to predict the 3D voxel flow and linearly synthesize the desired intermediate frame. Combined with neural network, PhaseNet [12] was designed to robustly handle challenging scenarios with larger motions. To handle multi-frame video interpolation, Jiang *et al.* [9] proposed Super-SloMo, which jointly solves interpolation and occlusion estimation. For better blending the two warped frames, Niklaus *et al.* [15] presented a context-aware synthesis approach that warps not only the input frames but also the pixel-wise contextual information. Based on the observation that closer objects should be preferably synthesized, the depth information [2] was exploited to further facilitate interpolation. The above works generally adopt the linear motion assumption, which is not a good approximation for real-world motions. To improve the accuracy of motion estimation, Xu *et al.* proposed a quadratic video interpolation method to exploit both the velocity and acceleration information. In this paper, we improve

[1] RQFP has no trainable parameters and only rectifies the formula of intermediate flow estimation. However, it requires matrix multiplication and costs more training time, so we adopt it after RCSN is equipped to speed up the entire learning process.

QVI in formulation, network architecture and training strategies, leading to a new state-of-the-art method EQVI.

Due to the limitation of using explicit optical flow, Choi *et al.* proposed CAIN [5], which employs channel attention mechanism and PixelShuffle [19] to directly learn the interpolated results. Recently, deformable convolutions [6] were introduced to this task. Instead of learning pixel-wise optical flow between two frames, FeatureFlow [7] explored feature flows in-between corresponding deep features. Moreover, Lee *et al.* [10] proposed a new warping module named Adaptive Collaboration of Flows (AdaCoF), which can estimate both kernel weights and offset vectors for each target pixel to synthesize the output frame.

3 Methodology

In the following, we first revisit the quadratic video interpolation (QVI) model [24] in Sect. 3.1. Then, in Sect. 3.2, 3.3 and 3.4, we describe the proposed Enhanced Quadratic Video Interpolation (EQVI) method, which includes three independent components to improve the original QVI model. Afterwards, the loss functions are introduced in Sect. 3.5. Finally, we depict the training strategies and protocols in Sect. 3.6.

3.1 Revisiting Quadratic Video Interpolation

Linear Video Interpolation. Given two adjacent frames I_0 and I_1, the aim of video interpolation is to generate an intermediate frame $\hat{I}_t$ at the temporal location t in between the two input frames. Existing algorithms usually assume that the motion between I_0 and I_1 is uniform and linear. That is, the displacement (optical flow) of the pixel from frame 0 to t can be written as a linear function:

$$f_{0\to t} = t f_{0\to 1}. \tag{1}$$

However, the objects in real scenarios do not necessarily move linearly.

Quadratic Video Interpolation. Different from most previous methods which assume the motion between two input frames is linear with uniform velocity, QVI [24] considers higher-dimentional information, i.e. the acceleration, to describe the in-between movement (refer to Fig. 3). Assume the acceleration is a constant, then the quadratic motion can be formulated as a uniform acceleration model:

$$f_{0\to t} = \frac{1}{2}at^2 + v_0 t, \tag{2}$$

where $f_{0\to t}$ denotes the displacement of the pixel from frame 0 to frame t, v_0 represents the velocity at frame 0, and a is the acceleration of the quadratic motion model.

Specifically, QVI takes four consecutive frames I_{-1}, I_0, I_1 and I_2 as inputs and predicts intermediate frame I_t between I_0 and I_1 for arbitrary $t \in (0, 1)$.

The prediction process can be summarized into four steps. (The overall solution pipeline of QVI can be depicted as Fig. 2, except that the quadratic flow prediction module takes two flow maps as input). First, four optical flows $f_{0\to-1}$, $f_{0\to1}$, $f_{1\to0}$ and $f_{1\to2}$ are estimated using PWC-Net [23]. Second, intermediate flows $f_{0\to t}$ and $f_{1\to t}$ are estimated by a quadratic flow prediction layer. Formally, if we put $t=-1$ and $t=1$ into Eq. (2), the follow equation is attained:

$$a^{qvi} = f_{0\to1} + f_{0\to-1}, \quad v_0^{qvi} = \frac{1}{2}(f_{0\to1} - f_{0\to-1}) \tag{3}$$

Third, the intermediate flows are transferred to backward flows $f_{t\to0}$ and $f_{t\to1}$ by a flow reversal layer, and are refined by an adaptive flow filtering with a UNet-like [18] refine network. Finally, the intermediate $\hat{I}_t$ is synthesized by fusing pixels in I_0 and I_1 warped by backward flows. More details can be found in the original QVI paper [24].

With the consideration of higher-dimensional information, the quadratic model can better fit the trajectories of real motions, thus predicting more accurate intermediate frames with state-of-the-art performance in benchmarks. In this work, we use QVI as a base model and enhance it by three proposed components.

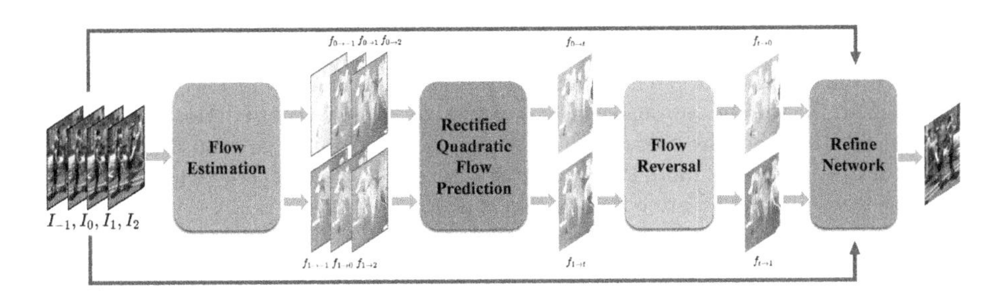

Fig. 2. Quadratic video interpolation model with rectified quadratic flow prediction (RQFP). In original QVI [24], the quadratic flow prediction is conducted with two flow maps ($f_{0\to-1}$ and $f_{0\to1}$), while the proposed EQVI employs three flow maps ($f_{0\to-1}$, $f_{0\to1}$ and $f_{0\to2}$) to predict the interpolated flow (forward direction).

3.2 Rectified Quadratic Flow Prediction

Based on the quadratic modelling [24], we utilize a least squares method to further improve the accuracy of the quadratic flow prediction. In Eq. (3), the acceleration is estimated by three frames (or two flow maps), which could lead to the situation where the accelerations calculated in the forward and backward directions are inconsistent (see Fig. 3(c)). Hence, we take a step further and use all four frames (or three flow maps) to estimate the quadratic model. If the

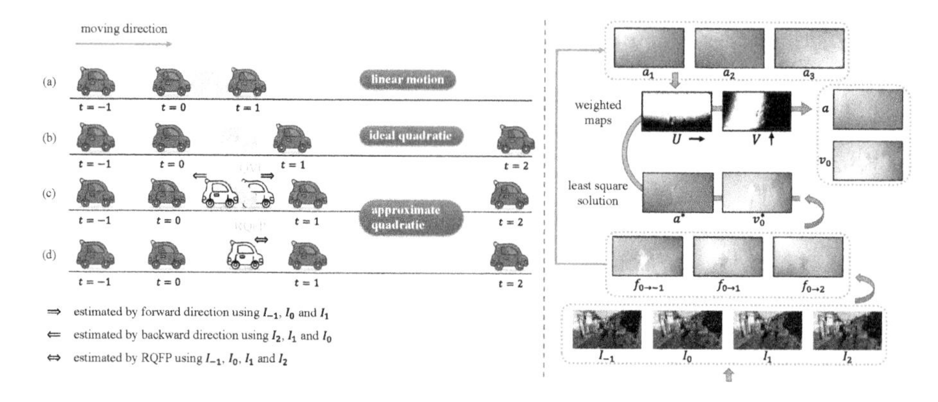

Fig. 3. **Left**: Illustration for linear, quadratic and rectified quadratic flow prediction. (a) A linear motion model where an object moves at a constant velocity. (b) An ideal quadratic motion model. The motion of four input adjacent frames perfectly satisfies a uniform acceleration model. (c) When the input four consecutive frames approximately satisfy the quadratic model, the estimation of motion information in forward and backward directions would differ, due to the existence of deviations. The translucent car and the sketched car represent the ground truth intermediate frame and the estimated output frame, respectively. (d) The proposed RQFP adopts four frames to estimate the motion by least squares method (see Eq. (6)). It can obtain the solution with minimum mean square error. **Right**: An example to illustrate the working mechanism of RQFP. Given four adjacent frames, firstly, the optical flows are estimated by the flow estimation network. Then, by Eq. (6), the least squares solutions of a^* and v_0^* are obtained. Meanwhile, through Eq. (7), the weighted maps α in U (Horizontal) and V (Vertical) directions are calculated. A brighter pixel in weighted map indicates the motion of this location is more likely to satisfy the quadratic model. Finally, the a and v_0 of the quadratic model are obtained by Eq. (8).

motion of the input four frames basically conforms to the uniform acceleration model, the following equations should approximately hold[2]:

$$f_{0\to-1} = -v_0 + \frac{1}{2}a, \quad f_{0\to1} = v_0 + \frac{1}{2}a, \quad f_{0\to2} = 2v_0 + 2a. \tag{4}$$

The matrix form of the above equations is:

$$\begin{bmatrix} -1 & 0.5 \\ 1 & 0.5 \\ 2 & 2 \end{bmatrix} \begin{bmatrix} v_0 \\ a \end{bmatrix} = \begin{bmatrix} f_{0\to-1} \\ f_{0\to1} \\ f_{0\to2} \end{bmatrix}. \tag{5}$$

We can obtain the least squares solution for above overdetermined equation:

$$x^* = [A^T A]^{-1} A^T b, \tag{6}$$

[2] Put $t = -1$, $t = 1$ and $t = 2$ into Eq. (2), respectively.

where $x^* = [v_0^* \quad a^*]^T$, $A = \begin{bmatrix} -1 & 0.5 \\ 1 & 0.5 \\ 2 & 2 \end{bmatrix}$, $b = \begin{bmatrix} f_{0\to-1} \\ f_{0\to1} \\ f_{0\to2} \end{bmatrix}$. When the motion of input four frames approximately satisfies the quadratic assumption, the least squares solution can make a better estimation of $f_{0\to t}$ with minimum mean square error, as shown in Fig. 3(d). Practically we adopt several simple rules to discriminate whether the motion satisfies the quadratic assumption. Picking any two equations from Eq. (4), we can compute an acceleration a_i as: $a_1 = f_{0\to-1} + f_{0\to1}$[3], $a_2 = \frac{2}{3}f_{0\to-1} + \frac{1}{3}f_{0\to2}$, and $a_3 = f_{0\to2} + 2f_{0\to2}$. Theoretically, for quadratic motion, a_1, a_2 and a_3 should be in the same direction (the dot product is positive) and approximately equal to each other. If the orientations of those accelerations are not consistent, we directly utilize the original a^{qvi} and v_0^{qvi} in Eq. (3); otherwise, we adopt the following weighting function (adapted from *tanh*) to fuse the rectified flow and the original QVI flow:

$$\alpha(z) = -\frac{1}{2}\left[\frac{e^{\omega(z-\gamma)} - e^{-\omega(z-\gamma)}}{e^{\omega(z-\gamma)} + e^{-\omega(z-\gamma)}}\right] + \frac{1}{2}, \tag{7}$$

where $z = |a_1 - a_2|$, ω is the axis of symmetry and γ is the stretc.hing factor. Empirically, we let $\omega = 5$ and $\gamma = 1$. A smaller z will lead to a larger weight $\alpha(z)$, indicating that there are more motions following the quadratic formula. Then, the final estimated v_0 and a are obtained by:

$$v_0 = \alpha v_0^* + (1-\alpha)v_0^{qvi}, \quad a = \alpha a^* + (1-\alpha)a^{qvi}. \tag{8}$$

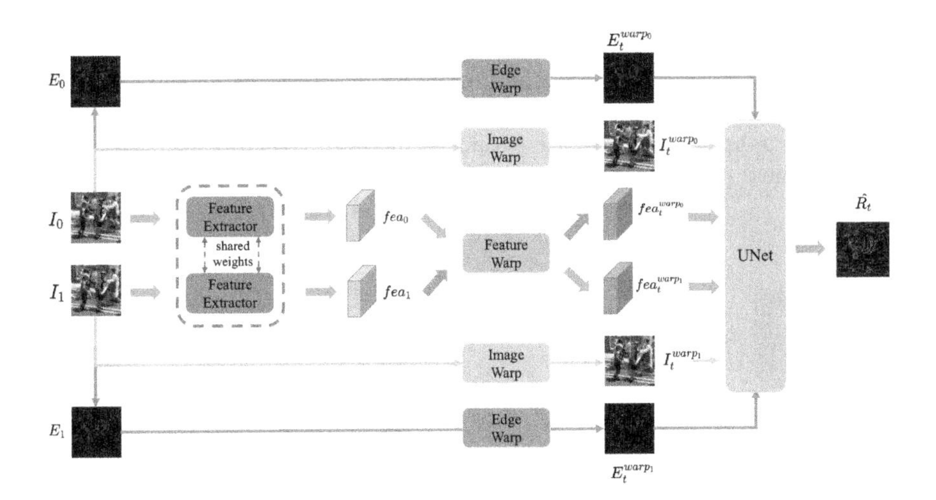

Fig. 4. Residual contextual synthesis network

[3] Derived from the first and second formulas of Eq. (4). Similar derivations for the others.

Figure 2 shows the quadratic video interpolation pipeline with rectified quadratic flow prediction module. An illustrative example of the working mechanism for RQFP is portrayed in the right part of Fig. 3.

3.3 Residual Contextual Synthesis Network

We introduce a residual contextual synthesis network (RCSN) to exploit the contextual information in high-dimensional feature space. The structure of RCSN is inspired by a recent video interpolation work [15]. As shown in Fig. 4, we adopt the *conv*1 layer of the pre-trained ResNet18 [8] to capture the contextual information of input frame I_0 and I_1. Then we apply backward warping on the extracted features with refined flow $f_{t\to 0}$ and $f_{t\to 1}$ to obtain pre-warped features. In addition, to preserve and leverage structural information, the edges of I_0 and I_1 are also extracted and warped. For simplicity, we calculate the gradient of each channel of the input frame as edge information. Afterwards, we feed the warped images, edges and features into a small network to synthesize a residual map $\hat{R}_t$, as depicted in Fig. 4. Finally, the refined output is obtained by $\hat{I}_t^{refined} = \hat{I}_t + \hat{R}_t$.

Instead of only blending pre-warped input frames, we take contextual features and edge information into account, which are complementary with the pixel-level information in image RGB space. These additional features can provide more robust and richer contextual information, which could help the model better deal with complex scenes and motion patterns. Different from [15], we let the context-aware synthesis network predict the residual image rather than the interpolated image. There are three main reasons for residual-learning: 1) The output of the baseline QVI has already achieved good result, so we want to predict a residual image to further refine the output, without violating the previous result. 2) Residual-learning can ease the training process and make the network converge faster. 3) By directly adding the learned residual, it is clearer to validate the performance gain obtained by the proposed module. Experiments have shown the effectiveness of the proposed component in Sect. 4.2.

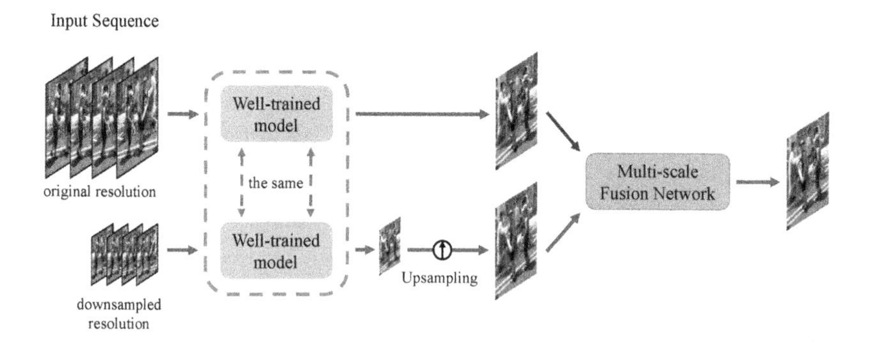

Fig. 5. Multi-scale fusion network

3.4 Multi-scale Fusion Network

We propose a novel multi-scale fusion network to capture various motions at different levels. This fusion network can be regarded as a learnable augmentation process at different resolutions. As shown in Fig. 5, once we have obtained a well-trained interpolation model from previous stages, we feed the model with an input sequence I_{in} and its downsampled counterpart I_{in}^{down}. Then a fusion network F is trained to predict a pixel-wise weighted map M to fuse the results.

$$M = F(Q(I_{in}), Up(Q(I_{in}^{down}))), \tag{9}$$

where Q represents the well-trained interpolation model and $Up(.)$ denotes bilinear upsampling operations. Then, the final output frame is obtained by

$$\hat{I}_t^{final} = M * Q(I_{in}) + (1 - M) * Up(Q(I_{in}^{down})). \tag{10}$$

3.5 Loss Functions

To reduce the difference between the predicted interpolated image $\hat{I}_t$ and the ground truth image I_t^{gt}, L_1 loss and the Laplacian pyramid loss L_{lap} are adopted in our method. The L_1 loss is as follows:

$$L_1 = \left\| \hat{I}_t - I_t^{gt} \right\|_1. \tag{11}$$

It is a commonly-used loss function which can directly measure the distortion/fidelity between the output and the ground truth in image pixel space.

Collaborative with L_1 loss, we additionally adopt a Laplacian pyramid loss which is introduced by [4] for optimizing generative networks.

$$L_{lap} = \sum_{i=1}^{n} 2^{i-1} \left\| L^i(\hat{I}_t) - L^i(I_t^{gt}) \right\|_1, \tag{12}$$

where $L^i(I)$ is the i-th level of the Laplacian pyramid representation of an image I. In our implementation, the number of Laplacian pyramid layer is 5 ($n = 5$). This loss is also adopted in [15]. As a supplement to L_1 loss, L_{lap} loss focuses more on the image edges and textures. Interestingly, experiments show that L_{lap} loss has excellent potential for optimizing PSNR (more details can be found in Sect. 4.5).

3.6 Training and Finetuning Protocols

Since the proposed three components can be adopted independently, we design a four-stage training strategy during the challenge. Such stage-wise training strategy can ease the overall optimization procedure and help verify the performance gain of each component. 1) Train a baseline model QVI. We use a new flow estimation method ScopeFlow [3] instead of the original PWC-Net [23]. The learning rate is initialized as 1×10^{-4} and decayed by a factor of 10 at 100th and 150th epoch. A total of 200 epochs were executed. 2) Based on the baseline model, we add the residual contextual synthesis network (RCSN) and finetune it. The learning rate is initialized as 1×10^{-4} and decayed by a factor of 10 at 15th and 25th epoch. A total of 35 epochs were executed. 3) We adopt rectified quadratic flow prediction formula (RQFP) to improve the model and finetune all the modules except optical flow estimation network. The learning rate is initialized as 1×10^{-5} and decayed by a factor of 10 at 15th and 25th epoch. A total of 30 epochs were executed. 4) We input the well-trained model into the multi-scale fusion network (MS-Fusion), and finetune it. The learning rate is initialized as 1×10^{-4}. A total of 5 epochs were executed. Note that the rectified quadratic flow prediction has no trainable parameters. We do not adopt it in the second stage, as it requires operations on matrix and would cost more training time. The overall training procedure is depicted in Fig. 6.

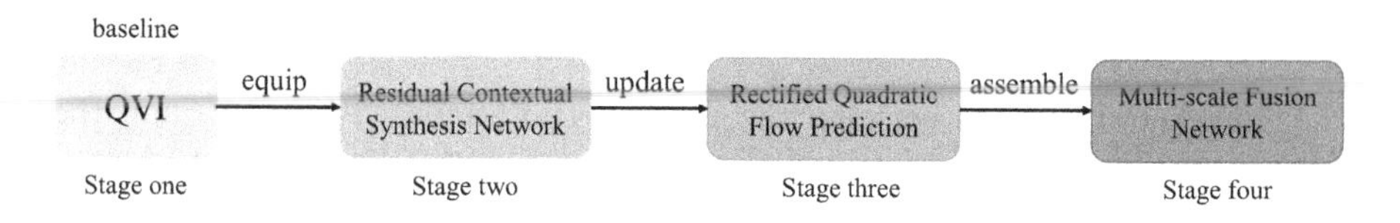

Fig. 6. The overall training procedure of the proposed EQVI method

4 Experiments

4.1 Training Settings and Details

Dataset and Metrics. Previous studies [7,9–12,15,17,24] on video frame interpolation have adopted many video datasets for training and evaluating, such as Middlebury [1], UCF101 [21], Vimeo-90k [25], Adobe 240 fps [22] and YouTube 960-fps [24]. Nevertheless, every algorithm utilizes different datasets, making it difficult to conduct a fair comparison among them. To tackle this problem, REDS_VTSR dataset is proposed for AIM 2019 VTSR Challenge [14] and AIM 2020 VTSR Challenge [20][4]. REDS_VTSR is derived from REDS dataset [13], which includes diverse scenes and motions. The input frame rate of REDS_VTSR is 15 fps and there are two-staged target frame rates of 30 fps (interpolate one frame) and 60 fps (interpolate three frames). REDS_VTSR consists of 240 training clips, 30 validation clips and 30 testing clips in total. We only use the 240 training clips of 60 fps for training. Since the ground truth of the test set is not available, we select five representative clips from the validation set to evaluate the performance, dubbed as REDS_VTSR5[5]. For simplicity, we only evaluate the

[4] https://data.vision.ee.ethz.ch/cvl/aim20/.

[5] The serial numbers are 002, 005, 010, 017 and 025.

performance of interpolating one frame (the ground truth frame rate is 30 fps). PSNR and SSIM are adopted for quantitative evaluation.

Implementation Details. The optical flow is estimated by ScopeFlow [3], a recent state-of-the-art algorithm. We fix the optical flow estimation module in all training stages. The adaptive flow filtering module is kept the same as QVI [24]. During training, a spatial patch size of 512×512 is cropped from the original 1280×720 frames. The mini-batch size is set to 12. We use $L_1 + 10 * L_{lap}$ loss to train the model. The training process is detailed in Sect. 3.6 and Fig. 6. We implement our models with PyTorch framework and train them using 4 NVIDIA GeForce RTX 2080 Ti GPUs. The entire training process lasts about 6 days.

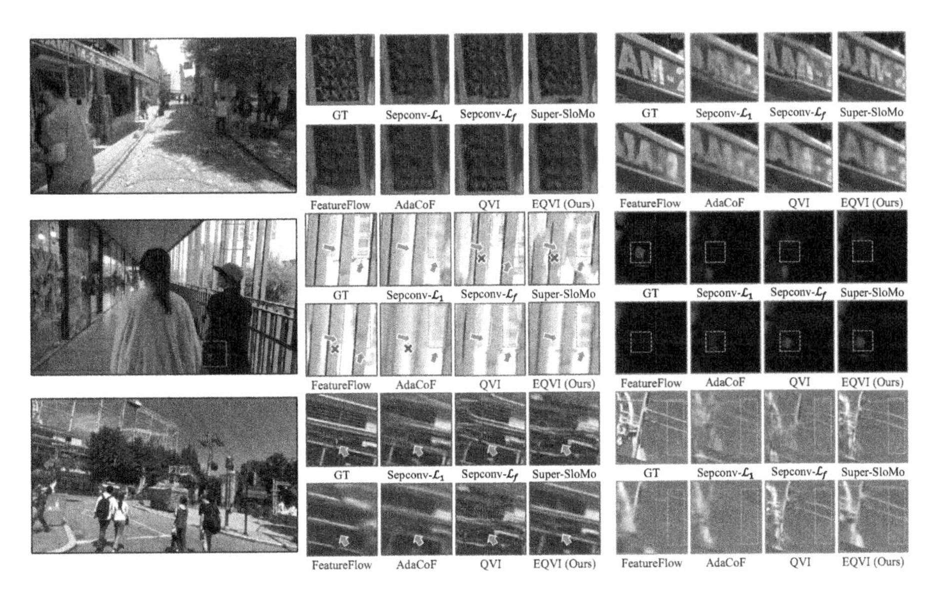

Fig. 7. Qualitative comparison with other state-of-the-art methods. For images with rich textures, EQVI can generate more accurate structures and edges; for challenging images with larger motion cases, the results produced by EQVI contain fewer ghosting and artifacts. Please see the highlighted marks in the figure.

4.2 Effectiveness of the Proposed Components

To facilitate the training process and demonstrate the effectiveness of each component, we introduce a four-stage training strategy. The evaluation results of each stage are listed in Table 1. The baseline model achieves 24.75 dB on the REDS_VTSR5 dataset. After equipping the residual contextual synthesis network, the performance reaches 24.93 dB, achieving an improvement of 0.18 dB, which reveals the great potential of exploiting the high-dimensional contextual information in feature space. Then, we update the model with the rectified quadratic flow prediction as described in Sect. 3.2, and the performance can be further boosted to 24.96 dB. In the fourth stage, the previously trained model is put into the multi-scale fusion network, after which the results can reach 24.97 dB. However, we find that the utilization of the multi-scale fusion network could decrease the performance of SSIM. In summary, the experimental results have shown that the proposed components have greatly improved the performance of the original QVI.

After the competition, we also train a model from scratch with RCSN and RQFP equipped. Interestingly, the results show that training from scratch can yield a better performance with 25.069 dB on REDS_VTSR5 dataset. However, it costs more training time (about 9 days). From the experiments, it can be concluded that the proposed method is very effective, flexible and easy to train.

Table 1. Test results on the REDS_VTSR5 dataset. Based on the standard QVI baseline, the proposed module improves the performance step by step.

	Baseline	RCSN	RQFP	MS-Fusion	PSNR	SSIM
Stage one	✓				24.746	0.717
Stage two	✓	✓			24.927	0.725
Stage three	✓	✓	✓		24.963	0.727
Stage four	✓	✓	✓	✓	24.971	0.726
From scratch	✓	✓	✓	×	25.069	0.729

4.3 Comparison with the State-of-the-Arts

We compare our EQVI model with several state-of-the-art methods on video interpolation: Sepconv [17], Super-SloMo [9], QVI [24], AdaCoF [10] and FeatureFlow [7]. However, most of these methods utilize different datasets for training and testing, making the comparison difficult. To be fair, we retrain QVI and AdaCoF with REDS_VTSR training set. For Sepconv, Super-SloMo and FeatureFlow, we just use their released pretrained models.

Quantitative results are shown in Table 2. Our EQVI outperforms all the state-of-the-art methods by a large margin. Specifically, compared with the standard QVI model which achieves the second-best result, we improve the performance by about 0.38 dB on the REDS_VTSR5 dataset. The visualization comparisons in Fig. 7 and Fig. 1 show that EQVI has its unique merits of dealing with large and complex motion. It can produce more accurate interpolated results with higher PSNR values while preserving the image textures and structures. To be specific, Sepconv, Super-SloMo, FeatureFlow and AdaCoF cannot model and estimate complex frame motions well, leading to inaccurate and blurry interpolated results with severe ghosting and artifacts. Compared with standard QVI, after equipping with three aforementioned components, the predicted images are further enhanced with more accurate textures and fewer ghosting artifacts.

Table 2. Quantitative comparison on REDS_VTSR5. EQVI achieves the best performance.

Method	PSNR	SSIM
Sepconv-$\mathcal{L}_1$ [17]	23.123	0.662
Sepconv-$\mathcal{L}_F$ [17]	22.770	0.641
Super-SloMo [9]	23.004	0.659
FeatureFlow [7]	22.990	0.666
AdaCoF [10]	23.462	0.675
QVI [24]	*24.591*	*0.710*
EQVI	**24.971**	**0.726**

Table 3. Top5 methods in the AIM2020 VTSR Challenge. Our method won the first place.

Method	15 fps → 30 fps		15 fps → 60 fps	
	PSNR	SSIM	PSNR	SSIM
Ours	**24.78**	*0.712*	**25.69**	*0.742*
2nd	*24.69*	**0.714**	*25.61*	**0.746**
3rd	24.40	0.697	25.19	0.727
4th	24.29	0.698	25.05	0.727
5th	23.59	0.672	24.36	0.700

4.4 Results of AIM2020 VTSR Challenge

We participated in the AIM2020 VTSR Challenge [20] with the proposed EQVI model and won the first place. The test results are ranked by PSNR in Table 3. In both 15 fps → 30 fps (interpolate one frame) and 15 fps → 60 fps (interpolate three frames) tracks, our EQVI achieves the highest PSNR values.

4.5 Ablation Study

Loss Combination. In the main experiments, we combine L_1 and L_{lap} loss functions together and achieve appealing performance in terms of PSNR and SSIM. To validate the effectiveness of different loss combinations, we conduct an ablation study using various losses to train the baseline model: 1) Only adopt L_1 loss. 2) Only adopt L_{lap} loss. 3) Combine L_1 and L_{lap} losses together. As shown in Table 4, adopting Laplacian loss L_{lap} can obtain better PSNR values than L_1 loss, but would decrease the performance of SSIM. Moreover, combining L_1 and L_{lap} can achieve higher PSNR, and to some extent, make up for the decrease of SSIM caused by adopting L_{lap} loss.

Effectiveness of RQFP. To directly demonstrate the effectiveness of the proposed rectified quadratic flow prediction formula, comparison experiments are conducted based on the well-trained model in stage two (see Fig. 6): 1) Finetune the model without RQFP. 2) Finetune the model with RQFP. The results are depicted in Table 5. The initial performance of the model in stage two is 24.93 dB. After equipping the proposed RQFP, the performance improves to 24.96 dB while the performance keeps nearly the same without RQFP.

Table 4. Ablation study on the loss combination.

Loss	PSNR	SSIM
L_1	24.646	**0.718**
L_{lap}	24.746	0.716
$L_{lap} + L_1$	**24.775**	0.717

Table 5. By finetuning with RQFP, the result is further boosted.

Model	PSNR	SSIM
Stage-two model	24.927	0.725
Finetune w/o RQFP	24.933	0.725
Finetune w/ RQFP	**24.963**	**0.727**

Exploration on RCSN. In Sect. 3.3, we discussed the proposed residual contextual synthesis network (RCSN), which integrates pre-warped features, images and edges together to exploit the contextual information as shown in Fig. 4. In this session, we further validate the effectiveness of such contextual information. As shown in Table 6, if we only feed pre-extracted features in RCSN, the PSNR and SSIM are 24.916 dB and 0.724, respectively. After adding pre-warped images into RCSN, the performance improves to 24.925 dB and 0.725. When edge information is further employed, the PSNR reaches 24.927 dB.

Table 6. Exploration on RCSN.

Contextual info	PSNR	SSIM
Features	24.916	0.724
Features + images	24.925	0.725
Features + images + edges	**24.927**	**0.725**

Table 7. Influence of optical flow.

Flow model	PSNR	SSIM
PWC-Net [23]	24.591	0.710
ScopeFlow [3]	**24.746**	**0.716**

Influence of Optical Flow Estimation. In flow-based methods, the accuracy of flow estimation is crucial. To validate the effects of different flow estimation algorithms in our method, we train the baseline model with two state-of-the-art algorithms - PWC-Net [23] and ScopeFlow [3]. In the optical flow benchmarks, ScopeFlow is superior to PWC-Net. As expected, as shown in Table 7, the baseline model equipped with ScopeFlow also outperforms that with PWC-Net, which shows the paramount importance of optical flow estimation algorithm in video frame interpolation task.

5 Conclusions

We have presented the winner solution of AIM2020 Video Temporal Super-Resolution Challenge. We propose an enhanced quadratic video interpolation (EQVI) model from the aspects of formulation, network architecture and training strategies. We adopt least squares method to rectify the estimation of motion flow. A residual contextual synthesis network is introduced to employ contextual information in high-dimensional feature space. In addition, we devise a novel learnable augmentation process – multi-scale fusion network to further boost the performance. The proposed EQVI model outperforms all recent state-of-the-art video interpolation methods quantitatively and qualitatively.

Acknowledgement. This work is partially supported by the National Natural Science Foundation of China (61906184), Science and Technology Service Network Initiative of Chinese Academy of Sciences (KFJ-STS-QYZX-092), Shenzhen Basic Research Program (JSGG20180507182100698, CXB201104220032A), the Joint Lab of CAS-HKShenzhen Institute of Artificial Intelligence and Robotics for Society.

References

1. Baker, S., Scharstein, D., Lewis, J., Roth, S., Black, M.J., Szeliski, R.: A database and evaluation methodology for optical flow. Int. J. Comput. Vision **92**(1), 1–31 (2011)
2. Bao, W., Lai, W.S., Ma, C., Zhang, X., Gao, Z., Yang, M.H.: Depth-aware video frame interpolation. In: Proceedings of the IEEE Conference on Computer Vision and Pattern Recognition, pp. 3703–3712 (2019)
3. Bar-Haim, A., Wolf, L.: ScopeFlow: dynamic scene scoping for optical flow. In: Proceedings of the IEEE/CVF Conference on Computer Vision and Pattern Recognition, pp. 7998–8007 (2020)
4. Bojanowski, P., Joulin, A., Lopez-Paz, D., Szlam, A.: Optimizing the latent space of generative networks. arXiv preprint arXiv:1707.05776 (2017)
5. Choi, M., Kim, H., Han, B., Xu, N., Lee, K.M.: Channel attention is all you need for video frame interpolation. In: AAAI, pp. 10663–10671 (2020)
6. Dai, J., et al.: Deformable convolutional networks. In: Proceedings of the IEEE International Conference on Computer Vision, pp. 764–773 (2017)
7. Gui, S., Wang, C., Chen, Q., Tao, D.: FeatureFlow: robust video interpolation via structure-to-texture generation. In: Proceedings of the IEEE/CVF Conference on Computer Vision and Pattern Recognition, pp. 14004–14013 (2020)
8. He, K., Zhang, X., Ren, S., Sun, J.: Deep residual learning for image recognition. In: Proceedings of the IEEE Conference on Computer Vision and Pattern Recognition, pp. 770–778 (2016)
9. Jiang, H., Sun, D., Jampani, V., Yang, M.H., Learned-Miller, E., Kautz, J.: Super SloMo: high quality estimation of multiple intermediate frames for video interpolation. In: Proceedings of the IEEE Conference on Computer Vision and Pattern Recognition, pp. 9000–9008 (2018)
10. Lee, H., Kim, T., Chung, T.y., Pak, D., Ban, Y., Lee, S.: AdaCoF: adaptive collaboration of flows for video frame interpolation. In: Proceedings of the IEEE/CVF Conference on Computer Vision and Pattern Recognition, pp. 5316–5325 (2020)
11. Liu, Z., Yeh, R.A., Tang, X., Liu, Y., Agarwala, A.: Video frame synthesis using deep voxel flow. In: Proceedings of the IEEE International Conference on Computer Vision, pp. 4463–4471 (2017)
12. Meyer, S., Djelouah, A., McWilliams, B., Sorkine-Hornung, A., Gross, M., Schroers, C.: PhaseNet for video frame interpolation. In: Proceedings of the IEEE Conference on Computer Vision and Pattern Recognition, pp. 498–507 (2018)

13. Nah, S., et al.: NTIRE 2019 challenges on video deblurring and super-resolution: Dataset and study. In: The IEEE Conference on Computer Vision and Pattern Recognition (CVPR) Workshops, June 2019
14. Nah, S., Son, S., Timofte, R., Lee, K.M., et al.: AIM 2019 challenge on video temporal super-resolution: methods and results. In: 2019 IEEE/CVF International Conference on Computer Vision (ICCV) Workshops, pp. 3388–3398. IEEE (2019)
15. Niklaus, S., Liu, F.: Context-aware synthesis for video frame interpolation. In: Proceedings of the IEEE Conference on Computer Vision and Pattern Recognition, pp. 1701–1710 (2018)
16. Niklaus, S., Mai, L., Liu, F.: Video frame interpolation via adaptive convolution. In: Proceedings of the IEEE Conference on Computer Vision and Pattern Recognition, pp. 670–679 (2017)
17. Niklaus, S., Mai, L., Liu, F.: Video frame interpolation via adaptive separable convolution. In: Proceedings of the IEEE International Conference on Computer Vision, pp. 261–270 (2017)
18. Ronneberger, O., Fischer, P., Brox, T.: U-Net: convolutional networks for biomedical image segmentation. In: Navab, N., Hornegger, J., Wells, W.M., Frangi, A.F. (eds.) MICCAI 2015. LNCS, vol. 9351, pp. 234–241. Springer, Cham (2015). https://doi.org/10.1007/978-3-319-24574-4_28
19. Shi, W., et al.: Real-time single image and video super-resolution using an efficient sub-pixel convolutional neural network. In: Proceedings of the IEEE Conference on Computer Vision and Pattern Recognition, pp. 1874–1883 (2016)
20. Son, S., Lee, J., Nah, S., Timofte, R., Lee, K.M., et al.: AIM 2020 challenge on video temporal super-resolution. In: European Conference on Computer Vision Workshops (2020)
21. Soomro, K., Zamir, A.R., Shah, M.: UCF101: a dataset of 101 human actions classes from videos in the wild. arXiv preprint arXiv:1212.0402 (2012)
22. Su, S., Delbracio, M., Wang, J., Sapiro, G., Heidrich, W., Wang, O.: Deep video deblurring for hand-held cameras. In: Proceedings of the IEEE Conference on Computer Vision and Pattern Recognition, pp. 1279–1288 (2017)
23. Sun, D., Yang, X., Liu, M.Y., Kautz, J.: PWC-Net: CNNs for optical flow using pyramid, warping, and cost volume. In: Proceedings of the IEEE Conference on Computer Vision and Pattern Recognition, pp. 8934–8943 (2018)
24. Xu, X., Siyao, L., Sun, W., Yin, Q., Yang, M.H.: Quadratic video interpolation. In: Advances in Neural Information Processing Systems, pp. 1647–1656 (2019)
25. Xue, T., Chen, B., Wu, J., Wei, D., Freeman, W.T.: Video enhancement with task-oriented flow. Int. J. Comput. Vision **127**(8), 1106–1125 (2019)

AIM 2020 Challenge on Video Extreme Super-Resolution: Methods and Results

Dario Fuoli[1(✉)], Zhiwu Huang[1], Shuhang Gu[2], Radu Timofte[1], Arnau Raventos[3], Aryan Esfandiari[3], Salah Karout[3], Xuan Xu[4,5], Xin Li[4,5], Xin Xiong[4,5], Jinge Wang[4,5], Pablo Navarrete Michelini[6], Wenhao Zhang[6], Dongyang Zhang[7,8], Hanwei Zhu[7,8], Dan Xia[7,8], Haoyu Chen[9,10], Jinjin Gu[9,10], Zhi Zhang[9,10], Tongtong Zhao[11,12], Shanshan Zhao[11,12], Kazutoshi Akita[13], Norimichi Ukita[13], P. S. Hrishikesh[14], Densen Puthussery[14], and C. V. Jiji[14]

[1] ETH Zürich, Zürich, Switzerland
dario.fuoli@vision.ee.ethz.ch
[2] University of Sydney, Sydney, Australia
[3] Huawei Technologies R&D UK, Ipswich, UK
[4] West Virginia University, Morgantown, USA
[5] Huazhong University of Science and Technology, Wuhan, China
[6] BOE Technology Group Co., Ltd., Beijing, China
[7] Jiangxi University of Finance and Economics, Nanchang, China
[8] National Key Laboratory for Remanufacturing, Army Academy of Armored Forces, Beijing, China
[9] Amazon Web Services, Seattle, USA
[10] The Chinese University of Hong Kong, Shenzhen, China
[11] Dalian Maritime University, Dalian, China
[12] China Everbright Bank Co., Ltd., Beijing, China
[13] Toyota Technological Institute (TTI), Nagoya, Japan
[14] College of Engineering, Trivandrum, Thiruvananthapuram, India
http://www.vision.ee.ethz.ch/aim20/

Abstract. This paper reviews the video extreme super-resolution challenge associated with the AIM 2020 workshop at ECCV 2020. Common scaling factors for learned video super-resolution (VSR) do not go beyond factor 4. Missing information can be restored well in this region, especially in HR videos, where the high-frequency content mostly consists of texture details. The task in this challenge is to upscale videos with an extreme factor of 16, which results in more serious degradations that

Dario Fuoli, Zhiwu Huang, Shuhang Gu and Radu Timofte are the AIM 2020 challenge organizers.
A. Raventos, A. Esfandiari, S. Karout, X. Xu, X. Li, X. Xiong, J. Wang, P. Navarrete Michelini, W. Zhang, D. Zhang, H. Zhu, D. Xia, H. Chen, J. Gu, Z. Zhang, T. Zhao, S. Zhao, K. Akita, N. Ukita, P. S. Hrishikesh, D. Puthussery and C. V. Jiji—Participants in the challenge. Appendix A contains the authors' teams and affiliations.

Electronic supplementary material The online version of this chapter (https://doi.org/10.1007/978-3-030-66823-5_4) contains supplementary material, which is available to authorized users.

A. Bartoli and A. Fusiello (Eds.): ECCV 2020 Workshops, LNCS 12538, pp. 57–81, 2020.
https://doi.org/10.1007/978-3-030-66823-5_4

also affect the structural integrity of the videos. A single pixel in the low-resolution (LR) domain corresponds to 256 pixels in the high-resolution (HR) domain. Due to this massive information loss, it is hard to accurately restore the missing information. Track 1 is set up to gauge the state-of-the-art for such a demanding task, where fidelity to the ground truth is measured by PSNR and SSIM. Perceptually higher quality can be achieved in trade-off for fidelity by generating plausible high-frequency content. Track 2 therefore aims at generating visually pleasing results, which are ranked according to human perception, evaluated by a user study. In contrast to single image super-resolution (SISR), VSR can benefit from additional information in the temporal domain. However, this also imposes an additional requirement, as the generated frames need to be consistent along time.

Keywords: Extreme super-resolution · Video restoration · Video enhancement · Challenge

1 Introduction

Super-resolution (SR) aims at reconstructing a high-resolution (HR) output from a given low-resolution (LR) input. Single image SR (SISR) generally focuses on restoring spatial details as the input consists of only one single image. By comparison, as the input of video SR (VSR) is usually composed of consecutive frames, it is expected to concentrate on the exploitation of the additional temporal correlations, which can help improving restoration quality over SISR methods. Making full use of the temporal associations among multiple frames and keeping the temporal consistency for VSR remain non-trivial problems. Furthermore, when moving towards more extreme settings, like higher scale factors that require to restore a large amount of pixels from severely limited information, the VSR problem will get much more challenging.

Fig. 1. Downscaled crops with the extreme factor of ×16 (top) and corresponding 96 × 96 HR crops (bottom).

Following our first AIM challenge [9], the goal of this challenge is to super resolve the given input videos with an extremely large zooming factor of 16, with searching for the current state-of-the-art and providing a standard benchmark protocol for future research in the field. Fig. 1 presents a few downscaled crops and their corresponding HR crops. In this challenge, track 1 aims at probing the state-of-the-art for the extreme VSR task, where fidelity to the ground truth is measured by peak signal-to-noise ratio (PSNR) and structural similarity index (SSIM). As a trade-off for the fidelity measurement, track 2 is designed for the production of visually pleasing videos, which are ranked by human perception opinions with a user study.

This challenge is one of the AIM 2020 associated challenges on: scene relighting and illumination estimation [7], image extreme inpainting [39], learned image signal processing pipeline [16], rendering realistic bokeh [17], real image super-resolution [50], efficient super-resolution [53], video temporal super-resolution [44] and video extreme super-resolution [11].

2 Related Work

For SR, deep learning based methods [4,6,21,25,32,40,41] have proven their superiority over traditional shallow learning methods. For example, [6] introduces convolution neural networks (CNN) to address the SISR problem. In particular, it proposes a very shallow network to deeply learn LR features, which are subsequently leveraged to generate HR images via non-linear mapping. To reduce the time complexity of the network operations in the HR space, [43] proposes an effective sub-pixel convolution network to extract and map features from the LR space to the HR space using convolutional layers instead of classical interpolations (e.g., bilinear and bicubic). [56] exploits a residual dense network block with direction connections for a more thorough extraction of local features from LR images. A comprehensive overview of SISR methods can be found in [47].

Compared with SISR, the VSR problem is considerably more complex due to the additional challenge of harnessing the temporal correlations among adjacent frames. To address this problem, a number of methods [5,26,34] are suggested to leverage temporal information by concatenating multiple LR frames to generate a single HR estimate. Following this strategy, [2] first warps consecutive frames towards the center frame, and then fuses the frames using a spatio-temporal network. [20] aggregates motion compensated adjacent frames, by computing optical flow and warping, followed by a few convolution layers for the processing on the fused frames. [29] calculates multiple HR estimates in parallel branches. In addition, it exploits an additional temporal modulation branch to balance the respective HR estimates for final aggregation. By contrast, [19] relies on implicit motion estimation. Dynamic upsampling filters and residuals are computed from adjacent LR frames with a single neural network. Finally, the dynamic upsampling filters are employed to process the center frame, which is then fused with the residuals. Similarly, [31] proposes a dynamic local filter network to perform implicit motion estimation and compensation. Besides, it suggests a global refinement neural network based on residual block and autoencoder structures to exploit non-local correlations and enhance the spatial consistency of the super-resolved frames. To address large motions, [48] devises an alignment module to align frames with deformable convolutions in a coarse-to-fine manner. Besides, it suggests a fusion module, where attention is applied both temporally and spatially, so as to emphasize important features for subsequent restoration.

In addition to the aggregation strategy, some other works suggest to make use of recurrent neural networks (RNN) for better VSR. Due to the better capacity of learning temporal information on input frames, they provide a potentially more powerful alternative to address the SR problem. For instance, [45] suggests an autoencoder style network as well as an intermediate convolutional long short-term memory (LSTM) layer. The whole network is capable of processing the preliminary HR estimate from a subpixel motion compensation layer, for better HR estimate. [15] proposes a bidirectional recurrent network, which exploits 2D and 3D convolutions with recurrent connections and combines a forward and a backward pass to produce the HR frames. To make use of temporal information, [42] designs a neural network that warps the previous HR output towards the current time step, by observing the optical flow in LR space. The warped output is concatenated with the current LR input frame and a SR network generates the HR estimate. [8] exploits a recurrent latent space propagation (RLSP) algorithm for more efficient VSR. Particularly, RLSP introduces high-dimensional latent states to propagate temporal information along frames in an implicit manner so that the efficiency can be highly improved.

Fig. 2. LR frame with extreme downscaling factor ×16 (left) and corresponding HR frame (right).

3 AIM 2020 Video Extreme SR Challenge Setup

3.1 Data

We use the Vid3oC [22] dataset for this challenge, which has been part of previous challenges [9,10]. The dataset is a collection of videos taken with three different cameras on a rig. This results in roughly aligned videos, which can be used for weak supervision. In this challenge, only the high-quality DSLR camera (Canon 5D Mark IV) is used to serve as HR ground truth. The corresponding LR source data is obtained by downscaling the ground truth by factor 16, using MATLAB's imresize function with standard settings, see Fig. 2. In order to retain proper pixel-alignment, the ground truth 1080 × 1920 FullHD frames are cropped to 1072 × 1920 before downscaling, to be dividable by 16. We provide 50 HR sequences to be used for training. To save bandwidth, these videos are provided as MP4 files together with scripts to extract and generate the LR source frames. Additionally, the dataset contains 16 paired sequences for validation and 16 paired sequences for testing, each consists of 120 frames in PNG format.

3.2 Challenge Phases

The challenge is hosted on CodaLab and is split up in a validation and a test phase. During the validation phase, only the validation source frames are provided and participants were asked to submit their super-resolved frames to the CodaLab servers to get feedback. Due to storage constraints on CodaLab, only a subset of frames could be submitted to the servers (every 20th frame in the sequence). In the following test phase, the final solutions had to be submitted to enter the challenge ranking. There was no feedback provided at this stage, in order to prevent overfitting to the test set. Additionally, the full set of frames had to be made accessible to the challenge organizers for the final rankings. After the submission deadline, the HR validation ground truth was released on CodaLab, for public use of our dataset.

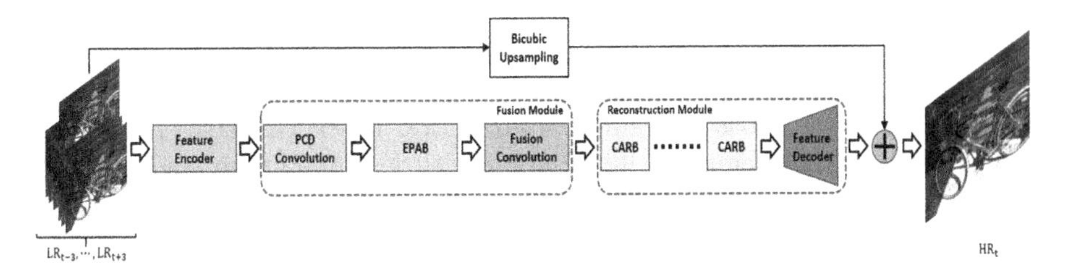

Fig. 3. Efficient Video Enhancement and Super-Resolution Net (EVESRNet) proposed by KirinUK.

3.3 Track 1 - Fidelity

This track aims at high fidelity restoration. For each team, the restored frames are compared to the ground truth in terms of PSNR and SSIM and can be objectively quantified by these pixel-level metrics. The focus is on restoring the data faithfully to the underlying ground truth. Commonly, methods for this task are trained with a pixel-level L1-loss or L2-loss. The final ranking among teams is determined by PSNR/SSIM exclusively, without visual assessment of the produced frames.

3.4 Track 2 - Perceptual

Super-resolution methods optimized for PSNR tend to oversmooth and often fail to restore the highest frequencies. Also, PSNR does not correlate well with human perception of quality. Therefore, the focus in the field has shifted towards generation of perceptually more pleasing results in trade-off for fidelity to the ground truth. Since the extreme scaling factor 16 and its associated large information loss prohibits high fidelity results, the only possibility to achieve realistically looking HR videos in this setting, is by hallucinating plausible high frequencies. Track 2 is aimed at upscaling the videos for highest perceptual quality. Quantitative assessment of perceptual quality is difficult and remains largely an open problem. We therefore resort to a user study in track 2, which is still the most reliable benchmark for perceptual quality evaluation.

4 Challenge Methods and Teams

4.1 KirinUK

Recent video super-resolution approaches [3] propose splitting the spatio-temporal attention operation in several dimensions. Their aim is to reduce the computational cost of a traditional 3D non-local block. Nevertheless, these methods still need to store the HWxHW attention matrices, which is challenging, especially when dealing with GPUs with limited amount of memory or when upscaling HR videos. To tackle this, the KirinUK team proposes to extend the VESRNet [3] architecture by replacing the Separate Non Local (SNL) module with an Efficient Point-Wise Temporal Attention Block (EPAB). This block aggregates the spatio-temporal information with less operations and memory

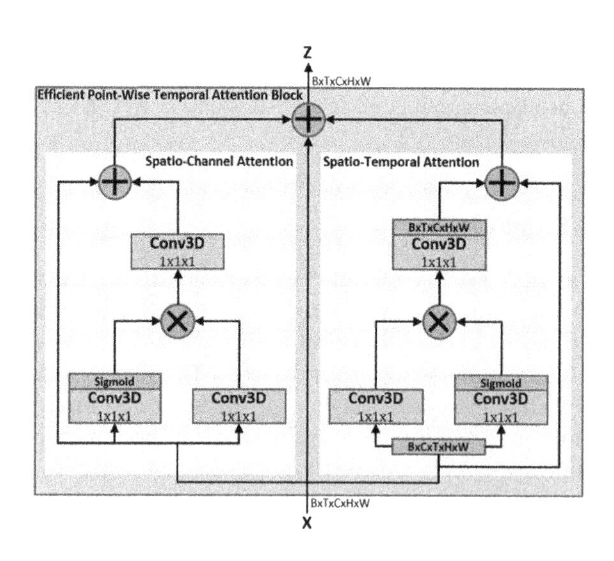

Fig. 4. Efficient Point-Wise Temporal Attention Block proposed by KirinUK.

consumption, while still keeping the same performance. The team names this new architecture Efficient Video Enhancement and Super-Resolution Net (EVESR-Net) and an overview of it can be seen in Fig. 3. It is mainly composed of Pyramid, Cascading and Deformable Convolutions (PCDs) [48], the EPAB, and Channel Attention Residual Blocks (CARBs) [3,55].

The EPAB module is illustrated in Fig. 4 and can be divided into two sub-blocks, the Spatio-Channel Attention (SCA) and the Spatio-Temporal Attention (STA). Both of them share the same structure, the only difference is the permutation operation at the beginning and at the end of the STA sub-block.

To perform Extreme Video SR, the team employs two 4x stages in cascade mode. Each stage was trained independently with an EVESRNet architecture. Moreover, their training does not start from scratch, they first pretrain a model with the REDS dataset [35] to initialize the networks. This helps preventing overfitting in the first stage where the amount of spatial data is limited. The REDS dataset was only utilized for pretraining. As a reference, the initialization model achieves 31.19 PSNR in the internal REDS validation set defined in [48], which corresponds to a +0.1 dB improvement with respect to EDVR [48].

Each track uses the same pipeline, the only differences are: 1) For the fidelity track, the two stages were trained using L2 loss. 2) For the perceptual track, the second stage was trained using the following combined loss:

$$L = \lambda_1 * L_{L1} + \lambda_2 * L_{VGG} + \lambda_3 * L_{RaGAN} \tag{1}$$

where λ_1 is 1e−3, λ_2 is 1 and λ_3 is 5e−3. They used a patch discriminator as in [18]. The rest of the hyperparameters are the same as in [49].

4.2 Team-WVU

Recently, deformable convolution [57] has been received increasingly more attention to solve low-level vision tasks such as video super-resolution. EDVR [48] and TDAN [46] have already successfully implemented deformable convolution to temporally align reference frame and its neighboring frames which can let networks better utilize both spatial and temporal information to enhance the final results.

Inspired by state-of-the-art video SR method EDVR [48], the team develops the novel Multi-Frame based Deformable Kernel Convolution Network [51] to temporally align the non-reference and reference frames with deformable kernel [12] convolution alignment module and enhance the edge and texture features via deformable kernel spatial attention module.

The overall diagram of proposed network is shown in Fig. 5. It mainly includes four parts, feature extractor, DKC_Align module (deformable kernel convolution alignment module), reconstruction module and upscale module. Different from PCD alignment module from EDVR, the team implemented stacked deformable kernel convolution layers instead of traditional convolution layer to extract offset. Deformable kernel can better adapt effective receptive fields than normal convolution [12] which can better enhance the offset extraction compared with the

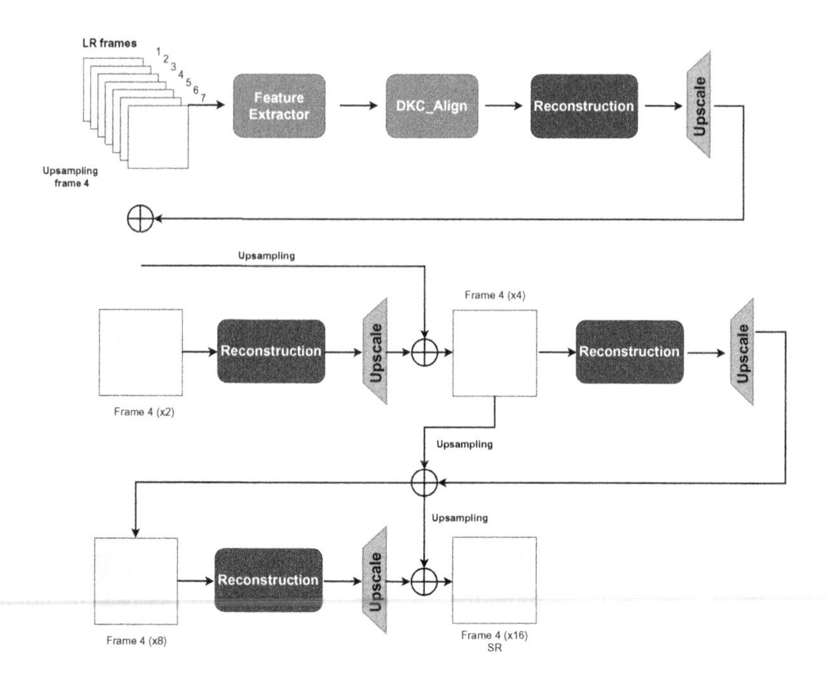

Fig. 5. Network proposed by Team-WVU.

normal convolution. On the step of reconstruction, to calibrate reconstructed feature maps before feeding into each upscaling module, the team proposes a Deformable Kernel Spatial Attention (DKSA) module (integrated to reconstruction module) to enhance the textures that can help the proposed network to reconstruct SR frames sharper and clearer. Because this challenge aims to super-resolve extreme LR videos with the scale factor of 16, to avoid generating undesired blurring and artifacts, the LR frames are super-resolved with a scale factor of 2 each time (see Fig. 5). Finally, the LR frames are super-resolved four times in total to upscale the LR frames with a magnification factor ×16. Charbonnier Loss [24,48] is used as the loss function for both track 1 and track 2 training, the loss can be expressed as follows:

$$L = \sqrt{||\hat{X}_r - X_r||^2 + \xi^2} \tag{2}$$

where $\xi = 1 \times 10^{-3}$, $\hat{X}_r$ is super-resolved frame and X_r is target frame (ground-truth).

4.3 BOE-IOT-AIBD

The team proposes 3D-MGBP, a fully 3D–convolutional architecture designed to scale efficiently for the difficult task of extreme video SR. 3D–MGBP is based on the Multi–Grid Back–Projection network introduced and studied in [36–38].

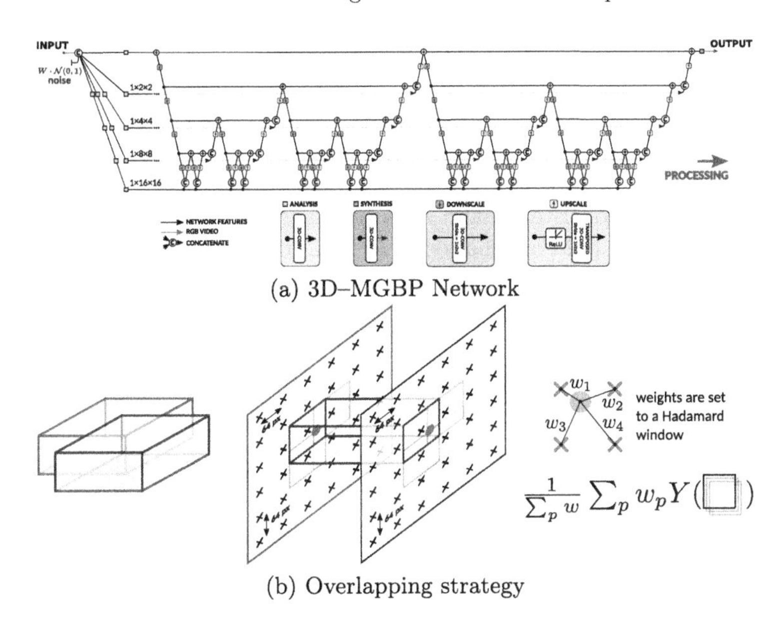

(a) 3D–MGBP Network

(b) Overlapping strategy

Fig. 6. 3D – MultiGrid BackProjection network (3D–MGBP) proposed by BOE-IOT-AIBD.

In particular, they extend the MGBPv2 network [36] that was designed to scale efficiently for the task of extreme image SR and was successfully used in the 2019–AIM Extreme Image SR competition [33] to win the Perceptual track of that challenge. For this challenge they redesigned the MGBPv2 network to use 3D–convolutions strided in space. The network works as a video enhancer that, ignoring memory constrains, can take a whole video stream and outputs a whole video stream with the same resolution and framerate. They input a 16× Bicubic upscaled video and the network enhances the quality of the video stream. The receptive field of the network extends in space as well as time by using 3D–convolutional kernels of size $3 \times 3 \times 3$. The overall architecture uses only 3D–convolutions and ReLU units. This is in contrast to general trends in video processing networks that often include attention, deformable convolutions, warping or other non–linear modules.

Figure 6 displays the diagram of the 3D–MGBP network used in the competition. In inference it is impossible for 3D–MGBP to process the whole video stream and so they extend the idea of overlapped patches used in MGBPv2 by using overlapped spatio–temporal patches (overlapping in space and time). More precisely, to upscale arbitrarily long video sequences they propose a patch based approach in which they average the output of overlapping video patches produced by the Bicubic upscaled input. First, they divide input streams into overlapping patches (of same size as training patches) as shown in Fig. 6; second, they multiply each output by weights set to a Hadamard window; and third, they average the results.

They trained the 3D–MGBP network starting from random parameters (no pre–trained models were used). For the Fidelity track they trained the model using L2 loss on the output spatio–temporal patch. For the Perceptual track they submitted the output of two different configurations of the same architecture. The first submission, labeled *Smooth*, was trained with L2 loss as they noticed better time–consistency and smooth edges. In their second submission, labeled *Texture*, they followed the loss and training strategy of G–MGBP [37], adding a noise input to activate and deactivate the generation of artificial details. The noise input consists of one channel of Gaussian noise concatenated to the Bicubic upscaled input. In this solution, although more noisy and farther from ground truth due to the perception–distortion trade–off [1], they noticed better perception of textures.

4.4 ZZX

In order to restore the high-frequency information of the video, the team designed the multi-scale aggregated upsampling based on high frequency attention (MAHA) network. The framework is illustrated in Fig. 7. The team inputs seven LR frames into the feature extraction module, and inspired by the EDVR [48], the Pyramid, Cascading and Deformable (PCD) alignment module was applied to address global motion. The non-local module was used to select valid inter-frame information. Then, the extracted reference frame features are concatenated with the alignment features that were utilized to perform spatial-temporal fusion in a progressive strategy, which helps to aggregate spatial-temporal information. Next, the team also proposed an attention-guided multi-level residual feature reconstruction module to fully improve feature representation. Finally, to generate a sharp structure HR video, the team computed the gradient map of the LR image to guide the spatial attention module.

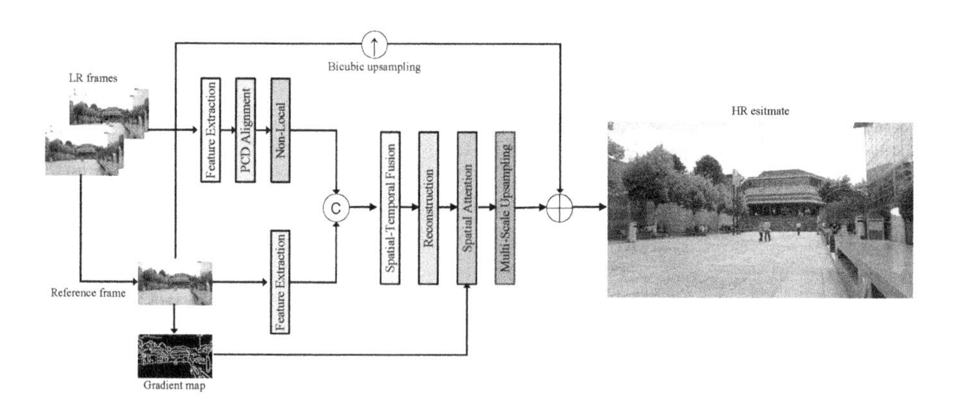

Fig. 7. Network proposed by ZZX.

The team divided the network training into two stages for both track 1 and track 2. For stage 1, the $L1$ loss was used. For stage 2, the team fine-tuned the results of stage 1 and using $0.15 * L_{L1} + 0.85 * L_{SSIM}$ to train the network. However, the team applied different testing strategies for track 1 & 2. For Track 1, they employed model fusion testing and test enhancement strategies to improve the PSNR value. For the track 2, the best validation performance was used to directly generate the test results.

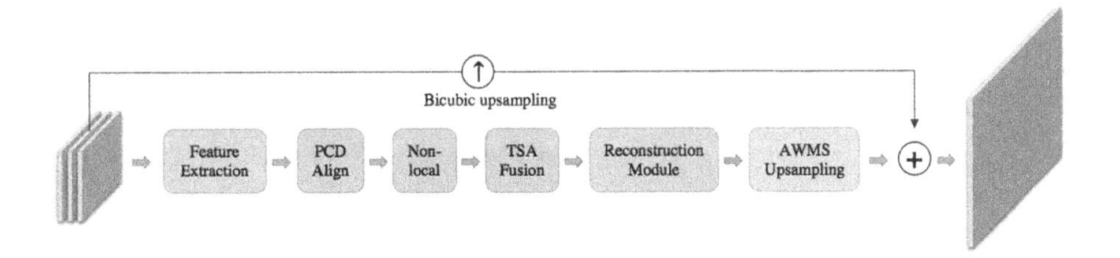

Fig. 8. Network proposed by sr_xxx.

4.5 sr_xxx

The team employs the high-level architecture design of EDVR [1], with improvements to accommodate large upscaling factors with up to 16. The used network is illustrated in Fig. 8. To explain the framework, they first start with EDVR as baseline. EDVR is a unified framework which can achieve good alignment and fusion quality in video restoration tasks. It proposed a Pyramid, Cascading and Deformable (PCD) alignment module, which for the first time uses deformable convolutions to align temporal frames. Besides, EDVR includes a Temporal and Spatial Attention (TSA) fusion module to emphasize important features.

Their proposed network takes 5 LR frames as input and generates one HR output image frame. They first conduct feature extraction, followed by PCD alignment module, Non-local module and TSA module to align and fuse multiple frames. Right after the reconstruction module, they use adaptive weighted multi-scale (AWMS) module as our upsampling layer. In the last module, they add the learned residual to a direct bicubic upsampled image to obtain the final HR outputs.

They trained two different reconstruction models and ensemble their results to obtain more stable texture reconstructions. To incorporate finer detail ensembling, they combine residual feature aggregation blocks [30] and residual channel attention blocks [55].

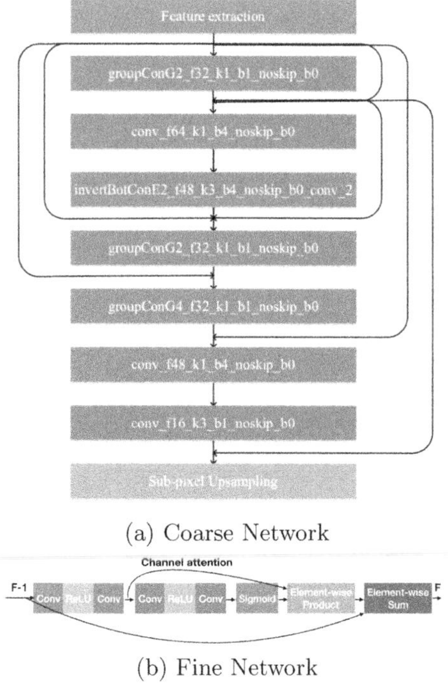

(a) Coarse Network

(b) Fine Network

Fig. 9. Network proposed by lyl.

4.6 lyl

As shown in Fig. 9, the team proposes a coarse to fine network for progressive super-resolution reconstruction. By using the suggested FineNet:lightweight upsampling module (LUM), they achieve competitive results with a modest number of parameters. Two requirements are contained in a coarse to fine network (CFN): (1) progressiveness and (2) merge the output of the LUM to correct the input in each level. Such a progressive cause-and-effect process helps to achieve the principle for image SR: high-level information can guide an LR image to recover a better SR image. In the proposed network, there are three indispensable parts to enforce the suggested CFN: (1) tying the loss at each level (2) using LUM structure and (3) providing a lower level extracted feature input to ensure the availability of low-level information.

They propose to construct their network based on the Laplacian pyramid framework, as shown in Fig. 9. Their model takes an LR image as input and progressively predicts residual images at $S_1, S_2 ... S_n$ levels where S is the scale factor. For example, the network consists of 4 sub-networks for super-reconstructing an LR image at a scale factor of 16, if the scale factor is 3, $S = S_1 \times S_2$, $S_1 = 1.5, S_2 = 2$. Their model has three branches: (1) feature extraction and (2) image reconstruction (3) loss function.

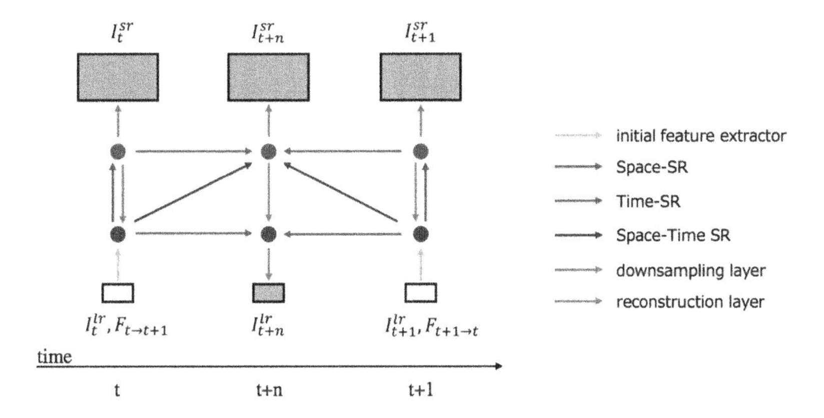

Fig. 10. Network proposed by TTI.

4.7 TTI

The team's base network for ×16 video SR is STARnet [13] shown in Fig. 10. With the idea that space and time are related, STARnet jointly optimizes three tasks (i.e., space SR, time SR, and space-time SR). In the experiments, STARnet was initially trained using three losses, i.e. space, time, and space-time losses, which evaluate the errors of images reconstructed through space SR paths (red arrows in the figure), time SR paths (blue arrows), and space-time SR paths (purple arrows), respectively. The network is then fine-tuned using only the space loss for optimizing the model, which is specialized for space SR. While space SR in STARnet is basically based on RBPN [14], this fine-tuning strategy allows them to be superior to RBPN trained only with the space loss. While the original STARnet employs pyflow [28] for optical flow computation, pyflow almost cannot estimate optical flows in LR frames in this challenge (i.e., 120 × 67 pixels). The optical flows are too small between subsequent frames. Based on an extensive survey, we chose sift-flow [27] that shows better performance on the LR images used in this challenge.

4.8 CET_CVLab

The architecture used by the team is inspired from the wide activation based network in [52] and channel attention network in [55]. As shown in Fig. 11, the network mainly consists of 3 blocks. A feature extraction block, a series of wide activation residual blocks and a set of progressive upsampling blocks (×2). Charbonnier loss is used for training the network as it better captures the edge information than with mean squared errorloss (MSE).

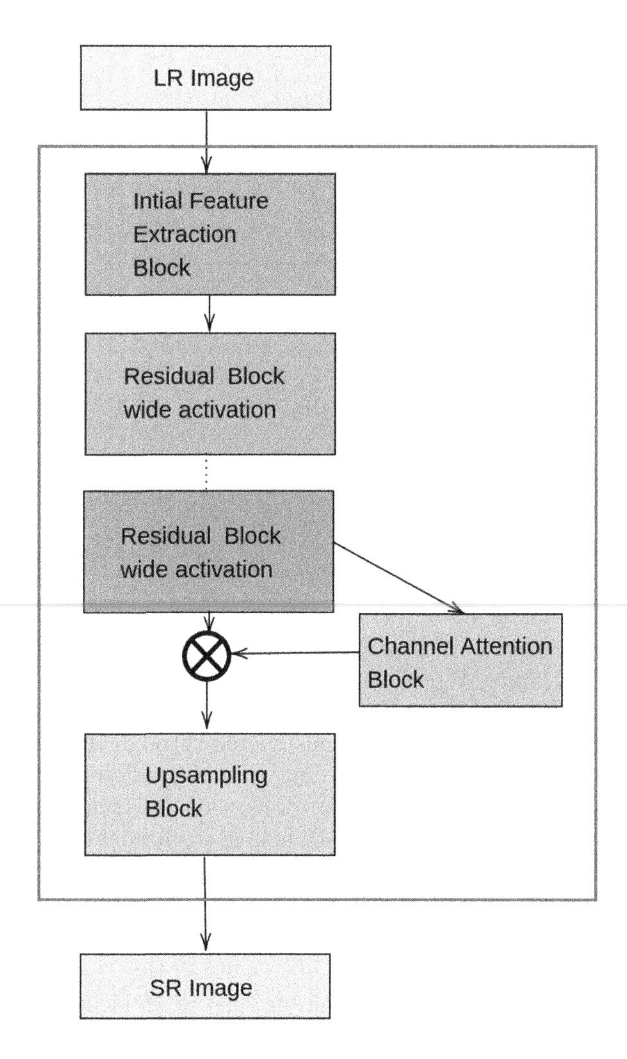

Fig. 11. Network proposed by CET-CVLab.

5 Challenge Results

5.1 Track 1 - Fidelity

This track aims at restoring the missing high frequencies that were lost during downsampling with the highest fidelity to the underlying ground truth. Challenge track 1 has 65 registered participants, from which 12 submitted solutions to the validation server and 8 teams entered the final ranking. The ranking, along with details about the training and testing are summarized in Table 1. Most provided solutions make use of large networks, as super-resolution with factor 16 is highly challenging and requires high-complexity networks in order to restore the details from learned priors. The top teams mostly employ window based

Table 1. Quantitative results for track 1. Train Time: days per model, Test Time: seconds per frame.

Method			↑PSNR	↑SSIM	Train req	Train time	Test req	Test time	Params	Extra data
Participants	1	KirinUK	**22.83**	**0.6450**	4 × V100	10d	1 × 2080Ti	6.1 s	45.29M	Yes
	2	Team-WVU	22.48	0.6378	4 × TitanXp	4d	1 × TitanXp	4.90 s	29.51M	No
		BOE-IOT-AIBD	22.48	0.6304	1 × V100	>30d	1 × 1080	4.83 s	53M	No
	4	sr_xxx	22.43	0.6353	8 × V100	2d	1×V100	4s	n/a	No
	5	ZZX	22.28	0.6321	6 × 1080Ti	4d	1 × 1080Ti	4 s	31,14M	No
	6	lyl	22.08	0.6256	1 × V100	2d	n/a	13 s	n/a	No
	7	TTI	21.91	0.6165	V100	n/a	n/a	0.249 s	n/a	No
	8	CET_CVLab	21.77	0.6112	1 × P100	6d	1 × P100	0.04 s	n/a	Yes
Bicubic (baseline)			20.69	0.5770						

approaches, attention modules and 3D-convolutions to additionally aggregate the temporal information. Team lyl and CET_CVLab do not process temporal information in their networks and instead rely only on a single frame for the upscaling process. They can not compete with the top teams, which shows the importance of temporal information for high-quality restoration. The winner in track 1 is team KirinUK with a PSNR score of 22.83 dB, followed by team Team-WVU and BOE-IOT-AIBD, which share the second place due to their identical PSNR scores.

Metrics. Since this track is about high fidelity restoration, we rank the teams according to PSNR, which is a pixel-level metric. Additionally, we compute SSIM scores which is a metric based on patch statistics and is considered to correlate better with human perception of image quality. PSNR does not explicitly enforce to retain smooth temporal dynamics. It is therefore possible, that a method can generate high image quality on frame level, but introduces temporal artifacts like flickering. Most window based and 3D-convolution approaches however manage to produce frames with only minimal flickering artifacts, as they have access to adjacent frames. On the other hand, the flickering is very prominent for the single frame enhancers in this challenge.

Visual Results. In addition to the metrics, we also provide visual examples in Fig. 12 for all competing methods in the challenge. We also show the Bicubic baseline (MATLAB's *imresize*) together with the ground truth frames for reference. All methods manage to clearly outperform the Bicubic baseline, which is also reflected in the PSNR and SSIM metrics in Table 1. The methods improve PSNR and SSIM by 1.08 dB to 2.14 dB and 0.0342 to 0.0680 respectively. As expected for such a challenging task, no method is capable of restoring all the fine details present in the ground truth. Predominantly, only sharp edges and smooth textures can be recovered even by the top teams, since the information loss is so extreme. Interestingly, teams KirinUK, Team-WVU, BOE-IOT-AIBD, sr_xxx, ZZX and TTI restore two windows (blue box highlight, second column) instead of a single one as present in the ground truth frame. This could indicate, the method's upscaling for such a large factor is highly dependent on image pri-

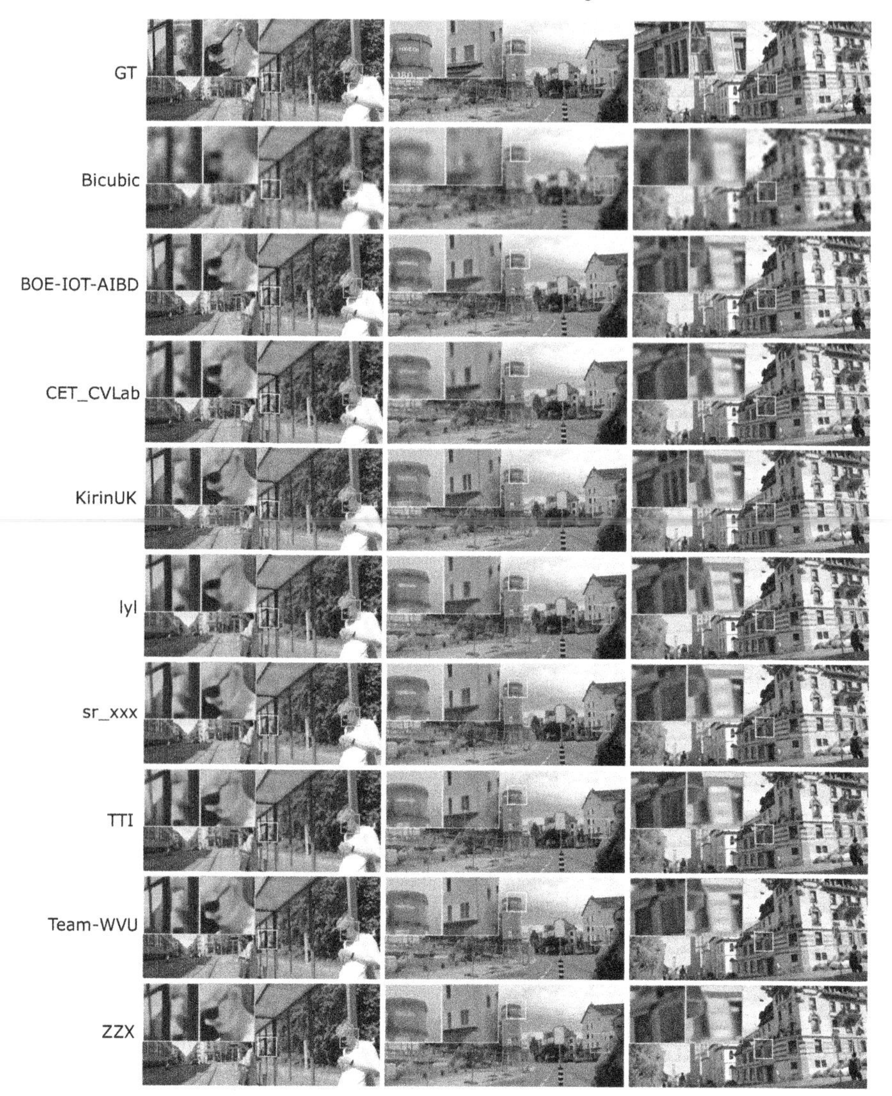

Fig. 12. Track 1 Visual results for all competing teams. Additionally, we show the ground truth (GT) and bicubic interpolation (Bicubic) for reference. To present the details more clearly and to fit all methods on a single page, the frames are cropped to 800×1920. Highlights from yellow and blue boxes are shown at the top left. (Color figure online)

Table 2. User study results for track 2. The results are obtained by a one vs. one user study on frame level and video level. Wins (tot) indicates absolute wins in all comparisons. Wins (%) reflects relative wins, which are normalized by the number of comparisons with other teams. Compared to absolute wins, the relative wins allow direct comparison between frame level and video level performance. The aggregated relative wins of both studies on frame and video level led to the final ranking. Additionally, we provide PSNR, SSIM and LPIPS scores for reference. Note, these metrics are not considered for ranking.

Method			K	T	Z	B (t)	s	B (s)	l	C	Wins (tot)	Wins (%)	↑PSNR	↑SSIM	↓LPIPS
Frame level	1	KirinUK	–	115	126	116	113	117	126	144	**857**	**76.52**	**22.79**	**0.6474**	**0.447**
	2	Team-WVU	45	–	85	97	97	110	109	132	675	60.27	22.48	0.6378	0.507
	3	ZZX	34	75	–	78	97	101	104	130	619	55.27	22.09	0.6268	0.505
	4	BOE-IOT-AIBD (t)	44	63	82	–	68	87	95	118	557	49.73	21.18	0.3633	0.514
	5	sr_xxx	47	63	63	92	–	95	109	131	600	53.57	22.43	0.6353	0.509
		BOE-IOT-AIBD (s)	43	50	59	73	65	–	85	113	488	43.57	22.48	0.6304	0.550
	7	lyl	34	51	56	65	51	75	–	119	451	40.27	22.08	0.6256	0.535
	8	CET_CVLab	16	28	30	42	29	47	41	–	233	20.80	21.77	0.6112	0.602
													Final scores		
													Frame	Video	Total
Video level	1	KirinUK	–	7	7	8	7	8	8	6	**51**	**72.86**	76.52	72.86	**149.38**
	2	Team-WVU	3	–	7	6	8	2	8	8	42	60.00	60.27	60.00	120.27
	3	ZZX	3	3	–	5	5	4	9	8	37	52.86	55.27	52.86	108.13
	4	BOE-IOT-AIBD (t)	2	4	5	–	8	8	5	8	40	57.14	49.73	57.14	106.87
	5	sr_xxx	3	2	5	2	–	4	7	6	29	41.43	53.57	41.43	95.00
		BOE-IOT-AIBD (s)	2	8	6	2	6	–	5	7	36	51.43	43.57	51.43	95.00
	7	lyl	2	2	1	5	3	5	–	7	25	35.71	40.27	35.71	75.98
	8	CET_CVLab	4	2	2	2	4	3	3	–	20	28.57	20.80	28.57	49.37

ors and having access to temporal information is not sufficient to achieve high restoration quality. Still, a top ranking is only achieved by teams that leverage temporal information, which shows its importance for video super-resolution, even in such extreme settings.

5.2 Track 2 - Perceptual

Due to the extreme information loss, it is hard to accurately restore the high-frequency content with respect to the ground truth. If deviations from the ground truth can be accepted and more visually pleasing results are desired, perceptual quality can be traded-off for fidelity. The results may not entirely reflect the underlying ground truth, but instead boost the perceptual quality considerably. We therefore do not rely on PSNR and SSIM for evaluation in track 2, but instead conduct a user study to asses human perceptual quality. Challenge track 2 has 54 registered participants, from which 7 submitted solutions to the validation server and 7 teams entered the final ranking

Metrics. Assessing perceptual quality quantitatively is difficult and remains largely an open problem. Attempts for such metrics have been made in the past and one of the most promising metrics is called Learned Perceptual Image Patch

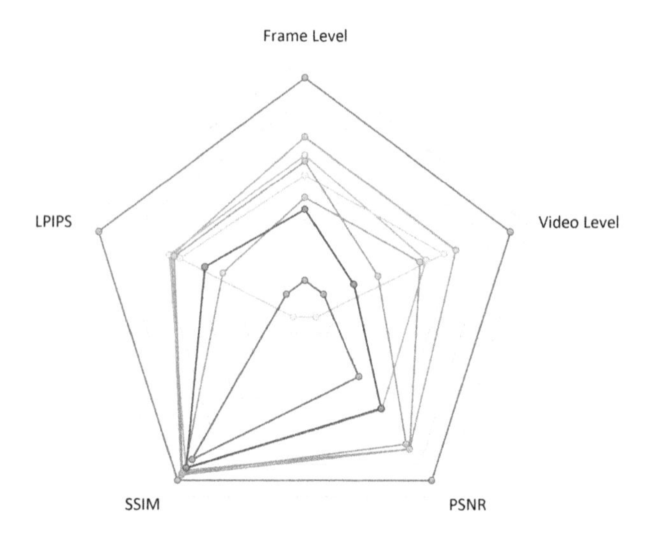

Fig. 13. Radar plot for track 2. All values per axis have been normalized to lie in [0,1]. The range for LPIPS has been reversed to indicate better values towards the outside in accordance with the other metrics.

Similarity (LPIPS), which is proposed in [54]. This metric measures similarity to the ground truth in feature space of popular architectures, e.g. [23]. While this metric is widely adopted for perceptual quality assessment, especially on images, it still fails in some cases to reflect human perception. Like PSNR and SSIM, it does not discriminate on temporal dynamics, which are crucial for high quality videos.

User Study. Since quantitative metrics are not reliable, we resort to a user study to rank the participating teams in track 2. For that matter we split the evaluation in two separate user studies, a frame level study and a video level study. The frame level study is meant to judge the image level quality and is performed on randomly subsampled frames from all 16 sequences in the test set. The competing method's frames are compared side-by-side in a one vs. one setting, resulting in 28 comparisons per frame. We asked 10 users to judge the frame level quality, which results in $16 \times 28 \times 10 = 4480$ total ratings. The detailed results are shown in a confusion matrix in Table 2. Each row shows the preference of the method in the first column against all other methods. Additionally, we show the total number of preferences (Wins (tot)) plus the relative preferences, which are normalized to the number of total comparisons $16 \times 7 \times 10 = 1120$ with all teams (Wins (%)). The video level study is meant for evaluating the temporal dynamics in the videos and the overall perceptual quality when watching the videos. Again, we generated side-by-side videos between all methods for com-

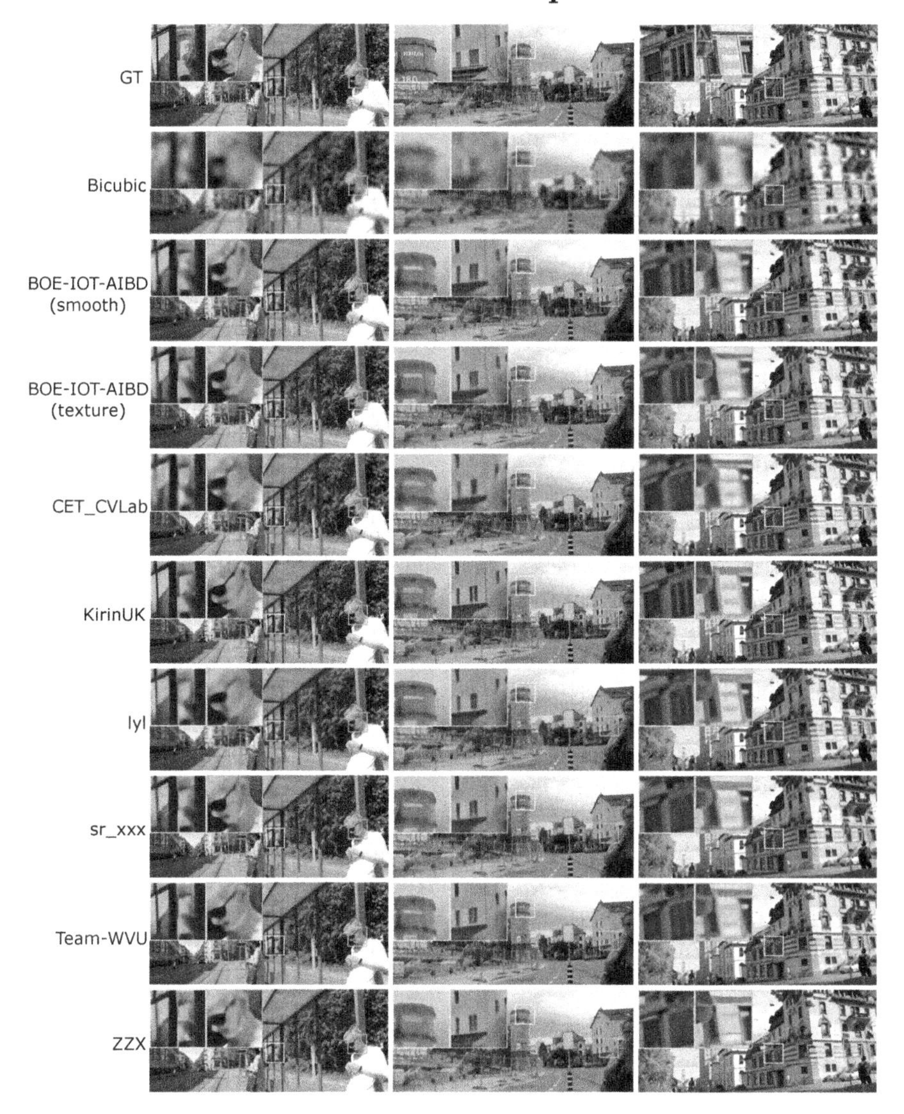

Fig. 14. Track 2 Visual results for all competing teams. Additionally, we show the ground truth (GT) and bicubic interpolation (Bicubic) for reference. To present the details more clearly and to fit all methods on a single page, the frames are cropped to 800×1920. Highlights from yellow and blue boxes are shown at the top left. Team BOE-IOT-AIBD provides two distinct solutions, which focus on high quality textures and temporal smoothness respectively. (Color figure online)

parison in a one vs. one setting. The videos are generated by compiling all 1920 frames from 16 sequences into a single video, showing two competing methods. This results in 28 short videos of ≈1 min. We also ask 10 users to perform the video level user study and get a total number of $28 \times 10 = 280$ ratings. In order to directly compare with the frame level study, we normalize the total wins by the total comparisons with all the teams ($10 \times 7 = 70$) to get the relative scores. We derive the ranking from the combined relative scores of both frame level and video level user studies (see Table 2, lower right). Team KirinUk is the clear winner in track 2, followed by Team-WVU and ZZX on the second and third place. We also provide PSNR, SSIM and LPIPS metrics for reference. A qualitative illustration of all track 2 results is presented in Fig. 13, including the user study results. Note, the final ranking is only derived from the user study.

Visual Results. To allow direct comparison to track 1, we provide the visual results on the same frames for track 2 in Fig. 14. Note that teams Team-WVU, sr_xxx, lyl and CET_CVLab submitted the same set of frames to both tracks, while KirinUK, BOE-IOT-AIBD and ZZX adapted their solutions to the specific requirements in both tracks. BOE-IOT-AIBD even provided two distinct solutions for track 2. One is optimized with emphasis on textures, the other is designed for temporal smoothness, abbreviated with (t) and (s) respectively in Table 2. Surprisingly, the texture based solution of BOE-IOT-AIBD also performs better in the video level user study. According to the users, the sharper texture details seem to have a higher impact on the quality than the flickering artifacts. The winning team KirinUK manages to not only outperform all other teams in both user studies, but also in the provided metrics PSNR, SSIM and LPIPS. However, it has to be considered, that the solution is optimized with L1 and VGG-loss, which are both closely related to these metrics. BOE-IOT-AIBD and KirinUK are the only teams that incorporate a GAN loss into their training strategy. On the other hand, Team-WVU trains its network only on the pixel-based Charbonnier Loss, and ZZX trains their perceptual solution on L1 and SSIM. Nevertheless, they outperform BOE-IOT-AIBD, which employs a GAN loss. Therefore, strong guidance from a pixel-based loss might be important for such an extreme scaling factor.

6 Conclusions

This paper presents the AIM 2020 challenge on Video Extreme Super-Resolution. We evaluate the performance in this challenging setting for both high fidelity restoration (track 1) and perceptual quality (track 2). The overall winner KirinUK manages to strike the best balance between restoration and perceptual quality. The participating teams provided innovative and diverse solutions to deal with the extreme upscaling factor of 16. Further improvements could be achieved by reducing and ideally removing the notorious flickering artifacts associated with video enhancement in general. On top of that, a more powerful generative setting could be designed for higher perceptual quality in track 2. Quantitative evaluation for perceptual quality still requires more research, especially in the video domain, where temporal consistency is important. We hope this challenge attracts more researchers to enter the area of extreme video super-resolution as it offers great opportunities for innovation.

Acknowledgements. We thank the AIM 2020 sponsors: Huawei, MediaTek, NVIDIA, Qualcomm, Google, and Computer Vision Lab (CVL), ETH Zurich.

Appendix A: Teams and Affiliations

AIM 2020 team

Title: AIM 2020 Challenge on Video Extreme Super-Resolution
Members: Dario Fuoli, Zhiwu Huang, Shuhang Gu, Radu Timofte
Affiliations:
Computer Vision Lab, ETH Zurich, Switzerland; The University of Sydney, Australia

KirinUK

Title: Efficient Video Enhancement and Super-Resolution Net (EVESRNet)
Members: Arnau Raventos, Aryan Esfandiari, Salah Karout
Affiliations:
Huawei Technologies R&D UK

Team-WVU

Title: Multi-Frame based Deformable Kernel Convolution Networks
Members: Xuan Xu, Xin Li, Xin Xiong, Jinge Wang
Affiliations:
West Virginia University, USA; Huazhong University of Science and Technology, China

BOE-IOT-AIBD

Title: Fully 3D–Convolutional MultiGrid–BackProjection Network
Members: Pablo Navarrete Michelini, Wenhao Zhang
Affiliations:
BOE Technology Group Co., Ltd.

ZZX

Title: Multi-Scale Aggregated Upsampling Extreme Video Based on High Frequency Attention
Members: Dongyang Zhang, Hanwei Zhu, Dan Xia
Affiliations:
Jiangxi University of Finance and Economics; National Key Laboratory for Remanufacturing, Army Academy of Armored Forces

sr_xxx

Title: Residual Receptive Attention for Video Super-Resolution

Members: Haoyu Chen, Jinjin Gu, Zhi Zhang
Affiliations:
Amazon Web Services; The Chinese University of Hong Kong, Shenzhen

lyl

Title: Coarse to Fine Pyramid Networks for Progressive Image Super-Resolution
Members: Tongtong Zhao, Shanshan Zhao
Affiliations:
Dalian Maritime University; China Everbright Bank Co., Ltd.

TTI

Title: STARnet
Members: Kazutoshi Akita, Norimichi Ukita
Affiliations:
Toyota Technological Institute (TTI)

CET_CVLab

Title: Video Extreme Super-Resolution using Progressive Wide Activation Net
Members: Hrishikesh P S, Densen Puthussery, Jiji C V
Affiliations:
College of Engineering, Trivandrum, India

References

1. Blau, Y., Michaeli, T.: The perception-distortion tradeoff. In: The IEEE Conference on Computer Vision and Pattern Recognition (CVPR), June 2018
2. Caballero, J., et al.: Real-time video super-resolution with spatio-temporal networks and motion compensation. In: The IEEE Conference on Computer Vision and Pattern Recognition (CVPR), July 2017
3. Chen, J., Tan, X., Shan, C., Liu, S., Chen, Z.: VESR-Net: the winning solution to Youku video enhancement and super-resolution challenge. arXiv preprint arXiv:2003.02115 (2020)
4. Dahl, R., Norouzi, M., Shlens, J.: Pixel recursive super resolution. In: The IEEE International Conference on Computer Vision (ICCV), October 2017
5. Dai, Q., Yoo, S., Kappeler, A., Katsaggelos, A.K.: Sparse representation-based multiple frame video super-resolution. IEEE Trans. Image Process. **26**(2), 765–781 (2017). https://doi.org/10.1109/TIP.2016.2631339
6. Dong, C., Loy, C.C., He, K., Tang, X.: Image super-resolution using deep convolutional networks. IEEE Trans. Pattern Anal. Mach. Intell. **38**(2), 295–307 (2016). https://doi.org/10.1109/TPAMI.2015.2439281
7. El Helou, M., Zhou, R., Süsstrunk, S., Timofte, R., et al.: AIM 2020: scene relighting and illumination estimation challenge. In: European Conference on Computer Vision Workshops (2020)
8. Fuoli, D., Gu, S., Timofte, R.: Efficient video super-resolution through recurrent latent space propagation. In: ICCV Workshops (2019)

9. Fuoli, D., Gu, S., Timofte, R., et al.: Aim 2019 challenge on video extreme super-resolution: methods and results. In: ICCV Workshops (2019)
10. Fuoli, D., Huang, Z., Danelljan, M., Timofte, R., et al.: NTIRE 2020 challenge on video quality mapping: Methods and results. In: The IEEE Conference on Computer Vision and Pattern Recognition (CVPR) Workshops, June 2020
11. Fuoli, D., Huang, Z., Gu, S., Timofte, R., et al.: AIM 2020 challenge on video extreme super-resolution: methods and results. In: European Conference on Computer Vision Workshops (2020)
12. Gao, H., Zhu, X., Lin, S., Dai, J.: Deformable kernels: adapting effective receptive fields for object deformation. In: International Conference on Learning Representations (2020). https://openreview.net/forum?id=SkxSv6VFvS
13. Haris, M., Shakhnarovich, G., Ukita, N.: Space-time-aware multi-resolution video enhancement. In: Proceedings of the IEEE/CVF Conference on Computer Vision and Pattern Recognition, pp. 2859–2868 (2020)
14. Haris, M., Shakhnarovich, G., Ukita, N.: Recurrent back-projection network for video super-resolution. In: Proceedings of the IEEE Conference on Computer Vision and Pattern Recognition, pp. 3897–3906 (2019)
15. Huang, Y., Wang, W., Wang, L.: Bidirectional recurrent convolutional networks for multi-frame super-resolution. In: Proceedings of the 28th International Conference on Neural Information Processing Systems, vol. 1, pp. 235–243. MIT Press, Cambridge (2015). http://dl.acm.org/citation.cfm?id=2969239.2969266
16. Ignatov, A., Timofte, R., et al.: AIM 2020 challenge on learned image signal processing pipeline. In: European Conference on Computer Vision Workshops (2020)
17. Ignatov, A., Timofte, R., et al.: AIM 2020 challenge on rendering realistic bokeh. In: European Conference on Computer Vision Workshops (2020)
18. Ji, X., Cao, Y., Tai, Y., Wang, C., Li, J., Huang, F.: Real-world super-resolution via kernel estimation and noise injection. In: Proceedings of the IEEE/CVF Conference on Computer Vision and Pattern Recognition Workshops, pp. 466–467 (2020)
19. Jo, Y., Wug Oh, S., Kang, J., Joo Kim, S.: Deep video super-resolution network using dynamic upsampling filters without explicit motion compensation. In: The IEEE Conference on Computer Vision and Pattern Recognition (CVPR), June 2018
20. Kappeler, A., Yoo, S., Dai, Q., Katsaggelos, A.: Video super-resolution with convolutional neural networks. IEEE Trans. Comput. Imaging **2**, 109–122 (2016)
21. Kim, J., Kwon Lee, J., Mu Lee, K.: Accurate image super-resolution using very deep convolutional networks. In: The IEEE Conference on Computer Vision and Pattern Recognition (CVPR), June 2016
22. Kim, S., Li, G., Fuoli, D., Danelljan, M., Huang, Z., Gu, S., Timofte, R.: The Vid3oC and IntVID datasets for video super resolution and quality mapping. In: ICCV Workshops (2019)
23. Krizhevsky, A., Sutskever, I.E., Hinton, G.: ImageNet classification with deep convolutional neural networks. Neural Inf. Process. Syst. **25** (2012). https://doi.org/10.1145/3065386
24. Lai, W.S., Huang, J.B., Ahuja, N., Yang, M.H.: Deep Laplacian pyramid networks for fast and accurate super-resolution. In: Proceedings of the IEEE Conference on Computer Vision and Pattern Recognition, pp. 624–632 (2017)
25. Ledig, C., et al.: Photo-realistic single image super-resolution using a generative adversarial network. In: The IEEE Conference on Computer Vision and Pattern Recognition (CVPR), July 2017

26. Liao, R., Tao, X., Li, R., Ma, Z., Jia, J.: Video super-resolution via deep draft-ensemble learning. In: The IEEE International Conference on Computer Vision (ICCV), December 2015
27. Liu, C., Yuen, J., Torralba, A.: SIFT Flow: dense correspondence across scenes and its applications. IEEE Trans. Pattern Anal. Mach. Intell. **33**(5), 978–994 (2010)
28. Liu, C., et al.: Beyond pixels: exploring new representations and applications for motion analysis. Ph.D. thesis, Massachusetts Institute of Technology (2009)
29. Liu, D., et al.: Robust video super-resolution with learned temporal dynamics. In: The IEEE International Conference on Computer Vision (ICCV), October 2017
30. Liu, J., Zhang, W., Tang, Y., Tang, J., Wu, G.: Residual feature aggregation network for image super-resolution. In: Proceedings of the IEEE/CVF Conference on Computer Vision and Pattern Recognition, pp. 2359–2368 (2020)
31. Liu, X., Kong, L., Zhou, Y., Zhao, J., Chen, J.: End-to-end trainable video super-resolution based on a new mechanism for implicit motion estimation and compensation. In: The IEEE Winter Conference on Applications of Computer Vision, pp. 2416–2425 (2020)
32. Lucas, A., Lopez Tapia, S., Molina, R., Katsaggelos, A.K.: Generative adversarial networks and perceptual losses for video super-resolution. arXiv e-prints, June 2018
33. Lugmayr, A., et al.: AIM 2019 challenge on real-world image super-resolution: methods and results. arXiv preprint arXiv:1911.07783 (2019)
34. Makansi, O., Ilg, E., Brox, T.: End-to-end learning of video super-resolution with motion compensation. arXiv e-prints, July 2017
35. Nah, S., et al.: NTIRE 2019 challenge on video deblurring and super-resolution: dataset and study. In: The IEEE Conference on Computer Vision and Pattern Recognition (CVPR) Workshops, June 2019
36. Navarrete Michelini, P., Chen, W., Liu, H., Zhu, D.: MGBPv2: scaling up multi-grid back-projection networks. In: The IEEE International Conference on Computer Vision Workshops (ICCVW), October 2019. https://arxiv.org/abs/1909.12983
37. Navarrete Michelini, P., Liu, H., Zhu, D.: Multi-scale recursive and perception-distortion controllable image super-resolution. In: The European Conference on Computer Vision Workshops (ECCVW), September 2018. http://arxiv.org/abs/1809.10711
38. Navarrete Michelini, P., Liu, H., Zhu, D.: Multigrid backprojection super-resolution and deep filter visualization. In: Proceedings of the Thirty-Third AAAI Conference on Artificial Intelligence (AAAI 2019). AAAI (2019)
39. Ntavelis, E., Romero, A., Bigdeli, S.A., Timofte, R., et al.: AIM 2020 challenge on image extreme inpainting. In: European Conference on Computer Vision Workshops (2020)
40. Pérez-Pellitero, E., Sajjadi, M.S.M., Hirsch, M., Schölkopf, B.: Photorealistic video super resolution. arXiv e-prints, July 2018
41. Sajjadi, M.S.M., Scholkopf, B., Hirsch, M.: EnhanceNet: single image super-resolution through automated texture synthesis. In: The IEEE International Conference on Computer Vision (ICCV), October 2017
42. Sajjadi, M.S.M., Vemulapalli, R., Brown, M.: Frame-recurrent video super-resolution. In: The IEEE Conference on Computer Vision and Pattern Recognition (CVPR), June 2018
43. Shi, W., et al.: Real-time single image and video super-resolution using an efficient sub-pixel convolutional neural network. In: CVPR (2016)
44. Son, S., Lee, J., Nah, S., Timofte, R., Lee, K.M., et al.: AIM 2020 challenge on video temporal super-resolution. In: European Conference on Computer Vision Workshops (2020)

45. Tao, X., Gao, H., Liao, R., Wang, J., Jia, J.: Detail-revealing deep video super-resolution. In: The IEEE International Conference on Computer Vision (ICCV), October 2017
46. Tian, Y., Zhang, Y., Fu, Y., Xu, C.: TDAN: temporally-deformable alignment network for video super-resolution. In: Proceedings of the IEEE/CVF Conference on Computer Vision and Pattern Recognition, pp. 3360–3369 (2020)
47. Timofte, R., Agustsson, E., Van Gool, L., Yang, M.H., Zhang, L.: NTIRE 2017 challenge on single image super-resolution: methods and results. In: The IEEE Conference on Computer Vision and Pattern Recognition (CVPR) Workshops, July 2017
48. Wang, X., Chan, K.C., Yu, K., Dong, C., Change Loy, C.: EDVR: video restoration with enhanced deformable convolutional networks. In: The IEEE Conference on Computer Vision and Pattern Recognition (CVPR) Workshops, June 2019
49. Wang, X., et al.: ESRGAN: enhanced super-resolution generative adversarial networks. In: Leal-Taixé, L., Roth, S. (eds.) ECCV 2018. LNCS, vol. 11133, pp. 63–79. Springer, Cham (2019). https://doi.org/10.1007/978-3-030-11021-5_5
50. Wei, P., Lu, H., Timofte, R., Lin, L., Zuo, W., et al.: AIM 2020 challenge on real image super-resolution. In: European Conference on Computer Vision Workshops (2020)
51. Xu, X., Xiong, X., Wang, J., Li, X.: Deformable kernel convolutional network for video extreme super-resolution. In: European Conference on Computer Vision Workshops (2020)
52. Yu, J., et al.: Wide activation for efficient and accurate image super-resolution. arXiv preprint arXiv:1808.08718 (2018)
53. Zhang, K., Danelljan, M., Li, Y., Timofte, R., et al.: AIM 2020 challenge on efficient super-resolution: Methods and results. In: European Conference on Computer Vision Workshops (2020)
54. Zhang, R., Isola, P., Efros, A.A., Shechtman, E., Wang, O.: The unreasonable effectiveness of deep features as a perceptual metric. In: CVPR (2018)
55. Zhang, Y., Li, K., Li, K., Wang, L., Zhong, B., Fu, Y.: Image super-resolution using very deep residual channel attention networks. In: Ferrari, V., Hebert, M., Sminchisescu, C., Weiss, Y. (eds.) ECCV 2018. LNCS, vol. 11211, pp. 294–310. Springer, Cham (2018). https://doi.org/10.1007/978-3-030-01234-2_18
56. Zhang, Y., Tian, Y., Kong, Y., Zhong, B., Fu, Y.: Residual dense network for image super-resolution. In: Proceedings of the IEEE Conference on Computer Vision and Pattern Recognition, pp. 2472–2481 (2018)
57. Zhu, X., Hu, H., Lin, S., Dai, J.: Deformable ConvNets v2: more deformable, better results. In: Proceedings of the IEEE Conference on Computer Vision and Pattern Recognition, pp. 9308–9316 (2019)

Deformable Kernel Convolutional Network for Video Extreme Super-Resolution

Xuan Xu[1], Xin Xiong[2], Jinge Wang[1], and Xin Li[1](✉)

[1] West Virginia University, Morgantown, WV 26505, USA
{xuxu,jnwang1}@mix.wvu.edu, xin.li@mail.wvu.edu

[2] Huazhong University of Science and Technology, Wuhan 430074, China
xiong_xin@hust.edu.cn

Abstract. Video super-resolution, which attempts to reconstruct high-resolution video frames from their corresponding low-resolution versions, has received increasingly more attention in recent years. Most existing approaches opt to use deformable convolution to temporally align neighboring frames and apply traditional spatial attention mechanism (convolution based) to enhance reconstructed features. However, such spatial-only strategies cannot fully utilize temporal dependency among video frames. In this paper, we propose a novel deep learning based VSR algorithm, named Deformable Kernel Spatial Attention Network (DKSAN). Thanks to newly designed Deformable Kernel Convolution Alignment (DKC_Align) and Deformable Kernel Spatial Attention (DKSA) modules, DKSAN can better exploit both spatial and temporal redundancies to facilitate the information propagation across different layers. We have tested DKSAN on AIM2020 Video Extreme Super-Resolution Challenge to super-resolve videos with a scale factor as large as 16. Experimental results demonstrate that our proposed DKSAN can achieve both better subjective and objective performance compared with the existing state-of-the-art EDVR on Vid3oC and IntVID datasets.

Keywords: Video super-resolution · Deep learning · Deformable kernels · Deformable convolution network · Attention mechanism

1 Introduction

Video Super-Resolution (VSR) refers to the task of reconstructing high-resolution (HR) video frames from their corresponding low-resolution (LR) observation data. Similar to image super-resolution, VSR aims at faithful recovery of important image structures (e.g., edges and textures) and has been widely used in practical applications from video surveillance [26] and high-definition Television (HDTV) [24] to video coding and streaming [31]. Existing VSR research can be mainly classified into two subfields, enhancing spatial super-resolution and enhancing temporal super-resolution. The former focuses

A. Bartoli and A. Fusiello (Eds.): ECCV 2020 Workshops, LNCS 12538, pp. 82–98, 2020.
https://doi.org/10.1007/978-3-030-66823-5_5

on super-resolving LR video frames to approximate HR video frames to improve visual quality of video; while the later refers to interpolate new frames between neighboring frames for the purpose of increasing video frame rate (e.g., from 30 fps to 60 fps). Different from Single Image Super-Resolution (SISR) which only needs to consider the information from spatial domain, both spatial and temporal dependencies have to be utilized by VSR algorithms in order to optimize their performance. In particular, how to effectively exploit temporal redundancy by motion compensation techniques has remained one of the key technical challenges in the task of VSR.

In order to explore the potential benefit from temporal information of VSR, several existing approaches [5,20,23,34] have used a sequence of consecutive LR frames (including one reference frame and several neighboring frames) as inputs to reconstruct the HR frame corresponding to the reference LR frame. To better exploit temporal dependency among multiple LR frames, the consecutive frames need to be aligned before the reconstruction of the HR frame. One of the most popular motion estimation methods, optical-flow estimation [11], is often considered and has been adopted by several VSR approaches [2,21,25]. However, VSR based on rigid motion estimation has to suffer from the potential problem arising from misalignment. For example, it is well known that there are two plagues with optical flow estimation: occlusion and aperture problems [29]. VSR based on incorrect motion estimation results may introduce undesired blurring and misregistration artifacts to the reconstructed HR frames.

In view of the weakness of rigid motion estimation approaches, alternative methods - namely deformable motion estimation - have been proposed as well. Recently, *deformable convolution* [4,42] has become more and more popular as a supplementary module to video frame alignment. Several VSR works such as [30,33,35] have already successfully applied varying forms of deformable convolution alignment module to temporally align neighboring frames with respect to the reference frame, which demonstrates improved motion compensation when compared with optical-flow-based methods. However, existing deformable alignment modules still learn the motion parameters via several standard convolution layers with fixed kernel configurations, which can not extract accurate motion information especially in the presence of large and deformable motion (e.g., in sport video). By contrast, deformable kernels [8] can adapt effective receptive fields [22] (i.e., the support of filters) by weighting the per-pixel contribution, which is expected to be capable of characterizing more sophisticated motion information.

In this paper, we propose a novel Multi-Frame based Deformable Kernel Spatial Attention Network (DKSAN) for video *extreme* super-resolution (the upscaling factor is as large as 16). Inspired by EDVR [35] which applies deformable convolution [42] to temporally align neighboring frames with reference frame, we have designed a new module not based on optical flow estimation, called Deformable Kernel Convolution Alignment (DKC_Align) module, to enhance deformable convolution [42]. The key idea is to combine deformable kernel with deformable convolution to extract and improve not only global but also local edge and texture features while aligning the neighboring frames with respect to the reference frame. Moreover, we have developed a Deformable Kernel Spatial Attention (DKSA) module to further enhance the spatial details of reconstructed feature maps, which extends the previous spatial attention works such as [13,35,38]. The novelty of DKSA module lies in that the deformable kernel [8] can better represent spatially-localized edge and texture features which are often important for the task of VSR than conventional convolution based spatial attention.

2 Related Works

Unlike image super-resolution which deals with reconstructing missing information in the spatial domain only, VSR has to not only reconstruct the missing high-frequency information in the spatial domain but also consider the motion-related consistency across different video frames in the temporal domain. In this section, we briefly review existing VSR approaches based on multi-frame such as [2,13,30,33,35,36], optical flow [11] alignment and deformable convolution [42] alignment.

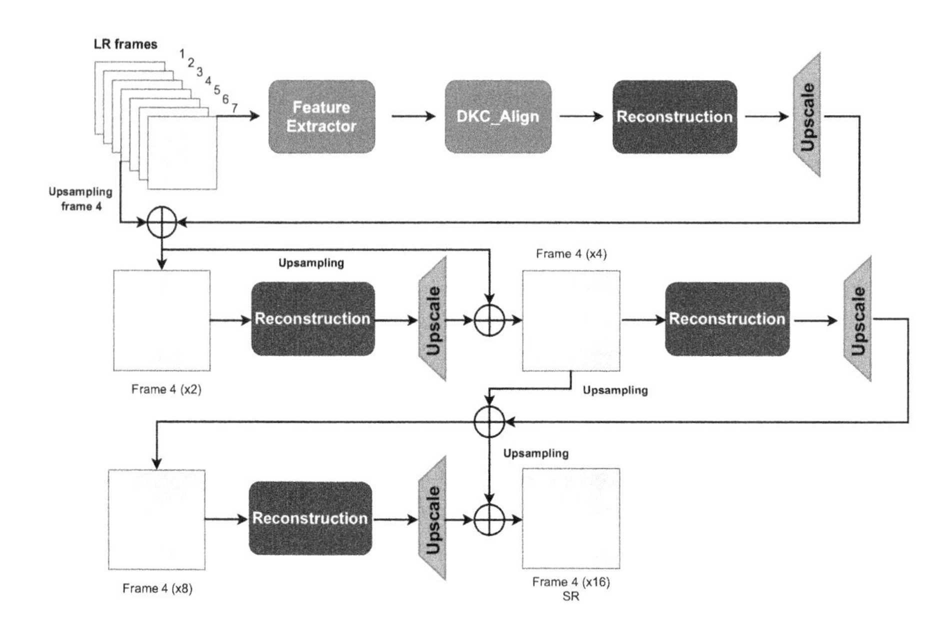

Fig. 1. Overview of DKSAN, $\oplus$ denotes element-wise sum.

2.1 Video Super-Resolution

One of the early works of applying optical-flow to VSR problems in order to utilize temporal and spatial information is [19]. In this work, a draft-ensemble strategy was introduced to use two robust optical flow methods: TV-l_1 flow and MDP flow to overcome the difficulty with large motion variation and then combine SR drafts via a deep convolutional neural network to generate the final SR result. Later, [15] proposed to use optical-flow to estimate motion compensation of consecutive LR frames and wrapped them as inputs of the CNN to generate SR frames. Those two-stage approaches are not optimal solution since they separate the motion compensation from frame reconstruction. To explore potential benefits of end-to-end learning architecture for VSR problem, a novel end-to-end deep CNN to joint train the estimation of optical flow and spatio-temporal networks called ESPCN was developed in [2]. In [28], a new layer called sub-pixel motion compensation (SPMC) was introduced to handle inter-frame motion alignment; it also applied a ConvLSTM [37] architecture for reconstruction and testing. Another work [9] proposed a recurrent back-projection network (RBPN) with encoder-decoder mechanism to extract spatial and temporal information. In [14], dynamic upsampling filters (DUF) was developed to avoid use the explicit motion compensation by computing pixels of local spatio-temporal neighbors of LR frames to learn implicit motion compensation. Most recently, a novel temporal group attention (TGA) framework [13] was proposed to group the input frames (7 frames) as three groups then generate temporal spatial attention maps to reconstruct the missing details in the reference frame. Another recent work [40] proposed to learn self-supervised motion representation, task-oriented flow (TOFlow), instead of fix optical flow as the motion compensation module for VSR problem.

2.2 Deformable Convolution

The inherent limitation with traditional CNNs is the capability of modeling geometric transformations because of their fixed kernel shape. Although dilated convolution can alleviate this limitation to some degree, it is still difficult for standard fixed-shape convolutional kernels to align the key points or salient features in the input images. To solve this problem, a deformable convolution network has been developed in [4,42] to improve the capability of modeling geometric transformations by adding flexible and learnable offsets. By acquiring information from other field rather than fixed local area, deformable convolution networks have been widely used by high-level vision tasks such as object detection [1] and segmentation [4]. Inspired by [42], a recent work [30] proposed a temporally-deformable alignment network (TDAN) to adapt deformable convolution to align the consecutive LR input frames at the feature level. Along this line of research, EDVR [35] designed a more aggressive alignment approach, PCD align module, to align the neighboring LR frames at different scale levels; also they proposed a temporal and spatial attention fusion module to future enhance important features. Another recent work [36] proposed a novel space-time video super-resolution framework to utilize deformable convolution and deformable ConvLSTM module to achieve temporal and spatial super-resolution at the same time. Most recently, [33] introduced another deformable convolution based VSR framework called deformable non-local network (DNLN) with non-local attention module and hierarchical feature fusion block to enhance the global details between neighboring frames and references. Those deformable alignment based methods have shown better performance than optical-flow based networks.

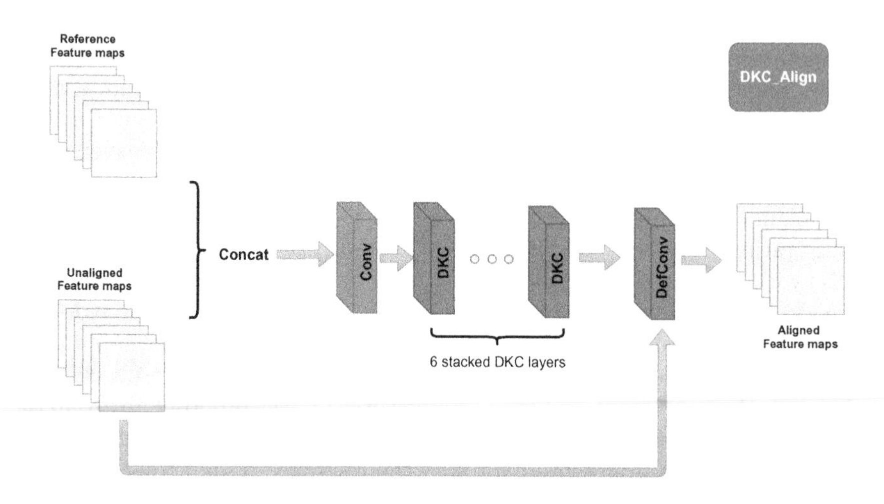

Fig. 2. Overview of DKC_Align module. Conv means convolution layer, DKC means deformable kernel convolution layer and DefConv stands for deformable convolution layer.

3 Proposed Methodology

The design of DKSAN network can be presented in the order of top-down hierarchy: DKSAN network (Fig. 1) $\rightarrow$ DKC_Align subnetwork (Fig. 2) $\rightarrow$ reconstruction module (Fig. 3).

3.1 Overview: Deformable Kernel Spatial Attention Network

For multi-frame based VSR, we are given a group of $2N+1$ consecutive LR frames $I_T^{LR} = \{I_{r-N}^{LR}, \ldots, I_{r-1}^{LR}, I_r^{LR}, I_{r+1}^{LR}, \ldots, I_{r+N}^{LR}\}$, where I_r^{LR} is denoted as frame at the center or reference frame and I_{r-N}^{LR} or I_{r+N}^{LR} are the neighboring frames of I_r^{LR}. The goal of multi-frame based VSR is to reconstruct a HR frame $\hat{Y}_r$ from the LR sequence of I_T^{LR} by exploiting both spatial and temporal redundancies in the sequence. The overall diagram of our proposed networks DKSAN is shown in Fig. 1. It mainly includes four parts: *feature extraction, DKC_Align module, reconstruct module*, and *upscale module*. Different from traditional deep learning based multi-frame VSR architectures, this work aims at super-resolving the LR videos at the extreme cases (e.g., with the scaling factor of 16). Due to large

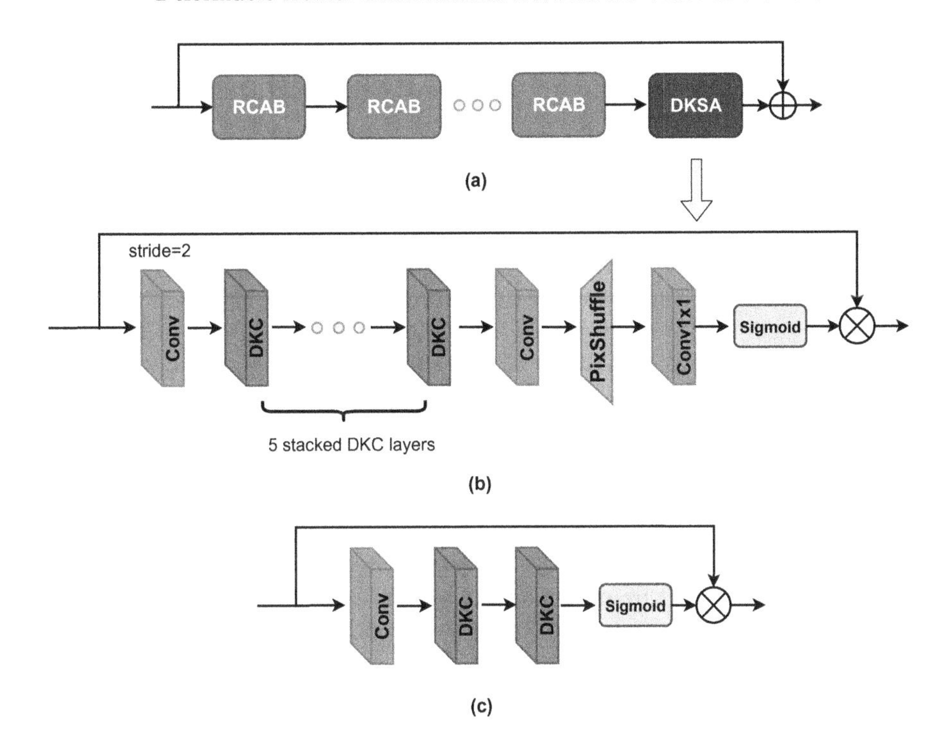

Fig. 3. Overview of reconstruction module; DKSA is deformable kernel spatial attention module shown in (b); a light version of DKSA is shown in (c); $\oplus$ and $\otimes$ denote element-wise sum and element-wise product, respectively.

scaling factor constraint, it is difficult to upscale the LR feature maps to the target HR ones directly. One-time upscaling approaches such as [13,30,35] tend to introduce undesired blurring and artifacts to super-resolved HR video frames.

To address this issue, we propose to construct a cascade of upscaling building blocks to iteratively super-resolve LR features several times (four times to reach the factor of $16 = 2^4$). Thanks to the cascade architecture, the LR frames can be super-resolved progressively to reconstruct the unknown HR frames more accurately than previous one-shot approaches. The whole problem of VSR can be formulated as follows:

$$\hat{Y}_r = \mathbb{F}(I_T^{LR}) \tag{1}$$

where I_T^{LR} denotes the consecutive LR frames and $\hat{Y}_r$ denotes the super-resolved reference frame I_r^{LR}. In particular, we extract the preliminary features of all input frames through the feature extraction which is stacked by several resblocks [35] without batch normalization layers. This procedure can be represented by:

$$F_{fea} = E_{res}(I_T^{LR}) \tag{2}$$

where E_{res} denotes the preliminary feature extraction, the output F_{fea} is the extracted feature maps for all input frames. Let define F_n is the neighboring feature, and F_r is the reference feature separated from F_{fea}. To align the neighboring feature and the reference feature with the proposed DKC_Align module E_{DKC_Align}, we have

$$F_{Align} = E_{DKC\_Align}(F_n, F_r) \tag{3}$$

$$F_{fusion} = \mathbf{W}_E(F_{Align}) \tag{4}$$

where $n \in [t-N, t+N]$ and $n \neq r$, F_{Align} is the concatenated aligned feature maps for each neighboring frame feature with reference frame feature. The details about this alignment module will be elaborated in Sect. 3.2; $\mathbf{W}_E \in \mathbb{R}^{1\times1\times C}$ is a 1×1 Conv layer. Conceptually similar to encoder-decoder configuration [3,9], the aligned feature F_{Align} (encoder outputs) will be fed to the reconstruction module and upscale module for the first-level upscaling (decoder) operation:

$$\hat{Y}_r^{level1} = U_1(E_{Recon1}(F_{fusion})) + B_{2\times}(I_r^{LR}) \tag{5}$$

where E_{Recon1} denotes the first level reconstruction module, U_1 is the first level upscaling module and $B_{2\times}$ stands for the Bicubic interpolation with scale factor of 2; $\hat{Y}_r^{level1}$ is the $2\times$ SR frame. Finally, to get the extreme super-resolved frame $\hat{Y}_r$, we repeat another 3 times of reconstruction operation:

$$\hat{Y}_r^{level2} = U_2(E_{Recon2}(E_2(\hat{Y}_r^{level1}))) + B_{2\times}(\hat{Y}_r^{level1}) \tag{6}$$

$$\hat{Y}_r^{level3} = U_3(E_{Recon3}(E_3(\hat{Y}_r^{level2}))) + B_{2\times}(\hat{Y}_r^{level2}) \tag{7}$$

$$\hat{Y}_r = U_4(E_{Recon4}(E_4(\hat{Y}_r^{level3}))) + B_{2\times}(\hat{Y}_r^{level3}) \tag{8}$$

where E_2, E_3, E_4 are the preliminary feature extractors for each level; U_2, U_3, U_4 denote the upscaling module for each corresponding level, respectively. The details about the reconstruction module are described in Sect. 3.3 including the DKSA module.

3.2 Deformable Kernel Alignment Module

Different from previous VSR works which applied optical flow to align neighboring frames with reference frame, [30] and [35] introduced to utilize modulated deformable convolution [42] to temporally align the given consecutive frames in order to add temporal information to VSR frameworks.

Deformable Convolution and Deformable Kernel. Inspired by [8,35], we propose a new alignment module, DKC_Align, to combine the deformable kernel [8] and deformable convolution [42] as shown in Fig. 2. First, let F_n and F_n^{align} denote the input and output feature maps (not the reference frame feature), $\mathbf{W}_k$ represents the weight kernel and p_k is the pre-specified offsets for the k-th location (K is the total sampling location), then the modulated deformable convolution can be described as follows:

$$F_n^{align}(p) = \sum_{k\in K} \mathbf{W}_k \cdot F_n(p + p_k + \Delta p_k) \cdot \Delta m_k \tag{9}$$

where $F_n^{align}(p)$ and $F_n(p)$ indicate the feature location p from F_n^{align} and F_n, Δp_k and Δm_k stand for the learnable offset and the modulation scalar, respectively. With Δp_k and Δm_k, the convolution will get the ability to be irregularly dilated to work with important feature points without the shape limitation of conventional convolution. Such process of deformable convolution can be regarded as a strategy of adapting the local receptive field to a support of arbitrary shape.

To get Δp_k and Δm_k and align the neighboring feature with reference feature in particular, we first concatenate the neighboring frame feature and the reference frame feature then fuse them with one Conv2D layer and fed them into several deformable kernel layers:

$$\Delta P_n, \Delta M_n = \mathbb{D}(f([F_n, F_r])), n \in [t-N, t+N], n \neq r \tag{10}$$

where f denotes the one Conv2d layer to fuse F_n and F_r, $\mathbb{D}$ represents the deformable kernel convolution layer. To formally express deformable kernel convolution layer $\mathbb{D}$, let Δk denote a learnable offset of the kernel $\mathbf{W}$, then deformable kernel convolution layer can be formulated as:

$$\mathbb{D} = \sum_{k \in K} \mathbf{W}_{k+\Delta k} \cdot f([F_n, F_r])(p + p_k + \Delta p_k), n \in [t-N, t+N], n \neq r \tag{11}$$

According to [8], deformable convolution can only adapt theoretical receptive fields by deforming the conventional convolution, but it cannot evaluate the contribution of each grid point. As a complementary tool to deformable convolution [4,42], deformable kernel [8] can weigh the contribution of each grid point to inform the network which point is more important (i.e., adaptive control of effective receptive fields). The advantage of combining deformable convolution with deformable kernel is to not only deform the convolution for extracting key grid points but also adaptively weigh the importance of each point (similar to the introduction of attention mechanism [32]). This way, the deformable convolution kernel layers will be more sensitive to the key feature points than traditional convolution layers and capable of extracting richer information to improve the alignment accuracy and reconstruction quality for our VSR task. Note that previous work such as EDVR [35] only studies the benefit of deformable convolution in VSR; while deformable kernel [8] was originally designed for high-level vision tasks such as object detection and classification. To the best of our knowledge, this work is the first to leverage of idea of combining deformable convolution with deformable kernel into the application of VSR.

3.3 Reconstruction Module

To get the super-resolved frame $\hat{Y}_r$, the output F_{fusion} from the DKC_Align module is fed into the reconstruction module. The reconstruction module includes several stacked RCAB blocks and the DKSA module (please refer to Fig. 3 (a)):

$$F_{recon} = E_{DKSA}(E_{RCABs}(F_{fusion})) + F_{fusion} \tag{12}$$

where F_{recon} is the final reconstruction features to be fed into upscale module (the architecture of upscale module is shown in Fig. 4 which includes several Conv layers, PixelShuffle and LeakyReLU), E_{RCABs} and E_{DKSA} denote the RCAB blocks and DKSA module. Note that RCAB module has the same structure as it proposed in RCAN [41] which includes resblock [10] and channel attention mechanism [12,39,41].

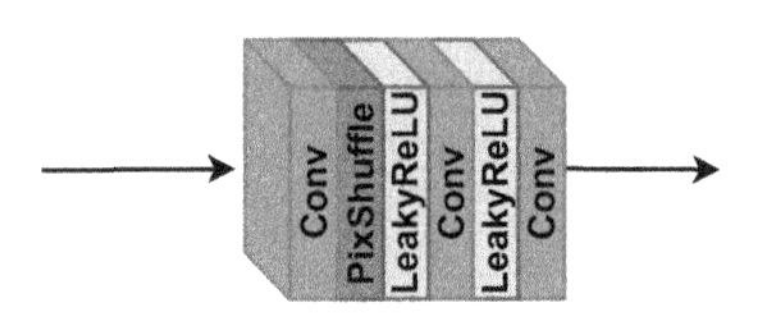

Fig. 4. The details of upscale module, the last Conv layer has only 3 feature maps output in order to generate RGB color frame.

Deformable Kernel Spatial Attention Module. In order to further calibrate output feature maps, we propose to construct a new Deformable Kernel based Spatial Attention (DKSA) module instead of traditional spatial attention mechanism. As shown in Fig. 3 (b), in DKSA, we first use one Conv layer to extract the output of the stacked RCAB blocks, then a couple of stacked Deformable Kernel Convolution (DKC) layers are placed to further extract key features from the naive feature map. As discussed in Sect. 3.2, deformable kernels can better measure the effective receptive field than standard convolution kernels. Therefore, DKSA can generate improved spatial attention maps to enforce networks pay more attention to important features such as edges and textures. Note that Fig. 3 (c) shows a light version of DKSA which is used by the level-1 reconstruction module.

4 Experimental Results

In this section, we demonstrate the training and test datasets, network setting, training details, experimental results and ablation study of proposed video extreme super-resolution approach.

4.1 Datasets

In this work, the training data we have used is Vid3oC [16] dataset provided by AIM2020 Video Extreme Super-Resolution Challenge. The Vid3oC dataset includes 50 videos for training, 16 sequences with 120 frames each for validation and 16 sequences with 120 frames each for testing. Note that the ground-truth of testing data are not released. Therefore, in this paper, we only show the validation results for Vid3oC dataset. In order to evaluate the validity of our

Table 1. Quantitative comparison on Vid3oC dataset for scaling factor of 16; s/f means seconds per frame. **Bold** font indicates the best result.

Video name	Scale	Bicubic	EDVR	DKSAN (ours)
		PSNR (dB)	PSNR (dB)	PSNR (dB)
050	x16	25.36	26.75	**29.17**
051	x16	23.20	23.76	**24.72**
052	x16	20.57	20.92	**21.61**
053	x16	21.61	22.15	**22.63**
054	x16	20.08	20.56	**21.15**
055	x16	20.01	20.36	**21.48**
056	x16	21.44	21.33	**22.54**
057	x16	20.22	20.33	**21.66**
058	x16	19.55	19.80	**21.45**
059	x16	20.22	20.92	**21.90**
060	x16	20.13	20.30	**21.38**
061	x16	21.08	21.58	**22.22**
062	x16	21.54	21.58	**23.12**
063	x16	21.54	22.00	**23.26**
064	x16	20.46	21.04	**21.94**
065	x16	22.53	23.41	**24.80**
Average	x16	21.22	21.67	**22.81**
Parameters	–	–	20.6M	29.5M
Runtime(s/f)	–	–	0.87	0.95

Table 2. Quantitative comparison on IntVID dataset for scaling factor of 16. **Bold** font indicates the best result.

Video name	Scale	Bicubic	EDVR	DKSAN (ours)
		PSNR (dB)	PSNR (dB)	PSNR (dB)
050	x16	21.56	21.81	**23.06**
051	x16	23.02	24.13	**24.92**
052	x16	29.56	29.33	**31.87**
053	x16	24.05	24.51	**25.09**
054	x16	31.34	33.15	**36.18**
055	x16	24.39	25.01	**26.88**
056	x16	31.16	31.93	**34.22**
057	x16	34.35	35.20	**39.75**
058	x16	36.00	37.36	**38.15**
059	x16	30.49	31.37	**34.17**
Average	x16	28.59	29.38	**31.43**

network, we choose 10 videos (050 to 059) from another dataset, IntVID [16], as a secondary test dataset. For each video, we extract 14 consecutive frames for testing.

4.2 Implementation Details

In the proposed DKSAN networks, to compare with EDVR, we set the kernel size as 3×3 with 128 filters for most of Conv layers, all deformable kernel layers and all deformable convolution layers. The kernel size of feature fusion layers is 1×1. The reduction ratio of channel attention module is still $r = 16$ as [41]. 5 resblocks are in feature extractor. The number of RCAB blocks are set to 30, 20, 15, 10 for each level (from 1 to 4) of reconstruction module. The PixelShuffle layer is the same as [27]. The last Conv layer filter is set to 3 in order to output color frames.

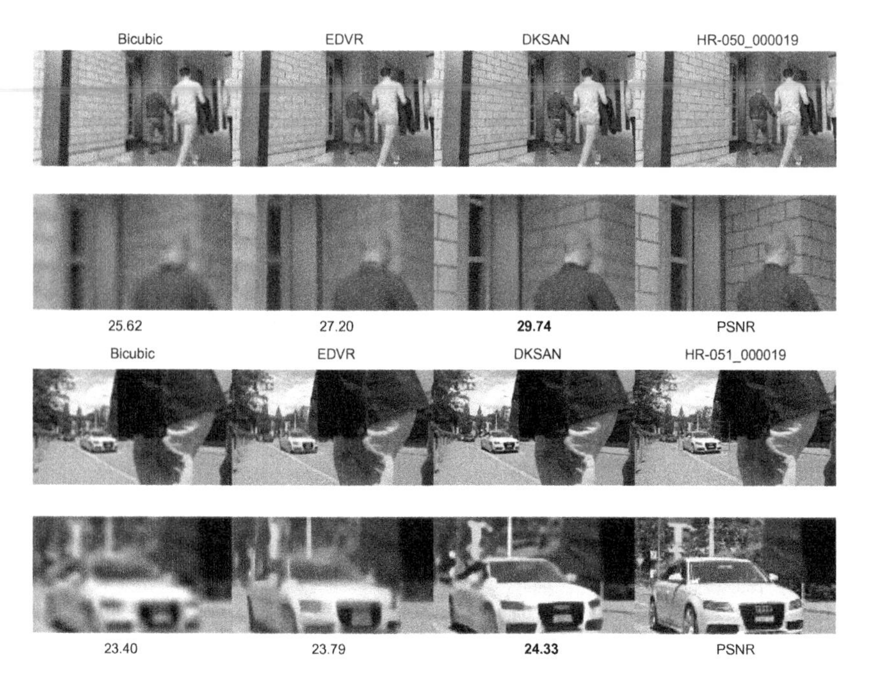

Fig. 5. Visual comparison results among competing approaches for Vid3oC dataset (video 050 and 051) at a scaling factor of 16.

In particular, we randomly crop the 7 low-res frames as small patches with the size of 32×32, and crop the corresponding 4th high-res frames with the size of 512×512. The batch size is 16. We augment the training set by random flips and rotations. The optimizer we used is ADAM [17] with $\beta_1 = 0.9$, $\beta_2 = 0.999$. The initial learning rate is set to 4×10^{-4}. The total training step is $115k$. The loss function we used is adapted Charbonnier penalty function [18]. The loss can be defined as Eq. 13 shown as follows:

$$Loss = \sqrt{||\hat{Y}_r - Y_r||^2 + \xi^2} \tag{13}$$

where $\xi = 1 \times 10^{-3}$, $\hat{Y}_r$ is super-resolved frame and Y_r is target frame (ground-truth). All experiments are trained on 4 NVIDIA Titan Xp GPUs with PyTorch framework Implementation.

Note that for fair comparison, we retrain EDVR with the same training dataset (Vid3oC) and keep most of EDVR setting as the same as the original implementation to run the experiment except setting the upscale module from factor 4 to factor 16 in order to make sure EDVR can generate extreme super-resolved frames.

4.3 Comparison Against State-of-the-Art

Because few existing works related to video extreme super-resolution (with a scale factor of 16), in this work, we have compared our proposed network against with Bicubic interpolation and state-of-the-art EDVR.

Table 1 shows the PSNR comparison results, number of parameters and running time (seconds per frame) of our approach with the competing methods, Bicubic interpolation and EDVR with the scaling factor of 16 on the validation set of Vid3oC [16]. From the Table, we can see that our DKSAN method has the best PSNR scores for all 16 testing videos. The significant PSNR gains (up to 2.4 dB) over previous state-of-the-art method EDVR. Since PSNR metrics cannot always evaluate the subjective quality of images, therefore, a qualitative result is shown in Fig. 5, we can easily observe that our proposed network DKSAN can better reconstruct the lines on the wall for "050_000019" and a clearer car for "051_000019" compared with EDVR.

To further verify the effectiveness of our proposed method, we selected another dataset, IntVid [16] as a secondary test dataset. From Table 2, we can easily find out that our proposed DKSAN has the best performance for all 10 testing videos compared with EDVR and bicubic interpolation. A qualitative result is shown in Fig. 6. For the subject "'050_0010", compared with EDVR, our DKSAN can better recover more details of the rear wing. Taking another example, in "054_0007", our DKSAN can reconstruct a much clearer face than EDVR does.

4.4 Ablation Studies

To investigate the effect of proposed DKC_Align module and DKSA module, we have conducted different strategies to remove the certain components from the final framework DKSAN. In particular, we have implemented four competing models for our ablation studies: 1) training with only resblocks, without channel attention, alignment and DKSA; 2) training without DKC_Align and DKSA

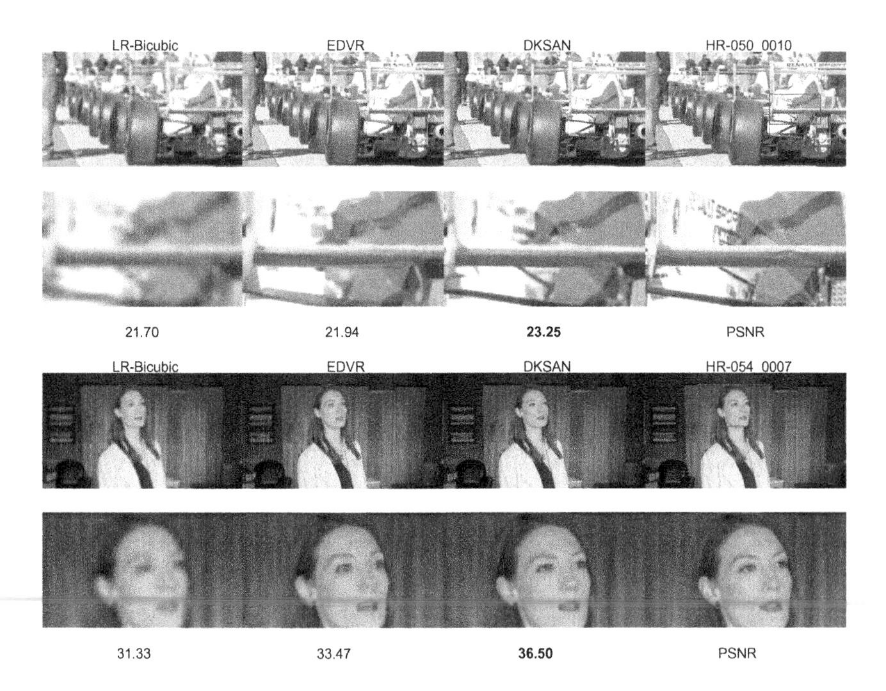

Fig. 6. Visual comparison results among competing approaches for IntVID dataset (video 050 and 054) at a scaling factor of 16.

module; 3) training with DKC_Align module but without DKSA module; 4) training with all modules (proposed DKSAN). Note that all experiments are trained under same dataset and conditions for fair comparison.

Table 3 shows the results, the number of parameters and running time (seconds per frame) of all four strategies mentioned previously with PSNR scores of each video and average. The backbone result is running only based on res-block, no channel attention, alignment and DKSA applied. From the results, we can see that the backbone has the worst performance; adding channel attention module but without DKC_Align and DKSA modules, the result is only 31.27 dB; after adding DKC_Align module, the result is improved to 31.32 dB; finally, we observe that after adding DKSA module (the full version of DKSAN), the result is further improved to 31.43 dB (**0.4 dB** and **0.16 dB** gained when compared with backbone and w/o Alignment & DKSA respectively) because of the effective module DKSA.

4.5 AIM 2020 Video Challenge

We have participated in the AIM2020 video extreme super-resolution challenges which is the second edition of AIM2019 challenges [6]. Our submissions won the **2nd place** for both track 1 and track 2 competitions. Note that track 1 is based

Table 3. Ablation Studies for DKSAN on IntVID dataset for scaling factor of 16. Backbone means only resblocks used; w/o Alignment & DKSA means DKC_Align and DKSA Module are not applied; w/o DKSA means only the DKSA module is not applied; s/f means seconds per frame. **Bold** font indicates the best result.

Video name	Backbone	w/o Alignment & DKSA	w/o DKSA	DKSAN (ours)
	PSNR (dB)	PSNR (dB)	PSNR (dB)	PSNR (dB)
050	22.80	22.98	22.87	**23.06**
051	24.72	24.89	24.89	**24.92**
052	31.65	31.75	31.85	**31.87**
053	25.05	25.07	**25.18**	25.09
054	35.30	35.73	35.86	**36.18**
055	26.52	**26.89**	26.69	26.88
056	33.79	33.92	**34.33**	34.22
057	38.97	39.13	39.27	**39.75**
058	38.09	38.21	**38.30**	38.15
059	33.38	34.15	34.16	**34.17**
Average	31.03	31.27	31.32	**31.43**
Parameters	26.1M	26.3M	27.0M	29.5M
Runtime(s/f)	0.83	0.89	0.92	0.97

on PSNR performance and track 2 is based on perceptual (see the AIM2020 challenge report [7] for more details).

5 Conclusions

In this work, we proposed a multi-frame based VSR networks DKSAN for extreme low-resolution videos. The novel temporal alignment module, DKC_Align, can help the networks to better learn and align the detailed features by improving both theoretical and effective receptive fields between reference frame and its neighboring frames. Furthermore, the DKSA module calibrated the reconstructed features to further enhance the edges and textures at the spatial domain. Thanks to the newly designed DKC_Align and DKSA modules, the proposed architecture can reconstruct high-quality HR frames from extreme LR frames and significantly improve both objective and subjective performance when compared with state-of-the-art approach EDVR [35].

Acknowledgment. This work is partially supported by the NSF under grants IIS-1908215 and OAC-1839909, the DoJ/NIJ under grant NIJ 2018-75-CX-0032, and the WV Higher Education Policy Commission Grant (HEPC.dsr.18.5).

References

1. Bertasius, G., Torresani, L., Shi, J.: Object detection in video with spatiotemporal sampling networks. In: Ferrari, V., Hebert, M., Sminchisescu, C., Weiss, Y. (eds.) ECCV 2018. LNCS, vol. 11216, pp. 342–357. Springer, Cham (2018). https://doi.org/10.1007/978-3-030-01258-8_21
2. Caballero, J., et al.: Real-time video super-resolution with spatio-temporal networks and motion compensation. In: Proceedings of the IEEE Conference on Computer Vision and Pattern Recognition, pp. 4778–4787 (2017)
3. Cheng, G., Matsune, A., Li, Q., Zhu, L., Zang, H., Zhan, S.: Encoder-decoder residual network for real super-resolution. In: Proceedings of the IEEE Conference on Computer Vision and Pattern Recognition Workshops (2019)
4. Dai, J., Qi, H., Xiong, Y., Li, Y., Zhang, G., Hu, H., Wei, Y.: Deformable convolutional networks. In: Proceedings of the IEEE International Conference on Computer Vision, pp. 764–773 (2017)
5. Farsiu, S., Robinson, M.D., Elad, M., Milanfar, P.: Fast and robust multiframe super resolution. IEEE Trans. Image Process. **13**(10), 1327–1344 (2004)
6. Fuoli, D., et al.: AIM 2019 challenge on video extreme super-resolution: methods and results. In: 2019 IEEE/CVF International Conference on Computer Vision Workshop (ICCVW), pp. 3467–3475 (2019)
7. Fuoli, D., Huang, Z., Gu, S., Timofte, R., et al.: AIM 2020 challenge on video extreme super-resolution: methods and results. In: Bartoli, A., Fusiello, A. (eds.) ECCV 2020, vol. 12538, pp. 57–81 (2020)
8. Gao, H., Zhu, X., Lin, S., Dai, J.: Deformable kernels: adapting effective receptive fields for object deformation. In: International Conference on Learning Representations (2020). https://openreview.net/forum?id=SkxSv6VFvS
9. Haris, M., Shakhnarovich, G., Ukita, N.: Recurrent back-projection network for video super-resolution. In: Proceedings of the IEEE Conference on Computer Vision and Pattern Recognition, pp. 3897–3906 (2019)
10. He, K., Zhang, X., Ren, S., Sun, J.: Deep residual learning for image recognition. In: Proceedings of the IEEE Conference on Computer Vision and Pattern Recognition, pp. 770–778 (2016)
11. Horn, B.K., Schunck, B.G.: Determining optical flow. In: Techniques and Applications of Image Understanding, vol. 281, pp. 319–331. International Society for Optics and Photonics (1981)
12. Hu, J., Shen, L., Sun, G.: Squeeze-and-excitation networks. In: Proceedings of the IEEE Conference on Computer Vision and Pattern Recognition, pp. 7132–7141 (2018)
13. Isobe, T., et a: Video super-resolution with temporal group attention. In: Proceedings of the IEEE/CVF Conference on Computer Vision and Pattern Recognition. pp. 8008–8017 (2020)
14. Jo, Y., Wug Oh, S., Kang, J., Joo Kim, S.: Deep video super-resolution network using dynamic upsampling filters without explicit motion compensation. In: Proceedings of the IEEE Conference on Computer Vision and Pattern Recognition, pp. 3224–3232 (2018)
15. Kappeler, A., Yoo, S., Dai, Q., Katsaggelos, A.K.: Video super-resolution with convolutional neural networks. IEEE Trans. Comput. Imaging **2**(2), 109–122 (2016)
16. Kim, S., Li, G., Fuoli, D., Danelljan, M., Huang, Z., Gu, S., Timofte, R.: The Vid3oC and IntVID datasets for video super resolution and quality mapping. In: 2019 IEEE/CVF International Conference on Computer Vision Workshop (ICCVW), pp. 3609–3616. IEEE (2019)

17. Kingma, D.P., Ba, J.: Adam: a method for stochastic optimization. arXiv preprint arXiv:1412.6980 (2014)
18. Lai, W.S., Huang, J.B., Ahuja, N., Yang, M.H.: Deep Laplacian pyramid networks for fast and accurate superresolution. In: IEEE Conference on Computer Vision and Pattern Recognition, vol. 2, p. 5 (2017)
19. Liao, R., Tao, X., Li, R., Ma, Z., Jia, J.: Video super-resolution via deep draft-ensemble learning. In: Proceedings of the IEEE International Conference on Computer Vision, pp. 531–539 (2015)
20. Liu, C., Sun, D.: On Bayesian adaptive video super resolution. IEEE Trans. Pattern Anal. Mach. Intell. **36**(2), 346–360 (2013)
21. Liu, D., et al.: Robust video super-resolution with learned temporal dynamics. In: Proceedings of the IEEE International Conference on Computer Vision, pp. 2507–2515 (2017)
22. Luo, W., Li, Y., Urtasun, R., Zemel, R.: Understanding the effective receptive field in deep convolutional neural networks. In: Advances in Neural Information Processing Systems, pp. 4898–4906 (2016)
23. Ma, Z., Liao, R., Tao, X., Xu, L., Jia, J., Wu, E.: Handling motion blur in multi-frame super-resolution. In: Proceedings of the IEEE Conference on Computer Vision and Pattern Recognition, pp. 5224–5232 (2015)
24. Matsuo, Y., Sakaida, S.: Super-resolution for 2k/8k television using wavelet-based image registration. In: IEEE Global Conference on Signal and Information Processing (GlobalSIP), pp. 378–382 (2017)
25. Sajjadi, M.S., Vemulapalli, R., Brown, M.: Frame-recurrent video super-resolution. In: Proceedings of the IEEE Conference on Computer Vision and Pattern Recognition, pp. 6626–6634 (2018)
26. Seibel, H., Goldenstein, S., Rocha, A.: Eyes on the target: super-resolution and license-plate recognition in low-quality surveillance videos. IEEE Access **5**, 20020–20035 (2017)
27. Shi, W., et al.: Real-time single image and video super-resolution using an efficient sub-pixel convolutional neural network. In: Proceedings of the IEEE Conference on Computer Vision and Pattern Recognition, pp. 1874–1883 (2016)
28. Tao, X., Gao, H., Liao, R., Wang, J., Jia, J.: Detail-revealing deep video super-resolution. In: Proceedings of the IEEE International Conference on Computer Vision, pp. 4472–4480 (2017)
29. Tekalp, A.M.: Digital Video Processing. Prentice Hall Press (2015)
30. Tian, Y., Zhang, Y., Fu, Y., Xu, C.: TDAN: Temporally-deformable alignment network for video super-resolution. In: Proceedings of the IEEE/CVF Conference on Computer Vision and Pattern Recognition, pp. 3360–3369 (2020)
31. Umeda, S., Yano, N., Watanabe, H., Ikai, T., Chujoh, T., Ito, N.: HDR video super-resolution for future video coding. In: 2018 International Workshop on Advanced Image Technology (IWAIT), pp. 1–4 (2018)
32. Vaswani, A., et al.: Attention is all you need. In: Advances in Neural Information Processing Systems, pp. 5998–6008 (2017)
33. Wang, H., Su, D., Liu, C., Jin, L., Sun, X., Peng, X.: Deformable non-local network for video super-resolution. IEEE Access **7**, 177734–177744 (2019)
34. Wang, W., Ren, C., He, X., Chen, H., Qing, L.: Video super-resolution via residual learning. IEEE Access **6**, 23767–23777 (2018)
35. Wang, X., Chan, K.C., Yu, K., Dong, C., Change Loy, C.: EDVR: video restoration with enhanced deformable convolutional networks. In: Proceedings of the IEEE Conference on Computer Vision and Pattern Recognition Workshops (2019)

36. Xiang, X., Tian, Y., Zhang, Y., Fu, Y., Allebach, J.P., Xu, C.: Zooming Slow-Mo: fast and accurate one-stage space-time video super-resolution. In: Proceedings of the IEEE/CVF Conference on Computer Vision and Pattern Recognition, pp. 3370–3379 (2020)
37. Xingjian, S., Chen, Z., Wang, H., Yeung, D.Y., Wong, W.K., Woo, W.c.: Convolutional LSTM network: a machine learning approach for precipitation nowcasting. In: Advances in Neural Information Processing Systems, pp. 802–810 (2015)
38. Xu, X., Li, X.: SCAN: spatial color attention networks for real single image super-resolution. In: 2019 IEEE/CVF Conference on Computer Vision and Pattern Recognition Workshops (CVPRW), pp. 2024–2032 (2019)
39. Xu, X., Ye, Y., Li, X.: Joint demosaicing and super-resolution (JDSR): network design and perceptual optimization. IEEE Trans. Comput. Imaging, 1 (2020)
40. Xue, T., Chen, B., Wu, J., Wei, D., Freeman, W.T.: Video enhancement with task-oriented flow. Int. J. Comput. Vis. **127**(8), 1106–1125 (2019)
41. Zhang, Y., Li, K., Li, K., Wang, L., Zhong, B., Fu, Y.: Image super-resolution using very deep residual channel attention networks. In: Ferrari, V., Hebert, M., Sminchisescu, C., Weiss, Y. (eds.) ECCV 2018. LNCS, vol. 11211, pp. 294–310. Springer, Cham (2018). https://doi.org/10.1007/978-3-030-01234-2_18
42. Zhu, X., Hu, H., Lin, S., Dai, J.: Deformable convnets V2: more deformable, better results. In: Proceedings of the IEEE Conference on Computer Vision and Pattern Recognition, pp. 9308–9316 (2019)

Multi-objective Reinforced Evolution in Mobile Neural Architecture Search

Xiangxiang Chu(✉), Bo Zhang, and Ruijun Xu

Xiaomi AI Lab, Beijing, China
{chuxiangxiang,zhangbo11,xuruijun}@xiaomi.com

Abstract. Fabricating neural models for a wide range of mobile devices is a challenging task due to highly constrained resources. Recent trends favor neural architecture search involving evolutionary algorithms (EA) and reinforcement learning (RL), however, they are separately used. In this paper, we present a novel multi-objective algorithm called MoreMNAS (**M**ulti-**O**bjective **R**einforced **E**volution in **M**obile **N**eural **A**rchitecture **S**earch) by leveraging good virtues from both sides. Particularly, we devise a variant of multi-objective genetic algorithm NSGA-II, where mutations are performed either by reinforcement learning or a natural mutating process. It maintains a delicate balance between *exploration* and *exploitation*. Not only does it prevent the search degradation, but it also makes use of the learned knowledge. Our experiments conducted in Super-resolution domain (SR) deliver rivaling models compared with some state-of-the-art methods at fewer FLOPS (Evaluation code can be found at https://github.com/xiaomi-automl/MoreMNAS).

Keywords: Image super-resolution · Neural Architecture Search

1 Introduction

Automated neural architecture search (NAS) has witnessed a victory versus human experts, confirming itself as the next generation paradigm of architectural engineering. The innovations are exhibited mainly in three parts: search space design, search strategy, and evaluation techniques.

Firstly, the search space of neural architectures is tailored in various forms regarding different search strategies. It was initially represented as a sequence of parameters to describe raw layers in NAS [42], followed by MetaQNN [2], ENAS [28], in which the selection of parameters is essentially finding subgraphs in a single directed acyclic graph (DAG). Inspired by successful modular design as in Inception [35], meta-architectures like predetermined placement of cells [22,43] or stacks of blocks [9] become a favourite. Each block or cell is a composition of layers, the search space is then outlined by operations within each module,

Electronic supplementary material The online version of this chapter (https://doi.org/10.1007/978-3-030-66823-5_6) contains supplementary material, which is available to authorized users.

A. Bartoli and A. Fusiello (Eds.): ECCV 2020 Workshops, LNCS 12538, pp. 99–113, 2020.
https://doi.org/10.1007/978-3-030-66823-5_6

like altering filter size and number, varying layer type, adding skip connections etc. Fine granular network description down to raw layers is more flexible but less tractable, while the coarse one with modular design is merely the opposite. Whereas in most genetic settings, a genotype representation such as an encoding of binary strings is preferred as in GeneticCNN [40], and in NSGA-Net [24].

Secondly, good search strategy can avoid abusive search in the vast space of neural architectures. Diverse investigations have been made with both evolutionary algorithms and reinforced learning, though mostly independent. Pure reinforced methods was initiated by NAS [42], later echoed by NASNet [43], ENAS [28], MetaQNN [2], MnasNet [37], MONAS [17] etc. Meanwhile, evolutionary approaches contain GeneticCNN [40], NEMO [19], LEMONADE [10], [29], NSGA-Net [24]. Other searching tactics like sequential model-based optimization (SMBO) are also exhibited in DeepArchitect [27], PNAS [22], and in PPP-Net [9]. Notice that multi-objective solutions exist in all categories. Nevertheless, the idea of putting RL and EA methods together is alluring since the former might degrade and the latter is sometimes less efficient, RENAS [3] attempted to mitigate this problem with integration of both methods, but unfortunately, they made the false claim since it still fails to converge.

Thirdly, typical evaluation involves training generated models and performing validation on a held-out set. Due to its computational cost, these models are normally not fully trained, based on the empirical conception that better models usually win at early stages. Less training data also facilitates the process but introduces biases. Other approaches seek to save time by initializing weights of newly morphed models with trained ones [10,28].

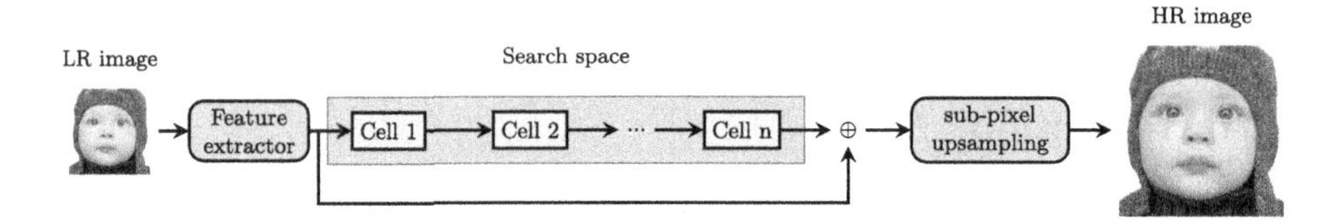

Fig. 1. A typical Super-Resolution network adapted for neural architecture search. The main body within the long-term residual block is constructed as searchable cells.

In this paper, we demonstrate a multi-objective reinforced evolutionary approach MoreMNAS to resolve inherent issues of each approach. Our main contributions are:

- applying multi-objective NAS in Super-resolution domain other than common classification tasks, our results dominate some of the state-of-the-art deep-learning based SR methods, with the best models aligned near the Pareto front in a single run,
- designing a cell-based search space (see Fig. 1) to allow for genetic crossover and mutation, particularly for the image super-resolution task, upon which we propose a hierarchical organization of reinforced mutator and natural mutation to speed up genetic selection,

- leveraging the advantages from both evolutionary algorithms and reinforcement learning to perform multi-objective NAS while overcoming the drawbacks from each method,
- involving minimum human expertise in model designing as an early guide, and imposing some practical constraints to obtain feasible solutions.

Experiments show that our generated models dominate some of the state-of-the-art methods: SRCNN [7], FSRCNN [8], and VDSR [18].

2 Related Work

2.1 Single-Objective Oriented NAS

Reinforcement Learning-Based Approaches. Excessive research works have applied reinforcement learning in neural architecture search as aforementioned. They can be loosely divided into two genres according to the type of RL techniques: Q-learning and Policy Gradient.

For Q-learning based methods like MetaQNN [2], a learning agent interacts with the environment by choosing CNN architectures from finite search space. It stores validation accuracy and architecture description in replay memory to be periodically sampled. The agent is enabled with ϵ-greedy strategy to balance exploration and exploitation.

In contrast, policy gradient-based methods feature a recurrent neural controller (RNN or LSTM) to generate models, with its parameters updated by a family of Policy Gradient algorithms: REINFORCEMENT [17,28,42], Proximal Policy Optimization [37,43]. The validation accuracy of models is constructed as a reward signal. The controller thus learns from experience and creates better models over time. Although some of them have produced superior models over human-designed ones on pilot tasks like CIFAR-10 and MNIST, these RL methods are subject to convergence problems, especially when scaling is involved. Besides, they are mainly single-objective, limiting its use in practice.

Evolution-Based Approaches. Pioneering studies on evolutionary neural architecture search form a subfield called Neuroevolution. For instance, NEAT [33] has given an in-depth discussion of this field, while itself evolves network topology along with weights to improve efficiency, but till recent works like GeneticCNN [40], [23], and [29], evolutionary approaches become substantially enhanced and practical to apply. Specifically, a population of neural models is initialized either randomly or non-trivially. It propagates itself through *crossover* and *mutation*, or *network morphism*, less competitive models are eliminated while the others continue to evolve. In this way, the reduction of traverses in search space is paramount. The selection of an individual model in each generation is based on its *fitness*, e.g. higher validation accuracy. In general, recent EA methods are proved to be on par with their RL counterparts and superior against human artistry, attested by AmoebaNet-A presented in [29].

Reinforced Evolution-Based Approaches. An attempt to resolve the gap between EA and RL goes to RENAS [3], where a reinforced mutation replaces random mutation in order to avoid degradation. The integration, however, fails to assure convergence in the stability test, because the selected genetic algorithm doesn't preserve advantages among generations. Better models once generated are possibly removed during its evolution process.

2.2 Multi-objective Oriented NAS

Deploying neural models in a wide range of settings naturally requires a multi-objective perspective. MONAS [17] extends NAS by scheming a linear combination of prediction accuracy and other objectives in this regard. But according to [6], a linear combination of objectives is suboptimal. Hence it is necessary to embed this search problem in a real multi-objective context, where a series of models is found along the *Pareto front* of multi-objectives like accuracy, computational cost or inference time, number of parameters, etc.

LEMONADE [10] utilizes Lamarckian inheritance mechanism which generates children with performance warm-started by their parents. It also makes the use of the fact that the performance of models (proxied by the number of parameters, multi-adds) are much easier to calculate than evaluation.

PPP-Net [9] is a multi-objective extension to [22], where Pareto-optimal models are picked in each generation. It also introduces a RNN regressor to speed up model evaluation.

NEMO [19] and NSGA-Net [24] adopt the classic non-dominated search algorithm (NSGA-II) to handle trade-off among multi-objectives. It groups models based on dominance, while measuring crowding distance gives priority to models within the same front. Besides, NSGA-Net uses Bayesian Optimization to profit from search history.

2.3 Image Super-Resolution and NAS

Neural architecture search has also attracted a little attention for Image super-resolution as in [32], in which it adopts an evolutionary algorithm to search from already well-designed efficient dense residual blocks. In contrast, we refrain ourselves from involving too many human priors and search from basic building blocks. Generally, NAS methods can profit better contrived search spaces. We instead employ a coarse-grained search space to evaluate the effectiveness and the generalization ability of our approach. This treatment also makes a fair comparison to the cited methods.

3 Multi-objective Reinforced Evolution

3.1 Building Single Image Super-Resolution as a CMOP

Single image super-resolution (SISR) is a classical low-level task in computer vision. Deep learning methods have obtained impressive results with a large margin than other methods. However, most of the studies concentrate on designing deeper and more complicated networks to boost PSNR (peak-signal-noise-ratio) and SSIM (structural similarity index), both of which are commonly used evaluation metrics [16,39].

In practice, the applications of SISR are inevitably constricted. For example, mobile devices usually have too limited resources to afford heavy computation of large neural models. Moreover, mobile users are so sensitive to responsiveness that the inference time spent on a forward calculation must be seriously taken into account. As a result, engineers are left with laborious model searching and tuning for devices of different configurations. In fact, It is a multi-objective problem (MOP). Natural thinking that transforms MOP into a single-objective problem by weighted summation is however impractical under such situation. In addition, some practical constraints such as minimal acceptable PSNR must be taken into account.

Here, we rephrase our single image super-resolution as a constrained multi-objective problem as follows,

$$\begin{aligned} \min_{m \in S} \quad & objs(m) \quad s.t. \quad psnr_{-} < psnr(m) < psnr_{+}, \\ & flops_{-} < flops(m) < flops_{+}, \quad params_{-} < params(m) < params_{+} \\ \text{where} \quad & objs(m) = \{(-psnr(m), flops(m), params(m)) | m \in S\}. \end{aligned} \tag{1}$$

In Eq. 1, m is a model from the search space S, and our objectives are to maximize the PSNR of a model evaluated on benchmarks while minimizing its FLOPS and number of parameters. The $_{-}$ subscript represents a lower bound and $_{+}$ means the corresponding upper constraint. In practice, $flops_{-}$ and $params_{-}$ can be set as zero to loosen the left side constraints while the $psnr_{+}$ can be set as $+\infty$ for the right side. Here, the model architecture is the *decision variable*. Generally speaking, the above objectives are competing or even conflicting. As a consequence, no model can achieve an optimum across all objectives.

3.2 Search Pipeline

As shown in Fig. 2, MoreMNAS contains three basic components: cell-based search space, a model generation **controller** regarding multi-objective based on NSGA-II and an **evaluator** to return multi-reward feedback for each child model. In our pipeline, the controller samples models from the search space, and it dispatches them to the evaluator to measure their performances, producing feedback for the controller in return.

In specific, the search space is composed of n cells, and each cell contains one or more basic operator blocks. Each block within a cell has m repeated basic operators so that it can represent various architectures. Moreover, all cells can be different from others. For simplicity, our cells share the same search space of repeated blocks.

As for the controller, we use an NSGA-II variation similar to NSGA-Net [24]. Unlike NSGA-Net, we perform crossovers on a cell-based architecture other than on binary encodings. Furthermore, our approach differs from NSGA-Net by a

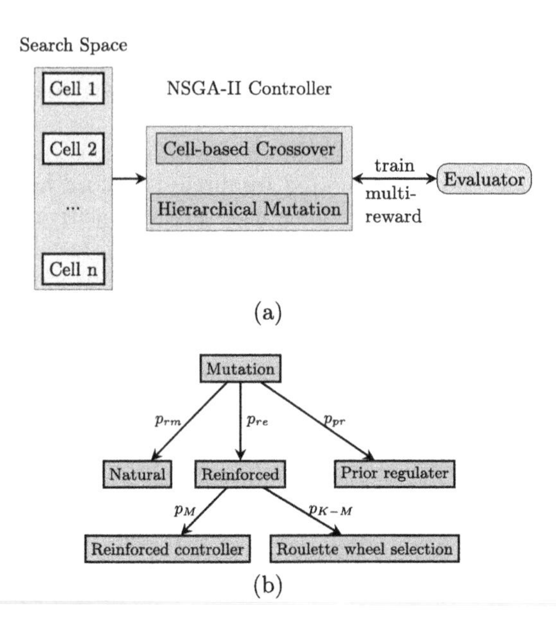

Fig. 2. (a) Our MoreMNAS searching pipeline. The search space is cell-based, from which we sample a model to train for some epochs for evaluation. In turn, the evaluated performance and network measures (no. of parameters, multi-adds) serve as multi-reward for reinforced evolutionary controller. Each cell is treated as a genotype for crossover and mutation. (b) The mutation scheme is set in a hierarchical manner, where each type of mutation has a preset probability p_x.

two-level hierarchical mutation process, which balances the trade-off between exploration and exploitation.

3.3 Search Space

Proper search space design is of great benefit to boost final performance. In fact, macro-level search [42] is harmful to mobile devices regarding some underlying hardware designs [25]. It suffers too much from the non-regularized combinations of arbitrary basic operators. Therefore, quite a few recent research no longer searches on the basis of wild basic operators but makes use of some good cells already discovered [37,43]. Partly motivated by MNAS [37], we also construct neural models on top of various cells. Unlike classification tasks, a super-resolution process can be divided into three consecutive sub-procedures: feature extraction, non-linear mapping, and reconstruction [7]. Most of the hottest research focuses on non-linear mapping process [1,7,12,20] where the final performance matters most. Therefore, we build our search space following this procedure (Fig. 1).

Specifically, our search space is composed of n_p cells, and each cell contains an amount of repeated basic operators. Different from [43], we don't place the same structure constraint for cells in one model. Hence, the operations for a single model can be represented by $s = (s_1, ..., s_n)$, where s_n is an element from the search space S_n for $cell_n$. Within S_i, we use the combinations to act as a basic element. Particularly, $cell_i$ contains the following basic operators for super-resolution,

- basic convolutions: 2D convolution, inverted bottleneck convolution with an expansion rate of 2, grouped convolution with groups in {2, 4}
- kernel sizes in {1, 3}
- filter numbers in {16, 32, 48, 64}
- whether or not to use skip connections
- the number of repeated blocks in {1, 2, 4}

For example, two repeated 3 × 3 2D convolutions with 16 channels is a basic operator. By this mean, each basic element can be presented by a unique index. The total search domain has the following complexity: $c(s) = \prod_{i=1}^{n} c(s_i)$. If we choose the same operator set with c elements for each cell, then the complexity can be simplified as $c(s) = c^n$, where $c = 4 \times 2 \times 4 \times 2 \times 3 = 192$ and $n = 7$ in our experiments. Moreover, this approach constructs a one-to-one mapping between the cell code and neural architectures, possessing an inherent advantage over NSGA-Net, whose space design involves a many-to-one mapping that further repetition removal steps must be taken to avoid meaningless training labors [24].[1]

Unlike classification task whose output distribution is often represented by a softmax function, single image super-resolution transforms a low-resolution picture to a high-resolution target. Its search space is composed of operators different from those of classification. If pooling and batch normalization are included, super-resolution results will deteriorate [20]. Besides, a down-sampling operation is excluded since it only blurs the features irreversibly.

3.4 Model Generation Based on NSGA-II with Cell-Level Crossover and Hierarchical Mutation

Population Initialization. A good initialization strategy usually involves good diversities, which is beneficial to speed up the evaluation. In this paper, we initialize each individual by randomly selecting basic operators for each cell from search space and repeating the cell for n times to build a model. We call this process *uniform initialization.* In such a way, we generate $2N$ individuals in total for the first generation.

Non-dominated Sorting. We use the same strategy as NSGA-II with an improved crowding distance to address the existing problem of original definition that doesn't differentiate individuals within the same cuboid [5].

[1] In fact, there are some many-to-one mappings within this design. However, their low probability makes them ignorable.

The improved crowding distance for individual j in n-th order for objective k is defined as follows,

$$dis^j = \sum_{k=1}^{K} \frac{f_{n+1}^k - f_n^k}{f_{max}^k - f_{min}^k} \tag{2}$$

where f_n^k is a fitness value for an individual in n-th order sorted by objective k.

Cell-Based Crossover. Unlike the encoding model with 0–1 bits in NSGA-Net, we take a more natural approach to perform crossover as the model consists of various cells by design. A crossover of two individuals (i.e., models) $x = (x^1, ..., x^i, ..., x^{n_p})$ and $y = (y^1, ..., y^i ..., y^{n_p})$ can result in a new child $z = (x^1, ..., x^{i-1}, y^i, x^{i+1}, ..., x^{n_p})$ when a single-point crossover is performed at position i. Other strategies such as *two-point* and k-*point* crossovers can also be applied [11,15,41].

Hierarchical Mutation. The top level of hierarchical mutation includes three mutually exclusive processes: *reinforced dominant*, *natural* and *prior-regularized* mutation. Whenever a new individual is prepared to be mutated, we sample from a category distribution over these three mutations, i.e. $p(mutation) = \{p_{re}, p_{rm}, p_{pr}\}$. The word reinforced dominant means that we use reinforcement learning to minimize objectives which are otherwise hard to predict. Actually, almost all objectives of multi-objective problems can be classified into two categories: those difficult to predict and the others not. Without loss of generality, we define K objectives, the first M ones are difficult to predict, and the remaining ones are not.

In super-resolution domain, for instance, the *mean square error* (MSE) is hard to obtain since it is computed between a ground truth high-resolution picture and the one generated by a deep neural model. In our optimization problem defined by Eq. 1, we have $K = 3$ objectives in total, where $M = 1$ objective is $-psnr$, and $K - M = 2$ objectives are respectively *multi-adds* and *number of parameters*. We take different steps to handle these two categories. For the former, we use M reinforced controllers, one for each objective. For the latter, we use *Roulette-wheel selection* to sample one cell in each step [21].

We also take advantage of learned knowledge from training generated models and their scores of objectives based on a reinforced mutation controller, which is designed to distill meaningful experience to evaluate model performance. This part of mutation can then be regarded as exploitation. In this procedure, the model generation is described as *Markov Decision Process* (MDP), which is represented by a LSTM controller [14] shown in Fig. 3. Initially, the zero state is reset for the LSTM controller, and a cell with index zero is injected into the LSTM controller after embedding. Here the zero index refers to a *null* element which is not related to any basic operator in S. Then the controller samples an action from S with a softmax function, i.e. category distribution. In turn, this sampled action serves as an input for the next step. This process is repeated until n cells are generated to build a complete model.

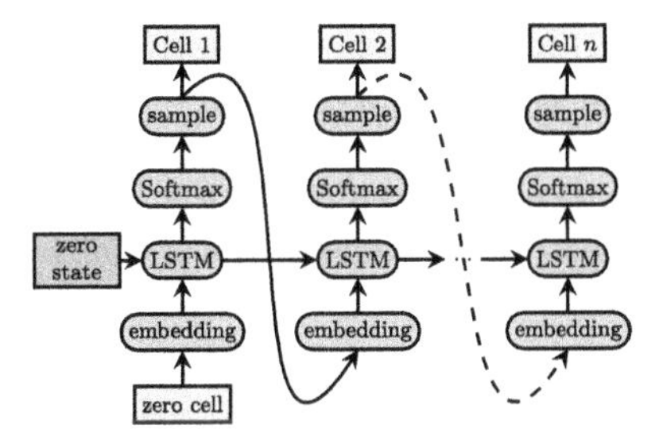

Fig. 3. Controller network for reinforced mutation.

From the perspective of reinforcement learning, each episode contains n steps. Note that the reward is delayed since it cannot be returned timely until the last cell is generated. In fact, PSNR is not suitable for a direct reward since it is measured in the logarithmic space. Instead, we define the reward based on the mean squared error between a generated high-resolution picture and the ground truth. In particular, the reward of a model m is defined as

$$reward(m) = \frac{0.001}{mse_m} - 0.5 \tag{3}$$

The minimization operator is a type of clipping trick to avoid instability from getting a too large value as a reward. Since the reward is obtained after model m is evaluated, which ends an episode, we adopt REINFORCEMENT to update this controller [34].

Under the assumption of MDP, given a policy π_θ represented by a neural network parameterized by θ, its gradient can be estimated in the form of mini-batch B by,

$$\begin{aligned} \nabla_\theta \pi_\theta &= \frac{1}{B} \sum_{b=1}^{B} \sum_{t=1}^{n} r_{t(b)} \log p_\theta(a_{t(b)} | s_{t(b)}) \\ &= \frac{1}{B} \sum_{b=1}^{B} \sum_{t=1}^{n} \gamma^t r_b \log p_\theta(a_{t(b)} | s_{t(b)}) \end{aligned} \tag{4}$$

where r_b is the reward defined by Eq. 3 for b-th model and γ is a discount coefficient. For simplicity, each cell has the same space configuration S.

Furthermore, natural mutation is applied to encourage exploration throughout the whole evolution. Since *elitist preservation mechanism* can afford no degradation during the model generation process, there is no need to take measures to weaken this natural mutation in the process of evolution. It comes with a byproduct of better exploration. Whereas for mutation controllers based on pure reinforcement learning, some strategies such as discouraging exploration gradually along with the training process are indispensable.[2]

[2] Taking DQN [26] for example, decreasing the hyper-parameter ϵ that allows for exploration during the learning process is a commonly used trick.

As per prior regularized mutation, we are partly motivated by the guidelines from ShuffleNet [25], in which some experimental suggestions regarding hardware implementation are proposed to decrease the running time costs, so did we draw ideas from various renowned neural architectures like ResNet [13] and MobileNet [30]. Apart from that, repeating the same operator many times to build a model is also proved advantageous. However, we introduce this prior with a small probability. To be more precise, we randomly choose a target position i for model $x = (x_1, ..., x_i, ..., x_n)$, and we mutate it to generate $x_{child} = (x_i, ..., x_i, ..., x_i)$.

Table 1. Hyperparameters for our search pipeline. P: Population, MR: mutation ratio

ITEM	VALUE	ITEM	VALUE
P	56	MR	0.8
p_{rm}	0.50	p_{re}	0.45
p_{pr}	0.05	p_M	0.75
p_{K-M}	0.25		

4 Experiments

Setup and Searching. In our experiment, the whole pipeline contains 200 generations, during each generation 56 individuals get spawned, i.e., there are 11200 models generated in total. Other hyper-parameters is listed in Table 1. In addition, all experiments are performed on a single Tesla V100 machine with 8 GPUs and a complete run takes about 7 days. Detailed hyper-parameters are shown in Tabel 1. Figure 5 plots the evolutionary process where the latest models push the Pareto front further.

Evaluator. Each evaluator trains a dispatched model on DIV2K dataset [38] across 200 epochs with a batch size of 16. In specific, the first 800 high-resolution (HR) pictures are used to construct the whole training set. The HR pictures are randomly cropped to 80 × 80, and then down-sampled to LR by bicubic interpolation. Furthermore, we use random flipping and rotation with $90°, 180°$ and $270°$ to perform data augmentation. Both feature extraction and restoration stage in Fig. 1 are composed of 32 3×3 2D convolution filters with a unit stride. Moreover, ReLU acts as the default activation function except for the last layer. We use Adam optimizer to train the model with $\beta_1 = 0.9$ and $\beta_2 = 0.999$. The initial learning rate is 1×10^{-4}, and decays by half every 100 epochs. In addition, we use L_1 loss between the generated HR images and the ground truth to guide the back-propagation.

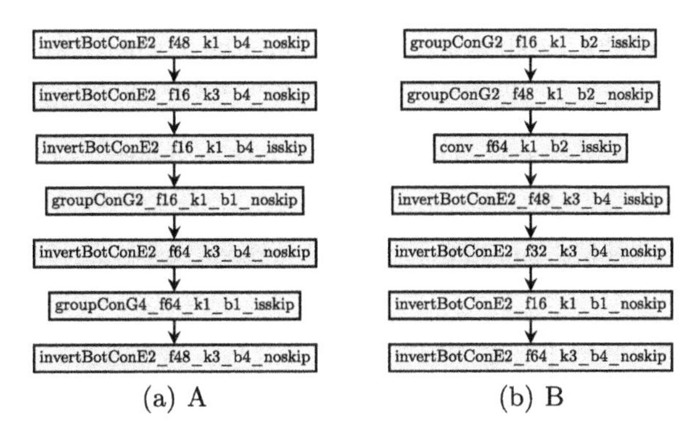

Fig. 4. Our searched models MoreMNAS-A and MoreMNAS-B. Note for instance, invertBotConEx_fy_kz_bu_isskip indicates it is an inverted bottleneck block with the first convolution of an expansion rate of x, a filter size (channel) of y, a kernel size of z and the block is repeated for u times. Whether to add the residual structure is decided by 'isskip' and 'noskip'.

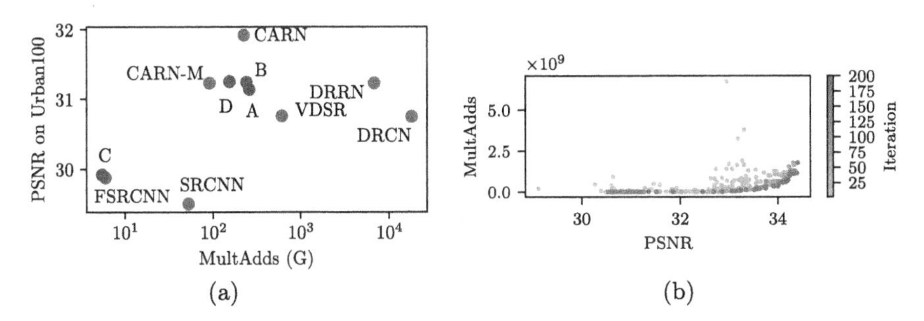

Fig. 5. (a) Our searched MoreMNAS-A, B, C, D (red dots) vs. other handcrafted models, (b) All the models explored in our reinforced evolution pipeline, near Pareto-front models are shown in magenta. (Color figure online)

Full Training. We select several models (see Fig. 4) located at our Pareto front for complete training using the same training set as well as data augmentation tricks. The only difference is that we use a larger cropping size (128×128) and a longer epoch 4800. In addition, we also use Adam optimizer to work on this task with $\beta_1 = 0.9$ and $\beta_2 = 0.999$. In specific, the initial learning rate is 10^{-3} for large models and 10^{-4} for small ones. The learning rate decays in half every 80000 back-propagations.

Comparison with Human Designed State-of-the-Art Models. Here we only consider those SR models with comparable parameters and multi-adds. In particular, we take SRCNN [7], FSRCNN [8] and VDSR [18], and we apply our models at $\times 2$ scale since it is one of the most commonly used tasks to draw comparisons. Figure 5 illustrates our results compared with other state-of-the-art models on Urban100 dataset.

The detailed results are shown in Table 2 and Fig. 6, where two models we call MOREMNAS-A and B hit higher PSNR and SSIM than VDSR across four evaluation sets with much fewer multi-adds. Another model called MOREMNAS-D dominates VDSR across three objectives with a quarter of its multi-adds.

Table 2. Comparisons with the state-of-the-art methods based on $\times 2$ super-resolution task†

Model	MultAdds	Params	SET5	SET14	B100	Urban100
			PSNR/SSIM	PSNR/SSIM	PSNR/SSIM	PSNR/SSIM
SRCNN [7]	52.7G	57K	36.66/0.9542	32.42/0.9063	31.36/0.8879	29.50/0.8946
FSRCNN [8]	6.0G	12K	37.00/0.9558	32.63/0.9088	31.53/0.8920	29.88/0.9020
VDSR [18]	612.6G	665K	37.53/0.9587	33.03/0.9124	31.90/0.8960	30.76/0.9140
DRRN [36]	6,796.9G	297K	37.74/0.9591	33.23/0.9136	32.05/0.8973	31.23/0.9188
MoreMNAS-A	238.6G	1039K	37.63/0.9584	33.23/0.9138	31.95/0.8961	31.24/0.9187
MoreMNAS-B	256.9G	1118K	37.58/0.9584	33.22/0.9135	31.91/0.8959	31.14/0.9175
MoreMNAS-C	5.5G	25K	37.06/0.9561	32.75/0.9094	31.50/0.8904	29.92/0.9023
MoreMNAS-D	152.4G	664K	37.57/0.9584	33.25/0.9142	31.94/0.8966	31.25/0.9191

† *The multi-adds are valued on* 480×480 *input resolution.*

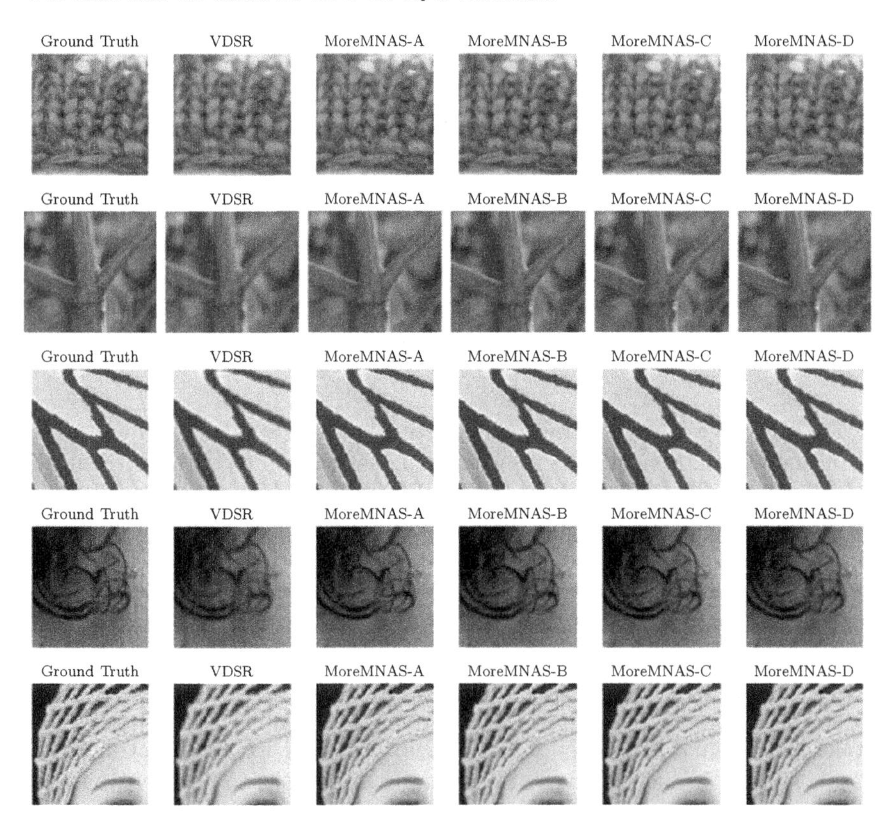

Fig. 6. Image super-resolution ($\times 2$) visualization of MoreMNAS models compared with VDSR [18] on Set5.

Besides, a light-weight model MOREMNAS-C matches FSRCNN with fewer multi-adds, which again dominates SRCNN across three aspects: higher score, fewer number of parameters and multi-adds. In addition, the non-linear mapping stage contains various cell blocks with a unit kernel, which shares some similarities with FSRCNN.

5 Conclusion

In this paper, we propose a multi-objective reinforced evolutionary algorithm in mobile neural architecture search, which seeks a better trade-off among various competing objectives. It has three obvious advantages: no recession of models during the whole process, good exploitation from reinforced mutation, a better balance of different objectives based on NSGA-II and Roulette-wheel selection. To our best knowledge, this is the first work to investigate multi-objective neural architecture search by combining NSGA-II and reinforcement learning.

Our method is evaluated in the super-resolution domain. We generate several light-weight models that are very competitive, sometimes dominating human expert designed ones such as SRCNN, VDSR across several critical objectives: PSNR, multi-adds, and the number of parameters. Finally, our algorithm can be applied in other situations, not only limited to a mobile setting. Noticeably, our method can benefit from SR-specific search spaces as in [32] to have enhanced performance.

Our future work will focus on accelerating the whole pipeline based on a regression evaluation from our sampled data. Second, fine-tuning hyper-parameters and replacing REINFORCEMENT with more powerful policy gradient algorithms such as PPO [31] and POP3D [4], which have more potential to push the Pareto front further.

References

1. Ahn, N., Kang, B., Sohn, K.A.: Fast, accurate, and lightweight super-resolution with cascading residual network. arXiv preprint arXiv:1803.08664 (2018)
2. Baker, B., Gupta, O., Naik, N., Raskar, R.: Designing neural network architectures using reinforcement learning. In: International Conference on Learning Representations (2017)
3. Chen, Y., Zhang, Q., Huang, C., Mu, L., Meng, G., Wang, X.: Reinforced evolutionary neural architecture search. arXiv preprint arXiv:1808.00193 (2018)
4. Chu, X.: Policy optimization with penalized point probability distance: an alternative to proximal policy optimization. arXiv preprint arXiv:1807.00442 (2018)
5. Chu, X., Yu, X.: Improved crowding distance for NSGA-II. arXiv preprint arXiv:1811.12667 (2018)
6. Deb, K., Pratap, A., Agarwal, S., Meyarivan, T.: A fast and elitist multiobjective genetic algorithm: NSGA-II. IEEE Trans. Evol. Comput. **6**(2), 182–197 (2002)
7. Dong, C., Loy, C.C., He, K., Tang, X.: Learning a deep convolutional network for image super-resolution. In: Fleet, D., Pajdla, T., Schiele, B., Tuytelaars, T. (eds.) ECCV 2014. LNCS, vol. 8692, pp. 184–199. Springer, Cham (2014). https://doi.org/10.1007/978-3-319-10593-2_13

8. Dong, C., Loy, C.C., Tang, X.: Accelerating the super-resolution convolutional neural network. In: Leibe, B., Matas, J., Sebe, N., Welling, M. (eds.) ECCV 2016. LNCS, vol. 9906, pp. 391–407. Springer, Cham (2016). https://doi.org/10.1007/978-3-319-46475-6_25
9. Dong, J.D., Cheng, A.C., Juan, D.C., Wei, W., Sun, M.: PPP-Net: platform-aware progressive search for pareto-optimal neural architectures. In: ICLR Workshop (2018). https://openreview.net/forum?id=B1NT3TAIM
10. Elsken, T., Metzen, J.H., Hutter, F.: Efficient multi-objective neural architecture search via Lamarckian evolution. In: ICLR (2018)
11. Gwiazda, T.D.: Crossover for Single-objective Numerical Optimization Problems, vol. 1. Tomasz Gwiazda (2006)
12. Haris, M., Shakhnarovich, G., Ukita, N.: Deep backprojection networks for super-resolution. In: Conference on Computer Vision and Pattern Recognition (2018)
13. He, K., Zhang, X., Ren, S., Sun, J.: Deep residual learning for image recognition. In: Proceedings of the IEEE Conference on Computer Vision and Pattern Recognition, pp. 770–778 (2016)
14. Hochreiter, S., Schmidhuber, J.: Long short-term memory. Neural Comput. **9**(8), 1735–1780 (1997)
15. Holland, J.H.: Adaptation in natural and artificial systems: an introductory analysis with applications to biology, control, and artificial intelligence (1975)
16. Hore, A., Ziou, D.: Image quality metrics: PSNR vs. SSIM. In: 2010 20th International Conference on Pattern Recognition (ICPR), pp. 2366–2369. IEEE (2010)
17. Hsu, C.H., et al.: MONAS: multi-objective neural architecture search using reinforcement learning. arXiv preprint arXiv:1806.10332 (2018)
18. Kim, J., Kwon Lee, J., Mu Lee, K.: Accurate image super-resolution using very deep convolutional networks. In: Proceedings of the IEEE Conference on Computer Vision and Pattern Recognition, pp. 1646–1654 (2016)
19. Kim, Y.H., Reddy, B., Yun, S., Seo, C.: NEMO: neuro-evolution with multiobjective optimization of deep neural network for speed and accuracy. In: JMLR: Workshop and Conference Proceedings (2017)
20. Lim, B., Son, S., Kim, H., Nah, S., Lee, K.M.: Enhanced deep residual networks for single image super-resolution. In: The IEEE Conference on Computer Vision and Pattern Recognition (CVPR) Workshops, vol. 1, p. 4 (2017)
21. Lipowski, A., Lipowska, D.: Roulette-wheel selection via stochastic acceptance. Phys. Stat. Mech. Appl. **391**(6), 2193–2196 (2012)
22. Liu, C., et al.: Progressive neural architecture search. arXiv preprint arXiv:1712.00559 (2017)
23. Liu, H., Simonyan, K., Vinyals, O., Fernando, C., Kavukcuoglu, K.: Hierarchical representations for efficient architecture search. arXiv preprint arXiv:1711.00436 (2017)
24. Lu, Z., Whalen, I., Boddeti, V., Dhebar, Y., Deb, K., Goodman, E., Banzhaf, W.: NSGA-Net: a multi-objective genetic algorithm for neural architecture search. arXiv preprint arXiv:1810.03522 (2018)
25. Ma, N., Zhang, X., Zheng, H.T., Sun, J.: ShuffleNet V2: practical guidelines for efficient CNN architecture design. arXiv preprint arXiv:1807.11164 1 (2018)
26. Mnih, V., et al.: Human-level control through deep reinforcement learning. Nature **518**(7540), 529 (2015)
27. Negrinho, R., Gordon, G.: Deeparchitect: automatically designing and training deep architectures. arXiv preprint arXiv:1704.08792 (2017)
28. Pham, H., Guan, M.Y., Zoph, B., Le, Q.V., Dean, J.: Efficient neural architecture search via parameter sharing. arXiv preprint arXiv:1802.03268 (2018)

29. Real, E., Aggarwal, A., Huang, Y., Le, Q.V.: Regularized evolution for image classifier architecture search. arXiv preprint arXiv:1802.01548 (2018)
30. Sandler, M., Howard, A., Zhu, M., Zhmoginov, A., Chen, L.C.: MobileNetV 2: inverted residuals and linear bottlenecks. In: Proceedings of the IEEE Conference on Computer Vision and Pattern Recognition, pp. 4510–4520 (2018)
31. Schulman, J., Wolski, F., Dhariwal, P., Radford, A., Klimov, O.: Proximal policy optimization algorithms. arXiv preprint arXiv:1707.06347 (2017)
32. Song, D., Xu, C., Jia, X., Chen, Y., Xu, C., Wang, Y.: Efficient residual dense block search for image super-resolution. In: Association for the Advancement of Artificial Intelligence (2020)
33. Stanley, K.O., Miikkulainen, R.: Evolving neural networks through augmenting topologies. Evol. Comput. **10**(2), 99–127 (2002)
34. Sutton, R.S., Barto, A.G.: Reinforcement Learning: An Introduction. MIT Press, Cambridge (2018)
35. Szegedy, C., et al.: Going deeper with convolutions. In: Proceedings of the IEEE Conference on Computer Vision and Pattern Recognition, pp. 1–9 (2015)
36. Tai, Y., Yang, J., Liu, X.: Image super-resolution via deep recursive residual network. In: Proceedings of the IEEE Conference on Computer Vision and Pattern Recognition, vol. 1, p. 5 (2017)
37. Tan, M., Chen, B., Pang, R., Vasudevan, V., Le, Q.V.: MnasNet: platform-aware neural architecture search for mobile. arXiv preprint arXiv:1807.11626 (2018)
38. Timofte, R., et al.: NTIRE 2017 challenge on single image super-resolution: methods and results. In: 2017 IEEE Conference on Computer Vision and Pattern Recognition Workshops (CVPRW), pp. 1110–1121. IEEE (2017)
39. Wang, Z., Bovik, A.C., Sheikh, H.R., Simoncelli, E.P.: Image quality assessment: from error visibility to structural similarity. IEEE Trans. Image Process. **13**(4), 600–612 (2004)
40. Xie, L., Yuille, A.: Genetic CNN. In: ICCV, pp. 1388–1397, October 2017. https://doi.org/10.1109/ICCV.2017.154
41. Yu, X., Gen, M.: Introduction to Evolutionary Algorithms. Springer, London (2010). https://doi.org/10.1007/978-1-84996-129-5
42. Zoph, B., Le, Q.V.: Neural architecture search with reinforcement learning. arXiv preprint arXiv:1611.01578 (2016)
43. Zoph, B., Vasudevan, V., Shlens, J., Le, Q.V.: Learning transferable architectures for scalable image recognition. arXiv preprint arXiv:1707.07012 2(6) (2017)

Deep Plug-and-Play Video Super-Resolution

Hannan Lu[1], Chaoyu Tong[1], Wei Lian[2], Dongwei Ren[3], and Wangmeng Zuo[1](✉)

[1] Harbin Institute of Technology, Harbin, China
{hannanlu,wmzuo}@hit.edu.cn, 19S003042@stu.hit.edu.cn
[2] Changzhi University, Changzhi, China
lianwei3@gmail.com
[3] Tianjin University, Tianjin, China
rendongweihit@gmail.com

Abstract. Video super-resolution (VSR) has been drawing increasing research attention due to its wide practical applications. Despite the unprecedented success of deep single image super-resolution (SISR), recent deep VSR methods devote much effort to designing modules for spatial alignment and feature fusion of multiple adjacent frames while failing to leverage the progress in SISR. In this paper, we propose a plug-and-play VSR framework, through which the state-of-the-art SISR models can be readily employed without re-training, and the proposed temporal consistency refinement network (TCRNet) can enhance the temporal consistency and visual quality. In particular, an SISR model is firstly adopted to super-resolve low-resolution video in a frame-by-frame manner. Instead of using multiple frames, our TCRNet only takes two adjacent frames as input. To alleviate the issue of spatial misalignments, we present an iterative residual refinement module for motion offset estimation. Furthermore, a deformable convolutional LSTM is proposed to exploit long-distance temporal information. The proposed TCRNet can be easily and stably trained using ℓ_2 loss function. Moreover, the VSR performance is further boosted by a bidirectional process. On popular benchmark datasets, our TCRNet can significantly enhance the temporal consistency when collaborating with various SISR models, and is superior to or at least on par with state-of-the-art VSR methods in terms of quantitative metrics and visually quality.

Keywords: Deep learning · Video super-resolution · Single image super-resolution · CNN · LSTM

Electronic supplementary material The online version of this chapter (https://doi.org/10.1007/978-3-030-66823-5_7) contains supplementary material, which is available to authorized users.

A. Bartoli and A. Fusiello (Eds.): ECCV 2020 Workshops, LNCS 12538, pp. 114–130, 2020.
https://doi.org/10.1007/978-3-030-66823-5_7

1 Introduction

Super-resolution(SR), aiming to generate high-resolution (HR) image or video from its low-resolution (LR) version, is a fundamental research topic in computer vision field. Due to the wide applications such as video surveillance and high definition display of images and videos, SR has drawn considerable research attention [1–3,7,10,11,14,34] in the past decades. Driven by the success of deep learning in low-level vision [8,26,32,33], single image super-resolution (SISR) has achieved unprecedented progress [4,19,21,30,35], where a variety of network architectures and training strategies have been developed to achieve quantitatively and qualitatively excellent performance. Video super-resolution (VSR) is more likely to achieve better visual quality due to the auxiliary information from adjacent frames, thus is drawing increasing research attention recently [3,9,11,23,25,28]. A natural solution is to super-resolve LR video using SISR method in a frame-by-frame manner. But the consecutive frames suffer from flicker artifacts as illustrated in Fig. 1, since the temporal information is not fully exploited. Therefore, existing VSR methods devote much effort to designing effective modules for spatial alignment and feature fusion of multiple adjacent frames. Optical flow is firstly adopted to predict motion offsets between adjacent frames [3,13,25], and the spatial alignment is explicitly accomplished in image level. Recently, implicit motion compensation in feature level is more preferred for the ability to better exploit temporal information. Examples include dynamic upsampling filter [12], sub-pixel motion compensation layer [24] and deformable convolution [27,29].

In most cases, existing VSR methods are individually studied without taking into account the progress of SISR, with only two exceptions which are VESPCN [3] and RBPN [11]. VESPCN prepends ESPCN [3] for SISR with an optical flow based motion compensation module, and RBPN integrates DBPN [10] to iteratively extract feature maps. However, only the SISR network architectures are incorporated into VSR and their parameters are end-to-end updated. Taking into account the fact that there are so many SISR models which have been well trained with state-of-the-art performance, it is unfortunate that they are overlooked when people study VSR models.

In this paper, we propose a plug-and-play VSR framework, through which the state-of-the-art SISR models can be readily plugged for tackling VSR. Generally speaking, a pre-trained SISR model can be adopted to obtain initial HR video in a frame-by-frame manner. We propose a simple yet effective temporal consistency refinement network (TCRNet) to predict temporally consistent and visually favorable HR video. As shown in Fig. 2, our TCRNet only takes two adjacent frames as input. We propose an iterative residual refinement of offset (IRRO) module to better estimate motion offset based on two adjacent LR frames, where the offset residual is gradually refined in multiple levels. Then, the HR frame $t-1$ is aligned to frame t using deformable convolution in feature level. To exploit long-distance temporal information, a deformable convolutional LSTM (DCLSTM) is proposed, where recurrent hidden states are also aligned to eliminate accumulated misalignments. Moreover, we propose a bidirectional

Fig. 1. A VSR example by different methods. HR frames by state-of-the-art SISR RRDB [30] suffer from flicker artifacts, based on which our proposed TCRNet can significantly enhance the temporal consistency to achieve superior or comparable results in comparison to state-of-the-art VSR methods FRVSR [23], DUF [12] and EDVR [29]. These results are demonstrated using animation figures, and please view them in Adobe Acrobat X or later versions.

VSR process to further boost the performance, where the video and its reverse version are successively forwarded through a TCRNet. As illustrated in Fig. 1, our TCRNet can significantly enhance the temporal consistency of HR frames super-resolved by SISR, and can achieve superior results than state-of-the-art VSR methods.

The proposed plug-and-play VSR has three merits: (1) Once TCRNet is trained, it can collaborate with any SISR model, and better SISR model would lead to better VSR performance. Also one interesting property is that future state-of-the-art SISR model is expected to be readily employed in our framework for better VSR quality. (2) Due to only taking two adjacent frames as input, spatial alignment and feature fusion are much easier than existing VSR methods. The training of TCRNet is also much easier and more stable. (3) The beginning and ending frames by TCRNet can also be well handled, whereas existing VSR methods simply discard them. In particular, through the bidirectional process, the beginning frames can be well handled using our method (Kindly refer to Figs. 4 and 5). Experimental results on popular VSR benchmark datasets have validated that the proposed TCRNet can effectively enhance the temporal consistency of super-resolved videos. In comparison to state-of-the-art VSR methods, the proposed plug-and-play VSR can achieve superior or at least comparable results.

Our contributions are summarized as follows:

- We propose a plug-and-play VSR framework, through which state-of-the-art SISR model can be readily utilized for tackling VSR.
- An iterative residual refinement of offset module is proposed for better estimating the motion offset.

- A deformable ConvLSTM is proposed to benefit from long-distance temporal information while at the same time suppressing accumulated misalignments.
- TCRNet is easy to train and can achieve superior or at least comparable results compared with state-of-the-art VSR methods.

2 Related Work

In this section, we briefly survey SISR and VSR methods.

2.1 Single Image Super-Resolution

As a pioneer work of deep learning-based SISR, SRCNN [7] is proposed to achieve descent performance in comparison with traditional SISR methods. Subsequently, a variety of effective network architectures have been studied for SISR, including very deep network [15], residual and dense blocks [18,19], recursive structure [14], channel attention [35]. Several acceleration strategies have also been developed to improve computational efficiency [5,17,24]. Besides, effective training losses such as GAN loss [18], perceptual loss [22] and texture matching loss [22] are exploited to achieve photo-realistic SISR. The state-of-the-art SISR models have achieved superior quantitative metrics and qualitative visual quality as well as high computational efficiency. However, most existing VSR models are individually studied without connection to SISR. In contrast, the proposed plug-and-play VSR framework provides an approach to benefit from existing and future state-of-the-art SISR models.

2.2 Video Super-Resolution

For VSR, it is critical to exploit temporal information, where HR frame can be predicted from several adjacent frames [12,23,25,27,29,31]. Pioneer VSR method adopts explicit motion estimation based on optical flow. The adjacent frames are warped for motion compensation, and the aligned frames are fed to CNN for predicting HR frames [13]. The image level alignment often yields artifacts. To address this issue, Caballero et al. [3] proposed the first end-to-end VSR method, where flow estimation and spatial alignment are performed in feature level. Subsequently, VSR methods devote much effort to designing modules for better spatial alignment of multiple adjacent frames. Instead of warping alignment in image level, temporal information can be better exploited in feature level by sub-pixel motion compensation layer [25], deformable convolution [27], dynamic upsampling filter [12] or pyramid and cascading deformable alignment [29]. However, the study on VSR seldom benefits from state-of-the-art SISR models. Only VESPCN [3] and RBPN [11] borrow SISR network architecture from ESPCN [11] and DBPN [10] respectively. It is wasteful that so many well trained SISR models do not contribute to tackling VSR. To address this issue, in this paper, the plug-and-play VSR provides a solution to incorporating SISR models into VSR. Besides, the proposed iterative residual refinement of offset module and deformable ConvLSTM can better exploit temporal information.

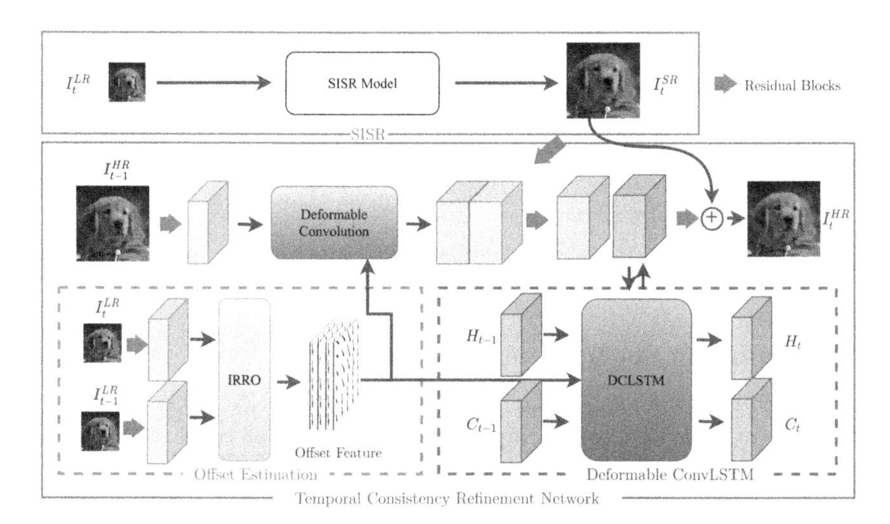

Fig. 2. Overview of plug-and-play VSR. The initial HR frames can be super-resolved by any SISR method, and the temporal consistency can be guaranteed by our TCRNet. In TCRNet, the motion offset is estimated by iterative residual refinement of offset (IRRO) module, then deformable convolution is adopted to align features in feature level, and finally Deformable ConvLSTM (DCLSTM) is further proposed for exploiting long-distance temporal information.

3 Deep Plug-and-Play Video Super-Resolution

The overview of our proposed plug-and-play VSR framework is illustrated in Fig. 2. Given a pre-trained SISR model, we can super-resolve LR video in a frame-by-frame manner. Let I_t^{LR} be LR frame t, and let I_t^{SR} be initial super-resolved HR frame by SISR. Without taking into account the temporal consistency, the initial HR video suffers from flicker artifacts. To address this issue, in the following, we will present the proposed temporal consistency refinement network (TCRNet).

3.1 Overview

For frame t, our goal is to predict final HR frame I_t^{HR}, where consecutive frames have good temporal consistency. The input is two adjacent frames $\{I_{t-1}^{HR}, I_t^{SR}\}$. We note that the two LR frames I_{t-1}^{LR} and I_t^{LR} are only used to estimate motion offset. The final HR frame I_t^{HR} is predicted by TCRNet $\mathcal{M}$,

$$I_t^{HR} = \mathcal{M}(I_t^{SR}, I_{t-1}^{HR}), \tag{1}$$

which can be accomplished through the following five stages.

(1) Feature Extraction. By forward through a feature extraction module including 3 Residual Blocks (ResBlocks), these HR frames I_{t-1}^{HR} and I_t^{SR} are mapped to feature tensors, which will be fused to predict HR frame I_t^{HR}. Besides, LR images I_{t-1}^{LR} and I_t^{LR} are fed to the shared 3-ResBlocks to extract features, which are then fed to IRRO to estimate motion offset.

(2) Iterative Residual Refinement of Offset. To alleviate adverse effect of misalignment between I_{t-1}^{HR} and I_t^{SR}, we propose an IRRO module to gradually refine motion offset. The initial HR frame I_t^{SR} may contain adverse disturbance introduced by SISR, yielding inaccurate offset estimation. To address this issue, IRRO takes the features of LR frames I_{t-1}^{LR} and I_t^{LR} as input. The output of IRRO module is the motion offset O. The details of IRRO module can be found in Sect. 3.2.

(3) Spatial Alignment. Given the estimated motion offset O, we adopt deformable convolution [28] to achieve spatial alignment. The deformable convolution is defined in Eq. (4), based on which the I_{t-1}^{HR} can be aligned to frame t. Then the features of I_{t-1}^{HR} and I_t^{SR} are concatenated and fused using a feature fusion module, which is a simple 3-ResBlocks.

(4) Deformable ConvLSTM. The IRRO and the following spatial alignment can only provide temporal information from one adjacent frame, thus we introduce a recurrent layer to better exploit the long-distance temporal information from more distant frames. However, the misalignments may be accumulated across multiple frames in the naive ConvLSTM. We hereby propose a Deformable ConvLSTM (DCLSTM), where the hidden states are also aligned to frame t. The details of DCLSTM can be found in Sect. 3.3.

(5) Reconstruction. The features are fed to a reconstruction module with 3-ResBlocks to generate the super-resolved HR residual. Finally, the HR frame I_t^{HR} is obtained by adding the predicted residual and initial I_t^{SR}.

In the following, we will present two key modules, IRRO and DCLSTM. The details of network architectures can be found in the supplementary material.

3.2 Iterative Residual Refinement of Offset

Figure 3 shows the architecture of IRRO module, where the motion offset is gradually refined across L scales. The features of I_t^{LR} and I_{t-1}^{LR} are downsampled to L scales using convolution with stride 2. On the coarsest level, the motion offset O^1 is initially estimated by feeding the concatenated features to the estimation function f_O,

$$O^1 = f_O(F_{t-1}^1, F_t^1), \tag{2}$$

where $O = \{\Delta p_k, \Delta m_k | k = 1, 2, ..., K\}$. Δm_k is the modulation scalar, and Δp_k is the offset of the convolution kernels. Taking a 3×3 kernel as an example, $K = 9$ and $p_k \in \{(-1,-1), (-1,0), ..., (0,1), (1,1)\}$ is k-th coordinate in 3×3 regular grid. The f_O is a 5-layer CNN with skip connections.

Then the motion offset is iteratively updated on finer levels by refining residual,

$$O^l = f_O(F_t^l, \hat{F}_{t-1}^l) + \hat{O}^l, \tag{3}$$

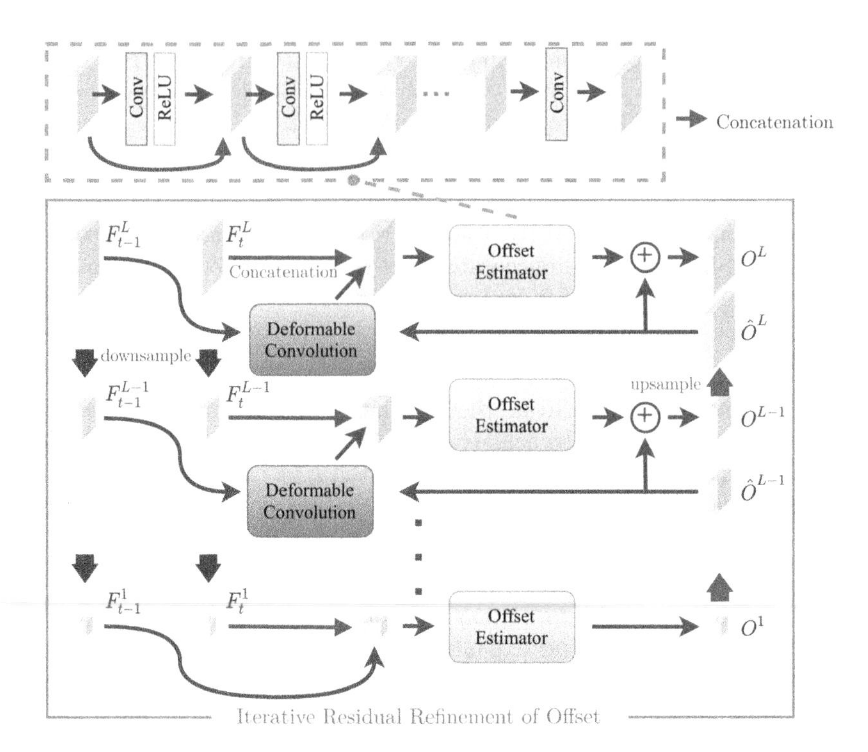

Fig. 3. Iterative Residual Refinement of Offset (IRRO) module. The motion offset is estimated across L levels. The initial offset is estimated on the coarsest level and is then gradually refined in upper levels until the finest level. The parameters of offset estimator are shared across different levels. For estimating offset residuals, the features are aligned using deformable convolution.

where $\hat{O}^l = 2 \cdot O^{l-1} \uparrow_{2\times}$ contains upsampled coordinate offsets Δp_k using $2\times$ bilinear upsampling. $\hat{F}^l_{t-1}$ is the feature aligned to frame t by

$$\hat{F}^l_{t-1} = f_D(F^l_{t-1}, \hat{O}^l), \tag{4}$$

where f_D is the deformable convolution[36] given motion offset $\hat{O}^l$. More specifically, a coordinate p_0 in feature map F_{t-1} can be aligned to frame t by

$$\hat{F}_{t-1}(\mathrm{p}_0) = \sum_{k=1}^{K} w_k \cdot F_{t-1}(\mathrm{p}_0 + \mathrm{p}_k + \Delta \mathrm{p}_k) \cdot \Delta m_k, \tag{5}$$

where w_k is the sampling weight. By iteratively performing Eq. (3), the motion offset can be gradually refined in finer levels. Until the finest level L, we obtain the motion offset $O_{t-1 \to t} = O^L$, based on which the features of I^{HR}_{t-1} are aligned to frame t using deformable convolution f_D Eq. (4).

In EDVR [29], PCD module adopts similar pyramid structure for motion offset estimation. Our IRRO differ from PCD in two aspects: (1) In PCD, misaligned features are directly concatenated to estimate offset in different levels.

In our IRRO module, the features are aligned using deformable convolution, and are then used to estimate residual of offset, resulting in gradually refined motion offset. (2) PCD adopts individual offset estimation networks for different levels, whereas f_O in our IRRO shares the same parameters across different levels.

3.3 Deformable ConvLSTM

The aligned features of I_{t-1}^{HR} are concatenated with the features of I_t^{SR}. The result is fed to a feature fusion module to obtain the current features F_t that will be fed into DCLSTM along with hidden states. Analogous to naive ConvLSTM, DCLSTM also includes an input gate P_t, a forget gate R_t, an output gate U_t, a cell state C_t and a hidden state H_t. To address the issue of misalignment with current frame t, the cell state C_t and hidden state H_t are firstly aligned based on estimated offset $O_{t-1\to t}$ using deformable convolution f_D in Eq. (4), and the deformable ConvLSTM can be formulated as,

$$\begin{aligned}
\hat{H}_t &= f_D(H_{t-1}, O_{t-1\to t}),\\
\hat{C}_t &= f_D(C_{t-1}, O_{t-1\to t}),\\
G_t &= \sigma(W_{gf} \otimes F_t + W_{gh} \otimes \hat{H}_t + b_g),\\
R_t &= \sigma(W_{rf} \otimes F_t + W_{rh} \otimes \hat{H}_t + b_r),\\
U_t &= \sigma(W_{uf} \otimes F_t + W_{uh} \otimes \hat{H}^t + b_u),\\
C_t &= R^t \odot \hat{C}_t + G_t \odot tanh(W_{cf} \otimes F_t + W_{ch} \otimes \hat{H}_t + b_c),\\
H_t &= U_t \odot tanh(C_t),
\end{aligned} \tag{6}$$

where $\otimes$ denotes 2D convolution, $\odot$ denotes entry-wise product and σ denotes *sigmoid* function. W and b are corresponding convolution matrix and bias vector.

3.4 Training and Beyond

It is easier to train our TCRNet than existing VSR methods. During training, ℓ_2 loss is simply adopted,

$$\mathcal{L} = ||I_t^{GT} - I_t^{HR}||^2, \tag{7}$$

where I_t^{GT} is the corresponding ground-truth frame t.

For the several beginning frames, there is no enough temporal information for DCLSTM, and thus flicker artifacts and temporal inconsistency may still exist. To address this issue, we suggest to employ TCRNet in a bidirectional manner, i.e., the video and its reverse version are successively fed to $\mathcal{M}$. The forward pass is exactly as Eq. (1), and the backward pass is formulated as,

$$\hat{I}_t^{HR} = \mathcal{M}(I_{t-1}^{HR}, I_t^{HR}), \tag{8}$$

where I_t^{SR} in Eq. (1) is substituted with I_t^{HR} generated in the forward pass. We experimentally found that the bidirectional strategy not only can improve the beginning frames, but also can further enhance the temporal consistency and visual quality of the whole HR video.

4 Experiments

In this section, we first describe our training and testing datasets as well as the training details. Second we conduct ablation study on TCRNet to analyze the contributions of IRRO module, deformable ConvLSTM and bidirectional processing. Third, we show that TCRNet can collaborate with various SISR models to achieve VSR. And Finally, we compare our method with state-of-the-art VSR methods on benchmark datasets.

4.1 Datasets and Training Details

Datasets. To train our plug-and-play TCRNet, we choose the diverse REDS video dataset [21], which is proposed in the NTIRE19 Competition. REDS dataset consists of 300 videos with resolution 720P, where diverse video contents are collected with various complicated real-world motions. Among the 300 videos, the ground-truth HR versions of 270 videos are provided for training and evaluation. Because the ground-truth images of REDS test set have not been released, we follow the experimental settings of EDVR [29], i.e., 266 videos for training and the other 4 videos for testing. To obtain the initial HR videos by SISR models, we choose two state-of-the-art SISR models, i.e., RRDB [30] and RCAN [35]. The pre-trained models of SISR methods are directly borrowed from the authors for training and testing on Vid4 dataset. We finetune RRDB, RCAN and EDSR on REDS when conducting evaluation on REDS4 dataset. Once our TCRNet is trained, it is evaluated on not only REDS dataset but also on another widely adopted testing dataset Vid4 [20].

Training Details. The IRRO in TCRNet is performed in $L = 6$ levels, which is sufficient to handle large and complicated inter-frame motions. Due to the introduction of DCLSTM in TCRNet, we randomly sample 11 consecutive frames as training batches. During training, patches with size 128×128 are randomly cropped from the consecutive clips and used as input. The mini-batch size is set as 16, and the training is done in 55,000 iterations. The initial learning rate is set as 4×10^{-4}, and is decayed after every $20,000$ iterations by multiplying with 0.5. As for the initialization, the hidden state of DCLSTM are initialized as zero tensor. All the other network parameters are randomly initialized, and are updated using the ADAM optimizer [16].

4.2 Ablation Study

To analyze the effectiveness of each module in TCRNet, we perform ablation studies on the testing dataset of Vid4. RCAN [35] is selected as the SISR model for generating initial HR frames. The models for ablation studies are trained with 7-frames clips for fast training. Besides our default TCRNet (i.e., #3 in Table 1), three variants of TCRNet (i.e., #1, #2 and #4 in Table 1) are trained.

- **#1**: The pyramid level $L = 1$ is set in IRRO module, and conventional ConvLSTM is adopted. That is, the motion offset is only estimated on the original scale, and the hidden states and cell states in ConvLSTM are not aligned according to the motion offset estimated by IRRO.
- **#2**: The pyramid level $L = 1$ is set in IRRO module, and DCLSTM is adopted.
- **#3**: The default TCRNet in which $L = 6$ is set in IRRO module, and DCLSTM is adopted.
- **#4**: This model shares the same settings with #3 model, except that the network parameters in IRRO are not shared across different levels.

Also, ablation studies on the iterative residual refinement mechanism of the proposed IRRO are also conducted, results are presented in supplementary material.

Table 1. Ablation studies on IRRO and DCLSTM.

	Iterative refine	Weight sharing	Align hidden	PSNR(dB) SSIM	Params(M)
#1	×	–	×	26.31/0.7849	2.332
#2	×	–	✓	26.53/0.7930	2.655
#3	✓	✓	✓	26.81/0.8062	3.186
#4	✓	×	✓	26.83/0.8071	7.429

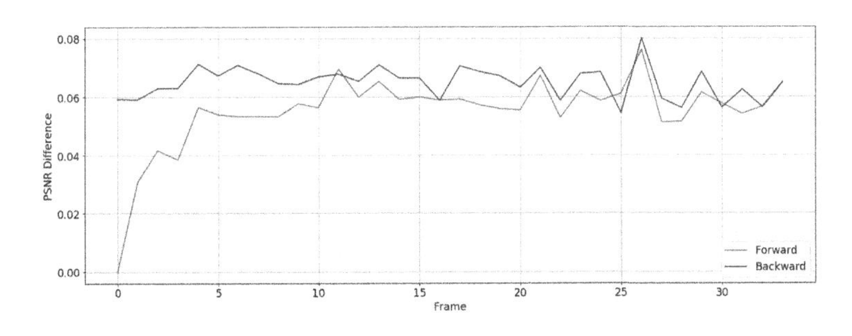

Fig. 4. Per-frame PSNR improvements between state-of-the-art SISR model RRDB [30] and our TCRNet by only forward pass and additional backward pass.

Iterative Residual Refinement of Offset. Comparing #2 model with #3 and #4 models in Table 1, one can see that IRRO with $L = 6$ performs better than #2 model which only estimates motion offset in original scale. The performance improvements of #3 and #4 models can then be attributed to the motion offset estimation in multi-scale of IRRO, which can better handle large and complicate inter-frame motions. Next, comparing #3 model with #4 model, we can see that more parameters without sharing parameters across different scales lead to higher PSNR and SSIM values. Taking into account the fact that #4 model has nearly 2 times parameters than #3 model and only brings such a small improvement margin, we choose #3 model as our default TCRNet.

Deformable ConvLSTM. Comparing #1 model with #2 model, one can easily find that for the model without aligning hidden states and cell states to the current frame in ConvLSTM, the performance would notably degrade. The extra parameters of DCLSTM only come from deformable convolution for spatial alignment. To sum up, the IRRO and DCLSTM can better handle the misalignments of adjacent frames, provide a better solution for TCRNet to exploit temporal information for VSR.

Bidirectional Processing. Since DCLSTM cannot accumulate sufficient recurrent information for several beginning frames, the VSR results of the beginning frames are usually not satisfactory. In Fig. 4, we show a VSR result of an example video from Vid4, where the Y-axis indicates the PSNR improvement in comparison to initial HR frames generated by SISR model. One can see that if we only use forward pass of TCRNet, the several beginning frames are not significantly better than the SISR results, and with increasing frames the temporal information can be gradually refined. But with the inclusion of backward pass of TCRNet, the beginning frames can be significantly improved, since the recurrent information has been well exploited in DCLSTM. From Tables 3 and 4, TCRNet with bidirectional processing can achieve superior VSR performance on both two testing datasets. Please refer to Fig. 5 to verify the visual improvements achieved by bidirectional processing.

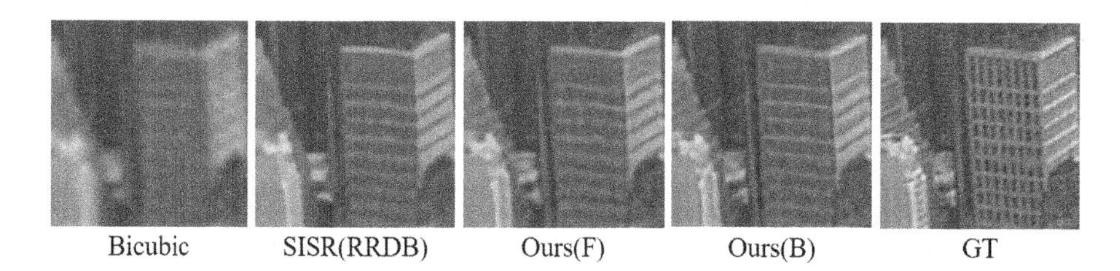

Fig. 5. Comparison of the results of the second frame by state-of-the-art SISR model RRDB [30], Ours(F) and Ours(B). With only forward pass, Ours(F) is similar with RRDB. Ours(B) benefits from bidirectional processing has better VSR visual quality.

4.3 Collaboration with Various SISR Models

In this subsection, we validate the performance of TCRNet when collaborating with various SISR models, including SRCNN [6], SRResNet [18], EDSR [19], RRDB [30] and RCAN [35]. The evaluation is performed on both the testing datasets of REDS and Vid4. Since TCRNet is trained on the super-resolved images by RRDB and RCAN, we first present the performance of TCRNet when collaborating with RRDB and RCAN. From Table 2, one can easily see that our TCRNet contributes to more than 2dB PSNR improvements in comparison to the initial results by RRDB and RCAN, indicating the effectiveness of the proposed plug-and-play VSR framework. From Fig. 1, the temporal consistency can be better improved by our TCRNet. Moreover, it is inspiring to see that our TCRNet can also make about 2dB improvements when collaborating with SRCNN, SRResNet and EDSR, indicating the good generalization ability to any other SISR models. Also, we can expect that our TCRNet will collaborate with future SISR models to achieve better VSR performance.

Due to the good performance in Table 2, in the following, we adopt RRDB plus TCRNet as the version of method. The single forward pass of TCRNet is denoted as Ours(F) and the bidirectional processing is denoted as Ours(B).

Table 2. Comparison of TCRNet when collaborating various SISR models, including SRCNN [6], SRResNet [18], EDSR [19], RRDB [30] and RCAN [35].

SISR method	Vid4		REDS4	
	Before	After	Before	After
SRCNN	24.76	26.98	27.32	30.30
SRResNet	25.38	27.23	28.33	30.79
EDSR	25.52	27.30	29.00	31.07
RRDB	25.65	27.33	29.08	31.08
RCAN	25.47	27.29	28.98	31.05

4.4 Comparison with State-of-the-Art Methods

In this subsection, we compare our method with state-of-the-art methods. On REDS testing dataset, our method is compared with RRDB [30], TOFlow [31], DUF [12] and EDVR [29]. For TOFlow [31], DUF [12] and EDVR [29], we use the pre-trained models provided by the authors of EDVR [29] to super-resolve LR frames and calculate quantitative metrics. We follow EDVR [29] to compute PSNR and SSIM values for the evaluation, which are reported in Table 3. Ours(B) with bidirectional processing performs better than all the competing methods. Ours(F) with only a single forward pass is also superior to the methods except EDVR. Figures 6 and 7 present the visual quality comparison

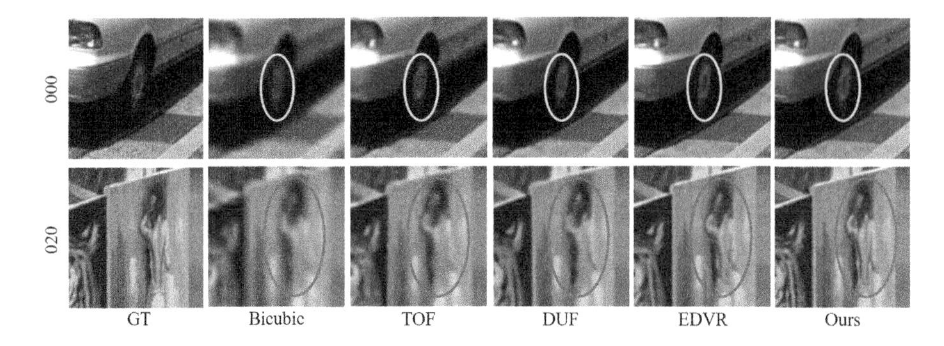

Fig. 6. Qualitative comparison on '000' and '020' of REDS4 dataset.

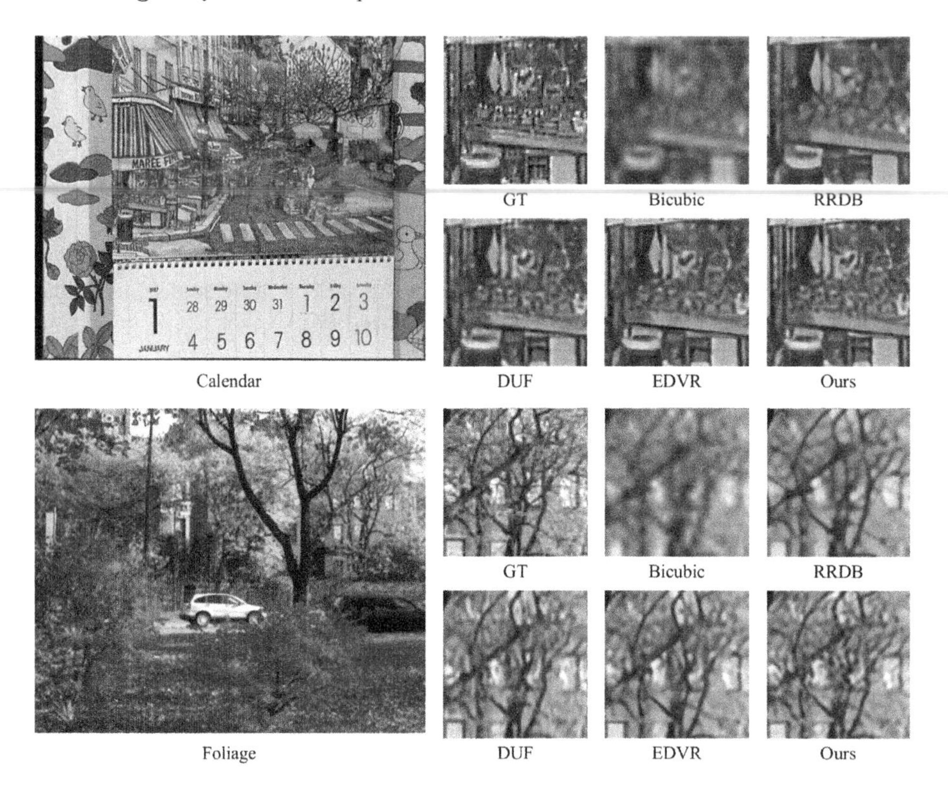

Fig. 7. Qualitative comparison on 'Calendar' and 'Foliage' of Vid4 dataset.

of super-resolved HR frames, and the corresponding videos are provided in the supplementary material.

We further conduct experiments on Vid4 testing dataset, where our method is compared with RRDB [30], TOFlow [31], FRVSR [23], DUF [12], RBPN [11] and EDVR [29]. We note that our TCRNet is finetuned on Vimeo90K dataset in 3,000 iterations when evaluating on Vid4, since EDVR [29] model for Vid4 is trained on Vimeo90K dataset. For FRVSR, we directly use the super-resolved HR frames provided by the authors, and for RBPN, these metrics are copied from their paper. The quantitative metrics are reported in Table 4. One can see that Ours(B) performs well in comparison to all these competing methods. Figures 6 and 7 present the visual quality comparison of super-resolved HR frames, from which one can find finer texture details by our methods. The corresponding videos are provided in the supplementary material.

Table 3. Comparison on REDS dataset [21], where bold and italic denotes the best and the second best performance. Our method is compared with RRDB [30], TOFlow [31], DUF [12] and EDVR [29].

	Bicubic	RRDB	TOFlow	DUF	EDVR	Ours(F)	Ours(B)
PSNR	26.13	29.08	27.96	28.64	*31.07*	30.38	**31.08**
SSIM	0.7290	0.8271	0.7977	0.8251	*0.8800*	0.8650	**0.8833**

Moreover, we provide the temporal file of one example video in supplementary material, which shows that Ours(F) and Ours(B) can significantly improve the temporal consistency of SISR methods while perform superior to or on par with state-of-the-art VSR methods.

To sum up, our plug-and-play VSR is an effective solution for tackling VSR based on pre-trained SISR models, and better SISR models can be directly incorporated to further boost VSR performance.

Table 4. Comparison on Vid4 dataset. Our method is compared with RRDB [30], TOFlow [31], FRVSR [23], DUF [12], RBPN [11] and EDVR [29].

	Bicubic	RRDB	TOFlow	FRVSR	DUF	RBPN	EDVR	Ours REDS	Ours Vimeo90K
PSNR	23.79	25.65	25.89	26.69	27.33	27.12	*27.39*	*27.39*	**27.41**
SSIM	0.6384	0.7472	0.7651	0.822	**0.8318**	0.818	0.8267	0.8161	*0.8280*

5 Conclusion

In this work, we proposed a plug-and-play VSR framework, through which any pre-trained SISR models can be incorporated to tackle the VSR task. To stabilize the temporal consistency of initial super-resolved HR frames, we propose a novel TCRNet. In TCRNet, spatial alignment is better achieved by iterative residual refinement of offset module, and Deformable ConvLSTM is proposed to exploit long-distance temporal information from more previous frames, while at the same time eliminating accumulated misalignments. Comprehensive experiments have demonstrated that the proposed TCRNet can collaborate with existing SISR models to perform superior to or on par with state-of-the-art VSR methods. Future state-of-the-art SISR model can also be incorporated into our plug-and-play VSR to further benefit VSR performance.

Acknowledgement. This project is partially supported by the National Natural Scientific Foundation of China (NSFC) under Grant No. 61671182, 61801326 and 61773002.

References

1. Belekos, S.P., Galatsanos, N.P., Katsaggelos, A.K.: Maximum a posteriori video super-resolution using a new multichannel image prior. IEEE Trans. Image Process. **19**(6), 1451–1464 (2010)
2. Ben-Ezra, M., Zomet, A., Nayar, S.K.: Video super-resolution using controlled subpixel detector shifts. IEEE Trans. Pattern Anal. Mach. Intell. **27**(6), 977–987 (2005)
3. Caballero, J., et al.: Real-time video super-resolution with spatio-temporal networks and motion compensation. In: Proceedings of the IEEE Conference on Computer Vision and Pattern Recognition, pp. 4778–4787 (2017)
4. Demir, U., Rawat, Y.S., Shah, M.: TinyVIRAT: low-resolution video action recognition. arXiv preprint arXiv:2007.07355 (2020)
5. Dong, C., Loy, C.C., Tang, X.: Accelerating the super-resolution convolutional neural network. In: Leibe, B., Matas, J., Sebe, N., Welling, M. (eds.) ECCV 2016. LNCS, vol. 9906, pp. 391–407. Springer, Cham (2016). https://doi.org/10.1007/978-3-319-46475-6_25
6. Dong, C., Loy, C.C., He, K., Tang, X.: Learning a deep convolutional network for image super-resolution. In: Fleet, D., Pajdla, T., Schiele, B., Tuytelaars, T. (eds.) ECCV 2014. LNCS, vol. 8692, pp. 184–199. Springer, Cham (2014). https://doi.org/10.1007/978-3-319-10593-2_13
7. Dong, C., Loy, C.C., He, K., Tang, X.: Image super-resolution using deep convolutional networks. IEEE Trans. Pattern Anal. Mach. Intell. **38**(2), 295–307 (2016)
8. Dong, W., Wang, P., Yin, W., Shi, G., Wu, F., Lu, X.: Denoising prior driven deep neural network for image restoration. IEEE Trans. Pattern Anal. Mach. Intell. **41**(10), 2305–2318 (2018)
9. Fuoli, D., Gu, S., Timofte, R.: Efficient video super-resolution through recurrent latent space propagation. arXiv preprint arXiv:1909.08080 (2019)
10. Haris, M., Shakhnarovich, G., Ukita, N.: Deep back-projection networks for super-resolution. In: Proceedings of the IEEE Conference on Computer Vision and Pattern Recognition, pp. 1664–1673 (2018)
11. Haris, M., Shakhnarovich, G., Ukita, N.: Recurrent back-projection network for video super-resolution. In: Proceedings of the IEEE Conference on Computer Vision and Pattern Recognition, pp. 3897–3906 (2019)
12. Jo, Y., Wug Oh, S., Kang, J., Joo Kim, S.: Deep video super-resolution network using dynamic upsampling filters without explicit motion compensation. In: Proceedings of the IEEE Conference on Computer Vision and Pattern Recognition, pp. 3224–3232 (2018)
13. Kappeler, A., Yoo, S., Dai, Q., Katsaggelos, A.K.: Video super-resolution with convolutional neural networks. IEEE Trans. Comput. Imaging **2**(2), 109–122 (2016)

14. Kim, J., Kwon Lee, J., Mu Lee, K.: Deeply-recursive convolutional network for image super-resolution. In: Proceedings of the IEEE Conference on Computer Vision and Pattern Recognition Workshops, pp. 1637–1645 (2016)
15. Kim, J., Lee, J.K., Lee, K.M.: Accurate image super-resolution using very deep convolutional networks. In: Proceedings of the IEEE Conference on Computer Vision and Pattern Recognition Workshops (2016)
16. Kingma, D.P., Ba, J.: Adam: a method for stochastic optimization. arXiv preprint arXiv:1412.6980 (2014)
17. Lai, W.S., Huang, J.B., Ahuja, N., Yang, M.H.: Deep Laplacian pyramid networks for fast and accurate super-resolution. In: Proceedings of the IEEE Conference on Computer Vision and Pattern Recognition Workshops (2017)
18. Ledig, C., et al.: Photo-realistic single image super-resolution using a generative adversarial network. In: Proceedings of the IEEE Conference on Computer Vision and Pattern Recognition, pp. 4681–4690 (2017)
19. Lim, B., Son, S., Kim, H., Nah, S., Mu Lee, K.: Enhanced deep residual networks for single image super-resolution. In: Proceedings of the IEEE Conference on Computer Vision and Pattern Recognition Workshops, pp. 136–144 (2017)
20. Liu, C., Sun, D.: On Bayesian adaptive video super resolution. IEEE Trans. Pattern Anal. Mach. Intell. **36**(2), 346–360 (2013)
21. Nah, S., et al.: NTIRE 2019 challenge on video deblurring and super-resolution: dataset and study. In: Proceedings of the IEEE Conference on Computer Vision and Pattern Recognition Workshops (2019)
22. Sajjadi, M.S.M., Scholkopf, B., Hirsch, M.: EnhanceNet: single image super-resolution through automated texture synthesis. In: Proceedings of the IEEE International Conference on Computer Vision (2017)
23. Sajjadi, M.S., Vemulapalli, R., Brown, M.: Frame-recurrent video super-resolution. In: Proceedings of the IEEE Conference on Computer Vision and Pattern Recognition, pp. 6626–6634 (2018)
24. Shi, W., et al.: Real-time single image and video super-resolution using an efficient sub-pixel convolutional neural network. In: Proceedings of the IEEE Conference on Computer Vision and Pattern Recognition, pp. 1874–1883 (2016)
25. Tao, X., Gao, H., Liao, R., Wang, J., Jia, J.: Detail-revealing deep video super-resolution. In: Proceedings of the IEEE International Conference on Computer Vision, pp. 4472–4480 (2017)
26. Tao, X., Gao, H., Shen, X., Wang, J., Jia, J.: Scale-recurrent network for deep image deblurring. In: Proceedings of the IEEE Conference on Computer Vision and Pattern Recognition, pp. 8174–8182 (2018)
27. Tian, Y., Zhang, Y., Fu, Y., Xu, C.: TDAN: temporally deformable alignment network for video super-resolution. arXiv preprint arXiv:1812.02898 (2018)
28. Wang, H., Su, D., Liu, C., Jin, L., Sun, X., Peng, X.: Deformable non-local network for video super-resolution. IEEE Access **7**, 177734–177744 (2019)
29. Wang, X., Chan, K.C., Yu, K., Dong, C., Change Loy, C.: EDVR: video restoration with enhanced deformable convolutional networks. In: Proceedings of the IEEE Conference on Computer Vision and Pattern Recognition Workshops (2019)
30. Wang, X., et al.: ESRGAN: enhanced super-resolution generative adversarial networks. In: Proceedings of the European Conference on Computer Vision (2018)
31. Xue, T., Chen, B., Wu, J., Wei, D., Freeman, W.T.: Video enhancement with task-oriented flow. Int. J. Comput. Vis. **127**(8), 1106–1125 (2019)
32. Zhang, J., et al.: Dynamic scene deblurring using spatially variant recurrent neural networks. In: Proceedings of the IEEE Conference on Computer Vision and Pattern Recognition (2018)

33. Zhang, K., Zuo, W., Chen, Y., Meng, D., Zhang, L.: Beyond a gaussian denoiser: residual learning of deep CNN for image denoising. IEEE Trans. Image Process. **26**(7), 3142–3155 (2017)
34. Zhang, K., Zuo, W., Zhang, L.: Deep plug-and-play super-resolution for arbitrary blur kernels. In: Proceedings of the IEEE Conference on Computer Vision and Pattern Recognition Workshops, pp. 1671–1681 (2019)
35. Zhang, Y., Li, K., Li, K., Wang, L., Zhong, B., Fu, Y.: Image super-resolution using very deep residual channel attention networks. In: Ferrari, V., Hebert, M., Sminchisescu, C., Weiss, Y. (eds.) ECCV 2018. LNCS, vol. 11211, pp. 294–310. Springer, Cham (2018). https://doi.org/10.1007/978-3-030-01234-2_18
36. Zhu, X., Hu, H., Lin, S., Dai, J.: Deformable ConvNets v2: more deformable, better results, pp. 9308–9316 (2019)

Deep Adaptive Inference Networks for Single Image Super-Resolution

Ming Liu[1], Zhilu Zhang[1], Liya Hou[1], Wangmeng Zuo[1](✉), and Lei Zhang[2]

[1] Harbin Institute of Technology, Harbin, China
csmliu@outlook.com, cszlzhang@outlook.com, h_liya@outlook.com,
wmzuo@hit.edu.cn

[2] The Hong Kong Polytechnic University, Hong Kong, China
cslzhang@comp.polyu.edu.hk

Abstract. Recent years have witnessed tremendous progress in single image super-resolution (SISR) owing to the deployment of deep convolutional neural networks (CNNs). For most existing methods, the computational cost of each SISR model is irrelevant to local image content, hardware platform and application scenario. Nonetheless, content and resource adaptive model is more preferred, and it is encouraging to apply simpler and efficient networks to the easier regions with less details and the scenarios with restricted efficiency constraints. In this paper, we take a step forward to address this issue by leveraging the adaptive inference networks for deep SISR (AdaDSR). In particular, our AdaDSR involves an SISR model as backbone and a lightweight adapter module which takes image features and resource constraint as input and predicts a map of local network depth. Adaptive inference can then be performed with the support of efficient sparse convolution, where only a fraction of the layers in the backbone is performed at a given position according to its predicted depth. The network learning can be formulated as joint optimization of reconstruction and network depth losses. In the inference stage, the average depth can be flexibly tuned to meet a range of efficiency constraints. Experiments demonstrate the effectiveness and adaptability of our AdaDSR in contrast to its counterparts (*e.g.*, EDSR and RCAN). Code is available at https://github.com/csmliu/AdaDSR.

Keywords: Single image super-resolution · Convolutional neural network · Adaptive inference

1 Introduction

Image super-resolution aims at recovering a high-resolution (HR) image from its low-resolution (LR) counterpart, which is a representative low-level vision task with many real-world applications such as medical imaging [29], surveillance [43]

Electronic supplementary material The online version of this chapter (https://doi.org/10.1007/978-3-030-66823-5_8) contains supplementary material, which is available to authorized users.

A. Bartoli and A. Fusiello (Eds.): ECCV 2020 Workshops, LNCS 12538, pp. 131–148, 2020.
https://doi.org/10.1007/978-3-030-66823-5_8

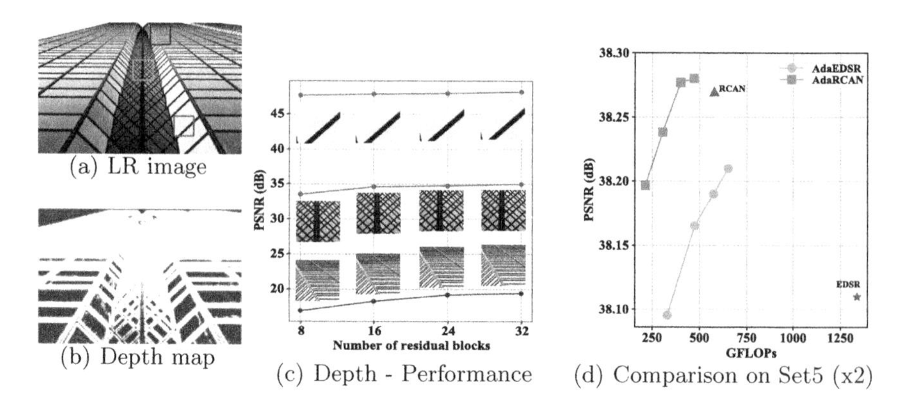

Fig. 1. Illustration of our motivation and performance. (a) and (b) show an LR image and the depth map predicted by AdaDSR, and three patches with various SISR difficulty are marked out. (c) explores the performance of EDSR models with different number of residual blocks on these patches. In (d), we compare two versions of our AdaDSR against their backbones on Set5 dataset.

and entertainment [39]. Recently, driven by the development of deep convolutional neural networks (CNNs), tremendous progress has been made in single image super-resolution (SISR). On the one hand, the quantitative performance of SISR has been continuously improved by many outstanding representative models such as SRCNN [6], VDSR [12], SRResNet [15], EDSR [17], RCAN [41], SAN [5], *etc.* On the other hand, considerable attention has also been given to handle several other issues in SISR, including visual quality [15], degradation model [38], and blind SISR [40].

Albeit their unprecedented success of SISR, for most existing networks, the computational cost of each model is still independent to image content and application scenarios. Given such an SISR model, once the training is finished, the inference process is deterministic and only depends on the model architecture and the input image size. Actually, instead of deterministic inference, it is inspiring to make the inference to be adaptive to local image content. To illustrate this point, Fig. 1(c) shows the SISR results of three image patches using EDSR [17] with different numbers of residual blocks. It can be seen that EDSR with 8 residual blocks is sufficient to super-resolve a smooth patch with less textures. In contrast, at least 24 residual blocks are required for the patch with rich details. Consequently, treating the whole image equally and processing all regions with identical number of residual blocks will certainly lead to the waste of computation resource. Thus, it is encouraging to develop the spatially adaptive inference method for better tradeoff between accuracy and efficiency.

Moreover, the SISR model may be deployed to diverse hardware platforms. Even for a given hardware device, the model can be run under different battery conditions or workloads, and has to meet various efficiency constraints. One natural solution is to design and train numerous deep SISR models in advance, and dynamically select the appropriate one according to the hardware platform and efficiency constraints. Nonetheless, both the training and storage of multiple deep SISR models are expensive, greatly limiting their practical applications to scenarios with highly dynamic efficiency constraints. Instead, we suggest to address this issue by making the inference method adaptive to efficiency constraints.

To make the learned model to adapt to local image content and efficiency constraints, this paper presents a kind of adaptive inference networks for deep SISR, *i.e.,* AdaDSR. Considering that stacked residual blocks have been widely adopted in the representative SISR models [15,17,41], the AdaDSR introduces a lightweight adapter module which takes image features as the input and produces a map of local network depth. Therefore, given a position with the local network depth d, only the first d blocks are required to be computed in the testing stage. Thus, our AdaDSR can apply shallower networks for the smooth regions, and exploit deeper ones for the regions with detailed textures, thereby benefiting the tradeoff between accuracy and efficiency. Taking all the positions into account, sparse convolution can be adopted to facilitate efficient and adaptive inference.

We further improve AdaDSR to be adaptive to efficiency constraints. Note that the average of depth map can be used as an indicator of inference efficiency. For simplicity, the efficiency constraint on hardware platform and application scenario can be represented as a specific desired depth. Thus, we also take the desired depth as an input of the adapter module, and require the average of predicted depth map to approximate the desired depth. And the learning of AdaDSR can then be formulated as the joint optimization of reconstruction and network depth loss. After training, we can dynamically set the desired depth values to accommodate various application scenarios, and then adopt our AdaDSR to meet the efficiency constraints.

Experiments are conducted to assess our AdaDSR. Without loss of generality, we adopt EDSR [17] model as the backbone of our AdaDSR (denoted by AdaEDSR). It can be observed from Fig. 1(b) that the predicted depth map has smaller depth values for the smooth regions and higher ones for the regions with rich small-scale details. As shown in Fig. 1(d), our AdaDSR can be flexibly tuned to meet various efficiency constraints (*e.g.,* FLOPs) by specifying proper desired depth values. In contrast, most existing SISR methods can only be performed with deterministic inference and fixed computational cost. Quantitative and qualitative results further show the effectiveness and adaptability of our AdaDSR in comparison to the state-of-the-art deep SISR methods. Furthermore, we also take another representative SISR model RCAN [41] as the backbone model (denoted by AdaRCAN), which illustrates the generality of our AdaDSR.

To sum up, the contributions of this work include:

- We present adaptive inference networks for deep SISR, *i.e.,* AdaDSR, which adds the backbone with a lightweight adapter module to produce local depth map for spatially adaptive inference.
- Both image features and desired depth are taken as the input of the adapter, and reconstruction loss is incorporated with depth loss for network learning,

thereby making AdaDSR equipped with sparse convolution to be adaptive to various efficiency constraints.
- Experiments show that our AdaDSR achieves better tradeoff between accuracy and efficiency than its counterparts (*i.e.,* EDSR and RCAN), and can adapt to different efficiency constraints without training from scratch.

2 Related Work

2.1 Deep Single Image Super-Resolution

Dong *et al.* introduce a three-layer convolutional network in their pioneer work SRCNN [6], since then, the quantitative performance of SISR has been continuously promoted. Kim *et al.* [12] further propose a deeper model named VDSR with residual blocks and adjustable gradient clipping. Liu *et al.* [19] propose MWCNN, which accelerates the running speed and enlarges the receptive field by deploying U-Net [27] like architecture, and multi-scale wavelet transformation is applied rather than down-up-sampling module to avoid information lost.

These methods take interpolated LR images as input, resulting in heavy computation burden, so many recent SISR methods choose to increase the spatial resolution via PixelShuffle [28] at the tail of the model. SRResNet [15], EDSR [17] and WDSR [35] follow this setting and have a deep main body by stacking several identical residual blocks [9] before the tail component, and they obtain better performance and efficiency by modifying the architecture of the residual blocks. Zhang *et al.* [41] build a very deep (>400 layers) yet narrow (64 channels *vs.* 256 channels in EDSR) RCAN model and learn a content-related weight for each feature channel inside the residual blocks. Dai *et al.* [5] propose SAN to obtain better feature representation via second-order attention model, and non-locally enhanced residual group is incorporated to capture long-distance features.

Apart from the fidelity track, considerable attention has also been given to handle several other issues in SISR. For example, SRGAN [15] incorporates adversarial loss to improve perceptual quality, DPSR [38] proposes a new degradation model and performs super-resolution and deblurring simultaneously, Zhang *et al.* [40] solve real image SISR problem in an unsupervised manner by taking advantage of generative adversarial networks. In addition, lightweight networks such as IDN [11] and CARN [1] are proposed, but most lightweight models are accelerated at the cost of quantitative performance. In this paper, we propose an AdaDSR model, which achieves better tradeoff between accuracy and efficiency.

2.2 Adaptive Inference

Traditional deterministic CNNs tend to be less flexible to meet various requirements in the applications. As a remedy, many adaptive inference methods have been explored in recent years. Inspired by [2], Upchurch *et al.* [32] propose to learn an interpolation of deep features extracted by a pre-trained model, and

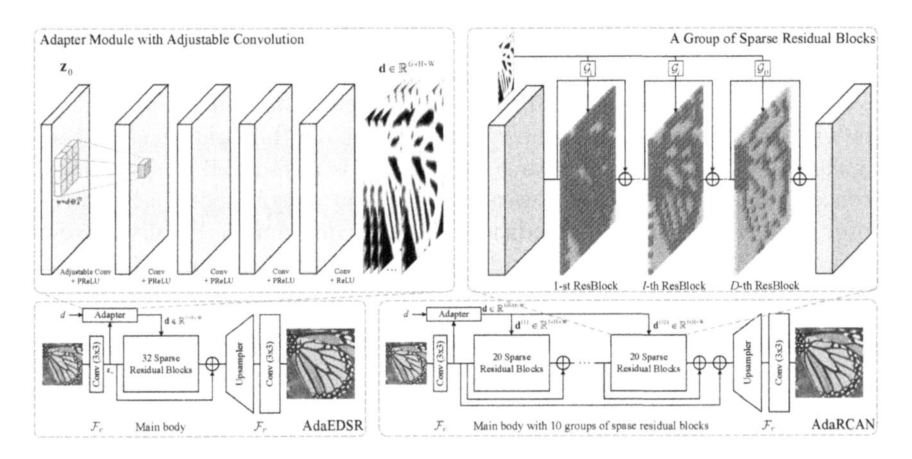

Fig. 2. Overall illustration of AdaDSR. On the top left, the adapter takes $\mathbf{z}_0$ as input and the weight of the first convolution is tuned by d on the fly. The adapter generats a depth map $\mathbf{d} \in \mathbb{R}^{G \times H \times W}$ (for AdaEDSR $G = 1$, while for AdaRCAN $G = 10$). Each channel of $\mathbf{d}$ is delivered to a group of sparse residual blocks (as shown on the top right). Only a fraction of the positions (marked by dark blue) require computation. (Color figure online)

manipulate the attributes of facial images. Shoshan *et al.* [30] further propose a DynamicNet by deploying tuning blocks alongside the backbone model, and linearly manipulate the features to learn an interpolation of two objectives, which can be tuned to explore the whole objective space during inference phase. Similarly, CFSNet [34] implements continuous transition of different objectives, and automatically learns the trade-off between perception and distortion for SISR.

Some methods also leverage adaptive inference to obtain computing efficient models. Li *et al.* [16] deploy multiple classifiers between the main blocks, and the last one performs as a teacher net to guide the previous ones. During the inference phase, the confidence score of a classifier indicates whether to perform the next block and the corresponding classifier. Figurnov *et al.* [7] predict a stop score for the patches, which determines whether to skip the subsequent layers, indicating different regions have unequal importance for detection tasks. Therefore, skipping layers at less important regions can save the inference time. Yu *et al.* [36] propose to build a denoising model with several multi-path blocks, and in each block, a path finder is deployed to select a proper path for each image patch. These methods are similar to our AdaDSR, however, they perform adaptive inference on patch-level, and the adaptation depends only on the features. In this paper, our AdaDSR implements pixel-wise adaptive inference via sparse convolution and is manually controllable to meet various resource constraints.

3 Proposed Method

This section presents our AdaDSR model for single image super-resolution. To begin with, we equip the backbone with a network depth map to facilitate spatially variant inference. Then, sparse convolution is introduced to speed up the inference by omitting the unnecessary computation. Furthermore, a lightweight adapter module is deployed to predict the network depth map. Finally, the overall network structure (see Fig. 2) and learning objective are provided.

3.1 AdaDSR with Spatially Variant Network Depth

SISR aims to learn a mapping to reconstruct high-resolution image $\hat{\mathbf{y}}$ from its low-resolution observation $\mathbf{x}$, and can be written as,

$$\hat{\mathbf{y}} = \mathcal{F}(\mathbf{x}; \Theta), \tag{1}$$

where $\mathcal{F}$ denotes the SISR network with the network parameters Θ. In this work, we consider a representative category of deep SISR networks that consist of three major modules, *i.e.*, feature extraction $\mathcal{F}_e$, residual blocks, and HR reconstruction $\mathcal{F}_r$. Several representative SISR models, *e.g.*, SRResNet [15], EDSR [17], and RCAN [41], belong to this category. Using EDSR as an example, we let $\mathbf{z}_0 = \mathcal{F}_e(\mathbf{x})$. The output of the residual blocks can then be formulated as,

$$\mathbf{z}^o = \mathbf{z}_0 + \sum_{l=1}^{D} \mathcal{F}_l(\mathbf{z}_{l-1}; \Theta_l), \tag{2}$$

where Θ_l is the network parameters associated with the l-th residual block. Given the output of the $(l-1)$-th residual block, the l-th residual block can be written as $\mathbf{z}_l = \mathbf{z}_{l-1} + \mathcal{F}_l(\mathbf{z}_{l-1}; \Theta_l)$. Finally, the reconstructed HR image can be obtained by $\hat{\mathbf{y}} = \mathcal{F}_r(\mathbf{z}^o; \Theta_r)$.

As shown in Fig. 1, the difficulty of super-resolution is spatially variant. For examples, it is not required to go through all the D residual blocks in Eq. (2) to reconstruct the smooth regions. As for the regions with rich and detailed textures, more residual blocks generally are required to fulfill high quality reconstruction. Therefore, we introduce a 2D network depth map $\mathbf{d}$ $(0 \leq d_{ij} \leq D)$ which has the same spatial size with $\mathbf{z}_0$. Intuitively, the network depth d_{ij} is smaller for the smooth region and larger for the region with rich details. To facilitate spatially adaptive inference, we modify Eq. (2) as,

$$\mathbf{z}^o = \mathbf{z}_0 + \sum_{l=1}^{D} \mathcal{G}_l(\mathbf{d}) \circ \mathcal{F}_l(\mathbf{z}_{l-1}; \Theta_l), \tag{3}$$

where $\circ$ denotes the entry-wise product. Here, $\mathcal{G}_l(d_{ij})$ is defined as,

$$\mathcal{G}_l(d_{ij}) = \begin{cases} 0, & d_{ij} < l-1 \\ 1, & d_{ij} > l \\ d_{ij} - (l-1), & otherwise \end{cases}. \tag{4}$$

Let $\lceil \cdot \rceil$ be the ceiling function, the last $D - \lceil d_{ij} \rceil$ residual blocks are not required to compute for a position with network depth d_{ij}. Given the 2D network depth map $\mathbf{d}$, we can exploit Eq. (3) to conduct spatially adaptive inference.

3.2 Sparse Convolution for Efficient Inference

Let $\mathbf{m}$ (*e.g.*, $\mathbf{m}_l = \mathcal{G}_l(\mathbf{d})$ for the l-th residual block) be a mask to indicate the positions where the convolution activations should be kept. As shown in Fig. 3, for some convolution implementations such as fast Fourier transform (FFT) [22,33] and Winograd [14] based algorithms, one should first perform a standard convolution to obtain the whole output feature by $\mathbf{o} = \mathbf{w} \star \mathbf{f}$. Here, $\mathbf{f}$, $\mathbf{w}$ and $\star$ denote input feature map, convolution kernel and convolution operation, respectively. Then the sparse results can be represented by $\mathbf{o}^* = \mathbf{m} \circ \mathbf{o}$. Nonetheless, such implementations meet the requirement of spatially adaptive inference while maintaining the same computational complexity with the standard convolution.

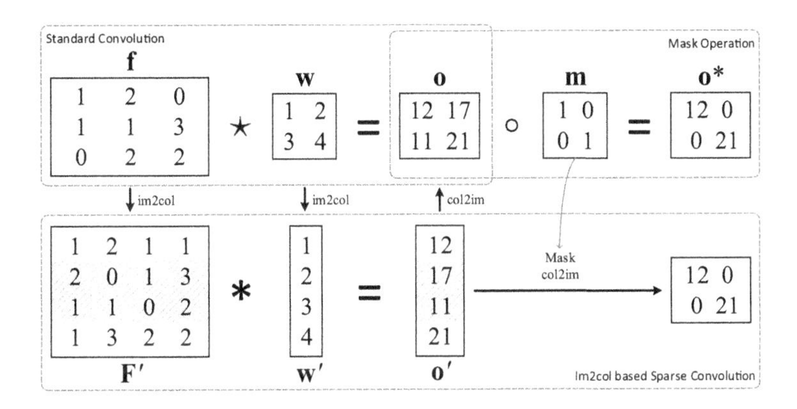

Fig. 3. An example to illustrate im2col [4] based sparse convolution. $\star$, $\circ$ and $*$ represent convolution, entry-wise product and matrix multiplication, respectively. $\mathbf{f}$, $\mathbf{w}$ and $\mathbf{o}$ are input feature, convolution kernel and output feature of standard convolution operation, which is implemented by arbitrary convolution implementation algorithms, while $\mathbf{F}'$ and $\mathbf{w}'$ are reorganized from $\mathbf{f}$ and $\mathbf{w}$ during the im2col procedure. Given the mask $\mathbf{m}$, the shaded rows do not exist in the im2col based sparse convolution, reducing computation amount comparing to standard convolution based sparse convolution.

Instead, we adopt the im2col [4] based sparse convolution for efficient adaptive inference. As shown in Fig. 3, the patch from $\mathbf{f}$ related to a point in $\mathbf{o}$ is organized as a row in matrix $\mathbf{F}'$, and the convolution kernel ($\mathbf{w}$) is also converted as vector $\mathbf{w}'$. Then the convolution operation is transformed into a matrix multiplication problem, which is highly optimized in many Basic Linear Algebra Subprograms (BLAS) libraries. Then, the result $\mathbf{o}'$ can be organized back to the output feature map. Given the mask $\mathbf{m}$, we can simply skip the corresponding row when constructing the reorganized input feature $\mathbf{F}'$ if it has zero mask value (see the shaded rows of $\mathbf{F}'$ in Fig. 3), and the computation is skipped as well. Note that we use a dilation operation to obtain the mask for the former convolution layer in the residual block. However, even without this operation, the performance degradation on most datasets will not exceed 0.02 dB. Thus, the spatially adaptive inference in Eq. (3) can be efficiently implemented via the im2col and col2im procedure. Moreover, the efficiency can be further improved when more rows are masked out, *i.e.*, when the average depth of $\mathbf{d}$ is smaller.

It is worth noting that sparse convolution has been suggested in many works and evaluated in image classification [20], object detection [18,26], model pruning [24] and 3D semantic segmentation [8] tasks. [18] and [26] are based on im2col and Winograd algorithm respectively, however, these methods implement patch-level sparse convolution. [8] designs new data structure for sparse convolution and constructs a whole CNN framework to suit the designed data structure, making it incompatible with standard methods. [20] incorporates sparsity into Winograd algorithm, which is not mathematically equivalent to the vanilla CNN. The most relevant work [24] skips unnecessary points when traversing all spatial positions and achieves pixel-level sparse convolution, which is implemented on serial devices (*e.g.*, CPUs) via for-loops. In this work, we use im2col based sparse convolution, which combines this intuitive thought and im2col algorithm, and deploy the proposed model on the parallel platforms (*e.g.*, GPUs). To the best of our knowledge, this is the first attempt to deploy pixel-wise sparse convolution on SISR task and achieves image content and resource adaptive inference.

3.3 Lightweight Adapter Module

In this subsection, we introduce a lightweight adapter module $\mathcal{P}$ to predict a 2D network depth map $\mathbf{d}$. In order to adapt to local image content, the adapter module $\mathcal{P}$ is required to produce lower network depth for smooth region and higher depth for detailed region. Let $\bar{d}$ be the average value of $\mathbf{d}$, and d be the desired network depth. To make the model to be adaptive to efficiency constraints, we also take the desired network depth d into account, and require that the decrease of d can result in smaller $\bar{d}$, *i.e.*, better inference efficiency.

As shown in Fig. 2, the adapter module $\mathcal{P}$ takes the feature map $\mathbf{z}_0$ as the input and is comprised of four convolution layers with PReLU nonlinearity followed by another convolution layer with ReLU nonlinearity. Let $\mathbf{d} = \mathcal{P}(\mathbf{z}_0; \Theta_a)$. We then use Eq. (4) to generate the mask $\mathbf{m}_l$ for each residual block. It is noted that $\mathbf{m}_l$ may not be a binary mask but contains many zeros. Thus, we can construct a sparse residual block which can omit the computation for the regions with zero mask values to facilitate efficient adaptive inference. To meet the efficiency constraint, we also take the desired network depth d as the input to the adapter, and predict the network depth map by

$$\mathbf{d} = \mathcal{P}(\mathbf{z}_0, d; \Theta_a), \tag{5}$$

where Θ_a denotes the network parameters of the adapter module. Denote the weight of the first convolution in the adapter as $\Theta_a^{(1)}$, we make the convolution adjustable by replacing the weight $\Theta_a^{(1)}$ with $d \cdot \Theta_a^{(1)}$ when the desired depth is d, then the adapter is able to meet the aforementioned d-oriented constraints.

3.4 Network Architecture and Learning Objective

Network Architecture. As shown in Fig. 2, our proposed AdaDSR is comprised of a backbone SISR network and a lightweight adapter module to facilitate image content and efficiency adaptive inference. Without loss of generality, in this section, we take EDSR [17] as the backbone to illustrate the network architecture, and it is feasible to apply our AdaDSR to other representative SISR models [15,35,41] with a number of residual blocks [9]. Following [17], the backbone involves 32 residual blocks, each of which has two 3×3 convolution layers with stride 1, padding 1 and 256 channels with ReLU nonlinearity. Another 3×3 convolution layer is deployed right behind the residual blocks. The feature extraction module $\mathcal{F}_e$ is a convolution layer, and the reconstruction module $\mathcal{F}_r$ is comprised of an upsampling unit to enlarge the features followed by a convolution layer which reconstructs the output image. The upsampling unit is composed by a series of Convolution-PixelShuffle [28] according to the super-resolution scale. Besides, the lightweight adapter module takes both the feature map $\mathbf{z}_0$ and the desired network depth d as the input, and consists of five convolution layers to produce an one-channel network depth map.

It is worth noting that, we implement two versions of AdaDSR. The first takes EDSR [17] as backbone, which is denoted by AdaEDSR. To further show the generality of proposed AdaDSR and compare against state-of-the-art methods, we also take RCAN [41] as backbone and implement an AdaRCAN model. The main difference is that, RCAN replaces the 32 residual blocks with 10 residual groups, and each residual groups is composed of 20 residual blocks equipped with channel attention. Therefore, we modify the adapter to generate 10 depth maps simultaneously, and each of which is deployed to a residual group.

Learning Objective. The learning objective of our AdaDSR includes a reconstruction loss term and a network depth loss term to achieve a proper tradeoff between SISR performance and efficiency. In terms of the SISR performance, we adopt the ℓ_1 reconstruction loss defined on the super-resolved output and the ground-truth high-resolution image,

$$\mathcal{L}_{rec} = \|\mathbf{y} - \hat{\mathbf{y}}\|_1, \tag{6}$$

where $\mathbf{y}$ and $\hat{\mathbf{y}}$ respectively represent the high-resolution ground-truth and the super-resolved image by our AdaDSR. Considering the efficiency constraint, we require the average $\bar{d}$ of the predicted network depth map to approximate the desired depth d, and then introduce the following network depth loss,

$$\mathcal{L}_{depth} = \max(0, \bar{d} - d). \tag{7}$$

To sum up, the overall learning objective of our AdaDSR is formulated as,

$$\mathcal{L} = \mathcal{L}_{rec} + \lambda \mathcal{L}_{depth}, \tag{8}$$

where λ is a tradeoff hyper-parameter and is set to 0.01 in all our experiments.

4 Experiments

4.1 Implementation Details

Model Training. For training our AdaDSR model, we use the 800 training images and the first five validation images from DIV2K dataset [31] as training and validation set, respectively. The input and output images are in RGB color space, and the input images are obtained by bicubic degradation model. Following previous works [15,17,41], during training we subtract the mean value of the DIV2K dataset on RGB channels and apply data augmentation on training images, including random horizontal flip, random vertical flip and 90° rotation. The AdaDSR model is optimized by the Adam [13] algorithm with $\beta_1 = 0.9$ and $\beta_2 = 0.999$ for 800 epochs. In each iteration, there are 16 LR patches of size 48×48. And the learning rate is initialized as 5×10^{-5} and decays to half after every 200 epochs. During training, the desired depth d is randomly sampled from $[0, D]$, where D is 32 and 20 for AdaEDSR and AdaRCAN, respectively. Note that, due to the data structure of the sparse convolution is identical with

Table 1. Quantitative results in comparison with the state-of-the-art methods. Best three methods are highlighted by red, blue and [illegible], respectively.

Method	Scale	Set5		Set14		B100		Urban100		Manga109	
		PSNR	SSIM	PSNR	SSIM	PSNR	SSIM	PSNR	SSIM	PSNR	SSIM
Bicubic	×2	33.66	0.9299	30.24	0.8688	29.56	0.8431	26.88	0.8403	30.80	0.9339
SRCNN [6]	×2	36.66	0.9542	32.45	0.9067	31.36	0.8879	29.50	0.8946	35.60	0.9663
VDSR [12]	×2	37.53	0.9590	33.05	0.9130	31.90	0.8960	30.77	0.9140	37.22	0.9750
EDSR [17]	×2	38.11	0.9602	33.92	0.9195	32.32	0.9013	32.93	0.9351	39.10	0.9773
AdaEDSR	×2	38.21	0.9611	33.97	0.9208	32.35	0.9017	32.91	0.9353	39.11	0.9778
RDN [42]	×2	38.24	0.9614	34.01	0.9212	32.34	0.9017	32.89	0.9353	39.18	0.9780
RCAN [41]	×2	[illegible]	[illegible]	34.12	0.9216	32.41	0.9027	33.34	0.9384	39.44	0.9786
SAN [5]	×2	38.31	0.9620	[illegible]	[illegible]	32.42	0.9028	[illegible]	[illegible]	[illegible]	0.9792
AdaRCAN	×2	38.28	0.9615	34.12	0.9216	32.41	[illegible]	33.29	0.9380	39.44	[illegible]
Bicubic	×3	30.39	0.8682	27.55	0.7742	27.21	0.7385	24.46	0.7349	26.95	0.8556
SRCNN [6]	×3	32.75	0.9090	29.30	0.8215	28.41	0.7863	26.24	0.7989	30.48	0.9117
VDSR [12]	×3	33.67	0.9210	29.78	0.8320	28.83	0.7990	27.14	0.8290	32.01	0.9340
EDSR [17]	×3	34.65	0.9280	30.52	0.8462	29.25	0.8093	28.80	0.8653	34.17	0.9476
AdaEDSR	×3	34.65	0.9288	30.57	0.8463	29.27	0.8091	28.78	0.8649	34.16	0.9482
RDN [42]	×3	34.71	0.9296	30.57	0.8468	29.26	0.8093	28.80	0.8653	34.13	0.9484
RCAN [41]	×3	[illegible]	[illegible]	30.65	0.8482	[illegible]	0.8111	29.09	0.8702	34.44	0.9499
SAN [5]	×3	34.75	0.9300	[illegible]	[illegible]	29.33	0.8112	[illegible]	[illegible]	[illegible]	[illegible]
AdaRCAN	×3	34.79	0.9302	30.65	0.8481	29.33	0.8111	29.03	0.8689	34.49	0.9498
Bicubic	×4	28.42	0.8104	26.00	0.7027	25.96	0.6675	23.14	0.6577	24.89	0.7866
SRCNN [6]	×4	30.48	0.8628	27.50	0.7513	26.90	0.7101	24.52	0.7221	27.58	0.8555
VDSR [12]	×4	31.35	0.8830	28.02	0.7680	27.29	0.0726	25.18	0.7540	28.83	0.8870
EDSR [17]	×4	32.46	0.8968	28.80	0.7876	27.71	0.7420	26.64	0.8033	31.02	0.9148
AdaEDSR	×4	32.49	0.8977	28.76	0.7865	27.71	0.7410	26.58	0.8011	30.96	0.9150
RDN [42]	×4	32.47	0.8990	28.81	0.7871	27.72	0.7419	26.61	0.8028	31.00	0.9151
RCAN [41]	×4	32.63	0.9002	[illegible]	0.7889	27.77	0.7436	26.82	0.8087	31.22	0.9173
SAN [5]	×4	32.64	0.9003	28.92	0.7888	27.78	0.7436	[illegible]	0.8068	[illegible]	[illegible]
AdaRCAN	×4	[illegible]	[illegible]	28.88	[illegible]	27.77	[illegible]	26.80	[illegible]	31.22	0.9172

Table 2. Inference efficiency in comparison with the state-of-the-art methods. Note that the GPU memory is not enough to run SAN [5] with scale ×2 on Urban100 and Manga109 datasets.

Method	Scale	Set5		Set14		B100		Urban100		Manga109	
		FLOPs (G)	Time (ms)	FLOPs (G)	Time (ms)	FLOPs (G)	Time (ms)	FLOPs (G)	Time (ms)	FLOPs (G)	Time (ms)
SRCNN [6]	×2	6.1	43.0	12.3	7.9	8.2	4.4	41.4	19.2	51.6	24.2
VDSR [12]	×2	70.5	86.9	143.0	88.7	95.4	58.0	481.6	301.0	599.5	368.8
EDSR [17]	×2	1338.8	395.7	2552.2	630.7	1776.9	469.9	8041.1	2163.8	9891.4	2554.5
AdaEDSR	×2	650.6	312.3	1397.3	489.8	965.3	371.5	4844.9	1655.2	5208.4	1864.5
RDN [42]	×2	801.1	345.9	1527.3	617.1	1063.3	407.7	4811.9	2198.5	5919.2	3417.1
RCAN [41]	×2	577.9	633.2	1101.8	813.6	767.0	607.0	3471.2	1955.0	4270.0	2342.9
SAN [5]	×2	3835.9	1276.0	17500.4	3314.0	3943.2	1637.8	372727.5	N/A	645359.4	N/A
AdaRCAN	×2	469.1	614.5	925.5	751.9	649.2	606.3	2907.2	1749.2	3300.7	2034.2
SRCNN [6]	×3	6.1	43.0	12.3	7.9	8.2	4.4	41.4	19.2	51.6	24.2
VDSR [12]	×3	70.5	86.9	143.0	88.7	95.4	58.0	481.6	301.0	599.5	368.8
EDSR [17]	×3	699.1	251.0	1305.7	341.7	924.1	259.6	3984.0	957.9	4904.0	1271.1
AdaEDSR	×3	504.8	231.4	1013.5	302.0	722.8	232.0	3314.2	858.4	3695.9	1023.3
RDN [42]	×3	437.1	195.0	816.3	290.3	577.8	190.2	2490.9	1045.3	3066.1	1404.0
RCAN [41]	×3	328.5	553.6	613.5	551.7	434.2	511.2	1872.0	1029.2	2304.2	1208.7
SAN [5]	×3	463.2	582.2	1930.1	992.9	517.5	600.7	36735.2	5416.2	61976.0	8194.4
AdaRCAN	×3	277.7	572.6	512.9	559.1	369.3	523.8	1596.3	968.1	1842.2	1107.1
SRCNN [6]	×4	6.1	43.0	12.3	7.9	8.2	4.4	41.4	19.2	51.6	24.2
VDSR [12]	×4	70.5	86.9	143.0	88.7	95.4	58.0	481.6	301.0	599.5	368.8
EDSR [17]	×4	501.9	214.6	908.8	239.8	655.7	240.8	2699.4	640.0	3297.6	762.2
AdaEDSR	×4	371.7	181.1	716.8	215.4	508.5	195.2	2265.8	563.3	2588.4	656.0
RDN [42]	×4	337.9	128.3	611.9	163.7	441.5	132.9	1817.3	512.4	2220.9	646.2
RCAN [41]	×4	270.1	546.9	489.0	505.7	352.8	490.0	1452.5	684.7	1774.4	843.8
SAN [5]	×4	159.4	482.7	522.2	568.5	190.9	445.2	7770.0	2258.0	12858.7	3174.3
AdaRCAN	×4	227.5	561.6	418.1	524.9	304.8	520.0	1263.0	659.8	1463.3	712.8

standard convolution, we can use the pretrained backbone model to initialize the AdaDSR model to improve the training stability and save training time.

Model Evaluation. Following previous works [15,17,41], we use PSNR and SSIM as model evaluation metrics, and five standard benchmark datasets (*i.e.*, Set5 [3], Set14 [37], B100 [21], Urban100 [10] and Manga109 [23]) are employed as test sets, and the PSNR and SSIM indices are calculated on the luminance channel (a.k.a. Y channel) of YCbCr color space with *scale* pixels on the boundary ignored. Furthermore, the computation efficiency is evaluated by FLOPs and inference time. For a fair comparison with the competing methods, when counting the running time, we implement all competing methods in our framework and replace the convolution layers of the main body with im2col [4] based convolutions. All evaluations are conducted in the PyTorch [25] environment running on a single Nvidia TITAN RTX GPU. The source code (with all comparison methods) and pre-trained models will be publicly available.

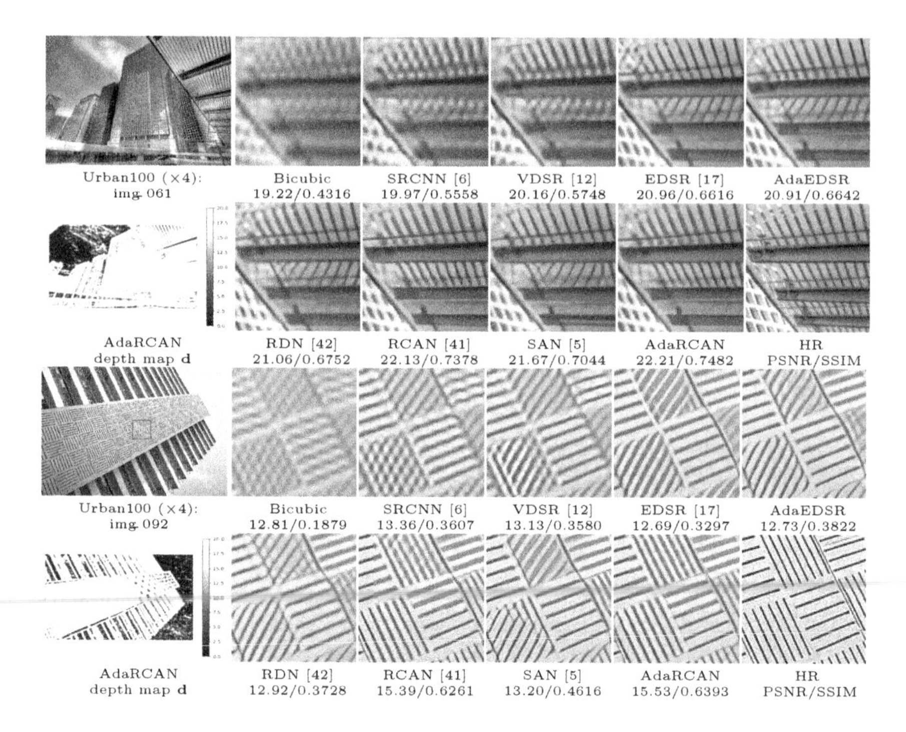

Fig. 4. Visual comparison for 4× SR on Urban100 dataset. Note that the depth map of AdaRCAN is an average of the 10 groups.

4.2 Comparison with State-of-the-Arts

To evaluate the effectiveness, we compare AdaDSR[1] with the backbone EDSR [17] and RCAN [41] models as well as four other state-of-the-art methods, *i.e.*, SRCNN [6], VDSR [12], RDN [42] and SAN [5]. Note that all visual results of other methods given in this section are generated by the officially released models, while the FLOPs and inference time are evaluated in our framework.

As shown in Table 1, both AdaEDSR and AdaRCAN perform favorably against their counterparts EDSR and RCAN in terms of quantitative PSNR and SSIM metrics. Besides, it can be seen from Table 2, although the adapter module introduces extra computation cost, it is very lightweight and efficient in comparison to the backbone super-resolution model, and the deployment of the lightweight adapter module greatly reduces computation amount of the whole model, resulting in lower FLOPs and faster inference, especially on large images (*e.g.*, Urban100 and Manga109). Note that SAN [5] has similar performance with RCAN and AdaRCAN, yet its computation cost is too heavy on large images.

Apart from the quantitative comparison, visual results are also given in Fig. 4. One can see that AdaEDSR and AdaRCAN are able to generate super-resolved

[1] The desired depth d is set to 32 and 20 for AdaEDSR and AdaRCAN in Tables 1 and 2, *i.e.*, the number of residual blocks in EDSR and that of each group in RCAN.

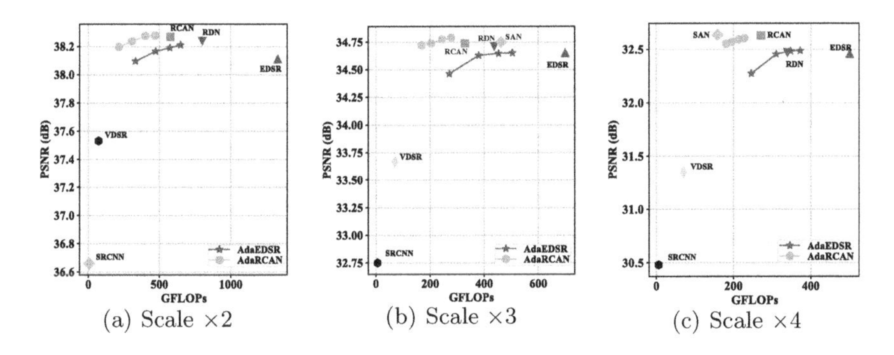

(a) Scale ×2 (b) Scale ×3 (c) Scale ×4

Fig. 5. Comparison against state-of-the-art methods in terms of FLOPs and PSNR on Set5. Note that SAN is not given in Scale ×2 due to that its computation cost is 3835.9 GFLOPs, which is much more than other methods.

images of similar or better visual quality to their counterparts. Kindly refer to the supplementary materials for more qualitative results. We also show the pixel-wise depth map $\mathbf{d}$ of AdaRCAN (due to space limit, we show the average of the depth maps for 10 groups of AdaRCAN) to discuss the relationship between the processed image and the depth map. As we can see from Fig. 4, greater depth is predicted for the regions with detailed textures, while most of the computation in smooth areas can be omitted for efficiency purpose, which is intuitive and verifies our discussions in Sect. 1.

Considering both quantitative and qualitative results, our AdaDSR can achieve comparable performance against state-of-the-art methods while greatly reducing the computation amount. Further analysis on the adaptive adjustment of d please refer to Sect. 4.3.

4.3 Adaptive Inference with Varying Depth

Taking both the feature map $\mathbf{z}_0$ and desired depth d as input, the adapter module is able to predict an image content adaptive network depth map while satisfying the computation efficiency constraints. Consequently, our AdaDSR can be flexibly tuned to meet various efficiency constraints on the fly. In comparison, the competing methods are based on deterministic inference and can only be performed with the fixed complexity. As shown in Fig. 5, we evaluate our AdaDSR model with different desired depth d (*i.e.*, 8, 16, 24, 32 for AdaEDSR and 5, 10, 15, 20 for AdaRCAN), and record the corresponding FLOPs and PSNR values on Set5. More results please refer to the supplementary materials.

From the figures, we can draw several conclusions. First, our AdaDSR can be tuned with the hyper-parameter d, and resulting in a curve in the figures, rather than a single point as the competing methods. With an increasing desired depth d, AdaDSR requires more computation resources and generates better super-resolved images. It is worth noting that, AdaDSR taps the potential of the backbone models, and can obtain comparable performance against the well-trained backbone model when higher d is set. Furthermore, AdaDSR reaches the saturation point with a relatively lower FLOPs, which indicates that a shallower model is sufficient for most regions. Experiments on both versions (*i.e.*, AdaEDSR and AdaRCAN) verify the effectiveness and generality of our adapter module.

5 Ablation Analysis

Considering training efficiency in the multi-GPU environment, without loss of generality, we conduct ablation analysis on AdaEDSR $\times 2$ model. Due to space limitation, more experiments please refer to the supplementary material.

EDSR Variants. To begin with, we train EDSR variants in our framework, *i.e.*, EDSR (8), EDSR (16), EDSR (24) and EDSR (32) by setting the number of residual blocks to 8, 16, 24 and 32, respectively. Note that EDSR (32) performs slightly better than released EDSR model, so we use this one for fair comparison. The quantitative results on Set5 are given in Table 3. Comparing all EDSR variants, generally one can observe performance gains as the model depth grows.

Besides, as previously illustrated in Fig. 1(c), a shallow model is sufficient for smooth areas, while regions with rich contexture usually require a deep model for better reconstruction of the details. Taking advantage of this phenomenon, with lightweight adapter, AdaDSR is able to predict suitable depth for various areas according to difficulty and resource constraints, and achieves better efficiency-performance tradeoff, resulting in the curve at the top left of their corresponding counterparts as shown in Figs. 1(d) and 5. Detailed data can be found in Table 3.

Table 3. Quantitative evaluation of EDSR and AdaEDSR variants on Set5 ($\times 2$).

Method	PSNR (dB)	SSIM	FLOPs (G)	Time (ms)	Method	PSNR (dB)	SSIM	FLOPs (G)	Time (ms)	Method	PSNR (dB)	SSIM	FLOPs (G)	Time (ms)
EDSR (8)	38.05	0.9607	408.23	147.2	FAdaEDSR (8)	38.17	0.9609	504.87	280.6	AdaEDSR (8)	38.10	0.9605	329.50	169.6
EDSR (16)	38.11	0.9610	718.41	230.7	FAdaEDSR (16)	38.21	0.9611	719.62	327.7	AdaEDSR (16)	38.17	0.9608	472.90	217.0
EDSR (24)	38.15	0.9612	1028.58	305.8	FAdaEDSR (24)	38.23	0.9613	997.95	366.8	AdaEDSR (24)	38.19	0.9610	574.85	243.8
EDSR (32)	38.16	0.9611	1338.76	395.7	FAdaEDSR (32)	38.24	0.9613	1358.30	402.9	AdaEDSR (32)	38.21	0.9611	650.65	312.3

AdaEDSR Variants. We further implement several AdaEDSR variants, *i.e.*, FAdaEDSR (8), FAdaEDSR (16), FAdaEDSR (24) and FAdaEDSR (32), which are trained with a fixed depth $d = 8$, 16, 24 and 32 respectively, and the adapter module takes only the image features as input. The models are trained under the same settings (except for the fixed d in the learning objective) with AdaEDSR. As shown in Table 3, with the per-pixel depth map, these models obtain much better quantitative results than EDSR variants with similar computation cost.

It is worth noting that FAdaEDSR (32) achieves comparable performance with RDN [42], which clearly shows the effectiveness of the predicted depth map. And with a deeper backbone, AdaDSR can potentialy achieve much higher performance with similar computation amount. We also show the performance of our AdaEDSR $\times 2$ model in Table 3. Specifically, AdaEDSR (n) means that desired depth $d = n$ at the test time. One can see that although the quantitative performance is slightly worse than FAdaEDSR, AdaEDSR is more computationally efficient and can be flexibly tuned in the testing phase, indicating that AdaDSR achieves adaptive inference with minor sacrifice of performance.

6 Conclusion

In this paper, we revisit the relationship between the model depth and quantitative performance on single image super-resolution task, and present an AdaDSR model by incorporating a lightweight adapter module and sparse convolution in deep SISR networks. The adapter module predicts an image content oriented network depth map, and the value is higher in regions with detailed textures and lower in smooth areas. According to the predicted depth, only a fraction of residual blocks are performed at each point by using im2col based sparse convolution. Furthermore, the parameters of the adapter module are adjustable on the fly according to the desired depth, so that the AdaDSR model can be tuned to meet various efficiency constraints in the inference phase. Experimental results show the effectiveness and adaptiveness of our AdaDSR model, and indicate that AdaDSR can obtain state-of-the-art performance while adaptive to a range of efficiency requirements.

Acknowledgement. This work is partially supported by the National Natural Scientific Foundation of China (NSFC) under Grant No.s 61671182, U19A2073.

References

1. Ahn, N., Kang, B., Sohn, K.-A.: Fast, accurate, and lightweight super-resolution with cascading residual network. In: Ferrari, V., Hebert, M., Sminchisescu, C., Weiss, Y. (eds.) ECCV 2018. LNCS, vol. 11214, pp. 256–272. Springer, Cham (2018). https://doi.org/10.1007/978-3-030-01249-6_16
2. Bengio, Y., Mesnil, G., Dauphin, Y., Rifai, S.: Better mixing via deep representations. In: International Conference on Machine Learning, pp. 552–560 (2013)
3. Bevilacqua, M., Roumy, A., Guillemot, C., Alberi-Morel, M.L.: Low-complexity single-image super-resolution based on nonnegative neighbor embedding (2012)
4. Chellapilla, K., Puri, S., Simard, P.: High performance convolutional neural networks for document processing. In: Tenth International Workshop on Frontiers in Handwriting Recognition. Suvisoft (2006)
5. Dai, T., Cai, J., Zhang, Y., Xia, S.T., Zhang, L.: Second-order attention network for single image super-resolution. In: Proceedings of the IEEE Conference on Computer Vision and Pattern Recognition, pp. 11065–11074 (2019)
6. Dong, C., Loy, C.C., He, K., Tang, X.: Image super-resolution using deep convolutional networks. IEEE Trans. Pattern Anal. Mach. Intell. **38**(2), 295–307 (2015)
7. Figurnov, M., et al.: Spatially adaptive computation time for residual networks. In: Proceedings of the IEEE Conference on Computer Vision and Pattern Recognition, pp. 1039–1048 (2017)

8. Graham, B., Engelcke, M., van der Maaten, L.: 3D semantic segmentation with submanifold sparse convolutional networks. In: Proceedings of the IEEE Conference on Computer Vision and Pattern Recognition, pp. 9224–9232 (2018)
9. He, K., Zhang, X., Ren, S., Sun, J.: Deep residual learning for image recognition. In: Proceedings of the IEEE Conference on Computer Vision and Pattern Recognition, pp. 770–778 (2016)
10. Huang, J.B., Singh, A., Ahuja, N.: Single image super-resolution from transformed self-exemplars. In: Proceedings of the IEEE Conference on Computer Vision and Pattern Recognition, pp. 5197–5206 (2015)
11. Hui, Z., Wang, X., Gao, X.: Fast and accurate single image super-resolution via information distillation network. In: Proceedings of the IEEE Conference on Computer Vision and Pattern Recognition, pp. 723–731 (2018)
12. Kim, J., Kwon Lee, J., Mu Lee, K.: Accurate image super-resolution using very deep convolutional networks. In: Proceedings of the IEEE Conference on Computer Vision and Pattern Recognition, pp. 1646–1654 (2016)
13. Kingma, D.P., Ba, J.: Adam: a method for stochastic optimization. arXiv preprint arXiv:1412.6980 (2014)
14. Lavin, A., Gray, S.: Fast algorithms for convolutional neural networks. In: Proceedings of the IEEE Conference on Computer Vision and Pattern Recognition, pp. 4013–4021 (2016)
15. Ledig, C., et al.: Photo-realistic single image super-resolution using a generative adversarial network. In: Proceedings of the IEEE Conference on Computer Vision and Pattern Recognition, pp. 4681–4690 (2017)
16. Li, H., Zhang, H., Qi, X., Yang, R., Huang, G.: Improved techniques for training adaptive deep networks. In: Proceedings of the IEEE International Conference on Computer Vision, pp. 1891–1900 (2019)
17. Lim, B., Son, S., Kim, H., Nah, S., Mu Lee, K.: Enhanced deep residual networks for single image super-resolution. In: The IEEE Conference on Computer Vision and Pattern Recognition Workshops, pp. 136–144 (2017)
18. Liu, B., Wang, M., Foroosh, H., Tappen, M., Pensky, M.: Sparse convolutional neural networks. In: Proceedings of the IEEE Conference on Computer Vision and Pattern Recognition, pp. 806–814 (2015)
19. Liu, P., Zhang, H., Zhang, K., Lin, L., Zuo, W.: Multi-level wavelet-CNN for image restoration. In: The IEEE Conference on Computer Vision and Pattern Recognition Workshops (2018)
20. Liu, X., Pool, J., Han, S., Dally, W.J.: Efficient sparse-winograd convolutional neural networks. arXiv preprint arXiv:1802.06367 (2018)
21. Martin, D., Fowlkes, C., Tal, D., Malik, J.: A database of human segmented natural images and its application to evaluating segmentation algorithms and measuring ecological statistics. In: Proceedings of the IEEE International Conference on Computer Vision, pp. 416–423 (2001)
22. Mathieu, M., Henaff, M., LeCun, Y.: Fast training of convolutional networks through FFTs. In: Proceedings of the International Conference on Learning Representations (2014)
23. Matsui, Y., et al.: Sketch-based manga retrieval using manga109 dataset. Multimedia Tools Appl. **76**(20), 21811–21838 (2017)
24. Park, J., et al.: Faster CNNs with direct sparse convolutions and guided pruning. In: Proceedings of the International Conference on Learning Representations (2017)
25. Paszke, A., et al.: PyTorch: an imperative style, high-performance deep learning library. In: Advances in Neural Information Processing Systems, pp. 8024–8035 (2019)

26. Ren, M., Pokrovsky, A., Yang, B., Urtasun, R.: SBNet: sparse blocks network for fast inference. In: Proceedings of the IEEE Conference on Computer Vision and Pattern Recognition, pp. 8711–8720 (2018)
27. Ronneberger, O., Fischer, P., Brox, T.: U-Net: convolutional networks for biomedical image segmentation. In: Navab, N., Hornegger, J., Wells, W.M., Frangi, A.F. (eds.) MICCAI 2015. LNCS, vol. 9351, pp. 234–241. Springer, Cham (2015). https://doi.org/10.1007/978-3-319-24574-4_28
28. Shi, W., et al.: Real-time single image and video super-resolution using an efficient sub-pixel convolutional neural network. In: Proceedings of the IEEE Conference on Computer Vision and Pattern Recognition, pp. 1874–1883 (2016)
29. Shi, W., et al.: Cardiac image super-resolution with global correspondence using multi-atlas patchmatch. In: Mori, K., Sakuma, I., Sato, Y., Barillot, C., Navab, N. (eds.) MICCAI 2013. LNCS, vol. 8151, pp. 9–16. Springer, Heidelberg (2013). https://doi.org/10.1007/978-3-642-40760-4_2
30. Shoshan, A., Mechrez, R., Zelnik-Manor, L.: Dynamic-Net: tuning the objective without re-training for synthesis tasks. In: Proceedings of the IEEE International Conference on Computer Vision, pp. 3215–3223 (2019)
31. Timofte, R., Agustsson, E., Van Gool, L., Yang, M.H., Zhang, L.: NTIRE 2017 challenge on single image super-resolution: methods and results. In: The IEEE Conference on Computer Vision and Pattern Recognition Workshops, pp. 114–125 (2017)
32. Upchurch, P., et al.: Deep feature interpolation for image content changes. In: Proceedings of the IEEE Conference on Computer Vision and Pattern Recognition, pp. 7064–7073 (2017)
33. Vasilache, N., Johnson, J., Mathieu, M., Chintala, S., Piantino, S., LeCun, Y.: Fast convolutional nets with fbfft: a GPU performance evaluation. In: Proceedings of the International Conference on Learning Representations (2015)
34. Wang, W., Guo, R., Tian, Y., Yang, W.: CFSNet: toward a controllable feature space for image restoration. In: Proceedings of the IEEE International Conference on Computer Vision, pp. 4140–4149 (2019)
35. Yu, J., et al.: Wide activation for efficient and accurate image super-resolution. arXiv preprint arXiv:1808.08718 (2018)
36. Yu, K., Wang, X., Dong, C., Tang, X., Loy, C.C.: Path-restore: learning network path selection for image restoration. arXiv preprint arXiv:1904.10343 (2019)
37. Zeyde, R., Elad, M., Protter, M.: On single image scale-up using sparse-representations. In: Boissonnat, J.-D., et al. (eds.) Curves and Surfaces 2010. LNCS, vol. 6920, pp. 711–730. Springer, Heidelberg (2012). https://doi.org/10.1007/978-3-642-27413-8_47
38. Zhang, K., Zuo, W., Zhang, L.: Deep plug-and-play super-resolution for arbitrary blur kernels. In: Proceedings of the IEEE Conference on Computer Vision and Pattern Recognition, pp. 1671–1681 (2019)
39. Zhang, Q., Ding, Y., Yu, B., Xu, M., Li, C.: Old film image enhancements based on sub-pixel convolutional network algorithm. In: Tenth International Conference on Graphics and Image Processing (2018)
40. Zhang, Y., Liu, S., Dong, C., Zhang, X., Yuan, Y.: Multiple cycle-in-cycle generative adversarial networks for unsupervised image super-resolution. IEEE Trans. Image Process. **29**, 1101–1112 (2019)
41. Zhang, Y., Li, K., Li, K., Wang, L., Zhong, B., Fu, Y.: Image super-resolution using very deep residual channel attention networks. In: Ferrari, V., Hebert, M., Sminchisescu, C., Weiss, Y. (eds.) ECCV 2018. LNCS, vol. 11211, pp. 294–310. Springer, Cham (2018). https://doi.org/10.1007/978-3-030-01234-2_18

42. Zhang, Y., Tian, Y., Kong, Y., Zhong, B., Fu, Y.: Residual dense network for image super-resolution. In: Proceedings of the IEEE Conference on Computer Vision and Pattern Recognition, pp. 2472–2481 (2018)
43. Zou, W.W., Yuen, P.C.: Very low resolution face recognition problem. IEEE Trans. Image Process. **21**(1), 327–340 (2011)

Densely Connecting Depth Maps for Monocular Depth Estimation

Jinqing Zhang[1], Haosong Yue[1](✉), Xingming Wu[1], Weihai Chen[1], and Changyun Wen[2]

[1] Beihang University, Beijing, China
{zhangmt0391,yuehaosong}@buaa.edu.cn, xmwubuaa@163.com, whchenbuaa@126.com

[2] Nanyang Technological University, Singapore, Singapore
ecywen@ntu.edu.sg

Abstract. Predicting depth map from a single RGB image is beneficial for many three-dimensional applications. Although recent monocular depth estimation methods have achieved impressive accuracy, the preference on high-level features or low-level features prevents them from balancing sharpness and fidelity of depth maps. In this work, we propose a dense connection mechanism that connects diverse sub-depth maps produced by the sub-predictors to the final depth map to contribute information from features at different levels. Besides, two kinds of diversity enhancement devices are proposed to increase the number and diversity of the sub-depth maps collected by the dense connection mechanism. Experimental results on KITTI and NYU Depth V2 datasets shows that, by fusing the dense connection mechanism and diversity enhancement devices, our proposed method achieves state-of-the-art accuracy and predicts sharp depth maps that restore reliable object structures.

Keywords: Monocular depth estimation · Dense connection mechanism · Diversity enhancement device

1 Introduction

Estimating depth from two-dimensional images acts as a bridge between computer vision and computer graphics. The depth values of each pixels in the image are quickly estimated and a point cloud is automatically generated, which can benefit applications like model reconstruction, point cloud based object detection and semantic segmentation. Different depth estimation methods deal with different image sources, including single RGB images [2,10,11], videos [14,41,42], stereo image pairs [13,21] and multi-view overlapping images [5,17,39]. Among them, monocular depth estimation not only requires the least information, but also avoids the constraints brought by traditional geometric processes like reprojection, which is often applied in the methods that use couples of images.

Recent deep learning based methods have greatly outperformed early traditional methods, which mainly rely on low-level features like edges, superpixels,

A. Bartoli and A. Fusiello (Eds.): ECCV 2020 Workshops, LNCS 12538, pp. 149–165, 2020.
https://doi.org/10.1007/978-3-030-66823-5_9

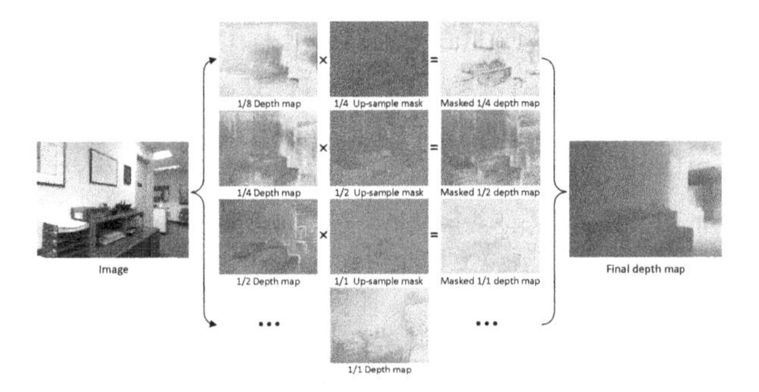

Fig. 1. Part of the sub-depth maps. The figure shows a part of the sub-depth maps and masks generated from a single image. The process of using masks to generate new sub-depth maps is also illustrated

or feature points [18,22,34]. However, few of them can generate perfect depth maps with both high sharpness and high fidelity due to the preference on features. Some methods [32,37,40] prefer high-level features like semantic information, making predicted depth maps restore reliable object structures but in low sharpness. Others [1,24,26] value low-level features and generate sharp depth maps with obvious distortion on objects.

Obtaining sharp depth maps with reliable object structures requires balanced utilization of features at different levels. There is a mechanism used in DenseNet [16] that builds dense connection between every features and bottlenecks in the block, which balances the components of relative high-level and low-level features in the output of the block. Inspired by this, we use multiple sub-predictors in the network to generate sub-depth maps and densely connect all the sub-depth maps and the sub-predictors to balance the proportion of the information of features at different levels in the final depth maps. Some other methods also generate sub-depth maps, but sparsely connect only one previous sub-depth map to the next predictor [24,31,42]. In other word, our method uses the DenseNet's way to connect the sub-depth maps, while the others use the ResNet's [15] way.

Dense connection mechanism gives the network the potential to balance sharpness and fidelity of depth maps, but it can not fully work if the sub-depth maps storage homogeneous information. In order to get more diverse sub-depth maps, two kinds of diversity enhancement devices are proposed. They are associated mask generators (AMG) and up-sampling mask generators (UMG). Both of them can generate masks by extracting information from features, but the features they enter and the occasion of using the masks they generate are different. These masks can directly multiply sub-depth maps at element level to create new diverse sub-depth maps. The overall structure of our network in Fig. 2 shows how the diversity enhancement devices participate in dense connection mechanism. The addition of AMGs and UMGs increases the total number of sub-depth maps from 4 to 30, and considerably enhances their diversity. A part of the sub-depth maps is shown in Fig. 1.

The combination of dense connection mechanism and diversity enhancement devices effectively balances the utilization of the features at different levels, and helps the network produces sharp depth maps with reliable object structures. Our method has been tested on NYU Depth V2 dataset [29] and KITTI dataset [12]. The experiment results show that our method achieves state-of-the-art accuracy and generates depth maps with both high sharpness and high fidelity. The reliable object structures in depth maps will also benefit potential applications like point cloud based object detection and semantic segmentation.

2 Related Work

Early monocular depth estimation methods usually utilize low-level features of images. For instance, Saxena et al. [34] use MRF to predict the depth of superpixels. Ladicky et al. [22] build a handcraft feature representation of images and generate depth maps through a classifier. These methods can only approximate the depth values and produce depth maps with insufficient information.

Later, methods using deep learning make a leap in the performance of monocular depth estimation. The first attempt of using CNN prototype is proposed in [10]. After that, improvements can be divided into two categories. One is to change the architecture of the network, and the other is to combine CNN with traditional image processing methods. Both routes can lead to higher accuracy and better performance, and some methods follow both of the routes.

2.1 Methods of Changing Network Structure

CNN based monocular depth estimation methods basically conform to the U-Net structure [33], which is originally designed for semantic segmentation, and typical CNN prototypes often act as the encoder of U-Net. It leads to the phenomenon that each time a better CNN prototype is proposed, better methods of monocular depth estimation follow up. AlexNet [20] in [10] is replaced by VGGNet [35] in [3,9], followed by ResNet [15] used in [11,26,37] and DenseNet [16] used in [1,24]. The decoder of the U-Net also includes unique network structures. Atrous Spatial Pyramid Pooling (ASPP) [4], a special structure which consists of multiple dilated convolutional layers, is widely used in semantic segmentation and also benefits monocular depth estimation methods [11,24,40]. Ren et al. [32] propose a two-stage robust depth estimation module that can fuse the knowledge of scene classification into depth estimation. Lee et al. [24] design a local planar guidance module that refines predicted depth map via spatial geometric information to get more precise results. Li et al. [27] apply channel-attention structure in the network to let it automatically decide which channels are reliable and perform well in both indoor and outdoor scenes. Li et al. [26] use two networks to predict the depth map and the gradient of the depth map respectively, and then combine them to get sharper depth maps.

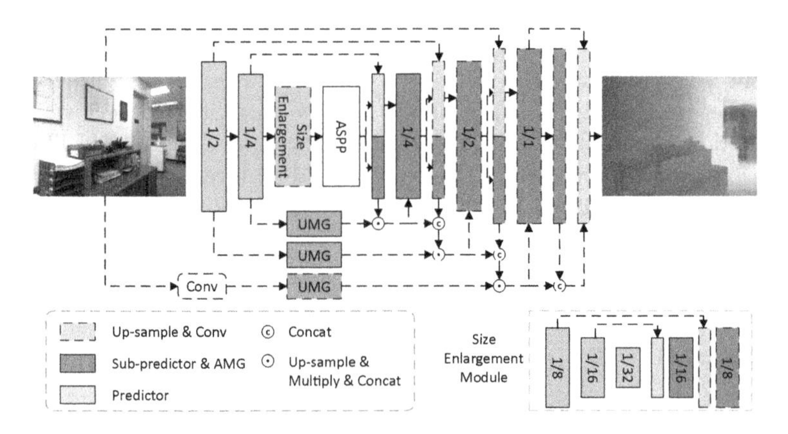

Fig. 2. The overall structure of our network. Our network follows the structure of U-Net. A version of DenseNet acts as the encoder of the network and the decoder consists of several sub-predictors, AMGs and UMGs. These components are densely connected via the dense connection mechanism

2.2 Methods of Combining CNN with Traditional Approach

Though deep learning have surpassed traditional image processing methods in many aspects, there are still some tasks that traditional approach can do better, such as edge detection and clustering. It is convincing that the combination with traditional approach can benefit monocular depth estimation methods. Conditional Random Field (CRF), a classical mathematic model that is powerful on tasks like semantic segmentation, are now applied in many CNN-based monocular depth estimation methods [28,36,37] to fuse the depth of superpixels or the depth map in different sizes. Fu et al. [11] utilize ordinal regression in depth estimation. The method changes the regression task into classification task by discretizing the continuous depth value and trains the network though a ordinal loss. Diaz et al. [8] follow the idea of ordinal regression, but change the depth label into soft probability distribution and generalize the method to other image processing tasks. Yin et al. [40] propose a method that calculates the loss of virtual surface normals, which enforces geometric constraints into the depth estimation and produced more reasonable point clouds.

3 Method

This section first introduces the overall architecture of our network, and then describes the two kinds of diversity enhancement devices in detail. After that, the proposed dense connection mechanism that collects and helps transport sub-depth maps is introduced, followed with the loss function.

3.1 Network Architecture

The network of our proposed method is shown in Fig. 2. It follows the structure of U-Net [33]. The encoder of the U-Net that is responsible for extracting features from image is DenseNet. Unlike ResNet that only inputs one output of previous bottleneck to the next bottleneck, DenseNet connects all of the outputs of previous bottlenecks in the same block to the next bottleneck. It lets the final bottleneck of a block in DenseNet can approach to both relative high-level and low-level features in this block at the same time, while the final bottleneck of a block in ResNet can only connect to the relative high-level features in this block. Since the skip connection only transports the output of the final bottleneck of a block, using DenseNet can more effectively transport the low-level features to the decoder.

Applying DenseNet helps balance the utilization of features at different levels, but the extension of the breadth of features makes the depth of features shallower because the low-level features will occupy the channels originally left for high-level features. To supplement the depth of features, the structure of ASPP [4] is applied in the decoder. The dilated convolutional layers in ASPP can significantly increase the receptive field of feature elements and make the level of features higher. In our method, a dense version of ASPP named DenseASPP [38] is used, which also densely connects the features and the dilated convolution layers.

In addition to DenseASPP, the decoder consists of sub-predictors, two kinds of mask generators and a final predictor. The four sub-predictors predict sub-depth maps in 1/8, 1/4, 1/2 and 1/1 resolution of original image. Associated mask generators (AMG) associate with the sub-predictors and generate associated masks (AM) that have the same size as their associated sub-depth maps and directly multiply them. The three up-sampling mask generators (UMG) generate up-sampling masks (UM) in 1/4, 1/2 and 1/1 resolution that multiply the up-sampled sub-depth maps. The final predictor at the end of the network is not associated with a AMG and combines the information in all sub-depth maps to generate the final depth map. All the mask generators and predictors are connected by our proposed dense connection mechanism.

3.2 Diversity Enhancement Device

The diversity enhancement devices in our network refer to two kinds of mask generators, AMGs and UMGs. They are designed to attach useful feature information to the sub-depth maps and effectively enhance their diversity. The structure of AMG and UMG and the function of the masks generated by them will be introduced respectively.

Associated Mask Generator. As the name implies, AMG is a network structure associated with sub-predictor. The structure of sub-predictor and AMG pair in Fig. 3 illustrates that they enter the same features and share several convolutional layers to reduce the dimension of features to 4. After that, each element in the sub-depth map corresponds with a 4 dimension vector represented

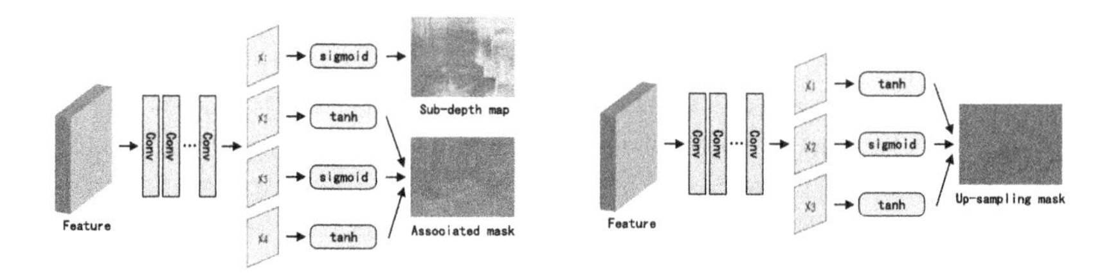

Fig. 3. Structure of sub-predictor & AMG pair (left) and UMG (right)

by (x_1, x_2, x_3, x_4). The sub-predictor uses x_1 to predict the sub-depth map, via a simple formula as

$$d = \text{sigmoid}(x_1) \times D_{max}. \tag{1}$$

The D_{max} in Eq. (1) is the theoretical maximum of the depth values of a specific dataset and act as a depth scale used to normalize the depth values. The sigmoid function maps x_1 to a number in [0,1], which can multiply D_{max} to get the predicted depth value.

The AMG takes the rest elements (x_2, x_3, x_4) to generate AM by

$$m_A = 2\left\{[1 + \alpha \tanh(x_2)] \, \text{sigmoid}(x_3) + \alpha \tanh(x_4)\right\}, \tag{2}$$

where α is a coefficient that limits the value in AM within a reasonable interval to guarantee the convergence of the network. We test several α values and find the network achieve the best performance when α is 0.1 (the results can be found in the section of ablation study). Eq (2) is inspire by the classical $y = wx + b$ equation and has $w = 1 + \alpha tanh(x_2)$, $x = sigmoid(x_3)$ and $b = \alpha tanh(x_4)$. The design of the mask generation equation gives the masks two special functions that will be introduce later and lets the masks be complicated enough to extract additional information from input features, while the simple combination of tanh and sigmoid function keeps the masks easy to train.

The AMs are used to multiply their associated sub-depth maps, so that each sub-predictor and AMG pair produces two different sub-depth maps at one time. The final predictor has the same structure as the sub-predictor except for the last convolutional layer that reduces the dimension of features to 1 instead of 4 because there is no AMG associated with it.

Up-Sampling Mask Generator. The structure of UMG shown in Fig. 3 is similar to AMG, but their input features are different. As shown in Fig. 2, the AMG takes the features transported between sub-predictors in the decoder. On the other hand, UMG takes the features transported by the skip connection from the encoder to the decoder. These features are reduced to 3 dimension via convolutional layers and each position in feature has a representation of vector (x_1, x_2, x_3). After that, UM is generated via Eq. (3), which follows the same idea as Eq. (2). The α in the formula also has the value of 0.1.

$$m_U = 2\{[1 + \alpha \tanh(x_1)] \text{sigmoid}(x_2) + \alpha \tanh(x_3)\}. \tag{3}$$

The occasion of using UMs is also different from AMs. They multiply smaller sub-depth maps after they are up-sampled into the same size as UMs. It is noteworthy that UMs can multiply sub-depth maps with much smaller sizes. For instance, a sub-depth map in 1/8 resolution can directly multiply the UM in 1/1 resolution after it has been up-sampled to 1/1 resolution.

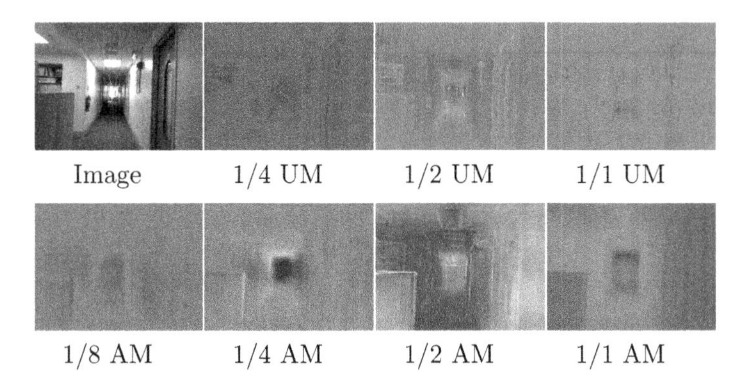

Fig. 4. Examples of AM and UM. Generally, AMs focus more on "adjustment" and UMs focus more on "selection"

Function of the Mask. As we know, the raw output of convolutional layer has a mean of 0. If the x_2, x_3, x_4 in Eq. (2) are replaced by 0, the value of m_A will be 1, which means the theoretical mean of elements in AMs is 1. The same is true for elements in UMs. When a sub-depth map is masked, the depth values in the sub-depth map directly multiply the corresponding elements in the mask and the overall distribution of the sub-depth map is changed. The changes can be divided into two cases.

In the first case, the mask elements have values around 1 and the depth values are slightly changed, which indicates that the elements in mask work as scale ratios and adjust the depth values to be more precise. We call this function "adjustment". In Fig. 4, it can be found that elements of plane regions in the masks have moderate intensity that is not too bright or too dark, telling the depth values of planes in masked sub-depth maps are gently adjusted to different but still trustworthy ones. The difference between masked and original sub-depth maps in adjusted regions indicates that the adjustment function of masks can use additional feature information to produce new and diverse sub-depth maps.

In the second case, the mask elements have values far from 1 and some parts of the depth values are dramatically changed, causing obvious steps between them and their neighbors. This case usually happens around object boundaries in low resolution sub-depth maps, where the original depth values are not trustworthy. We call this function "selection". It can also be found in Fig. 4 that elements around object boundaries in these masks are brighter or darker than their neighbors. After multiplying these elements, the depth values are changed into extremely large or small ones, informing the following sub-predictors that the depth values in these regions are wrong. Then, the sub-predictors will use the information of additional low-level features or the sub-depth maps in larger resolution to make a more reliable prediction. This can lead to clearer boundaries in the final depth maps.

Both AMs and UMs can do adjustment and selection. However, the diverse intensity of plane regions in AMs indicates that AMs focus more on adjustment because AMs and their associated sub-depth maps carry homology information, while the similar intensity of plane regions in UMs indicates that UMs focus more on selection because UMs always carry the information of the relative low-level features. Through the two functions, the information of features will be efficiently attached to the sub-depth maps to increase their number and diversity.

3.3 Dense Connection Mechanism

AMG and UMG are powerful tools to attach feature information to sub-depth maps. Our proposed dense connection mechanism then regularly reuses the masks generated by them to create many diverse sub-depth maps, and enters current sub-depth maps into the next predictor to provide rich information of features at different levels for the next prediction. Figure 2 gives a full view of our proposed dense connection mechanism.

We make a sub-depth map set to store all the obtained sub-depth maps. It starts from the raw and masked sub-depth maps generated by sub-predictor and AMG pair in 1/8 resolution. Each time the set meets a new sub-predictor and AMG pair, all the sub-depth maps in the set will be concatenated with the input features of the pair. After prediction, the generated raw and masked sub-depth maps will be concatenated with the map set and the total number of sub-depth maps plus 2. If the next sub-predictor and AMG pair requires the input in larger size, all the sub-depth maps will be up-sampled to the required size and immediately multiply the UM in that size. For instance, if the sub-depth maps in the set with a 1/8 resolution meet a sub-predictor that input features in 1/4 resolution, the sub-depth maps will be up-sampled to the 1/4 resolution and multiply the UM in 1/4 resolution. Multiplying UM generates as many new sub-depth maps as the current map set has. The new sub-depth maps will then be concatenated with the set and the number of sub-depth maps is doubled. There are 4 sub-predictor and AMG pairs and 3 UMGs in the network distributed alternately, which means the total number of the sub-depth maps that will be entered to the final predictor is $[(2 \times 2 + 2) \times 2 + 2] \times 2 + 2 = 30$.

With the dense connection mechanism, all of the predicted depth maps will be kept in the sub-depth maps set, even the raw sub-depth map in 1/8 resolution that carries the least information will be kept and connected with the final predictor. The predictors in the network can access to a set of sub-depth maps with much more diversity than the normal connecting method and generate sharp depth maps with reliable object structures by balancing the utilization of features at different levels.

Table 1. Performance on KITTI dataset. The "cap" refers to the effective depth range for calculation. The best results in each metrics are in **bold**, the second best result are underlined. Our method get the best or second best results on all of the metrics

Method	Cap	Higher is better			Lower is better			
		δ_1	δ_2	δ_3	AbsRel	SqRel	RMSE	RMSE_{log}
Eigen et al. [10]	0–80 m	0.692	0.899	0.967	0.190	1.515	7.156	0.270
Liu et al. [28]	0–80 m	0.656	0.881	0.958	0.217	–	7.046	–
Zhou et al. [42]	0–80 m	0.678	0.885	0.957	0.208	1.768	6.856	0.283
Kuznietsov et al. [21]	0–80 m	0.862	0.960	0.986	0.113	0.741	4.621	0.189
Fu et al. [11]	0–80 m	0.932	0.984	0.994	0.072	0.307	**2.727**	0.120
Godard et al. [14]	0–80 m	0.876	0.958	0.980	0.106	0.806	4.630	0.193
Lee et al. [24]	0–80 m	0.950	**0.993**	**0.999**	**0.064**	0.266	3.105	0.134
Yin et al. [40]	0–80 m	0.938	0.990	0.998	0.072	-	3.258	0.117
Ours	0–80 m	**0.952**	**0.993**	0.998	**0.064**	**0.265**	3.084	**0.106**
Zhou et al. [42]	0–50 m	0.696	0.900	0.966	0.201	1.391	5.181	0.264
Kuznietsov et al. [21]	0–50 m	0.875	0.964	0.988	0.108	0.595	3.518	0.179
Fu et al. [11]	0–50 m	0.936	0.985	0.995	0.071	0.268	**2.271**	0.116
Lee et al. [24]	0–50 m	0.959	0.994	**0.999**	**0.060**	**0.182**	2.292	0.127
Ours	0–50 m	**0.960**	**0.995**	**0.999**	0.061	0.200	2.283	**0.099**

3.4 Loss Function

The task of monocular depth estimation requires a relative accuracy, which can tolerate a large absolute error when the depth value is large. In order to fit this, we use the scale-invariant error (SIE) proposed by Eigen et al. in [10] as the loss function. The weighted formula of SIE can be simplified into

$$l = \frac{1}{n}\sum_i \sigma_i^2 - \frac{\lambda}{n^2}(\sum_i \sigma_i)^2. \tag{4}$$

The formula has $\sigma = \log d - \log \hat{d}$, where d and $\hat{d}$ represent predicted depth value and ground truth respectively.

SIE uses logarithmic function to lessen the error in far distances and cares the relative positions of all pixel pairs in the predicted depth maps to remain the object structures in ground truth. Although SIE fits the requirement of relative accuracy well, the absolute accuracy is still important for accuracy. Therefore, the λ in Eq. (4), a constant between 0 and 1, controls the proportion of SIE in the loss function. When λ is 1, the equation calculates the pure SIE. When λ is 0, the equation calculates the absolute logarithmic error. According to experiments, the best performance is achieved when λ is 0.75.

Table 2. Performance on NYU Depth V2 dataset. The best results in each metrics are in **bold**, the second best result are underlined. Our method get best or second best results on all of the metrics

Method	Higher is better			Lower is better		
	δ_1	δ_2	δ_3	AbsRel	RMSE	log10
Eigen et al. [10]	0.611	0.887	0.971	0.215	0.907	–
Liu et al. [28]	0.650	0.906	0.976	0.213	0.759	0.087
Li et al. [26]	0.789	0.955	0.988	0.152	0.611	0.064
Laina et al. [23]	0.811	0.953	0.988	0.127	0.573	0.055
Xu et al. [36]	0.811	0.954	0.987	0.121	0.586	0.052
Fu et al. [11]	0.828	0.965	0.992	0.115	0.509	0.051
Alhashim et al. [1]	0.846	0.974	0.994	0.123	0.465	0.053
Lee et al. [24]	0.878	**0.979**	0.995	0.113	0.461	0.048
Yin et al. [40]	0.875	0.976	0.994	**0.109**	0.500	**0.047**
Ours	**0.881**	**0.979**	**0.996**	0.112	**0.447**	0.048

4 Experiments

In this section, we evaluate our method on two challenging depth estimation datasets, i.e. KITTI dataset and NYU Depth V2 dataset, and compare its performance with several state-of-the-art methods. The results of ablation study will also be shown to demonstrate the effectiveness of our proposed mechanisms.

4.1 Dataset

KITTI Dataset. KITTI dataset is a famous autopilot dataset that contains image and depth map pairs from 61 outdoor scenes. The depth maps are captured through Lidar, making it is possible to get precise depth values in far positions and have a $D_{max} = 80$ m. The training set and test set we use follow a split made by Eigen [10]. The image and depth map pairs used for evaluating have a number of 697 and come from 29 scenes. The training set contains 23K pairs that come from the rest 32 scenes. The original size of the images in KITTI dataset is 375×1241. We centrally crop the images to 352×1216 to fit the structure of U-Net.

NYU Depth V2 Dataset. NYU Depth V2 dataset contains image and depth map pairs from 464 indoor scenes. The depth maps are captured by depth camera, which can produce dense depth maps and have a $D_{max} = 10m$. We randomly choose 50K pairs from the raw dataset as the training set and evaluate our method on 654 pairs chosen by Eigen in official annotated dataset. The raw depth maps in training set are inpainted referencing the methods described in [25] and [6] to fill the vacuum. We train our network using the original image size of 480×640 to achieve high resolution.

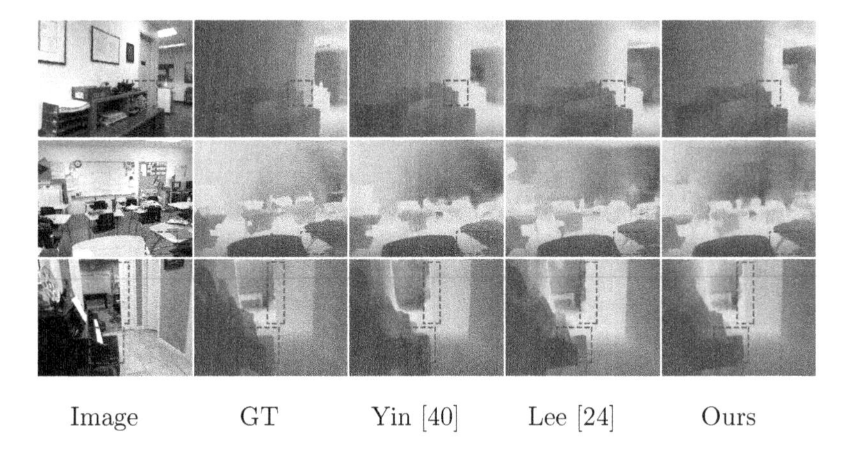

Fig. 5. Predicted depth maps on NYU Depth V2 dataset. The framed regions emphasize that our depth maps restore more reliable object structures

4.2 Implementation Details

We implement our method on a powerful deep learning platform PyTorch [30]. The encoder we use is a specific version of DenseNet named DenseNet-161 [16], which is official release and has been pre-trained on ImageNet [7]. During the training process, we use an Adam optimizer [19] with a weight decay of 10^{-4}. An exponential decay is applied on the learning rate of the optimizer, which starts from 10^{-4} and times 0.97 per one tenth epoch. We use a single TITAN X GPU for training. The total epoch is 10 with a batch size of 3 for both KITTI and NYU Depth V2 dataset. The training on KITTI Dataset costs about 15 h and 11.7GiB GPU memory and the training on NYU Depth V2 Dataset costs about 23 h and 8.9GiB GPU memory.

The data augmentation applied on training set follows the method used in [10] and [1], including a random left-right flip with a probability of 0.5, a random color channel swap with a probability of 0.5, a random scaling with a ratio limited in $[0.8, 1.25]$ and a random rotation with an angle limited in $[-5, 5]$ degrees.

4.3 Evaluation Result

We compare the performance of our method with state-of-the-art methods on both KITTI and NYU Depth V2 datasets. The metrics used to evaluate the accuracy including: mean absolute relative error (AbsRel), mean square relative error (SqRel), mean $\log_{10}$ error (log10), root mean square error (RMSE), root mean square log error (RMSE_{log}) and the accuracy under threshold calculated by $\delta_i = P[\max(d/\hat{d}, \hat{d}/d) < 1.25^i], i = 1, 2, 3$.

The evaluation results on KITTI dataset and NYU Depth V2 dataset are listed in Table 1 and Table 2, respectively. Since there are not many supervised methods that have been evaluated on KITTI dataset, we also compare our

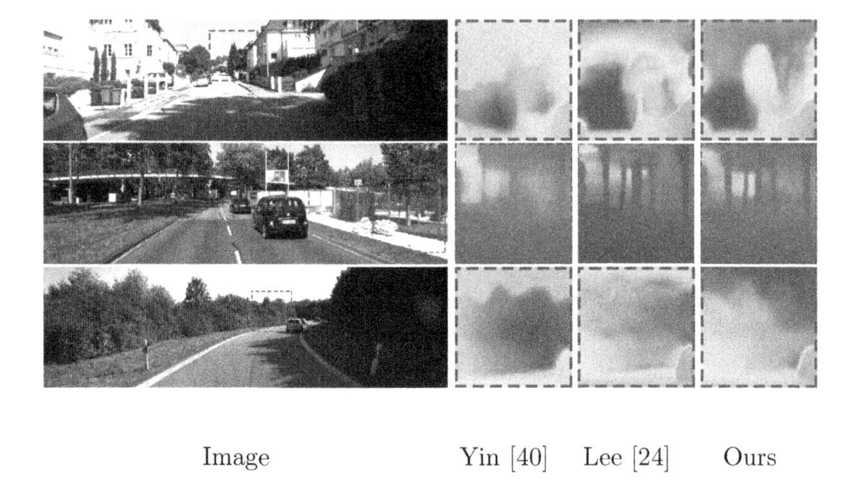

Fig. 6. Predicted depth maps on KITTI dataset. The figures in the last three columns show the depth maps of the framed region in the first column images. Our depth maps are sharp and restore more reliable object structures

Table 3. Metrics of different network versions on NYU V2 Depth dataset. The "params" is the total number of the parameters in the network and "maps" is the total number of sub-depth maps

Version	Params	Maps	Higher is better			Lower is better		
			δ_1	δ_2	δ_3	AbsRel	RMSE	log10
Baseline	37.42M	0	0.830	0.975	0.995	0.137	0.493	0.057
ASPP	41.20M	0	0.854	0.978	0.996	0.125	0.469	0.052
AMG	41.22M	8	0.858	**0.980**	**0.997**	0.121	0.468	0.052
UMG	41.27M	14	0.862	0.980	0.996	0.119	0.466	0.051
Nomask	41.22M	4	0.860	0.979	0.996	0.120	0.469	0.051
Ours	41.27M	30	**0.881**	0.979	0.996	**0.112**	**0.447**	**0.048**

method with some self-supervised methods that use videos to train the networks [14,21,42]. From the comparison, it can be found that our method achieves state-of-the-art accuracy on both two datasets and get the best or second best results on all of the metrics. Besides, our method performs better especially on accuracy under thresholds that reflect the overall accuracy of the prediction results.

Our method not only achieves better performance on metrics, but also generates depth maps with balanced sharpness and fidelity. Some predicted depth maps on NYU Depth V2 dataset are shown in Fig. 5. And Fig. 6 shows some parts of the predicted depth maps on KITTI dataset. It can be found that our depth maps achieve high sharpness comparable to the method in Lee et al. [24], which generates sharp depth maps but with distorted object structures, and restore more reliable object structures than other methods. The dotted frame

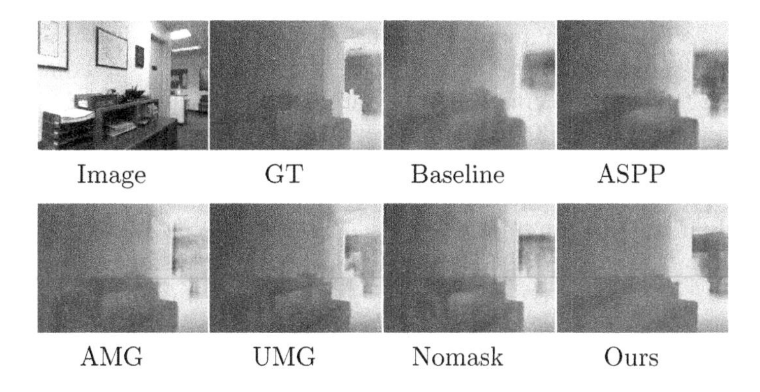

Fig. 7. Predicted depth maps in ablation study on NYU V2 Depth dataset. The network closer to "Ours" generate sharper depth maps with more reliable details

emphasize our improvement on performance. For example, the Lee et al. [24]'s door edge framed by the blue frame in Fig. 5 is bended, while ours is straight. The reliable object structures in our depth maps will benefit the potential applications like point cloud based object detection and semantic segmentation.

4.4 Ablation Study

To confirm the effectiveness of our proposed mechanisms, an ablation study is carried out. We train different versions of network on the NYU Depth V2 dataset using the same hyperparameters, and record their performance in Table 3. The "Baseline" in Table 3 only takes DenseNet-161 as the encoder and several upsampling layers as the decoder. Then we gradually add DenseASPP [38], AMG and UMG to the "Baseline" and get network versions called "ASPP", "AMG" and "UMG". We also apply dense connection mechanism to the "ASPP" and "UMG" to get "Nomask" and the complete network "Ours". The depth maps predicted by above network versions are shown in Fig. 7.

The change of metrics indicates that all of the proposed components gradually promote the performance of the network and the accuracy increases with

Table 4. Metrics of different α on NYU V2 Depth dataset. The network achieves the best performance when α is 0.1

α	Higher is better			Lower is better		
	δ_1	δ_2	δ_3	AbsRel	RMSE	log10
0	0.872	**0.980**	0.996	0.116	0.448	0.049
0.05	0.876	0.979	0.995	0.113	0.452	0.049
0.1	**0.881**	0.979	**0.996**	**0.112**	**0.447**	**0.048**
0.2	0.875	0.979	0.996	0.114	0.450	0.049

the number of sub-depth maps (ignoring the influence of dense connection mechanism). It should also be noticed that the AMGs and UMGs make obvious progress and only increase 0.07M parameters, which is much less than DenseASPP that increase 3.78M parameters. The improvement from "ASPP" to "Nomask" and the improvement from "UMG" to "Ours" highlight the importance of our dense connection mechanism which brings considerable progress while hardly increasing parameters. In Fig. 7, it can also be found that the network closer to "Ours" predicts sharper depth maps that restore more details in images.

We also test the network accuracy when α in Eq. (2) and Eq. (3) has different value and list the results in Table 4. It can be found that the α slightly influences the performance on metrics, and network achieves best performance when α is 0.1. It also noteworthy that when α is 0, which means mask generators only use 1 channel to generate masks instead of 3, the network can not achieve the best performance.

5 Conclusions

In this paper, we propose the idea of densely connecting sub-depth maps with high diversity to the predictors. The sub-depth maps with sufficient diversity have been attached with information of features at different levels, making the network balance the preference on features and predict depth map with both high sharpness and high fidelity. To implement this idea into the network, a dense connection mechanism is proposed to effectively connect the sub-depth maps and the predictors. Two kinds of diversity enhancement devices, AMG and UMG, are also proposed to generate more diverse sub-depth maps restoring information of features at different levels, which further improve the ability of dense connection mechanism.

The proposed method has been evaluated on KITTI and NYU Depth V2 dataset and compared with existing methods on several metrics. The quantitative results show that our method achieve state-of-the-art accuracy. The comparison between predicted depth maps demonstrates that the depth maps predicted by our method are sharper and restore more reliable object structures.

Acknowledgement. This research was funded by National Natural Science Foundation of China (No. 61603020, No. 61620106012), and the Fundamental Research Funds for the Central Universities (No. YWF-20-BJ-J-923, No. YWF-19-BJ-J-355).

References

1. Alhashim, I., Wonka, P.: High quality monocular depth estimation via transfer learning. arXiv preprint arXiv:1812.11941 (2018)
2. Cao, Y., Wu, Z., Shen, C.: Estimating depth from monocular images as classification using deep fully convolutional residual networks. IEEE Trans. Circ. Syst. Video Technol. **28**(11), 3174–3182 (2017)

3. Chakrabarti, A., Shao, J., Shakhnarovich, G.: Depth from a single image by harmonizing overcomplete local network predictions. In: Advances in Neural Information Processing Systems, pp. 2658–2666 (2016)
4. Chen, L.C., Papandreou, G., Kokkinos, I., Murphy, K., Yuille, A.L.: DeepLab: semantic image segmentation with deep convolutional Nets, Atrous convolution, and fully connected CRFs. IEEE Trans. Pattern Anal. Mach. Intell. **40**(4), 834–848 (2017)
5. Chen, R., Han, S., Xu, J., Su, H.: Point-based multi-view stereo network. In: Proceedings of the IEEE International Conference on Computer Vision, pp. 1538–1547 (2019)
6. Chen, W., Yue, H., Wang, J., Wu, X.: An improved edge detection algorithm for depth map inpainting. Optics Lasers Eng. **55**, 69–77 (2014)
7. Deng, J., Dong, W., Socher, R., Li, L.J., Li, K., Fei-Fei, L.: ImageNet: a large-scale hierarchical image database. In: 2009 IEEE Conference on Computer Vision and Pattern Recognition, pp. 248–255. IEEE (2009)
8. Diaz, R., Marathe, A.: Soft labels for ordinal regression. In: Proceedings of the IEEE Conference on Computer Vision and Pattern Recognition, pp. 4738–4747 (2019)
9. Eigen, D., Fergus, R.: Predicting depth, surface normals and semantic labels with a common multi-scale convolutional architecture. In: Proceedings of the IEEE International Conference on Computer Vision, pp. 2650–2658 (2015)
10. Eigen, D., Puhrsch, C., Fergus, R.: Depth map prediction from a single image using a multi-scale deep network. In: Advances in Neural Information Processing Systems, pp. 2366–2374 (2014)
11. Fu, H., Gong, M., Wang, C., Batmanghelich, K., Tao, D.: Deep ordinal regression network for monocular depth estimation. In: Proceedings of the IEEE Conference on Computer Vision and Pattern Recognition, pp. 2002–2011 (2018)
12. Geiger, A., Lenz, P., Stiller, C., Urtasun, R.: Vision meets robotics: the KITTI dataset. Int. J. Robot. Res. (IJRR) **32**, 1231–1237 (2013)
13. Godard, C., Mac Aodha, O., Brostow, G.J.: Unsupervised monocular depth estimation with left-right consistency. In: Proceedings of the IEEE Conference on Computer Vision and Pattern Recognition, pp. 270–279 (2017)
14. Godard, C., Mac Aodha, O., Firman, M., Brostow, G.J.: Digging into self-supervised monocular depth estimation. In: Proceedings of the IEEE International Conference on Computer Vision, pp. 3828–3838 (2019)
15. He, K., Zhang, X., Ren, S., Sun, J.: Deep residual learning for image recognition. In: Proceedings of the IEEE Conference on Computer Vision and Pattern Recognition, pp. 770–778 (2016)
16. Huang, G., Liu, Z., Van Der Maaten, L., Weinberger, K.Q.: Densely connected convolutional networks. In: Proceedings of the IEEE Conference on Computer Vision and Pattern Recognition, pp. 4700–4708 (2017)
17. Huang, P.H., Matzen, K., Kopf, J., Ahuja, N., Huang, J.B.: DeepMVS: learning multi-view stereopsis. In: Proceedings of the IEEE Conference on Computer Vision and Pattern Recognition, pp. 2821–2830 (2018)
18. Karsch, K., Liu, C., Kang, S.B.: Depth transfer: depth extraction from video using non-parametric sampling. IEEE Trans. Pattern Anal. Mach. Intell. **36**(11), 2144–2158 (2014)
19. Kingma, D.P., Ba, J.: Adam: a method for stochastic optimization. arXiv preprint arXiv:1412.6980 (2014)

20. Krizhevsky, A., Sutskever, I., Hinton, G.E.: ImageNet classification with deep convolutional neural networks. In: Advances in Neural Information Processing Systems, pp. 1097–1105 (2012)
21. Kuznietsov, Y., Stuckler, J., Leibe, B.: Semi-supervised deep learning for monocular depth map prediction. In: Proceedings of the IEEE Conference on Computer Vision and Pattern Recognition, pp. 6647–6655 (2017)
22. Ladicky, L., Shi, J., Pollefeys, M.: Pulling things out of perspective. In: Proceedings of the IEEE Conference on Computer Vision and Pattern Recognition, pp. 89–96 (2014)
23. Laina, I., Rupprecht, C., Belagiannis, V., Tombari, F., Navab, N.: Deeper depth prediction with fully convolutional residual networks. In: 2016 Fourth International Conference on 3D Vision (3DV), pp. 239–248. IEEE (2016)
24. Lee, J.H., Han, M.K., Ko, D.W., Suh, I.H.: From big to small: multi-scale local planar guidance for monocular depth estimation. arXiv preprint arXiv:1907.10326 (2019)
25. Levin, A., Lischinski, D., Weiss, Y.: Colorization using optimization. In: ACM SIGGRAPH 2004 Papers, pp. 689–694 (2004)
26. Li, J., Klein, R., Yao, A.: A two-streamed network for estimating fine-scaled depth maps from single RGB images. In: Proceedings of the IEEE International Conference on Computer Vision, pp. 3372–3380 (2017)
27. Li, R., Xian, K., Shen, C., Cao, Z., Lu, H., Hang, L.: Deep attention-based classification network for robust depth prediction. In: Jawahar, C.V., Li, H., Mori, G., Schindler, K. (eds.) ACCV 2018. LNCS, vol. 11364, pp. 663–678. Springer, Cham (2019). https://doi.org/10.1007/978-3-030-20870-7_41
28. Liu, F., Shen, C., Lin, G., Reid, I.: Learning depth from single monocular images using deep convolutional neural fields. IEEE Trans. Pattern Anal. Mach. Intell. **38**(10), 2024–2039 (2015)
29. Silberman, N., Hoiem, D., Kohli, P., Fergus, R.: Indoor segmentation and support inference from RGBD images. In: Fitzgibbon, A., Lazebnik, S., Perona, P., Sato, Y., Schmid, C. (eds.) ECCV 2012. LNCS, vol. 7576, pp. 746–760. Springer, Heidelberg (2012). https://doi.org/10.1007/978-3-642-33715-4_54
30. Paszke, A., et al.: Pytorch: an imperative style, high-performance deep learning library. In: Advances in Neural Information Processing Systems, pp. 8024–8035 (2019)
31. Pillai, S., Ambruş, R., Gaidon, A.: SuperDepth: self-supervised, super-resolved monocular depth estimation. In: 2019 International Conference on Robotics and Automation (ICRA), pp. 9250–9256. IEEE (2019)
32. Ren, H., El-khamy, M., Lee, J.: Deep robust single image depth estimation neural network using scene understanding. arXiv preprint arXiv:1906.03279 (2019)
33. Ronneberger, O., Fischer, P., Brox, T.: U-Net: convolutional networks for biomedical image segmentation. In: Navab, N., Hornegger, J., Wells, W.M., Frangi, A.F. (eds.) MICCAI 2015. LNCS, vol. 9351, pp. 234–241. Springer, Cham (2015). https://doi.org/10.1007/978-3-319-24574-4_28
34. Saxena, A., Sun, M., Ng, A.Y.: Make3D: learning 3D scene structure from a single still image. IEEE Trans. Pattern Anal. Mach. Intell. **31**(5), 824–840 (2008)
35. Simonyan, K., Zisserman, A.: Very deep convolutional networks for large-scale image recognition. arXiv preprint arXiv:1409.1556 (2014)
36. Xu, D., Ricci, E., Ouyang, W., Wang, X., Sebe, N.: Multi-scale continuous CRFs as sequential deep networks for monocular depth estimation. In: Proceedings of the IEEE Conference on Computer Vision and Pattern Recognition, pp. 5354–5362 (2017)

37. Xu, D., Wang, W., Tang, H., Liu, H., Sebe, N., Ricci, E.: Structured attention guided convolutional neural fields for monocular depth estimation. In: Proceedings of the IEEE Conference on Computer Vision and Pattern Recognition, pp. 3917–3925 (2018)
38. Yang, M., Yu, K., Zhang, C., Li, Z., Yang, K.: DenseASPP for semantic segmentation in street scenes. In: Proceedings of the IEEE Conference on Computer Vision and Pattern Recognition, pp. 3684–3692 (2018)
39. Yao, Y., Luo, Z., Li, S., Shen, T., Fang, T., Quan, L.: Recurrent MVSNet for high-resolution multi-view stereo depth inference. In: Proceedings of the IEEE Conference on Computer Vision and Pattern Recognition, pp. 5525–5534 (2019)
40. Yin, W., Liu, Y., Shen, C., Yan, Y.: Enforcing geometric constraints of virtual normal for depth prediction. In: Proceedings of the IEEE International Conference on Computer Vision, pp. 5684–5693 (2019)
41. Zhan, H., Garg, R., Saroj Weerasekera, C., Li, K., Agarwal, H., Reid, I.: Unsupervised learning of monocular depth estimation and visual odometry with deep feature reconstruction. In: Proceedings of the IEEE Conference on Computer Vision and Pattern Recognition, pp. 340–349 (2018)
42. Zhou, T., Brown, M., Snavely, N., Lowe, D.G.: Unsupervised learning of depth and ego-motion from video. In: Proceedings of the IEEE Conference on Computer Vision and Pattern Recognition, pp. 1851–1858 (2017)

Single Image Dehazing for a Variety of Haze Scenarios Using Back Projected Pyramid Network

Ayush Singh[1], Ajay Bhave[1], and Dilip K. Prasad[2(✉)]

[1] Indian Institute of Technology (ISM), Dhanbad 826004, India
ayush.s.18je0204@cse.iitism.ac.in
[2] UiT The Arctic University of Norway, 9019 Tromsø, Norway
dilip.prasad@uit.no

Abstract. Learning to dehaze single hazy images, especially using a small training dataset is quite challenging. We propose a novel generative adversarial network architecture for this problem, namely back projected pyramid network (BPPNet), that gives good performance for a variety of challenging haze conditions, including dense haze and inhomogeneous haze. Our architecture incorporates learning of multiple levels of complexities while retaining spatial context through iterative blocks of UNets and structural information of multiple scales through a novel pyramidal convolution block. These blocks together for the generator and are amenable to learning through back projection. We have shown that our network can be trained without over-fitting using as few as 20 image pairs of hazy and non-hazy images. We report the state of the art performances on NTIRE 2018 homogeneous haze datasets for indoor and outdoor images, NTIRE 2019 denseHaze dataset, and NTIRE 2020 non-homogeneous haze dataset.

Keywords: Single image dehazing · Generative adversarial network · Back projection · Deep learning

1 Introduction

The quality of images of scenes in our daily life is greatly affected by the particles suspended in the environment, such as due to dust, smoke, mist, fog, smog, etc. Bad weather also contributes to this. Beside significantly higher and non-uniform noise in the images, the usual effects are reduced visibility, reduced sharpness, and contrast of the objects within the visibility and obscuring of other objects. Therefore, performing computer vision tasks like object detection, object recognition, tracking and segmentation becomes complicated for such images. Therefore, the true potential of computer vision empowered automated and remote surveillance systems such as drones and robots cannot be realized under hazy conditions. Thus, it is of interest to enhance the quality of images taken under homogeneous and non-homogeneous hazy conditions and recover the

A. Bartoli and A. Fusiello (Eds.): ECCV 2020 Workshops, LNCS 12538, pp. 166–181, 2020.
https://doi.org/10.1007/978-3-030-66823-5_10

details of the scene. Haze removal or dehazing algorithms address this problem (Fig. 1).

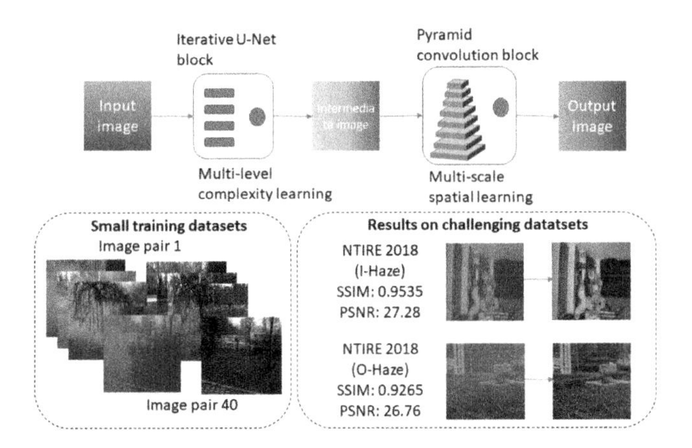

Fig. 1. A compact representation of our novel generator and its important features.

There has been a significant activity in the topic of dehazing in recent years. New algorithms ranging from physics-based solvers, image processing based algorithms, as well as deep learning-based approaches, are being proposed. Furthermore, newer challenges are being undertaken, including dehazing in the presence of dense haze, non-homogenous haze, and using a single RGB image of a scene. It is being recognized that deep learning architectures provide better performance than the other approaches for diverse and challenging dehazing scenarios if suitably designed large datasets are available. However, dehazing images through deep learning on a small dataset using a single RGB image is quite challenging and of significant practical interest. For example, in the situation of fire management or natural disaster management, a suitable dehazing model characteristic of the situation needs to be learned quickly using a small number of images in haze and corresponding pre-disaster images.

We propose a novel deep learning architecture that is amenable to reliable learning of dehazing model using a small dataset. Our novel generative adversarial network (GAN) architecture includes iterative blocks of UNets to model haze features of different complexities and a pyramid convolution block to preserve and restore spatial features of different scales. The key contributions of this paper are as follows:

- A novel technique named pyramid convolution is introduced for dehazing to obtain spatial features of multiple scales structural information.
- We have used iterative UNet block for image dehazing tasks to make the generator learn different and complex features of haze without the loss of local and global structural information or without making the network too deep to result into loss of spatial features.

- The model used is end-to-end trainable with hazy image as input and haze-free image as the desired output. Therefore the conventional practice of using the atmospheric scattering model is obviated, and the problems encountered in inverse reconstruction are circumvented. It also makes the approach more versatile and applicable to haze scenarios where the conventional simplified atmospheric model may not apply.
- Extensive experimentation is done on four contemporary challenging datasets, namely I-Haze and O-Haze datasets of NTIRE 2018 challenge, Dense-haze dataset of NTIRE 2019 challenge, and non-homogeneous dehazing dataset of NTIRE 2020 challenge.

The outline of the paper is as follows. Section 2 presents related work, and Sect. 3 introduces our architecture and learning approach. Section 4 presents numerical experiments and results. Section 5 includes an ablation study on the proposed method. Section 6 concludes the paper.

2 Related Work

Since this paper's focus is single image dehazing, we exclude a discussion on studies that required multiple images, for example, exploiting polarization, to perform dehazing. Single image dehazing is an ill posed problem because the number of measurements is not sufficient for learning the haze model, and the non-linearity of the haze model implies higher sensitivity to noise. Single image based dehazing exploits polarization-independent atmospheric scattering model proposed by Koschmieder [16] and its characteristics such as dark channel, color attenuation and haze-free priors. According to this model, the hazy image is specified by the atmospheric light (generally assumed uniform), the albedo of the objects in the scene, and the transmission map of the hazy medium. More details can be found in [16] and its subsequent citations, including recent ones [7,23]. We have to predict the unknown transmission map and global atmospheric light. In the past, many methods have been proposed for this task. The methods can be divided into two categories, namely (i) Traditional handcrafted prior based methods and (ii) Learning based methods.

Traditional Handcrafted Prior Based Methods: Fattal [9] proposed a physically grounded method by estimating the albedo of the scene. Tan [22] proposed the use of the Markov random field to maximize the local contrast of the image. He et al. [12] proposed a dark channel prior for the estimation of the transmission map. Fattal [10] proposed a color-line method based on the observation that small image patches typically exhibit a one-dimensional distribution in the RGB color space. Traditional handcrafted prior methods give good results for certain cases but are not robust for all the cases.

Learning Based Approaches: In recent years, many learning based methods have been proposed for single image dehazing that encash the success of deep learning in image processing tasks, availability of large datasets, and better computation resources. Some examples are briefly mentioned here. Cai et al. [6]

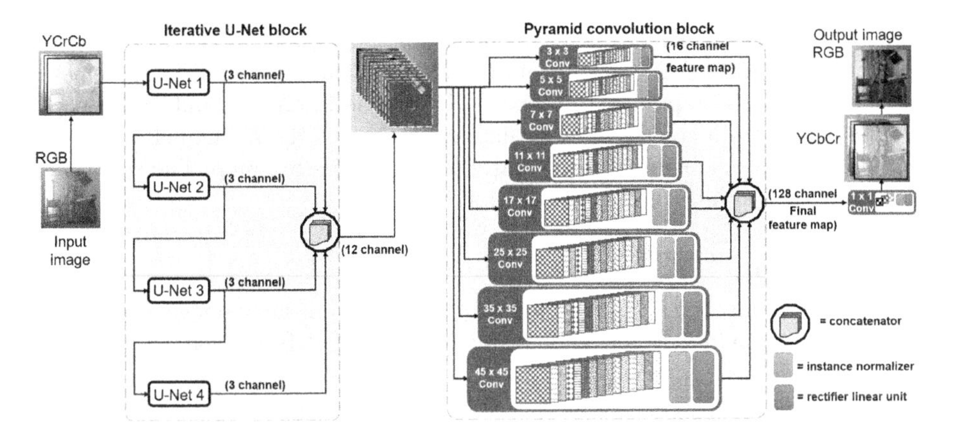

Fig. 2. The architecture of our generator.

proposed an end-to-end CNN based deep architecture to estimate the transmission map. Ren et al. [20] proposed a multi-scale deep architecture, which also estimates the transmission map from the haze image. Zhang et al. in [25] proposed a deep network architecture that estimates the transmission map and the atmospheric light. These estimates are then used together with the atmospheric scattering model to generate the haze-free image.

Our Approach in Context: In contrast to these approaches, our approach is an end-to-end learning based approach in which the learnt model directly predicts the haze-free image without needing to reconstruct the transmission map and the atmospheric light, or using the atmospheric scattering model. It is therefore more versatile to be trained for situations where the atmospheric scattering model of [16] may not apply or may be too simple. Example includes non-uniform haze. It also circumvents the numerical errors and artifacts associated with the use of inverse approaches of reconstructing the haze-free image from the transmission map and atmospheric light.

3 Proposed Method

In this section we present our model, namely back projected pyramid network (BPPNet). The overall architecture is based on generative adversarial network [11], where a generator generates a haze-free image from a hazy image, and a discriminator tells whether the image provided to it is real or not.

3.1 Generator

The architecture of the generator is shown in Fig. 2. It comprises of two blocks in series, namely (i) iterative UNet block, (ii) pyramid convolution block, which we describe next.

Iterative UNet Block (IUB): This block consists of multiple UNet [21] units connected in series i.e. the output of one UNet (architecture in the supplementary) is fed as the input to the next UNet. In addition, the output of each UNet is passed to a concatenator, which concatenates the 3 channel output of all the UNets, providing an intermediate 12 channel feature map. The equations describing the working of IUB are the following.

$$I_1 = \text{UNET}_1(I_{\text{haze}}); I_i = \text{UNET}_i(I_{i-1}) \quad \text{for} \quad i > 1, \tag{1}$$

where I_i is the output of ith UNet unit, I_{haze} is the input hazy image after being transformed to YCbCr space, and the output $\hat{I}_{\text{IUB}}$ of IUB is given as

$$\hat{I}_{\text{IUB}} = I_1 \oplus I_2 \oplus \ldots I_M \tag{2}$$

where $\oplus$ indicates concatenation operator and M is the number of UNet unit. We have used $M = 4$. An ablation study on the value of M is presented later in Sect. 5. Here, we discuss the need of more than one UNet.

In principle, a single UNet may be able to support dehazing to some extent. However, it may not be able to extract complex features and generate an output with fine details. One way to tackle this problem is to increase the number of layers in the encoder block so that more complex features can be learned. But the layers in the encoder block are arranged in feed forward fashion, and the height and the width of layers decreases upon moving further. This causes loss in spatial information and reduces the possibility of extraction of spatial features of high complexity. Therefore, we take an alternate approach of creating sequence of the multiple UNets. The sequence of UNets may be interpreted as a sequence of multiple encoder-decoder pairs with skip connections. The encoder in each UNet extracts the features from input tensor in the downsample feature map and decoder uses those features and projects them into an upsample latent space with same height and width as input tensor. In this way, each generator helps in learning increasingly complex features of haze while the decoder helps in retaining the spatial information in the image. Lastly, the concatenation step ensures that complexity of all the levels are available for subsequent reconstruction.

We illustrate the effect of using multiple UNets in Fig. 3. Histogram equalized 3-channel output of each UNet is shown as an RGB image. It is seen that the spatial context is preserved, and at the same time haze introduced blur of different complexities are present in the outputs of different images. The haze in the last UNet output is flatter across the image and shows large scale blurs while the haze in the first UNet is local and introduces small scale blurs. Therefore, most dehazing is accomplished in UNet1, although the subsequent UNets pick the dehazing components that the previous UNets did not. Figure 3 also explains our choice of only four UNet blocks even though more blocks could be used in principle. We explain our choice in two parts. First, there is a trade-off involved between accuracy and speed when choosing the number of UNet blocks. Second, as seen in the histograms in Fig. 3, the dynamic range of channels decreases with every

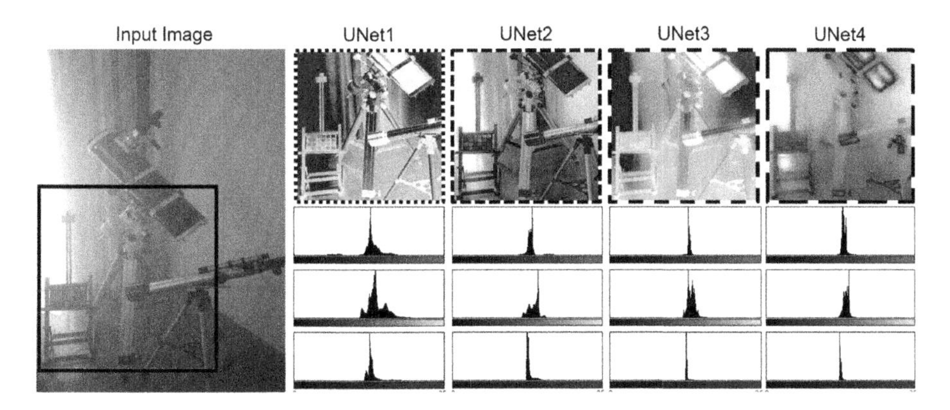

Fig. 3. The effect of the successive UNet units is illustrated. Images are histogram equalized for better visualization. The histograms of the channels becomes narrower after passing more number of UNet units, indicating that adding more UNet units may cease to create more value after a certain limit.

subsequent UNet block, thereby indicating the reduction in the usable information content. The standard deviation of the intensity values in the 3 channels after UNet4 is ~12.2. Adding more blocks would further reduce this value, and therefore not provide significantly exploitable data for dehazing.

Pyramid Convolution (PyCon) Block: Although the iterative UNet block does provide global and local structural information, the output lacks the global and local structural information for different sized objects. An underlying reason is that the structural information from different scales are not directly used to generate the output. To overcome this issue, we have used a novel pyramid convolution technique. Earlier pyramid pooling has been used in [25] to leverage the "global structural information". However, since the pooling layers are not learnable, we instead employ learnable convolution layers that can easily outperform the pooling layers in leveraging the information.

We employ many convolution layers of different kernel sizes in parallel on the input map (the 12 channel output of iterative UNet block). Corresponding to different kernel sizes used for convolution, different output maps are generated with structural information of different spatial scales. The kernel sizes are chosen as 3, 5, 7, 11, 17, 25, 35, and 45, as shown in Fig. 2. Odd sized kernels are used since pixels in the intermediate feature map are assumed to be symmetrical around output pixel. We observed introduction of distortion over layers upon using even-sized kernels, indicating the importance of exploiting the symmetry of the features around the output pixel. Zero padding is used to ensure that the features at the edges are not lost. After the generation of feature map from corresponding kernels, all the generated maps are concatenated to make an output feature map of 128 channels, which is subsequently used for the final image construction by applying a convolution layer of kernel size 3×3 with zero

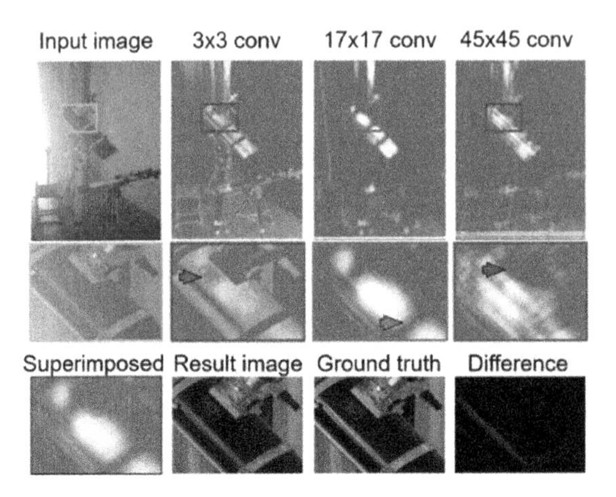

Fig. 4. Feature maps corresponding to one of the channels of 3×3, 17×17, and 45×45 convolution layer respectively. The figure shows that smaller kernel size generates smaller scale features such as edges while large kernel size generates large scale features such as big patches.

padding. In this manner, the local to global information is directly used for the final image reconstruction.

The effect of using pyramid convolution is shown in Fig. 4. In the zoom-ins shown in the middle panel, the arrows indicate some features of the size of the convolution filter used for generating that particular feature map. The illustrated 3 channels are superimposed as a hypothetical RGB image in the bottom left of Fig. 4 to demonstrate the types of details present in a subset of the output feature map. Since we have used 8 convolution filters that operate on 12 channel input, we generate a total of 128 channels in the output feature map with a large variety of spatial features of multiple scales learned and restored. Therefore, the result image shown in the bottom panel has spatial features closely matching the ground truth, resulting in a low difference map (shown in the bottom right).

One may consider using the 12 channel output of the iterative UNet block for generating the dehazed image directly, without employing the PyCon block. To indicate the importance of including the PyCon block, we include an ablation study in Sect. 5.

3.2 Discriminator

We have used a patch discriminator to determine whether a particular patch is realistic or not. The patches overlap in order to eliminate the problem of low performance on the edges. We have used 4×4 convolution layers in discriminator. After every convolution layer, we have added an activation layer with an activation leaky rectified linear unit (Leaky ReLu) except the last layer where the activation function is sigmoid. The size of the convolution kernel used is

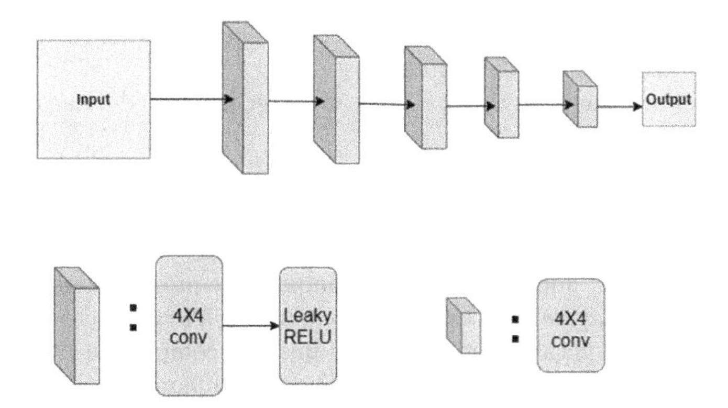

Fig. 5. Architecture of the discriminator.

4×4, and the output map size is 62×62 for an input of size 512×512. The architecture of our discriminator is shown in Fig. 5.

3.3 Loss Functions

Most dehazing models use the mean squares error (MSE) as the loss function [26]. However, MSE is known to be only weakly correlated with human perception of image quality [13]. Hence, we employ additional loss functions that are closer to human perception. To this end, we have used a combination of MSE (L_2 loss), adversarial loss L_{adv}, content loss L_{con}, and structural similarity loss L_{SSIM}. We define the remaining loss functions below.

The adversarial loss for the generator L_{adv} and the discriminator L_{dis} is defined as:

$$L_{\text{adv}} = \langle \log(D(I_{\text{pred}})) \rangle, \tag{3}$$

$$L_{\text{dis}} = \langle \log(D(I_{\text{GT}})) \rangle + \langle \log(1 - D(I_{\text{pred}})) \rangle, \tag{4}$$

where $(I_{\text{haze}}, I_{\text{GT}})$ are the supervised pair of the hazy image and the corresponding ground truth. $D(I)$ is the discriminator's estimate of the probability that data instance I provided to it is indeed real. Similarly, $G(I)$ is the generator output for the input instance I. Further, $I_{\text{pred}} = G(I_{\text{haze}})$. The notation $<>$ indicates the expected value over all the supervised pairs.

The MSE, also referred to as the L_2 loss, is defined as the average norm 2 distance between I_{GT} and I_{pred}:

$$L_2 = \langle I_{\text{GT}} - I_{\text{pred}} \rangle \tag{5}$$

Our content loss is the VGG based perceptual loss [14], defined as:

$$L_{\text{con}} = \Big\langle \sum_i \frac{1}{N_i} || \, \phi_i(I_{\text{GT}}) - \phi_i(I_{\text{pred}}) \, || \Big\rangle, \tag{6}$$

where N_i is the number of elements in the ith activation of VGG-19 and ϕ_i represents i^{th} activation of VGG-19.

The structural similarity loss L_{SSIM} over reconstructed image I_{pred} and ground truth I_{GT} is defined as:

$$L_{\text{SSIM}} = 1 - \langle \text{SSIM}(I_{\text{GT}}, I_{\text{pred}}) \rangle, \tag{7}$$

where $\text{SSIM}(I, I')$ is the SSIM [24] between the images I and I'. We note that although the losses L_2 and L_{SSIM} directly compare the predicted and the ground truth images, the nature of comparison is quite different across them. L_2 is insensitive to the structural details but retains the comparison of the general energy and dynamic range of the two images being compared. L_{SSIM} on the other hand compared the structural content in the images with less sensitivity to the contrast. Therefore, including these two loss functions provide complementary aspects of comparison between the predicted and the ground truth images.

The overall generator loss L_{G} and discriminator loss L_{D} are given as

$$L_{\text{G}} = A_1 L_{\text{adv}} + A_2 L_{\text{con}} + A_3 L_2 + A_4 L_{\text{SSIM}} \tag{8}$$

$$L_{\text{D}} = B_1 L_{\text{D}_{\text{adv}}}. \tag{9}$$

We have heuristically chosen the values of the constant weights in the above equation as $A_1 = 0.7$, $A_2 = 0.5$, $A_3 = 1.0$, $A_4 = 1.0$, and $B_1 = 1.0$.

4 Experimental Results

4.1 Datasets

We have trained and tested our model on the following four datasets, namely NTIRE 2018 image dehazing indoor dataset (abbreviated as I-Haze), NTIRE 2018 image dehazing outdoor dataset (O-Haze), Dense-Haze dataset of NTIRE 2019, and NTIRE 2020 dataset for non-homogeneous dehazing challenge (NTIRE 2020).

I-Haze [4] ***and O-Haze*** [2]: These datasets contains 25 and 35 hazy images (size 2833×4657 pixels) respectively for training. Both datasets contain 5 hazy images for validation along with their corresponding ground truth images. For both of these datasets, the training was done on training data and validation images were used for testing because although 5 hazy images for testing were given but their ground truths were not available to make the quantitative comparison.

Dense-Haze [1]: This dataset contains 45 hazy images (size 1200×1600 pixels) for training and 5 hazy images for validation and 5 more for testing with their corresponding ground truth images. We have done training on training data and tested our model with test data.

NTIRE 2020 [8]: This dataset contains 45 hazy images (size 1200×1600 pixels) for training with their corresponding ground truth. It is the first dataset of

Table 1. Quantitative comparison of various state of the art methods with our model on I-Haze, O-Haze and Dense-Haze datasets. Our model does the dehazing task in real-time at an average running time of 0.0311 s i.e. 31.1 ms per image.

I-Haze dataset									
Metric	Input	CVPR '09 [12]	TIP '15 [27]	ECCV '16 [20]	CVPR '16 [5]	ICCV '17 [17]	CVPRW '18 [26]	Our model	
SSIM	0.7302	0.7516	0.6065	0.7545	0.6537	0.7323	0.8705	**0.8994**	
PSNR	13.80	14.43	12.24	15.22	14.12	13.98	22.53	**22.56**	
O-Haze dataset									
Metric	Input	CVPR '09 [12]	TIP '15 [27]	ECCV '16 [20]	CVPR '16 [5]	ICCV '17 [17]	CVPRW '18 [26]	Our model	
SSIM	0.5907	0.6532	0.5965	0.6495	0.5849	0.5385	0.7205	**0.8919**	
PSNR	13.56	16.78	16.08	17.56	15.98	15.03	24.24	**24.27**	
Dense-Haze dataset									
Metric	CVPR '09 [12]	Meng et al. [18]	Fattal [10]	Cai et al. [6]	Ancuti et al. [3]	CVPR '16 [5]	ECCV '16 [20]	Morales et al. [19]	Our model
SSIM	0.398	0.352	0.326	0.374	0.306	0.358	0.369	0.569	**0.613**
PSNR	14.56	14.62	12.11	11.36	13.67	13.18	12.52	16.37	**17.01**

non-homogeneous haze in our knowledge. As ground truth for validation was not given, hence we used only 40 image pairs for training and calculated our results on the rest of the 5 images.

4.2 Training Details

The optimizer used for the training was Adam [15] with the initial learning rate for 0.001 and 0.001 for generator and discriminator respectively. We have randomly cropped large square patches from the training images. The crop size was 1024×1024 for NTIRE 2020 and Dense-Haze. Leveraging on the even large sizes of images in I-Haze and O-Haze datasets, we created four crops each of two different sizes, 1024×1024 and 2048×2048. We then resized all the patches to 512×512 using bicubic interpolation. These patches are randomly cropped for each epoch i.e. these patches are not same for every epoch. This has created an effectively larger dataset out of the small dataset available for training for each of the considered datasets. For datasets named I-Haze, O-Haze and NTIRE 2020, we converted these randomly cropped resize patches from RGB space to YCbCr space and then used them for training. For Dense-Haze dataset we directly used RGB patches for training.

We decreased the learning rate of the generator whenever the loss became stable. We stopped training when the learning rate of the generator reached 0.00001 and the loss stabilized. We also tried to decrease the learning rate of discriminator but found that doing so did not give the best results.

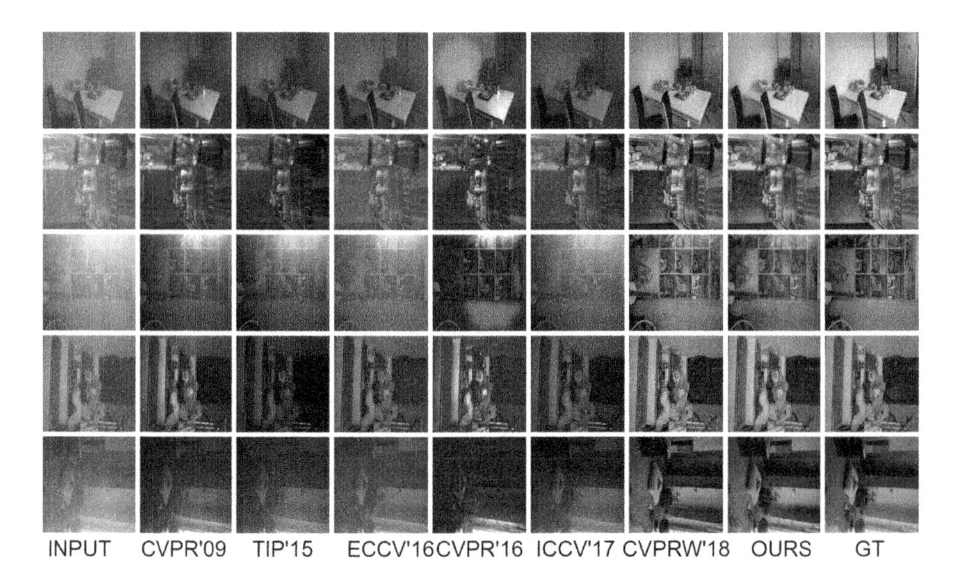

Fig. 6. Qualitative comparison of various benchmark with our model on I-Haze dataset

4.3 Results

Here, we present our results and their comparison with the results of other models available in the literature. We note that we converted the test input image size to 512×512 for all the datasets for generating our results in view of our hardware (GPU) memory constraints. Quantitative evaluation is performed using the SSIM metric and the peak signal to noise ratio (PSNR). The metrics are computed in the RGB space even if the training was done in YCbCr space. The quantitative results are compared with earlier state-of-the-art in Table 1. The metrics for the other methods are reproduced from [26] for the I-Haze and O-Haze dataset. The benchmark for Dense-haze was provided in [1]. We further include the results of Morales et al. [19] for comparison.

I-Haze: The average PSNR and SSIM of our method for this dataset on validation data were 22.56 and 0.8994 respectively. It is evident from Table 1 that our model outperforms the state-of-the-art in both SSIM and PSNR index by a good margin. Qualitative comparison results on the test data are shown in Fig. 6. It is evident that only CVPRW'18 [26] competes with our method in the quality of dehazed image and match with the ground truth.

O-Haze: The average PSNR and SSIM for this dataset on validation data were 24.27 and 0.8919 respectively on validation data, see Table 1. Our model clearly outperforms all the state-of-the-art in both PSNR and SSIM index by a large margin. For SSIM, the closest performing method was CVPRW'18 [26] with SSIM of 0.7205, which is significantly lower than ours i.e 0.8919. It is notable from the results of all the methods that this dataset is more challenging than

Fig. 7. Qualitative comparison of various methods with our model on O-Haze dataset.

I-Haze. Nonetheless, our method provides comparable performance over both I-Haze and O-Haze datasets. The qualitative comparison of results on the test data are shown in Fig. 7. Similar to the I-Haze dataset, only CVPRW'18 [26] and our method generate dehazed images of good quality.

As compared to I-Haze results in Fig. 6, it is more strongly evident in Fig. 7 that the color cast of our method is slightly mismatched with the ground truth, where CVPRW'18 performs better than our method. However, CVPRW'18 shows poorer structural continuity than our method, as evident more strongly in Fig. 6.

Dense-Haze: From Table 1, it is evident that this dataset is significantly more challenging that the I-Haze and O-Haze datasets. All methods perform quite poorer for this dataset, as compared to the numbers reported for I-Haze and O-Haze dataset. Even though the performance of our method is also poorer for this dataset as compared to the other datasets, its SSIM and PSNR values are significantly better than the other 8 methods whose results are included in Table 1 for this dataset. Qualitative comparison with select methods is shown in Fig. 8. The results clearly illustrate the challenge of this dataset as the features and details of the scene are almost imperceptible through the dense haze. Only our method is capable of dehazing the image effectively and bringing forth the details of the scene. Nonetheless, the color cast is non-uniformly compensated and different from the ground truth in regions.

NTIRE 2020: As the ground truth for test data is not given, we randomly chose 5 images for testing and used the rest of the 40 image pair for training. The average SSIM and PSNR are 0.8726 and 19.40 respectively. This SSIM value

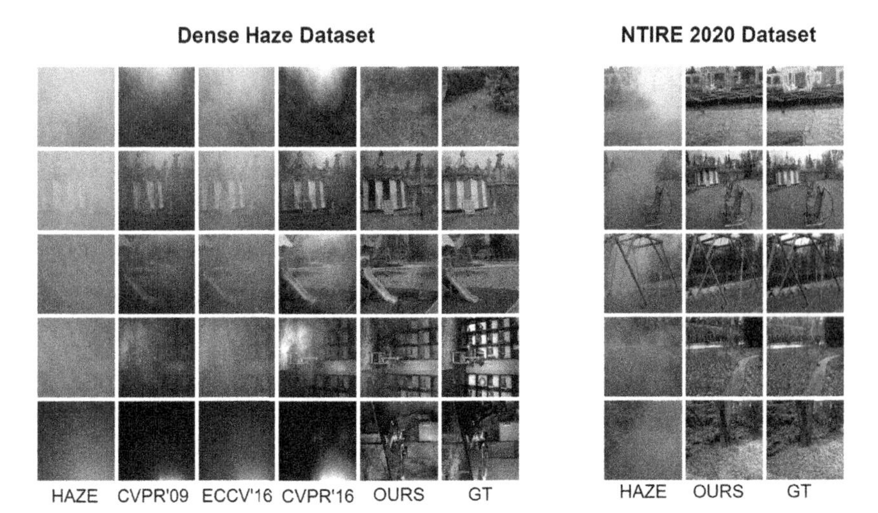

Fig. 8. Qualitative results for Dense-Haze and NTIRE 2020 dataset.

is better than the best SSIM observed in the competition and informed to the participants in a personal email after the test phase. The qualitative results are shown in Fig. 8. The observations are generally similar to the observations for the Dense-Haze dataset. Our results are qualitatively quite close to the ground truth and show the ability of our method to recover the details lost in haze, despite the non-homogeneity of the haze. Second, we observe a little bit of mismatch in the color reproduction and in-homogeneity in the color cast, which needs further work. We expect that the problem of color cast inhomogeneity may be related to the inhomogeneity in the haze itself, which may have been present in the Dense-Haze data as well but may not have been perceptible due to the generally high density of haze.

5 Ablation Study

We conduct ablation study using I-Haze and O-Haze datasets. We consider the ablation associated with the architectural elements in Sect. 5.1, loss components in Sect. 5.2, and the image space used in training in Sect. 5.3.

5.1 Architecture Ablation

Here, we consider ablation study relating to the number of UNet units used in the iterative UNet block and the absence or presence of the pyramid convolution block. The results are shown in Table 2. It is evident that decreasing or increasing the number of UNet blocks degrades the performance and the use of $M = 4$ UNet blocks is optimal for the architecture. This is in agreement in the observations derived from Fig. 3. Similarly, dropping the PyCon block also degrades the performance.

Table 2. The results of ablation study are presented here. The reference indicates the use of 4 UNet blocks, inclusion of PyCon block with layer configuration as shown in Fig. 2. All loss functions discussed in Sect. 3.3 are used and the entire architecture uses YCbCr space, such as shown in Fig. 2.

(a) Ablation study on architectural units

Dataset	Metric	Reference	1 UNet	2 UNets	3 UNets	5 UNets	No PyCon
I-Haze	SSIM	**0.8994**	0.8572	0.8679	0.8820	0.8932	0.8878
	PSNR	**22.57**	18.54	19.94	20.92	21.62	21.17
O-Haze	SSIM	**0.8927**	0.8500	0.8639	0.8795	0.8639	0.8768
	PSNR	**24.30**	21.34	22.36	23.06	22.36	23.13

(b) Ablation study on losses and the image space

		Reference	The loss function dropped				Direct use of RGB, not the YCbCr space
			L_{adv}	L_{con}	L_2	L_{SSIM}	
I-Haze	SSIM	**0.8994**	0.8620	0.8372	0.8648	0.8343	0.8944
	PSNR	**22.57**	19.52	18.99	20.02	19.58	20.94
O-Haze	SSIM	**0.8927**	0.8608	0.8271	0.8650	0.8568	0.8712
	PSNR	**24.30**	22.26	20.66	22.78	22.44	22.54

5.2 Loss Ablation

We proposed in Sect. 3.3 to use four types of loss functions for the training of the generator. Here, we consider the effect of dropping one loss function at a time. The results are presented in the bottom panel of Table 2. It is seen than dropping any of the loss function results into significant deterioration of performance. This indicates the different and complementary roles each of these loss functions is playing. Our observation of the qualitative results, discussed in Sect. 4, we might need to introduce another loss function related to the preservation of the color cast or color constancy.

5.3 Use of RGB Versus YCbCr Space

If we used RGB space instead of YCbCr space for training, we observe a degraded performance in terms of SSIM as reported in Sect. 2(b). However, we note that this observation is not consistent over all the datasets. Specifically, we noted that for Dense-Haze, the YCbCr conversion gave little poorer results than RGB based training. Hence, we have used RGB patches for training on Dense-Haze dataset.

6 Conclusion

The presented single image dehazing method is an end-to-end trainable architecture that is applicable in diverse situations such as indoor, outdoor, dense, and non-homogeneous haze even though training datasets used are small in each of these cases. It beats the state-of-the-art results in terms of SSIM and PSNR for all the three datasets whose results are available. Qualitative results for indoor images indicate preservation of colors in the reconstructed image in the I-Haze dataset while a poorer color reconstruction is observed in the results of other datasets. In the future, we will improve our model to inherently include color preservation and seamless color cast as well. Source code, results, and trained model are shared at our project page (https://github.com/ayu-22/BPPNet-Back-Projected-Pyramid-Network).

References

1. Ancuti, C.O., Ancuti, C., Sbert, M., Timofte, R.: Dense-Haze: a benchmark for image dehazing with dense-haze and haze-free images. In: IEEE International Conference on Image Processing, pp. 1014–1018 (2019)
2. Ancuti, C.O., Ancuti, C., Timofte, R., De Vleeschouwer, C.: O-HAZE: a dehazing benchmark with real hazy and haze-free outdoor images. In: IEEE Conference on Computer Vision and Pattern Recognition Workshops, pp. 754–762 (2018)
3. Ancuti, C., Ancuti, C.O., De Vleeschouwer, C., Bovik, A.C.: Night-time dehazing by fusion. In: IEEE International Conference on Image Processing, pp. 2256–2260 (2016)
4. Ancuti, C., Ancuti, C.O., Timofte, R., De Vleeschouwer, C.: I-HAZE: a dehazing benchmark with real hazy and haze-free indoor images. In: Blanc-Talon, J., Helbert, D., Philips, W., Popescu, D., Scheunders, P. (eds.) ACIVS 2018. LNCS, vol. 11182, pp. 620–631. Springer, Cham (2018). https://doi.org/10.1007/978-3-030-01449-0_52
5. Berman, D., Avidan, S., et al.: Non-local image dehazing. In: IEEE Conference on Computer Vision and Pattern Recognition, pp. 1674–1682 (2016)
6. Cai, B., Xu, X., Jia, K., Qing, C., Tao, D.: DehazeNet: an end-to-end system for single image haze removal. IEEE Trans. Image Process. **25**(11), 5187–5198 (2016)
7. Chen, S., Chen, Y., Qu, Y., Huang, J., Hong, M.: Multi-scale adaptive dehazing network. In: IEEE Conference on Computer Vision and Pattern Recognition Workshops (2019)
8. Cosmin Ancuti, Codruta O. Ancuti, R.T.: NTIRE 2020 Non Homogeneous Dehazing Challenge (2020). https://competitions.codalab.org/competitions/22236
9. Fattal, R.: Single image dehazing. ACM Trans. Graph. **27**(3), 1–9 (2008)
10. Fattal, R.: Dehazing using color-lines. ACM Trans. Graph. **34**(1), 1–14 (2014)
11. Goodfellow, I., et al.: Generative adversarial nets. In: Advances in Neural Information Processing Systems, pp. 2672–2680 (2014)
12. He, K., Sun, J., Tang, X.: Single image haze removal using dark channel prior. IEEE Trans. Pattern Anal. Mach. Intell. **33**(12), 2341–2353 (2010)
13. Huang, S.C., Chen, B.H., Wang, W.J.: Visibility restoration of single hazy images captured in real-world weather conditions. IEEE Trans. Circ. Syst. Video Technol. **24**(10), 1814–1824 (2014)
14. Johnson, J., Alahi, A., Fei-Fei, L.: Perceptual losses for real-time style transfer and super-resolution. In: Leibe, B., Matas, J., Sebe, N., Welling, M. (eds.) ECCV 2016. LNCS, vol. 9906, pp. 694–711. Springer, Cham (2016). https://doi.org/10.1007/978-3-319-46475-6_43
15. Kingma, D.P., Ba, J.: Adam: a method for stochastic optimization. arXiv preprint arXiv:1412.6980 (2014)
16. Koschmieder, H.: Theorie der horizontalen Sichtweite. Keim & Nemnich, Munich (1925)

17. Li, B., Peng, X., Wang, Z., Xu, J., Feng, D.: AOD-Net: all-in-one dehazing network. In: The IEEE International Conference on Computer Vision (2017)
18. Meng, G., Wang, Y., Duan, J., Xiang, S., Pan, C.: Efficient image dehazing with boundary constraint and contextual regularization. In: IEEE International Conference on Computer Vision, pp. 617–624 (2013)
19. Morales, P., Klinghoffer, T., Jae Lee, S.: Feature forwarding for efficient single image dehazing. In: IEEE Conference on Computer Vision and Pattern Recognition Workshops (2019)
20. Ren, W., Liu, S., Zhang, H., Pan, J., Cao, X., Yang, M.-H.: Single image dehazing via multi-scale convolutional neural networks. In: Leibe, B., Matas, J., Sebe, N., Welling, M. (eds.) ECCV 2016. LNCS, vol. 9906, pp. 154–169. Springer, Cham (2016). https://doi.org/10.1007/978-3-319-46475-6_10
21. Ronneberger, O., Fischer, P., Brox, T.: U-Net: convolutional networks for biomedical image segmentation. In: Navab, N., Hornegger, J., Wells, W.M., Frangi, A.F. (eds.) MICCAI 2015. LNCS, vol. 9351, pp. 234–241. Springer, Cham (2015). https://doi.org/10.1007/978-3-319-24574-4_28
22. Tan, R.T.: Visibility in bad weather from a single image. In: IEEE Conference on Computer Vision and Pattern Recognition, pp. 1–8 (2008)
23. Vazquez-Corral, J., Finlayson, G.D., Bertalmío, M.: Physical-based optimization for non-physical image dehazing methods. Opt. Express **28**(7), 9327–9339 (2020)
24. Wang, Z., Bovik, A.C., Sheikh, H.R., Simoncelli, E.P.: Image quality assessment: from error visibility to structural similarity. IEEE Trans. Image Process. **13**(4), 600–612 (2004)
25. Zhang, H., Patel, V.M.: Densely connected pyramid dehazing network. In: IEEE Conference on Computer Vision and Pattern Recognition, pp. 3194–3203 (2018)
26. Zhang, H., Sindagi, V., Patel, V.M.: Multi-scale single image dehazing using perceptual pyramid deep network. In: IEEE Conference on Computer Vision and Pattern Recognition Workshops, pp. 902–911 (2018)
27. Zhu, Q., Mai, J., Shao, L.: A fast single image haze removal algorithm using color attenuation prior. IEEE Trans. Image Process. **24**(11), 3522–3533 (2015)

A Benchmark for Inpainting of Clothing Images with Irregular Holes

Furkan Kınlı(✉), Barış Özcan, and Furkan Kıraç

Özyeğin University, İstanbul, Turkey
{furkan.kinli,furkan.kirac}@ozyegin.edu.tr,
baris.ozcan.10097@ozu.edu.tr

Fig. 1. Masked & generated images of our model employing dilated partial convolutions for clothing image inpainting.

Abstract. Fashion image understanding is an active research field with a large number of practical applications for the industry. Despite its practical impacts on intelligent fashion analysis systems, clothing image inpainting has not been extensively examined yet. For that matter, we present an extensive benchmark of clothing image inpainting on well-known fashion datasets. Furthermore, we introduce the use of a dilated version of partial convolutions, which efficiently derive the mask update step, and empirically show that the proposed method reduces the required number of layers to form fully-transparent masks. Experiments show that dilated partial convolutions (DPConv) improve the quantitative inpainting performance when compared to the other inpainting strategies, especially it performs better when the mask size is 20% or more of the image.

Keywords: Image inpainting · Fashion image understanding · Dilated convolutions · Partial convolutions

A. Bartoli and A. Fusiello (Eds.): ECCV 2020 Workshops, LNCS 12538, pp. 182–199, 2020.
https://doi.org/10.1007/978-3-030-66823-5_11

1 Introduction

Image inpainting is the task of filling the holes in a particular image with some missing regions in such a way that the generated image should be visually plausible and semantically coherent. There are numerous vision applications including object removal, regional editing, super-resolution, stitching and many others that can be practicable by employing different image inpainting strategies. In the literature, the most prominent studies [2,19,33,38,41,57,58] proposing different strategies for solving inpainting problems have focused on mostly natural scene understanding, street view understanding and face completion tasks.

Fashion image understanding is an active research field that has enormous potential of practical applications in the industry. With the achievements of deep learning-based solutions [8,15,16,43] and increasing the number of datasets representing more real-world-like cases [7,36,44,62], different solutions have been proposed for various image-related vision tasks in fashion domain such as clothing category classification [7,20,31,36,50], attribute recognition [7,17,36,53,59], clothing segmentation [7,23,29,32,37], clothing image retrieval [5,18,22,27,39,51] and clothing generation [11–13,17,21,30,35,46,55,61]. At this point, the recent breakthroughs in deep generative learning solutions [10,24,25] lead to arise some opportunities for the applications in fashion domain including designing new clothing items for recommendation systems [17,46], virtual try-on systems [12,21,30] and fashion synthesis [11,13,35,55,61]. Despite these extensive studies on different tasks in fashion domain, image inpainting has not been extensively examined yet. Considering the practical impacts of achieving this task on real-world applications of fashion analysis, we present an extensive benchmark for clothing image inpainting by employing a generative approach that recovers the irregular missing regions in the set of clothing items (Fig. 1).

To extend the practical advantages of this solution for fashion domain, we need further investigation of the drawbacks of the current fashion analysis systems working on real-world cases. Due to possible deformations and occlusions in real-world scenarios, understanding fashion images is a challenging task to make successful inferences. For instance, first, attribute recognition models trained with commercial images often fail to successfully infer the attributes of social media images (in other words, consumer images), due to the occlusions that appears on the parts representing the actual attribute of a particular clothing. Next, overlapping items in a single clothing image builds natural occlusion scenarios for both items when segmentation is applied to the image. Although such segmentation models [4,9,52] have the ability to infer the possible locations of the occluded region of an object, filling this region in a visually plausible and semantically correct way in order to analyze the segmented clothing items is yet to be achieved. Moreover, for a visual recommendation system working with real-world clothing images, any deformation on the extracted clothing items can change the possible combinations of it with the other items. Based on such drawbacks on fashion analysis systems, applying different inpainting strategies may be practical for understanding fashion images better. Therefore, we want to contribute to the fundamental research effort of solving inpainting problems by redirecting it to fashion domain (Fig. 2).

Fig. 2. Difficult scenarios for fashion image understanding due to the occlusions (natural, multiple persons, multiple clothing items)

In image inpainting literature, there are several strategies that use image statistics or deep learning-based solutions. PatchMatch [2] is one of the most prominent strategies relying on the image statistics, and it basically searches for the best fitting patch to fill the rectangular-formed missing parts in the images. Although it is possible to generate visually plausible results by using this inpainting strategy, it cannot achieve semantic coherence in hard scenarios since it only depends on the statistics of available parts in the image. On the other hand, deep learning-based solutions are more suitable for image inpainting since deep neural networks can inherently learn the hidden representations by preserving semantic priors. Earlier studies [19,41,57] try to solve the problem of initialization of the pixels in the missing parts to condition the output by assigning a fixed value for these pixels. In these studies, the results have the visual artifacts, and need some additional post-processing methods (*e.g.* fast marching [48], Poisson image blending [42] and the refinement network [57]) to refine them. Partial convolutional mechanism [33] addresses this limitation by conditioning the only valid pixels. To achieve this, convolutional layers are masked and re-weighted, and then the mask is updated in such a way that progressively filling up the hole pixels. Recently, Nazeri *et al.* [38] considers inpainting problem as image completion by predicting the structure (*i.e.* edge maps) of the main content in the image. Apart from these studies, we focus on increasing the receptive field size of low-resolution feature maps of partial convolutions, and thus, we propose dilated partial convolutions whose the kernel of partial convolution [33] is dilated by a given parameter, and capable of gathering more information from near-to-hole regions in order to update the masks more efficiently without requiring to decrease their spatial dimensionality up to very low-resolution.

In summary, our main contributions are as follows:

- We introduce an extensive benchmark of clothing image inpainting on a variety of challenging datasets including FashionGen [44], FashionAI [62], DeepFashion [36] and DeepFashion2 [7], and attempt to redirect the fundamental research efforts on image inpainting problems to fashion domain.
- To the best of our knowledge, this is the first study to mention the practical usefulness of achieving visually plausible and semantically coherent inpainting of fashion images for industrial applications.
- We present the idea of using dilated version of partial convolutions for image inpainting tasks, where the dilated input window for partial convolutions derives more efficient mask update step.
- We empirically show that dilated partial convolutions reduce the number of layers that requires to lead to a mask without any holes, and thus, it makes possible to achieve better inpainting quality without reducing the spatial dimensionality of encoder output.

2 Related Work

Image inpainting is a challenging task, and it has been extensively studied for a decade in different vision-related research fields, especially on natural scene understanding and face recognition. Several traditional inpainting strategies such as [1,3,6,14,48] try to synthesize texture information in the holes by employing available image statistics. However, these methods often fail to achieve inpainting the images without any artifacts since using only available image statistics may mislead the results about the texture information between holes and non-hole regions when it varies. Barnes *et al.* [2] proposes a fast algorithm, namely *PatchMatch*, which iteratively searches for the best fitting patches to the missing parts of the image. Although this method can produce visually plausible results in a faster way, it still lacks of producing semantically coherent results, and far from being suitable for real-time image processing pipeline.

In deep learning-based solutions, earlier approaches use a constant value for initializing the holes before passing throughout the network. First, *Context Encoders* [41] employs an Auto-encoder architecture with adversarial training strategy in order to fill a central large hole in the images. Yang *et al.* [54] extends the idea in [41] with post-processing the output that considers only the available image statistics. Song *et al.* [47] proposes a robust training scheme in coarse-to-fine manner. Iizuka *et al.* [19] introduces a different adversarial training strategy to provide both local and global consistency in the generated images. Also, in [19], it is demonstrated that sufficiently larger receptive fields work well in image inpainting tasks, and dilated convolution is adopted for increasing the size of the receptive fields. Yu *et al.* [57] improves the idea of using dilated convolutions in inpainting task by adding contextual attention mechanism on top of low-resolution feature maps for explicitly matching and attending to relevant background patches. Liu *et al.* [33] presents partial convolutions where the convolution is masked and re-weighted to be conditioned on only valid pixels. Yu *et al.* [58] proposes gated convolutions that have the ability to learn the features dynamically at each spatial location across all layers, and this mechanism improves the color consistency and semantic coherence in the generated images. Liu *et al.* [34] employs coherent semantic attention and consistency loss to the refinement network to construct the correlation between the features of hole regions, even in deeper layers. Kınlı *et al.* [28] observes the effect of collaborating with textual features extracted by image descriptions on inpainting performance. Nazeri *et al.* [38] proposes a novel method that predicts the image structure of the missing region in the form of edge maps, and these predicted edge maps are passed to the second stage to guide the inpainting.

3 Methodology

Our proposed model enhances the capability of partial convolutions [33], which alters this mechanism by adding a dilation factor to its convolutional filters, and employing self-attention to the decoder part of our model. In this section, we first explain the reasoning behind dilated partial convolutions and present its formulation. Then, we describe our architecture, and lastly discuss the loss functions of proposed model.

3.1 Dilated Partial Convolutions

Unlike in natural images where spatially-near pixels yield a larger correlation, for clothing image inpainting, the correlated pixels may be far apart in a particular image (*e.g.* a pattern on a shirt can cover a larger area of the image, and more importantly, this pattern may not be spatially continuous in such scenarios with occlusions or deformations). Even though both partial and dilated convolutional layers prove that they can produce visually plausible results in inpainting tasks, they have their own particular shortcomings. When partial convolutions are employed, the layers cannot gather the information from the correlation of far apart pixels as in dilated convolutions, on the other hand, the network does not only focus on non-hole regions of the images by using only dilated convolutions [56]. We address both these shortcomings by introducing the use of dilated partial convolutions where the input window is dilated by a given parameter, and the mask of partial convolution is applied afterwards. This allows the network to focus on non-hole regions, and it has larger receptive fields to utilize correlations of far apart pixels. More importantly, as stated in [33], the consecutive layers of partial convolutions will eventually lead to a mask without any holes depending on the input mask, and we have empirically proven that dilated partial convolutions reduce the number of required layers to achieve this, and the input mask covers up a larger percentage of the input image in earlier part of the network. The discussion and detailed empirical results can be found in Discussion part. The formal definition of dilated partial convolutions can be formulated as follows:

$$\mathcal{M} = \sum_{m=-M}^{M} \sum_{n=-N}^{N} m(x - ln, y - lm) \tag{1}$$

$$Z = \frac{(2M + 1) \times (2N + 1)}{\mathcal{M}} \tag{2}$$

$$(f_m \circledast g_l)(x, y) = \frac{\sum_{m=-M}^{M} \sum_{n=-N}^{N} f(x - ln, y - lm) g(n, m) m(x - ln, y - lm)}{Z} \tag{3}$$

where f, g and m represents the input, the convolutional kernel with a size of $(2M+1) \times (2N+1)$ and the corresponding mask, respectively. Equation (2) has essentially the same scaling factor as in [33], but modified to be applicable to dilated partial convolutions. Note that when the dilation factor $l = 1$, Eq. (3) becomes the partial convolution without any dilation. Mask update is calculated in the same manner as in partial convolutions, and can be formulated as follows:

$$m' = \begin{cases} 1, & \text{if}(\mathcal{M} > 1) \\ 0, & \text{otherwise} \end{cases} \tag{4}$$

where the updated mask m' gets closer to become fully-transparent (*i.e.* all pixels become ones) after a certain number of layers. Dilated partial convolutions allows m' to become fully-transparent in earlier layers when compared to partial convolutions.

3.2 Model Architecture

We have designed a similar *U-Net-like* [43] architecture as in [33], where all low resolution layers are replaced with dilated partial convolutions while the first four layers are left as partial convolutions. The binary mask in the first layer is defined as the input corruption mask. Moreover, in the decoder stage, we have used self-attention module [49] to make use of spatially distant but related features. All of the skip connections are concatenated with the layer on corresponding level, instead of adding to them, and also dilated partial convolutions are residually-connected. The dilation rate is progressively increased up to the last dilated convolutional layer (*i.e.* multi-scale context aggregation), as in [56]. As working on the fashion datasets, the input sizes for all models are picked as 256×256. Overall architecture design can be seen in Fig. 3.

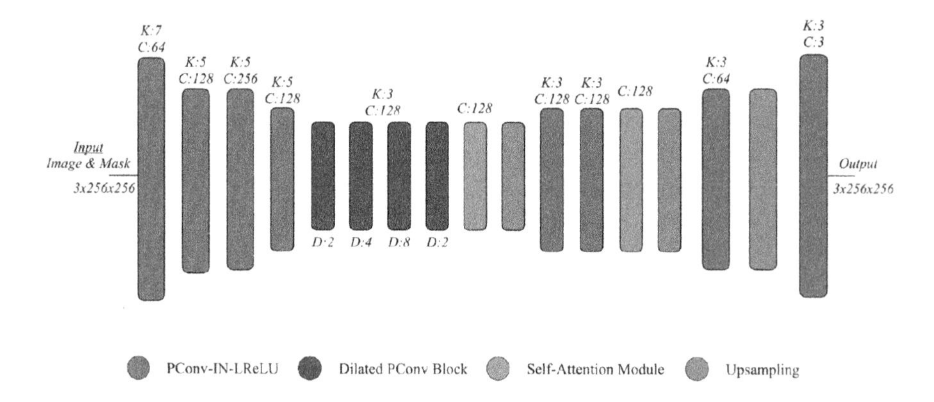

Fig. 3. Overview of our architecture design.

3.3 Loss Functions

We have followed a similar loss function scheme to [33], but with slight differences. The total loss function is introduced in Eq. (5), where $\mathcal{L}_{pixel}$ is the pixel loss, $\mathcal{L}_{style}$ is the style loss, $\mathcal{L}_{adv}$ is the adversarial loss and $\mathcal{L}_{tv}$ is the total variation (TV) loss. We found that the coefficients in Eq. (5) work better than the ones in [33] in our experimental settings. Moreover, the perceptual loss is substituted with an adversarial loss since we experimentally found that the benefits of the perceptual loss do not afford its computational cost, and also we want to take the advantage of adversarial training for our model.

$$\mathcal{L}_{total} = 10\mathcal{L}_{pixel} + 120\mathcal{L}_{style} + 10^{-3}\mathcal{L}_{adv} + 10^{-4}\mathcal{L}_{tv} \tag{5}$$

Pixel loss $\mathcal{L}_{pixel}$ is the sum of ℓ_1 losses of the hole and non-hole regions between the output image $\mathbf{I}_{out}$ and the ground truth image $\mathbf{I}_{gt}$, as given in (6), where $N_{\mathbf{I}_{gt}} = C \times H \times W$ as C, H and W are the input dimensions, and $\hat{m}$ denotes the initial mask on the input.

$$\mathcal{L}_{pixel} = \frac{1}{N_{\mathbf{I}_{gt}}}\left(\left\|(1-\hat{m}) \odot (\mathbf{I}_{out} - \mathbf{I}_{gt})\right\|_1 + \left\|\hat{m} \odot (\mathbf{I}_{out} - \mathbf{I}_{gt})\right\|_1\right) \tag{6}$$

The total style loss $\mathcal{L}_{style}$ is calculated by summing the style losses for $\mathbf{I}_{out}$ and $\mathbf{I}_{comp}$, denoted as $\mathcal{L}_{style_{out}}$ and $\mathcal{L}_{style_{comp}}$, respectively (Eq. (7)). To obtain the composite image $\mathbf{I}_{comp}$, all non-hole pixels in the output image are replaced with the ground truth pixels. Style loss requires to calculate the activation maps $\Psi^{\mathbf{I}_*}{}_p$ of pre-trained VGG-16 with respect to the input image $\mathbf{I}_*$, where p denotes the layer index. For a given layer index p, the normalization factor is defined as $K_p = \frac{1}{C_p H_p W_p}$.

$$\mathcal{L}_{style} = \sum_{p=0}^{P-1} \frac{1}{C_p^2} \left\| K_p((\Psi^{\mathbf{I}_{out}}{}_p)^T (\Psi^{\mathbf{I}_{out}}{}_p) - (\Psi^{\mathbf{I}_{gt}}{}_p)^T (\Psi^{\mathbf{I}_{gt}}{}_p)) \right\|_1 + \sum_{p=0}^{P-1} \frac{1}{C_p^2} \left\| K_p((\Psi^{\mathbf{I}_{comp}}{}_p)^T (\Psi^{\mathbf{I}_{comp}}{}_p) - (\Psi^{\mathbf{I}_{gt}}{}_p)^T (\Psi^{\mathbf{I}_{gt}}{}_p)) \right\|_1 \tag{7}$$

The adversarial loss given in Eq. (8) is adopted from [10], where $\mathcal{G}$ is the generator network, $\mathcal{D}$ is the discriminator network, and $\mathbf{I}^M$ is the masked input image. The discriminator $\mathcal{D}$ network is a simple CNN with a depth of 5, and trained to optimize the loss function given in Eq. (9). To train the discriminator better, we flipped the labels with the probability of 0.1 to make the labels noisy for the discriminator, and also we applied label smoothing where, for each instance, if it is real, then replace its label with a random number between 0.7 and 1.2, otherwise, replace it with 0.0 and 0.3. Lastly, we tried to apply random dropout for the decoder part of our model.

$$\mathcal{L}_{adv_G} := \mathbb{E}\left[\left(\mathcal{D}\left(\mathcal{G}\left(\mathbf{I}^M \right) - 1 \right) \right)^2 \right] \tag{8}$$

$$\mathcal{L}_{adv_D} := \mathbb{E}\left[\mathcal{D}(\hat{\mathbf{I}})^2 \right] + \mathbb{E}\left[\left(\mathcal{D}\left(\mathbf{I} \right) - 1 \right)^2 \right] \tag{9}$$

Lastly, total variation loss L_{tv}, as given in (10), enforces the spatial continuity on the generated images, where R is the region after applying a 1-pixel dilation to a hole region in the input image.

$$\mathcal{L}_{tv} = \sum_{(i,j)\in R,(i,j+1)\in R} \frac{\left\| \mathbf{I}_{comp}^{i,j+1} - \mathbf{I}_{comp}^{i,j} \right\|_1}{N_{\mathbf{I}_{comp}}} + \sum_{(i,j)\in R,(i+1,j)\in R} \frac{\left\| \mathbf{I}_{comp}^{i+1,j} - \mathbf{I}_{comp}^{i,j} \right\|_1}{N_{\mathbf{I}_{comp}}} \tag{10}$$

4 Experiments

In this study, we introduce an extensive benchmark on fashion image inpainting, and also propose enhanced version of partial convolutions, namely *dilated partial convolutions*. We have conducted our experiments on four different well-known fashion datasets, which are FashionGen [44], FashionAI [62], DeepFashion [36] and DeepFashion2 [7].

4.1 Experimental Setup

Training Details. In our experiments, we used Adam optimizer [26] with $\beta_1 = 0.5$ and $\beta_2 = 0.9$ for both generator and discriminator networks. The initial learning rate for generator network is 2×10^{-4}, and for discriminator network, is 10^{-4}. We trained our model for ∼120.000 steps for each dataset with batch size of 64. We only applied horizontal flipping to the input images, no other data augmentation technique is applied. For DeepFashion and DeepFashion2 datasets, the size of images varies, so we use random cropping (central cropping for testing) and resizing to obtain the size of 256×256 for training images. We implemented our framework with PyTorch library [40], and used 2x NVIDIA Tesla V100 GPUs for our training. Source code: https://github.com/birdortyedi/fashion-image-inpainting

Datasets. To create an extensive benchmark for fashion image inpainting, we picked four common fashion datasets to use in our experiments. Before training, we prepared the datasets to the image inpainting setup by applying inpainting masks[1] to the images, which erase some parts randomly. To make sure that the images have some holes on clothing parts, we generated the masks with a heuristic, where at least a small portion of the clothing parts has been erased by the mask. Then, we run our experiments for each method on each dataset separately.

FashionGen (FG). [44] contains several different fashion products collected from an online platform selling the luxury goods from independent designers. Each product is represented by an image, the description, its attributes, and relational information defined by professional designers. We extract the instances belonging to these categories from the dataset. At this point, training set has 200K clothing images from 9 different categories (*top, sweater, pant, jean, shirt, dress, short, skirt, coat*), and validation set has 30K clothing images.

FashionAI (FAI). [62] is introduced in FashionAI Global Challenge 2018 by The Vision & Beauty Team of Alibaba Group and Institute of Textile & Clothing of The Hong Kong Polytechnic University. For this dataset, there are 180K consumer images of different clothing categories for training, and 10K for testing.

DeepFashion (DF). [36] is a dataset containing ∼800K diverse fashion images with their rich annotations (*i.e.* 46 categories, 1.000 descriptive attributes, bounding boxes and landmark information) ranging from well-posed product images to real-world-like consumer photos. In our experiments on this dataset, we follow the same procedure as in [36] to split training and testing sets, so we have 207K clothing images for training, and 40K for testing.

DeepFashion2 (DF2). [7] is one of the largest open-source fashion database that contains 491K high-resolution clothing images with 13 different categories and numerous attributes. Similar to DeepFashion, we again follow the procedure in the paper of dataset [7] to split the published version of the data.

[1] https://github.com/karfly/qd-imd.

Table 1. Quantitative comparison between our proposed model and the state-of-the-art methods on four well-known fashion datasets.

Mask ratios	[0.0:0.2]				[0.2:0.4]				[0.4:]			
Datasets	FG	FAI	DF	DF2	FG	FAI	DF	DF2	FG	FAI	DF	DF2
ℓ_1 (PM) %	**0.66**	1.16	1.34	1.32	1.34	2.11	2.03	1.96	3.70	6.12	5.74	5.43
ℓ_1 (PConv) %	0.73	0.92	0.91	0.94	1.18	1.48	1.34	1.47	2.82	4.86	3.76	3.70
ℓ_1 (GConv) %	0.76	0.95	0.93	0.95	1.22	1.58	1.23	1.36	2.99	5.20	3.65	3.62
ℓ_1 (**DPConv**) %	0.70	**0.91**	**0.87**	**0.92**	**1.14**	**1.39**	**1.20**	**1.33**	**2.61**	**4.03**	**3.36**	**3.38**
PSNR (PM)	16.14	12.21	12.30	12.55	14.71	11.23	11.37	11.49	13.36	9.74	9.96	10.09
PSNR (PConv)	16.04	**12.33**	12.43	12.73	15.04	11.49	11.66	11.94	13.91	11.11	11.47	11.89
PSNR (GConv)	15.98	12.20	12.34	12.66	14.86	11.42	11.61	11.84	13.79	11.02	11.39	11.77
PSNR (**DPConv**)	**16.23**	12.27	**12.56**	**12.89**	**15.15**	**11.81**	**11.88**	**12.12**	**14.16**	**11.63**	**11.84**	**12.28**
SSIM (PM)	**0.891**	0.773	0.784	0.804	0.816	0.737	0.744	0.769	0.738	0.622	0.671	0.677
SSIM (PConv)	0.851	**0.784**	0.777	0.804	0.770	0.735	0.729	0.733	0.721	0.649	0.689	0.688
SSIM (GConv)	0.845	0.751	0.768	0.796	0.783	0.723	0.731	0.741	0.717	0.624	0.670	0.681
SSIM (**DPConv**)	0.855	0.782	**0.794**	**0.811**	**0.824**	**0.739**	**0.750**	**0.777**	**0.766**	**0.683**	**0.717**	**0.719**
FID (PM)	**2.99**	5.44	**4.36**	5.78	6.21	11.98	11.65	12.30	25.92	28.91	27.94	29.74
FID (PConv)	3.76	6.04	5.93	6.22	5.56	11.05	10.36	11.28	23.47	27.42	24.39	27.52
FID (GConv)	3.57	5.86	5.63	6.18	5.28	10.84	10.12	11.01	23.18	27.19	24.21	26.95
FID (**DPConv**)	3.18	**5.24**	4.66	**5.48**	**5.20**	**10.08**	**9.26**	**10.21**	**22.82**	**25.55**	**23.94**	**25.62**

4.2 Quantitative Results

As is the case with the previous studies [33,34,57,58], we evaluate the performance of our model with four different metrics, which are ℓ_1 error percentage, SSIM [60], PSNR and FID [45]. Then, we compare the quantitative performance of DPConv with PatchMatch (PM) [2] as a non-learning method, Partial Convolutions (PConv) [33] and Gated Convolutions (GConv) [58], as learning-based methods. We used the third-party implementations for PM[2] and GConv[3], but re-implemented PConv according to the layer implementation[4] and the architecture details referred in their paper [33]. The quantitative results of both our model and the state-of-the-art methods/models are shown in Table 1. The observations are as follows: (1) Dilated partial convolutions are more robust to the changes in the mask ratios. (2) According to SSIM metric, although PM performs well in the cases where the mask ratio is smaller, the performances on all measurements significantly decrease for more complex cases. (3) The performances of all methods are very similar on less complex dataset (*i.e.* FashionGen), whereas the impact of our proposed model can be clearly observed as mask ratio increases in the settings of more complex datasets (*i.e.* DeepFashion and DeepFashion2). (4) Using dilated version of partial convolutions in image inpainting improves the overall performance without reducing the spatial dimensionality of the encoder output. (5) Overall, our model, namely *DPConv*, outperforms the current inpainting strategies on four well-known fashion datasets.

[2] https://github.com/MingtaoGuo/PatchMatch.
[3] https://github.com/avalonstrel/GatedConvolution_pytorch.
[4] https://github.com/NVIDIA/partialconv.

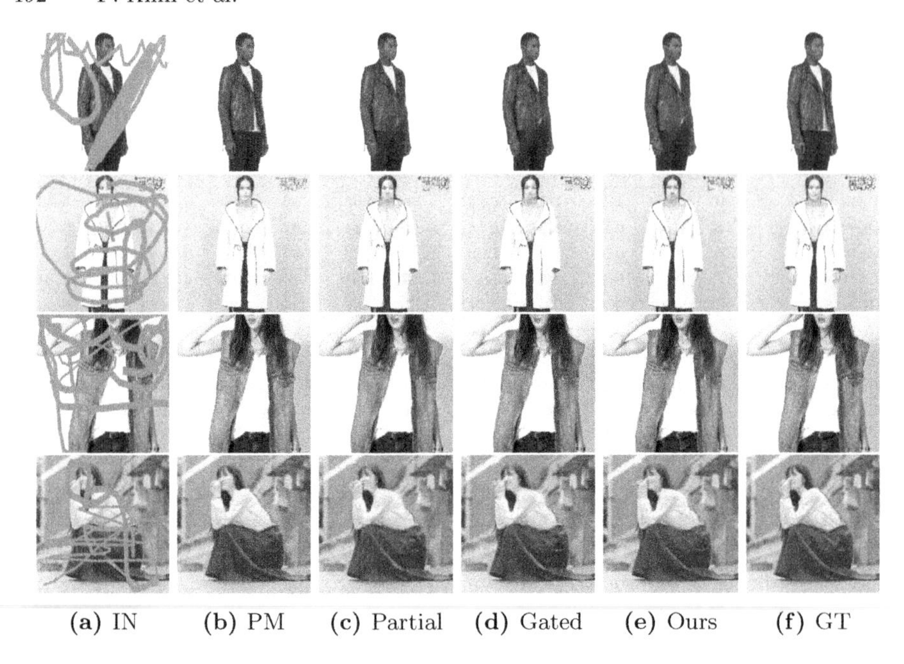

Fig. 4. Comparison the results of our proposed model and the state-of-the-art methods.

4.3 Qualitative Results

Next, we compare the visual compatibility of the results of our model with the other state-of-the-art methods [2,33,58]. Figure 4 shows the results on four well-known fashion datasets [7,36,44,62]. We apply the same settings with ours to the training of all methods, and do not perform any post-processing for the outputs. The results demonstrate that all inpainting strategies can produce visually plausible and semantically coherent clothing images, and the visual outputs of these inpainting strategies can be utilized in order to increase the effectiveness of fashion image understanding solutions (*e.g.* removing the disrupted regions and inpainting them). However, DPConv accomplish it by employing a shallower network architecture, and taking advantage of the efficient mask update step of dilated partial convolutions. Furthermore, to compare the strategies, we can say that (1) DPConv and PConv seem to produce very similar outputs, but when looking into the details, DPConv preserves the visual coherence better (*e.g.* In Fig. 4, the collar of the coat in row 2 & the head of woman in row 4). (2) PM cannot produce smooth outputs in contrast to the other methods. (3) GConv cannot fill the holes by preserving the semantic coherence, and residue of the input mask can be still seen in GConv outputs.

Ablation Study: We conduct additional experiments to observe the effect of using dilated convolutions at certain stages of the inpainting networks. The first one is called *DPConv*† whose layers before dilated partial convolutional block also have dilation of 2 in their kernels. The latter, namely *DGConv*, is the same architecture with our model, but employs gated convolutions, instead of partial convolutions. Table 2 demonstrates the quantitative evaluation of these models and ours. The observations are concluded as follows: (1) Using dilation in every stage of the network without applying multi-scale context aggregation strategy has negative effect on the performance of our model. (2) Due to the computational burden, applying multi-scale context aggregation to each stage is not feasible for this task. (3) Gated convolutions with dilation on the kernels leading to the lower-level feature maps shows a similar impact on the qualitative results, which DPConv does it on PConv. (4) However, DPConv still mostly outperforms DGConv on different mask ratios on four well-known fashion datasets (for more, see Fig. 7).

Table 2. The evaluation of the usage of dilation in certain stages of the networks for different inpainting strategies.

Mask ratios	[0.0:0.2]				[0.2:0.4]				[0.4:]			
Datasets	FG	FAI	DF	DF2	FG	FAI	DF	DF2	FG	FAI	DF	DF2
ℓ_1 (DGConv) %	0.72	0.92	0.90	**0.92**	1.17	1.43	1.21	**1.33**	2.68	4.21	3.41	3.47
ℓ_1 (DPConv†) %	0.62	1.02	0.94	1.20	1.21	1.72	1.86	2.04	3.18	5.16	5.66	5.39
ℓ_1 (**DPConv**) %	0.70	**0.91**	**0.87**	**0.92**	**1.14**	**1.39**	**1.20**	**1.33**	**2.61**	**4.03**	**3.36**	**3.38**
PSNR (DGConv)	16.11	**12.31**	12.54	**12.91**	15.09	11.76	**11.92**	12.10	14.12	11.56	11.70	11.92
PSNR (DPConv†)	15.96	12.14	12.32	12.61	14.75	11.27	11.39	11.61	13.40	10.97	11.04	11.19
PSNR (**DPConv**)	**16.23**	12.27	**12.56**	12.89	**15.15**	**11.81**	11.88	**12.12**	**14.16**	**11.63**	**11.84**	**12.28**
SSIM (DGConv)	**0.858**	0.779	0.788	0.802	0.818	0.736	**0.754**	0.774	0.750	**0.688**	0.712	0.704
SSIM (DPConv†)	0.869	0.773	0.762	0.788	0.791	0.717	0.728	0.735	0.719	0.626	0.664	0.679
SSIM (**DPConv**)	0.855	**0.782**	**0.794**	**0.811**	**0.824**	**0.739**	0.750	**0.777**	**0.766**	0.683	**0.717**	**0.719**
FID (DGConv)	3.38	5.49	4.90	5.89	**5.06**	10.36	9.53	10.46	22.87	25.71	24.02	25.88
FID (DPConv†)	3.96	6.01	6.73	5.82	6.16	11.70	11.99	11.38	24.02	26.89	24.54	26.24
FID (**DPConv**)	**3.18**	**5.24**	**4.66**	**5.48**	5.20	**10.08**	**9.26**	**10.21**	**22.82**	**25.55**	**23.94**	**25.62**

4.4 The Effect of Dilation in Partial Convolutions

Dilated partial convolutions complete the masks to become fully-transparent (*i.e.* masks without any zeros) throughout the network with less number of consecutive layers, when compared to partial convolutions. To empirically prove this, we designed an experiment to analyze the behaviour of both layers given different input masks. In this experiment, we used 12.000 random masks with ~2.000 masks for each 10% range of mask ratios from 0% up to 60%. For each range of mask ratios, the average number of layers required to obtain a fully-transparent mask is calculated, and the maximum number of layers that can be stacked is limited to 20. As illustrated in Fig. 5a, dilated partial convolutions can reach fully-transparent masks 2.1 layers earlier on average (*i.e.* ~15% less layers). Figure 5b emphasizes the exponential growth of the required number of layers with respect to all masks in the case of using partial convolutions or dilated partial convolutions. The result of this experiment shows that the dilated

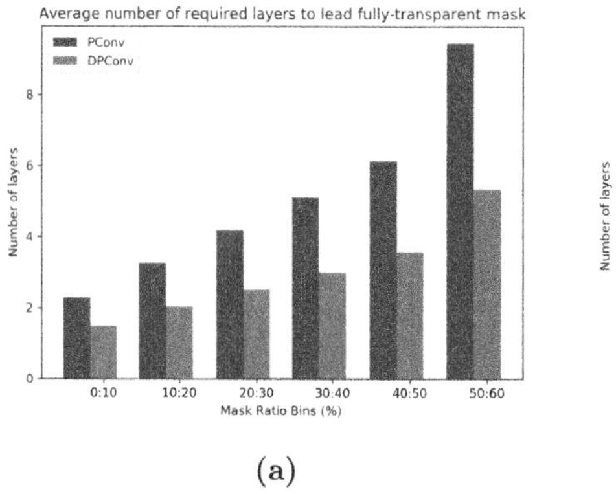

(a)

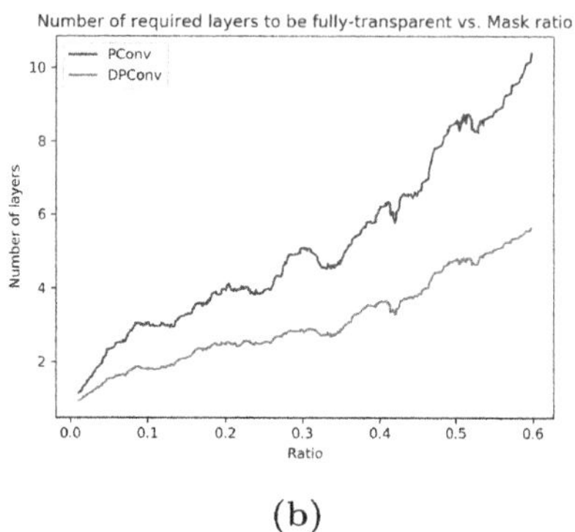

(b)

Fig. 5. Experimental results of analyzing the behavior of partial convolutions and dilated partial convolutions given random masks.

version of partial convolutions has a practical impact of *leading fully-transparent masks in higher resolution without requiring to go deeper throughout the network*, and thus leads to a more efficient mask update step and faster learning process. Note that PConv decreases the feature map to 2×2 in its decoder part while DPConv starts to upsampling at the feature maps sized 32×32.

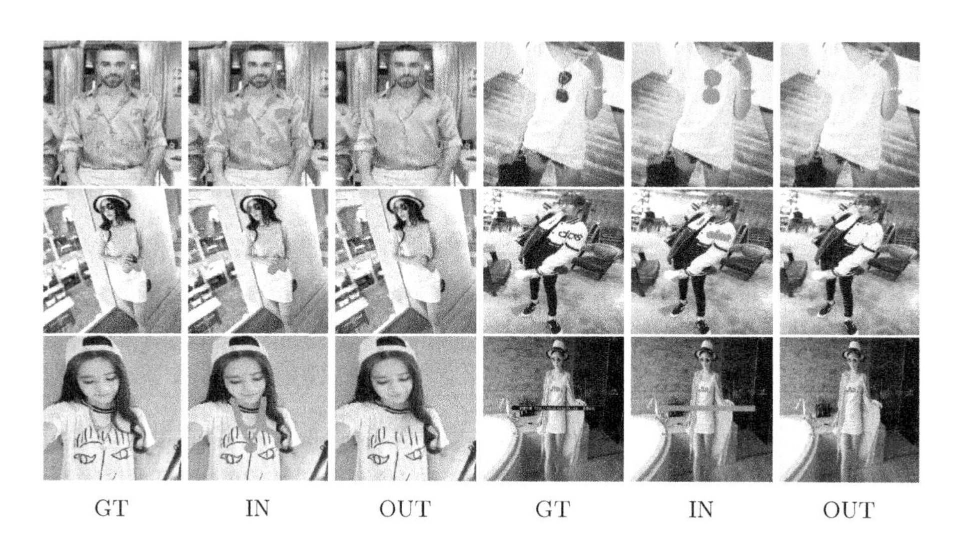

Fig. 6. Example inpainting results of the images containing some overlapping or disruptive parts on clothing items.

4.5 Practical Usage in Fashion Domain

Image inpainting can improve the performance of fashion image understanding solutions when applied as pre-process or post-process. To give a brief idea about how it can be useful for such systems, we picked a number of images that contain different items overlapping the clothing items or that have some disruptive parts (*e.g.* banner, logo). We have manually created the masks that remove these items/parts from the images. Figure 6 demonstrates the inpainting results of picked images where DPConv trained on DeepFashion2 dataset is employed for testing. At this point, we showed that clothing image inpainting can be useful for fashion editing (*e.g.* removing accessories like eye-glasses and necklace, logos, banners or non-clothing items, even changing the design of clothing), and it makes the main-stream fashion image understanding solutions work better.

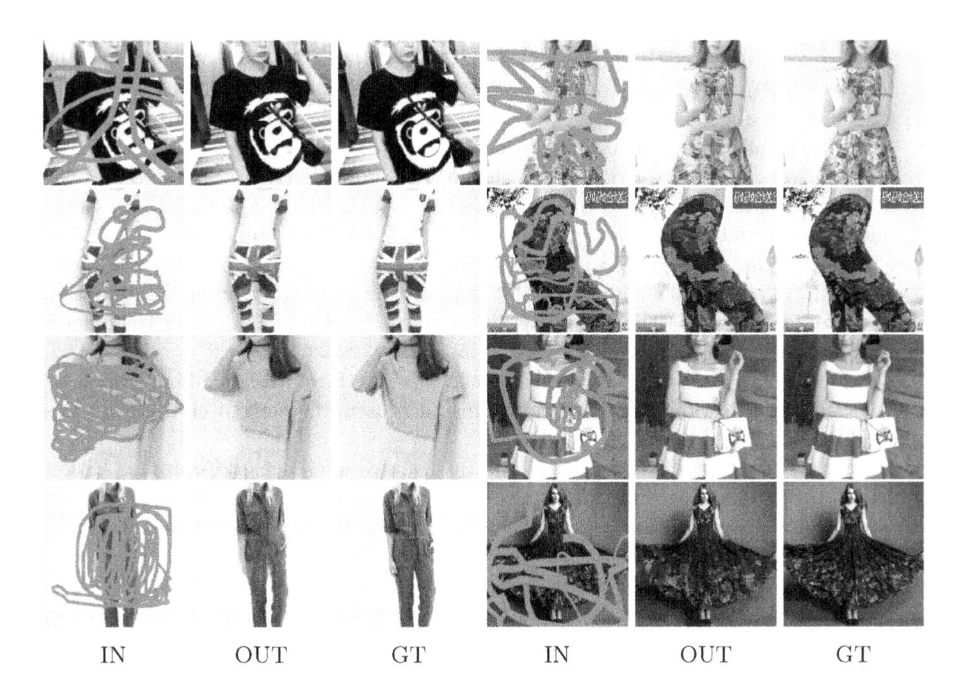

Fig. 7. More inpainting results of DPConv on four different datasets. Zoom in for better view.

5 Conclusion

In this study, we present an extensive benchmark for clothing image inpainting, which may be practical for industrial applications in fashion domain. Qualitative and quantitative comparisons demonstrate that proposed method improves image inpainting performance when compared to the previous state-of-the-art methods, and produce visually plausible and semantically coherent results for clothing images. Overall performances of inpainting strategies proves that AI-based fashion image understanding solutions can employ inpainting to their pipeline in order to improve the general performance.

References

1. Ballester, C., Bertalmio, M., Caselles, V., Sapiro, G., Verdera, J.: Filling-in by joint interpolation of vector fields and gray levels. IEEE Trans. Image Process. **10**(8), 1200–1211 (2001). https://doi.org/10.1109/83.935036
2. Barnes, C., Shechtman, E., Finkelstein, A., Goldman, D.B.: PatchMatch: a randomized correspondence algorithm for structural image editing. ACM Trans. Graph. **28**(3), 24 (2009)
3. Bertalmio, M., Sapiro, G., Caselles, V., Ballester, C.: Image inpainting. In: Proceedings of the 27th Annual Conference on Computer Graphics and Interactive Techniques, SIGGRAPH 2000, pp. 417–424. ACM Press/Addison-Wesley Publishing Co. (2000). https://doi.org/10.1145/344779.344972
4. Chen, K., et al.: MMDetection: open MMLab detection toolbox and benchmark. arXiv preprint arXiv:1906.07155 (2019)
5. Di, W., Wah, C., Bhardwaj, A., Piramuthu, R., Sundaresan, N.: Style finder: fine-grained clothing style detection and retrieval. In: The IEEE Conference on Computer Vision and Pattern Recognition (CVPR) Workshops, June 2013
6. Efros, A.A., Freeman, W.T.: Image quilting for texture synthesis and transfer. In: Proceedings of the 28th Annual Conference on Computer Graphics and Interactive Techniques, SIGGRAPH 2001, pp. 341–346. Association for Computing Machinery, New York (2001). https://doi.org/10.1145/383259.383296
7. Ge, Y., Zhang, R., Wu, L., Wang, X., Tang, X., Luo, P.: A versatile benchmark for detection, pose estimation, segmentation and re-identification of clothing images (2019)
8. Girshick, R.: Fast R-CNN. In: IEEE International Conference on Computer Vision (ICCV), pp. 1440–1448, December 2015. https://doi.org/10.1109/ICCV.2015.169
9. Girshick, R., Radosavovic, I., Gkioxari, G., Dollár, P., He, K.: Detectron (2018). https://github.com/facebookresearch/detectron
10. Goodfellow, I., et al.: Generative adversarial nets. In: Ghahramani, Z., Welling, M., Cortes, C., Lawrence, N.D., Weinberger, K.Q. (eds.) Advances in Neural Information Processing Systems, vol. 27, pp. 2672–2680. Curran Associates, Inc. (2014). http://papers.nips.cc/paper/5423-generative-adversarial-nets.pdf
11. Gunel, M., Erdem, E., Erdem, A.: Language guided fashion image manipulation with feature-wise transformations. In: First Workshop on Computer Vision in Art, Fashion and Design - in conjunction with ECCV 2018 (2018)
12. Han, X., Wu, Z., Wu, Z., Yu, R., Davis, L.S.: VITON: an image-based virtual try-on network. In: IEEE/CVF Conference on Computer Vision and Pattern Recognition, pp. 7543–7552, June 2018. https://doi.org/10.1109/CVPR.2018.00787
13. Han, X., Wu, Z., Huang, W., Scott, M.R., Davis, L.S.: FiNet: compatible and diverse fashion image inpainting. In: Proceedings of the IEEE International Conference on Computer Vision, pp. 4481–4491 (2019)
14. Hays, J., Efros, A.A.: Scene completion using millions of photographs. Commun. ACM **51**(10), 87–94 (2008). https://doi.org/10.1145/1400181.1400202

15. He, K., Gkioxari, G., Dollár, P., Girshick, R.: Mask R-CNN. In: IEEE International Conference on Computer Vision (ICCV), pp. 2980–2988, October 2017. https://doi.org/10.1109/ICCV.2017.322
16. He, K., Zhang, X., Ren, S., Sun, J.: Deep residual learning for image recognition. In: IEEE Conference on Computer Vision and Pattern Recognition (CVPR), pp. 770–778, June 2016. https://doi.org/10.1109/CVPR.2016.90
17. Hsiao, W.L., Grauman, K.: Creating capsule wardrobes from fashion images. In: 2018 IEEE/CVF Conference on Computer Vision and Pattern Recognition, June 2018. https://doi.org/10.1109/cvpr.2018.00748
18. Huang, J., Feris, R.S., Chen, Q., Yan, S.: Cross-domain image retrieval with a dual attribute-aware ranking network. In: Proceedings of the IEEE International Conference on Computer Vision, pp. 1062–1070 (2015)
19. Iizuka, S., Simo-Serra, E., Ishikawa, H.: Globally and locally consistent image completion. ACM Trans. Graph. **36**(4) (2017). https://doi.org/10.1145/3072959.3073659
20. Inoue, N., Simo-Serra, E., Yamasaki, T., Ishikawa, H.: Multi-label fashion image classification with minimal human supervision. In: The IEEE International Conference on Computer Vision (ICCV) Workshops, October 2017
21. Jae Lee, H., Lee, R., Kang, M., Cho, M., Park, G.: LA-VITON: a network for looking-attractive virtual try-on. In: The IEEE International Conference on Computer Vision (ICCV) Workshops, October 2019
22. Jagadeesh, V., Piramuthu, R., Bhardwaj, A., Di, W., Sundaresan, N.: Large scale visual recommendations from street fashion images. In: Proceedings of the 20th ACM SIGKDD International Conference on Knowledge Discovery and Data Mining, KDD 2014, pp. 1925–1934. Association for Computing Machinery, New York (2014). https://doi.org/10.1145/2623330.2623332
23. Ji, W., et al.: Semantic locality-aware deformable network for clothing segmentation. In: IJCAI, pp. 764–770 (2018)
24. Karras, T., Aila, T., Laine, S., Lehtinen, J.: Progressive growing of GANs for improved quality, stability, and variation. In: International Conference on Learning Representations (2018). https://openreview.net/forum?id=Hk99zCeAb
25. Karras, T., Laine, S., Aila, T.: A style-based generator architecture for generative adversarial networks. In: 2019 IEEE/CVF Conference on Computer Vision and Pattern Recognition (CVPR), June 2019. https://doi.org/10.1109/cvpr.2019.00453
26. Kingma, D., Ba, J.: Adam: a method for stochastic optimization. In: International Conference on Learning Representations, December 2014
27. Kinli, F., Ozcan, B., Kirac, F.: Fashion image retrieval with capsule networks. In: Proceedings of the IEEE International Conference on Computer Vision Workshops (2019)
28. Kınlı, F., Özcan, B., Kıraç, F.: Description-aware fashion image inpainting with convolutional neural networks in coarse-to-fine manner. In: Proceedings of the 2020 6th International Conference on Computer and Technology Applications, ICCTA 2020, pp. 74–79 (2020). https://doi.org/10.1145/3397125.3397155
29. Korneliusson, M., Martinsson, J., Mogren, O.: Generative modelling of semantic segmentation data in the fashion domain. In: The IEEE International Conference on Computer Vision (ICCV) Workshops, October 2019
30. Kubo, S., Iwasawa, Y., Suzuki, M., Matsuo, Y.: UVTON: UV mapping to consider the 3D structure of a human in image-based virtual try-on network. In: The IEEE International Conference on Computer Vision (ICCV) Workshops, October 2019
31. Kınlı, F., Kıraç, F.: FashionCapsNet: clothing classification with capsule networks. J. Inf. Technol. **13**, 87–96 (2020). https://doi.org/10.17671/gazibtd.580222

32. Liang, X., Lin, L., Yang, W., Luo, P., Huang, J., Yan, S.: Clothes co-parsing via joint image segmentation and labeling with application to clothing retrieval. IEEE Trans. Multimedia **18**(6), 1175–1186 (2016)
33. Liu, G., Reda, F.A., Shih, K.J., Wang, T.-C., Tao, A., Catanzaro, B.: Image inpainting for irregular holes using partial convolutions. In: Ferrari, V., Hebert, M., Sminchisescu, C., Weiss, Y. (eds.) ECCV 2018. LNCS, vol. 11215, pp. 89–105. Springer, Cham (2018). https://doi.org/10.1007/978-3-030-01252-6_6
34. Liu, H., Jiang, B., Xiao, Y., Yang, C.: Coherent semantic attention for image inpainting. In: The IEEE International Conference on Computer Vision (ICCV), October 2019
35. Liu, L., Zhang, H., Ji, Y., Wu, Q.M.J.: Toward AI fashion design: an attribute-GAN model for clothing match. Neurocomputing **341**, 156–167 (2019)
36. Liu, Z., Luo, P., Qiu, S., Wang, X., Tang, X.: DeepFashion: powering robust clothes recognition and retrieval with rich annotations. In: Proceedings of IEEE Conference on Computer Vision and Pattern Recognition (CVPR), June 2016
37. Martinsson, J., Mogren, O.: Semantic segmentation of fashion images using feature pyramid networks. In: Proceedings of the IEEE International Conference on Computer Vision Workshops (2019)
38. Nazeri, K., Ng, E., Joseph, T., Qureshi, F., Ebrahimi, M.: EdgeConnect: structure guided image inpainting using edge prediction. In: Proceedings of the IEEE/CVF International Conference on Computer Vision (ICCV) Workshops, October 2019
39. Opitz, M., Waltner, G., Possegger, H., Bischof, H.: Bier - boosting independent embeddings robustly. In: 2017 IEEE International Conference on Computer Vision (ICCV), pp. 5199–5208, October 2017. https://doi.org/10.1109/ICCV.2017.555
40. Paszke, A., et al.: Pytorch: an imperative style, high-performance deep learning library. In: Advances in Neural Information Processing Systems, vol. 32, pp. 8024–8035. Curran Associates, Inc. (2019). http://papers.neurips.cc/paper/9015-pytorch-an-imperative-style-high-performance-deep-learning-library.pdf
41. Pathak, D., Krähenbühl, P., Donahue, J., Darrell, T., Efros, A.: Context encoders: feature learning by inpainting. In: Computer Vision and Pattern Recognition (CVPR) (2016)
42. Pérez, P., Gangnet, M., Blake, A.: Poisson image editing. ACM Trans. Graph. **22**(3), 313–318 (2003). https://doi.org/10.1145/882262.882269
43. Ronneberger, O., Fischer, P., Brox, T.: U-Net: convolutional networks for biomedical image segmentation. In: Navab, N., Hornegger, J., Wells, W.M., Frangi, A.F. (eds.) MICCAI 2015. LNCS, vol. 9351, pp. 234–241. Springer, Cham (2015). https://doi.org/10.1007/978-3-319-24574-4_28
44. Rostamzadeh, N., et al.: Fashion-Gen: the Generative Fashion Dataset and Challenge. ArXiv e-prints, June 2018
45. Salimans, T., Goodfellow, I., Zaremba, W., Cheung, V., Radford, A., Chen, X.: Improved techniques for training GANs. In: Advances in Neural Information Processing Systems, pp. 2234–2242 (2016)
46. Sbai, O., Elhoseiny, M., Bordes, A., LeCun, Y., Couprie, C.: DesIGN: design inspiration from generative networks. In: Leal-Taixé, L., Roth, S. (eds.) ECCV 2018. LNCS, vol. 11131, pp. 37–44. Springer, Cham (2019). https://doi.org/10.1007/978-3-030-11015-4_5
47. Song, Y., et al.: Contextual-based image inpainting: infer, match, and translate. In: Ferrari, V., Hebert, M., Sminchisescu, C., Weiss, Y. (eds.) ECCV 2018. LNCS, vol. 11206, pp. 3–18. Springer, Cham (2018). https://doi.org/10.1007/978-3-030-01216-8_1

48. Telea, A.: An image inpainting technique based on the fast marching method. J. Graph. GPU Game Tools **9**, 23–34 (2004)
49. Vaswani, A., et al.: Attention is all you need. In: Guyon, I., Luxburg, U.V., Bengio, S., Wallach, H., Fergus, R., Vishwanathan, S., Garnett, R. (eds.) Advances in Neural Information Processing Systems, vol. 30, pp. 5998–6008. Curran Associates, Inc. (2017). http://papers.nips.cc/paper/7181-attention-is-all-you-need.pdf
50. Wang, W., Xu, Y., Shen, J., Zhu, S.C.: Attentive fashion grammar network for fashion landmark detection and clothing category classification. In: The IEEE Conference on Computer Vision and Pattern Recognition (CVPR), June 2018
51. Wang, Z., Gu, Y., Zhang, Y., Zhou, J., Gu, X.: Clothing retrieval with visual attention model. In: 2017 IEEE Visual Communications and Image Processing (VCIP), December 2017. https://doi.org/10.1109/vcip.2017.8305144
52. Wu, Y., Kirillov, A., Massa, F., Lo, W.Y., Girshick, R.: Detectron2 (2019). https://github.com/facebookresearch/detectron2
53. Yamaguchi, K., Okatani, T., Sudo, K., Murasaki, K., Taniguchi, Y.: Mix and match: joint model for clothing and attribute recognition. In: BMVC, vol. 1, p. 4 (2015)
54. Yang, C., Lu, X., Lin, Z., Shechtman, E., Wang, O., Li, H.: High-resolution image inpainting using multi-scale neural patch synthesis. In: IEEE Conference on Computer Vision and Pattern Recognition (CVPR), pp. 4076–4084, July 2017. https://doi.org/10.1109/CVPR.2017.434
55. Yildirim, G., Jetchev, N., Vollgraf, R., Bergmann, U.: Generating high-resolution fashion model images wearing custom outfits. In: The IEEE International Conference on Computer Vision (ICCV) Workshops, October 2019
56. Yu, F., Koltun, V.: Multi-scale context aggregation by dilated convolutions. In: ICLR (2016)
57. Yu, J., Lin, Z., Yang, J., Shen, X., Lu, X., Huang, T.S.: Generative image inpainting with contextual attention. In: IEEE/CVF Conference on Computer Vision and Pattern Recognition, June 2018. https://doi.org/10.1109/cvpr.2018.00577
58. Yu, J., Lin, Z., Yang, J., Shen, X., Lu, X., Huang, T.S.: Free-form image inpainting with gated convolution. In: The IEEE International Conference on Computer Vision (ICCV), October 2019
59. Zhang, S., Song, Z., Cao, X., Zhang, H., Zhou, J.: Task-aware attention model for clothing attribute prediction. IEEE Trans. Circ. Syst. Video Technol. **30**, 1051–1064 (2019)
60. Wang, Z., Bovik, A.C., Sheikh, H.R., Simoncelli, E.P.: Image quality assessment: from error visibility to structural similarity. IEEE Trans. Image Process. **13**(4), 600–612 (2004). https://doi.org/10.1109/TIP.2003.819861
61. Zhu, S., Fidler, S., Urtasun, R., Lin, D., Loy, C.C.: Be your own Prada: fashion synthesis with structural coherence. In: 2017 IEEE International Conference on Computer Vision (ICCV), October 2017. https://doi.org/10.1109/iccv.2017.186
62. Zou, X., Kong, X., Wong, W., Wang, C., Liu, Y., Cao, Y.: FashionAI: a hierarchical dataset for fashion understanding. In: The IEEE Conference on Computer Vision and Pattern Recognition (CVPR) Workshops, June 2019

Learning to Improve Image Compression Without Changing the Standard Decoder

Yannick Strümpler(✉), Ren Yang, and Radu Timofte

ETH Zurich, Rämistrasse 101, 8092 Zurich, Switzerland
styannic@ethz.ch, {ren.yang,timofter}@vision.ee.ethz.ch

Abstract. In recent years we have witnessed an increasing interest in applying Deep Neural Networks (DNNs) to improve the rate-distortion performance in image compression. However, the existing approaches either train a post-processing DNN on the decoder side, or propose learning for image compression in an end-to-end manner. This way, the trained DNNs are required in the decoder, leading to the incompatibility to the standard image decoders (*e.g.*, JPEG) in personal computers and mobiles. Therefore, we propose learning to improve the encoding performance with the standard decoder. In this paper, We work on JPEG as an example. Specifically, a frequency-domain pre-editing method is proposed to optimize the distribution of DCT coefficients, aiming at facilitating the JPEG compression. Moreover, we propose learning the JPEG quantization table jointly with the pre-editing network. Most importantly, we do not modify the JPEG decoder and therefore our approach is applicable when viewing images with the widely used standard JPEG decoder. The experiments validate that our approach successfully improves the rate-distortion performance of JPEG in terms of various quality metrics, such as PSNR, MS-SSIM and LPIPS. Visually, this translates to better overall color retention especially when strong compression is applied.

Keywords: DNN · Image compression · Decoder compatibility

1 Introduction

The past decades have witnessed the increasing popularity of transmitting images over the Internet, while also the typical image resolution has become larger. Therefore, improving the performance of image compression is essential for the efficient transmission of images over the band-limited Internet. In recent years, there has been an increasing interest in employing Deep Neural Networks (DNNs) into image compression frameworks to improve the rate-distortion performance. Specifically, some works, *e.g.*, [12,28,36,40,42], apply DNNs to reduce the compression artifacts on the decoder side. Other works, *e.g.*, [4,5,15,19,22,32], proposed learning for image compression with end-to-end DNNs and advance the state-of-the-art performance of image compression.

A. Bartoli and A. Fusiello (Eds.): ECCV 2020 Workshops, LNCS 12538, pp. 200–216, 2020.
https://doi.org/10.1007/978-3-030-66823-5_12

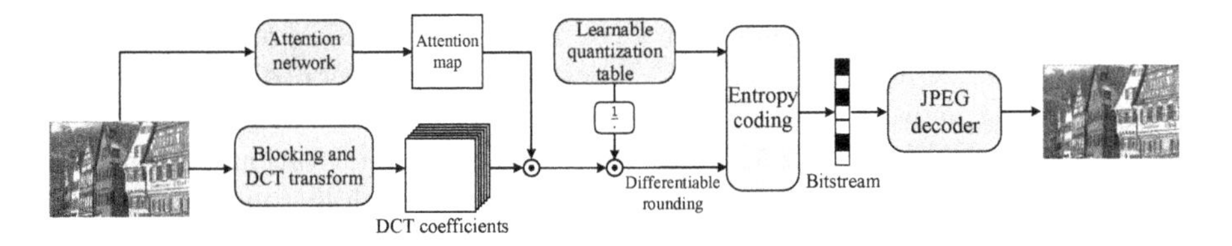

Fig. 1. The overview of the proposed approach. In this paper, ⊙ indicates element-wise multiplication.

However, each of these approaches requires a specifically trained decoder (post-processing DNNs or the DNN-based decoder), and therefore cannot be supported by the commonly used image viewers in computers and mobiles. Such incompatibility reduces the practicability of these approaches.

To overcome this shortcoming, this paper proposes adopting deep learning strategies to optimize the handcrafted image encoder without modification in the decoder side. We work on the most commonly used image compression standard JPEG [34] as an example. Our approach improves the rate-distortion performance of JPEG while ensuring that the bitstreams are decodable by the standard JPEG decoder. As such, it is compatible with all image viewers in personal computers and mobiles. To be specific, as shown in Fig. 1, we propose pre-editing the input image in the frequency domain by a learned attention map. The attention map learns to weight the DCT coefficients to facilitate the compression of the input image. Moreover, we propose learning the quantization table in the JPEG encoder. Unlike the standard JPEG that uses hand-crafted quantization tables, we propose jointly optimizing them with the attention network for rate-distortion performance. Note that, since the DCT transform is differentiable, we build a differentiable JPEG pipeline during training, and thus the proposed attention network and the learnable quantization table can be jointly trained in an end-to-end manner.

The contribution of this paper can be summarized as:

1. We propose improving the rate-distortion performance by optimizing the JPEG encoder in the frequency domain and keep the standard JPEG decoder.
2. We propose an attention network which learns to facilitate the JPEG compression by editing the DCT coefficients.
3. We propose a learnable quantization table which can be jointly optimized with the attention network towards the rate-distortion performance in an end-to-end manner.

2 Related Work

During the past a few years, plenty of works apply DNNs to improve the performance of image compression. Among them, [10,12,13,20,28,36–40] proposed post-processing DNNs to enhance the compressed image without bit-rate overhead, thus improving the rate-distortion performance. For example,

Dong *et al.*[12] proposed a four-layer DNN to reduce the compression artifacts of JPEG. Later, Guo *et al.* [13] and Wang *et al.* [36] designed the advanced post-processing network based on the prior knowledge of the JPEG algorithm. Afterwards, DnCNN [40] and Memnet [28] were proposed for various restoration tasks, including the enhancement of JPEG images. Meanwhile, some approaches aim at the post-processing of the images compressed by the HEVC [27] intra mode, *e.g.*, Dai *et al.* [10], DS-CNN-I [39] and QE-CNN-I [38]. Most recently, Xing *et al.* [37] proposed a dynamic DNN to blindly enhance the images compressed with different quality, and showed the effectiveness on both JPEG- and HEVC-compressed images.

Besides, there is more and more interest in training end-to-end DNNs for learned image compression [2,4,5,15,16,19,21–23,30,32,33]. For instance, a compressive auto-encoder is proposed in [30], which achieves comparable performance with JPEG 2000 [26]. Then, in [4] and [7], Balle *et al.* proposed jointly training the auto-encoder with the factorized and hyperpior entropy model, respectively. Meanwhile, Fabian *et al.* [22] adopted a 3D-CNN to learn the conditional probability of the elements in latent representations. Later, the hierarchical prior [23] and the context adaptive [19] entropy models were proposed, and successfully outperform the latest image compression standard BPG [8]. Most recently, the coarse-to-fine entropy model was proposed in [15] to fully explore the spatial redundancy and achieves the state-of-the-art learned image compression performance. Moreover, several approaches [16,32,33] proposed recurrently encoding the residual to compress images at various bit-rates with a single learned model.

However, all aforementioned approaches utilize the trained DNNs at the decoder side, and therefore they are incompatible with the standard image decoders which are widely used in personal computers and mobiles. This limits their applicability in practical scenarios. To overcome this shortcoming, this paper proposes improving rate-distortion performance without modifying the standard JPEG decoder. As far as we know, Talebi *et al.* [29] is the only work on pre-editing before JPEG compression, which trains a DNN in the pixel domain before the JPEG encoder to pre-edit input images. Different from [29], we propose learning to improve the JPEG encoder in the frequency domain, *i.e.*, learning an attention map to apply spatial weighting to the DCT coefficients and learning the quantization tables to optimize rate-distortion performance.

3 The Proposed Approach

3.1 The JPEG Algorithm

We first briefly introduce the JPEG algorithm. The first step in JPEG compression is to convert the input image from the RGB color space to the YCbCr colorspace. Next, the image is divided into blocks of 8×8 pixels which we index by $(n, m) \in [1, N] \times [1, M]$. Each block is then transformed through the (forward) discrete cosine transform ((F)DCT) into frequency space. We denote the DCT coefficients of block (n, m) for the luminance channel Y with $\mathrm{F}^{(Y)}[n, m] \in \mathbb{R}^{8 \times 8}$

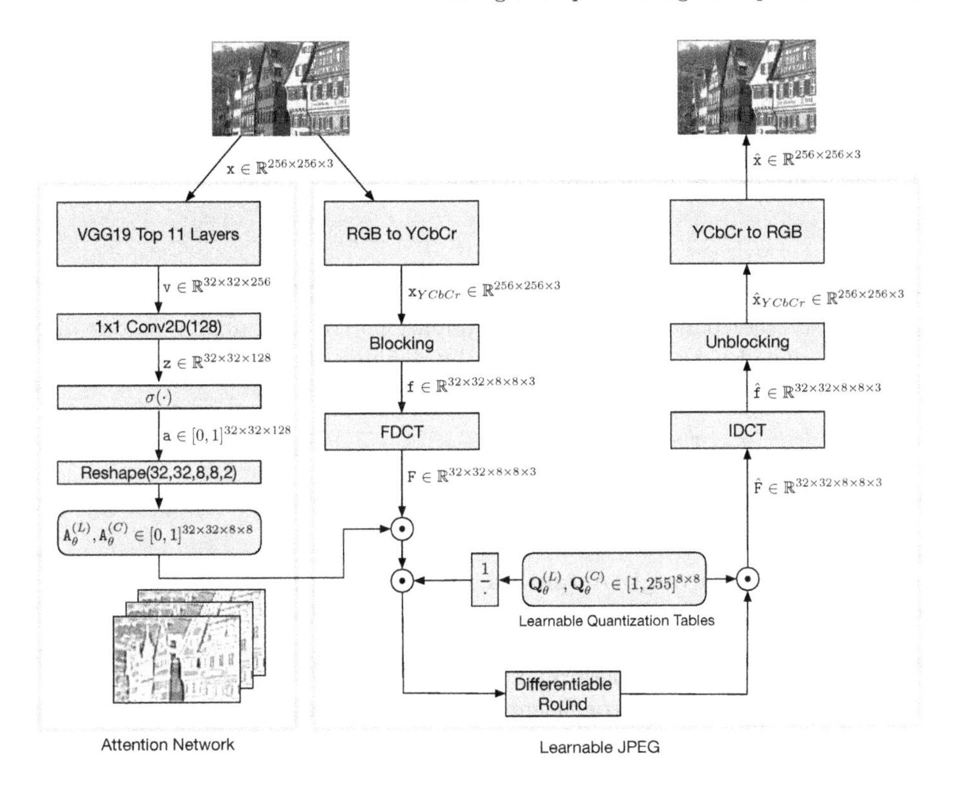

Fig. 2. Network architecture of the proposed solution

and accordingly for channels Cb, Cr. Subsequently, the DCT coefficients are quantized using two quantization tables: $\mathbf{Q}^{(L)}$ for the luminance channel Y and $\mathbf{Q}^{(C)}$ for the chrominance channels Cb, Cr. Quantization is applied through elementwise division by the quantization table followed by the rounding function, i.e. for a block in the Y channel:

$$\hat{Z}_{u,v}^{(Y)} = \left\lfloor \frac{F_{u,v}^{(Y)}}{Q_{u,v}^{(L)}} \right\rceil \text{ for } u, v \in [1, 8] \tag{1}$$

Finally, the quantized DCT coefficients are encoded by lossless entropy coding resulting in the compressed image file that also stores the quantization tables.

3.2 Proposed Network Architecture

In Fig. 2, we show a graphical representation summarizing the input-output flow of our proposed architecture. As Fig. 2 shows, we propose an attention network to pre-edit the input image in the frequency domain by weighting the DCT coefficients, and propose learning the quantization table jointly with the attention network. To optimize our DNN-based approach on the JPEG encoder, we adopt the differentiable JPEG encoder-decoder pipeline as introduced by Shin *et al.* [24]. In particular this pipeline leaves out the non-differentiable entropy coding and decoding because it is lossless and does not impact the reconstruction loss. Additionally, the rounding operation in the quantization step is replaced by a differentiable 3rd order approximation:

$$\lfloor x \rceil_{approx} = \lfloor x \rceil + (\lfloor x \rceil - x)^3. \tag{2}$$

In our implementation we use this approximation to backpropagate the gradient and use true rounding in forward evaluation. We can summarize our differentiable encoder-decoder architecture by defining the encoder and decoder functions as:

$$\hat{\mathtt{Z}} = E_\theta(\mathtt{x}), \quad \hat{\mathtt{x}} = D_\theta(\hat{\mathtt{Z}}) \implies \quad \hat{\mathtt{x}} = D_\theta(E_\theta(\mathtt{x})). \tag{3}$$

where $\mathtt{x}$ is the RGB input image, $\hat{\mathtt{x}}$ is the reconstructed RGB output image and $\hat{\mathtt{Z}}$ are the quantized DCT coefficients.

Image Editing Through Attention. We propose a novel approach to pre-editing the image before quantization to improve the compression quality. [29] has shown that an image smoothing network before compressing the image improves the compression performance. We also employ a smoothing mechanism that acts on the DCT coefficients directly. We use a parallel branch in the architecture shown in Fig. 2 that extracts image features from the original image using a pre-trained version of VGG-19, which is a 19 layer variant of the VGG network [25]. In particular we use the output of the third 2×2 max-pooling layer that has 256 channels. Note that after three max-pooling operations the spatial resolution of the image is reduced by a factor of 8. Recalling that each 8×8 block after the FDCT is represented by 64 DCT coefficients, we can just interpret the DCT coefficients as subchannels to each Y, Cr, Cb channel and get a feature map with the same spatial dimensions as the third layer VGG feature map. We then use a 1×1 convolutional layer to reduce the channel dimension to 128. By using a sigmoid activation we limit the outputs to the range $[0, 1] \subset \mathbb{R}$. Now we reshape the 128 output channels to a $8 \times 8 \times 2$ tensor for each block. We split this tensor into 2 giving us the attention tensors for luminance and chrominance for all $N \cdot M$ blocks:

$$\begin{aligned} \mathtt{A}^{(L)} &\in \{x \in \mathbb{R} \mid 0 \leq x \leq 1\}^{N \times M \times 8 \times 8}, \\ \mathtt{A}^{(C)} &\in \{x \in \mathbb{R} \mid 0 \leq x \leq 1\}^{N \times M \times 8 \times 8}. \end{aligned} \tag{4}$$

We now multiply each DCT coefficient by its importance score in Eq. 5 before we apply the learnable quantization table.

Learnable Quantization Table. In the following, we use the differentiable JPEG pipeline to learn the quantization tables by introducing $\mathbf{Q}_\theta^{(L)}$ and $\mathbf{Q}_\theta^{(C)}$ as optimization variables. We use the subscript θ to indicate that this quantity is learned. We limit the value range of the quantization tables to $[1, 255]$ by specifying a clipping function that adjusts the range after an optimization step.

Using the Hadamard product $\odot$ for ease of notation, the proposed approach can be summarized as:

$$\begin{aligned}
\hat{\mathtt{Z}}^{(Y)}[n,m] &= \left\lfloor \mathtt{F}^{(Y)}[n,m] \odot \mathtt{A}^{(L)}[n,m] \odot \bar{\mathbf{Q}}_{\theta}^{(L)} \right\rceil_{approx} \\
\hat{\mathtt{Z}}^{(Cr)}[n,m] &= \left\lfloor \mathtt{F}^{(Cr)}[n,m] \odot \mathtt{A}^{(C)}[n,m] \odot \bar{\mathbf{Q}}_{\theta}^{(C)} \right\rceil_{approx} \\
\hat{\mathtt{Z}}^{(Cb)}[n,m] &= \left\lfloor \mathtt{F}^{(Cb)}[n,m] \odot \mathtt{A}^{(C)}[n,m] \odot \bar{\mathbf{Q}}_{\theta}^{(C)} \right\rceil_{approx} \\
&\text{for } n \in [1,N], m \in [1,M], \\
\text{with } \bar{Q}_{u,v}^{(L)} &= \frac{1}{Q_{u,v}^{(L)}}, \quad \bar{Q}_{u,v}^{(C)} = \frac{1}{Q_{u,v}^{(C)}}, \quad \text{for } u,v \in [1,8].
\end{aligned} \tag{5}$$

It is important to understand that this modification is not recoverable in the decoder. Multiplying the DCT coefficients by a number smaller or equal to 1 acts like a frequency filter. Typically, higher frequencies get suppressed more so we get a low-pass filter. By distributing the attention weights across the spatial dimension and the DCT-coefficient dimension we get a smoothing filter that is adaptive spatially and across different frequencies. The big advantage of using such an attention mechanism instead of a feedforward smoothing network is that we can control the flow of information by limiting the norm of the attention maps. Note that, to get the final JPEG image for evaluation, we perform the entropy coding step with the default Huffman tables used for JPEG.

3.3 Evaluation Metrics

A widely used metric for measuring image similarity is the Mean Squared Error (MSE) that can be converted to the Peak Signal to Noise Ratio (PSNR) on a logarithmic scale. Similarly to [9] we define the MSE and PSNR for the tensors $\mathtt{x}, \hat{\mathtt{x}} \in \mathcal{X}$ of arbitrary dimension:

$$\begin{aligned}
\text{MSE}(\mathtt{x}, \hat{\mathtt{x}}) &= \frac{1}{|\mathcal{P}|} \sum_{\mathbf{p} \in \mathcal{P}} (x_{\mathbf{p}} - \hat{x}_{\mathbf{p}})^2 \\
\text{PSNR}(\mathtt{x}, \hat{\mathtt{x}}) &= 10 \log_{10} \left(\frac{255^2}{\text{MSE}(\mathtt{x}, \hat{\mathtt{x}})} \right)
\end{aligned} \tag{6}$$

where $\mathcal{P}$ is the set of pixel indices and $x_{\mathbf{p}}, \hat{x}_{\mathbf{p}} \in [0, 255], \quad \forall \mathbf{p} \in \mathcal{P}$.

To better represent local statistics we also use the Multi-Scale Structural Similarity (MS-SSIM) [35] with its default implementation *tf.image.ssim_multiscale* in Tensorflow. With deep neural networks dominating computer vision tasks, Zhang *et al.* [41] have developed the Learned Perceptual Image Patch Similarity (LPIPS) metric that leverages the power of deep features to judge image similarity. We use the version of this metric that is based on AlexNet [18] features.

For measuring the strength of the compression we define the bit rate in bits per pixel as:

$$\mathrm{BPP}\left[\frac{\mathrm{bit}}{\mathrm{px}}\right] = \frac{\text{file size [bit]}}{\text{total number of pixels [px]}}. \tag{7}$$

3.4 Formulation of the Loss

Optimizing a compression task has to tackle the fundamental balance between quality and file size that we introduced as the rate-distortion tradeoff. For the RGB input image $\mathtt{x}$ and the reconstructed RGB image $\hat{\mathtt{x}}$, given the learned parameters θ, our loss function has the general form:

$$\begin{aligned} \mathcal{L}(\mathtt{x}, \hat{\mathtt{x}}; \theta) &= \lambda \cdot \mathrm{d}(\mathtt{x}, \hat{\mathtt{x}}) + \mathrm{r}(\mathtt{x}, \hat{\mathtt{x}}; \theta), \\ \text{with } \mathtt{x}, \hat{\mathtt{x}} &\in \{t \in \mathbb{R} \mid 0 \le t \le 255\}^{8N \times 8M \times 3}, \quad \lambda \in \mathbb{R}^+, \end{aligned} \tag{8}$$

where we call $\mathrm{r}(\mathtt{x}, \hat{\mathtt{x}}; \theta)$ the rate loss and $\mathrm{d}(\mathtt{x}, \hat{\mathtt{x}})$ the distortion loss. The parameter λ determines the ratio of the two major loss terms and hence balances distortion and rate. Usually the network is trained for several values of λ to obtain a rate-distortion curve.

Distortion Loss. The distortion loss measures how close the reconstructed image is to the original image. We use the combination of the fidelity loss MSE Eq. 11 and the perceptual loss LPIPS:

$$\mathrm{d}(\mathtt{x}, \hat{\mathtt{x}}) = \mathrm{MSE}(\mathtt{x}, \hat{\mathtt{x}}) + \gamma \cdot \mathrm{LPIPS}(\mathtt{x}, \hat{\mathtt{x}}) \tag{9}$$

where we introduce γ as the LPIPS loss weight.

Rate Loss. The final bit per pixel rate that determines the file size of the JPEG file is proportional to the discrete entropy of the quantized DCT coefficients, given that the consecutive Huffman encoding is optimal. Because of the previously mentioned differentiability issue, we cannot directly add the entropy as a loss term to be minimized. Instead, [29] uses the differentiable entropy estimator proposed in [6] that relies on a continuous density estimator explained in the appendix of [7]. Using this estimator to optimize our learned JPEG architecture did not result in improved performance over the standard JPEG. We hence propose a novel formulation of the rate loss that is entirely regularization based:

$$\mathrm{r}(\mathtt{x}; \theta) = \alpha(\|\bar{\mathbf{Q}}_\theta^{(L)}\|_1 + \|\bar{\mathbf{Q}}_\theta^{(C)}\|_1) + \beta(\mathrm{mean}(\mathtt{A}_\theta^{(L)}(\mathtt{x})) + \mathrm{mean}(\mathtt{A}_\theta^{(C)}(\mathtt{x}))) \tag{10}$$

with the mean function:

$$\mathrm{mean}(\mathtt{A}) = \frac{1}{|\mathcal{P}|} \sum_{\mathbf{p} \in \mathcal{P}} A_{\mathbf{p}}, \tag{11}$$

where $\mathcal{P}$ is the index set over all entries in the tensor $\mathtt{A}$. The intuition behind this is that large quantization values reduce the entropy monotonically. It is important to note that these loss terms are only related to the reduction in entropy and not the entropy itself. For optimization with gradient descent, a constant offset of the true entropy does not affect the gradient as it vanishes in the derivative. We denote the attention maps as functions of $\mathbf{x}$ here to point out that they are input dependent, the quantization tables on the other hand are not a function of the input.

Based on Eq. 3 and the loss definitions above we can formulate the optimization objective as:

$$\min_{\theta} \mathcal{L}(\mathbf{x}, D_{\theta}(E_{\theta}(\mathbf{x})); \theta). \tag{12}$$

4 Experiments

4.1 Datasets

The learned JPEG network is trained on the dataset provided by Hasinoff *et al.* [14]. It consists of 3640 HDR image bursts with a resolution of 12 megapixels. For training we use the merged HDR images provided in the dataset. We extract image patches of size 256 that are obtained from randomly cropping. We evaluate our model on the Kodak dataset [1], consisting of 24 uncompressed images of size 768×512. Additionally, we evaluate on the DIV2K [3,31] validation set that contains 100 high quality images with 2040 pixels along the long edge.

4.2 Training Procedure

The Tensorflow implementation of our model is trained using gradient descent with the Adam [17] optimizer. The learning rate was set to 10^{-6}. We use a batch size of 8 throughout the experiments. During training, the rate distortion parameter λ is varied in the interval $[10^{-4}, 10^{-1}]$. For any choice of λ, the network is trained for 20000 steps. We used GPU based training on the NVIDIA TITAN X as well as the NVIDIA GTX 1080Ti for training in most experiments. For the single image training we used the NVIDIA TESLA P100.

The VGG layers of the attention network are initialized using weights from pretraining on ImageNet [11] and refined during training. The 1×1 convolutional layer is initialized with the Glorot uniform initializer. The quantization tables optimization variables are initialized uniformly in the interval $[1s, 2s]$ and are limited to the range $[1s, 255s]$ where $s > 0$ is a scaling factor, which we set to $s = 10^{-5}$. The final quantization tables are calculated by multiplying with s^{-1}. This choice was made because the quantization tables are trained jointly with the neural network weights which generally have a much smaller magnitude. The introduced scaling factor makes up for that difference in magnitude and allows all parameters to be updated with the same optimizer and learning rate. Alternatively, one could use a separate optimizer with an adjusted learning rate to only train the quantization tables.

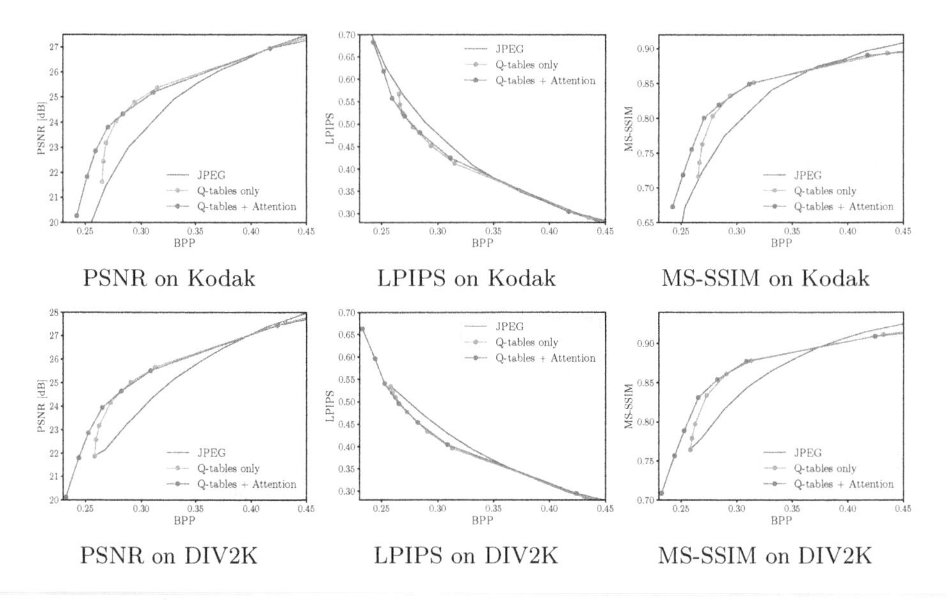

Fig. 3. Metrics comparison for learned quantization tables (Q-tables only) against learned quantization tables together with attention (Q-tables + Attention) and the JPEG baseline.

4.3 Comparison of the Proposed Solution to JPEG

We use the Python package Pillow to interface the standard JPEG encoder libjpeg We compute the JPEG baseline by compressing the test images for quality factors in $[1, 90]$. We do not use any chroma subsampling and use default Huffman tables for entropy coding.

It is important to note that the plots used for comparison are averaged as follows: The model trained for a certain rate-distortion parameter λ is evaluated on the whole test set. We average the bpp as well as the results for each metric over the test set. This gives us one data point per λ which we present in the plot. In Fig. 3, we show the performance of learning the quantization tables with and without the attention based editing. We set the hyperparameters to $\alpha = 10$, $\beta = 1$, $\gamma = 500$ which achieve the best results in our experiments.

Generally, we see a clear performance increase over the JPEG baseline for both configurations at bpp <0.4. The advantage in PSNR when using the attention network shows in the lowest bit per pixel range where the smoothing allows to further reduce the file size. In Fig. 4, we can see a side-by-side comparison of multiple images compressed with either the standard JPEG encoder, our proposed encoder with trained quantization tables or our proposed encoder with trained quantization tables and attention based editing. The most notable difference is that JPEG performs clearly worse in terms of color accuracy. The addition of the attention network helps retain better color, especially at lower bpp with only slightly sacrificing detail. At higher bpp, *i.e.*, in the third image, the visual quality is similar as without attention but the achieved bitrate is lower.

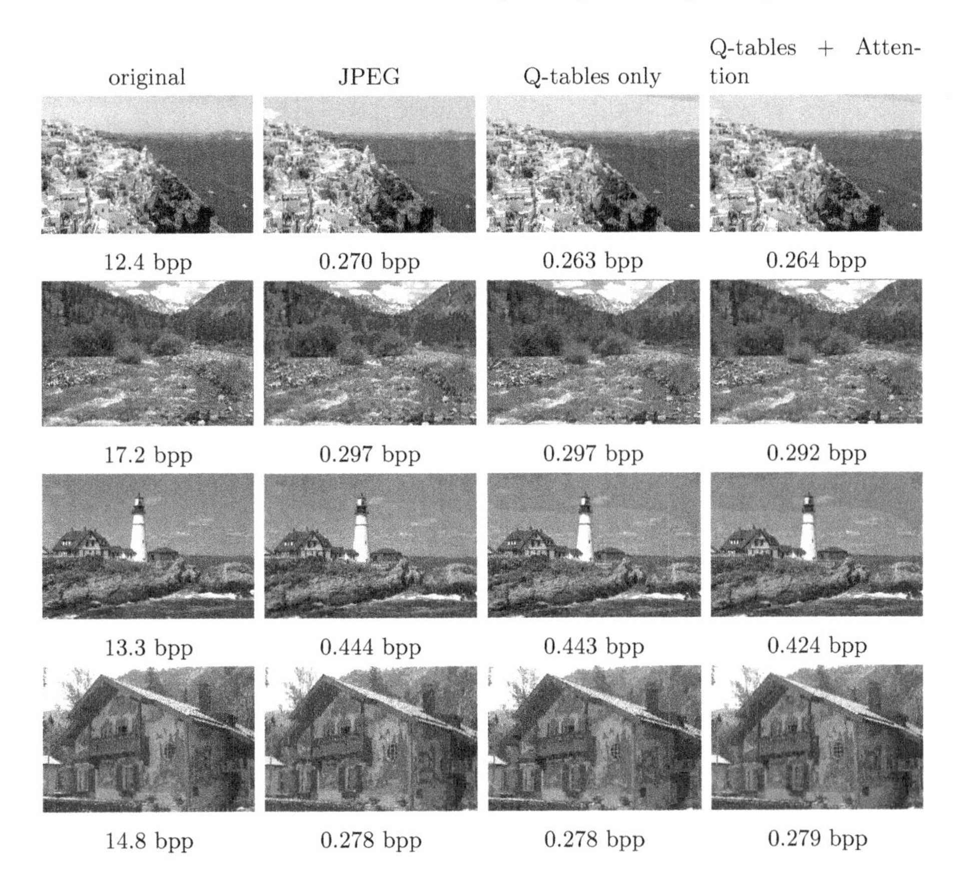

Fig. 4. Comparison of compressed image output from the JPEG baseline and our proposed solution for learning only the quantization tables (Q-tables only) or learning the quantization tables jointly with the attention map (Q-tables + Attention).

4.4 Evaluation of Our Rate Loss as a Proxy of BPP

Since we cannot use the true bits per pixel of an image in the optimization we propose an alternative rate loss in Eq. 10. For the evaluation we use a model trained with the hyperparameters $\alpha = 10$, $\beta = 1$, $\gamma = 500$. In Fig. 5, we show the rate loss components and the combination of both for the DIV2K and Kodak dataset. Except for one outlier in the attention map loss, both components show a monotonically increasing curve, hence they are suitable to be used as a proxy to the true bit rate. Combining both loss terms with the weights $\alpha = 10$, $\beta = 1$ it is evident that the final rate loss used for optimization is dominated by the quantization table loss. We choose so because the majority of the entropy reduction should be achieved through the quantization tables.

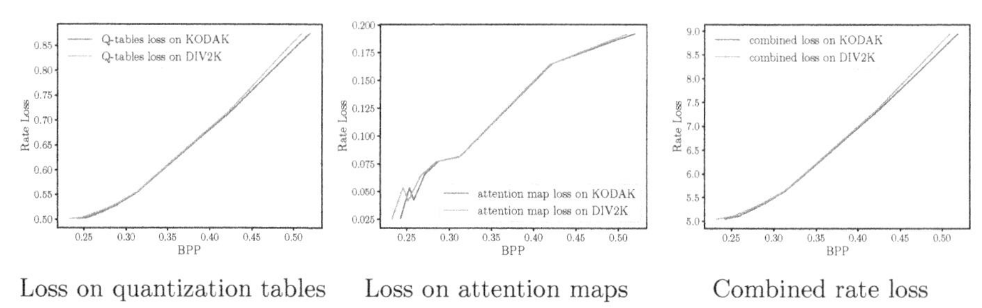

Loss on quantization tables Loss on attention maps Combined rate loss

Fig. 5. Comparing our proposed rate loss to the true bpp. The loss terms are combined with the weights $\alpha = 10$ (Q-tables), $\beta = 1$ (attention maps).

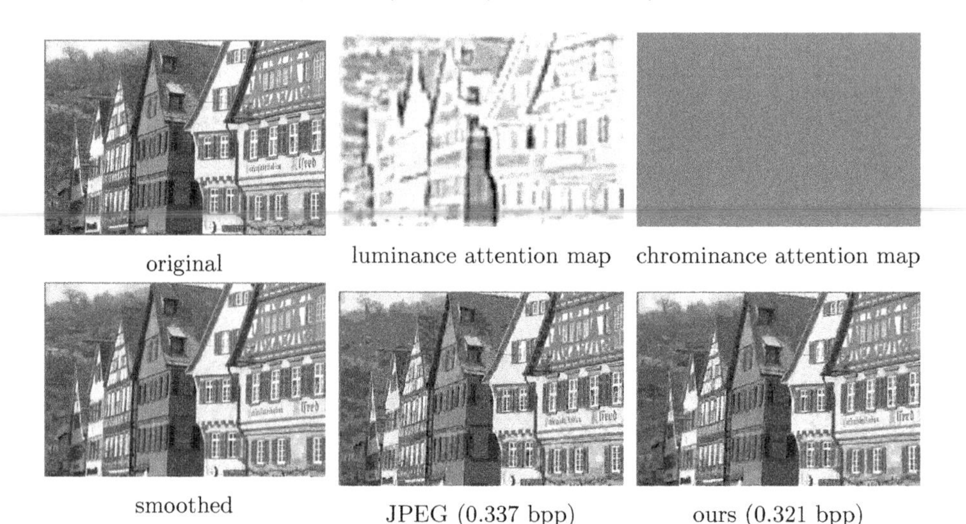

original luminance attention map chrominance attention map

smoothed JPEG (0.337 bpp) ours (0.321 bpp)

Fig. 6. Visualizing attention maps and smoothing output.

4.5 Impact of the Attention Network

In this section we want to visually show the effect of attention based editing to interpret its benefits. We use the network with attention trained with $\alpha = 10, \beta = 1, \gamma = 500$ and evaluate without rounding in the quantization step to only show the modifications through the attention maps. For visualization of the attention maps we average over all 8×8 DCT frequencies and show the spatial dimensions. For better visibility we also normalize the attention maps.

Looking at Fig. 6d shows the smoothed output of the attention based editing. We can generally see that fine details are reduced while strong edges are retained. Figure 6b shows the attention map for the luminance channel and we can see that prominent edges (*e.g.* roof line) and textures (*e.g.* text on the building) get a higher attention score. Areas in the background and planes with few contrast edges get assigned a lower importance and are smoothed more. In the chrominance attention map in Fig. 6c we do not see any spatial adaptation.

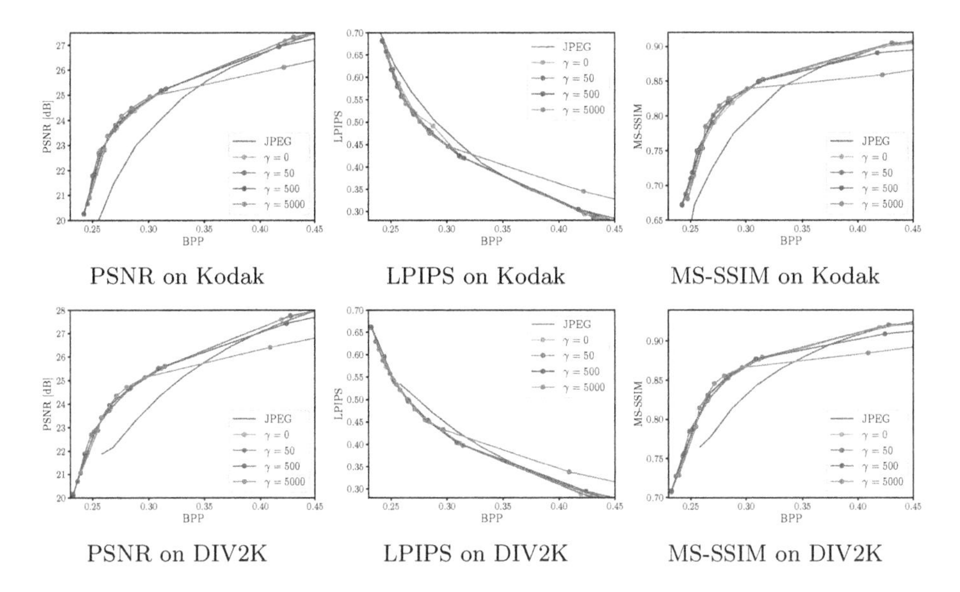

Fig. 7. Ablation study for varying the LPIPS loss weight $\gamma \in \{0, 50, 500, 5000\}$.

Comparing the final compression result of our network to the JPEG baseline shows that there is slightly less detail in the letter on the building, however overall the image looks more pleasing with significantly better color retention. The benefit of the smoothing is that the quantization table can contain smaller values while still achieving similar bit rates. Smaller quantization table values translate to more levels in the image that overall leads to better colors.

4.6 Impact of the Perceptual Loss

In this experiment we show how the addition of a perceptual loss term effects the compression performance. Setting the hyperparameter of the perceptual loss to $\gamma = 500$ leads to an approximately even loss contribution of MSE and LPIPS. Setting $\gamma = 50, 5000$ we put less or more emphasis on the LPIPS loss term. The other weight hyperparameters stay as before: $\alpha = 10$, $\beta = 1$.

When looking at Fig. 7 we do not see a big difference in performance between setting the LPIPS weight to $\gamma = 0$, $\gamma = 50$ or $\gamma = 500$. The general trend we see is that higher γ show a sweet spot between 0.25-0.3 bpp, but have slightly worse performance on the low and high end of the reported range. This is especially visible for $\gamma = 5000$. When comparing the image output in Fig. 8, we see the best detail retention and color reproduction in the wheel for $\gamma = 50$ and $\gamma = 500$. Both have clearly less color artifacts than the JPEG baseline. This is especially visible around the tires where the JPEG baseline produces purple color artifacts. For $\gamma = 5000$ we see a slightly worse performance than for a lower settings of γ. Although the image looks slightly sharper, the blockiness is also more apparent.

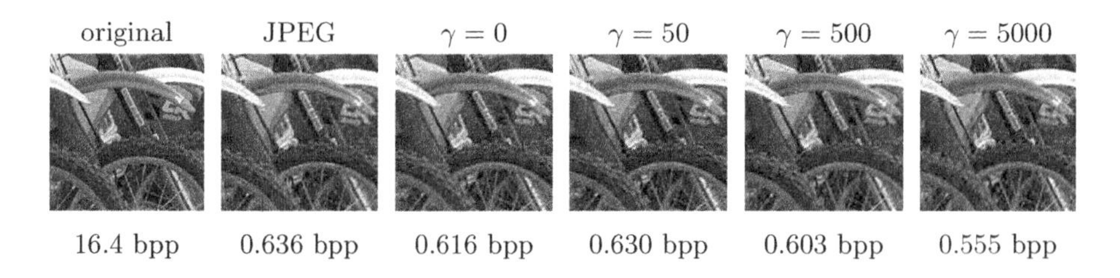

Fig. 8. Visual quality at different LPIPS weights (cropped view).

4.7 Refinement on Kodak

In a further experiment we started off with the network with attention that was already trained on the Hasinoff [14] dataset with the hyperparameters set to $\alpha = 10$, $\beta = 1$, $\gamma = 500$. We use images 13–24 from the Kodak [1] dataset to train the network for another 10000 steps, resulting in 30000 steps in total. This is done for each setting of the tradeoff parameter λ. The testing is performed on images 1–12 in the dataset.

The goal of refining the network on part of the test set, is to see how much the compression performance benefits when the distribution of train and test images is closer. In other words, large improvements suggest that the network does not generalize well from the Hasinoff [14] to the Kodak [1] dataset. Comparing the metrics shown in Fig. 9 we see almost no improvement at low bit rates. At higher bit rates, the network benefits increasingly from the refinement. This is especially visible in the MS-SSIM metric. After the refinement, the network achieves similarity metrics that are roughly equal to the JPEG baseline above 0.4 bpp, whereas previously the performance was inferior for these higher bitrates.

4.8 Single Image Optimization

In this experiment we explore the limit of the achievable compression of our network with attention by optimizing it for each individual test image. We do so by following the same training procedure as usual but since we only have a single image in the training set, we use a batch size of 1 and train directly on the full image (without random cropping as before). We keep all hyperparameters the same as for the refinement. In total we train our model for 6 values of the tradeoff parameter $\lambda \in [10^{-3}, 10^{-1}]$ for each of the 24 images in the Kodak dataset. This results in a total of 144 models that we then evaluate for the particular image that each model was trained on. As before we average the measured metrics over each setting of λ to get an overall rate-distorion curve.

Looking at the result shown in Fig. 9 shows that the networks trained on single images outperform the refined and unrefined network at bit rates larger than 0.3 bpp as to be expected. For very low bit rates all three approaches perform roughly the same. These findings support the belief that the worse performance of the normal model at higher bit rates, stems from the lack of

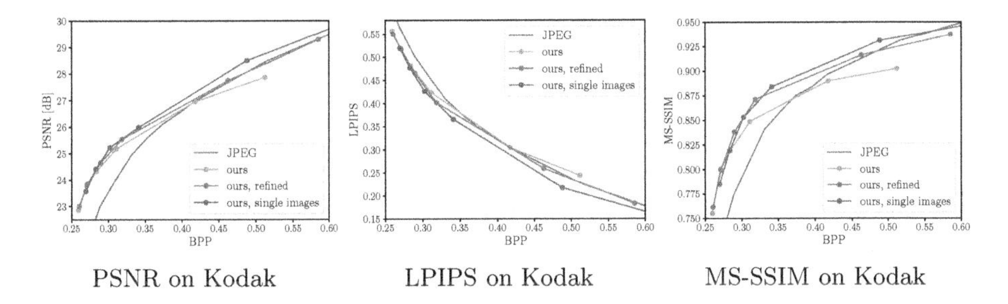

Fig. 9. Comparing the rate-distortion performance of our solution when refined on part of the Kodak test set and when trained for single images.

generalization. In fact, training and testing on single images does not require any generalization at all, explaining the higher scores.

5 Conclusion

We have shown in our experiments that our approach to improving JPEG encoding through attention based smoothing and learned quantization tables leads to better compression for low bit rates in terms of absolute deviation (MSE, PSNR) and perceptual metrics (LPIPS, MS-SSIM). Most importantly, the improved encoder remains entirely compatible with any standard JPEG decoder. Working at low bit rates, our solution still produces significant blocking artifacts that are simply inherent to the JPEG algorithm and cannot be avoided. We were however able to retain significantly better colors than the standard JPEG implementation. We have also shown that we can achieve this improvements without directly estimating the entropy of the DCT coefficients, but by regularizing the attention maps and the learned quantization tables. Our experiments on optimizing for single images have shown that individually learned quantization tables perform better at bit rates higher than 0.35. As a possible extension of our approach we suggest to additionally learn a network predicting the quantization tables for each image individually instead of learning global quantization tables that work for all images. This would make the quantization tables adaptive to each image, while the compression can still be done in a single forward evaluation.

Improving compression quality at low bit rates is relevant to all areas that require very small image files. A possible application are websites where the content should be loaded as fast as possible even when the internet connection is slow. The proposed solution can also be used to render previews that are updated with a higher quality version once they are fully loaded.

References

1. Kodak lossless true color image suite. http://r0k.us/graphics/kodak/

2. Agustsson, E., et al.: Soft-to-hard vector quantization for end-to-end learning compressible representations. In: Advances in Neural Information Processing Systems (NeurIPS), pp. 1141–1151 (2017)
3. Agustsson, E., Timofte, R.: NTIRE 2017 challenge on single image super-resolution: dataset and study, pp. 1122–1131, July 2017. https://doi.org/10.1109/CVPRW.2017.150
4. Ballé, J., Laparra, V., Simoncelli, E.P.: End-to-end optimized image compression. In: Proceedings of the International Conference on Learning Representations (ICLR) (2017)
5. Ballé, J., Minnen, D., Singh, S., Hwang, S.J., Johnston, N.: Variational image compression with a scale hyperprior. In: Proceedings of the International Conference on Learning Representations (ICLR) (2018)
6. Ballé, J., Laparra, V., Simoncelli, E.P.: End-to-end optimized image compression (2016)
7. Ballé, J., Minnen, D., Singh, S., Hwang, S.J., Johnston, N.: Variational image compression with a scale hyperprior (2018)
8. Bellard, F.: BPG image format. https://bellard.org/bpg/
9. Cavigelli, L., Hager, P., Benini, L.: CAS-CNN: a deep convolutional neural network for image compression artifact suppression. In: International Joint Conference on Neural Networks (IJCNN), May 2017. https://doi.org/10.1109/ijcnn.2017.7965927, http://dx.doi.org/10.1109/IJCNN.2017.7965927
10. Dai, Y., Liu, D., Wu, F.: A convolutional neural network approach for post-processing in HEVC intra coding. In: Amsaleg, L., Guðmundsson, G.Þ., Gurrin, C., Jónsson, B.Þ., Satoh, S. (eds.) MMM 2017. LNCS, vol. 10132, pp. 28–39. Springer, Cham (2017). https://doi.org/10.1007/978-3-319-51811-4_3
11. Deng, J., Dong, W., Socher, R., Li, L.J., Li, K., Fei-Fei, L.: ImageNet: a large-scale hierarchical image database. In: CVPR 2009 (2009)
12. Dong, C., Deng, Y., Change Loy, C., Tang, X.: Compression artifacts reduction by a deep convolutional network. In: Proceedings of the IEEE International Conference on Computer Vision, pp. 576–584 (2015)
13. Guo, J., Chao, H.: Building dual-domain representations for compression artifacts reduction. In: Leibe, B., Matas, J., Sebe, N., Welling, M. (eds.) ECCV 2016. LNCS, vol. 9905, pp. 628–644. Springer, Cham (2016). https://doi.org/10.1007/978-3-319-46448-0_38
14. Hasinoff, S., et al.: Burst photography for high dynamic range and low-light imaging on mobile cameras. In: SIGGRAPH Asia (2016). http://www.hdrplusdata.org/hdrplus.pdf
15. Hu, Y., Yang, W., Liu, J.: Coarse-to-fine hyper-prior modeling for learned image compression. In: Proceedings of the AAAI Conference on Artificial Intelligence (2020)
16. Johnston, N., et al.: Improved lossy image compression with priming and spatially adaptive bit rates for recurrent networks. In: Proceedings of the IEEE Conference on Computer Vision and Pattern Recognition (CVPR), pp. 4385–4393 (2018)
17. Kingma, D.P., Ba, J.: Adam: A method for stochastic optimization. In: Proceedings of the International Conference on Learning Representations (ICLR) (2015)
18. Krizhevsky, A., Sutskever, I., Hinton, G.E.: ImageNet classification with deep convolutional neural networks. In: Advances in Neural Information Processing Systems (2012)
19. Lee, J., Cho, S., Beack, S.K.: Context-adaptive entropy model for end-to-end optimized image compression. In: Proceedings of the International Conference on Learning Representations (ICLR) (2019)

20. Li, K., Bare, B., Yan, B.: An efficient deep convolutional neural networks model for compressed image deblocking. In: IEEE International Conference on Multimedia and Expo (ICME), pp. 1320–1325. IEEE (2017)
21. Li, M., Zuo, W., Gu, S., Zhao, D., Zhang, D.: Learning convolutional networks for content-weighted image compression. In: Proceedings of the IEEE Conference on Computer Vision and Pattern Recognition (CVPR), pp. 3214–3223 (2018)
22. Mentzer, F., Agustsson, E., Tschannen, M., Timofte, R., Van Gool, L.: Conditional probability models for deep image compression. In: Proceedings of the IEEE Conference on Computer Vision and Pattern Recognition (CVPR), pp. 4394–4402 (2018)
23. Minnen, D., Ballé, J., Toderici, G.D.: Joint autoregressive and hierarchical priors for learned image compression. In: Advances in Neural Information Processing Systems (NeurIPS), pp. 10771–10780 (2018)
24. Shin, R.: JPEG-resistant adversarial images (2017)
25. Simonyan, K., Zisserman, A.: Very deep convolutional networks for large-scale image recognition (2014)
26. Skodras, A., Christopoulos, C., Ebrahimi, T.: The JPEG 2000 still image compression standard. IEEE Signal Process. Mag. **18**(5), 36–58 (2001)
27. Sullivan, G.J., Ohm, J.R., Han, W.J., Wiegand, T.: Overview of the high efficiency video coding (HEVC) standard. IEEE Trans. Circ. Syst. Video Technol. **22**(12), 1649–1668 (2012)
28. Tai, Y., Yang, J., Liu, X., Xu, C.: MemNet: a persistent memory network for image restoration. In: Proceedings of the IEEE International Conference on Computer Vision, pp. 4539–4547 (2017)
29. Talebi, H., et al.: Better compression with deep pre-editing (2020)
30. Theis, L., Shi, W., Cunningham, A., Huszár, F.: Lossy image compression with compressive autoencoders. In: Proceedings of the International Conference on Learning Representations (ICLR) (2017)
31. Timofte, R., et al.: NTIRE 2017 challenge on single image super-resolution: methods and results. In: The IEEE Conference on Computer Vision and Pattern Recognition (CVPR) Workshops, July 2017
32. Toderici, G., et al.: Variable rate image compression with recurrent neural networks. In: Proceedings of the International Conference on Learning Representations (ICLR) (2016)
33. Toderici, G., et al.: Full resolution image compression with recurrent neural networks. In: Proceedings of the IEEE Conference on Computer Vision and Pattern Recognition (CVPR), pp. 5306–5314 (2017)
34. Wallace, G.K.: The JPEG still picture compression standard. IEEE Trans. Consum. Electron. **38**(1), xviii-xxxiv (1992)
35. Wang, Z., Simoncelli, E.P., Bovik, A.C.: Multiscale structural similarity for image quality assessment. In: The Thrity-Seventh Asilomar Conference on Signals, Systems Computers, vol. 2, pp. 1398–1402 (2003)
36. Wang, Z., Liu, D., Chang, S., Ling, Q., Yang, Y., Huang, T.S.: D3: deep dual-domain based fast restoration of JPEG-compressed images. In: Proceedings of the IEEE Conference on Computer Vision and Pattern Recognition, pp. 2764–2772 (2016)
37. Xing, Q., Xu, M., Li, T., Guan, Z.: Early exit or not: resource-efficient blind quality enhancement for compressed images. In: Vedaldi, A., Bischof, H., Brox, T., Frahm, J.-M. (eds.) ECCV 2020. LNCS, vol. 12361, pp. 275–292. Springer, Cham (2020). https://doi.org/10.1007/978-3-030-58517-4_17

38. Yang, R., Xu, M., Liu, T., Wang, Z., Guan, Z.: Enhancing quality for HEVC compressed videos. IEEE Trans. Circ. Syst. Video Technol. **29**, 2039–2054 (2018)
39. Yang, R., Xu, M., Wang, Z.: Decoder-side HEVC quality enhancement with scalable convolutional neural network. In: Proceedings of the IEEE International Conference on Multimedia and Expo (ICME), pp. 817–822. IEEE (2017)
40. Zhang, K., Zuo, W., Chen, Y., Meng, D., Zhang, L.: Beyond a gaussian denoiser: residual learning of deep CNN for image denoising. IEEE Trans. Image Process. **26**(7), 3142–3155 (2017)
41. Zhang, R., Isola, P., Efros, A.A., Shechtman, E., Wang, O.: The unreasonable effectiveness of deep features as a perceptual metric (2018)
42. Zhang, Y., Tian, Y., Kong, Y., Zhong, B., Fu, Y.: Residual dense network for image restoration. IEEE Transactions on Pattern Analysis and Machine Intelligence (2020)

Conditional Adversarial Camera Model Anonymization

Jerone T. A. Andrews(✉), Yidan Zhang, and Lewis D. Griffin

Department of Computer Science, University College London, London, UK
jerone.andrews@cs.ucl.ac.uk

Abstract. The model of camera that was used to capture a particular photographic image (model attribution) is typically inferred from high-frequency model-specific artifacts present within the image. Model anonymization is the process of transforming these artifacts such that the apparent capture model is changed. We propose a conditional adversarial approach for learning such transformations. In contrast to previous works, we cast model anonymization as the process of transforming both high and low spatial frequency information. We augment the objective with the loss from a pre-trained dual-stream model attribution classifier, which constrains the generative network to transform the full range of artifacts. Quantitative comparisons demonstrate the efficacy of our framework in a restrictive non-interactive black-box setting.

Keywords: Camera model anonymization · Conditional generative adversarial nets · Adversarial training · Non-interactive black-box attacks · Image editing/manipulation · Camera model attribution/identification

1 Introduction

Photographic images can be attributed to the specific camera model used for capture [28]. Attribution is facilitated by inferring model-specific digital acquisition and processing artifacts present within high-frequency pixel patterns [4,24,29,51,60,61]. While such artifacts have been used to verify the origin and integrity of images, attribution evidently raises concerns about unjustifiable misuse. This is particularly pertinent to individuals such as human rights' activists, photojournalists and whistle-blowers, that reserve the right to privacy and anonymity [22].

In this work, we are not concerned with attribution per se, but the challenging problem of camera model anonymization [14,40,48]. Model anonymization is the process of transforming model-specific artifacts s.t. the apparent capture model is changed. Namely, the goal is to learn a function that transforms the innate

Electronic supplementary material The online version of this chapter (https://doi.org/10.1007/978-3-030-66823-5_13) contains supplementary material, which is available to authorized users.

A. Bartoli and A. Fusiello (Eds.): ECCV 2020 Workshops, LNCS 12538, pp. 217–235, 2020.
https://doi.org/10.1007/978-3-030-66823-5_13

model-specific artifacts of an image to those of a disparate target model. Such a system could then be used to preserve privacy, or conversely for validating the robustness and reliability of attribution methods, particularly when attribution results are admitted as forensic evidence in civil or criminal cases.

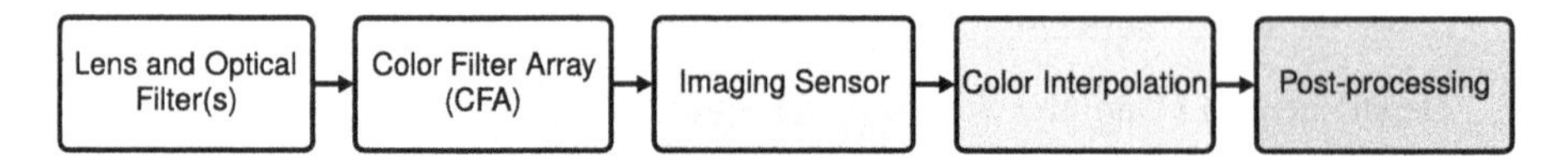

Light enters the imaging device via a system of lenses and optical filters. The CFA mosaic projects each pixel to a color: R, G or B. The sensor output is a single-channel mosaic representation. In order to obtain an RGB color image, color interpolation (a.k.a. demosaicing) estimates the missing color information in each channel based on neighboring pixels. Preceding digital storage, the image undergoes various post-processing (e.g. white balancing, color correction, gamma correction, compression).

Fig. 1. Simplified digital image acquisition pipeline

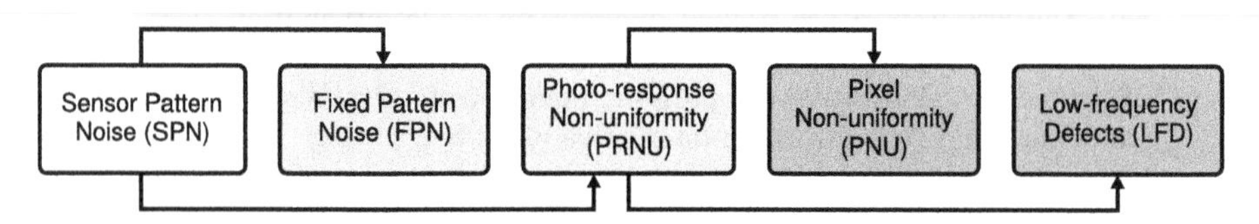

Imaging SPN is defined as any noise component that survives frame averaging [36]—i.e. systematic distortions—and primarily comprises of FPN and PRNU. FPN is defined as the pixel-to-pixel differences when the sensor is not exposed to light [47], whereas PRNU is the dominant component of SPN. PRNU is largely caused by PNU, which is defined as stochastic variations in the sensitivity of individual pixel sensors to light. Finally, LFD are artifacts that change slowly in intensity over long spatial distances (e.g. optical vignetting). LFD are a result of the lens and camera optics, as opposed to the imaging sensor.

Fig. 2. Imaging sensor pattern noise sources

Broadly, previous work on model anonymization tend to view anonymization as *solely* necessitating the attenuation [10,48] or transformation [14,15,20] of the device-specific pixel non-uniformity imaging sensor noise [25], which is defined as slight variations in the sensitivity of individual pixel sensors. Although initially device-specific (prior to color interpolation), these variations propagate nonlinearly through the processing steps (Fig. 1) that result in the final image and thus end up also depending on model-specific aspects, such as color interpolation, on-sensor signal transfer, sensor design and compression [26].

Pixel non-uniformity is the dominant noise component of what is termed photo-response non-uniformity (Fig. 2). Photo-response non-uniformity noise, however, also contains contributions from low spatial frequency artifacts (independent of the imaging sensor) caused by light refraction on dust particles, optical surfaces and properties of the camera model optics [46]. Such artifacts include optical vignetting [44], which corresponds to the fall-off in light intensity towards the corners of an image.

Model anonymization approaches based on pixel non-uniformity invariably suppress the noise-free imaging sensor response (image content), via a denoising filter, and instead work with the *noise residual* (the observed image minus its estimated noise-free image content). This is premised on improving the signal-to-noise ratio between the photo-response non-uniformity noise (signal of interest) and the observable image. However, this precludes the anonymization process from attending to discriminative low spatial frequency model-specific artifacts, since they no longer exist within the high-frequency noise residual.

Despite model anonymization loosely falling under the remit of image editing [55,62,65], the targeted anonymizing transformations that we seek should not alter the image content of an input. Minimal distortion is easily achieved by formulating the problem as a lossy reconstruction task, however low distortion is often at odds with high perceptual quality [7]. We therefore absorb the lossy reconstruction task into a simple adversarial training procedure, taking inspiration from conditional generative adversarial networks [54]. The gist of the procedure is that the generator transforms (with low distortion) an input image conditioned on a target camera model label, and tries to fool the discriminator into thinking the transformed image's prediction error features (low-level high-frequency pixel value dependency features) are real and coherent with the condition.

In contrast to previous work, we cast model anonymization as the process of transforming both high and low spatial frequency model-specific artifacts. With this in mind, our conditional adversarial camera model anonymizing (Cama) framework also includes a fixed (w.r.t. its parameters) dual-stream discriminative decision-making component (evaluator) that decides whether a transformed image belongs to its target class. Each stream captures a different aspect of the input data. Specifically, we decompose an input into its high and low spatial frequency components and assign each to its own stream. The intuition is that this allows the evaluator to *independently* reason over specific information present in each. Augmenting the adversarial objective with the evaluator's *discriminative* objective reinforces the transformation process and ensures that both high and low spatial frequency model-specific artifacts are attended to by the generator.

Quantitative results underscore the efficacy of our framework when attacking a variety of non-interactive black-box target classifiers, irrespective of whether an input image was captured by a camera model *known* to our framework.

1.1 Related Work

Camera Model Attribution. Classical approaches to camera model attribution typically construct parametric models of particular physical or algorithmic in-camera processes [6,13,18,19,24,60]. Others operate *blindly*, viewing attribution as a texture classification problem, and derive a set of heuristically designed features irrespective of their physical meaning [29,42,52,63]. Recent methods based on deep learning take a data-driven approach, obviating the need for explicit prior domain knowledge [4,5,9,45,61]. A common thread unifying most classical and contemporary approaches is image content suppression (employed as a preprocessing step) [4,24,29,51,60,61]. That is, it is *a priori* assumed that model-specific image acquisition and processing artifacts are wholly contained in the high-frequency pixel non-uniformity noise (corrupted by in-camera processes) as opposed to the low-frequency image content. Notwithstanding, it has been empirically shown [58] that this preprocessing step is not strictly necessary when the model attribution classifier is a sufficiently deep convolutional neural network (convnet), i.e. training may be performed directly in the spatial domain. Here we focus on deceiving convnet classifiers—with and without image content suppression—which can be considered state-of-the-art.

Camera Model Anonymization. As a direct consequence of the fixation of model attribution methods on model-specific artifacts contained within the pixel non-uniformity noise, model anonymization methods have mostly focused on attenuating or misaligning these high-frequency micro-patterns. Notable approaches include flat-fielding [8,31], pixel non-uniformity estimation and subtraction [10,21,40], irreversible forced seam-carving [22], image patch replacement [23] and image inpainting [48]. However, a principal issue with the aforementioned approaches is the detectable absence of model-specific artifacts within the anonymized images [11]. In contrast, we *transform* the underlying model-specific artifacts of images rather than distorting them.

Adversarial Examples. In image classification, adversarial examples refer to misclassified inputs obtained by applying imperceptible non-random perturbations [59]. Such attacks are well studied and are typically categorized based on the knowledge available to the adversary (as well as whether the attack causes an untargeted or targeted misclassification). Broadly, white-box attacks require complete knowledge (architecture and parameters) of the classifier to be attacked, whereas black-box attacks require only partial knowledge (obtained by querying the targeted classifier). Similar to other image classification tasks, recent research has shown that model attribution convnets are also extremely vulnerable to adversarial examples, particularly in white-box scenarios [15,20,34,50]. Nevertheless, in the challenging black-box setting, there are clear issues w.r.t. the transferability of adversarial examples [20,50]. In fact, the apparent lack of transferability has been echoed in other image forensic classification tasks [2,33], such as median filtering and resizing detection, which is in stark contrast to what has been observed in classical object-centric classification tasks.

Generative Adversarial Networks. Generative adversarial networks (GANs) [32] offer a viable framework for training generative models and are increasingly being used for tasks such as image generation [41], image editing [65] and representation learning [53]. Extending this framework to conditional image generation applications, conditional GANs (cGANs) [54] have been successfully applied to image-to-image translation [38,66,67] and modifying image attributes [55,64]. Most relevant to our work are the approaches of MISL [14,15] and SpoC [20], which both propose *multiple* camera model anonymizing (unconditional) GANs with access to a single fixed evaluator. That is, for each target camera model, a separate (generator, discriminator) tuple is trained, thus markedly increasing the computational cost. Moreover, MISL only considers the transformation of high-frequency artifacts contained within noise residuals. Differently, SpoC *implicitly* aims to transform image- and noise residual-based artifacts. Using a fixed pre-processor, the discriminative networks (including the evaluator) receive as input the original image concatenated channel-wise with its noise residual. The main problem with concatenating these modalities is that it does not necessarily constrain the discriminative networks to reason over the specific information present in each. Notably, neural networks are known to take *shortcuts* [27], and we posit that the discriminative networks learn to rely almost entirely on the noise residual components, as this high-frequency information is not obfuscated by the non-discriminative image content (leading to faster convergence).

2 Method

In this section, we outline the attack setting and desiderata, and explain the motivation and core components of our conditional adversarial camera model anonymizer (Cama). Refer to Appendix A for an extended motivation.

2.1 Attack Setting and Desiderata

We denote by $x \in \mathbb{R}^d$ and $y \in \mathbb{N}_c = \{1, \ldots, c\}$ an image and its ground truth (source) camera model label, respectively, sampled from a dataset p_{data}. Consider a *target* (i.e. to be attacked) convnet classifier F with c classes trained over input-output tuples $(x, y) \sim p_{\text{data}}(x, y)$. Given x, F outputs a prediction vector of class probabilities $F : x \mapsto F(x) \in [0, 1]^c$.

In this work, we operate in a *non-interactive black-box setting*: we do not assume to have knowledge of the parameters, architecture or training randomness of F, nor can we interact with it. We do, however, assume that we can sample from a dataset similar to p_{data}, which we denote by q_{data}. Precisely, we can sample tuples of the following form: $(x, y) \sim q_{\text{data}}(x, y)$ s.t. $y \in \mathbb{N}_{c'}$, where $c' \leq c$. That is, the set of possible image class labels in p_{data} is a superset of the set of possible image class labels in q_{data}, i.e. $\mathbb{N}_c \supseteq \mathbb{N}_{c'}$.

Suppose $(x, y) \sim q_{\text{data}}(x, y)$ and $y' \in \mathbb{N}_{c'}$, where $y' \neq y$ is a target label. Our aim is to learn a function $G : (x, y') \mapsto x' \approx x$ s.t. the maximum probability satisfies $\arg\max_i F(x')_i = y'$. This is known as a *targeted* attack, whereas the maximum probability of an *untargeted* attack must satisfy $\arg\max_i F(x')_i \neq y$. This work focuses on targeted attacks.

2.2 Cama: Conditional Adversarial Camera Model Anonymizer

In this framework, our model has two class conditional components: a generator G that transforms an image x conditioned on a target class label y', and a

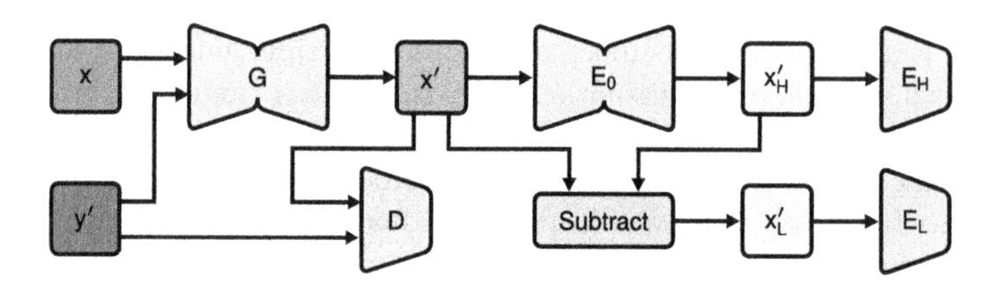

Fig. 3. Flow diagram of our method Cama given an input tuple (x, y')

discriminator D that predicts whether the low-level high-frequency pixel value dependency features of any given image conditioned on a label are real or fake. In addition, our model has a fixed (w.r.t. its parameters) dual-stream discriminative decision-making component E (evaluator) that decides whether a transformed image x' belongs to its target class y'. In essence, E serves as a *surrogate* for the non-interactive black-box F. W.r.t. E, a transformed image x' is decomposed into its high and low spatial frequency components (x'_{H} and x'_{L}, respectively), via E_0, with each assigned to a separate stream (E_{H} and E_{L}, respectively). The evaluator then reasons over the information present in x'_{H} and x'_{L} separately (via E_{H} and E_{L}, respectively). This reinforces the transformation process, as G is constrained to transform both high and low spatial frequency camera model-specific artifacts used by the evaluator for discrimination. Cama is illustrated in Fig. 3.

Our objective contains three types of terms: an *adversarial loss* for matching the distribution of transformed images to the data distribution q_{data}; a *pixel-wise loss* to incentivize the preservation of image content; and a *classification loss* to encourage G to apply transformations that result in transformed images lying in their respective target classes.

Adversarial Loss. To learn plausible conditional transformations, we apply an adversarial loss to G. For G and D, the training process alternates between G minimizing

$$L_{\mathrm{adv}} = \mathbb{E}_{\substack{x \sim q_{\mathrm{data}}(x) \\ y' \sim q_{\mathrm{data}}(y)}} \left[\left(D\left(x', y'\right) - 1 \right)^2 \right], \tag{1}$$

and D minimizing

$$\begin{aligned} L_{\mathrm{dis}} = \; & \mathbb{E}_{(x,y) \sim q_{\mathrm{data}}(x,y)} \left[\left(D\left(x, y\right) - 1 \right)^2 \right] \\ & + \frac{1}{2} \left[\mathbb{E}_{\substack{x \sim q_{\mathrm{data}}(x) \\ y' \sim q_{\mathrm{data}}(y)}} \left[D\left(x', y'\right)^2 \right] + \mathbb{E}_{\substack{x \sim q_{\mathrm{data}}(x) \\ y' \sim q_{\mathrm{data}}(y)}} \left[D\left(x, y'\right)^2 \right] \right]. \end{aligned} \tag{2}$$

Note that this is the matching-aware [56] least squares formulation [49] of the generative adversarial objective [32], which offers increased learning stability, generates higher quality results and encourages G to output images *aligned* with their target labels.

In this work, the first layer of the discriminator is a *constrained* convolutional layer [3] (originally proposed for image manipulation detection), which learns a set of prediction error filters. Each filter's central value is constrained to equal -1, whereas its remaining elements are constrained to sum to unity. In other words, the constrained layer extracts prediction error features (low-level high-frequency pixel value dependency features) by learning a normalized linear combination of the central pixel value based on its local neighborhood [4]. This concurrently serves to suppress image content.

Pixel-Wise Loss. Although D constrains G to learn credible transformations, it does not ensure that an input image's content is preserved. To incentivize this, we incorporate a simple pixel-wise L^1 norm loss between x' and x:

$$L_{\text{pix}} = \mathbb{E}_{\substack{x \sim q_{\text{data}}(x) \\ y' \sim q_{\text{data}}(y)}} \left[\|x - x'\|_1 \right]. \tag{3}$$

The addition of this loss tasks G with producing images that are close to the ground truth input image, i.e. $x' \approx x$. We prefer an L^1 norm loss over an L^2 norm loss since it has been observed to encourage less blurring [66].

Classification Loss. Ideally, x' should possess all relevant target model-specific artifacts that are the result of in-camera processes, i.e. artifacts associated with y'. However, the previously introduced adversarial and pixel-wise losses give no guarantees as to whether x' lies in class y' according to an attribution classifier such as F. In particular, as G is guided by D, the adversarial game may fixate on discovering and fixing peculiarities in x' such as abnormal interpolation patterns [1,16]. This could result in a failure to transform the *range* of salient model-specific (high and low spatial frequency) artifacts learned by a discriminative classifier, i.e. $\arg\max_i F(x')_i \neq y'$. To evaluate and reinforce the transformation process, we adopt a pre-trained fixed dual-stream camera model attribution convnet classifier E, where E is employed as a *proxy* for F. To decompose an input image x' into its high and low spatial frequency components, the evaluator is prefixed with a preprocessor (convnet) $E_0 : x \mapsto E_0(x') = x'_{\text{H}}$, where x'_{H} is the high-frequency noise residual of x'. The low spatial frequency components are computed as $x'_{\text{L}} = x' - x'_{\text{H}}$. Formally, given x' we propose to reduce the expected negative log-likelihood w.r.t. y' by minimizing

$$L_{\text{clf}} = -\frac{1}{2} \mathbb{E}_{\substack{x \sim q_{\text{data}}(x) \\ y' \sim q_{\text{data}}(y)}} \left[\log \left(E_{\text{H}}(x'_{\text{H}})_{y'} E_{\text{L}}(x'_{\text{L}})_{y'} \right) \right], \tag{4}$$

where E_{H} and E_{L} denote the high and low spatial frequency streams of E, respectively. Incorporating this loss into the generative objective encourages G to update its parameters s.t. the predicted class of x' is y' according to both streams of E.

Full Objective. The full generative objective is as follows:

$$L_{\text{gen}} = L_{\text{adv}} + \lambda_{\text{pix}} L_{\text{pix}} + \lambda_{\text{clf}} L_{\text{clf}}, \tag{5}$$

where λ_{pix} and λ_{clf} are weights that control the contribution of the three objectives. The discriminative objective is as outlined in Eq. (2).

3 Implementation

To facilitate reproducibility, we make our code publicly available.[1]

Table 1. Dataset itemization. Shown are the number of images per class within each set

y	Camera model	Set		
		q_{data}	p_{data}	p_{test}
1	Kodak M1063	760	765	100
2	Casio EX-Z150	324	335	100
3	Nikon CoolPixS710	352	330	100
4	Praktica DCZ5.9	345	358	100
5	Olympus mju-1050SW	374	379	100
6	Ricoh GX100	353	296	100
7	Rollei RCP-7325XS	-	339	100
8	Panasonic DMC-FZ50	-	677	100
9	Samsung NV15	-	396	100
10	Samsung L74wide	-	432	100
11	Fujifilm FinePixJ50	-	397	100
12	Canon Ixus70	-	333	100
Total		2508	5037	1200

Dataset. To provide a comparison on a prevalent digital image forensics benchmark, we use the Dresden image database [30] of RGB `.jpeg` images. Specifically, we use a subset of images from 12 camera models centrally cropped to a common resolution of 512×512. We partition the images into three disjoint sets (Table 1), which are disjoint w.r.t. the specific devices used to capture the images. Throughout, sets q_{data}, p_{data}, and p_{test} are used for constructing non-interactive black-box attacks, target classifier training, and evaluating attack methods, respectively.

Network Architecture. The generative network is adapted from [39] and contains two residual blocks. Following [14], prior to being fed to the network, an input image $x \in \mathbb{R}^d$ is preprocessed (using the Bayer 'RGGB' pattern) s.t. it is projected to a color filter array mosaic pattern and then back to a demosaiced RGB image. This serves to remove an input image's original demosaicing traces and forces the generator to *re-demosaic* its input w.r.t. the target camera model condition. The preprocessed image and target label condition $y' \mapsto \{0,1\}^d$ are then concatenated in the channel dimension, where the y'-th channel of a target condition is filled with ones with the remaining channels filled with zeros. The discriminative network is adapted from [38] and operates at patch-level by classifying 34×34 overlapping input patches as real or fake. The constrained convolutional layer has three 5×5 filters with stride and zero-padding equal to 1 and 2, respectively, s.t. the output of this layer retains the spatial size of its input. Similar in principle to [55], label conditions are reshaped and concatenated in the filter dimension of the output of the first *standard* convolutional layer. The evaluator's preprocessor E_0 employs the same underlying architecture as G, whereas each stream (i.e. E_{H} and E_{L}) uses a ResNet-18 architecture [35].

[1] https://github.com/jeroneandrews/cama.

Training Details. For training, we perform data augmentation by extracting non-overlapping 64×64 patches from 512×512 images $\sim q_{\text{data}}$ and using dihedral group Dih_4 transformations. We first train E_0 to approximate ground truth noise residuals obtained through wavelet-based Wiener filtering.[2] We minimize an L^2 norm loss for 90 epochs using Adam [43] with default parameters, learning rate 1e–4, weight decay 5e–4 and batch size 128. Fixing E_0, we separately train E_{H} and E_{L} to minimize a negative log-likelihood loss. We train both for 90 epochs using SGD with momentum 0.9, initial learning rate 0.1, weight decay 5e–4 and batch size 128. Fixing the modules of E, we empirically set $\lambda_{\text{pix}} = 10$ and $\lambda_{\text{clf}} = 0.01$ in Eq. (5). We optimize Cama for 200 epochs using the Adam solver with learning rate 2e–4, momentum parameter $\beta_1 = 0.5$ and batch size 32.

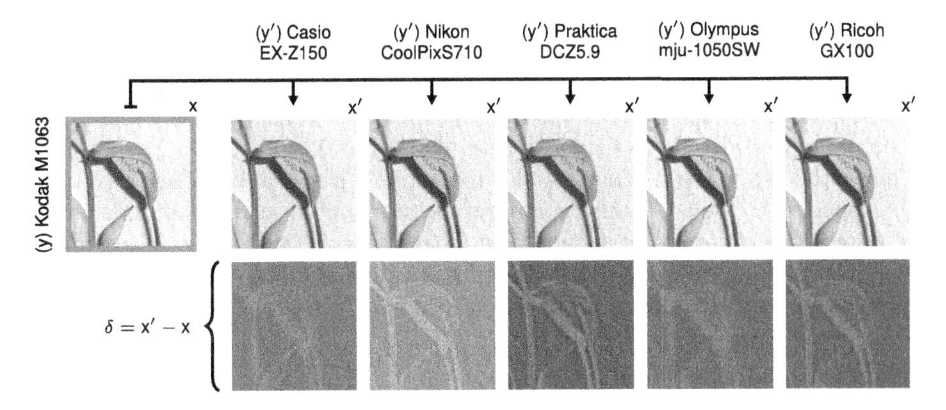

Fig. 4. Example of Cama transformed images x' with different target label conditions y' given an in-distribution input image x. The applied transformations (amplified for visualization purposes) are shown as δ

4 Experiments

In this section, we provide evidence supporting our main claims: (i) model anonymization requires the transformation of both high and low spatial frequency artifacts, and (ii) better anonymization performance can be obtained by employing an adversarial evaluator that reasons over specific information present in a transformed image's high and low spatial frequency components *separately*.

[2] We refer the reader to [25] for details on this denoising filter.

4.1 Experimental Setup

All networks used in this work are fully-convolutional, therefore we can both classify and anonymize images of arbitrary size. Evaluation is always performed on 512×512 images $\sim p_{\text{test}}$. Figure 4 illustrates Cama's ability to transform an input image using different target label conditions. The applied transformations do not alter the image content and are (largely) imperceptible. See Appendix B for additional qualitative results on both in-distribution and out-of-distribution images.

Baselines. To contextualize our approach, we compare against several baselines quantitatively. From the GAN literature (Sect. 1.1), we recast the unconditional approaches MISL [14,15] and SpoC [20] as cGANs. The evaluator of MISL has a single-stream and is prefixed with a constrained convolutional layer. The discriminator and single-stream evaluator of SpoC use a fixed preprocessor, which concatenates an RGB image and its third-order finite differences channel-wise. To ensure a fair comparison, we implement the baselines using the same architecture and details as Cama (where appropriate), and set the parameters of all methods s.t. the mean peak signal-to-noise ratio is approximately 35 dB. See Appendix C for additional comparisons to targeted attack methods from the adversarial examples literature (and mean peak signal-to-noise ratios when anonymizing images $\sim p_{\text{test}}$). We omit these from the main body due to space limitations and their poor performance in contrast to the selected baselines.

Target Classifiers. To validate the efficacy of Cama, we vary the architecture of a target camera model attribution classifier F, its prefixed preprocessor and its training data ($\sim p_{\text{data}}$). Namely, we consider ResNet-18 [35], ResNet-50 [35], DenseNet-100 [37] and VGG-16 [57] architectures. Regarding the preprocessors, image content suppression is performed by fixed hand-crafted high-pass filtering (HP) [61], third-order finite differencing (FD) [20], constrained convolution filtering (CC) [3] or wavelet-based Wiener filtering (WW) [25]. We also consider the channel-wise concatenation of an RGB image with its third-order finite differences (RGB+FD) [20]. No preprocessing is simply denoted by RGB. For training, F is trained over tuples $\sim p_{\text{data}}(x, \mathbb{N}_6)$ or $p_{\text{data}}(x, \mathbb{N}_{12})$. We refer to the former as a *complete overlap* and the latter as a *partial overlap* of known camera model classes.[3] The training details (e.g. training epochs, optimizer, data augmentation, etc.) are the same as used for our evaluator streams (described in Sect. 3). To confirm that the classifiers F can perform accurate camera model attribution, we compute their classification accuracy on relevant non-anonymized images $\sim p_{\text{test}}$. We obtain a mean accuracy of 99.5%, which varies by 1.5% from one target classifier to the next.

[3] Recall that all anonymization methods are trained on $(x, y) \sim q_{\text{data}}(x, \mathbb{N}_6)$.

Evaluation Measures. We consider the targeted success rate (TSR) for evaluating the anonymization ability of a non-interactive black-box attack, which is defined as the fraction of all possible x' that satisfy $\arg\max_i F(x')_i = y'$. For completeness, we also report the untargeted success rate (USR), i.e. the fraction of x' that satisfy $\arg\max_i F(x')_i \neq y$. We report these rates separately for in-distribution images (i.e. captured by camera models *known* to an attack framework) and out-of-distribution images (i.e. captured by camera models *unknown* to an attack framework). The best scores are always shown in boldface.

4.2 Results

Same Architecture Complete Overlap. We first analyze the success rates against ResNet-18 target classifiers F trained over $(x, y) \sim p_{\text{data}}(x, \mathbb{N}_6)$. Recall that the evaluators used during adversarial training employ the same underlying architecture. Table 2a shows the results when anonymizing in-distribution images $\sim p_{\text{test}}(x, \mathbb{N}_6)$. Examining the results based on a target classifier's preprocessor, our approach has the highest TSR and USR in 5/6 and 4/6 cases, respectively. Table 2b shows that our approach has the highest TSR in all cases when anonymizing out-of-distribution images $\sim p_{\text{test}}(x, \mathbb{N}_{12} \setminus \mathbb{N}_6)$.[4]

Table 2. TSR (USR) in the *same architecture complete overlap* setting when anonymizing (a) in-distribution images and (b) out-of-distribution images

(a)

Attack	Preprocessor						
	RGB	RGB+FD	FD	WW	CC	HP	Mean
MISL	21.8 (60.0)	42.6 (79.2)	65.3 (87.9)	49.3 (83.1)	80.1 (85.5)	52.7 (86.7)	52.0 (80.4)
SpoC	64.8 (81.0)	**94.4 (95.4)**	84.3 (92.6)	90.8 (94.0)	75.4 (78.9)	83.5 (**94.4**)	82.2 (89.4)
Cama	**91.2 (96.3)**	86.8 (87.7)	**94.2 (97.4)**	**97.3 (98.0)**	**88.1 (88.6)**	**89.7** (92.6)	**91.2 (93.4)**

(b)

Attack	Preprocessor						
	RGB	RGB+FD	FD	WW	CC	HP	Mean
MISL	29.2	46.4	73.0	56.1	86.9	56.2	58.0
SpoC	80.1	97.4	90.7	96.6	91.9	92.9	91.6
Cama	**98.5**	**97.7**	**94.2**	**98.1**	**98.8**	**97.8**	**97.5**

Same Architecture Partial Overlap. In this setting, an attack method is unaware of all possible classes of image learned by target classifiers F, which are trained over $(x, y) \sim p_{\text{data}}(x, \mathbb{N}_{12})$. This represents a slightly more realistic scenario. Table 3a shows the results when anonymizing in-distribution images. We achieve the highest TSR and USR in 6/6 and 5/6 cases, respectively. W.r.t. anonymizing out-of-distribution, Table 3b shows that Cama achieves the highest TSR and USR in all cases.

[4] Note that $\mathbb{N}_{12} \setminus \mathbb{N}_6 = \{7, \ldots, 12\}$.

Table 3. TSR (USR) in the *same architecture partial overlap* setting when anonymizing (a) in-distribution images and (b) out-of-distribution images

(a)

Attack	Preprocessor						Mean
	RGB	RGB+FD	FD	WW	CC	HP	
MISL	17.7 (70.5)	12.5 (88.4)	47.4 (90.5)	26.2 (93.4)	20.9 (90.4)	35.4 (91.6)	26.7 (87.5)
SpoC	60.9 (80.6)	79.6 (**90.9**)	83.0 (96.0)	80.4 (96.0)	49.0 (86.4)	47.6 (89.5)	66.8 (89.9)
Cama	**81.8 (91.3)**	**80.7** (85.3)	**94.9 (98.3)**	**94.4 (97.1)**	**82.0 (93.5)**	**82.5 (95.3)**	**86.0 (93.5)**

(b)

Attack	Preprocessor						Mean
	RGB	RGB+FD	FD	WW	CC	HP	
MISL	16.9 (86.6)	15.4 (81.3)	57.8 (91.4)	30.9 (85.7)	21.1 (84.0)	40.9 (92.9)	30.5 (87.0)
SpoC	70.6 (92.2)	89.9 (95.1)	88.6 (95.6)	89.1 (96.0)	61.8 (90.6)	53.7 (86.9)	75.6 (92.7)
Cama	**86.6 (95.0)**	**96.5 (98.2)**	**95.5 (98.8)**	**95.6 (99.1)**	**83.6 (95.5)**	**92.1 (97.0)**	**91.7 (97.3)**

Architecture Transfer Complete Overlap. To investigate whether the attacks transfer to other architectures—i.e. architectures that are distinct from an attack method's ResNet-18-based evaluator—we ran experiments using different target classifier architectures: ResNet-50 (R-50), DenseNet-100 (D-100) and VGG-16 (V-16). Table 4a shows the results when anonymizing in-distribution images: the success rates and trend are similar to what we observed when attacking ResNet-18 target classifiers (Table 2b). In particular, Cama has the highest TSR and USR in 18/18 and 14/18 cases, respectively. On out-of-distribution images, we attain the highest success rate in all cases (Table 4b).

Architecture Transfer Partial Overlap. This represents the most realistic and interesting setting for assessing the performance of a non-interactive black-box attack. Notably, on in-distribution images we achieve the highest TSR and USR in 18/18 and 15/18 cases (Table 5a), respectively. Moreover, as shown in Table 5b, we attain the best success rates in every case when anonymizing out-of-distribution images.

4.3 Discussion

The experimental results validate that our approach (Cama) is able to reliably perform targeted transformations. Importantly, not only can we successfully perform targeted transformations on in-distribution images captured by camera models *known* to our framework, but also on out-of-distribution images captured by camera models *unknown* to our framework. Most significantly, our attack methodology transfers across different target classifiers, i.e. as we vary a target classifier's architecture, preprocessing module and training data. Our results are non-trivial: for instance, there is no reason that the feature space of a VGG-16-based target classifier should *behave* in the same manner as an adversarial evaluator's ResNet-18. This shows that our method has good generalization ability and that the applied transformations go above and beyond mere

Table 4. TSR (USR) in the *architecture transfer complete overlap* setting when anonymizing (a) in-distribution images and (b) out-of-distribution images

(a)

	Attack	Preprocessor						
		RGB	RGB+FD	FD	WW	CC	HP	Mean
R-50	MISL	20.8 (57.2)	20.8 (76.0)	71.3 (89.6)	55.1 (86.4)	45.5 (67.8)	69.7 (**90.6**)	47.2 (77.9)
	SpoC	60.0 (75.6)	78.7 (**91.6**)	90.5 (**94.5**)	88.1 (91.6)	73.1 (81.0)	77.2 (89.3)	77.9 (87.3)
	Cama	**87.9 (92.3)**	**79.8** (83.7)	**92.0** (92.5)	**95.1 (95.9)**	**93.9 (95.9)**	**89.3** (89.9)	**89.7 (91.7)**
D-100	MISL	28.7 (59.1)	57.4 (75.4)	62.0 (86.8)	54.9 (83.6)	96.8 (**99.0**)	62.9 (88.7)	60.4 (82.1)
	SpoC	62.0 (77.0)	86.9 (91.5)	88.7 (94.2)	86.1 (92.9)	78.9 (88.5)	87.0 (95.9)	81.6 (90.0)
	Cama	**97.8 (98.7)**	**97.4 (97.8)**	**98.8 (99.3)**	**99.5 (99.8)**	**98.6** (98.9)	**96.9 (99.5)**	**98.2 (99.0)**
V-16	MISL	16.6 (83.2)	60.8 (84.3)	50.3 (83.9)	42.0 (83.7)	56.8 (77.8)	66.2 (86.2)	48.8 (83.2)
	SpoC	46.0 (80.1)	95.5 (96.8)	90.3 (95.3)	69.6 (88.0)	89.2 (93.3)	84.5 (91.3)	79.2 (90.8)
	Cama	**88.0 (95.9)**	**99.0 (99.3)**	**98.1 (98.8)**	**87.4 (96.6)**	**98.7 (99.3)**	**98.4 (98.8)**	**94.9 (98.1)**

(b)

	Attack	Preprocessor						
		RGB	RGB+FD	FD	WW	CC	HP	Mean
R-50	MISL	30.3	27.8	77.1	64.9	51.4	74.8	54.4
	SpoC	75.2	83.8	95.5	96.6	87.7	86.7	87.6
	Cama	**95.9**	**96.7**	**99.2**	**99.4**	**98.9**	**99.1**	**98.2**
D-100	MISL	34.9	73.3	67.6	61.4	97.5	66.9	66.9
	SpoC	80.9	93.6	94.2	90.9	84.4	92.5	89.4
	Cama	**99.1**	**99.6**	**99.6**	**99.7**	**99.3**	**98.6**	**99.3**
V-16	MISL	17.1	71.9	50.6	50.8	68.3	71.6	55.0
	SpoC	55.8	98.2	92.3	77.4	96.3	93.8	85.6
	Cama	**91.6**	**99.8**	**98.2**	**91.7**	**99.3**	**99.3**	**96.6**

adversarial noise. This last point is especially apparent when one considers the results attained using adversarial example methods (Appendix C).

As hypothesized, it is critical to transform both high and low spatial frequency artifacts. This can be readily seen by the poor performance of MISL, which wholly focuses on the transformation of high-frequency artifacts, and therefore cannot attend to lower-frequency model-specific artifacts used by RGB-based target classifiers (i.e. without image content suppression). Moreover, while the method of prediction-error filtering, using a constrained convolutional layer, has been successfully used for camera model attribution, it was originally proposed for image manipulation detection [3]. It is unclear how these features relate to model-specific artifacts extracted by other methods that suppress image content as a preprocessing step, i.e. other than being useful for image manipulation detection. Principally, when employed by an evaluator (as is the case for MISL), the learned generator is incapable of reliably causing targeted misclassifications when faced by target classifiers that employ a dissimilar image content suppressor and/or architecture.

During optimization, the generator of SpoC is guided by an evaluator and discriminator that concatenate an RGB input image channel-wise with its image residual. We posit that concatenating these input modalities does not effectively force the generative model to update its parameters s.t. lower-frequency artifacts contained in the RGB channels of an input are attended to. As evidenced by the results, the evaluator and discriminator of SpoC pay more attention to the

Table 5. TSR (USR) in the *architecture transfer partial overlap* setting when anonymizing (a) in-distribution images and (b) out-of-distribution images

(a)

	Attack	Preprocessor						Mean
		RGB	RGB+FD	FD	WW	CC	HP	
R-50	MISL	8.4 (81.5)	8.2 (89.5)	20.8 (91.7)	19.1 (96.2)	33.8 (83.0)	36.5 (**91.6**)	21.1 (88.9)
	SpoC	62.7 (90.3)	66.4 (**95.7**)	73.9 (95.0)	71.2 (96.6)	70.9 (94.7)	51.2 (**91.6**)	66.0 (94.0)
	Cama	**92.6 (97.6)**	**73.3** (88.2)	**83.3 (98.2)**	**91.9 (99.3)**	**81.9 (96.4)**	**75.3** (88.4)	**83.0 (94.7)**
D-100	MISL	11.5 (71.9)	31.0 (91.8)	37.6 (96.9)	20.6 (94.7)	16.2 (93.1)	24.1 (98.3)	23.5 (91.1)
	SpoC	52.6 (81.9)	89.2 (96.2)	85.9 (97.5)	79.8 (94.5)	76.3 (96.5)	60.9 (94.5)	74.1 (93.5)
	Cama	**88.3 (95.0)**	**96.7 (98.6)**	**96.3 (99.5)**	**96.9 (99.2)**	**88.9 (97.4)**	**88.9 (99.2)**	**92.7 (98.2)**
V-16	MISL	30.1 (86.5)	21.5 (96.2)	46.6 (85.8)	27.5 (97.2)	96.9 (**99.9**)	28.6 (96.4)	41.9 (93.7)
	SpoC	81.7 (93.3)	81.7 (96.3)	76.3 (92.4)	80.5 (95.1)	64.1 (90.6)	61.3 (96.2)	74.3 (94.0)
	Cama	**98.3 (99.5)**	**95.1 (98.4)**	**94.5 (99.0)**	**94.3 (99.4)**	**97.3** (98.9)	**92.9 (99.2)**	**95.4 (99.1)**

(b)

	Attack	Preprocessor						Mean
		RGB	RGB+FD	FD	WW	CC	HP	
R-50	MISL	7.6 (79.3)	9.8 (85.9)	27.1 (82.2)	22.8 (85.4)	39.2 (86.0)	42.0 (91.9)	24.8 (85.1)
	SpoC	70.7 (93.0)	77.0 (92.8)	82.8 (94.6)	80.3 (94.3)	72.5 (91.3)	56.7 (94.6)	73.3 (93.4)
	Cama	**96.1 (97.9)**	**88.4 (94.4)**	**88.7 (96.0)**	**97.5 (99.1)**	**81.3 (93.6)**	**86.8 (97.1)**	**89.8 (96.4)**
D-100	MISL	10.6 (78.4)	34.7 (85.5)	40.5 (82.9)	20.7 (85.6)	15.1 (80.0)	29.4 (86.9)	25.2 (83.2)
	SpoC	66.5 (89.7)	93.4 (97.0)	91.1 (96.5)	86.9 (96.2)	82.1 (91.3)	75.5 (94.2)	82.6 (94.1)
	Cama	**92.3 (95.5)**	**97.8 (98.5)**	**95.2 (98.3)**	**97.3 (98.8)**	**91.2 (95.6)**	**94.7 (98.0)**	**94.8 (97.4)**
V-16	MISL	33.9 (87.7)	31.4 (84.6)	49.8 (88.0)	23.9 (83.6)	98.0 (99.6)	30.6 (83.2)	44.6 (87.8)
	SpoC	77.7 (95.7)	85.8 (94.9)	79.0 (92.0)	84.9 (94.8)	74.1 (90.6)	70.0 (87.0)	78.6 (92.5)
	Cama	**98.9 (99.6)**	**97.6 (98.8)**	**93.0 (97.4)**	**95.4 (97.5)**	**96.4 (98.8)**	**95.9 (97.9)**	**96.2 (98.3)**

image residual channels. In contradistinction to SpoC, Cama *constrains* its generator using a dual-stream evaluator that independently reasons over high and low spatial frequency artifacts. This consistently improves performance, since the generator is tasked with fooling both streams such that it cannot easily take a *shortcut*—i.e. predominantly focus on the salient high-frequency information. Patently, anonymization methods that are capable of transforming model-specific artifacts of a high and low spatial frequency are better able to deceive unknown non-interactive black-box target classifiers of varying types. This is particularly useful, since we do not know *a priori* in which space a target classifier operates.

5 Conclusion

The method proposed in this paper, Cama, offers a way to preserve privacy by transforming an image's ground truth camera model-specific artifacts to those of a disparate target camera model. By formulating the learning procedure as necessitating the transformation of both high and low spatial frequency artifacts, we proposed to incorporate a fixed pre-trained dual-stream evaluator into the generative objective. The evaluator serves to reinforce the transformation process by *independently* reasoning over information present in the high and low spatial frequencies. Experimental results demonstrate that our approach (i) can successfully anonymize images captured by camera models *known* and *unknown* to our framework, and (ii) results in targeted transformations that are non-interactive black-box target classifier-agnostic (i.e. as we vary the architecture, training data and preprocessing module of a target classifier).

While the preservation of privacy is evidently beneficial to certain vulnerable individuals, anonymization could equally be open to misuse [17]. As society is at present afflicted by the deliberate dissemination of misinformation, we require more robust and reliable digital forensic methods for authenticating the origin and integrity of images. In particular, when faced by synthesized photo-realistic *deepfakes* [12], which additionally mimic the intrinsic artifacts of a target camera model (and/or device) as made possible by Cama.

Acknowledgments. JTAA is supported by the Royal Academy of Engineering (RAEng) and the Office of the Chief Science Adviser for National Security under the UK Intelligence Community Postdoctoral Fellowship Programme.

References

1. Arandjelović, R., Zisserman, A.: Object discovery with a copy-pasting GAN. arXiv preprint arXiv:1905.11369 (2019)
2. Barni, M., Kallas, K., Nowroozi, E., Tondi, B.: On the transferability of adversarial examples against CNN-based image forensics. In: ICASSP 2019 – 2019 IEEE International Conference on Acoustics, Speech and Signal Processing (ICASSP), pp. 8286–8290. IEEE (2019)
3. Bayar, B., Stamm, M.C.: A deep learning approach to universal image manipulation detection using a new convolutional layer. In: Proceedings of the 4th ACM Workshop on Information Hiding and Multimedia Security, pp. 5–10. ACM (2016)
4. Bayar, B., Stamm, M.C.: Design principles of convolutional neural networks for multimedia forensics. Electron. Imaging **2017**(7), 77–86 (2017)
5. Bayar, B., Stamm, M.C.: Towards open set camera model identification using a deep learning framework. In: 2018 IEEE International Conference on Acoustics, Speech and Signal Processing (ICASSP), pp. 2007–2011. IEEE (2018)
6. Bayram, S., Sencar, H., Memon, N., Avcibas, I.: Source camera identification based on CFA interpolation. In: IEEE International Conference on Image Processing 2005, vol. 3, p. III-69. IEEE (2005)
7. Blau, Y., Michaeli, T.: The perception-distortion tradeoff. In: Proceedings of the IEEE Conference on Computer Vision and Pattern Recognition, pp. 6228–6237 (2018)
8. Böhme, R., Kirchner, M.: Counter-forensics: attacking image forensics. In: Sencar, H., Memon, N. (eds.) Digital Image Forensics, pp. 327–366. Springer, New York (2013). https://doi.org/10.1007/978-1-4614-0757-7_12
9. Bondi, L., Baroffio, L., Güera, D., Bestagini, P., Delp, E.J., Tubaro, S.: First steps toward camera model identification with convolutional neural networks. IEEE Signal Process. Lett. **24**(3), 259–263 (2016)
10. Bonettini, N., et al.: Fooling PRNU-based detectors through convolutional neural networks. In: 2018 26th European Signal Processing Conference (EUSIPCO), pp. 957–961. IEEE (2018)
11. Bonettini, N., Güera, D., Bondi, L., Bestagini, P., Delp, E.J., Tubaro, S.: Image anonymization detection with deep handcrafted features. In: 2019 IEEE International Conference on Image Processing (ICIP), pp. 2304–2308. IEEE (2019)

12. Caldwell, M., Andrews, J.T.A., Tanay, T., Griffin, L.D.: AI-enabled future crime. Crime Sci. **9**(1), 1–13 (2020). https://doi.org/10.1186/s40163-020-00123-8
13. Chen, C., Stamm, M.C.: Camera model identification framework using an ensemble of demosaicing features. In: 2015 IEEE International Workshop on Information Forensics and Security (WIFS), pp. 1–6. IEEE (2015)
14. Chen, C., Zhao, X., Stamm, M.C.: MISLGAN: an anti-forensic camera model falsification framework using a generative adversarial network. In: 2018 25th IEEE International Conference on Image Processing (ICIP), pp. 535–539. IEEE (2018)
15. Chen, C., Zhao, X., Stamm, M.C.: Generative adversarial attacks against deep-learning-based camera model identification. IEEE Trans. Inf. Forensics Secur. (2019)
16. Chen, T., Zhai, X., Ritter, M., Lucic, M., Houlsby, N.: Self-supervised GANs via auxiliary rotation loss. In: Proceedings of the IEEE Conference on Computer Vision and Pattern Recognition, pp. 12154–12163 (2019)
17. Chesney, B., Citron, D.: Deep fakes: a looming challenge for privacy, democracy, and national security. Calif. L. Rev. **107**, 1753 (2019)
18. Choi, K.S., Lam, E.Y., Wong, K.K.: Source camera identification by JPEG compression statistics for image forensics. In: TENCON 2006 – 2006 IEEE Region 10 Conference, pp. 1–4. IEEE (2006)
19. Choi, K.S., Lam, E.Y., Wong, K.K.: Source camera identification using footprints from lens aberration. In: Digital Photography II, vol. 6069, pp. 172–179 (2006)
20. Cozzolino, D., Thies, J., Rössler, A., Nießner, M., Verdoliva, L.: SpoC: spoofing camera fingerprints. arXiv preprint arXiv:1911.12069 (2019)
21. Dirik, A.E., Karaküçük, A.: Forensic use of photo response non-uniformity of imaging sensors and a counter method. Opt. Express **22**(1), 470–482 (2014)
22. Dirik, A.E., Sencar, H.T., Memon, N.: Analysis of seam-carving-based anonymization of images against PRNU noise pattern-based source attribution. IEEE Trans. Inf. Forensics Secur. **9**(12), 2277–2290 (2014)
23. Entrieri, J., Kirchner, M.: Patch-based desynchronization of digital camera sensor fingerprints. Electron. Imaging **2016**(8), 1–9 (2016)
24. Filler, T., Fridrich, J., Goljan, M.: Using sensor pattern noise for camera model identification. In: 2008 15th IEEE International Conference on Image Processing, pp. 1296–1299. IEEE (2008)
25. Fridrich, J.: Digital image forensics. IEEE Signal Process. Mag. **26**(2), 26–37 (2009)
26. Fridrich, J.: Sensor defects in digital image forensic. In: Sencar, H., Memon, N. (eds.) Digital Image Forensics, pp. 179–218. Springer, New York (2013). https://doi.org/10.1007/978-1-4614-0757-7_6
27. Geirhos, R., et al.: Shortcut learning in deep neural networks. arXiv preprint arXiv:2004.07780 (2020)
28. Geradts, Z.J., Bijhold, J., Kieft, M., Kurosawa, K., Kuroki, K., Saitoh, N.: Methods for identification of images acquired with digital cameras. In: Enabling Technologies for Law Enforcement and Security, vol. 4232, pp. 505–512. International Society for Optics and Photonics (2001)
29. Gloe, T.: Feature-based forensic camera model identification. In: Shi, Y.Q., Katzenbeisser, S. (eds.) Transactions on Data Hiding and Multimedia Security VIII. LNCS, vol. 7228, pp. 42–62. Springer, Heidelberg (2012). https://doi.org/10.1007/978-3-642-31971-6_3
30. Gloe, T., Böhme, R.: The 'Dresden Image Database' for benchmarking digital image forensics. In: Proceedings of the 2010 ACM Symposium on Applied Computing, pp. 1584–1590. ACM (2010)

31. Gloe, T., Kirchner, M., Winkler, A., Böhme, R.: Can we trust digital image forensics? In: Proceedings of the 15th ACM International Conference on Multimedia, pp. 78–86. ACM (2007)
32. Goodfellow, I., et al.: Generative adversarial nets. In: Advances in Neural Information Processing Systems, pp. 2672–2680 (2014)
33. Gragnaniello, D., Marra, F., Poggi, G., Verdoliva, L.: Analysis of adversarial attacks against CNN-based image forgery detectors. In: 2018 26th European Signal Processing Conference (EUSIPCO), pp. 967–971. IEEE (2018)
34. Güera, D., Wang, Y., Bondi, L., Bestagini, P., Tubaro, S., Delp, E.J.: A counter-forensic method for CNN-based camera model identification. In: Proceedings of the IEEE Conference on Computer Vision and Pattern Recognition Workshops, pp. 28–35 (2017)
35. He, K., Zhang, X., Ren, S., Sun, J.: Deep residual learning for image recognition. In: Proceedings of the IEEE Conference on Computer Vision and Pattern Recognition, pp. 770–778 (2016)
36. Holst, G.C.: CCD arrays, cameras, and displays, 2nd edn. Society of Photo Optical (1998)
37. Huang, G., Liu, Z., Van Der Maaten, L., Weinberger, K.Q.: Densely connected convolutional networks. In: Proceedings of the IEEE Conference on Computer Vision and Pattern Recognition, pp. 4700–4708 (2017)
38. Isola, P., Zhu, J.Y., Zhou, T., Efros, A.A.: Image-to-image translation with conditional adversarial networks. In: Proceedings of the IEEE Conference on Computer Vision and Pattern Recognition, pp. 1125–1134 (2017)
39. Johnson, J., Alahi, A., Fei-Fei, L.: Perceptual losses for real-time style transfer and super-resolution. In: Leibe, B., Matas, J., Sebe, N., Welling, M. (eds.) ECCV 2016. LNCS, vol. 9906, pp. 694–711. Springer, Cham (2016). https://doi.org/10.1007/978-3-319-46475-6_43
40. Karaküçük, A., Dirik, A.E.: Adaptive photo-response non-uniformity noise removal against image source attribution. Digit. Investig. **12**, 66–76 (2015)
41. Karras, T., Aila, T., Laine, S., Lehtinen, J.: Progressive growing of GANs for improved quality, stability, and variation. arXiv preprint arXiv:1710.10196 (2017)
42. Kharrazi, M., Sencar, H.T., Memon, N.: Blind source camera identification. In: 2004 International Conference on Image Processing, ICIP 2004, vol. 1, pp. 709–712. IEEE (2004)
43. Kingma, D.P., Ba, J.: Adam: a method for stochastic optimization. arXiv preprint arXiv:1412.6980 (2014)
44. Kirchner, M., Gloe, T.: Forensic camera model identification. In: Handbook of Digital Forensics of Multimedia Data and Devices, pp. 329–374 (2015)
45. Kuzin, A., Fattakhov, A., Kibardin, I., Iglovikov, V.I., Dautov, R.: Camera model identification using convolutional neural networks. In: 2018 IEEE International Conference on Big Data (Big Data), pp. 3107–3110. IEEE (2018)
46. Lukáš, J., Fridrich, J., Goljan, M.: Detecting digital image forgeries using sensor pattern noise. In: Security, Steganography, and Watermarking of Multimedia Contents VIII, vol. 6072, p. 60720Y. International Society for Optics and Photonics (2006)
47. Lukáš, J., Fridrich, J., Goljan, M.: Digital camera identification from sensor pattern noise. IEEE Trans. Inf. Forensics Secur. **1**(2), 205–214 (2006)
48. Mandelli, S., Bondi, L., Lameri, S., Lipari, V., Bestagini, P., Tubaro, S.: Inpainting-based camera anonymization. In: 2017 IEEE International Conference on Image Processing (ICIP), pp. 1522–1526. IEEE (2017)

49. Mao, X., Li, Q., Xie, H., Lau, R.Y., Wang, Z., Paul Smolley, S.: Least squares generative adversarial networks. In: Proceedings of the IEEE International Conference on Computer Vision, pp. 2794–2802 (2017)
50. Marra, F., Gragnaniello, D., Verdoliva, L.: On the vulnerability of deep learning to adversarial attacks for camera model identification. Sig. Process. Image Commun. **65**, 240–248 (2018)
51. Marra, F., Poggi, G., Sansone, C., Verdoliva, L.: Evaluation of residual-based local features for camera model identification. In: Murino, V., Puppo, E., Sona, D., Cristani, M., Sansone, C. (eds.) ICIAP 2015. LNCS, vol. 9281, pp. 11–18. Springer, Cham (2015). https://doi.org/10.1007/978-3-319-23222-5_2
52. Marra, F., Poggi, G., Sansone, C., Verdoliva, L.: A study of co-occurrence based local features for camera model identification. Multimed. Tools Appl. **76**(4), 4765–4781 (2016). https://doi.org/10.1007/s11042-016-3663-0
53. Mathieu, M.F., Zhao, J.J., Zhao, J., Ramesh, A., Sprechmann, P., LeCun, Y.: Disentangling factors of variation in deep representation using adversarial training. In: Advances in Neural Information Processing Systems, pp. 5040–5048 (2016)
54. Mirza, M., Osindero, S.: Conditional generative adversarial nets. arXiv preprint arXiv:1411.1784 (2014)
55. Perarnau, G., Van De Weijer, J., Raducanu, B., Álvarez, J.M.: Invertible conditional GANs for image editing. arXiv preprint arXiv:1611.06355 (2016)
56. Reed, S., Akata, Z., Yan, X., Logeswaran, L., Schiele, B., Lee, H.: Generative adversarial text to image synthesis. arXiv preprint arXiv:1605.05396 (2016)
57. Simonyan, K., Zisserman, A.: Very deep convolutional networks for large-scale image recognition. arXiv preprint arXiv:1409.1556 (2014)
58. The IEEE Signal Processing Society: IEEE Signal Processing Cup 2018: Forensic Camera Model Identification Challenge (2018). https://www.kaggle.com/c/sp-society-camera-model-identification
59. Szegedy, C., et al.: Intriguing properties of neural networks. arXiv preprint arXiv:1312.6199 (2013)
60. Tuama, A., Comby, F., Chaumont, M.: Source camera model identification using features from contaminated sensor noise. In: Shi, Y.-Q., Kim, H.J., Pérez-González, F., Echizen, I. (eds.) IWDW 2015. LNCS, vol. 9569, pp. 83–93. Springer, Cham (2016). https://doi.org/10.1007/978-3-319-31960-5_8
61. Tuama, A., Comby, F., Chaumont, M.: Camera model identification with the use of deep convolutional neural networks. In: 2016 IEEE International Workshop on Information Forensics and Security (WIFS), pp. 1–6. IEEE (2016)
62. Wang, Z., Tang, X., Luo, W., Gao, S.: Face aging with identity-preserved conditional generative adversarial networks. In: Proceedings of the IEEE Conference on Computer Vision and Pattern Recognition, pp. 7939–7947 (2018)
63. Xu, G., Shi, Y.Q.: Camera model identification using local binary patterns. In: 2012 IEEE International Conference on Multimedia and Expo, pp. 392–397. IEEE (2012)
64. Yang, H., Huang, D., Wang, Y., Jain, A.K.: Learning face age progression: a pyramid architecture of GANs. In: Proceedings of the IEEE Conference on Computer Vision and Pattern Recognition, pp. 31–39 (2018)
65. Zhu, J.-Y., Krähenbühl, P., Shechtman, E., Efros, A.A.: Generative visual manipulation on the natural image manifold. In: Leibe, B., Matas, J., Sebe, N., Welling, M. (eds.) ECCV 2016. LNCS, vol. 9909, pp. 597–613. Springer, Cham (2016). https://doi.org/10.1007/978-3-319-46454-1_36

66. Zhu, J.Y., Park, T., Isola, P., Efros, A.A.: Unpaired image-to-image translation using cycle-consistent adversarial networks. In: Proceedings of the IEEE International Conference on Computer Vision, pp. 2223–2232 (2017)
67. Zhu, J.Y., et al.: Toward multimodal image-to-image translation. In: Advances in Neural Information Processing Systems, pp. 465–476 (2017)

Disrupting Deepfakes: Adversarial Attacks Against Conditional Image Translation Networks and Facial Manipulation Systems

Nataniel Ruiz(✉), Sarah Adel Bargal, and Stan Sclaroff

Boston University, Boston, MA, USA
{nruiz9,sbargal,sclaroff}@bu.edu

Abstract. Face modification systems using deep learning have become increasingly powerful and accessible. Given images of a person's face, such systems can generate new images of that same person under different expressions and poses. Some systems can also modify targeted attributes such as hair color or age. This type of manipulated images and video have been coined Deepfakes. In order to prevent a malicious user from generating modified images of a person without their consent we tackle the new problem of generating adversarial attacks against such image translation systems, which disrupt the resulting output image. We call this problem *disrupting deepfakes*. Most image translation architectures are generative models conditioned on an attribute (e.g. put a smile on this person's face). We are first to propose and successfully apply (1) class transferable adversarial attacks that generalize to different classes, which means that the attacker does not need to have knowledge about the conditioning class, and (2) adversarial training for generative adversarial networks (GANs) as a first step towards robust image translation networks. Finally, in our scenario, the deepfaker can adaptively blur the image and potentially mount a successful defense against disruption. We present a spread-spectrum adversarial attack, which evades blur defenses. We open-source our code.

Keywords: Adversarial attacks · Image translation · Face modification · Deepfake · Generative models · GAN · Privacy

1 Introduction

Advances in image translation using generative adversarial networks (GANs) have allowed the rise of face manipulation systems that achieve impressive realism. Some face manipulation systems can create new images of a person's face under different expressions and poses [21,30]. Other face manipulation systems modify the age, hair color, gender or other attributes of the person [6,7].

Electronic supplementary material The online version of this chapter (https://doi.org/10.1007/978-3-030-66823-5_14) contains supplementary material, which is available to authorized users.

A. Bartoli and A. Fusiello (Eds.): ECCV 2020 Workshops, LNCS 12538, pp. 236–251, 2020.
https://doi.org/10.1007/978-3-030-66823-5_14

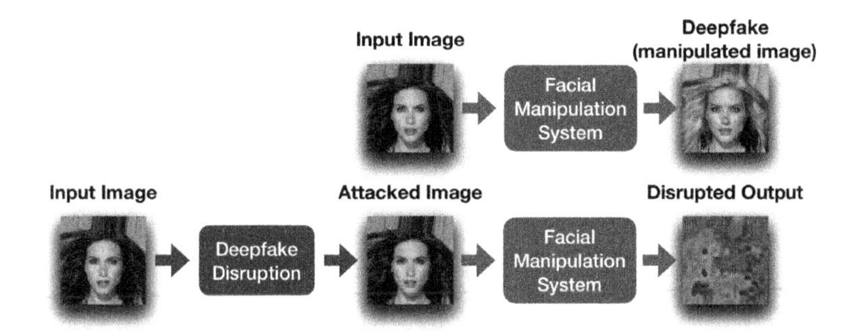

Fig. 1. An illustration of deepfake disruption. After applying an imperceptible disruption filter on the image the output of StarGAN [6] is successfully disrupted.

Given the widespread availability of these systems, malicious actors can modify images of a person without their consent. There have been occasions where faces of celebrities have been transferred to videos with explicit content without their consent [1] and companies such as Facebook have banned uploading modified pictures and video of people [2].

One way of mitigating this risk is to develop systems that can detect whether an image or video has been modified using one of these systems. There have been recent efforts in this direction, with varying levels of success [26,27].

There is work showing that deep neural networks are vulnerable to adversarial attacks [5,12,20,23], where an attacker applies imperceptible perturbations to an image causing it to be incorrectly classified. We distinguish different attack scenarios. In a *white-box* scenario the attacker has perfect knowledge of the architecture, model parameters and defenses in place. In a *black-box* scenario, the attacker is only able to query the target model for output labels for chosen inputs. There are several different definitions of *gray-box* scenarios. In this work, a gray-box scenario denotes perfect knowledge of the model and parameters, but ignorance of the pre-processing defense mechanisms in place (such as blurring). We focus on white-box and gray-box settings.

Another way of combating malicious actors is by *disrupting the deepfaker's ability to generate a deepfake*. In this work we propose a solution by adapting traditional adversarial attacks that are imperceptible to the human eye in the source image, but interfere with translation of this image using image translation networks. A successful disruption corresponds to the generated image being sufficiently deteriorated such that it has to be discarded or such that the modification is perceptually evident (Fig. 1). We present a formal and quantifiable definition of disruption success in Sect. 3.

Most facial manipulation architectures are conditioned both on the input image and on a target conditioning class. One example, is to define the target expression of the generated face using this attribute class (e.g. put a smile on the person's face). In this example, if we want to prevent a malicious actor from putting a smile on the person's face in the image, we need to know that

Fig. 2. An example of our deepfake disruptions on StarGAN [6] and GANimation [21]. Some image translation networks are more prone to disruption.

the malicious actor has selected the smile attribute instead of, for instance, eye closing. In this work, we are first to formalize the problem of disrupting class conditional image translation, and present two variants of class transferable disruptions that improve generalization to different conditioning attributes.

We explore defenses of image translation systems and are the first to propose adversarial training for GANs, where the inputs to the generator and discriminator are adversarially attacked during training. We show that this is a strong defense that alleviates disruptions in the white-box scenario.

Finally, in our proposed scenario the disruptor plays first by providing the perturbed image. The deepfaker can adaptively blur the image to deter the attack on their deepfake generation system. In this scenario the efficacy of naive disruptions drop. We present a novel spread-spectrum disruption that evades a variety of blur defenses in this gray-box setting.

In summary:

- We present baseline methods for disrupting deepfakes by adapting adversarial attack methods to image translation networks.
- We are the first to address disruptions on conditional image translation networks. We propose and evaluate novel disruption methods that transfer from one conditioning class to another.
- We are the first to propose and evaluate adversarial training for generative adversarial networks. Our novel *G+D adversarial training* alleviates disruptions in a white-box setting.
- We propose a novel spread-spectrum disruption that evades blur defenses in a gray-box scenario.

2 Related Work

There are several works exploring image translation using deep neural networks [6,7,13,21,28,30,31]. Some of these works apply image translation to face images in order to generate new images of individuals with modified expression or attributes [6,7,21,30].

There is a large amount of work that explores adversarial attacks on deep neural networks for classification [5,12,17–20,23]. Fast Gradient Sign Method (FGSM), a one-step gradient attack was proposed by Goodfellow *et al.* [12]. Stronger iterative attacks such as iterative FGSM (I-FGSM) [15] and Projected Gradient Descent (PGD) [16] have been proposed. Sabour *et al.* [22] explore feature-space attacks on deep neural network classifiers using L-BFGS.

Tabacof *et al.* [24] and Kos *et al.* [14] explore adversarial attacks against Variational Autoencoders (VAE) and VAE-GANs, where an adversarial image is compressed into a latent space and instead of being reconstructed into the original image is reconstructed into an image of a different semantic class. In contrast, our work focuses on attacks against image translation systems. Additionally, our objective is to disrupt deepfake generation as opposed to changing the output image to a different semantic class.

Chu et al. [8] show that information hiding occurs during CycleGAN training and is similar in spirit to an adversarial attack [23]. Bashkirova et al. [4] explore self-adversarial attacks in cycle-consistent image translation networks. [4] proposes two methods for defending against such attacks leading to more honest translations by attenuating the self-adversarial hidden embedding. While [4] addresses self-adversarial attacks [8], which lead to higher translation quality, our work addresses adversarial attacks [23], which seek to disrupt the image translation process.

Wang *et al.* [25] adapt adversarial attacks to the image translation scenario for traffic scenes on the pix2pixHD and CycleGAN networks. Yeh *et al.* [29], is concurrent work to ours, and proposes adapting PGD to attack pix2pixHD and CycleGAN networks in the face domain. Most face manipulation networks are conditional image translation networks, [25,29] do not address this scenario and do not explore defenses for such attacks. We are the first to explore attacks against conditional image translation GANs as well as attacks that transfer to different conditioning classes. We are also the first to propose adversarial training [16] for image translation GANs. Madry *et al.* [16] propose adversarial training using strong adversaries to alleviate adversarial attacks against deep neural network classifiers. In this work, we propose two adaptations of this technique for GANs, as a first step towards robust image translation networks.

A version of spread-spectrum watermarking for images was proposed by Cox *et al.* [10]. Athalye *et al.* [3] proposes the expectation over transformation (EoT) method for synthesizing adversarial examples robust to pre-processing transformations. However, Athalye *et al.* [3] demonstrate their method on affine transformations, noise and others, but do not consider blur. In this work, we propose a faster heuristic iterative spread-spectrum disruption for evading blur defenses.

3 Method

We describe methods for image translation disruption (Sect. 3.1), our proposed conditional image translation disruption techniques (Sect. 3.2), our proposed adversarial training techniques for GANs (Sect. 3.3) and our proposed spread-spectrum disruption (Sect. 3.4).

3.1 Image Translation Disruption

Similar to an adversarial example, we want to generate a disruption by adding a human-imperceptible perturbation $\boldsymbol{\eta}$ to the input image:

$$\tilde{\boldsymbol{x}} = \boldsymbol{x} + \boldsymbol{\eta}, \tag{1}$$

where $\tilde{\boldsymbol{x}}$ is the generated disrupted input image and $\boldsymbol{x}$ is the input image. By feeding the original image or the disrupted input image to a generator we have the mappings $\boldsymbol{G}(\boldsymbol{x}) = \boldsymbol{y}$ and $\boldsymbol{G}(\tilde{\boldsymbol{x}}) = \tilde{\boldsymbol{y}}$, respectively, where $\boldsymbol{y}$ and $\tilde{\boldsymbol{y}}$ are the translated output images and $\boldsymbol{G}$ is an image translation network (e.g. the generator of an image translation GAN, or a VAE).

We consider a disruption successful when it introduces perceptible corruptions or modifications onto the output $\tilde{\boldsymbol{y}}$ of the network leading a human observer to notice that the image has been altered and therefore distrust its source.

We operationalize this phenomenon. Adversarial attack research has focused on attacks showing low distortions using the L^0, L^2 and L^∞ distance metrics. The logic behind using attacks with low distortion is that the larger the distance, the more apparent the alteration of the image, such that an observer could detect it. In contrast, we seek to *maximize* the distortion of our output, with respect to a well-chosen reference $\boldsymbol{r}$.

$$\max_{\boldsymbol{\eta}} L(\boldsymbol{G}(\boldsymbol{x} + \boldsymbol{\eta}), \boldsymbol{r}), \quad \text{subject to } ||\boldsymbol{\eta}||_\infty \leq \epsilon, \tag{2}$$

where ϵ is the maximum magnitude of the perturbation and L is a distance function. If we pick r to be the ground-truth output, $\boldsymbol{r} = \boldsymbol{G}(\boldsymbol{x})$, we get the *ideal* disruption which aims to maximize the distortion of the output.

We can also formulate a *targeted* disruption, which pushes the output $\tilde{\boldsymbol{y}}$ to be close to $\boldsymbol{r}$:

$$\boldsymbol{\eta} = \arg\min_{\boldsymbol{\eta}} L(\boldsymbol{G}(\boldsymbol{x} + \boldsymbol{\eta}), \boldsymbol{r}), \quad \text{subject to } ||\boldsymbol{\eta}||_\infty \leq \epsilon. \tag{3}$$

Note that the ideal disruption is a special case of the targeted disruption where we minimize the negative distortion instead and select $\boldsymbol{r} = \boldsymbol{G}(\boldsymbol{x})$. We can thus disrupt an image *towards* a target or *away from* a target.

We can generate a targeted disruption by adapting well-established adversarial attacks: FGSM, I-FGSM, and PGD. Fast Gradient Sign Method (FGSM) [12] generates an attack in one forward-backward step, and is adapted as follows:

$$\boldsymbol{\eta} = \epsilon \ \text{sign}[\nabla_{\boldsymbol{x}} L(\boldsymbol{G}(\boldsymbol{x}), \boldsymbol{r})], \tag{4}$$

where ϵ is the size of the FGSM step. Iterative Fast Gradient Sign Method (I-FGSM) [15] generates a stronger adversarial attack in multiple forward-backward steps. We adapt this method for the targeted disruption scenario as follows:

$$\tilde{\boldsymbol{x}}_t = \text{clip}(\tilde{\boldsymbol{x}}_{t-1} - \boldsymbol{a} \ \text{sign}[\nabla_{\tilde{\boldsymbol{x}}} L(\boldsymbol{G}(\tilde{\boldsymbol{x}}_{t-1}), \boldsymbol{r})]), \tag{5}$$

where $\boldsymbol{a}$ is the step size and the constraint $||\tilde{\boldsymbol{x}} - \boldsymbol{x}||_\infty \leq \epsilon$ is enforced by the clip function. For disruptions *away from* the target $\boldsymbol{r}$ instead of *towards* $\boldsymbol{r}$, using the negative gradient of the loss in the equations above is sufficient. For an adapted Projected Gradient Descent (PGD) [16], we initialize the disrupted image $\tilde{\boldsymbol{x}}_0$ randomly inside the ϵ-ball around $\boldsymbol{x}$ and use the I-FGSM update function.

3.2 Conditional Image Translation Disruption

Many image translation systems are conditioned not only on the input image, but on a target class as well:

$$\boldsymbol{y} = \boldsymbol{G}(\boldsymbol{x}, \boldsymbol{c}), \tag{6}$$

where $\boldsymbol{x}$ is the input image, $\boldsymbol{c}$ is the target class and $\boldsymbol{y}$ is the output image. A target class can be an attribute of a dataset, for example blond or brown-haired.

A disruption for the data/class pair $(\boldsymbol{x}, \boldsymbol{c}_i)$ is not guaranteed to transfer to the data/class pair $(\boldsymbol{x}, \boldsymbol{c}_j)$ when $i \neq j$. We can define the problem of looking for a class transferable disruption as follows:

$$\boldsymbol{\eta} = \arg\min_{\boldsymbol{\eta}} \mathbb{E}_c[L(\boldsymbol{G}(\boldsymbol{x} + \boldsymbol{\eta}, \boldsymbol{c}), \boldsymbol{r})], \quad \text{subject to } ||\boldsymbol{\eta}||_\infty \leq \epsilon. \tag{7}$$

We can write this empirically as an optimization problem:

$$\boldsymbol{\eta} = \arg\min_{\boldsymbol{\eta}} \sum_{c} [L(\boldsymbol{G}(\boldsymbol{x} + \boldsymbol{\eta}, \boldsymbol{c}), \boldsymbol{r})], \quad \text{subject to } ||\boldsymbol{\eta}||_\infty \leq \epsilon. \tag{8}$$

Iterative Class Transferable Disruption. In order to solve this problem, we present a novel disruption on class conditional image translation systems that increases the transferability of our disruption to different classes. We perform a modified I-FGSM disruption:

$$\tilde{\boldsymbol{x}}_t = \text{clip}(\tilde{\boldsymbol{x}}_{t-1} - \boldsymbol{a}\ \text{sign}[\nabla_{\tilde{\boldsymbol{x}}} L(\boldsymbol{G}(\tilde{\boldsymbol{x}}_{t-1}, \boldsymbol{c}_k), \boldsymbol{r})]). \tag{9}$$

We initialize $k = 1$ and increment k at every iteration, until we reach $k = K$ where K is the number of classes. We then reset $k = 1$.

Joint Class Transferable Disruption. We propose a disruption which seeks to minimize the expected value of the distance to the target $\boldsymbol{r}$ at every step t. For this, we compute this loss term at every step of an I-FGSM disruption and use it to inform our update step:

$$\tilde{\boldsymbol{x}}_t = \text{clip}(\tilde{\boldsymbol{x}}_{t-1} - \boldsymbol{a}\ \text{sign}[\nabla_{\tilde{\boldsymbol{x}}} \sum_{c} L(\boldsymbol{G}(\tilde{\boldsymbol{x}}_{t-1}, \boldsymbol{c}), \boldsymbol{r})]). \tag{10}$$

3.3 GAN Adversarial Training

Adversarial training for classifier deep neural networks was proposed by Madry *et al.* [16]. It incorporates strong PGD attacks on the training data for the classifier. We propose the first adaptations of adversarial training for generative adversarial networks. Our methods, described below, are a first step in attempting to defend against image translation disruption.

Generator Adversarial Training. A conditional image translation GAN uses the following adversarial loss:

$$\mathcal{L} = \mathbb{E}_{x} \left[\log \boldsymbol{D}(\boldsymbol{x})\right] \ + \mathbb{E}_{x,c}\left[\log \left(1 - \boldsymbol{D}(\boldsymbol{G}(\boldsymbol{x}, \boldsymbol{c}))\right)\right], \tag{11}$$

where $\boldsymbol{D}$ is the discriminator. In order to make the generator resistant to adversarial examples, we train the GAN using the modified loss:

$$\mathcal{L} = \mathbb{E}_{x} \left[\log \boldsymbol{D}(\boldsymbol{x})\right] \ + \mathbb{E}_{x,c,\eta}\left[\log \left(1 - \boldsymbol{D}(\boldsymbol{G}(\boldsymbol{x} + \boldsymbol{\eta}, \boldsymbol{c}))\right)\right]. \tag{12}$$

We hypothesize that this adversarial perturbation η leads the generator to be more robust to adversarial attacks. In other words, we expect that the generator of the image-translation GAN is harder to disrupt since it has learned to produce correct image translations with images that have been attacked.

Generator+Discriminator (G+D) Adversarial Training. Instead of only training the generator to be indifferent to adversarial examples, we also train the discriminator on adversarial examples:

$$\mathcal{L} = \mathbb{E}_{x,\eta_1} \left[\log \boldsymbol{D}(\boldsymbol{x} + \boldsymbol{\eta}_1)\right] \ + \mathbb{E}_{x,c,\eta_2,\eta_3}\left[\log \left(1 - \boldsymbol{D}(\boldsymbol{G}(\boldsymbol{x} + \boldsymbol{\eta}_2, \boldsymbol{c}) + \boldsymbol{\eta}_3)\right)\right]. \tag{13}$$

Here, we add adversarial perturbations to (1) the generator input (η_2) (2) the real image fed into the discriminator (η_1) and (3) the fake image fed into the discriminator (η_3). Both the generator and the discriminator learn to be invariant to adversarial attacks and play the GAN game with adversarial examples. We expect this to yield a more robust image translation network since the discriminator is being trained using the same distribution as the generator (as opposed to G. adversarial training. where the distributions differ), and the newly robust discriminator should in turn push the generator to be more resistant to adversarial attacks.

3.4 Spread-Spectrum Evasion of Blur Defenses

In our scenario blurring can be an effective defense against disruptions because the disruptor might ignore the type or magnitude of blur being used. Adaptive attacks are hard to put in place since the disruptor plays first and the defender can vary their preprocessing according to the attack. In order to successfully disrupt a network in this scenario, we propose a spread-spectrum evasion of blur defenses that transfers to different types of blur. We perform a modified I-FGSM update

$$\tilde{\boldsymbol{x}}_t = \text{clip}(\tilde{\boldsymbol{x}}_{t-1} - \epsilon \ \text{sign}[\nabla_{\tilde{\boldsymbol{x}}} L(\boldsymbol{f}_k(\boldsymbol{G}(\tilde{\boldsymbol{x}}_{t-1})), \boldsymbol{r})]), \tag{14}$$

where $\boldsymbol{f}_k$ is a blurring convolution operation, and we have K different blurring methods with different magnitudes and types. We initialize $k = 1$ and increment k at every iteration of the algorithm, until we reach $k = K$ where K is the total number of blur types and magnitudes. We then reset $k = 1$.

4 Experiments

In this section we demonstrate that our proposed image-level FGSM, I-FGSM and PGD-based disruptions are able to disrupt different recent image translation architectures such as GANimation [21], StarGAN [6], pix2pixHD [28] and CycleGAN [31]. In Sect. 4.1, we show that the ideal formulation of an image-level disruption presented in Sect. 3.1, is the most effective at producing large distortions in the output. In Sect. 4.2, we demonstrate that both our *iterative class transferable disruption* and *joint class transferable disruption* are able to transfer to different conditioning classes. In Sect. 4.3, we test our disruptions against two defenses in a white-box setting. We show that our proposed *G+D adversarial training* is most effective at alleviating disruptions, although strong disruptions are able to overcome this defense. Finally, in Sect. 4.4 we show that blurring is an effective defense against disruptions in a gray-box setting, in which the disruptor does not know the type or magnitude of the pre-processing blur. We then demonstrate that our proposed *spread-spectrum adversarial disruption* evades different blur defenses in this scenario. All disruptions in our experiments use $L = L^2$.

Architectures and Datasets. We use the GANimation [21], StarGAN [7], pix2pixHD [28] and CycleGAN [31] image translation architectures. We use an open-source implementation of GANimation trained for 37 epochs on the CelebA dataset for 80 action units (AU) from the Facial Action Unit Coding System (FACS) [11]. We test GANimation on 50 random images from the CelebA dataset (4,000 disruptions). We use the official open-source implementation of StarGAN, trained on the CelebA dataset for the five attributes black hair, blond hair, brown hair, gender and aged. We test StarGAN on 50 random images from the CelebA dataset (250 disruptions). For pix2pixHD we use the official open-source implementation, which was trained for label-to-street view translation on the Cityscapes dataset [9]. We test pix2pixHD on 50 random images from the Cityscapes test set. For CycleGAN we use the official open-source implementation for both the zebra-to-horses and photograph-to-Monet painting translations. We disrupt 100 images from both datasets. We use the pre-trained models provided in the open-source implementations, unless specifically noted.

4.1 Image Translation Disruption

Success Scenario. In order to develop intuition on the relationship between our main L^2 and L^1 distortion metrics and the qualitative distortion caused on image translations, we display in Fig. 3 a scale that shows qualitative examples of disrupted outputs and their respective distortion metrics. We can see that when the L^2 and L^1 metric becomes larger than 0.05 we have very noticeable distortions in the output images. Throughout the experiments section, we report the percentage of successfully disrupted images (% dis.), which correspond to the percentage of outputs presenting a distortion $L^2 \geq 0.05$.

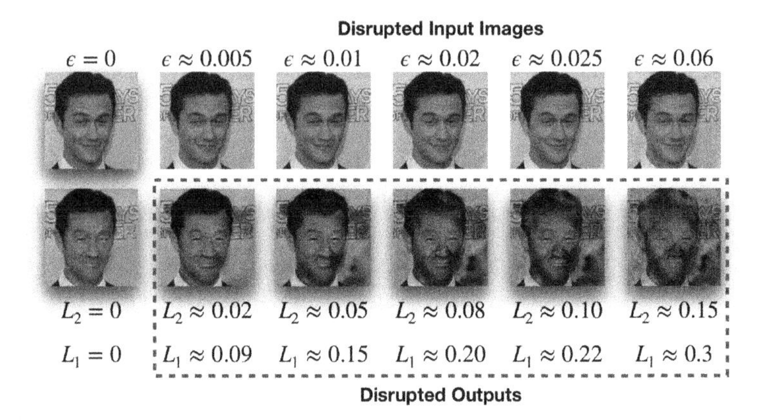

Fig. 3. Equivalence scale between L_2 and L_1 distances and qualitative distortion on disrupted StarGAN images. We also show the original image and output with no disruption. Images with $L_2 \geq 0.05$ have very noticeable distortions.

Table 1. Comparison of L^1 and L^2 pixel-wise errors, as well as the percentage of disrupted images (% dis.) for different disruption methods on different facial manipulation architectures and datasets. All disruptions use $\epsilon = 0.05$ unless noted. We notice that strong disruptions are successful on all tested architectures.

Architecture (Dataset)	FGSM			I-FGSM			PGD		
	L^1	L^2	% dis	L^1	L^2	% dis	L^1	L^2	% dis
StarGAN (CelebA)	0.462	0.332	100%	1.134	1.525	100%	1.119	1.479	100%
GANimation (CelebA)	0.090	0.017	0%	0.142	0.046	34.9%	0.139	0.044	30.4%
GANimation (CelebA, $\epsilon = 0.1$)	0.121	0.024	1.5%	0.212	0.098	93.9%	0.190	0.077	83.7%
pix2pixHD (Cityscapes)	0.240	0.118	96%	0.935	1.110	100%	0.922	1.084	100%
CycleGAN (Horse)	0.133	0.040	21%	0.385	0.242	100%	0.402	0.253	100%
CycleGAN (Monet)	0.155	0.039	22%	0.817	0.802	100%	0.881	0.898	100%

Vulnerable Image Translation Architectures. We show that we are able to disrupt the StarGAN, pix2pixHD and CycleGAN architectures with very successful results using either I-FGSM or PGD in Table 1. Our white-box disruptions are effective on several recent image translation architectures and several different translation domains. GANimation reveals itself to be more robust to disruptions of magnitude $\epsilon = 0.05$ than StarGAN, although it can be successfully disrupted with stronger disruptions ($\epsilon = 0.1$). The metrics reported in Table 1 are the average of the L^1 and L^2 errors on all dataset samples, where we compute the error for each sample by comparing the ground-truth output $\boldsymbol{G}(\boldsymbol{x})$ with the disrupted output $\boldsymbol{G}(\tilde{\boldsymbol{x}})$, using the following formulas $L^1 = ||\boldsymbol{G}(\tilde{\boldsymbol{x}}) - \boldsymbol{G}(\boldsymbol{x})||_1$ and $L^2 = ||\boldsymbol{G}(\tilde{\boldsymbol{x}}) - \boldsymbol{G}(\boldsymbol{x})||_2$. For I-FGSM and PGD we use 20 steps with step size of 0.01. We use our ideal formulation for all disruptions.

Table 2. Comparison of efficacy of FGSM, I-FGSM and PGD methods with different disruption targets for the StarGAN generator and the CelebA dataset.

Target	FGSM		I-FGSM		PGD	
	L^1	L^2	L^1	L^2	L^1	L^2
Towards black	0.494	0.336	0.494	0.335	0.465	0.304
Towards white	0.471	0.362	0.711	0.694	0.699	0.666
Towards random noise	**0.509**	**0.409**	0.607	0.532	0.594	0.511
Away from input	0.449	0.319	1.086	1.444	1.054	1.354
Away from output	0.465	0.335	**1.156**	**1.574**	**1.119**	**1.480**

We show examples of successfully disrupted image translations on GANimation and StarGAN in Fig. 2 using I-FGSM. We observe different qualitative behaviors for disruptions on different architectures. Nevertheless, all of our disruptions successfully make the modifications in the image obvious for any observer, thus avoiding any type of undetected manipulation of an image.

Ideal Disruption. In Sect. 3.1, we derived an ideal disruption for our success metric. In order to execute this disruption we first need to obtain the ground-truth output of the image translation network $\boldsymbol{G}(\boldsymbol{x})$ for the image $\boldsymbol{x}$ being disrupted. We push the disrupted output $\boldsymbol{G}(\tilde{\boldsymbol{x}})$ to be maximally different from $\boldsymbol{G}(\boldsymbol{x})$. We compare this ideal disruption (designated as *Away From* Output in Table 2) to targeted disruptions with different targets such as a black image, a white image and random noise. We also compare it to a less computationally intensive disruption called *Away From* Input, which seeks to maximize the distortion between our disrupted output $\boldsymbol{G}(\tilde{\boldsymbol{x}})$ and our original input $\boldsymbol{x}$.

We display the results for the StarGAN architecture on the CelebA dataset in Table 2. As expected, the *Away From* Output disruption is the most effective using I-FGSM and PGD. All disruptions show similar effectiveness when using one-step FGSM. *Away From* Input seems similarly effective to the *Away From* Output for I-FGSM and PGD, yet it does not have to compute $\boldsymbol{G}(\boldsymbol{x})$, thus saving one forward pass of the generator.

Finally, we show in Table 3 comparisons of our image-level *Away From* Output disruption to the feature-level attack for Variational Autoencoders (VAE) presented in Kos *et al.* [14]. Although in Kos *et al.* [14] attacks are only targeted on the latent vector of a VAE, here we attack every possible intermediate feature map of the image translation network using this attack. The other two attacks presented in Kos *et al.* [14] cannot be applied to the image-translation scenario. We disrupt the StarGAN architecture on the CelebA dataset. Both disruptions use the 10-step PGD optimization formulation with $\epsilon = 0.05$. We notice that while both disruptions are successful, our image-level formulation obtains stronger distortions on average.

Table 3. Comparison of our image-level PGD disruption to an adapted feature-level disruption from Kos *et al.* [14] on the StarGAN architecture.

Layer	Kos *et al.* [14]											
	1	2	3	4	5	6	7	8	9	10	11	Ours
L^1	0.367	0.406	0.583	0.671	0.661	0.622	0.573	0.554	0.512	0.489	0.778	**1.066**
L^2	0.218	0.269	0.503	0.656	0.621	0.558	0.478	0.443	0.384	0.331	0.817	**1.365**

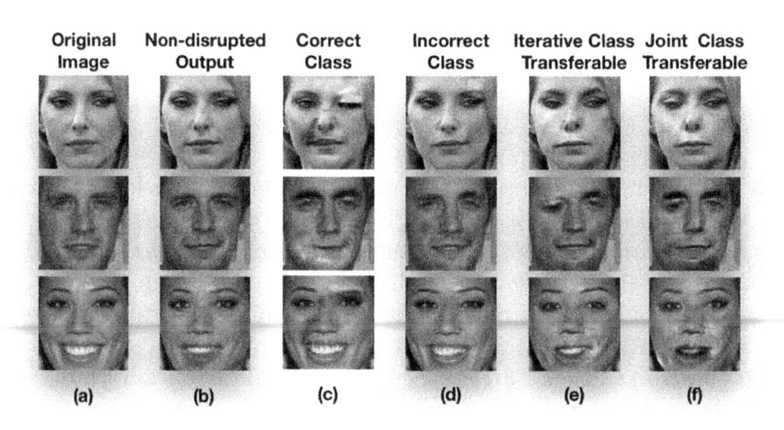

Fig. 4. Examples of our class transferable disruptions. (a) Input image. (b) The ground truth GANimation output without disruption. (c) A disruption using the correct action unit correctly is successful. (d) A disruption with a incorrect target AU is not successful. (e) Our iterative class transferable disruption and (f) joint class transferable disruption are able to transfer across different action units and successfully disrupt the deepfake generation.

4.2 Class Transferable Adversarial Disruption

Class Conditional Image Translation Systems such as GANimation and StarGAN are conditional GANs. Both are conditioned on an input image. Additionally, GANimation is conditioned on the target AU intensities and StarGAN is conditioned on a target attribute. As the disruptor we *do know* which image the malicious actor wants to modify (our image), and in some scenarios we *might know* the architecture and weights that they are using (white-box disruption), yet in almost all cases we *do not know* whether they want to put a smile on the person's face or close their eyes, for example. Since this non-perfect knowledge scenario is probable, we want a disruption that transfers to all of the classes in a class conditional image translation network.

In our experiments we have noticed that attention-driven face manipulation systems such as GANimation present an issue with class transfer. GANimation generates a color mask as well as an attention mask designating the parts of the image that should be replaced with the color mask.

Table 4. Class transferability results for our proposed disruptions. This disruption seeks maximal disruption in the output image. We present the distance between the ground-truth non-disrupted output and the disrupted output images, *higher distance* is better.

	L^1	L^2	% dis.
Incorrect class	0.144	0.053	45.7%
Iterative class transferable	**0.171**	**0.075**	**75.6%**
Joint class transferable	0.157	0.062	53.8%
Correct class	0.166	0.071	68.7%

Table 5. Class transferability results for our proposed disruptions. This disruption seeks minimal change in the input image. We present the distance between the input and output images, *lower distance* is better.

	L^1	L^2
Incorrect class	1.69×10^{-3}	3.09×10^{-4}
Iterative class transferable	6.07×10^{-4}	8.02×10^{-5}
Joint class transferable	3.86×10^{-4}	$\mathbf{1.67 \times 10^{-5}}$
Correct class	$\mathbf{9.88 \times 10^{-5}}$	4.73×10^{-5}
No disruption	9.10×10^{-2}	2.15×10^{-2}

In Fig. 4, we present qualitative examples of our proposed iterative class transferable disruption and joint class transferable disruption. The goal of these disruptions is to transfer to all action unit inputs for GANimation. We compare this to the unsuccessful disruption transfer case where the disruption is targeted to the incorrect AU. Columns (e) and (f) of Fig. 4 show our iterative class transferable disruption and our joint class transferable disruption successfully disrupting the deepfakes, whereas attempting to disrupt the system using the incorrect AU is not effective (column (c)).

Quantitative results demonstrating the superiority of our proposed methods can be found in Table 4. For our disruptions, we use 80 iterations of PGD, magnitude $\epsilon = 0.05$ and a step of 0.01.

For our second experiment, presented in Table 5, instead of disrupting the input image such that the output is visibly distorted, we disrupt the input image such that the output is the identity. In other words, we want the input image to be untouched by the image translation network. We use 80 iterations of I-FGSM, magnitude $\epsilon = 0.05$ and a step of 0.005.

4.3 GAN Adversarial Training and Other Defenses

We present results for our *generator adversarial training* and *G+D adversarial training* proposed in Sect. 3.3. In Table 6, we can see that *generator adversarial training* is somewhat effective at alleviating a strong 10-step PGD disruption.

Table 6. Image translation disruptions on StarGAN with different defenses.

Defense	FGSM			I-FGSM			PGD		
	L^1	L^2	% dis.	L^1	L^2	% dis.	L^1	L^2	% dis.
No defense	0.489	0.377	100	0.877	1.011	100	0.863	0.981	100
Blur	0.160	0.048	37.6	0.285	0.138	89.6	0.279	0.133	89.2
Adv. G. Training	0.125	0.032	15.6	0.317	0.183	96	0.319	0.186	95.2
Adv. G+D Training	0.141	0.036	17.2	0.283	0.138	87.6	0.281	0.136	87.6
Adv. G. Train. + Blur	0.138	0.039	21.6	0.225	0.100	63.2	0.224	0.099	61.2
Adv. G+D Train. + Blur	**0.116**	**0.026**	**10.4**	**0.184**	**0.062**	**36.8**	**0.184**	**0.062**	**37.2**

G+D adversarial training proves to be even more effective than *generator adversarial training*.

Additionally, in the same Table 6, we present results for a Gaussian blur test-time defense ($\sigma = 1.5$). We disrupt this blur defense in a white-box manner. With perfect knowledge of the pre-processing, we can simply backpropagate through that step and obtain a disruption. We achieve the biggest resistance to disruption by combining blurring and *G+D adversarial training*, although strong PGD disruptions are still relatively successful. Nevertheless, this is a first step towards robust image translation networks.

We use a 10-step PGD ($\epsilon = 0.1$) for both *generator adversarial training* and *G+D adversarial training*. We trained StarGAN for 50,000 iterations using a batch size of 14. We use an FGSM disruption $\epsilon = 0.05$, a 10-step I-FGSM disruption $\epsilon = 0.05$ with step size 0.01 and a 10-step PGD disruption $\epsilon = 0.05$ with step size 0.01.

4.4 Spread-Spectrum Evasion of Blur Defenses

In this Section, we evaluate our proposed *spread-spectrum adversarial disruption* which seeks to evade blur defenses in a gray-box scenario, with high transferability between types and magnitudes of blur. In Fig. 5 we present the proportion of test images successfully disrupted ($L^2 \geq 0.05$) for our spread-spectrum method, a white-box perfect knowledge disruption, an adaptation of EoT [3] to the blur scenario and a disruption which does not use any evasion method. We notice that both our method and EoT defeat diverse magnitudes and types of blur and achieve relatively similar performance. Our method achieves better performance on the Gaussian blur scenarios with high magnitude of blur, whereas EoT outperforms our method on the box blur cases, on average. Our iterative spread-spectrum method is roughly K times faster than EoT since it only has to perform one forward-backward pass per iteration of I-FGSM instead of K to compute the loss. Additionally, in Fig. 6, we present random qualitative samples, which show the effectiveness of our method over a naive disruption.

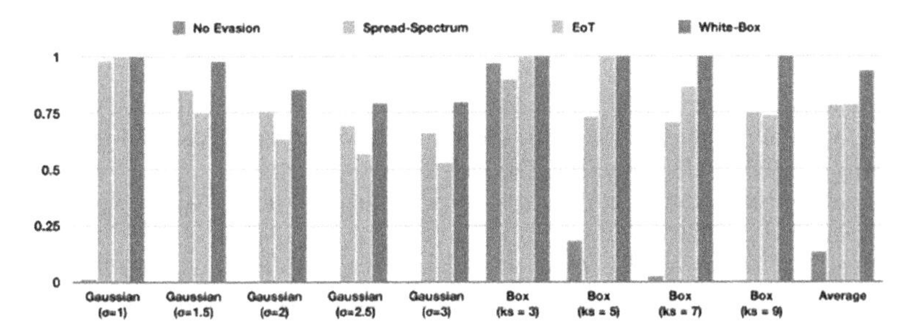

Fig. 5. Proportion of disrupted images ($L^2 \geq 0.05$) for different blur evasions under different blur defenses.

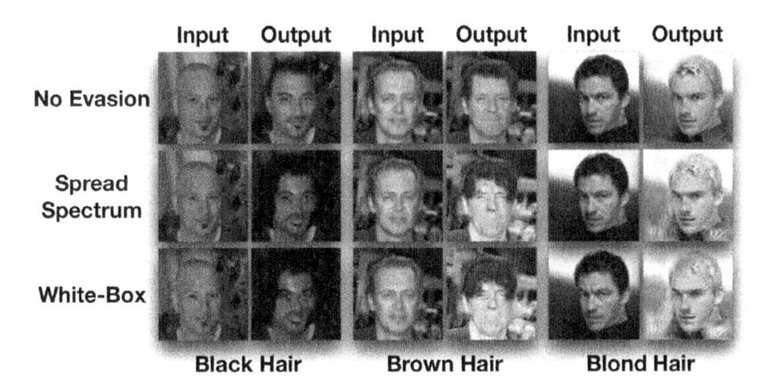

Fig. 6. An example of our spread-spectrum evasion of blur defenses for a Gaussian blur ($\sigma = 1.5$). The first row shows a naive disruption, the second row shows our spread-spectrum evasion and the last row shows a white-box disruption.

5 Conclusion

In this paper we presented a novel approach to defend against image translation-based deepfake generation. Instead of trying to detect whether an image has been modified after the fact, we defend against the non-authorized manipulation by disrupting conditional image translation facial manipulation networks using adapted adversarial attacks.

We operationalized our definition of a successful disruption, which allowed us to formulate an ideal disruption that can be undertaken using traditional adversarial attack methods such as FGSM, I-FGSM and PGD. We demonstrated that this disruption is superior to other alternatives. Since many face modification networks are conditioned on a target attribute, we proposed two disruptions which transfer from one attribute to another and showed their effectiveness over naive disruptions. In addition, we proposed adversarial training for GANs, which is a first step towards image translation networks that are resistant to disruption. Finally, blurring is an effective defense against naive disruptions in a gray-box scenario and can allow a malicious actor to bypass the disruption and modify the image. We presented a spread-spectrum disruption which evades a wide range of blur defenses.

References

1. Deepfakes, revenge porn, and the impact on women. https://www.forbes.com/sites/chenxiwang/2019/11/01/deepfakes-revenge-porn-and-the-impact-on-women/#7dfb95bf1f53. Accessed 10 Dec 2019
2. Facebook to ban 'deepfakes'. https://www.bbc.com/news/technology-51018758. Accessed 10 Jan 2020
3. Athalye, A., Engstrom, L., Ilyas, A., Kwok, K.: Synthesizing robust adversarial examples. In: International Conference on Machine Learning, pp. 284–293 (2018)
4. Bashkirova, D., Usman, B., Saenko, K.: Adversarial self-defense for cycle-consistent GANs. In: Advances in Neural Information Processing Systems, pp. 635–645 (2019)
5. Carlini, N., Wagner, D.: Towards evaluating the robustness of neural networks. In: 2017 IEEE Symposium on Security and Privacy (SP), pp. 39–57. IEEE (2017)
6. Choi, Y., Choi, M., Kim, M., Ha, J.W., Kim, S., Choo, J.: StarGAN: unified generative adversarial networks for multi-domain image-to-image translation. In: Proceedings of the IEEE Conference on Computer Vision and Pattern Recognition, pp. 8789–8797 (2018)
7. Choi, Y., Uh, Y., Yoo, J., Ha, J.W.: StarGAN v2: diverse image synthesis for multiple domains. arXiv preprint. arXiv:1912.01865 (2019)
8. Chu, C., Zhmoginov, A., Sandler, M.: CycleGAN, a master of steganography. arXiv preprint. arXiv:1712.02950 (2017)
9. Cordts, M., et al.: The cityscapes dataset for semantic urban scene understanding. In: Proceedings of the IEEE Conference on Computer Vision and Pattern Recognition, pp. 3213–3223 (2016)
10. Cox, I.J., Kilian, J., Leighton, F.T., Shamoon, T.: Secure spread spectrum watermarking for multimedia. IEEE Trans. Image Process. **6**(12), 1673–1687 (1997)
11. Ekman, R.: What the Face Reveals: Basic and Applied Studies of Spontaneous Expression Using the Facial Action Coding System (FACS). Oxford University Press, Oxford (1997)
12. Goodfellow, I., Shlens, J., Szegedy, C.: Explaining and harnessing adversarial examples. In: International Conference on Learning Representations (2015). http://arxiv.org/abs/1412.6572
13. Isola, P., Zhu, J.Y., Zhou, T., Efros, A.A.: Image-to-image translation with conditional adversarial networks. In: Proceedings of the IEEE Conference on Computer Vision and Pattern Recognition, pp. 1125–1134 (2017)
14. Kos, J., Fischer, I., Song, D.: Adversarial examples for generative models. In: 2018 IEEE Security and Privacy Workshops (SPW), pp. 36–42. IEEE (2018)
15. Kurakin, A., Goodfellow, I., Bengio, S.: Adversarial examples in the physical world. arXiv preprint. arXiv:1607.02533 (2016)
16. Madry, A., Makelov, A., Schmidt, L., Tsipras, D., Vladu, A.: Towards deep learning models resistant to adversarial attacks. In: International Conference on Learning Representations (2018). https://openreview.net/forum?id=rJzIBfZAb
17. Moosavi-Dezfooli, S.M., Fawzi, A., Fawzi, O., Frossard, P.: Universal adversarial perturbations. In: Proceedings of the IEEE Conference on Computer Vision and Pattern Recognition, pp. 1765–1773 (2017)
18. Moosavi-Dezfooli, S.M., Fawzi, A., Frossard, P.: DeepFool: a simple and accurate method to fool deep neural networks. In: Proceedings of the IEEE Conference on Computer Vision and Pattern Recognition, pp. 2574–2582 (2016)
19. Nguyen, A., Yosinski, J., Clune, J.: Deep neural networks are easily fooled: high confidence predictions for unrecognizable images. In: Proceedings of the IEEE Conference on Computer Vision and Pattern Recognition, pp. 427–436 (2015)

20. Papernot, N., McDaniel, P., Jha, S., Fredrikson, M., Celik, Z.B., Swami, A.: The limitations of deep learning in adversarial settings. In: 2016 IEEE European Symposium on Security and Privacy (EuroS&P), pp. 372–387. IEEE (2016)
21. Pumarola, A., Agudo, A., Martinez, A.M., Sanfeliu, A., Moreno-Noguer, F.: GANimation: anatomically-aware facial animation from a single image. In: Proceedings of the European Conference on Computer Vision (ECCV), pp. 818–833 (2018)
22. Sabour, S., Cao, Y., Faghri, F., Fleet, D.J.: Adversarial manipulation of deep representations. arXiv preprint. arXiv:1511.05122 (2015)
23. Szegedy, C., et al.: Intriguing properties of neural networks. arXiv preprint arXiv:1312.6199 (2013)
24. Tabacof, P., Tavares, J., Valle, E.: Adversarial images for variational autoencoders. arXiv preprint. arXiv:1612.00155 (2016)
25. Wang, L., Cho, W., Yoon, K.J.: Deceiving image-to-image translation networks for autonomous driving with adversarial perturbations. IEEE Robot. Autom. Lett. **PP**, 1 (2020). https://doi.org/10.1109/LRA.2020.2967289
26. Wang, R., Ma, L., Juefei-Xu, F., Xie, X., Wang, J., Liu, Y.: FakeSpotter: a simple baseline for spotting AI-synthesized fake faces. arXiv preprint. arXiv:1909.06122 (2019)
27. Wang, S.Y., Wang, O., Zhang, R., Owens, A., Efros, A.A.: CNN-generated images are surprisingly easy to spot...for now. In: CVPR (2020)
28. Wang, T.C., Liu, M.Y., Zhu, J.Y., Tao, A., Kautz, J., Catanzaro, B.: High-resolution image synthesis and semantic manipulation with conditional GANs. In: Proceedings of the IEEE Conference on Computer Vision and Pattern Recognition, pp. 8798–8807 (2018)
29. Yeh, C.Y., Chen, H.W., Tsai, S.L., Wang, S.D.: Disrupting image-translation-based DeepFake algorithms with adversarial attacks. In: The IEEE Winter Conference on Applications of Computer Vision (WACV) Workshops, March 2020
30. Zakharov, E., Shysheya, A., Burkov, E., Lempitsky, V.: Few-shot adversarial learning of realistic neural talking head models. In: Proceedings of the IEEE International Conference on Computer Vision, pp. 9459–9468 (2019)
31. Zhu, J.Y., Park, T., Isola, P., Efros, A.A.: Unpaired image-to-image translation using cycle-consistent adversarial networks. In: Proceedings of the IEEE International Conference on Computer Vision, pp. 2223–2232 (2017)

Efficiently Detecting Plausible Locations for Object Placement Using Masked Convolutions

Anna Volokitin(✉), Igor Susmelj, Eirikur Agustsson, Luc Van Gool, and Radu Timofte

Computer Vision Lab, ETH Zürich, Zürich, Switzerland
{voanna,aeirikur,vangool,timofter}@vision.ee.ethz.ch,
isusmelj@gmail.com

Abstract. Being able to insert new objects into images is an important problem for both artistic image editing and for data augmentation. For a successful image manipulation, the plausible placement and the blending of the new objects in the image are critical. In this paper, we propose a fast method for the automatic selection of plausible locations for object insertion into images. Like previous work, we approach the object placement problem as a detection problem – given a bounding box, we evaluate whether an object is present inside the box based only on the neighborhood of the box. However, previous work requires a forward pass for each potential bounding box location. We propose instead to make use of *masked convolutions* to compute featuremaps for left, right, top and bottom contexts just once per image. Combining these features in such a way that no information from inside a bounding box is propagated to the final classifier allows the model to evaluate a grid of proposals on the featuremaps rather than on the image, speeding up inference dramatically. We validate that our model can generate plausible placements using experiments on the COCO dataset and on a user study. Our method trades off speed for performance, as compared to a patch based approach.

1 Introduction

Much progress has been made in the area of automatic editing of images. Notable examples of editing whole existing images are style transfer [22], image-to-image translation [6], and super-resolution [8].

Another line of research has focused on making small modifications to existing images. This area has greatly benefited from deep learning approaches and includes applications such as inpainting [11], image blending [9,21], and synthesizing new objects into images [14]. Object placement systems also belong to this line of work. Many methods have been developed for data augmentation [1,3,4,15] and robotics applications, with methods mostly making use of

A. Volokitin and I. Susmelj—denotes equal contribution

A. Bartoli and A. Fusiello (Eds.): ECCV 2020 Workshops, LNCS 12538, pp. 252–266, 2020.
https://doi.org/10.1007/978-3-030-66823-5_15

flat surface estimations [7,17]. We continue this line of research and focus on improving the inference speed of placement systems. We take inspiration from the approach of Dvornik *et al.* [3], which frames object placement as a context detection problem, in which the detector only has access to the context of a proposed object location, but does not see the pixels inside the bounding box.

We propose a model for image placement which takes advantage of masked convolutions to lift the sliding window of proposals from the image level to the featuremap level, and speeds up the inference over 10×. Our model uses four sub-networks where the receptive fields do not grow in the left, right, top and bottom directions, respectively, to compute these direction-specific context features once per image and uses them to evaluate all proposals at all locations of the image efficiently. We also characterize this model in terms of plausibility of placed objects in realistic locations of images, both in quantitative results on the COCO dataset, and a user study. Our evaluation indicates that the masked convolution model sacrifices from 2% to 7% in accuracy in exchange for a 10× speed up in inference per proposal, which may be an appropriate tradeoff for when on-the-fly data augmentation is required, such as in a robotics application, or in online learning (Fig. 1).

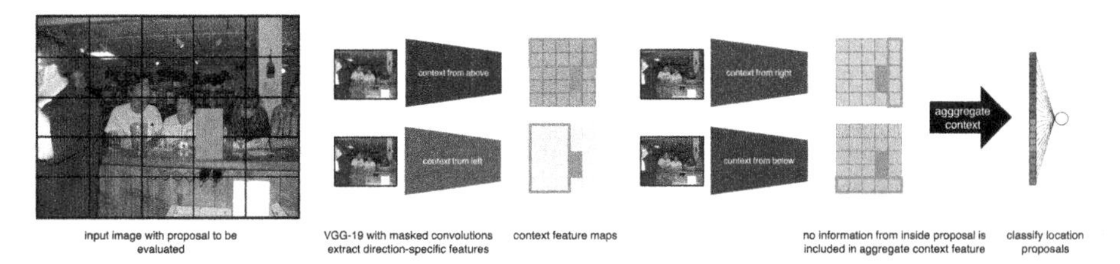

Fig. 1. Overview of masked convolution method. We show how one proposal (the pink filled rectangle) is used to train the network. Four networks with masked convolutions compute featuremaps, whose receptive fields do not grow in a certain direction. The proposal is then classified based on context features which correspond to the regions to the top, bottom, left and right of the proposal. At inference time, the featuremaps are computed once for the image, and we classify a sliding window of proposals on the spatial dimensions of the featuremaps. (Color figure online)

2 Related Work

The object placement approach most similar to ours is from Dvornik *et al.* [3]. Dvornik *et al.* propose to train a model to predict whether a given bounding box contains an object or not, based only on a margin of pixels around the box. In particular, patches are classified into one of $C + 1$ categories, corresponding to either one of C object classes or a negative example where no object can be placed. These predictions are then used to paste in object instances with Poisson blending, and the images are later used to augment training data for object detection, semantic and instance segmentation networks.

Also related is the Spatial Transformer GAN [9] which uses adversarial training for iterative object placement. The generator predicts a sequence of geometric warps of the foreground image, whereas the discriminator judges the realism of the final composition. The approach is validated for constrained settings, *i.e.* indoor furniture placement and eyeglasses placement. This work is complementary to our own, as the iterative warping steps learned by this model are small, and make fine-grained corrections to the initial placement. The model proposed in this work rather proposes a global placement that we validate for both unconstrained outdoor and indoor conditions, which could be used as an initialization for [9].

Surface Estimation Approaches. Many methods for object placement rely on surface estimation and have mainly been developed for robotics. [17] uses a combination of an estimated surface from a point cloud with a segmentation of a scene to detect free locations on surfaces where a robot might place an object. Similarly, [7] learns to place objects in new environments in a robotics setting, based on point clouds.

Data Augmentation. Another line of work has explored inserting objects into images to create more training data for supervised learning tasks. [4] learn to place objects in indoor scenes to create more training data for object detection algorithms. Their method relies on supporting surface estimation and segmentation maps, using only the semantic categories of *table*, *desk* and *counter* as valid areas for placing objects. The segmentation model is jointly trained for semantic segmentation and depth estimation, which is used for scaling objects. [15] use object placement as a data augmentation method for instance segmentation. They use the fact that most objects in Cityscapes [2] occur on similar horizontal scan-lines, and insert new objects onto horizontal scan-lines of existing objects. [14] synthesize pedestrians into existing Cityscapes [2] images and show that additional synthetic training data can be useful for object detection. The locations for synthesizing pedestrians are randomly sampled, and then manually cleaned by removing implausible locations, such as in walls or on cars. Neither [14] nor [15] take into account the scale of placed objects. [1] also create more data for object detection algorithms by synthesizing pedestrians into scenes by learning a Spawn Probability Map for clusters of images in a dataset, estimating camera parameters, and then rendering 3d models of pedestrians into the scene.

3 Masked Convolution Network

Our goal is to train a model which, for any given bounding box on an image, will output the likelihood of this box containing an object of a given category, without having access to the contents of the bounding box. Since we would like to evaluate this model at all possible image locations, we propose to make use of masked convolutions to compute image features only once per image. We can then evaluate a sliding window for placement proposals on the featuremaps of our masked convolution networks instead of evaluating them on the input image.

Our approach is inspired by the use of causal convolutions in the WaveNet model [19] to ensure that only information about the past is used for predicting the future. In our model, we instead restrict the dependence of output convolutional featuremaps on certain spatial directions in the input image.

We compute four featuremaps, with information flow restricted in the up, down, left and right directions respectively. When computing, for example, the featuremap that is restricted from the right, we would like to obtain a featuremap in which each element of the output featuremap $\mathbf{b}_{\nrightarrow}$ is independent of elements to the right of it in the preceding featuremap $\mathbf{a}_{\nrightarrow}$. Using python indexing notation to indicate indexing of the featuremaps $\mathbf{a}_{\nrightarrow}$ and $\mathbf{b}_{\nrightarrow}$, and f indicating functional dependence, we can write that:

$$\mathbf{b}_{\nrightarrow}[i,j] = f(\mathbf{a}_{\nrightarrow}[:,0:j+1]) \tag{1}$$

By restricting information from the right from propagating through each successive network layer, we obtain a final featuremap whose elements are functionally independent of the corresponding input image locations to their right.

Assuming we have the final featuremap $\mathbf{b}_{\nrightarrow}$ that is restricted from the right, which has a downsampling factor D, and a possible bounding box with image coordinates of the corners as x_{min}, y_{min} and x_{max}, y_{max}, we can encode the context corresponding to the region to the left of this bounding box by using the information in $\mathbf{b}_{\nrightarrow}[:, 0 : \lfloor \frac{x_{min}}{D} \rfloor]$. We can analogously compute context features for the three remaining directions, changing the indexing and featuremaps accordingly. PixelCNN [20] has also adapted the causal convolutions from Wavenet to 2D image inputs. A conceptually similar approach using partial convolutions has been applied to the inpainting of irregular holes by [12].

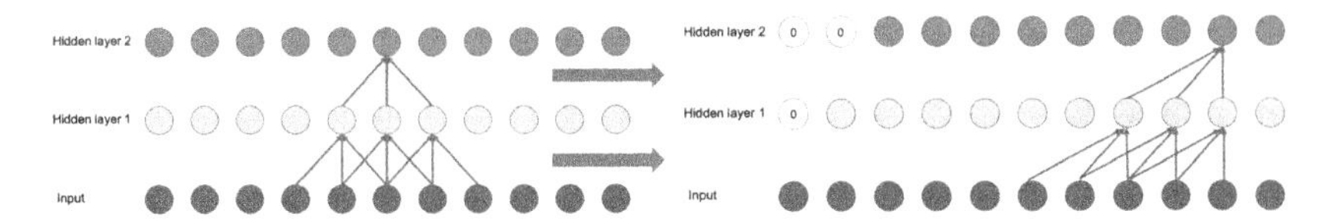

Fig. 2. A 1-D illustration of our masked convolution implementation with shifting. Shifting the output of a regular 1×3 convolution to the right is equivalent to convolving with a 1×5 kernel where the two right elements are set to 0, thus restricting information from the right spatial direction from propagating up the feature hierarchy.

3.1 Using a Shift Layer to Construct Masked Convolutions

We implement masked convolutions by applying standard 3×3 convolutions, and then shifting the output relative to the input. Figure 2 shows how this achieves the desired effect of restricting the functional dependence of the output featuremap in a given spatial direction. Here, we can see that the original 1×3 kernels behave like 1×5 kernels, where the two right most elements are masked (*i.e.* set to zero) after shifting. This insight allows us to convert an existing feature extraction network (in our case VGG-19 [18]) into a masked convolution network, without directly masking any of its kernels and instead introducing shift layers appropriately after convolution and pooling layers.

The convolution of a standard 3 × 3 kernel with a 1 × 3 shift kernel in 2D is written below:

$$\begin{bmatrix} w_{0,0} & w_{1,0} & w_{2,0} \\ w_{0,1} & w_{1,1} & w_{2,1} \\ w_{0,2} & w_{1,2} & w_{2,2} \end{bmatrix} * \begin{bmatrix} 1 & 0 & 0 \end{bmatrix} = \begin{bmatrix} w_{0,0} & w_{1,0} & w_{2,0} & 0 & 0 \\ w_{0,1} & w_{1,1} & w_{2,1} & 0 & 0 \\ w_{0,2} & w_{1,2} & w_{2,2} & 0 & 0 \end{bmatrix}$$

Context features which correspond to the regions to the top, bottom, left and right of the proposal are combined and fed to a fully connected layer, which predicts whether an object could be in the proposal or not.

Table 1. Size of anchors and number of samples per object category for training and testing. There are an equal number of positive and negative samples in the test set.

	Small				Medium				Large			
		Num samples				Num samples				Num samples		
	size	train +	train −	test	size	train +	train −	test	size	train +	train −	test
Apples	22 × 19	705	2167	367	64 × 51	1146	3541	603	119 × 92	482	1475	249
Bottles	11 × 31	2832	8531	1408	27 × 62	5833	17614	2917	49 × 104	2619	7818	1280
Books	12 × 23	1656	5084	859	44 × 24	2444	7367	1217	90 × 43	4048	12013	1946
Cows	32 × 26	1080	3360	577	100 × 76	1415	4369	744	187 × 140	228	485	130

4 Experimental Setup

4.1 Dataset

For our placement experiments, we used the COCO [10] (Common Objects in Context) dataset.

In particular, we use the categories *apples, books, bottles* and *cows*, as there were many instances of these objects in the dataset, and these objects are (mostly) rigid. Bottles and books are especially difficult categories since they can occur in different shapes and orientations and also often appear in interaction with humans.

We cluster the bounding boxes of the objects into three sizes, and use only these sizes, which we call anchors, as in the Faster R-CNN method [16]. This means a bounding box gets assigned to the smallest anchor it will fit inside. The anchor sizes per category are shown in Table 1. One difference between placement detection and object detection is that real-world images contain only positive examples for good placements, and no negative ones (*i.e.* we never learn that a particular place *cannot* contain an object). To obtain negative samples, we sample random locations from the same image and assume that these are on average poor locations for objects.

We create a data set which has three times as many negatives as positives, as in Faster R-CNN. The data set is split into 67% for training 33% for testing. The test set additionally gets processed such that it has an equal number of positive and negative samples. For each of the methods described below, we train a binary classifier for each size-category combination (3 sizes $\times$ 4 categories = 12 models per method in total).

4.2 Compared Methods

Location Prior. We train a random forest model to classify anchors based only on the bounding box coordinates. This model, therefore, learns to capture the non-uniform distribution of bounding boxes of the respective class and size.

Patch Classifier. As a reference point, we implement an approach similar to Dvornik *et al.* [3], which we call the *patch classifier*. This network is trained to classify image patches with zeroed out proposals as being plausible object placement or not. As the feature extractor, we make use of the convolutional part of VGG-19, or `conv5_4`, which corresponds to a $512 \times 7 \times 7$ featuremap given a $3 \times 224 \times 224$ input. The downsampling factor for `conv5_4` is 32. A fully connected layer, acting as a binary classifier, is added on top to classify each patch as containing an object or not. For inference, we use a sliding window and evaluate placements at many locations in the image. Such an image classification approach requires us to compute forward passes for a crop around each proposed location.

All inputs for the patch classifier model are $3 \times 224 \times 224$ pixels. The procedure for generating these is shown in Fig. 3.

Fig. 3. Examples of context patch extraction. First, we zero out the anchor corresponding to the bounding box, and then crop out a region around it containing a context border of the same size as the anchor, and resize this patch to 224×224 using bilinear interpolation.

Masked Convolutions Network. As in the patch classifier model, we use VGG-19 as the base architecture into which we inject our shift layers, following either convolutional or MaxPool layers. Here, we use features from `conv5_4` without the final MaxPool2D, which corresponds to a downsampling factor of 16. We have chosen to use VGG-19 as our base architecture, as our shift-layer approach cannot be applied to create masked convolutions in architectures with skip-connections, such as ResNet [5].

As a result of using only the convolutional part of VGG-19 to extract features, the resulting Masked Convolution model is fully-convolutional, and can be applied to inputs of any size. This means that for the masked convolution experiments, we can use the raw image and bounding box data, with 64-pixel zero-padding.

Using masked convolutions gives us access to four featuremaps, whereas each one is forced to not "see" in one of the directions (left, top, right, down). They contain information about a proposed anchor's context. There are several possible ways to aggregate this information, which we explore, and which are shown in Fig. 4.

Cross Each featuremap is indexed at one location to provide information about *e.g.* context to the right. The feature is indexed at the location closest to the middle of the edge of the corresponding side. All four features are concatenated and used as input to a dense classification layer.

Star Same as above, but we use three spatial locations for indexing each featuremap – at the beginning, middle and end of each edge. The resulting 12 features are concatenated for the classification layer.

Pooling Features are max-pooled from a context area of a given size down to a smaller resolution. For instance, we can pool a 16×16 grid of features into a 2×2 grid which is used for the final classification.

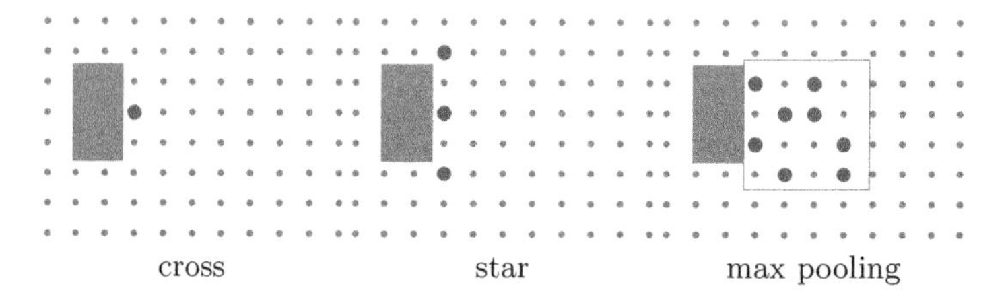

Fig. 4. Different ways to aggregate context featuremaps based on a proposal (shown downscaled and aligned to the spatial dimensions of the featuremap in pink). We show operations only on the right context featuremap. Max pooling shown from 4×4 to 2×2. (Color figure online)

4.3 Optimization

All models are trained with stochastic gradient descent, with a learning rate of 0.0001 and momentum of 0.9 for 5 epochs. Models are trained with a binary cross-entropy loss. The masked convolution network requires four networks to be trained, so the batch size for training is 1. We also train the patch classifier model with batch size 1 for ease of comparison.

We also experiment with finetuning the VGG-19 network or using it as a fixed feature extractor. When finetuning, we train in total for 15 epochs.

For inference, we evaluate all locations on a grid with a stride of 16 pixels, and take the highest scoring location as the predicted anchor. This results in around 1000 evaluations per COCO image.

5 Experiments

5.1 Quantitative Evaluations

It is difficult to evaluate a model which proposes locations for objects, as in any image there can be many plausible locations where an object *could* be placed. However, we do not have ground truth data for the implausibility of a given placement. In other words, we only have positive examples in our training set, and no negative ones. To evaluate the masked convolution model quantitatively, we propose multiple experiments and metrics.

Accuracy. We compute all models' accuracy on a test set with 50% positive and 50% negative examples, which are randomly sampled image locations, (see Table 1 for number of samples), and compare training with and without finetuning, as well as the the best settings for context aggregation for the masked convolution models in Table 4. We threshold the models' predictions at 0.5. The best setup for masked convolution models was to aggregate context by pooling down from a 16×16 grid to an 8×8 grid, which we call MC Pool 16, 8 in the table (Table 2).

Table 2. Test set accuracy for three models. Finetuning improves performance for both patch classifier and masked convolution setups. We also see that accuracy decreases with size, likely due to a smaller amount of training data available for the large anchors.

		Apples			Books			Bottles			Cows			Avg
Model	Finetuning	S	M	L	S	M	L	S	M	L	S	M	L	
Location prior	–	55	54	54	59	58	56	56	54	57	58	53	52	56
Patch classifier	✗	83	66	58	92	82	79	83	77	72	85	71	**66**	76
Patch classifier	✓	82	**71**	**61**	**93**	**84**	**81**	**87**	**79**	**74**	85	73	63	**78**
MC Pool 16, 8	✗	84	66	56	82	76	74	80	70	66	87	73	60	73
MC Pool 16, 8	✓	**85**	**71**	59	86	78	78	81	76	69	**88**	**74**	62	76

The location prior method performs at chance level, whereas all methods using context features perform significantly better. Finetuning the VGG-19 feature extraction network improves performance slightly, both for the patch classifier model and for the masked convolution network.

Both patch classifier models have slightly higher accuracy than their respective masked convolution models. One possible reason for this could be that the masked convolution network is more difficult to optimize due to the shift layers compared to the patch classifier. However, the masked convolution model is significantly above the location prior level.

We also see that performance decreases with size across all categories and setups. This could be because the number of samples available for training is the lowest for this. Another difficulty for the models trained on large anchors is that the true object could be significantly smaller than the proposed anchor, making the context of the anchor less useful than the context around a tighter box.

Speed Comparison Between Patch Classifier and Masked Convolutions. We define efficiency as the number of placement proposals a model can evaluate per second and compare the efficiency of the patch classifier and masked convolution models for different numbers of evaluated proposals. An image of a resolution of 320×277 results around 240 proposals whereas a 1984×1345 pixels contain 10000 proposals. The patch classifier model is much slower, as it runs a forward pass through the whole network for every proposal, whereas the masked convolution model only computes context features once per image. Figure 5 shows that the masked convolution model is faster by an order of magnitude for more than 4000 proposals. Times are shown for GPU computation.

User Study. To evaluate the visual quality of the models, we evaluate placements with a user study on Amazon Mechanical Turk[1]. We consider the location prior model, the finetuned patch classifier model and the finetuned masked convolution model. In each case, we show an image with two boxes on it: an anchor which truly contains the object as well as a placement proposed by our method. The user is instructed to choose the box behind which the given object is more likely to be. A perfect model would be indistinguishable to from ground truth placements and thus be chosen 50% of the time. Each pairwise comparison was evaluated by 30 users. Eight images from each combination of size and object were evaluated per method, totaling that 96 pairs of boxes. We show user study results in Table 3.

A score of 0.28 for the patch classifier model means placements produced by this model are preferred to ground truth placements by users around one in four times. We also see that although the masked convolution model does not perform as well as the patch classifier model, it is still preferred over ground truth placements about one in five times.

Comparison with a Surface Estimation Method. We also compare the patch classifier and masked convolution models against a model for placement

[1] https://www.mturk.com/.

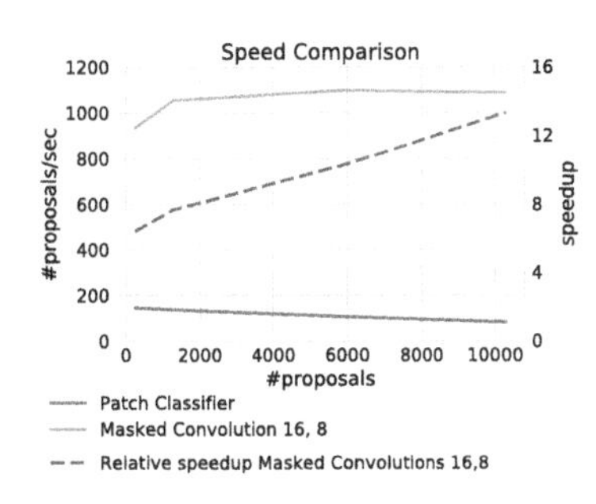

Fig. 5. Using the masked convolution model allows us to evaluate many more proposals than the patch classifier per second.

Table 3. We show the proportion of users that chose a method's placement as being more plausible than COCO ground truth placement and than the synthetic set placements of [4]. The patch classifier's proposals have the highest visual quality, whereas the rightmost column shows the superior performance of the MC Pool 16, 8 in terms of proposals per second.

	User pref. of method over		
Method	GT COCO	Synth. Set	props/s
GT (theoretically)	0.5	0.5	–
Location prior	0.09	–	–
Patch classifier	0.28	0.40	137
MC Pool 16, 8	0.18	0.22	1054

Fig. 6. Comparison of random placements on kitchens from the NYU Depth Dataset V2. blue: MC Pool 16, 8, yellow: patch classifier, pink: synthetic set [4]. Both the patch classifier and synthetic set placements tend to be in reasonable locations. (Color figure online)

proposed by Georgakis *et al.* [4], as the authors have provided synthetic placements on the kitchens from the NYU Depth Dataset V2 [13]. We refer to placements from this model as *synthetic set.*

This model is constrained using predicted depth and segmentation maps to place objects on horizontal surfaces with the semantic segmentation label of *counter, desk* or *table*, which should prevent the model from making very unreasonable placements.

The dataset contains 71 kitchen scenes, and we manually annotated 269 images with bounding boxes that represented plausible bottle locations. Three times as many boxes were randomly sampled as negatives. The bounding boxes were chosen to correspond to the medium size bottle anchor. We finetuned both the patch classifier and masked convolution models for placing medium sized bottles on these annotations for 3 epochs using SGD with a learning rate of 0.0001. The annotations were done on images from 36 sequences, while the methods were compared on a held-out set of images from the 35 other scenes.

As images from [4] contained multiple placed objects, we selected the two images per test sequence which contain the largest bounding box which fit entirely inside our anchor, totaling 70 images. Each pair of images was evaluated by 30 users.

To compare against the model of Georgakis *et al.*, we ran a user study with 70 pairs of images, in which one image in the pair came from [4] and the other from either the patch classifier or masked convolution model. Evaluators were instructed to pick the image which had a more plausible placement for the bottle. We show the results of the user study in Table 3. Each pair of images was evaluated by 30 users.

The patch classifier model is preferred by users 40% of the time, which means that this method has a good understanding of scene semantics, as the model proposed by [4] is constrained to never place objects in the air. Sample placements from both the patch classifier, masked convolution model and from [4] are shown in Fig. 6. The masked convolution method is preferred 22% of the time over a more constrained method.

Masked Convolution Model Study. We now further explore the properties of the masked convolution model.

Context Aggregation. We investigate the best way to make use of context when predicting the feasibility of an object placement. We compare the *cross*, *star* and *pooling* methods, and also explore how large of a region we should pool over. Results are shown in Table 4. We see that pooling is superior over the fixed indexing approaches, and that results are best when pooling over a larger area. These experiments are done without finetuning.

Table 4. Test set accuracy for different context aggregation settings without finetuning. We see that using a larger area to pool context from helps our model and that having a finer-grained spatial resolution is also better. Using pooling outperforms the fixed indexing setups.

	Pool dims		Apples			Books			Bottles			Cows			Avg
Model	In	Out	S	M	L	S	M	L	S	M	L	S	M	L	
MC Cross			68	57	50	65	57	55	62	57	56	75	62	56	60
MC Star			72	63	52	73	66	59	67	56	54	79	67	54	64
MC Pool	4×4	2×2	74	63	53	77	62	64	69	59	61	81	73	56	66
MC Pool	8×8	2×2	81	69	56	81	66	63	72	61	62	83	72	58	69
MC Pool	8×8	4×4	81	67	54	**84**	71	68	77	66	64	84	70	60	71
MC Pool	16×16	2×2	**85**	**72**	**58**	81	64	69	67	65	56	**88**	**78**	**61**	70
MC Pool	16×16	4×4	84	71	57	83	69	69	77	64	62	85	65	**61**	71
MC Pool	16×16	8×8	84	66	56	82	**76**	**74**	**80**	**70**	**66**	87	73	60	**73**

Table 5. Proposal recall of the finetuned patch classifier and masked convolution models, showing how well our models are able to predict true object locations in the test set. When we evaluate our models at a denser grid of proposals, we are able to recall a higher number of true object placements.

		Apples			Books			Bottles			Cows			Avg
Model	Grid	S	M	L	S	M	L	S	M	L	S	M	L	
MC Pool	4	59	45	35	53	33	53	49	53	49	90	86	86	58
MC Pool	8	52	39	32	40	28	38	29	38	46	87	84	80	49
MC Pool	16	21	31	27	15	18	28	12	28	39	65	80	75	37
MC Pool	32	5	20	21	4	7	15	3	10	23	17	74	64	22
Patch classifier	32	8	17	16	4	2	4	4	9	15	14	71	64	19

Runtime-Recall Tradeoff. The advantage of the masked convolution model over the patch classifier model is speed. As we evaluate the sliding window on the context featuremaps, which have low spatial resolution, we do not have access to fine-grained spatial information. In fact the masked convolution model can only evaluate proposals with a stride of 16 pixels, because this is the downsampling factor of the final featuremap. We can, however, decrease this stride, and increase the density of the evaluated grid of proposals, by shifting the input image. To create a grid of proposals with a stride of 8 pixels, we can evaluate the masked convolution model on the original image, the image shifted right by 8 pixels, the image shifted down by 8 pixels, and on the image shifted both right by 8 pixels and down by 8 pixels, as shown in Fig. 7. Also evaluating our model on the shifted image gives us four times as many evaluated object placements. A denser grid of proposals allows us to recall more locations of true objects in our test set. However, with each doubling of the grid density, the amount of computation is quadrupled.

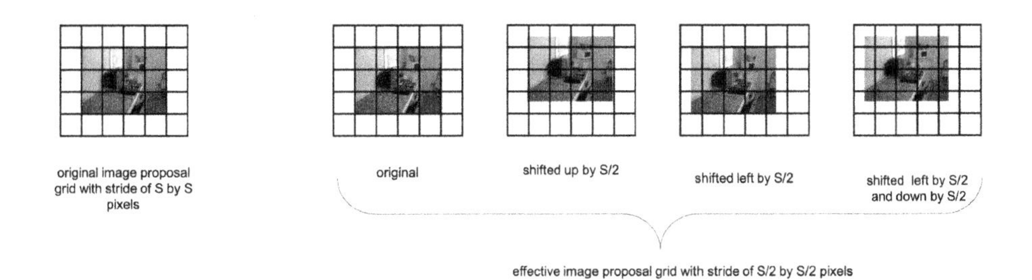

Fig. 7. By shifting the input image relative to the downsampling grid of the Masked Convolution model, we can effectively evaluate placement recall at a finer-spaced grid on the input image.

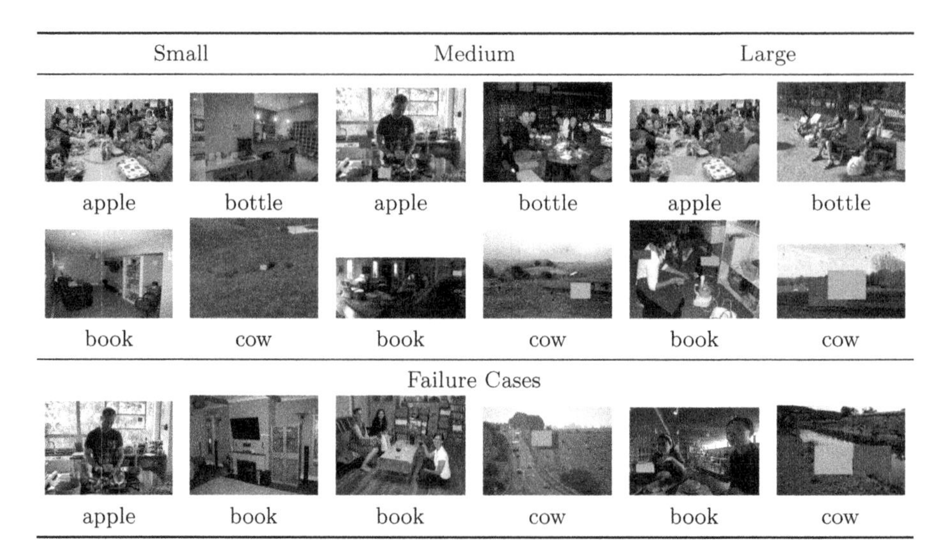

Fig. 8. Sample placements for masked convolution (blue) and patch classified (yellow) models. First two rows show good placements. The bottom row shows failure cases. (Color figure online)

We compute recall for different densities of a grid of proposals. We quantify recall by computing the proportion of predicted locations which have an Intersection over Union (IoU) score of 0.5 or greater with the ground truth anchor.

Table 5 shows that the model manages to recall a greater proportion of the true placements in the dataset when evaluating proposals at a higher grid resolution. It is worth bearing in mind that the recall metric penalizes our model for not learning all possible placement modes in the dataset, which is very challenging in a dataset like COCO, which contains a large amount of variation. This could contribute to the reason that models for placing cows have high recall – there are not as many different modes for the placements of cows as for books in a cluttered house.

We additionally show the recall for the patch classifier, for a grid resolution of 32×32, which has been evaluated on a subset of 100 samples of the test set, which performs similarly to the masked convolution model. All other lines in the table are evaluated on the full test set.

We see that there is a tradeoff between speed and spatial precision in predictions, and the operating point on this tradeoff can be chosen by users of our model.

5.2 Sample Placements

Figure 8 shows both successful and unsuccessful placements by the patch classifier and masked convolution model, in a variety of challenging locations, such as cluttered indoor scenes. We see that both the masked convolution model and the patch classifier model learn to place books on bookshelves or apples and bottles on cluttered surfaces, for example, which is not possible for models constrained to use flat surfaces. However, the model also can fail when there is a relatively large uniform area, where it is difficult to distinguish a wall from a table.

6 Conclusion

In this work, we have proposed a masked convolution approach for fast object placement that makes use of the object detection framework. Although our proposed approach is 2%–7% less accurate, it runs ten times faster per proposal as compared with a patch-based method, and does significantly outperform the location prior. We have shown that users also have the option of using the model at higher proposal grid resolutions for improved placement quality. Our system may be of use in proposing many candidate placement locations for generating on-the-fly augmented data.

References

1. Cheung, E.C., Wong, T.K., Bera, A., Manocha, D.: MixedPeds: pedestrian detection in unannotated videos using synthetically generated human-agents for training. arXiv preprint. arXiv:1707.09100 (2017)
2. Cordts, M., et al.: The cityscapes dataset for semantic urban scene understanding. CoRR, abs/1604.01685 (2016)
3. Dvornik, N., Mairal, J., Schmid, C.: On the importance of visual context for data augmentation in scene understanding. arXiv:1809.02492, September 2019
4. Dwibedi, D., Misra, I., Hebert, M.: Cut, paste and learn: surprisingly easy synthesis for instance detection. In: Proceedings of the IEEE International Conference on Computer Vision, pp. 1301–1310 (2017)
5. He, K., Zhang, X., Ren, S., Sun, J.: Deep residual learning for image recognition. In: Proceedings of the IEEE Conference on Computer Vision and Pattern Recognition (CVPR), June 2016
6. Isola, P., Zhu, J.-Y., Zhou, T., Efros, A.A.: Image-to-image translation with conditional adversarial networks. arXiv preprint (2017)
7. Jiang, Y., Lim, M., Zheng, C., Saxena, A.: Learning to place new objects in a scene. CoRR, abs/1202.1694 (2012)
8. Ledig, C., et al.: Photo-realistic single image super-resolution using a generative adversarial network. In Proceedings of the IEEE Conference on Computer Vision and Pattern Recognition (CVPR), July 2017
9. Lin, C.-H., Yumer, E., Wang, O., Shechtman, E., Lucey, S.: St-GAN: spatial transformer generative adversarial networks for image compositing. In: Proceedings of the IEEE Conference on Computer Vision and Pattern Recognition, pp. 9455–9464 (2018)
10. Lin, T., et al.: Microsoft COCO: common objects in context. CoRR, abs/1405.0312 (2014)
11. Liu, G., Reda, F.A., Shih, K.J., Wang, T.-C., Tao, A., Catanzaro, B.: Image inpainting for irregular holes using partial convolutions. arXiv preprint arXiv:1804.07723 (2018)

12. Liu, G., Reda, F.A., Shih, K.J., Wang, T.-C., Tao, A., Catanzaro, B.: Image inpainting for irregular holes using partial convolutions. In: Proceedings of the European Conference on Computer Vision (ECCV), September 2018
13. Silberman, N., Hoiem, D., Kohli, P., Fergus, R.: Indoor segmentation and support inference from RGBD images. In: Fitzgibbon, A., Lazebnik, S., Perona, P., Sato, Y., Schmid, C. (eds.) ECCV 2012. LNCS, vol. 7576, pp. 746–760. Springer, Heidelberg (2012). https://doi.org/10.1007/978-3-642-33715-4_54
14. Ouyang, X., Cheng, Y., Jiang, Y., Li, C., Zhou, P.: Pedestrian-synthesis-GAN: generating pedestrian data in real scene and beyond. CoRR, abs/1804.02047 (2018)
15. Remez, T., Huang, J., Brown, M.: Learning to segment via cut-and-paste. In: Proceedings of the European Conference on Computer Vision (ECCV), pp. 37–52 (2018)
16. Ren, S., He, K., Girshick, R.B., Sun, J.: Faster R-CNN: towards real-time object detection with region proposal networks. CoRR, abs/1506.01497 (2015)
17. Schuster, M.J., Okerman, J., Nguyen, H., Rehg, J.M., Kemp, C.C.: Perceiving clutter and surfaces for object placement in indoor environments. In: 2010 10th IEEE-RAS International Conference on Humanoid Robots, pp. 152–159 (2010)
18. Simonyan, K., Zisserman, A.: Very deep convolutional networks for large-scale image recognition. CoRR, abs/1409.1556 (2014)
19. van den Oord, A., et al.: WaveNet: a generative model for raw audio. CoRR, abs/1609.03499 (2016)
20. van den Oord, A., Kalchbrenner, N., Espeholt, L., Vinyals, O., Graves, A., et al.: Conditional image generation with PixelCNN decoders. In: Advances in Neural Information Processing Systems, pp. 4790–4798 (2016)
21. Zhao, H., Shen, X., Lin, Z., Sunkavalli, B. Price, K., Jia, J.: Compositing-aware image search. In: Proceedings of the European Conference on Computer Vision (ECCV), pp. 502–516 (2018)
22. Zhu, J., Park, T., Isola, P., Efros, A.A.: Unpaired image-to-image translation using cycle-consistent adversarial networks. CoRR, abs/1703.10593 (2017)

L2-Constrained RemNet for Camera Model Identification and Image Manipulation Detection

Abdul Muntakim Rafi[1], Jonathan Wu[1], and Md. Kamrul Hasan[2(✉)]

[1] University of Windsor, 401 Sunset Avenue, Windsor, ON N9B 3P4, Canada
{rafi11,jwu}@uwindsor.ca

[2] Bangladesh University of Engineering and Technology, Dhaka 1205, Bangladesh
khasan@eee.buet.ac.bd

Abstract. Source camera model identification (CMI) and image manipulation detection are of paramount importance in image forensics. In this paper, we propose an L2-constrained Remnant Convolutional Neural Network (L2-constrained RemNet) for performing these two crucial tasks. The proposed network architecture consists of a dynamic preprocessor block and a classification block. An L2 loss is applied to the output of the preprocessor block, and categorical crossentropy loss is calculated based on the output of the classification block. The whole network is trained in an end-to-end manner by minimizing the total loss, which is a combination of the L2 loss and the categorical crossentropy loss. Aided by the L2 loss, the data-adaptive preprocessor learns to suppress the unnecessary image contents and assists the classification block in extracting robust image forensics features. We train and test the network on the Dresden database and achieve an overall accuracy of 98.15%, where all the test images are from devices and scenes not used during training to replicate practical applications. The network also outperforms other state-of-the-art CNNs even when the images are manipulated. Furthermore, we attain an overall accuracy of 99.68% in image manipulation detection, which implies that it can be used as a general-purpose network for image forensic tasks.

Keywords: Image forensics · Camera model identification · Image manipulation detection · Convolutional Neural Networks

1 Introduction

Camera model identification (CMI) and image manipulation detection are crucial tasks in image forensics with applications in criminal investigations, authenticating evidence, detecting forgery, etc. Digital images go through various camera-internal processing before being saved in the device [28]. Moreover, they are often manipulated after they leave the device that has been used to capture them. Nowadays, professional image editing tools like Adobe Photoshop, ACDsee, and

A. Bartoli and A. Fusiello (Eds.): ECCV 2020 Workshops, LNCS 12538, pp. 267–282, 2020.
https://doi.org/10.1007/978-3-030-66823-5_16

Hornil Stylepix are readily available, consequently making image manipulation a common phenomenon [14]. Also, images undergo different kinds of manipulations when they are shared online. We have observed a proliferation of digitally altered images with the advent of modern technologies. When the authenticity of such images is questioned, a forensic analyst has to answer two questions first, what is the source of the image under question and whether the image has been manipulated. The image metadata cannot be trusted as a reliable source, as this data can be forged. Therefore, a forensic analyst resorts to different image forensics techniques to answer these questions.

Image forensics is an active research area, and several methods exist in the literature for finding out the source camera model and detecting image-processing operations of a questioned image. But researches are conducted discretely for finding out the source and manipulation history of an image. In [33,41], we can find a brief overview of the approaches proposed over the last two decades. We see that initial research in CMI has focused on merging image-markers, such as watermarks, device-specific code, etc. [33]. However, using separate external features for each camera model is an unmanageable task [18]. Consequently, researchers have focused on utilizing the intrinsic features, such as the Color Filter Array (CFA) pattern [6], interpolation algorithms [26], and Image Quality Metrics (IQM) [22]. Utilizing Photo Response Non-Uniformity (PRNU) noise patterns have been proposed for device-level identification [16,21]. Although sensor noise carries device-specific noise artifacts, researchers have developed methods to perform CMI using sensor noise patterns [29,43]. Most of these approaches attempt to extract camera model-specific features and compare the features with a pre-calculated reference for the corresponding camera model [10]. In the case of image manipulation, traces are found in the image according to the type of processing it has gone through [5]. Following this theory, researchers have used distinct forensic approaches for identifying different kinds of image manipulation, such as resizing [19,34], contrast enhancement [40,51], and multiple jpeg compression [7,31], etc. The drawback of using the above-mentioned statistical feature-based approaches is that the performance degrades sharply, when new cases arise that have not been considered during feature vector selection [14]. For that reason, more recent researches have focused on becoming data-driven, such as utilizing local pixel dependencies used in steganalysis [20,32] to perform CMI [11,30] and detect image manipulation [35]. In [17], the authors propose a Gaussian mixture model for image manipulation detection. Though these approaches provide good results, extracting features for different manipulations requires substantial computational resources, and the performance degrades severely depending on the size of the questioned image [14].

Recently, researchers have started applying Convolutional Neural Networks (CNNs) for image forensic tasks [49]. It is expected as CNNs have performed extremely well in different image classification tasks [38]. Usually, CNNs tend to learn features related to the content of an image, whereas, for image forensics, we need to refrain CNNs from learning image contents [4]. As a result, a common practice while using CNNs in digital image forensics is adding a preprocessing layer at the beginning of the CNN architecture. Chen et al. [12] have proposed using a median filter, whereas Tuama et al. [45] have used a high-pass filter before feeding images in their respective CNNs. However, such crude filtering is not supported by the literature as the artifacts introduced by different camera-internal processing and manipulations can lie in both low and high frequency domain [29]. Therefore, fixed filters as preprocessor may lose forensics-related features. Bayar and Stamm [4] have proposed a data-driven constrained convolutional layer which has performed better than the above-mentioned fixed filters. Rafi et al. [37] have used a completely data-driven preprocessor block followed by a classification block to perform CMI. Bayar and Stamm [4] have also used their constrained CNN for image manipulation detection. However, some CNN based approaches do not use any preprocessing scheme. Yang et al. use the idea of multi-scale receptive fields on an input image to perform CMI [50]. In [8], the authors use CNN and support vector machine (SVM) for CMI, where they use the CNN part as a feature extractor. In [36], explores the performance of DenseNet [25] in both CMI and image manipulation detection. In [14], the authors investigate the performance of densely connected CNNs in image manipulation detection. Owing to the performance of the data-driven preprocessing schemes, it can be inferred that further researches need to be conducted to make the preprocessing operations more robust for image forensic tasks. Several researches exist in the literature that use auxiliary loss function to enhance the discrimination between learned features [42,46,47]. There is a scope of utilizing such auxiliary loss functions in the modular CNN architectures for image forensics.

Despite the numerous researches conducted in this field, most researchers have explored CMI and image manipulation detection problems discretely. Bayar and Stamm [4] show that it is possible to use the same approach for both tasks. Therefore, research for coming up with a general-purpose neural network suitable for both CMI and image manipulation detection requires more attention. Also, strict measures should be followed while conducting experiments so that the proposed methods can be applied in real-life scenarios. Kirchner and Gloe suggest that the test set should always consist of images captured by devices that have not been used during training or validation [28]. Also, the scenes in the test set should be different from those used during training and validation. Here, *scene* refers to a combination of a location and a specific viewpoint. Keeping separate devices and scenes in the test set is compulsory for replicating real-life conditions and making the result reliable for practical applications. These evaluation criteria will ensure that the neural network is free from *data leakage* [1] during testing and can not overperform by learning features specific to the device or scene. Besides, the performances of CMI and image manipulation detection should be measured using images manipulated at different intensities. We strictly follow the above-mentioned points in our experiments.

In this paper, we propose a general-purpose novel CNN architecture, called L2-constrained Remnant Convolutional Neural Network (L2-constrained RemNet) for performing two crucial tasks in image forensics, CMI and image manipulation detection. Our proposed CNN has two parts, a preprocessor block and a classification block. The preprocessor architecture consists of several data-driven remnant blocks, and an L2 loss is applied to the output of the preprocessor block. A CNN based classification block follows the preprocessor block, and categorical crossentropy loss is calculated based on its output. The total loss function is a combination of the L2 loss and the categorical crossentropy loss. The whole network is trained end-to-end while minimizing the total loss. The L2-constrained preprocessor learns to suppress image contents making it easier for the classification block to extract image forensics features. Our experiments show that the proposed method can outperform other state-of-the-art networks in both image forensic tasks.

We organize the rest of the paper as follows. A brief overview of the general pipeline of digital images is provided in Sect. 2. Section 3 contains a description of our proposed CNN and loss function. We discuss our training and evaluation criteria, along with the experimental results in Sect. 4. Finally, we conclude in Sect. 5.

2 Image Acquisition Pipeline

Digital images go through multiple operations from being captured by a digital camera to being available in different online platforms [28]. We first describe the image acquisition pipeline of digital cameras, as depicted in Fig. 1. In a typical digital camera, the light of a scene passes through a system of lenses and optical filters, which is then collected by an optical sensor. A color filter array (CFA) is used before the sensor to obtain RGB color images so that the individual sensor element records light of a certain color. The remaining color information is estimated from surrounding pixels through a process called CFA interpolation or demosaicing. After demosaicing, the image goes through a number of post-processing operations (e.g., color correction, edge enhancement, and compression) before it is saved on a storage device. As described in [28], most of these components leave certain 'fingerprints' in the images, which can be utilized in different image forensic tasks. Manufacturers generally employ different lens systems in their different camera models, which causes lens distortion artifacts, such as radial lens distortion, chromatic aberration, and vignetting. The CFA layout and demosaicing process vary widely among different models and

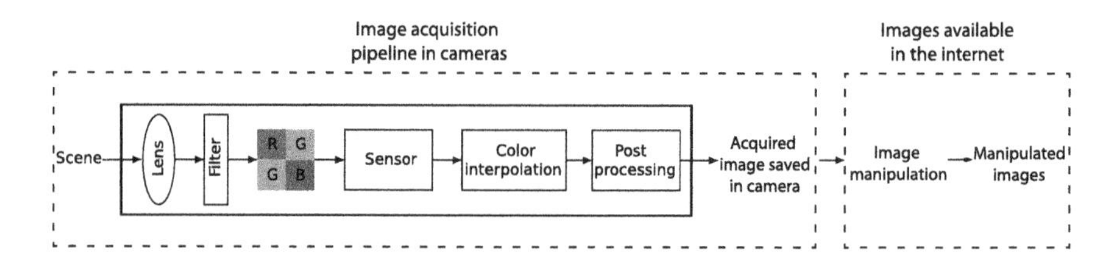

Fig. 1. General pipeline of available digital images.

are generally considered one of the most distinctive model-specific signatures. The sensor pattern noise (SPN) is the most unique characteristic of a digital camera, and it is used excessively in the literature for source identification. In addition to the camera-internal processing operations, digital images face different manipulations when they are edited by different image editing softwares. Moreover, they are resized or re-compressed when uploaded to photo-sharing websites or social media applications [3]. Therefore, image forensics techniques should be made robust to these common manipulation operations.

3 Proposed Method

In this paper, we propose a CNN-based patch-level method for CMI and image manipulation detection. A schematic representation of our proposed method is shown in Fig. 2.

As shown, we first extract high quality clusters of size 256×256 from an input image. From each cluster, patches of size 64×64 are taken and fed to the L2-constrained RemNet. It then generates a class probability map for each patch. We assign a camera model or image manipulation type label to each cluster by averaging the class probability maps of its patches. The final prediction is made based on the majority voting on the labels of the clusters of an image.

As well known, CNNs in their standard form tend to learn content-specific features from the training images. In designing CNNs for image forensic tasks, it has been, therefore, a common practice to use a preprocessing scheme to suppress the image contents and intensify the minute signatures induced by the image acquisition pipeline or image manipulation operation. Unlike the conventional approaches, the benefit of designing a dynamic preprocessing block is that it can adapt itself optimally to perform different image forensic tasks. To this end, we propose a general-purpose novel CNN architecture, called L2-constrained

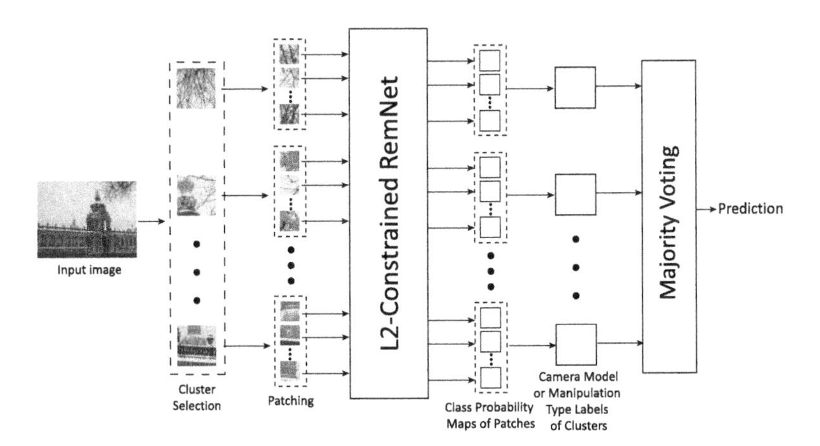

Fig. 2. Schematic representation of the proposed method for CMI and image manipulation detection.

RemNet. A data-driven preprocessing block coupled with L2 loss is used at the beginning of the network, which is followed by a classification block. A line diagram of the proposed network model is shown in Fig. 3. The details of our proposed model are presented in the following.

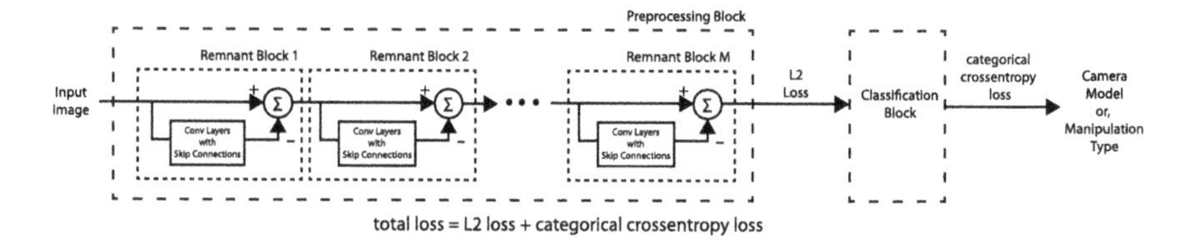

Fig. 3. Line diagram of proposed modular network.

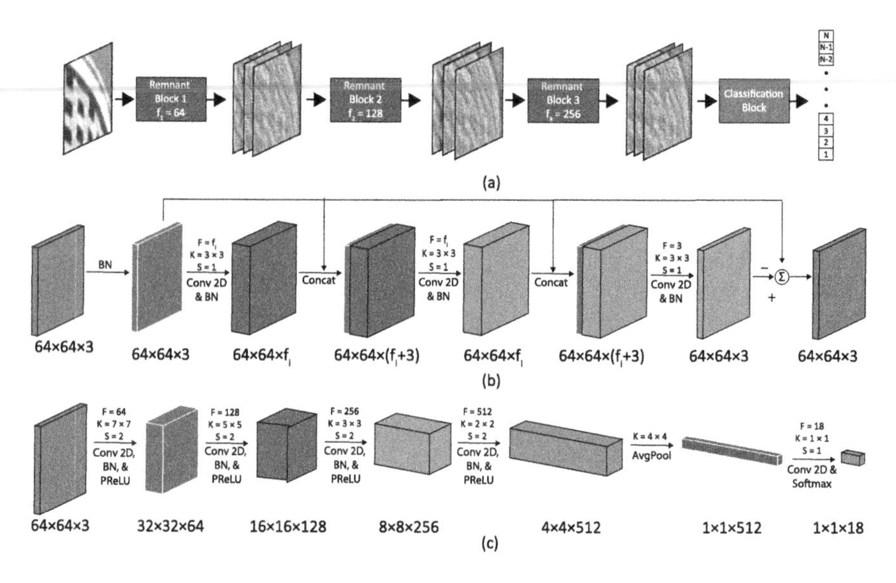

Fig. 4. Architecture of our proposed L2-constrained RemNet. The overall architecture with preprocessor block and the classification block is illustrated in (a). The structure of the remnant blocks and the classification blocks are shown in (b) and (c), respectively. Here, AvgPool and Conv2D stand for average pooling and 2D convolution, respectively. The letters F, K, and S represent the number of filters, their kernel size, and strides, respectively.

3.1 Preprocessing Block

We use the remnant blocks proposed in [37] as our preprocessing block. The architecture is influenced by the highway networks [39]. The inherent camera model-specific features are very subtle and minute features of the image

[13,29,41]. The problem of diminishing minute model-specific features is alleviated in the remnant block through the use of skip connections. Also, it refrains the minute features from being lost in a layer. Moreover, it resolves the vanishing gradient problem [44] during training. The use of activation is avoided in the remnant blocks as per the design requirements in [4]. Also, it is motivated by our wish to make them perform as optimal digital filters. The final layer of a remnant block is subtracted from its input in a pixelwise manner. This subtraction helps regulate information flow. While choosing the depth of a remnant block, the number of filters in each convolutional layer, and kernel size– we use the hyperparameters proposed in [37]. The architecture of the remnant block is illustrated in Fig. 4.

Each remnant block has three convolutional layers. The kernel size is chosen as 3×3. Each layer is followed by BN (batch normalization). The feature space is widened from $64 \times 64 \times 3$ to $64 \times 64 \times f_i$ in the first two convolutional layers and then reduced to the original dimension again. The output of the last convolutional layer is subtracted from the input. As the convolutional layers are followed by BN, the input to the block is batch normalized as well. To preserve input information throughout a block, the input is propagated to every convolutional layer inside the block. We use three remnant blocks in total. 64, 128, and 256 are chosen as f_i for the consecutive remnant blocks.

Table 1. Architecture of our proposed L2-constrained RemNet

Layers	Output size	Kernels*
Remnant block 1	64×64×3	$f_1 = 64$
Remnant block 2	64×64×3	$f_2 = 128$
Remnant block 3	64×64×3	$f_3 = 256$
Classification block		
Conv 2D, BN, & PReLU	32×32×64	F = 64, K = 7×7, S = 2
Conv 2D, BN, & PReLU	16×16×128	F = 128, K = 5×5, S = 2
Conv 2D, BN, & PReLU	8×8×256	F = 256, K = 3×3, S = 2
Conv 2D, BN, & PReLU	4×4×512	F = 512, K = 2×2, S = 2
Average pool	1×1×512	K = 4×4
Conv 2D	1×1×N	F = N, K = 1×1, S = 1
Softmax	N	–

*Here, F, K, and S represent the number of filters, kernel size, and strides, respectively. N represents the number of class.

3.2 Classification Block

We use the classifier proposed in [37] as our classification block as well. The output of the preprocessing block is passed a classification block. The architecture of the classification block is provided in Table 1. It extracts higher-level image forensics features by gradually reducing the dimensions of the feature space, and finally provide a class probability of the source camera model or the manipulation type of the input image.

The classification block starts with four convolutional layers. Each of the first four convolutional layers is followed by a BN layer and a PReLU activation. The output of the fourth convolutional layer is followed by an average-pooling operation. Lastly, a final convolutional layer with softmax activation is used to generate a probability for the final prediction. The design choices for the classification block are the same as the hyperparameters proposed in [37]. There are no fully connected layers in the classification block, which keeps the number of parameters less. Consequently, the network is less prone to overfitting and trains within a shorter time.

3.3 Loss Function

The preprocessing block contains M remnant blocks. The i-th remnant block applies a transformation H_i on its input $\mathbf{x_i}$ (which is also the output of the $(i-1)$-th remnant block) and subtracts it from its input to produce the output $\mathbf{y_{P_i}}$:

$$\mathbf{y_{P_i}} = \mathbf{x_i} - H\left(\mathbf{x_i}, \mathbf{W_{P_i}}\right), \tag{1}$$

The output of the last remnant block is $\mathbf{y_{PM}}$. A loss is calculated based on a flattened version of this output:

$$L_2 = \sum_{l=1}^{N_{param}} y_{PM_l}^2. \tag{2}$$

Here, $\mathbf{y_{PM_l}}$ is the l-th element of $\mathbf{y_{PM}}$ and N_{param} is the total number of elements in $\mathbf{y_{PM}}$. Afterwards, $\mathbf{y_{PM}}$ is fed the classifier block that applies a transformation G to generate the final output $\mathbf{y_c}$:

$$\mathbf{y_c} = G\left(\mathbf{y_{PM}}, \mathbf{W_c}\right). \tag{3}$$

We calculate categorical crossentropy loss between this output and the ground truth using:

$$L_{xent} = \sum_{k=1}^{N_{class}} y_{c_i}^{*(k)} \log\left(y_{c_i}^{(k)}\right). \tag{4}$$

where $y_{c_i}^{*(k)}$ and $y_{c_i}^{(k)}$ are the true label and the network output of the i-th image at the k-th class among the N_{class} classes, respectively. The total loss L is defined using the following equation:

$$L = \alpha * L_2 + L_{xent}. \tag{5}$$

Here, α indicates how much weight we want to put in the suppression of the residue from the preprocessor block. A larger choice for α may cause the vanishing gradient problem for the classifier [44]. We empirically set the value of α as 0.5. During backpropagation, the gradient of L_2 is used to update the weights of the preprocessing block. The gradient of L_{xent} is used to update the weights of both the preprocessing block and the classifier block. The whole network is trained in an end-to-end manner. The preprocessing block outputs a residue of the input, and L_2 attempts to minimize this output, which results in suppression of image contents. Simultaneously, the classifier tries to extract useful features from this residue for accurate predictions to minimize L_{xent}. Minimization of L results in rich image forensics features in the residue for the classifier block.

4 Experimental Results

We perform a number of experiments to prove the efficacy of our proposed method. We discuss the experiments and the results in this section.

4.1 Camera Model Identification

We evaluate our L2-constrained RemNet on Dresden Dataset [23]. The dataset contains images captured with 73 devices of 27 different camera models. Multiple snaps have been captured from different scenes for each device. We discard eight camera models to choose the specific camera models, which have images captured using more than one device. Our goal is to keep one device excluded during training and use it only for testing. Also, we consider Nikon D70 and Nikon D70s, as a single camera model, according to [28]. We end up with 18 camera models. We split the dataset into train, validation, and test sets following strict criteria that the camera device and scenes used during testing are not used for training or validation. This results in 7938, 1353, and 540 images in the train, validation, and test set, respectively. These criteria, proposed in [28] and used in [8], is quite necessary to make sure that the evaluation is not biased owing to device-specific features and scene-specific features.

Data augmentation is a commonly used method in deep learning to reduce overfitting. Recently, researchers have started using it for CMI as well [36,37,49]. Also, our goal is to perform CMI from both unaltered and manipulated images. Therefore, we choose different image manipulations as our data augmentations, as proposed in [37]. The types of augmentation that we use in this work are:

- JPEG-Compression with a quality factor of 70%, 80%, and 90%
- Resizing by a factor of 0.5, 0.8, 1.5, and 2.0
- Gamma-Correction with a factor of $\gamma = 0.8$ and 1.2

This increases our data by nine folds. Afterward, we extract clusters of 256×256 size from the images. However, saturated and flat regions inside an image are not less likely to contain features related to CMI. Therefore, we follow the selection strategy proposed in [8,37] to extract high quality image clusters. For every cluster $\mathcal{P}$ in an image, its quality $Q(\mathcal{P})$ is computed as

$$Q(\mathcal{P}) = \frac{1}{3} \sum_{c \in [R,G,B]} \left[\alpha \cdot \beta \cdot (\mu_c - \mu_c^2) + (1-\alpha) \cdot (1 - e^{\gamma \sigma_c})\right] \tag{6}$$

where α, β, and γ are empirically set constants (set to 0.7, 4 and ln(0.01), respectively), μ_c and σ_c, $c \in [R,G,B]$ are the mean and standard deviation of the red, green, and blue components of cluster $\mathcal{P}$, respectively.

Although we extract 256×256 sized high quality clusters, we use 64×64 input size for our network according to [8,37,50,52]. Patches of 64×64 are randomly selected from a cluster of 256×256 during training. This strategy introduces statistical variations during training which is discussed in detail in [37]. We extract 20 clusters of size 256×256 from each image and this results in 1587600 and 270600 train and validation clusters. We use our custom loss function (see Subsect. 3.3) and Adam [27] optimizer with exponential decay rate factors $\beta_1 = 0.9$ and $\beta_2 = 0.999$. The choice for our batch size is 64. The learning rate starts with 10^{-3} and we decrease it with a factor of 0.5 if the softmax classification loss (L_{xent}) does not decrease in three successive epochs. We train our network for a maximum of 70 epochs and save the weight with the least validation softmax classification loss for evaluation.

The test set contains 10800 clusters of size 256×256 from 540 full images. During testing, we average the predictions on all non-overlapping patches of size 64×64 to make a prediction for a cluster and assign a camera model label $\hat{L_n}$ to it. We use majority voting to make the final prediction $\hat{L}$ for the full image. The metric we use for evaluating the performance of our models is provided in the following equation:

$$Accuracy = \frac{N_C}{N_T}. \tag{7}$$

Here, N_C is the number of correct prediction and N_T is the total number of test images. We also compare our results with four other state-of-art CNNs in

Table 2. Accuracy (in %) of different methods in CMI for unaltered test dataset

Method	Dresden dataset
Yang et al. [50]	95.19
Bayar and Stamm [4]	93.89
Bondi et al. [8]	92.59
DenseNet [25]	95.05
ResNet [24]	95.19
ResNeXt [48]	95.55
RemNet without preprocessing block [37]	95.74
RemNet [37]	97.59
Proposed method	**98.15**

CMI [4], fusion residual networks [8,50], and [37]. Moreover, we provide comparison with two other popular deep CNNs, ResNet [24] and DenseNet [25] as they have been used for image forensics as well [2,9,15,36]. We use the same input size for all the networks for fair comparison.

Table 3. Accuracy (in %) of different methods in CMI for manipulated test dataset

Method	Gamma correction				JPEG compression				Resize scale			
	0.5	0.75	1.25	1.5	95	90	85	80	0.8	0.9	1.1	1.2
Yang et al. [50]	94.26	95.37	95.00	92.78	94.07	94.07	92.59	92.59	94.26	92.59	90.93	90.56
Bayar and Stamm [4]	93.52	94.44	94.44	94.63	92.59	94.81	88.15	85.74	88.15	87.04	64.44	59.07
Bondi et al. [8]	85.92	91.85	89.07	92.03	84.07	85.92	91.48	90.74	92.56	92.77	91.48	89.44
DenseNet [25]	91.66	95.18	92.03	94.62	92.77	92.96	94.26	94.81	95.00	94.81	94.44	94.26
ResNet [24]	91.85	95.18	92.77	94.81	93.88	**94.82**	95.55	95.00	95.18	95.18	95.00	95.18
ResNeXt [48]	94.25	95.55	93.88	95.18	95.18	**94.82**	94.25	94.07	95.00	95.00	**96.11**	**95.55**
RemNet [37]	96.11	97.22	96.11	95.56	**97.59**	**94.82**	92.59	92.78	95.00	93.33	92.04	92.41
Proposed method	**96.29**	**98.14**	**97.59**	**97.96**	92.96	93.33	**96.11**	**97.03**	**96.67**	**96.67**	90.74	91.66

At first, we evaluate the performance of the models on the unaltered test dataset. L2-constrained RemNet achieves an overall accuracy of 98.15%, which is better than all other approaches we compare with (see Table 2). It should be noted that we set the value for α in our custom loss function (5) empirically. We have achieved accuracy of 97.77%, 98.15%, and 97.77%, when α is chosen as 0.1, 0.5, and 1, respectively. Therefore, we propose using $\alpha = 0.5$.

We perform several experiments to justify the use of the L2-constrained preprocessing block in our network. First, we train the RemNet without any preprocessing block at the beginning of the network, that is, we only train the classification block. Then, we train the RemNet without any auxiliary L2 loss at the output of the preprocessing block. Afterward, we experiment with replacing the L2 loss with the L1 loss. The lower accuracy of the RemNet without the preprocessing block justifies the use of the preprocessing step. Similarly, the lower accuracy of RemNet without any additional loss justifies the use of the auxiliary loss. When we use the L1 loss in our custom loss function, the total loss oscillates throughout the training and does not converge. After a complete run, the L1-constrained RemNet attains an accuracy of 58.88%. The L1 loss enforces sparsity on the output of the preprocessing block, whereas the image forensics features, in this case, are non-sparse and present throughout the image. The L2 loss forces the output of the preprocessing block to be small and provides a non-sparse solution.

Furthermore, we apply various manipulations on the test set and evaluate the performance of our method. To make sure that the network has not overfitted on the manipulation factors used during training, we also manipulate the test images with factors that are not used during training. The test images are created using gamma correction with $\gamma = 0.5$, 0.75, 1.25, and 1.5; JPEG compression quality factors (QFs) 95%, 90%, 85%, and 80%; and resize scaling factor of 0.8, 0.9, 1.1, 1.2. The highest result for each manipulation factor is made bold (see Table 3).

We can see that our proposed method has substantial improvement over other methods for Gamma Correction. In the case of JPEG Compression, our network achieves better performance for two factors, and RemNet [37] achieves better performance in two. For Resize manipulation, we see that ResNeXt [48] gains higher accuracy for two manipulation factors, whereas our proposed method gains higher accuracy in the other two factors. We can conclude that our proposed method proves to be most robust to external manipulation. Also, deep CNNs perform better than shallow networks in the face of manipulated images.

4.2 Image Manipulation Detection

Now, we show the use of our network in a completely different image forensic task. We use it to identify the kind of image-manipulation done on an image. The same network is used here except the number of output classes, which is four– unaltered, rescale, JPEG compression, and gamma correction. The input size for all the networks is also maintained at (64×64). We use the same train and validation set from our experiments with CMI and sub-divide it into the four manipulation classes. The L2-constrained RemNet is then trained to detect the type of manipulation applied to an image. It is to be mentioned that, during training, our dataset consisting of 1587600 train and 270600 validation clusters has been reduced in order to make the training data evenly distributed among four classes. Since the number of unaltered train and validation clusters are 158760 and 27060, respectively, we select 158760 train and 27060 validation clusters randomly for each type of manipulation.

In testing, we have used the test images from the Dresden dataset and generated a total of $540 \times 12 = 6480$ test images, which include 540 unaltered images; $540 \times 4 = 2160$ gamma-corrected images with $\gamma = 0.5, 0.75, 1.25$, and 1.5; $540 \times 3 = 1620$ JPEG compressed images compressed with factors of 85%, 90%, and 95%; and $540 \times 4 = 2160$ resized images images with scaling factor of 0.8, 0.9, 1.1, and 1.2. Details of the results are given in Table 5. We achieve an overall accuracy of 99.68% in this task whereas RemNet [37], Bayar and Stamm [4], and Yang et al. [50] achieve 98.27%, 87.28% and 91.74%, respectively (see Table 4). We demonstrate the detection accuracy for different factors of manipulation in Table 5. For gamma-corrected images, the performances of [50], RemNet [37] and our proposed method are substantially better than that of [4]. In the case of JPEG compression, all four networks perform almost the same except at the

Table 4. Accuracy (in %) of different methods in image manipulation detection

Method	Dresden dataset
Yang et al. [50]	91.74
Bayar and Stamm [4]	87.28
RemNet [37]	98.27
Proposed method	**99.68**

Table 5. Accuracy (in %) of image manipulation detection for different manipulation factors

Method	Gamma correction				JPEG compression			Rescale			
	0.5	0.75	1.25	1.5	95	90	85	0.8	0.9	1.1	1.2
Yang et al. [50]	99.07	98.52	97.04	98.70	49.44	100	100	100	97.40	60.74	100
Bayar and Stamm [4]	94.44	83.33	77.22	90.56	11.30	100	100	100	100	90.93	99.63
RemNet [37]	100	**99.81**	**99.63**	**100**	81.48	98.33	100	100	100	100	100
Proposed method	**100**	99.63	99.26	98.7	**100**	**98.7**	**100**	**100**	**100**	**100**	**100**

compression factor of 95, where [4] and [50] fail miserably by misclassifying most of the compressed images as unaltered images. There is a significant drop in the detection accuracy for RemNet [37] as well. This is expected since there is very little difference between the original image and JPEG compressed image with factor 95. However, our proposed method achieves 100% accuracy even at this factor, which indicates that the network can detect even minute manipulation artifacts introduced during manipulation operation. When detecting rescaled images, our network and RemNet [37] performs the same by attaining a 100% accuracy. Of the other two networks, [4] performs better than [50].

5 Conclusion

In this paper, we have proposed an L2 loss constrained RemNet for performing two important image forensics tasks, namely, CMI and image manipulation detection. The proposed modular CNN model comprises of a dynamic preprocessor and a classification block in series. The L2-constrained preprocessor, while trained end-to-end along with the classification block, suppresses unnecessary image contents dynamically and generates a residue of the image from where the classification block can easily extract image forensics features. We have comprehensively conducted multiple experiments on the Dresden dataset to demonstrate the efficacy of such a preprocessing scheme assisted by the L2 loss in CMI. During testing, we use images captured by devices not seen during training to replicate practical applications. The results of the experiments have shown that our proposed method can be successfully used in real-world scenarios. Additionally, we have used our proposed method for image manipulation detection. The satisfactory performances of our network on both classification tasks prove that it can be used for a general-purpose network for image forensics.

References

1. Alneyadi, S., Sithirasenan, E., Muthukkumarasamy, V.: A survey on data leakage prevention systems. J. Netw. Comput. Appl. **62**, 137–152 (2016)

2. Barni, M., Costanzo, A., Nowroozi, E., Tondi, B.: CNN-based detection of generic contrast adjustment with jpeg post-processing. In: Proceedings of the IEEE International Conference on Image Process (ICIP), pp. 3803–3807, October 2018. https://doi.org/10.1109/ICIP.2018.8451698
3. Bayar, B., Stamm, M.C.: Augmented convolutional feature maps for robust CNN-based camera model identification. In: Proceedings of the IEEE International Conference on Image Processing, (ICIP), pp. 4098–4102. IEEE (2017)
4. Bayar, B., Stamm, M.C.: Design principles of convolutional neural networks for multimedia forensics. Electron. Imaging **2017**(7), 77–86 (2017)
5. Bayar, B., Stamm, M.C.: Constrained convolutional neural networks: A new approach towards general purpose image manipulation detection. IEEE Trans. Inf. Forensics Secur. **13**(11), 2691–2706 (2018)
6. Bayram, S., Sencar, H., Memon, N., Avcibas, I.: Source camera identification based on CFA interpolation. In: Proceedings of the IEEE International Conference on Image Processing, (ICIP), vol. 3, pp. III–69. IEEE (2005)
7. Bianchi, T., Piva, A.: Image forgery localization via block-grained analysis of JPEG artifacts. IEEE Trans. Inf. Forensics Secur. **7**(3), 1003–1017 (2012)
8. Bondi, L., Baroffio, L., Güera, D., Bestagini, P., Delp, E.J., Tubaro, S.: First steps toward camera model identification with convolutional neural networks. IEEE Signal Process. Lett. **24**(3), 259–263 (2017)
9. Boroumand, M., Fridrich, J.: Deep learning for detecting processing history of images. Electron. Imaging **2018**, 213 (2018)
10. Cao, H., Kot, A.C.: Accurate detection of demosaicing regularity for digital image forensics. IEEE Trans. Inf. Forensics Secur. **4**(4), 899–910 (2009)
11. Chen, C., Stamm, M.C.: Camera model identification framework using an ensemble of demosaicing features. In: Proceedings of the IEEE International Workshop on Information Forensics and Security (WIFS), pp. 1–6. IEEE (2015)
12. Chen, J., Kang, X., Liu, Y., Wang, Z.J.: Median filtering forensics based on convolutional neural networks. IEEE Signal Process. Lett. **22**(11), 1849–1853 (2015)
13. Chen, M., Fridrich, J., Goljan, M., Lukás, J.: Determining image origin and integrity using sensor noise. IEEE Trans. Inf. Forensics Secur. **3**(1), 74–90 (2008)
14. Chen, Y., Kang, X., Shi, Y.Q., Wang, Z.J.: A multi-purpose image forensic method using densely connected convolutional neural networks. J. Real-Time Image Process. **16**(3), 725–740 (2019)
15. Chen, Y., Kang, X., Wang, Z.J., Zhang, Q.: Densely connected convolutional neural network for multi-purpose image forensics under anti-forensic attacks. In: Proceedings of the 6th ACM Workshop on Information Hiding Multimedia Security, pp. 91–96. ACM, New York (2018). https://doi.org/10.1145/3206004.3206013
16. Dirik, A.E., Sencar, H.T., Memon, N.: Source camera identification based on sensor dust characteristics. In: Proceedings of the IEEE Workshop on Signal Processing Applications for Public Security and Forensics, pp. 1–6. IEEE (2007)
17. Fan, W., Wang, K., Cayre, F.: General-purpose image forensics using patch likelihood under image statistical models. In: 2015 IEEE International Workshop on Information Forensics and Security (WIFS), pp. 1–6. IEEE (2015)
18. Farid, H.: Image forgery detection. IEEE Signal Process. Mag. **26**(2), 16–25 (2009)
19. Feng, X., Cox, I.J., Doerr, G.: Normalized energy density-based forensic detection of resampled images. IEEE Trans. Multimedia **14**(3), 536–545 (2012)
20. Fridrich, J., Kodovsky, J.: Rich models for steganalysis of digital images. IEEE Trans. Inf. Forensics Secur. **7**(3), 868–882 (2012)
21. Fridrich, J., Lukas, J., Goljan, M.: Digital camera identification from sensor noise. IEEE Trans. Inf. Forensics Secur. **1**(2), 205–214 (2006)

22. Gloe, T.: Feature-based forensic camera model identification. In: Shi, Y.Q., Katzenbeisser, S. (eds.) Transactions on Data Hiding and Multimedia Security VIII. LNCS, vol. 7228, pp. 42–62. Springer, Heidelberg (2012). https://doi.org/10.1007/978-3-642-31971-6_3
23. Gloe, T., Böhme, R.: The Dresden image database for benchmarking digital image forensics. J. Digit. Forensic Pract. **3**, 150–159 (2010)
24. He, K., Zhang, X., Ren, S., Sun, J.: Deep residual learning for image recognition. In: Proceedings of the IEEE Conference on Computer Vision and Pattern Recognition (CVPR), pp. 770–778 (2016)
25. Huang, G., Liu, Z., van der Maaten, L., Weinberger, K.Q.: Densely connected convolutional networks. In: Proceedings of the IEEE Conference on Computer Vision and Pattern Recognition (CVPR), pp. 2261–2269, July 2017. https://doi.org/10.1109/CVPR.2017.243
26. Kharrazi, M., Sencar, H.T., Memon, N.: Blind source camera identification. In: Proceedings of the IEEE International Conference on Image Processing, (ICIP), vol. 1, pp. 709–712. IEEE (2004)
27. Kingma, D.P., Ba, J.: Adam: a method for stochastic optimization. arXiv preprint arXiv:1412.6980 (2014)
28. Kirchner, M., Gloe, T.: Forensic camera model identification. In: Proceedings of the WOL Handbook of Digital Forensics of Multimedia Data and Devices, pp. 329–374 (2015)
29. Lukas, J., Fridrich, J., Goljan, M.: Digital camera identification from sensor pattern noise. IEEE Trans. Inf. Forensics Secur. **1**(2), 205–214 (2006)
30. Marra, F., Poggi, G., Sansone, C., Verdoliva, L.: A study of co-occurrence based local features for camera model identification. Multimedia Tools Appl. **76**(4), 4765–4781 (2017). https://doi.org/10.1007/s11042-016-3663-0
31. Neelamani, R., De Queiroz, R., Fan, Z., Dash, S., Baraniuk, R.G.: JPEG compression history estimation for color images. IEEE Trans. Image Process. **15**(6), 1365–1378 (2006)
32. Pevny, T., Bas, P., Fridrich, J.: Steganalysis by subtractive pixel adjacency matrix. IEEE Trans. Inf. Forensics Secur. **5**(2), 215–224 (2010)
33. Piva, A.: An overview on image forensics. Proc. ISRN Signal Process. **2013**, 496701 (2013)
34. Popescu, A.C., Farid, H.: Exposing digital forgeries by detecting traces of resampling. IEEE Trans. Signal Process. **53**(2), 758–767 (2005)
35. Qiu, X., Li, H., Luo, W., Huang, J.: A universal image forensic strategy based on steganalytic model. In: Proceedings of the 2nd ACM Workshop on Information Hiding and Multimedia Security, pp. 165–170 (2014)
36. Rafi, A.M., et al.: Application of DenseNet in camera model identification and post-processing detection. In: CVPR Workshops, pp. 19–28 (2019)
37. Rafi, A.M., Tonmoy, T.I., Kamal, U., Hoque, R., Hasan, M., et al.: RemNet: remnant convolutional neural network for camera model identification. arXiv preprint arXiv:1902.00694 (2019)
38. Schmidhuber, J.: Deep learning in neural networks: an overview. Neural Netw. **61**, 85–117 (2015)
39. Srivastava, R.K., Greff, K., Schmidhuber, J.: Highway networks. arXiv preprint arXiv:1505.00387 (2015)
40. Stamm, M.C., Liu, K.R.: Forensic detection of image manipulation using statistical intrinsic fingerprints. IEEE Trans. Inf. Forensics Secur. **5**(3), 492–506 (2010)
41. Stamm, M.C., Wu, M., Liu, K.R.: Information forensics: an overview of the first decade. IEEE Access **1**, 167–200 (2013)

42. Sun, Y., Wang, X., Tang, X.: Deeply learned face representations are sparse, selective, and robust. In: Proceedings of the IEEE Conference on Computer Vision and Pattern Recognition, pp. 2892–2900 (2015)
43. Thai, T.H., Cogranne, R., Retraint, F.: Camera model identification based on the heteroscedastic noise model. IEEE Trans. Image Process. **23**(1), 250–263 (2014)
44. Tong, T., Li, G., Liu, X., Gao, Q.: Image super-resolution using dense skip connections. In: Proceedings of the IEEE International Conference on Computer Vision, pp. 4799–4807 (2017)
45. Tuama, A., Comby, F., Chaumont, M.: Camera model identification with the use of deep convolutional neural networks. In: Proceedings of the IEEE International Workshop on Information Forensics and Security (WIFS), pp. 1–6. IEEE (2016)
46. Wen, Y., Li, Z., Qiao, Y.: Latent factor guided convolutional neural networks for age-invariant face recognition. In: Proceedings of the IEEE Conference on Computer Vision and Pattern Recognition, pp. 4893–4901 (2016)
47. Wen, Y., Zhang, K., Li, Z., Qiao, Yu.: A discriminative feature learning approach for deep face recognition. In: Leibe, B., Matas, J., Sebe, N., Welling, M. (eds.) ECCV 2016. LNCS, vol. 9911, pp. 499–515. Springer, Cham (2016). https://doi.org/10.1007/978-3-319-46478-7_31
48. Xie, S., Girshick, R., Dollár, P., Tu, Z., He, K.: Aggregated residual transformations for deep neural networks. In: Proceedings of the IEEE Conference on Computer Vision and Pattern Recognition, pp. 1492–1500 (2017)
49. Yang, P., Baracchi, D., Ni, R., Zhao, Y., Argenti, F., Piva, A.: A survey of deep learning-based source image forensics. J. Imaging **6**(3), 9 (2020)
50. Yang, P., Zhao, W., Ni, R., Zhao, Y.: Source camera identification based on content-adaptive fusion network. Pattern Recogn. Lett. **119**, 195–204 (2019)
51. Yao, H., Wang, S., Zhang, X.: Detect piecewise linear contrast enhancement and estimate parameters using spectral analysis of image histogram (2009)
52. Yao, H., Qiao, T., Xu, M., Zheng, N.: Robust multi-classifier for camera model identification based on convolution neural network. IEEE Access **6**, 24973–24982 (2018)

W25 - Assistive Computer Vision and Robotics

W25 - Assistive Computer Vision and Robotics

The Eighth International Workshop on Assistive Computer Vision and Robotics (ACVR2020) was held as a virtual meeting alongside the 16th European Conference on Computer Vision (ECCV2020) on August 28, 2020. This workshop followed previous editions held in conjunction with past ECCV and ICCV conferences. With respect to previous editions which mainly focused on health-related assistive technologies, this year's edition has extended the set of topics to cover a broader spectrum of assistive applications to support daily life, safety, and industrial processes. The new topics have been introduced to account for the huge growth of Computer Vision and Robotics in recent years, which makes possible many more assistive applications with an impact on our lives. The workshop received 39 submissions, which were peer reviewed by experts in the field following a double-blind review scheme. Based on the reviewers' comments, we accepted 15 papers, which corresponds to a 38% acceptance rate. Authors were asked to prepare a 10-minute video presentation of their work and upload it to the "Tutorials and Workshops" platform. Authors also provided a 1-minute video summary, which was played back live during the workshop to encourage participation in the live sessions and discussion. Live Q&A sessions followed the playback of the 1-minute videos to allow authors and attendees to interact. The workshop program was completed by four invited speeches given by Dhruv Batra (Georgia Tech, USA), Walterio Mayol-Cuevas (University of Bristol, UK), Marc Pollefeys (ETH Zurich, CH), and Jim Rehg, (Georgia Tech, USA). The organizers would like to take this opportunity to thank all the members of the technical program committee, who delivered quality reviews of the submitted papers, the CVPL association, who endorsed the event, as well as all the attendees, who, with their participation and feedback, contributed to the discussion on how assistive technologies for health, safety, and industrial processes are evolving to have a significant impact on our lives.

August 2020

Giovanni M. Farinella
Antonino Furnari
Marco Leo
Gerard G. Medioni
Mohan Trivedi

We Learn Better Road Pothole Detection: From Attention Aggregation to Adversarial Domain Adaptation

Rui Fan[1], Hengli Wang[2], Mohammud J. Bocus[3], and Ming Liu[2](✉)

[1] UC San Diego, La Jolla, USA
rui.fan@ieee.org
[2] HKUST Robotics Institute, Kowloon, Hong Kong
{hwangdf,eelium}@ust.hk
[3] University of Bristol, Bristol, UK
junaid.bocus@bristol.ac.uk

Abstract. Manual visual inspection performed by certified inspectors is still the main form of road pothole detection. This process is, however, not only tedious, time-consuming and costly, but also dangerous for the inspectors. Furthermore, the road pothole detection results are always subjective, because they depend entirely on the individual experience. Our recently introduced disparity (or inverse depth) transformation algorithm allows better discrimination between damaged and undamaged road areas, and it can be easily deployed to any semantic segmentation network for better road pothole detection results. To boost the performance, we propose a novel attention aggregation (AA) framework, which takes the advantages of different types of attention modules. In addition, we develop an effective training set augmentation technique based on adversarial domain adaptation, where the synthetic road RGB images and transformed road disparity (or inverse depth) images are generated to enhance the training of semantic segmentation networks. The experimental results demonstrate that, firstly, the transformed disparity (or inverse depth) images become more informative; secondly, AA-UNet and AA-RTFNet, our best performing implementations, respectively outperform all other state-of-the-art single-modal and data-fusion networks for road pothole detection; and finally, the training set augmentation technique based on adversarial domain adaptation not only improves the accuracy of the state-of-the-art semantic segmentation networks, but also accelerates their convergence.

Source Code and Dataset:
http://sites.google.com/view/pothole-600

R. Fan and H. Wang—These authors contributed equally to this work and are therefore joint first authors.

A. Bartoli and A. Fusiello (Eds.): ECCV 2020 Workshops, LNCS 12538, pp. 285–300, 2020.
https://doi.org/10.1007/978-3-030-66823-5_17

1 Introduction

Potholes are small concave depressions on the road surface [1]. They arise due to a number of environmental factors, such as water permeating into the ground under the asphalt road surface [2]. The affected road areas are further deteriorated due to the vibration of tires, making the road surface impracticable for driving. Furthermore, vehicular traffic can cause the subsurface materials to move, and this generates a weak spot under the street. With time, the road damage worsens due to the frequent movement of vehicles over the surface and this causes new road potholes to emerge [3].

Road pothole is not just an inconvenience, but also poses a safety risk, because it can severely affect vehicle condition, driving comfort, and traffic safety [2]. It was reported in 2015 that Danielle Rowe, an Olympic gold medalist as well as three-time world champion, had eight fractured ribs resulting in a punctured lung, after hitting a pothole during a race [4]. Therefore, it is crucial and necessary to regularly inspect road potholes and repair them in time.

Currently, manual visual inspection performed by certified inspectors is still the main form of road pothole detection [5]. However, this process is not only time-consuming, exhausting and expensive, but also hazardous for the inspectors [3]. For example, the city of San Diego repairs more than 30K potholes per year using hot patches compound and bagged asphalt, and they have been requesting residents to report potholes so as to relieve the burden on the local road maintenance department [6]. Elsewhere, the UK government is set to pledge billions of pounds for filling potholes across the country [7]. Additionally, the pothole detection results are always subjective, as the decisions depend entirely on the inspector's experience and judgment [8]. Hence, there has been a strong demand for automated road condition assessment systems, which can not only acquire 2D/3D road data, but also detect and predict road potholes accurately, robustly and objectively [9].

Specifically, automated road pothole detection has been considered as more than an infrastructure maintenance problem in recent years, as many self-driving car companies have included road pothole detection into their autonomous car perception modules. For instance, Jaguar Land Rover announced their recent research achievements on road pothole detection/prediction [10], where the vehicles can not only gather the location and severity data of the road potholes, but also send driver warnings to slow down the car. Ford also claimed that they were experimenting with data-driven technologies to warn drivers of the pothole locations [11]. Furthermore, during the Consumer Electronics Show (CES) 2020, Mobileye demonstrated their solutions[1] for road pothole detection, which are based on machine vision and intelligence. With recent advances in image analysis and deep learning, especially for 3D vision data, depth/disparity image analysis and convolutional neural networks (CNNs) have become the mainstream techniques for road pothole detection [8].

---

[1] http://s21.q4cdn.com/600692695/files/doc_presentations/2020/1/Mobileye-CES-2020-presentation.pdf.

Given the 3D road data, image segmentation algorithms are typically performed to detect potholes. For example, Jahanshahi *et al.* [12] employed Otsu's thresholding method [13] to segment depth images for road pothole detection. In [2], we proposed a disparity image transformation algorithm, which can better distinguish between damaged and undamaged road areas. The road potholes were then detected using a surface modeling approach. Subsequently, we minimized the computational complexity of our algorithm and successfully embedded it in a drone for real-time road inspection [8]. Recently, the aforementioned algorithm was proved to have a numeric solution [5], which allows it to be easily deployed to any existing semantic segmentation networks for end-to-end road pothole detection.

In this paper, we first briefly introduce the disparity (or inverse depth, as disparity is in inverse proportion to depth) transformation (DT) algorithm proposed in [5]. We then exploit the aggregation of different types of attention modules (AMs) so as to improve the semantic segmentation networks for better road pothole detection. Furthermore, we develop a novel adversarial domain adaptation framework for training set augmentation. Moreover, we publish our road pothole detection dataset, named *Pothole-600*, at http://sites.google.com/view/pothole-600 for research purposes. According to our experimental results presented in Sect. 6, training CNNs with augmented road data yields better semantic segmentation results, where convergence is achieved with fewer iterations at the same time.

2 Related Works

2.1 Semantic Segmentation

Fully convolutional network (FCN) [14] was the first end-to-end single-modal CNN designed for semantic segmentation. Based on FCN, U-Net [15] adopts an encoder-decoder architecture. It also adds skip connections between the encoder and decoder to help smooth the gradient flow and restore the locations of objects. Additionally, PSPNet [16], DeepLabv3+ [17] and DenseASPP [18] leverage a pyramid pooling module to extract context information for better segmentation performance. Furthermore, GSCNN [19] employs a two-branch framework consisting of a shape branch and a regular branch, which can effectively improve the semantic predictions on the boundaries.

Different from the above-mentioned single-modal networks, many data-fusion networks have also been proposed to improve semantic segmentation accuracy by extracting and fusing the features from multi-modalities of visual information [20,21]. For instance, FuseNet [22] and depth-aware CNN [23] adopt the popular encoder-decoder architecture, but employ different operations to fuse the feature maps obtained from the RGB and depth branches. Moreover, RTFNet [24] was developed to improve semantic segmentation performance by fusing the features extracted from RGB images and thermal images. It also adopts an encoder-decoder architecture and an element-wise addition fusion strategy.

2.2 Attention Module

Due to their simplicity and effectiveness, AMs have been widely used in various computer vision tasks. AMs typically learn the weight distribution (WD) of an input feature map and output an updated feature map based on the learned WD [25]. Specifically, Squeeze-and-Excitation Network (SENet) [26] employs a channel-wise AM to improve image classification accuracy. Furthermore, Wang *et al.* [27] presented a non-local module to capture long-range dependencies for video classification. OCNet [28] and DANet [29] proposed different self-attention modules that are capable of using contextual information for semantic segmentation. Moreover, CCNet [30] adopts a criss-cross AM to obtain dense contextual information in a more efficient way. Different from the aforementioned studies, we propose an attention aggregation (AA) framework that focuses on the combination of different AMs. Based on this idea, our proposed AA-UNet and AA-RTFNet can take advantage of different AMs and yield accurate results for road pothole detection.

2.3 Adversarial Domain Adaptation

Since the concept of "generative adversarial network (GAN)" [31] was first introduced in 2014, great efforts have been made in this research area to improve the existing computer vision algorithms. The recipe for their success is the use of an adversarial loss, which makes the generated synthetic images become indistinguishable from the real images when minimized [32].

Recent image-to-image translation approaches typically utilize a dataset, which contains paired source and target images, to learn a parametric translation using CNNs. One of the most well-known work is the "pix2pix" framework [33] proposed by Isola *et al.*, which employs a conditional GAN to learn the mapping from source images to target images.

In addition to the paired image-to-image translation approaches mentioned above, many unsupervised approaches have also been proposed in recent years to tackle unpaired image-to-image translation problem, where the primary goal is to learn a mapping $G : \mathcal{S} \rightarrow \mathcal{T}$ from source domain $\mathcal{S}$ to target domain $\mathcal{T}$, so that the distribution of images from $G(\mathcal{S})$ is indistinguishable from the distribution $\mathcal{T}$. CycleGAN [32] is a representative work handling unpaired image-to-image translation, where an inverse mapping $F : \mathcal{T} \rightarrow \mathcal{S}$ and a cycle-consistency loss (aiming at forcing $F(G(\mathcal{S})) \approxeq \mathcal{S}$) were coupled with $G : \mathcal{S} \rightarrow \mathcal{T}$. Our proposed training set augmentation technique is developed based on CycleGAN [32], but it performs paired image-to-image translation.

3 Disparity (or Inverse Depth) Transformation

DT aims at transforming a disparity or inverse depth image $\mathbf{G}$ into a quasi bird's eye view, whereby the pixels in the undamaged road areas possess similar values, while they differ significantly from those of the pothole pixels.

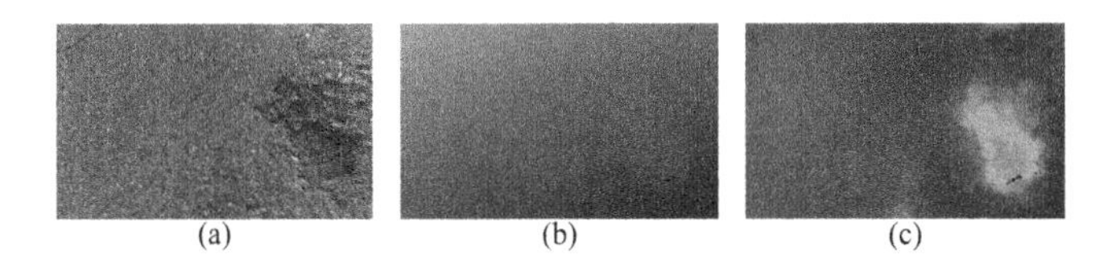

Fig. 1. Disparity transformation: (a) RGB image; (b) disparity image produced by PT-SRP [34]; and (c) transformed disparity image.

Since the concept of "v-disparity domain" was introduced in [35], disparity image analysis has become a common technique used for 3D driving scene understanding [8]. The projections of the on-road disparity (or inverse depth) pixels in the v-disparity domain can be represented by a non-linear model as follows:

$$\tilde{\mathbf{q}} = \mathbf{M}\tilde{\mathbf{p}} = \varkappa \begin{bmatrix} -\sin\Phi & \cos\Phi & \kappa \\ 0 & 1/\varkappa & 0 \\ 0 & 0 & 1/\varkappa \end{bmatrix} \tilde{\mathbf{p}}, \tag{1}$$

where $\tilde{\mathbf{p}} = [u, v, 1]^\top$ is the homogeneous coordinates of a pixel in the disparity (or inverse depth) image, and $\tilde{\mathbf{q}} = [g, v, 1]^\top$ is the homogeneous coordinates of its projection in the v-disparity domain. Φ can be estimated via [8]:

$$\arg\min_{\Phi} \mathbf{g}^\top\mathbf{g} - \mathbf{g}^\top\mathbf{T}(\Phi)\big(\mathbf{T}(\Phi)^\top\mathbf{T}(\Phi)\big)^{-1}\mathbf{T}(\Phi)^\top\mathbf{g}, \tag{2}$$

where $\mathbf{g}$ is a k-entry vector of disparity (or inverse depth) values, $\mathbf{1}_k$ is a k-entry vector of ones, $\mathbf{u}$ and $\mathbf{v}$ are two k-entry vectors storing the horizontal and vertical coordinates of the observed pixels, respectively, and $\mathbf{T}(\Phi) = [\mathbf{1}_k, \cos\Phi\mathbf{v} - \sin\Phi\mathbf{u}]$. (2) has a closed-form solution as follows [5]:

$$\Phi = \arctan\frac{\omega_4\omega_0 - \omega_3\omega_1 + q\sqrt{\Delta}}{\omega_3\omega_2 + \omega_5\omega_1 - \omega_5\omega_0 - \omega_4\omega_2} \quad \text{s.t. } q \in \{-1, 1\}, \tag{3}$$

where

$$\Delta = (\omega_4\omega_0 - \omega_3\omega_1)^2 + (\omega_3\omega_2 - \omega_5\omega_0)^2 - (\omega_4\omega_2 - \omega_5\omega_1)^2. \tag{4}$$

The expressions of ω_0-ω_5 are given in [5]. κ and $\varkappa$ can then be obtained using:

$$\mathbf{x} = \varkappa \begin{bmatrix} \kappa \\ 1 \end{bmatrix} = \big(\mathbf{T}(\Phi)^\top\mathbf{T}(\Phi)\big)^{-1}\mathbf{T}(\Phi)^\top\mathbf{g}. \tag{5}$$

DT can therefore be realized using [5]:

$$\mathbf{G}'(\mathbf{p}) = \mathbf{G}(\mathbf{p}) - \varkappa\big(\cos\Phi v - \sin\Phi u\big) - \varkappa\kappa + \Lambda, \tag{6}$$

where Λ is a constant used to ensure that the values in the transformed disparity (or depth inverse) image $\mathbf{G}'$ are non-negative. An example of the transformed disparity (or inverse depth) image is shown in Fig. 1, where it can be observed that the damaged road area becomes highly distinguishable. The effectiveness of DT on improving semantic segmentation is discussed in Sect. 6.4.

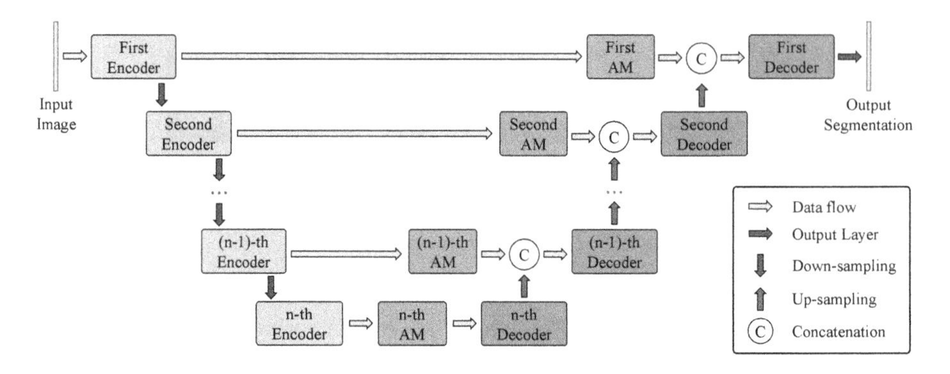

Fig. 2. The architecture of the proposed attention aggregation framework for our AA-UNet and AA-RTFNet.

4 Attention Aggregation Framework

The architecture of our proposed attention aggregation framework is illustrated in Fig. 2. We add different AMs into the existing CNNs that adopt the popular encoder-decoder architecture. Firstly, U-Net [15] has demonstrated the effectiveness of employing skip connections, which concatenate the same-scale feature maps produced by the encoder and decoder. However, these two feature maps can present large difference because of the different numbers of transformations undergone, which can result in significant performance degradation. To alleviate this drawback, we add an AM for the encoder feature map before the concatenation in each skip connection, as shown in Fig. 2 (from the 1st to $(n-1)$-th AMs), where n denotes the number of network levels. These AMs enable the encoder feature maps to focus on the potholes, which can shorten the gap between the same-scale feature maps produced by the encoder and decoder. This further improves pothole detection performance. Secondly, many studies [29,30] have already demonstrated that adding an AM for a high-level feature map can significantly improve the overall performance. Therefore, we follow this paradigm and add an AM at the highest level, as shown in Fig. 2 (n-th AM).

We use three AMs in our attention aggregation framework: 1) Channel Attention Module (CAM), 2) Position Attention Module (PAM) and 3) Dual Attention Module (DAM) [29], as illustrated in Fig. 3. Similar to SENet [26], our CAM is designed to assign each channel with a weight since some channels are more important. It first employs a global average pooling layer to squeeze spatial information, and then utilizes fully connected (FC) layers to generate the WD, which is finally combined with the input feature map by element-wise multiplication operation to generate the output feature map. Different from CAM, our PAM focuses on spatial information. It first generates the spatial WD and applies it on the input feature map to generate the output feature map. DAM [29] is composed of a channel attention submodule and a position attention submodule. Different from our CAM and PAM, these two submodules adopt the self-attention scheme to generate the WD, which can achieve better performance at the expense of

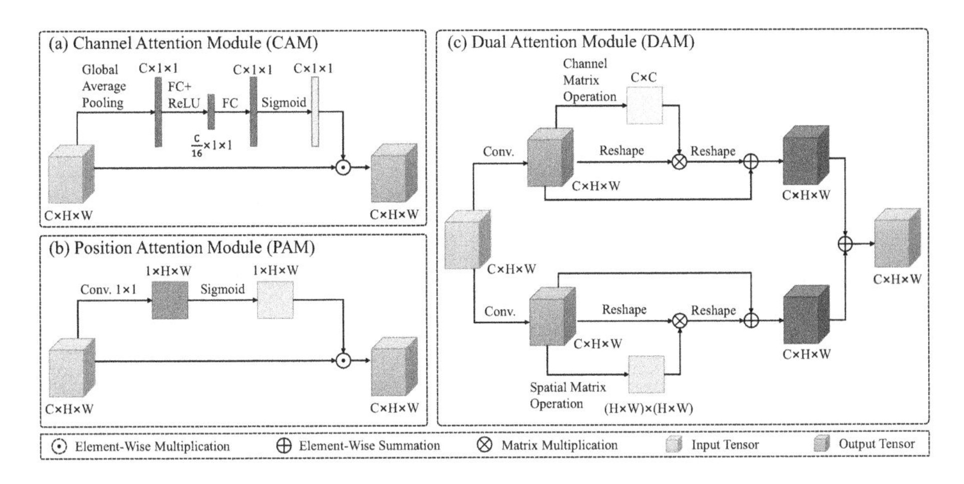

Fig. 3. The illustrations of the three AMs used in our attention aggregation framework.

a higher computational complexity. Since the memory consumed by DAM will grow significantly with the increase of feature map size, we only use it at the highest level (n-th AM) so as to ensure computational efficiency.

To demonstrate the effectiveness of our framework, we employ it in a single-modal network (U-Net) and a data-fusion network (RTFNet), and dub them as AA-UNet and AA-RTFNet, respectively. The specific architectures (the selection of each AM) of our AA-UNet and AA-RTFNet are discussed in Sect. 6.3.

5 Adversarial Domain Adaptation for Training Set Augmentation

In this paper, adversarial domain adaptation is utilized to augment training set so that the semantic segmentation networks can perform more robustly. Our proposed training set augmentation framework is illustrated in Fig. 4, where $F_1 : \mathcal{S}_1 \rightarrow \mathcal{T}$ translates RGB images $s_{1_i} \in \mathcal{S}_1$ to pothole detection ground truth $t_i \in \mathcal{T}$; $G_1 : \mathcal{T} \rightarrow \mathcal{S}_1$ translates pothole detection ground truth $t_i \in \mathcal{T}$ back to RGB images $s_{1_i} \in \mathcal{S}_1$; $F_2 : \mathcal{S}_2 \rightarrow \mathcal{T}$ translates our transformed disparity images $s_{2_i} \in \mathcal{S}_2$ to pothole detection ground truth $t_i \in \mathcal{T}$; and $G_2 : \mathcal{T} \rightarrow \mathcal{S}_2$ translates pothole detection ground truth $t_i \in \mathcal{T}$ back to our transformed disparity images $s_{2_i} \in \mathcal{S}_2$. The learning of G_1 and G_2 is guided by the intra-class means. Our full objective is:

$$\begin{aligned}\mathcal{L}(G_1, G_2, F_1, F_2, D_{\mathcal{S}_1}, D_{\mathcal{S}_2}, D_{\mathcal{T}}) = & \mathcal{L}_{\text{GAN}}(G_1, D_{\mathcal{S}_1}, \mathcal{T}, \mathcal{S}_1) + \mathcal{L}_{\text{GAN}}(F_1, D_{\mathcal{T}}, \mathcal{S}_1, \mathcal{T}) \\ & + \mathcal{L}_{\text{GAN}}(G_2, D_{\mathcal{S}_2}, \mathcal{T}, \mathcal{S}_2) + \mathcal{L}_{\text{GAN}}(F_2, D_{\mathcal{T}}, \mathcal{S}_2, \mathcal{T}) \\ & + \mathcal{L}_{\text{cyc}}(G_1, F_1) + \mathcal{L}_{\text{cyc}}(G_2, F_2),\end{aligned} \tag{7}$$

where

$$\mathcal{L}_{GAN}(G, D_{\mathcal{S}}, \mathcal{T}, \mathcal{S}) = \mathbb{E}_{s \sim p_{\text{data}}(s)}[\log D_{\mathcal{S}}(s)] + \mathbb{E}_{t \sim p_{\text{data}}(t)}[\log(1 - D_{\mathcal{S}}(G(t)))], \tag{8}$$

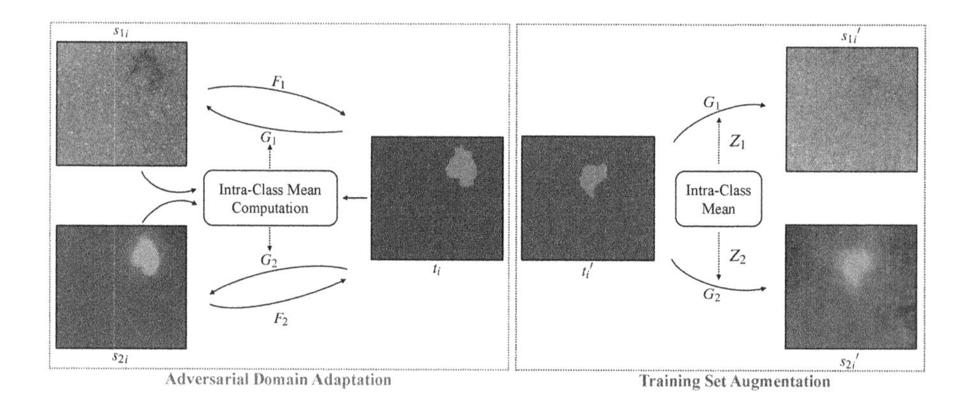

Fig. 4. Adversarial domain adaptation for training set augmentation.

$$\mathcal{L}_{GAN}(F, D_{\mathcal{T}}, \mathcal{S}, \mathcal{T}) = \mathbb{E}_{t \sim p_{\text{data}}(t)}[\log D_{\mathcal{T}}(t)] + \mathbb{E}_{s \sim p_{\text{data}}(s)}[\log(1 - D_{\mathcal{T}}(F(s)))], \quad (9)$$

$$\mathcal{L}_{cyc}(G, F) = \mathbb{E}_{s \sim p_{\text{data}}(s)}[G(F(s)) - s] + \mathbb{E}_{t \sim p_{\text{data}}(t)}[F(G(t)) - t], \quad (10)$$

$D_{\mathcal{S}}$ and $D_{\mathcal{T}}$ are two adversarial discriminators: $D_{\mathcal{S}}$ aims to distinguish between images $\{s\}$ and the translated images $\{G(t)\}$, while $D_{\mathcal{T}}$ aims to distinguish between images $\{t\}$ and the translated images $\{F(s)\}$; $s \sim p_{\text{data}}(s)$ and $t \sim p_{\text{data}}(t)$ denote the data distributions of the source and target domains, respectively.

With well-learned mapping functions G_1 and G_2, we can generate an infinite number of synthetic RGB images ${s_1}_i' \in \mathcal{S}_1'$ and their corresponding synthetic transformed disparity images ${s_2}_i' \in \mathcal{S}_2'$ from a randomly generated pothole detection ground truth $t_i' \in \mathcal{T}'$. In order to expand the distributions of the two domains $s_1' \sim p_{\text{data}}(s_1')$ and $s_2' \sim p_{\text{data}}(s_2')$, we add random Gaussian noises Z_1 and Z_2 into G_1 and G_2 when generating ${s_1}_i'$ and ${s_2}_i'$, as shown in Fig. 4. Some examples in the augmented training set are shown in Fig. 5. The benefits of our proposed training set augmentation technique for semantic segmentation are discussed in Sect. 6.4.

6 Experiments

6.1 Datasets

Pothole-600. In our experiments, we utilized a stereo camera to capture stereo road images. These images are then split into a training set, a validation set and a testing set, which contains 240, 180 and 180 pairs of RGB images and transformed disparity images, respectively.

Augmented Training Set. We use adversarial domain adaptation to produce an augmented training set, which contains 2,400 pairs of RGB images and transformed disparity images. The performance comparison between using the original training set and using the augmented training set is presented in Sect. 6.4.

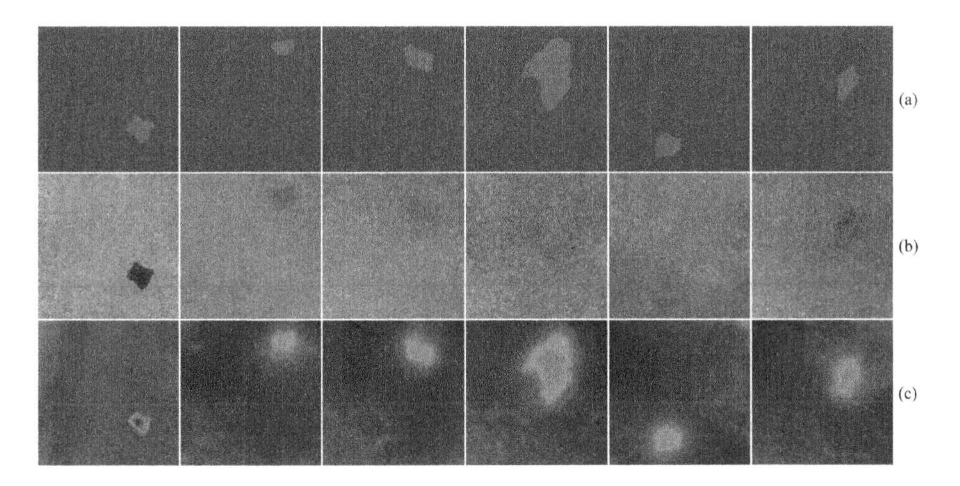

Fig. 5. Examples of training set augmentation results: (a) randomly created pothole detection ground truth; (b) generated RGB images; and (c) generated transformed disparity images.

6.2 Experimental Setup

In our experiments, we first select the architecture of our AA-UNet and AA-RTFNet, as presented in Sect. 6.3. Then, we compare our AA-UNet and AA-RTFNet with eight state-of-the-art (SoA) CNNs (five single-modal ones and three data-fusion ones) for road pothole detection. Each single-modal CNN is trained using RGB images (RGB) and transformed disparity images (T-Disp), respectively; while each data-fusion CNN is trained using RGB and transformed disparity images (RGB+T-Disp). Furthermore, we also select different numbers of RGB images and transformed disparity images from our augmented training set to train the CNNs. The experimental results are presented in Sect. 6.4.

To quantify the performance of these CNNs, we adopt the commonly used F-score (Fsc) and intersection over union (IoU) metrics, and compute their mean values across the testing set, denoted as mFsc and mIoU, respectively. Moreover, the stochastic gradient descent with momentum (SGDM) [36] is used to optimize the CNNs.

6.3 Architecture Selection of AA-UNet and AA-RTFNet

In this subsection, we conduct experiments to select the best architecture for our AA-UNet and AA-RTFNet. All the AA-UNet variants use the same training setups, so do all the AA-RTFNet variants. It should be noted here that $n = 5$ is for both AA-UNet and AA-RTFNet. We also record the inference time of each variant on an NVIDIA GTX 1080Ti graphics card for comparison. (B)–(L) in Table 1 present the effects of a single AM at different network levels. We can see that an AM can bring in better performance improvement when it is added at a higher level, as this can influence the subsequent processes. Moreover, DAM

Table 1. Performances of different AA-UNet variants on the Pothole-600 validation set, where (A) is the U-Net baseline; and (B)–(T) are different variants. Best Results are shown in bold type.

No.	Attention aggregation scheme					Evaluation metrics		
	1st	2nd	3rd	4th	5th	mFsc (%)	mIoU (%)	Runtime (ms)
(A)	–	–	–	–	–	75.9	61.2	**31.3**
(B)	–	–	–	–	DAM	79.7	66.3	33.7
(C)	–	–	–	–	CAM	79.5	66.0	31.5
(D)	–	–	–	–	PAM	79.4	65.9	31.7
(E)	–	–	–	CAM	–	78.7	64.9	31.6
(F)	–	–	–	PAM	–	78.5	64.6	31.9
(G)	–	–	CAM	–	–	78.0	63.9	31.8
(H)	–	–	PAM	–	–	77.7	63.5	32.0
(I)	–	CAM	–	–	–	77.8	63.6	32.1
(J)	–	PAM	–	–	–	77.5	63.2	32.6
(K)	CAM	–	–	–	–	77.6	63.4	32.3
(L)	PAM	–	–	–	–	77.8	63.7	33.5
(M)	–	–	–	CAM	DAM	80.2	66.9	33.8
(N)	–	–	–	PAM	DAM	77.1	62.7	34.0
(O)	–	–	CAM	CAM	DAM	80.7	67.6	33.9
(P)	–	–	PAM	CAM	DAM	77.8	63.6	34.2
(Q)	–	CAM	CAM	CAM	DAM	81.0	68.0	34.1
(R)	–	PAM	CAM	CAM	DAM	79.7	66.2	34.5
(S)	CAM	CAM	CAM	CAM	DAM	81.3	68.5	34.3
(T)	PAM	CAM	CAM	CAM	DAM	**82.6**	**70.3**	34.7

outperforms CAM and PAM at the highest level, since DAM adopts the self-attention scheme, which can achieve better performance, as mentioned above. Furthermore, our CAM performs better than our PAM at higher levels, since feature maps at higher levels have more channels but limited spatial sizes and it is more useful to apply weights on channels. Conversely, feature maps at lower levels have larger spatial sizes but limited channels, and thus it is more useful to adopt our PAM.

Based on these observations, we test the performance of different attention aggregation schemes for our AA-UNet and AA-RTFNet on the validation set, as shown on (M)–(T) in Table 1 and (B)–(J) in Table 2, respectively. We can see that adopting PAM at the lowest network level, adopting DAM at the highest network level, and adopting CAM at other network levels can achieve the best performance for both AA-UNet and AA-RTFNet. Compared with the baseline models, our AA-UNet and AA-RTFNet can increase the mIoU by 9.1% and 5.4%,

Table 2. Performances of different AA-RTFNet variants on the Pothole-600 validation set, where (A) is the RTFNet baseline; and (B)–(J) are different variants. Best Results are shown in bold type.

No.	Attention aggregation scheme					Evaluation metrics		
	1st	2nd	3rd	4th	5th	mFsc (%)	mIoU (%)	Runtime (ms)
(A)	–	–	–	–	–	81.3	68.5	**46.7**
(B)	–	–	–	–	DAM	82.5	70.2	49.1
(C)	–	–	–	CAM	DAM	82.6	70.4	49.2
(D)	–	–	–	PAM	DAM	81.7	69.0	49.4
(E)	–	–	CAM	CAM	DAM	82.8	70.7	49.3
(F)	–	–	PAM	CAM	DAM	81.9	69.3	49.7
(G)	–	CAM	CAM	CAM	DAM	83.4	71.6	49.6
(H)	–	PAM	CAM	CAM	DAM	83.1	71.1	49.9
(I)	CAM	CAM	CAM	CAM	DAM	84.1	72.5	50.0
(J)	PAM	CAM	CAM	CAM	DAM	**85.0**	**73.9**	50.2

respectively, with acceptable extra runtime, which demonstrates the effectiveness and efficiency of our attention aggregation framework.

6.4 Performance Evaluation of Road Pothole Detection

In this subsection, we evaluate the performance of our AA-UNet and AA-RTFNet both qualitatively and quantitatively on the testing set. As mentioned previously, we use different numbers of images selected from the augmented training set to train each CNN. λ denotes the number of samples used in the augmented training set versus the number of samples in the original training set. For example, $\lambda = 2$ means that we train the CNN with $240 \times 2 = 480$ samples randomly selected from the augmented training set. In addition, we introduce a new evaluation metric δ for better comparison. For a given training setup, δ is defined as ratio of the number of iterations for the network to converge using the augmented training set to that of the original training set. $\delta < 1$ means that the training setup converges faster than the baseline setup.

The quantitative results are shown in Fig. 6, where we can clearly observe that the single-modal CNNs with our transformed disparity images as inputs generally perform better than they do with RGB images as inputs, and the mIoU increases by about 17–31%. This is because our transformed disparity images can make the road potholes become highly distinguishable, and can thus benefit all CNNs for road pothole detection. Moreover, we can see that when $\lambda \geq 4$, the CNNs trained with the augmented training set generally outperform themselves when trained with the original training set, and $\delta < 1$ holds in most cases, which demonstrates that adversarial domain adaptation can not only significantly improve pothole detection accuracy but can also accelerate the network convergence. Compared

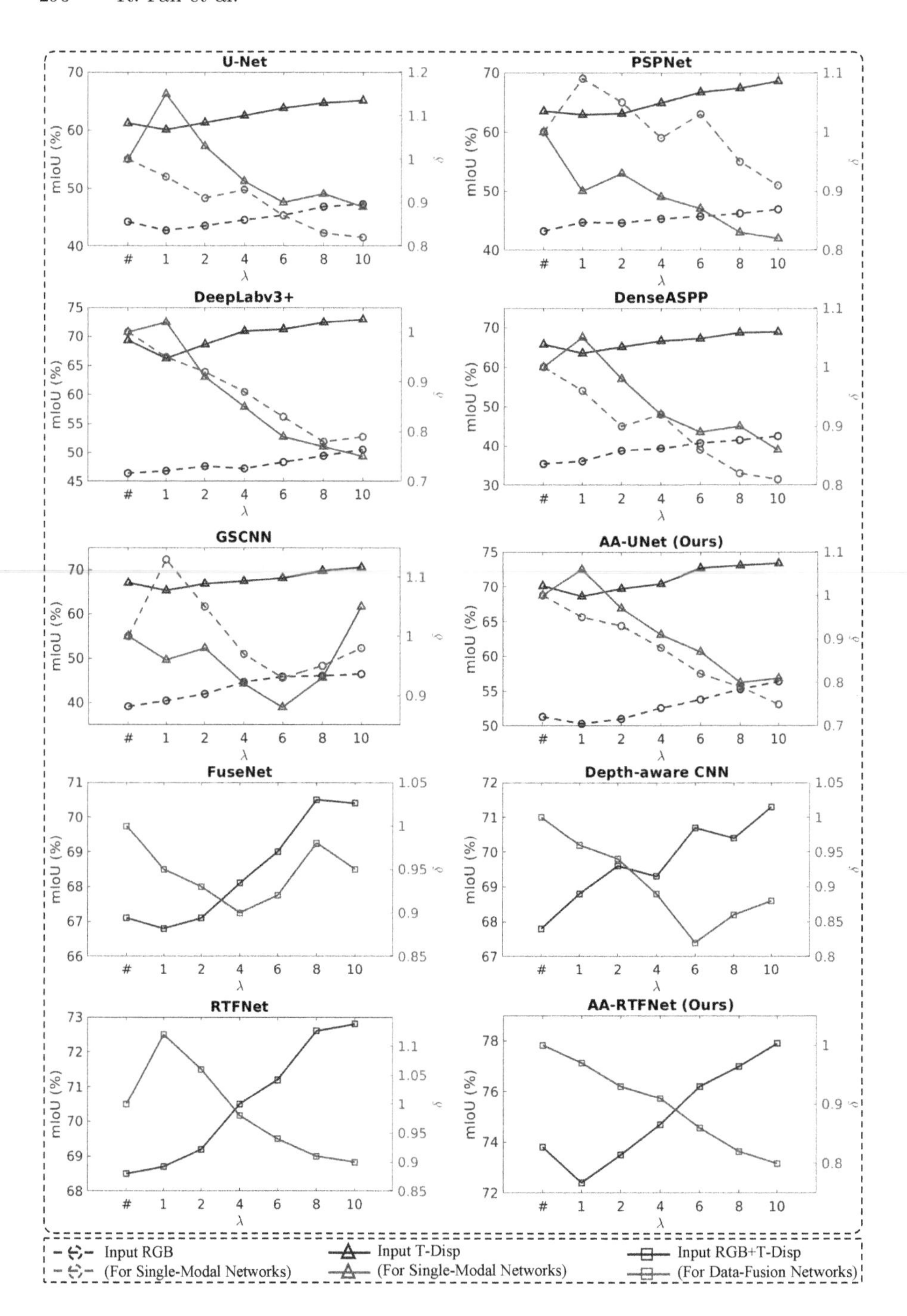

Fig. 6. Performance comparison among eight SoA CNNs, AA-UNet and AA-RTFNet on the Pothole-600 testing set, where the symbol "#" in the λ axis means that we use the original training set in the CNN.

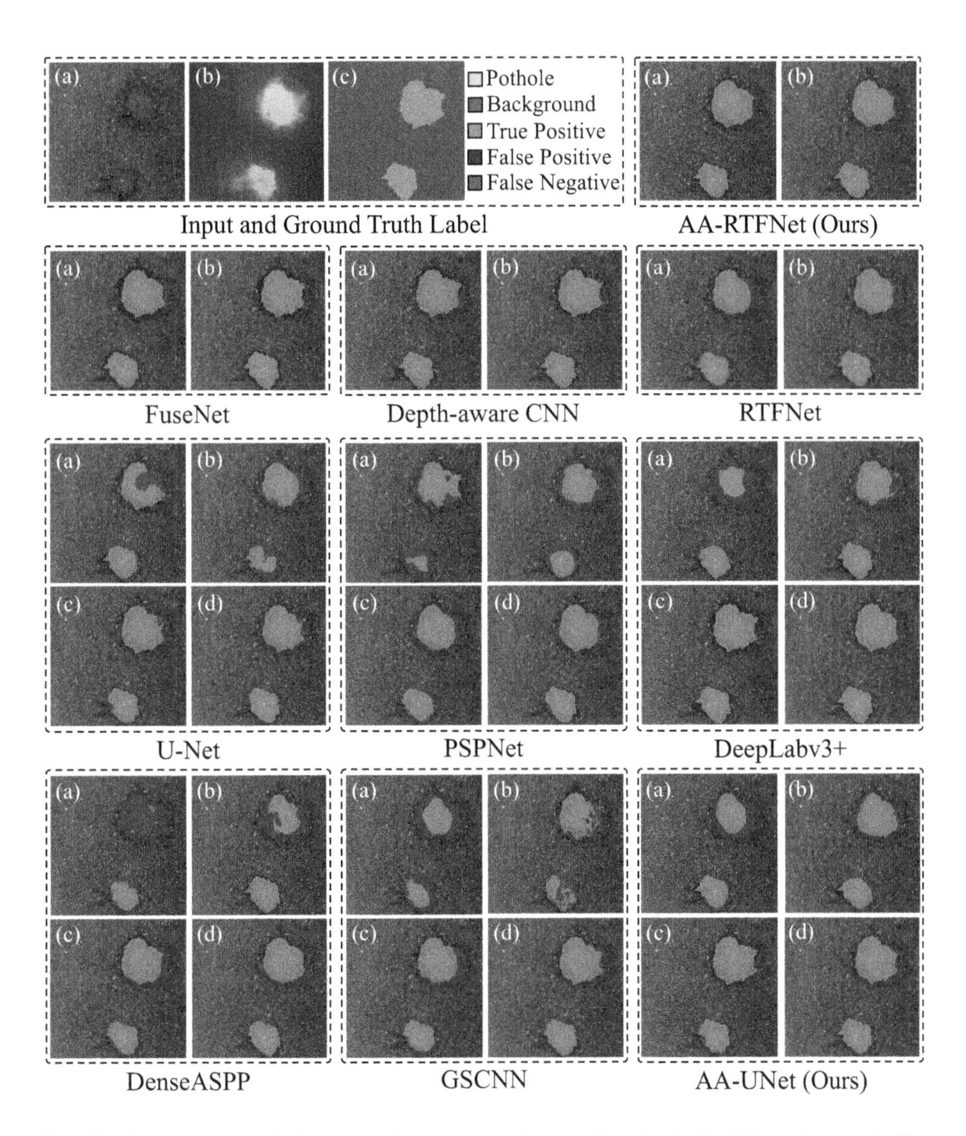

Fig. 7. An example of the experimental results on the Pothole-600 testing set. For the input and ground truth label block: (a) RGB, (b) T-Disp, and (c) ground truth label; For the single-modal network (including U-Net [15], PSPNet [16], DeepLabv3+ [17], DenseASPP [18], GSCNN [19] and our AA-UNet) blocks: (a) input RGB from the original training set, (b) input RGB from the whole augmented training set, (c) input T-Disp from the original training set, and (d) input T-Disp from the whole augmented training set; For the data-fusion network (including FuseNet [22], Depth-aware CNN [23], RTFNet [24] and our AA-RTFNet) blocks: (a) input RGB+T-Disp from the original training set, and (b) input RGB+T-Disp from the whole augmented training set.

with the training setup using the original training set, an increase of around 3–8% is witnessed on the mIoU for the training setup using the whole augmented training set. This is because these two sets share very similar distributions, and our augmented training set possesses an expanded distribution, which can improve road pothole detection performance. In addition, our AA-UNet and AA-RTFNet outperform all other SoA single-modal and data-fusion networks for road pothole detection, respectively, which strongly validates the effectiveness and efficiency of our attention aggregation framework. Readers can see that our AA-UNet can increase the mIoU by approximately 3–14% compared with the SoA single-modal networks, and our AA-RTFNet can increase the mIoU by about 5–8% compared with the SoA data-fusion networks. The qualitative results shown in Fig. 7 can also confirm the superiority of our proposed approaches.

7 Conclusion

The major contributions of this paper include: a) a novel attention aggregation framework, which can help the CNNs focus more on salient objects, such as road potholes, so as to improve semantic segmentation for better pothole detection results; b) a novel training set augmentation technique developed based on adversarial domain adaptation, which can produce more synthetic road RGB images and their corresponding transformed road disparity (or inverse depth) images to improve both the efficiency and accuracy of CNN training; c) a large-scale road pothole detection dataset, publicly available at http://sites.google.com/view/pothole-600 for research purposes. The experimental results validated the effectiveness and feasibility of our proposed attention aggregation framework and the training set augmentation technique for enhancing road pothole detection. Moreover, we believe our proposed techniques can also be used for many other semantic segmentation applications, such as freespace detection.

Acknowledgements. This work was supported by the National Natural Science Foundation of China, under grant No. U1713211, Collaborative Research Fund by Research Grants Council Hong Kong, under Project No. C4063-18G, and the Research Grant Council of Hong Kong SAR Government, China, under Project No. 11210017, awarded to Prof. Ming Liu.

References

1. Mathavan, S., Kamal, K., Rahman, M.: A review of three-dimensional imaging technologies for pavement distress detection and measurements. IEEE Trans. Intell. Transp. Syst. **16**(5), 2353–2362 (2015)
2. Fan, R., Ozgunalp, U., Hosking, B., Liu, M., Pitas, I.: Pothole detection based on disparity transformation and road surface modeling. IEEE Trans. Image Process. **29**, 897–908 (2019)
3. Koch, C., Georgieva, K., Kasireddy, V., Akinci, B., Fieguth, P.: A review on computer vision based defect detection and condition assessment of concrete and asphalt civil infrastructure. Adv. Eng. Inf. **29**(2), 196–210 (2015)

4. Majendie, M.: Dani king: 'it was just a freak accident but I thought I was going to die'. Technical report, Independent, June 2015
5. Fan, R., Liu, M.: Road damage detection based on unsupervised disparity map segmentation. IEEE Trans. Intell. Transp. Syst. **21**, 4906–4911 (2019)
6. Devine, R.: City of San Diego asking residents to report potholes. Technical report, NBC San Diego, January 2017
7. News, B.: Government to pledge billions for filling potholes. Technical report, BBC News, March 2020
8. Fan, R., Jiao, J., Pan, J., Huang, H., Shen, S., Liu, M.: Real-time dense stereo embedded in a UAV for road inspection. In: Proceedings of the IEEE Conference on Computer Vision and Pattern Recognition (CVPR) Workshops, pp. 535–543. IEEE (2019)
9. Leo, M., Furnari, A., Medioni, G.G., Trivedi, M., Farinella, G.M.: Deep learning for assistive computer vision. In: Proceedings of the European Conference on Computer Vision (ECCV), pp. 0–0, (2018)
10. Rover, J.L.: Pothole detection technology research announced by jaguar land rover
11. Baraniuk, C.: Ford developing pothole alert system for drivers, February 2017
12. Jahanshahi, M.R., Jazizadeh, F., Masri, S.F., Becerik-Gerber, B.: Unsupervised approach for autonomous pavement-defect detection and quantification using an inexpensive depth sensor. J. Comput. Civ. Eng. **27**(6), 743–754 (2013)
13. Otsu, N.: A threshold selection method from gray-level histograms. IEEE Trans. Syst. Man Cybern. **9**(1), 62–66 (1979)
14. Long, J., Shelhamer, E., Darrell, T.: Fully convolutional networks for semantic segmentation. In: Proceedings of the IEEE Conference on Computer Vision and Pattern Recognition, pp. 3431–3440 (2015)
15. Ronneberger, O., Fischer, P., Brox, T.: U-Net: convolutional networks for biomedical image segmentation. In: Navab, N., Hornegger, J., Wells, W.M., Frangi, A.F. (eds.) MICCAI 2015. LNCS, vol. 9351, pp. 234–241. Springer, Cham (2015). https://doi.org/10.1007/978-3-319-24574-4_28
16. Zhao, H., Shi, J., Qi, X., Wang, X., Jia, J.: Pyramid scene parsing network. In: Proceedings of the IEEE Conference on Computer Vision and Pattern Recognition, pp. 2881–2890 (2017)
17. Chen, L.C., Zhu, Y., Papandreou, G., Schroff, F., Adam, H.: Encoder-decoder with atrous separable convolution for semantic image segmentation. In: Proceedings of the European Conference on Computer Vision (ECCV), pp. 801–818 (2018)
18. Yang, M., Yu, K., Zhang, C., Li, Z., Yang, K.: DenseASPP for semantic segmentation in street scenes. In: Proceedings of the IEEE Conference on Computer Vision and Pattern Recognition, 3684–3692 (2018)
19. Takikawa, T., Acuna, D., Jampani, V., Fidler, S.: Gated-SCNN: gated shape CNNs for semantic segmentation. In: Proceedings of the IEEE International Conference on Computer Vision, pp. 5229–5238 (2019)
20. Wang, H., Fan, R., Sun, Y., Liu, M.: Applying surface normal information in drivable area and road anomaly detection for ground mobile robots. In: 2020 IEEE/RSJ International Conference on Intelligent Robots and Systems (IROS) (2020). (To be published)
21. Fan, R., Wang, H., Cai, P., Liu, M.: SNE-RoadSeg: incorporating surface normal information into semantic segmentation for accurate freespace detection. In: Vedaldi, A., Bischof, H., Brox, T., Frahm, J.-M. (eds.) ECCV 2020. LNCS, vol. 12375, pp. 340–356. Springer, Cham (2020). https://doi.org/10.1007/978-3-030-58577-8_21

22. Hazirbas, C., Ma, L., Domokos, C., Cremers, D.: FuseNet: incorporating depth into semantic segmentation via fusion-based CNN architecture. In: Lai, S.-H., Lepetit, V., Nishino, K., Sato, Y. (eds.) ACCV 2016. LNCS, vol. 10111, pp. 213–228. Springer, Cham (2017). https://doi.org/10.1007/978-3-319-54181-5_14
23. Wang, W., Neumann, U.: Depth-aware CNN for RGB-D segmentation. In: Proceedings of the European Conference on Computer Vision (ECCV), pp. 135–150 (2018)
24. Sun, Y., Zuo, W., Liu, M.: RTFNet: RGB-thermal fusion network for semantic segmentation of urban scenes. IEEE Robot. Autom. Lett. **4**(3), 2576–2583 (2019)
25. Vaswani, A., et al.: Attention is all you need. In: Advances in Neural Information Processing Systems, pp. 5998–6008 (2017)
26. Hu, J., Shen, L., Sun, G.: Squeeze-and-excitation networks. In: Proceedings of the IEEE Conference on Computer Vision and Pattern Recognition, pp. 7132–7141 (2018)
27. Wang, X., Girshick, R., Gupta, A., He, K.: Non-local neural networks. In: Proceedings of the IEEE Conference on Computer Vision and Pattern Recognition, pp. 7794–7803 (2018)
28. Yuan, Y., Wang, J.: OCNet: object context network for scene parsing. arXiv preprint arXiv:1809.00916 (2018)
29. Fu, J., et al.: Dual attention network for scene segmentation. In: Proceedings of the IEEE Conference on Computer Vision and Pattern Recognition, pp. 3146–3154 (2019)
30. Huang, Z., Wang, X., Huang, L., Huang, C., Wei, Y., Liu, W.: CCNet: criss-cross attention for semantic segmentation. In: Proceedings of the IEEE International Conference on Computer Vision, pp. 603–612 (2019)
31. Goodfellow, I., et al.: Generative adversarial nets. In: Advances in Neural Information Processing Systems, pp. 2672–2680 (2014)
32. Zhu, J.Y., Park, T., Isola, P., Efros, A.A.: Unpaired image-to-image translation using cycle-consistent adversarial networks. In: Proceedings of the IEEE International Conference on Computer Vision, pp. 2223–2232 (2017)
33. Isola, P., Zhu, J.Y., Zhou, T., Efros, A.A.: Image-to-image translation with conditional adversarial networks. In: Proceedings of the IEEE Conference on Computer Vision and Pattern Recognition, pp. 1125–1134 (2017)
34. Fan, R., Ai, X., Dahnoun, N.: Road surface 3D reconstruction based on dense subpixel disparity map estimation. IEEE Trans. Image Process. **27**(6), 3025–3035 (2018)
35. Labayrade, R., Aubert, D.: A single framework for vehicle roll, pitch, yaw estimation and obstacles detection by stereovision. In: IEEE IV2003 Intelligent Vehicles Symposium. Proceedings (Cat. No. 03TH8683), pp. 31–36. IEEE (2003)
36. LeCun, Y., Bengio, Y., Hinton, G.: Deep learning. Nature **521**(7553), 436–444 (2015)

Multi-channel Transformers for Multi-articulatory Sign Language Translation

Necati Cihan Camgoz[1(✉)], Oscar Koller[2], Simon Hadfield[1], and Richard Bowden[1]

[1] CVSSP, University of Surrey, Guildford, UK
{n.camgoz,s.hadfield,r.bowden}@surrey.ac.uk
[2] Microsoft, Munich, Germany
oscar.koller@microsoft.com

Abstract. Sign languages use multiple asynchronous information channels (articulators), not just the hands but also the face and body, which computational approaches often ignore. In this paper we tackle the multi-articulatory sign language translation task and propose a novel multi-channel transformer architecture. The proposed architecture allows both the inter and intra contextual relationships between different sign articulators to be modelled within the transformer network itself, while also maintaining channel specific information. We evaluate our approach on the RWTH-PHOENIX-Weather-2014T dataset and report competitive translation performance. Importantly, we overcome the reliance on gloss annotations which underpin other state-of-the-art approaches, thereby removing the need for expensive curated datasets.

Keywords: Sign language translation · Multi-channel · Sequence-to-sequence

1 Introduction

Sign languages are the main medium of communication of the Deaf. Every country typically has its own sign language and although some grammatical structures are shared, as are signs that rely upon heavy iconicity, different sign language have unique vocabularies [57,59]. Contrary to spoken and written languages, sign languages are visual. This makes automatic sign language understanding a novel research field where computer vision and natural language processing meet with a view to understanding and translating the spatio-temporal linguistic constructs of sign [7].

Signers use multiple channels to convey information [61]. These channels, also known as articulators in linguistics [46], can be grouped under two main

Electronic supplementary material The online version of this chapter (https://doi.org/10.1007/978-3-030-66823-5_18) contains supplementary material, which is available to authorized users.

A. Bartoli and A. Fusiello (Eds.): ECCV 2020 Workshops, LNCS 12538, pp. 301–319, 2020.
https://doi.org/10.1007/978-3-030-66823-5_18

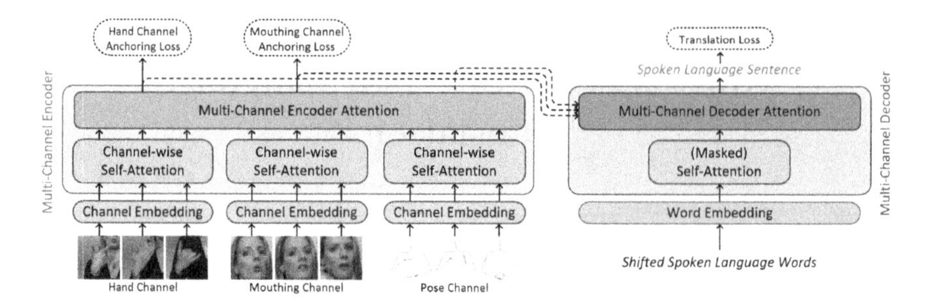

Fig. 1. An overview of the proposed Multi-channel Transformer architecture applied to the multi-articulatory SLT task.

categories with respect to their role in conveying information, namely manual and non-manual features [8]. Manual features include the hand shape and its motion. Although manual features can be considered as the dominant part of the sign morphology, they alone do not encapsulate the full context of the conveyed information. To give clarity, emphasis and additional meaning, signers use non-manual features, such as facial expressions, mouth gestures, mouthings[1] and body pose. Furthermore, both manual and non-manual features effect each other's meaning when used together.

To date, the literature in the field has predominantly focused on using the manual features to realize sign language recognition and translation [13,39,50], thus ignoring the rich and essential information contained in the non-manual features. This focus on the manual features is partially responsible for the common misconception that sign language recognition and translation problems are special sub-tasks of the gesture recognition field [54]. Sign language is as rich and complex as any spoken language. However, the multi-channel nature adds additional complexity as channels are not synchronised.

In contrast to much of the existing literature, in this paper we model sign language by incorporating both manual and non-manual features into SLT. To achieve this, we utilize multiple channels which correspond to the articulatory subunits of the sign, namely hand shape, upper body pose and mouthings. We explore several approaches to combine the information present in these channels using both early and late fusion in a transformer architecture. Based on these findings we then introduce a novel deep learning architecture, the Multi-channel Transformer. This approach incorporates both inter and intra channel contextual relationships to learn meaningful spatio-temporal representations of asynchronous sign articulators, while also preserving channel specific information by using anchoring losses. Although this approach was designed specifically for SLT, we believe it can also be used to tackle other multi-channel sequence-to-sequence learning tasks, such as audio-visual speech recognition [1]. An overview of the Multi-channel Transformer in the context of SLT can be seen in Fig. 1.

[1] Mouthings are lip patterns that accompany a sign.

We evaluate our approach on the challenging RWTH-PHOENIX-Weather-2014**T** (PHOENIX14**T**) dataset which provides both sign gloss[2] annotations and spoken language translations. Previous approaches [12,13] on PHOENIX14**T** heavily relied upon sign gloss annotations, which are labor intensive to obtain. We aim to remove this dependency on gloss annotation, by utilizing channel specific features obtained from related tasks, such as human pose estimation approaches [14,27] to represent upper body pose channel or lip reading features [3,19] to represent mouthings [37,38]. Removing the dependency on manual annotation allows our approach to be scaled beyond what is possible with previous techniques, potentially using huge collections of un-annotated data. We empirically show that by integrating multiple articulator channels into our multi-channel transformer, it is possible to achieve competitive SLT performance which is *on par* with models trained using additional gloss annotation.

The contributions of the paper can be summarized as: (1) We overcome the need for expensive gloss-level supervision by combining multiple articulatory channels with anchoring losses to achieve competitive continuous SLT performance on the PHOENIX14**T** dataset. (2) We propose a novel multi-channel transformer architecture that supports multi-channel, asynchronous, sequence-to-sequence learning and (3) We use this to introduce the first successful approach to multi-articulatory SLT, which models the inter and intra relationship of manual and non-manual channels.

2 Literature Review

Computational processing of sign languages is an important field and expected to have tremendous impact on language deprivation of Deaf children, accessibility, sign linguistics and human-computer interaction in general. Its first attempt dates back more than thirty years: A patent describing a hardwired electronic glove that recognized American Sign Language (ASL) finger spelling from hand configurations [24]. In the early days the field moved slowly, focusing first on isolated [60], then continuous sign language recognition [55]. With the rise of deep learning, enthusiasm was revived and accelerated the field [10,39,41]. The recognition of limited domain but continuous real-life sign language became feasible [11,20,21,28,40]. Driven by linguistic evidence [5,52,65], the field realized that sign language recognition needs to focus on more than just the hands. Earlier works looked at several modalities separately, such as the face in general [36,63], head pose [45], the mouth [2,37,38], eye-gaze [15], and body pose [17,53]. More recently, multi-stream architectures showed strong performance [35,68].

Nevertheless, sign recognition only addresses part of the communication barrier between Deaf and hearing people. Sign languages follow a distinct grammar and are not word by word translations of spoken languages. After successful recognition, reordering and mapping into the target spoken language complete the communication pipeline. In early works, recognition and translation were

[2] Glosses can be considered as the minimal lexical items of the sign languages.

treated as two independent processing steps. Isolated single signs were recognized and subsequently translated [16]. Often, existing work exclusively considered the problem as a text-to-text translation problem [9,56], despite the visual nature of sign language and the lack of a written representation.

Generally speaking, much of the available sign translation literature falsely declares sign recognition as sign translation [22,25,26,64]. Camgoz *et al.* [13] were the first to release a joint recognition and translation corpus with videos, glosses and translations to spoken language, covering real-life sign language, recorded from the broadcast news. They proposed to tackle the task based on a Neural Machine Translation (NMT) framework relying on input tokenization of the videos and subsequent sequence-to-sequence networks with attention. Their best performing tokenization method was based on strong sign recognition models trained using gloss annotations with full video frames and achieved an 18.1 BLEU-4 score, while a simple tokenization scheme (not trained with glosses) only reached 9.6 BLEU-4 on the test set of PHOENIX14**T**. Orbay and Akarun [48] investigated different tokenization methods on the same corpus and showed again that a pretrained hand shape recognizer [39] outperforms simpler approaches and reaches 14.6 BLEU-4. While they also investigated transformer architectures and multiple hands as input, the results underperformed. Ko *et al.* [34] describe a non-public dataset covering sign language videos, gloss annotation and translation. Their method relies on detected body keypoints only. It hence misses the important appearance based characteristics of sign. More recently, Camgoz *et al.* [12] proposed Sign Language Transformers, a joint end-to-end sign language recognition and translation approach. They used pre-trained gloss representations as inputs to their networks and trained transformer encoders using gloss annotations to learn meaningful spatio-temporal representations for SLT. Their approach is the current state-of-the-art on PHOENIX14**T**. They report 20.2 BLEU-4 for pre-trained gloss features to spoken language translation, and 21.3 BLEU-4 with the additional gloss recognition supervision.

Overall, previous work in the space of SLT has two major short-comings, which we intend to address with this paper: (1) The beauty of translations is the abundance of available training data, as they can be created in real-time by interpreters. Glosses are expensive to create and limit data availability. No previous work was able to achieve competitive performance while not relying on glosses. (2) So far SLT has never considered multiple articulators.

3 Background on Neural Machine Translation

The objective of machine translation is to learn the conditional probability $p(\mathcal{Y}|\mathcal{X})$ where $\mathcal{X} = (x_1, ..., x_T)$ is a sentence from the source language with T tokens and $\mathcal{Y} = (y_1, ..., y_U)$ is the desired corresponding translation of said sentence in the target language. To learn this mapping using neural networks, Kalchbrenner *et al.* [32] proposed using an *encoder-decoder* architecture, where the source sentence is encoded into a fixed sized "context" vector which is then used to decode the target sentence. Cho *et al.* [18] and Sutzeker *et al.* [58] further improved this approach by assigning the encoding and decoding stages of translation to individual specialized Recurrent Neural Networks (RNNs).

The main drawback of RNN-based approaches are long term dependency issues. Although there have been practical solutions to this, such as source sentence reversing [58], the context vector is still of fixed size, and thus cannot perfectly encode arbitrarily long input sequences. To overcome the information bottleneck imposed by using the last hidden state of the RNN as the context vector, Bahdanau *et al.* [4] proposed an attention mechanism, which was a breakthrough in the field of NMT. The idea behind the attention mechanism is to use a soft-search over the encoder outputs at each step of target sentence decoding. This was realized by conditioning target word prediction on a context vector which is a weighted sum of the source sentence representations. The weighting in turn is done by a learnt scoring function which measures the relevance of the decoders current hidden state and the encoder outputs. Luong *et al.* [44] further improved this approach by proposing a dot product attention (scoring) function as:

$$\text{context} = \text{softmax}\left(QK^T\right)V \tag{1}$$

where queries, Q, correspond to the hidden state of the decoder at a given time step, and keys, K, and values, V, represent the encoder outputs.

More recently, Vaswani *et al.* [62] introduced self-attention mechanisms, which refine the source and target token representations by looking at the context they have been used in. Combining encoder and decoder self-attention layers with encoder-decoder attention, Vaswani *et al.* proposed Transformer networks, a fully connected network (as opposed to being RNN-based) which has revolutionized the field of machine translation. In contrast to RNN-based models, transformers obtain Q, K and V values by using individually learnt linear projection matrices at each attention layer. Vaswani *et al.* also introduced "scaled" dot-product attention as:

$$\text{context} = \text{softmax}\left(\frac{QK^T}{\sqrt{d_m}}\right)V \tag{2}$$

where d_m is the number of hidden units of the model. The motivation behind the scaling operation is to counteract the effect of gradients becoming extremely small in cases where the number of hidden units is high and in-turn, the dot products grow large [62].

In this work we extend the transformer network architecture and adapt it to the task of multi-channel sequence-to-sequence learning. We propose a multi-channel attention layer to refine the representations of each source channel in the context of other source channels, while maintaining channel specific information using anchoring losses. We also adapt the encoder-decoder attention layer to be able to use multiple source channel representations.

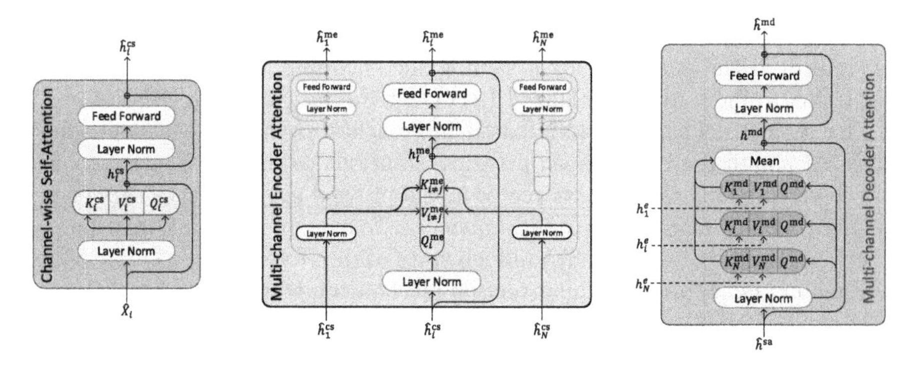

Fig. 2. A detailed overview of the introduced attention modules: (left) Channel-wise Self-Attention, (middle) Multi-channel Encoder Attention, and (right) Mult-channel Decoder Attention

4 Multi-channel Transformers

In this section we introduce Multi-channel Transformers, a novel architecture for sequence-to-sequence learning problems where the source information is embedded across several asynchronous channels. Given source sequences $\mathcal{X} = (X_1, ..., X_N)$, where X_i is the ith source channel with a cardinality of T_i, our objective is to learn the conditional probability $p(\mathcal{Y}|\mathcal{X})$, where $\mathcal{Y} = (y_1, ..., y_U)$ is the target sequence with U tokens. In the application domain of SLT, these channels correspond to representations of the manual and non-manual features of the sign. An overview of the multi-channel transformer can be seen in Fig. 1, while individual attention modules introduced in this paper are visualized in Fig. 2. To keep the formulation simple, and to focus the readers attention on the differentiating factors of our architecture, we omit the multi-headed attention, layer normalization and residual connections from our equations, which are the same as the original transformer networks [62].

4.1 Channel and Word Embeddings

As with other machine translation tasks, we start by projecting both the source channel features and the one-hot word vectors into a denser embedding where similar inputs lie close to one-another. To achieve this we use linear layers. We employ normalization and activation layers to change the scale of the embedded channel features and give additional non-linear representational capability to the model. The transformer networks do not have an implicit structure to model the position of a token within the sequence. To overcome this, we employ positional encoding [62] to add temporal ordering to the embedded representations. The embedding process for an input feature $x_{i,t}$ coming from the ith channel at time t can be formalized as:

$$\hat{x}_{i,t} = \text{Activ}\left(\text{Norm}\left(x_{i,t}W_i^{\text{ce}} + b_i^{\text{ce}}\right)\right) + \text{PosEnc}(t) \tag{3}$$

where W_i^{ce} and b_i^{ce} are channel specific learnt parameters of the linear projection layers. Similarly, the word embedding is as follows:

$$\hat{y}_u = y_u W^{\text{we}} + b^{\text{we}} + \text{PosEnc}(u) \tag{4}$$

where W^{we} and b^{we} are the weights of a linear layer which are either learned from scratch or pretrained on a large corpus [6,31].

4.2 Multi-channel Encoder Layer

Channel-Wise Self Attention (cs): Each multi-channel encoder layer starts by learning the contextual relationships within a single channel by utilizing individual self-attention layers (See Fig. 2 (left)). As per the original transformer implementation, we use the scaled dot product scoring function in the attention mechanisms. Given embedded source channel representations, $\hat{X}_i$, we obtain Queries, Keys and Values for the channel i as[3]:

$$\begin{aligned} Q_i^{\text{cs}} &= \hat{X}_i W_i^{\text{cs,q}} + b_i^{\text{cs,q}} \\ K_i^{\text{cs}} &= \hat{X}_i W_i^{\text{cs,k}} + b_i^{\text{cs,k}} \\ V_i^{\text{cs}} &= \hat{X}_i W_i^{\text{cs,v}} + b_i^{\text{cs,v}} \end{aligned} \tag{5}$$

which are then passed to the channel-wise self attention function to have their intra channel contextual relationship modeled as:

$$h_i^{\text{cs}} = \text{softmax}\left(\frac{Q_i^{\text{cs}} \left(K_i^{\text{cs}}\right)^T}{\sqrt{d_m}}\right) V_i^{\text{cs}} \tag{6}$$

where h_i^{CS} is the spatio-temporal representation of the ith source channel and d_m is the hidden size of the model. We also utilize individual feed forward layers as described in [62] for each channel as:

$$\text{FF}(x) = \max\left(0, xW_1^{\text{ff}} + b_1\right) W_2^{\text{ff}} + b_2 \tag{7}$$

By feeding the contextually modeled channel representations through feed forward layers, we obtain the final outputs of the channel-wise attention layer of our multi-channel encoder layer as:

$$\hat{h}_i^{\text{cs}} = \text{FF}_i^{\text{cs}}(h_i^{\text{cs}}) \tag{8}$$

Multi-channel Encoder Attention (me): We now introduce the multi-channel encoder attention, which learns the contextual relationship between the

[3] Note that we use a vectorized formulation in our equations. All softmax and bias addition operations are done row-wise.

self-attended channel representations (See Fig. 2 (middle)). As we are using dot product attention, we start by obtaining Q, K and V for each source as:

$$\begin{aligned} Q_i^{\mathrm{me}} &= \hat{h}_i^{\mathrm{cs}} W_i^{\mathrm{me,q}} + b_i^{\mathrm{me,q}} \\ K_i^{\mathrm{me}} &= \hat{h}_i^{\mathrm{cs}} W_i^{\mathrm{me,k}} + b_i^{\mathrm{me,k}} \\ V_i^{\mathrm{me}} &= \hat{h}_i^{\mathrm{cs}} W_i^{\mathrm{me,v}} + b_i^{\mathrm{me,v}} \end{aligned} \tag{9}$$

These values are then passed to the multi-channel attention layers where the Queries of each channel are used to estimate the scores over the concatenated Keys of the other channels. These scores are then used to calculate the channel-fused representations by taking a weighted sum over the other channels' concatenated Values. More formally, multi-channel attention can be defined as:

$$h_i^{\mathrm{me}} = \mathrm{softmax}\left(\frac{Q_i^{\mathrm{me}}\left(\left[\forall K_j^{\mathrm{me}} \text{ where } j \neq i\right]\right)^T}{\sqrt{d_m}}\right)\left[\forall V_j^{\mathrm{me}} \text{ where } j \neq i\right] \tag{10}$$

We would like to note that, the concatenation operation ($[\quad]$) is performed over the time axis, thus making our approach applicable to tasks where the source channels have a different numbers of tokens. We then pass multi-channel attention outputs to individual feed forward layers to obtain the final outputs of the multi-channel encoder layer as:

$$\hat{h}_i^{\mathrm{me}} = \mathrm{FF}_i^{\mathrm{me}}(h_i^{\mathrm{me}}) \tag{11}$$

Several multi-channel encoder layers can be stacked to form the encoder network with the aim of learning more complex multi-channel contextual representations, $h^e = (h_1^e, ..., h_N^e)$, where h_i^e is the output corresponding to the ith source channel.

4.3 Multi-channel Decoder Layer

Transformer networks utilize a masked self attention and an encoder-decoder attention in each decoder layer. The subsequent masking on self-attention is essential, as the target tokens' successors will not be available at inference time. In our approach, we also employ the masked self-attention to model the contextual relationship between target tokens' and its predecessors. However, we replace encoder-decoder attention with multi-channel decoder attention, which is modified to work with multiple source channel representations (See Fig. 2 (right)). Given the word embeddings $\hat{\mathcal{Y}}$ of a sentence $\mathcal{Y}$, we first obtain the masked self-attention (sa) outputs $\hat{h}^{\mathrm{sa}}$ using the generic approach [62], which are then in turn passed to our multi-channel decoder attention.

Multi-channel Decoder Attention (md): In generic transformers, encoder-decoder attention Queries are obtained from the decoder self-attention estimates, $\hat{h}^{\mathrm{sa}}$, while Keys and Values are calculated from the final encoder layer outputs, h_e. In order to incorporate information coming from multiple channels using transformer models, we propose the multi-channel decoder attention module. We first obtain the Q, K and V as:

$$\begin{aligned} Q^{\mathrm{md}} &= \hat{h}^{\mathrm{sa}} W^{\mathrm{md,q}} + b^{\mathrm{md,q}} \\ K_i^{\mathrm{md}} &= h_i^{e} W_i^{\mathrm{md,k}} + b_i^{\mathrm{md,k}} \\ V_i^{\mathrm{md}} &= h_i^{e} W_i^{\mathrm{md,v}} + b_i^{\mathrm{md,v}} \end{aligned} \tag{12}$$

Note that each source channel i has their own learned Key and Value matrices, $W_i^{md,k}$ and $W_i^{md,v}$ respectively.

These are then passed to the multi-channel decoder attention module where the Queries of each target token are scored against all channel Keys. Channel scores are then used to calculate the weighted average of their respective Values. Individual channel outputs are averaged to obtain the final output of the multi-channel decoder attention module. This process can be formalized as:

$$h^{\mathrm{md}} = \frac{1}{N} \sum_{i=1}^{N} \left(\mathrm{softmax} \left(\frac{Q^{\mathrm{md}} \left(K_i^{\mathrm{md}} \right)^T}{\sqrt{d_m}} \right) V_i^{\mathrm{md}} \right) \tag{13}$$

The attention module outputs are then passed through a feed forward layer to obtain the final representations of the multi-channel decoder layer as:

$$\hat{h}^{\mathrm{md}} = \mathrm{FF}^{\mathrm{md}}(h^{\mathrm{md}}) \tag{14}$$

Like the multi-channel encoder layer, multiple decoder layers can be stacked to improve representation capabilities of the decoder network. The output of the stacked decoder is denoted as $h^d = (h_1, ..., h_U)$ which is used to condition target token generation.

4.4 Loss Functions

We propose training multi-channel transformers using two types of loss function, namely a Translation Loss, which is commonly used in machine translation, and a Channel Anchoring Loss, which aims to preserve channel specific information during encoding.

Translation Loss: Although different loss functions have been used to train translation models, such as a mixture-of-softmaxes [66], token level cross-entropy loss is the most common approach to learn network parameters. Given a source-target pair, the translation loss, $\mathcal{L}_T$, is calculated as the accumulation of the error at each decoding step u, which is estimated using a classification loss over the target vocabulary as:

$$\mathcal{L}_T = 1 - \prod_{u=1}^{U} \sum_{g=1}^{G} p(y_u^g) p(\hat{y}_u^g) \tag{15}$$

where $p(y_u^g)$ and $p(\hat{y}_u^g)$ represent the ground truth and the generation probabilities of the target y^g at decoding step u, respectively, and G is the target language vocabulary size. In our networks, the probability of generating target token y_u at the decoding step u is conditioned on the hidden state of the decoder network h^d_u at the corresponding time step, $p(\hat{y}_u) = p(\hat{y}_u|h^d_u)$. Softmaxed linear projection of h^d_u is used to model the probability of producing tokens over the whole target vocabulary as:

$$p(\hat{y}_u|h^d_u) = \text{softmax}(h^d_u W^o + b^o) \tag{16}$$

where W^o and b^o are the trainable parameters of a linear layer.

Channel Anchoring Loss: For source channels, where we have access to a relevant classifier, we use an anchoring loss to preserve channel specific information. Predictions of these classifiers are used as ground truth to calculate token level cross entropy losses in the same manner as the translation loss. Given the classifier outputs corresponding to the ith channel, $C_i = (c_{i,1}, ..., c_{i,T_i})$, and the hidden state of the encoder, h^e, we first calculate the prediction probabilities over the target channel classes as:

$$p(\hat{c}_{i,t}|h^e) = \text{softmax}(h^e W^o_i + b^o_i) \tag{17}$$

where $p(\hat{c}_{i,t}|h^e)$ represent the prediction probabilities over the ith channel's classifier vocabulary, while W^o_i and b^o_i are the weights and biases of the linear layer used for the ith channel, respectively. We then use a modified version of Eq. 15 to calculate the ith channel's anchoring loss, $\mathcal{L}_{A,i}$, as:

$$\mathcal{L}_{A,i} = 1 - \prod_{t=1}^{T} \sum_{g=1}^{G_i} p(c^g_{i,t}) p(\hat{c}^g_{i,t}|h^e_t) \tag{18}$$

where $p(c^g_{i,t})$ and $p(\hat{c}^g_{i,t})$ represent the classifier output and the predicted probabilities of the class c^d_i at the encoders t^{th} step, respectively, while G_i is the number of target classes of the classifier corresponding to channel i. For example, one can use a hand shape classifier's convolutional layer as input channel and the same classifier's predictions as ground truth for hand channel anchoring loss to preserve the hand shape information, as well as to regularize the translation loss.

Total Loss: We use a weighed combination of Translation loss, $\mathcal{L}_T$, and Anchoring losses, $\mathcal{L}_A = (\mathcal{L}_{A,1}, ..., \mathcal{L}_{A,N})$, during training as:

$$\mathcal{L} = \lambda_T \mathcal{L}_T + \lambda_A \sum_{i=1}^{N} \mathcal{L}_{A,i} \tag{19}$$

where λ_T and λ_A decide the importance of each loss function during training.

5 Implementation and Evaluation Details

Dataset: We evaluate our model on the challenging PHOENIX14T [13] dataset, which is currently the only publicly available large vocabulary continuous SLT dataset aimed at vision based sign language research.

Sign Channels: We use three different articulators/channels to represent the manual and non-manual features of the sign, namely hand shapes, mouthings and upper body pose. We employ the models proposed and used in [35] and from these networks extracted 1024 dimensional Convolutional Neural Network (CNN) features for each frame (last layer before the fully connected layer) for hand shape and mouthing channels. We use the class prediction from the same network to anchor the channel representations. Although these networks were trained on 61 and 40 hand shapes and mouthing classes respectively (including transition/silence class), the predictions only contained 52 and 36 classes. Hence, our anchoring losses are calculated over the predicted number of classes.

To represent the upper body pose of the signers, we extract 2D skeletal pose information using the OpenPose library [14]. We then employ a 2D-to-3D lifting approach designed specifically for sign language production to obtain the final 3D joint positions of 50 upper body pose joints [67]. As there were no prior subunit classes for the upper body pose on PHOENIX14**T**, we do not utilize an anchoring loss on the pose channel.

Training and Network Details: Our networks are trained using the PyTorch framework [51] with a modified version of the JoeyNMT library [42]. We use Adam [33] optimizer with a batch size of 32, a learning rate of 10^{-3} ($\beta_1 = 0.9, \beta_2 = 0.998$) and a weight decay of 10^{-3}. We utilize Xavier [23] initialization and train all networks from scratch. We do not apply dropout and only use a single headed scaled dot-product attention to reduce the number of hyper-parameters in our experiments.

Decoding: During training we use a greedy search to evaluate development set translation performance. At inference, we employ beam search decoding with the beam width ranging from 0 to 10. We also employ a length penalty as proposed by [30] with α values ranging from 0 to 5. We use the development set to find the best performing beam width and α, and use these during test set inference for final results.

Performance Metrics: We use BLEU [49] and ROUGE [43] scores to measure the translation performance. To give the reader a better understanding of the networks behaviour, we repeat each experiment 10 times and report mean and standard deviation of BLEU-4 and ROUGE scores on both development and test sets. We also report our best result for every setup based on the BLEU-4 score as per the development set. BLEU-4 score is also used as the validation score for our learning scheduler and for early stopping.

6 Experiment Results

In this section we propose several multi-channel SLT experimental setups and report our quantitative results. We start by sharing single channel SLT performance using different network architectures, varying the number of hidden units, both to set a baseline for our multi-channel approaches and to find the optimal network size. After that, we propose two naive channel fusion approaches,

Table 1. Single channel SLT baselines using different network architectures.

		Dev set				Test set			
		BLEU-4		ROUGE		BLEU-4		ROUGE	
Channel	HS × FF	Best	mean ± std	–	mean ± std	–	mean ± std	–	mean ± std
Hand	32 × 64	14.54	14.05 ± 0.42	38.47	38.49 ± 0.45	13.88	13.80 ± 0.63	38.05	38.04 ± 0.49
Hand	64 × 128	**16.44**	15.70 ± 0.41	40.79	40.45 ± 0.64	16.18	15.63 ± 0.65	40.62	40.07 ± 0.80
Hand	128 × 256	16.32	**15.91** ± 0.43	**41.87**	**41.08** ± 0.73	**16.76**	**16.02** ± 0.88	**41.85**	**40.67** ± 0.93
Hand	256 × 512	16.06	15.41 ± 0.46	41.46	40.13 ± 0.83	15.43	15.57 ± 0.60	40.48	39.88 ± 0.71
Mouthing	32 × 64	11.70	11.24 ± 0.34	33.26	32.84 ± 0.60	10.77	11.01 ± 0.34	33.51	33.05 ± 0.34
Mouthing	64 × 128	12.91	12.55 ± 0.30	36.22	35.17 ± 0.71	12.83	12.62 ± 0.50	35.30	35.04 ± 0.72
Mouthing	128 × 256	**13.74**	**13.08** ± 0.41	**37.20**	**35.96** ± 0.79	**13.77**	**13.50** ± 0.37	**37.24**	**36.60** ± 0.72
Mouthing	256 × 512	12.86	12.40 ± 0.42	34.13	34.63 ± 0.64	13.25	12.34 ± 0.51	35.53	34.83 ± 0.47
Pose	32 × 64	9.64	8.91 ± 0.42	30.27	29.92 ± 0.67	9.64	8.55 ± 0.62	30.03	29.13 ± 0.79
Pose	64 × 128	10.64	10.28 ± 0.26	31.06	31.31 ± 0.47	9.97	9.88 ± 0.18	29.63	30.44 ± 0.60
Pose	128 × 256	**11.02**	**10.52** ± 0.31	**32.22**	**31.85** ± 0.42	**10.26**	**10.03** ± 0.46	**30.44**	**30.79** ± 0.82
Pose	256 × 512	10.06	9.50 ± 0.51	31.03	30.11 ± 0.75	9.51	8.62 ± 0.70	29.92	28.65 ± 0.99
Gloss	32 × 64	17.21	16.03 ± 0.49	42.20	41.26 ± 0.64	15.45	15.68 ± 0.43	41.21	40.71 ± 0.42
Gloss	64 × 128	18.50	18.16 ± 0.23	44.99	43.87 ± 0.68	18.14	17.89 ± 0.56	43.57	43.02 ± 0.69
Gloss	128 × 256	19.43	**19.14** ± 0.36	**46.10**	**45.17** ± 0.63	19.52	**19.08** ± 0.48	**45.32**	**44.52** ± 0.80
Gloss	256 × 512	**19.52**	18.36 ± 0.50	45.97	44.16 ± 0.79	**19.61**	18.60 ± 0.63	45.29	43.92 ± 0.98

namely early fusion and late fusion, to set a fusion benchmark for our novel *Multi-channel Transformer* architecture. Finally, we report the performance of the multi-channel transformer approach with and without the channel anchoring losses and compare our results against the state-of-the-art.

We apply batch normalization [29] and a soft-sign activation function [47] to input channel embeddings before passing them to our networks. See supplementary material for empirical justification for this choice.

6.1 Single Channel Baselines

In the first set of experiments, we train single channel SLT models. The main objective of these experiments is to set translation baselines for all future multi-channel fusion models. However, we would also like to examine the relative information presented in each channel by comparing their translation performance against one another. In addition, we wish to identify the optimal network setup for each channel to guide the future experiments. Therefore, we conduct experiments with four network setups for all three articulators with sizes varying from 32 × 64 to 256 × 512 (hidden size (HS) × number of feed forward (FF) units). All networks were built using two encoder and decoder layers.

As can be seen in Table 1, **H**and is the best performing channel in all network setups. Furthermore, using a network setup of 128 × 256 outperforms all of the alternatives. We believe this is closely related to the limited number of training samples we have and the over-fitting issues that come with it. Therefore, for the rest of our experiments we use 128 × 256 parameters for each channel.

We further train a **G**loss single channel network to set a baseline for our multi-channel approaches to compare against. As shown in Table 1, using CNN

Table 2. SLT performance of early and late channel fusion approaches.

			Dev Set				Test Set			
			BLEU-4		ROUGE		BLEU-4		ROUGE	
Fusion	Channels	HSxFF	Best	mean ± std	-	mean ± std	-	mean ± std	-	mean ± std
Early	**H + M**	2*(128x256)	**17.25**	**16.73** ± 0.57	**42.04**	**41.72** ± 0.61	**17.37**	**16.73** ± 0.82	**42.35**	**41.76** ± 0.80
Early	**H + P**	2*(128x256)	16.17	15.70 ± 0.32	40.51	40.28 ± 0.46	15.75	15.83 ± 0.30	40.54	40.34 ± 0.70
Early	**M + P**	2*(128x256)	13.57	12.91 ± 0.30	36.43	35.88 ± 0.40	13.23	13.02 ± 0.42	35.66	36.01 ± 0.72
Early	**H + M + P**	3*(128x256)	15.69	15.08 ± 0.55	39.78	39.38 ± 0.73	15.19	15.19 ± 0.49	39.99	39.43 ± 0.80
Late	**H + M**	2*(128x256)	**17.03**	**16.36** ± 0.48	41.69	**41.58** ± 0.79	16.81	**16.67** ± 0.49	41.69	**41.69** ± 0.54
Late	**H + P**	2*(128x256)	16.61	16.16 ± 0.33	41.54	41.18 ± 0.58	15.90	16.07 ± 0.76	40.81	40.50 ± 0.99
Late	**M + P**	2*(128x256)	14.22	13.55 ± 0.31	36.44	37.03 ± 0.64	14.11	13.65 ± 0.49	36.25	36.95 ± 0.65
Late	**H + M + P**	3*(128x256)	17.00	16.35 ± 0.38	**42.09**	41.29 ± 0.42	**16.95**	16.50 ± 0.47	**42.12**	41.53 ± 0.52

features that were trained using gloss level annotations outperforms all single sign articular based models (19.52 *vs.* 16.44 dev BLEU-4 score). Although the 256×512 network setup obtained the best individual development and test set translation performances, the mean performance of the 128×256 network was better, encouraging us to utilize this setup going forward.

6.2 Early and Late Fusion of Sign Channels

To set another benchmark for our multi-channel transformer, we propose two naive multi-channel fusion approaches, namely early and late fusion. In the early fusion setup, features from different channels are concatenated to create a fused representation of each frame. These representations are then used to train SLT models, as if they were features coming from a single channel. Hence, the contextual relationship is performed in an implicit manner by the transformer architecture. In our second, late fusion setup, individual SLT models are built which are then fused at the decoder output level, *i.e.* h^d, by concatenation. The fused representation is then used to generate target tokens using a linear projection layer. Compared to early fusion, this approach's capability to learn more abstract relationships is limited as the fusion is only done by a single linear layer. We examine all four possible fusions of the three channels. Network setup is set to linearly scale with respect to the number of channels that are fused together with a factor of 128×256 per channel.

As can be seen in Table 2, fusion of **H**ands and **M**outh yields slightly better results than single channel translation models (excluding gloss). However, unlike late fusion, which saw improvement in all scenarios, early fusion's performance gets worse as more features are added to the network. As this means having more parameters in our networks, we believe this is due to the natural propensity of the transformers to over-fit on small training datasets, like ours.

6.3 Multi-channel Transformers

In this set of experiments we examine the translation performance of the proposed multi-channel transformer architecture for multi-articulatory SLT. We first start by investigating the effects of the anchoring loss. We then compare our best performing method against other fusion options, gloss based translation

Table 3. Multi-channel Transformer based multi-articulatory SLT results.

			Dev Set				Test Set			
			BLEU-4		ROUGE		BLEU-4		ROUGE	
Channels	Anchoring Loss	HSxFF	Best	mean ± std	-	mean ± std	-	mean ± std	-	mean ± std
H + M	✗	2*(128x256)	17.71	16.97 ± 0.53	43.43	42.02 ± 0.86	17.72	17.19 ± 0.73	42.70	41.95 ± 0.85
H + P	✗	2*(128x256)	17.20	16.36 ± 0.58	42.15	41.23 ± 0.46	16.41	16.25 ± 0.66	40.56	40.87 ± 0.67
M + P	✗	2*(128x256)	14.17	13.50 ± 0.40	36.82	36.62 ± 0.52	13.43	13.93 ± 0.44	37.03	37.43 ± 0.65
H + M + P	✗	3*(128x256)	17.98	16.89 ± 0.59	44.01	41.85 ± 0.93	17.15	16.85 ± 0.65	42.38	41.83 ± 0.85
H + M	✗	2*(256x512)	15.95	15.46 ± 0.34	41.01	40.31 ± 0.50	15.87	15.80 ± 0.36	40.30	40.40 ± 0.70
H + P	✗	2*(256x512)	15.41	14.95 ± 0.33	40.10	39.22 ± 0.53	15.91	15.14 ± 0.59	40.24	39.11 ± 0.69
M + P	✗	2*(256x512)	13.39	12.60 ± 0.49	35.70	35.01 ± 0.73	13.38	12.74 ± 0.48	36.89	35.39 ± 0.79
H + M + P	✗	3*(256x512)	15.87	14.97 ± 0.51	40.53	39.65 ± 0.79	16.02	15.17 ± 0.81	40.15	39.61 ± 1.12
H + M	✓	2*(128x256)	18.52	17.93 ± 0.39	44.56	43.25 ± 0.57	17.93	17.76 ± 0.49	43.21	42.91 ± 0.49
H + P	✓	2*(128x256)	17.70	16.53 ± 0.67	43.19	41.26 ± 0.99	16.93	16.41 ± 0.48	42.62	40.80 ± 0.82
M + P	✓	2*(128x256)	15.14	14.60 ± 0.32	38.57	38.45 ± 0.45	15.32	15.05 ± 0.72	38.47	38.66 ± 0.76
H + M + P	✓	3*(128x256)	18.80	17.81 ± 0.68	44.24	43.17 ± 0.81	18.30	17.75 ± 0.58	43.65	42.90 ± 0.62
H + M	✓	2*(256x512)	19.05	18.07 ± 0.44	45.04	43.76 ± 0.82	**19.21**	17.71 ± 0.72	**45.05**	43.29 ± 0.99
H + P	✓	2*(256x512)	16.80	16.29 ± 0.36	41.15	40.86 ± 0.42	16.68	16.29 ± 0.48	41.34	40.68 ± 0.76
M + P	✓	2*(256x512)	15.14	14.60 ± 0.26	39.45	38.47 ± 0.62	15.36	15.13 ± 0.44	40.06	39.09 ± 0.62
H + M + P	✓	**3*(256x512)**	**19.51**	**18.66** ± 0.52	**45.90**	**44.30** ± 0.92	18.51	**18.31** ± 0.57	43.57	**43.75** ± 0.63
Gloss	–	128x256	19.43	**19.14** ± 0.36	**46.10**	**45.17** ± 0.63	19.52	**19.08** ± 0.48	**45.32**	**44.52** ± 0.80
Gloss	–	**256x512**	**19.52**	18.36 ± 0.50	45.97	44.16 ± 0.79	**19.61**	18.60 ± 0.63	45.29	43.92 ± 0.98
Orbay *et al.* [48]	–	–	–	–	–	–	14.56	–	38.05	–
Sign2Gloss→Gloss2Text [13]	–	–	17.89	–	43.76	–	17.79	–	43.45	–
Sign2Gloss2Text [13]	–	–	18.40	–	44.14	–	18.13	–	43.80	–
(Gloss) Sign2Text [12]	–	**3x(512x2048)**	**20.69**	–	–	–	**20.17**	–	–	–
(Gloss) Sign2(Gloss+Text) [12]	–	**3x(512x2048)**	**22.38**	–	–	–	**21.32**	–	–	–

and other state-of-the-art methods. As with other fusion experiments, we examine all possible fusion combinations. In addition to using the 128 × 256 network setup, we also evaluate having a larger network to see if the additional anchoring losses help with over-fitting by regularizing the translation loss.

As can be seen in the first row of Table 3, while using the same number of parameters as the early and late fusion setups, our proposed *Multi-Channel Transformer* approach outperforms both configurations. However, doubling the network size does effect the direct application of multi-channel attention negatively. To counteract this issue and to examine the effects of the anchoring loss, we run experiments with both 128 × 256 and 256 × 512 setups. We normalize our losses on the sequence level instead of token level and we set the anchoring loss weight, λ_A, to 0.15 to counteract different source (video) and target (sentence) sequence lengths. Using the anchoring losses not only improves the performance of the 128 × 256 models but also allows the 256 × 512 networks to achieve similar translation performance to using gloss features. We believe this is due to two main factors. Firstly, the anchoring loss forces the encoder channels to preserve the channel specific information while being contextually modeled against other articulators. Secondly, it acts as a regularizer for the translation loss and counteracts the over-fitting previously discussed.

Compared to the state-of-the-art, our best multi-channel transformer model surpasses the performance of several previous models [13,48], some of which are heavily reliant on gloss annotation. Furthermore, our multi-channel models perform *on par* with our single gloss channel model, and yields competitive translation performance to the state-of-the-art transformer based approaches [12], which utilize larger models and uses gloss supervision on several levels (pretrained Gloss CNN features and transformer encoder supervision). However, due to their dependence on gloss annotations, such models [12,13] can not be scaled to larger un-annotated datasets, which is not a limiting factor for the proposed multi-channel transformer approach. See supplementary material for qualitative translation examples from our best multi-articulatory translation model.

7 Conclusion

This paper presented a novel approach to Neural Machine Translation in the context of sign language. Our novel multi-channel transformer architecture allows both the inter and intra contextual relationship between different asynchronous channels to be modelled within the transformer network itself. Experiments on RWTH-PHOENIX-Weather-2014T dataset demonstrate the approach achieves *on par* or competitive performance against the state-of-the-art. More importantly, we overcome the reliance on gloss information which underpins other state-of-the-art approaches. Now we have broken the dependency upon gloss information, future work will be to scale learning to larger dataset where gloss information is not available, such as broadcast footage.

Acknowledgements. This work received funding from the SNSF Sinergia project 'SMILE' (CRSII2_160811), the European Union's Horizon2020 research and innovation programme under grant agreement no. 762021 'Content4All' and the EPSRC project 'ExTOL' (EP/R03298X/1). This work reflects only the author's view and the Commission is not responsible for any use that may be made of the information it contains. We would also like to thank NVIDIA Corporation for their GPU grant.

References

1. Afouras, T., Chung, J.S., Senior, A., Vinyals, O., Zisserman, A.: Deep audio-visual speech recognition. IEEE Trans. Pattern Anal. Mach. Intell. (TPAMI) (2018)
2. Antonakos, E., Pitsikalis, V., Rodomagoulakis, I., Maragos, P.: Unsupervised classification of extreme facial events using active appearance models tracking for sign language videos. In: Proceedings of the IEEE International Conference on Image Processing (ICIP) (2012)
3. Assael, Y.M., Shillingford, B., Whiteson, S., De Freitas, N.: LipNet: end-to-end sentence-level lipreading. In: GPU Technology Conference (2017)
4. Bahdanau, D., Cho, K., Bengio, Y.: Neural machine translation by jointly learning to align and translate. In: Proceedings of the International Conference on Learning Representations (ICLR) (2015)
5. Bellugi, U., Fischer, S.: A comparison of sign language and spoken language. Cognition **1**(2), 173–200 (1972)
6. Bojanowski, P., Grave, E., Joulin, A., Mikolov, T.: Enriching word vectors with subword information. Trans. Assoc. Comput. Linguis. (ACL) **5**, 135–146 (2017)
7. Bragg, D., et al.: Sign language recognition, generation, and translation: an interdisciplinary perspective. In: Proceedings of the International ACM SIGACCESS Conference on Computers and Accessibility (ASSETS) (2019)
8. Brentari, D.: Sign Language Phonology. Cambridge University Press, Cambridge (2019)
9. Bungeroth, J., Ney, H.: Statistical Sign Language Translation. In: Proceedings of the Workshop on Representation and Processing of Sign Languages at International Conference on Language Resources and Evaluation (LREC) (2004)

10. Camgoz, N.C., Hadfield, S., Koller, O., Bowden, R.: Using convolutional 3D neural networks for user-independent continuous gesture recognition. In: Proceedings of the IEEE International Conference on Pattern Recognition Workshops (ICPRW) (2016)
11. Camgoz, N.C., Hadfield, S., Koller, O., Bowden, R.: SubUNets: end-to-end hand shape and continuous sign language recognition. In: Proceedings of the IEEE International Conference on Computer Vision (ICCV) (2017)
12. Camgoz, N.C., Hadfield, S., Koller, O., Bowden, R.: Sign language transformers: joint end-to-end sign language recognition and translation. In: Proceedings of the IEEE Conference on Computer Vision and Pattern Recognition (CVPR) (2020)
13. Camgoz, N.C., Hadfield, S., Koller, O., Ney, H., Bowden, R.: Neural sign language translation. In: Proceedings of the IEEE Conference on Computer Vision and Pattern Recognition (CVPR) (2018)
14. Cao, Z., Hidalgo, G.M., Simon, T., Wei, S., Sheikh, Y.: OpenPose: realtime multi-person 2D pose estimation using part affinity fields. IEEE Trans. Pattern Anal. Mach. Intell. (TPAMI) (2019)
15. Caridakis, G., Asteriadis, S., Karpouzis, K.: Non-manual cues in automatic sign language recognition. Pers. Ubiquit. Comput. **18**(1), 37–46 (2014)
16. Chai, X., et al.: Sign language recognition and translation with Kinect. In: Proceedings of the International Conference on Automatic Face and Gesture Recognition (FG) (2013)
17. Charles, J., Pfister, T., Everingham, M., Zisserman, A.: Automatic and efficient human pose estimation for sign language videos. Int. J. Comput. Vision (IJCV) **110**(1), 70–90 (2014)
18. Cho, K., et al.: Learning phrase representations using RNN encoder-decoder for statistical machine translation. In: Proceedings of the Conference on Empirical Methods in Natural Language Processing (EMNLP) (2014)
19. Chung, J.S., Senior, A., Vinyals, O., Zisserman, A.: Lip reading sentences in the wild. In: Proceedings of the IEEE Conference on Computer Vision and Pattern Recognition (CVPR) (2017)
20. Cui, R., Liu, H., Zhang, C.: Recurrent convolutional neural networks for continuous sign language recognition by staged optimization. In: Proceedings of the IEEE Conference on Computer Vision and Pattern Recognition (CVPR) (2017)
21. Cui, R., Liu, H., Zhang, C.: A deep neural framework for continuous sign language recognition by iterative training. IEEE Trans. Multimedia **21**, 1880–1891 (2019)
22. Fang, B. Co, J., Zhang, M.: DeepASL: enabling ubiquitous and non-intrusive word and sentence-level sign language translation. In: Proceedings of the ACM Conference on Embedded Networked Sensor Systems (SenSys) (2017)
23. Glorot, X., Bengio, Y.: Understanding the difficulty of training deep feedforward neural networks. In: Proceedings of the International Conference on Artificial Intelligence and Statistics (2010)
24. Grimes, G.J.: Digital Data Entry Glove Interface Device, US Patent 4,414,537 (1983)
25. Guo, D., Wang, S., Tian, Q., Wang, M.: Dense temporal convolution network for sign language translation. In: Proceedings of the AAAI Conference on Artificial Intelligence (2019)
26. Guo, D., Zhou, W., Li, H., Wang, M.: Hierarchical LSTM for sign language translation. In: Proceedings of the AAAI Conference on Artificial Intelligence (2018)
27. Hidalgo, G., et al.: Single-network whole-body pose estimation. In: IEEE International Conference on Computer Vision (ICCV) (2019)

28. Huang, J., Zhou, W., Zhang, Q., Li, H., Li, W.: Video-based sign language recognition without temporal segmentation. In: Proceedings of the AAAI Conference on Artificial Intelligence (2018)
29. Ioffe, S., Szegedy, C.: Batch normalization: accelerating deep network training by reducing internal covariate shift. In: Proceedings of the International Conference on Machine Learning (ICML) (2015)
30. Johnson, M., et al.: Google's multilingual neural machine translation system: enabling zero-shot translation. Trans. Assoc. Comput. Linguist. **5**, 339–351 (2017)
31. Joulin, A., Grave, E., Bojanowski, P., Mikolov, T.: Bag of tricks for efficient text classification. In: Proceedings of the 15th Conference of the European Chapter of the Association for Computational Linguistics (ACL) (2017)
32. Kalchbrenner, N., Blunsom, P.: Recurrent continuous translation models. In: Proceedings of the Conference on Empirical Methods in Natural Language Processing (EMNLP) (2013)
33. Kingma, D.P., Ba, J.: Adam: a method for stochastic optimization. In: Proceedings of the International Conference on Learning Representations (ICLR) (2014)
34. Ko, S.K., Kim, C.J., Jung, H., Cho, C.: Neural sign language translation based on human keypoint estimation. Appl. Sci. **9**(13), 2683 (2019)
35. Koller, O., Camgoz, N.C., Bowden, R., Ney, H.: Weakly supervised learning with multi-stream CNN-LSTM-HMMs to discover sequential parallelism in sign language videos. IEEE Trans. Pattern Anal. Mach. Intell. (TPAMI) (2019)
36. Koller, O., Forster, J., Ney, H.: Continuous sign language recognition: towards large vocabulary statistical recognition systems handling multiple signers. Comput. Vis. Image Underst. (CVIU) **141**, 108–125 (2015)
37. Koller, O., Ney, H., Bowden, R.: Read my lips: continuous signer independent weakly supervised viseme recognition. In: European Conference on Computer Vision (ECCV) (2014)
38. Koller, O., Ney, H., Bowden, R.: Deep learning of mouth shapes for sign language. In: Proceedings of the IEEE International Conference on Computer Vision Workshops (ICCVW) (2015)
39. Koller, O., Ney, H., Bowden, R.: Deep hand: how to train a CNN on 1 million hand images when your data is continuous and weakly labelled. In: Proceedings of the IEEE Conference on Computer Vision and Pattern Recognition (CVPR) (2016)
40. Koller, O., Zargaran, S., Ney, H.: Re-sign: re-aligned end-to-end sequence modelling with deep recurrent CNN-HMMs. In: Proceedings of the IEEE Conference on Computer Vision and Pattern Recognition (CVPR) (2017)
41. Koller, O., Zargaran, S., Ney, H., Bowden, R.: Deep sign: hybrid CNN-HMM for continuous sign language recognition. In: Proceedings of the British Machine Vision Conference (BMVC) (2016)
42. Kreutzer, J., Bastings, J., Riezler, S.: Joey NMT: a minimalist NMT toolkit for novices. In: Proceedings of the Conference on Empirical Methods in Natural Language Processing (EMNLP): System Demonstrations (2019)
43. Lin, C.Y.: ROUGE: a package for automatic evaluation of summaries. In: Proceedings of the Annual Meeting of the Association for Computational Linguistics, Text Summarization Branches Out Workshop (2004)
44. Luong, M.T., Pham, H., Manning, C.D.: Effective approaches to attention-based neural machine translation. In: Proceedings of the Conference on Empirical Methods in Natural Language Processing (EMNLP) (2015)

45. Luzardo, M., Karppa, M., Laaksonen, J., Jantunen, T.: Head pose estimation for sign language video. In: Kämäräinen, J.-K., Koskela, M. (eds.) SCIA 2013. LNCS, vol. 7944, pp. 349–360. Springer, Heidelberg (2013). https://doi.org/10.1007/978-3-642-38886-6_34
46. Malaia, E., Borneman, J.D., Wilbur, R.B.: Information transfer capacity of articulators in American sign language. Lang. Speech **61**(1), 97–112 (2018)
47. Nwankpa, C., Ijomah, W., Gachagan, A., Marshall, S.: Activation functions: comparison of trends in practice and research for deep learning. arXiv:1811.03378 (2018)
48. Orbay, A., Akarun, L.: Neural sign language translation by learning tokenization. arXiv:2002.00479 (2020)
49. Papineni, K., Roukos, S., Ward, T., Zhu, W.J.: BLEU: a method for automatic evaluation of machine translation. In: Proceedings of the Annual Meeting of the Association for Computational Linguistics (ACL) (2002)
50. Parton, B.S.: Sign language recognition and translation: a multidisciplined approach from the field of artificial intelligence. J. Deaf Stud. Deaf Educ. **11**(1), 94–101 (2005)
51. Paszke, A., et al.: Automatic differentiation in PyTorch. In: Proceedings of the Advances in Neural Information Processing Systems Workshops (NIPSW) (2017)
52. Pfau, R., Quer, J.: Nonmanuals: their grammatical and prosodic roles. In: Sign Languages. Cambridge University Press (2010)
53. Pfister, T., Charles, J., Everingham, M., Zisserman, A.: Automatic and efficient long term arm and hand tracking for continuous sign language TV broadcasts. In: Proceedings of the British Machine Vision Conference (BMVC) (2012)
54. Pigou, L., Dieleman, S., Kindermans, P.J., Schrauwen, B.: Sign language recognition using convolutional neural networks. In: European Conference on Computer Vision Workshops (ECCVW) (2014)
55. Starner, T., Weaver, J., Pentland, A.: Real-time American sign language recognition using desk and wearable computer based video. IEEE Trans. Pattern Anal. Mach. Intell. (TPAMI) **20**(12), 1371–1375 (1998)
56. Stein, D., Schmidt, C., Ney, H.: Sign language machine translation overkill. In: International Workshop on Spoken Language Translation (2010)
57. Stokoe, W.C.: Sign language structure. Ann. Rev. Anthropol. **9**(1) (1980)
58. Sutskever, I., Vinyals, O., Le, Q.V.: Sequence to sequence learning with neural networks. In: Proceedings of the Advances in Neural Information Processing Systems (NIPS) (2014)
59. Sutton-Spence, R., Woll, B.: The Linguistics of British Sign Language: An Introduction. Cambridge University Press, Cambridge (1999)
60. Tamura, S., Kawasaki, S.: Recognition of sign language motion images. Pattern Recogn. **21**(4), 343–353 (1988)
61. Valli, C., Lucas, C.: Linguistics of American Sign Language: An Introduction. Gallaudet University Press, Washington, D.C. (2000)
62. Vaswani, A., et al.: Attention is all you need. In: Proceedings of the Advances in Neural Information Processing Systems (NIPS) (2017)
63. Vogler, C., Goldenstein, S.: Facial movement analysis in ASL. Univ. Access Inf. Soc. **6**(4) (2008)
64. Wang, S., Guo, D., Zhou, W.g., Zha, Z.J., Wang, M.: Connectionist temporal fusion for sign language translation. In: Proceedings of the ACM International Conference on Multimedia (2018)

65. Wilbur, R.B.: Phonological and Prosodic Layering of Nonmanuals in American Sign Language. The Signs of Language Revisited: An Anthology to Honor Ursula Bellugi and Edward Klima (2000)
66. Yang, Z., Dai, Z., Salakhutdinov, R., Cohen, W.W.: Breaking the softmax bottleneck: a high-rank RNN language model. In: Proceedings of the International Conference on Learning Representations (ICLR) (2018)
67. Zelinka, J., Kanis, J., Salajka, P.: NN-based Czech sign language synthesis. In: Salah, A.A., Karpov, A., Potapova, R. (eds.) SPECOM 2019. LNCS (LNAI), vol. 11658, pp. 559–568. Springer, Cham (2019). https://doi.org/10.1007/978-3-030-26061-3_57
68. Zhou, H., Zhou, W., Zhou, Y., Li, H.: Spatial-temporal multi-cue network for continuous sign language recognition. In: Proceedings of the AAAI Conference on Artificial Intelligence (2020)

Synthetic Convolutional Features for Improved Semantic Segmentation

Yang He[1,2](✉), Bernt Schiele[2], and Mario Fritz[1]

[1] CISPA Helmholtz Center for Information Security, Saarbrücken, Germany
{yang.he,fritz}@cispa.saarland
[2] Max Planck Institute for Informatics, Saarland Informatics Campus, Saarbrücken, Germany
schiele@mpi-inf.mpg.de

Abstract. Recently, learning-based image synthesis has enabled to generate high resolution images, either applying popular adversarial training or a powerful perceptual loss. However, it remains challenging to successfully leverage synthetic data for improving semantic segmentation with additional synthetic images. Therefore, we suggest to generate intermediate convolutional features and propose the first synthesis approach that is catered to such intermediate convolutional features. This allows us to generate new features from label masks and include them successfully into the training procedure in order to improve the performance of semantic segmentation. Experimental results and analysis on two challenging datasets *Cityscapes* and *ADE20K* show that our generated feature improves performance on segmentation tasks.

1 Introduction

Semantic image segmentation is a fundamental problem in computer vision, and has many applications in scene understanding, perception, robotics and in the medical area. To achieve robust segmentation performance, models usually are trained with data augmentation like flipping and re-scaling to make full use of expensively annotated data.

Recent work leverages synthetic images as data augmentation for computer vision tasks benefiting from capable graphic engines and development of generative modeling [9], e.g. for gaze estimation [24] and hand pose estimation [15], etc. However, using synthesized images for semantic segmentation remains challenging, because of the complexity of scenes and exponential combinations of different elements. Previous work on semantic segmentation with synthetic data [10,22,23,28,31] focuses on domain adaptation problems that aim to reduce the distribution gap between synthetic images and real images, instead of improving a segmentation model trained with fully annotated real data. Besides, even

Electronic supplementary material The online version of this chapter (https://doi.org/10.1007/978-3-030-66823-5_19) contains supplementary material, which is available to authorized users.

A. Bartoli and A. Fusiello (Eds.): ECCV 2020 Workshops, LNCS 12538, pp. 320–336, 2020.
https://doi.org/10.1007/978-3-030-66823-5_19

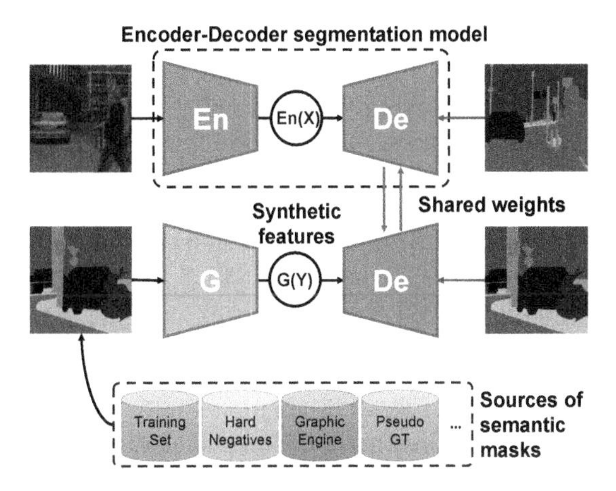

Fig. 1. Our pipeline for semantic segmentation. A generator is learned to synthesize convolution features and our semantic segmentation model is trained with synthetic features and real images.

though high resolution realistic generated images [4,17,19,29] are able to acquire, better segmentation results has not been shown by training with those generated images, comparing to training with real images. When inspecting these generated images visually [19,29], there are still some visual artifacts, which affect the low-level convolutional layers significantly. Learning with those regions, low-level representations are probably degenerated, and thus high-level representations are also hard to effectively build on top of them. As a result, it becomes hard to train a model with such images, which might lead to decreased segmentation performance.

Because of the difficulty of image synthesis, we question if we really need to generate high-quality images for boosting the training of semantic segmentation models. Instead, we present a feature synthesis-based data augmentation approach for semantic segmentation, as shown in Fig. 1. We aim to learn a semantic segmentation model with a mixture of real images and synthetic data from a generator, allowing to sample paired data from semantic layout masks, which assign categories for each pixel. Modern semantic image segmentation models built on fully convolutional architectures, as a result, we are able to synthesize convolutional features to approximate the distribution of real features with a similar generation pipeline to images [37]. Hence, those synthetic features are able to used as additional training data to improve semantic segmentation.

The capacity and quality of data is the key to success for data augmentation. Different to image synthesis, the designed synthetic features have lower spatial dimension but a larger number of channels. Consequently, it is hard to directly apply existing image synthesis architectures for the feature generation task, and thus a new effective architecture is needed. An ideal feature generator is supposed to meet the following requirements. (1) It allows us sample multiple diverse features from a semantic mask input, and thus provides us numerous training pairs. (2) Those synthetic features should follow a similar distribution as extracted features from real images. In other words, the synthetic features are able to be segmented by a trained model, with comparable performance to real features. (3) The final challenge is that raw images contain many detailed information which are compressed in the feature domain. The generator should be powerful enough to represent those important details. To achieve above goals, we design a generator under the framework of multi-modal translation [37] with a network architecture catered to the convolutional feature synthesis task.

The main contributions of this work are: (1) We propose to synthesize convolution features for data augmentation for semantic image segmentation, leading to improved results. (2) We present a feature generative model, and analyze its effectiveness according to a series of ablation studies. (3) Several techniques are proposed to leverage the synthetic features, including online hard negative mining, generation from additional masks, and label smoothing regularization.

2 Related Work

Generative Adversarial Networks (GAN) and Image Translation: GAN [9] was proposed to capture arbitrary data distribution with learning a discriminator and a generator in an adversarial way. It was extended to conditional version [14] that feeds extra information like class labels into a generator, which allows to model the conditional distribution of data and develop many interesting applications. For example, Reed *et al.* [20] applied adversarial training to generate different types of flowers from labeled attribute. Odena *et al.* [16] added an auxiliary classifier to the generator that is supposed to recognize the generated data as the input class of generator, and synthesized 1000 ImageNet [6] classes. Particularly, Isola *et al.* [12] proposed the *pix2pix* framework that performs image translation, such as generating colorful shoe images from skeleton images, or generating realistic street scenes from semantic layouts. Besides, Zhu *et al.* [36] added a cycle-consistency constraint to achieve unpaired image translation. Recently, Zhu *et al.* [37] model the distribution of latent representation under the *pix2pix* framework and generate multiple output images from different modalities, which is called *BicycleGAN*. Different to [37], we aim to generate dense features, which has totally different dimension in spatial and channel.

Semantic Segmentation with GAN: GAN has been applied to semantic segmentation with respective to providing additional loss term [13] or leveraging unlabeled training data [11,26]. Luc *et al.* first applied GAN into semantic segmentation area, which learned a discriminator taking posteriors from a segmentation model as the input, and tried to fool the discriminator with the posteriors. Thus the discriminator provides additional loss term to semantic segmentation model and it is updated within the adversarial training. Besides, people made efforts in leveraging unlabeled data and then lead to semi-supervised learning settings. On one hand, unlabeled data provide a real distribution of natural images for adversarial training, and they might be helpful to successful train a

GAN model. One the other hand, the discriminator is able to provide penalty gradients for those unlabeled data, thus it is possible to utilize more data to improve the performance.

Data Augmentation with GAN: There are several works utilizing generated data with GAN in computer vision tasks [1,2,7,15,18,24,25,32,34]. Xian *et al.* [32] proposed to generate embedding visual features from attributes with GAN for zero-shot learning. They successful trained a classifier with mixing synthetic features for unseen classes and real features of seen classes. As a result, significant improvement was achieved in generalized zero shot image classification problem. Besides, Sixt *et al.* [25] generated large amounts of realistic labeled images by combining a 3D model. Peng *et al.* [18] applied adversarial training to generate many hard occlusion and rotation patterns for augmentation in human pose estimation task. Zheng *et al.* [34] leveraged large amounts of unlabeled generated images to improve person re-identification task. GAN was also applied to generate image/label pairs in semantic segmentation. Bowles *et al.* [2] regard label image as an additional channel, and generate four-channel outputs from a noise, and augment the training set in semantic segmentation. Finally, [24] and [15] address the problem of gaze estimation and hand pose estimation by utilizing the data from rendering system and training a GAN to eliminate the distribution gap between synthetic data and real data.

We highlight that our approach is the first to synthesize dense features, and improve complex task of semantic segmentation by providing synthetic features. Different to previous work [12,37], features should be successfully used as training data to improve a model, instead of only examples with good visual quality. Besides, to generate dense features instead of a vector in [32], we also need more comprehensive architectures.

3 Semantic Segmentation with Dense Feature Synthesis

Motivated by the great importance of data augmentation, we depict how to generate dense convolutional features and then leverage the generated features as additional training data to improve semantic segmentation, as illustrated in Fig. 1. Generation and classification are reverse problems, which translate between images and labels each other. With a paired training set $\mathcal{T} = \{(\mathrm{X}^i, \mathrm{Y}^i)\}_{i=1}^{n}$, we can learn a segmentation model $\mathrm{Y}^i = S(\mathrm{X}^i)$ as well as an image generator $\mathrm{X}^i = G_{img}(\mathrm{Y}^i)$. Naturally, it is able to train a segmentation model with the augmented dataset $\mathcal{T} \cup \{G_{img}(\mathrm{Y}^i), \mathrm{Y}^i\}_{i=1}^{m}$.

Alternative to generating images, we generate convolutional features for providing more data, and our pipeline is presented in Fig. 1. Semantic segmentation model S consists of encoder E and decoder D, as a result, we can extract features for an image by $E(\mathrm{X})$ and segment the image by $D(E(\mathrm{X}))$. Hence, our goal is: (1) to learn a generator G_{feat} which is able to produce realistic features, formally $p(E(\mathrm{X})|\mathrm{Y}) \sim p(G_{feat}(\mathrm{Y})|\mathrm{Y})$; (2) to learn the parameters for the segmentation model with the mixture of synthetic and real pairs $\{G_{feat}(\mathrm{Y}^i), \mathrm{Y}^i\}_{i=1}^{m} \cup \{E(\mathrm{X}^i), \mathrm{Y}^i\}_{i=1}^{n}$.

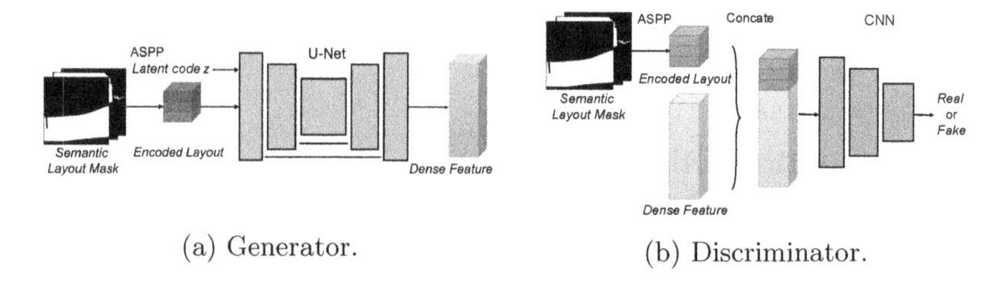

(a) Generator. (b) Discriminator.

Fig. 2. The illustration of network architectures in our adversarial feature generator.

In detail, the encoder E is updated with real images only, which captures low-level local features. The decoder D is shared by synthetic and real features, therefore, both steams contribute to the updating of the decoder. Therefore, the loss function of model training is formulated as

$$\mathcal{L} = \mathbb{E}[-\log D(E(\mathrm{X})|\mathrm{Y})] + \mathbb{E}[-\log D(G_{feat}(\mathrm{Y})|\mathrm{Y})], \tag{1}$$

where the output of D is per class probabilities for each pixel normalized with softmax, and the feature generator G_{feat} is first trained with $\mathcal{T}$ and E.

3.1 Convolutional Feature Synthesis

We present our model for effective feature synthesis, which is crucial for our goal, as discussed in Sect. 1. Generating multiple diverse features is challenging, because they encode information of large areas as well as details, which cannot be ignored. Also, the synthetic features should follow a similar distribution as extracted real features. We formulate our generator by using the BicycleGAN objective [37], which is shown to be successful in one-to-many image translation tasks. In the objective, the generation is driven by a conditional input and a random latent vector, allowing us to sample multiple different examples. Reconstructions on latent vectors and input features guarantee the quality of sampled features. Finally, an adversarial loss helps to generate features with useful details.

Architecture. It turns out that previous synthesis approaches are not directly applicable to feature synthesis, as they emit an output with the same dimension of the input. Convolutional features are compressed from images with smaller spatial dimension but much larger channel number, encoding location information and many useful details [8]. Hence, representing such information correctly plays an important role to successful feature synthesis. As shown in Fig. 2, our generator takes a high resolution semantic layout mask to produce low resolution feature maps. Our discriminator takes a layout/feature pair as input, to judge if the feature is compatible with the layout or not.

Preserving Resolution by Atrous Pooling. Atrous spatial pyramid pooling (ASPP) is an effective module in semantic segmentation used to aggregate multi-scale context information [3]. Here, we take advantages of ASPP in multi-scale representation capability, and effectively encode a high resolution semantic layout mask. In our ASPP module, there are three convolution layers with dilation 1,2, and 4, to capture neighboring information and wider context. Stride operation is followed after ASPP, leading to downsampled resolution. After applying several ASPP modules, the encoded semantic layout reaches to the same spatial dimension as the features. We feed encoded semantic layout and a latent code into a U-Net [21] and sample a convolutional feature. In our discriminator for adversarial training, we also apply ASPP module to encode high resolution semantic layout masks. We concatenate the encoded layout and its corresponding real/fake feature together to classify the feature/mask pair as real or fake.

3.2 Regularization on Synthetic Features

Label smoothing regularization (LSR) has been shown to reduce the influence of noisy labels and improve generalization [27]. Because not all sampled synthetic features are perfect, we apply LSR on the synthetic features.

In Eq. (1), the per class probabilities for X are expressively described as $p_i(k|\mathrm{X}) = \frac{\exp(r_i^k)}{\sum_{k=1}^{K}\exp(r_i^k)}$ for each label $k \in \{1, \cdots, K\}$, where r_i^k is the unnormalized log probabilities for k-th class, indexing at i-th location. Similarly, the per class probabilities for synthetic features $G_{feat}(\mathrm{Y})$ are $p_i(k|G_{feat}(\mathrm{Y})) = \frac{\exp(s_i^k)}{\sum_{k=1}^{K}\exp(s_i^k)}$. The negative log likelihood in Eq. (1) can be rewritten as

$$\mathcal{L} = \mathbb{E}(-\sum_i \log p_i(k|\mathrm{X}) q_{real}(k)) + \mathbb{E}(-\sum_i \log p_i(k|G_{feat}(\mathrm{Y})) q_{syn}(k)) \quad (2)$$

with weighting functions $q_{real}(k)$ and $q_{syn}(k)$ for the branches using real images and synthetic features, respectively. For cross entropy loss, it only takes the probability for designed label; for the version with LSR, it takes all the probabilities to compute a loss. They can be formulated in an unified formulation, i.e.,

$$q_\epsilon(k) = \begin{cases} 1 - \frac{K-1}{K}\epsilon, & k = y \\ \frac{\epsilon}{K}, & k \neq y, \end{cases} \quad (3)$$

where ϵ is a small value in the range of (0,1) for label smoothing regularization. It will become cross entropy when $\epsilon = 0$. As a result, we set $q_{real} = q_0$ and $q_{syn} = q_\epsilon$ $(\epsilon > 0)$ in Eq. (2).

3.3 Online Hard Negative Mining

Except generating features by selecting layouts randomly, we can search hard examples during training, which have a large loss value. We do online hard negative mining and feature generation alternatively. We randomly sample some image patches and compute their loss value, the top ranking patches are used to generate the features for the next several training iterations.

3.4 Additional Semantic Masks

Since we generate paired data from semantic layout masks, it is possible to acquire more data than augmenting a training set only, by providing novel mask. For example, in traffic scenarios, the environment is fixed, but everyday road users are different. It is interesting to know if segmentation model can be further improved by seeing more combination of road users and still objects. We present more semantic masks from different sources in Table 5 of Sect. 4.2.

4 Experiments and Analysis

4.1 Experimental Settings

We evaluate our data augmentation scheme using PSPNet [33] on the *Cityscapes* [5] and *ADE20K* [35] datasets. *Cityscapes* captures traffic scenes in various cities under different weather and illumination conditions containing 2975 training image pairs with detailed annotations, and 19998 extra images with coarse annotations. Except still frames, it also provides a short video for each frame. *ADE20K* has 20210 training images at different image resolutions including a variety of indoor and outdoor scenes. We evaluate our approach with widely used measurements for semantic segmentation for all the datasets including pixel accuracy (PixelAcc), class accuracy (ClassAcc), mean intersection over union (mIoU) and frequent weighted intersection over union (fwIoU).

Implementation Details. We implement our generator with the modification of [37] using the PyTorch framework. We implement our segmentation model with synthetic features under the official PSPNet implementation [33], and apply released ResNet-101 and ResNet-50 based PSPNet as our baselines for *Cityscapes* and *ADE20K*.

To begin with, we extract 3000 and 40000 patches on the `conv4_12` and `conv4_3` layers from official released models to train the generator for *Cityscapes* and *ADE20K*, respectively. We set 60 and 20 epochs for those datasets. We set three ASPP stages with 96, 192, 384 output feature maps and encode the masks by 8. Besides, we follow the BicycleGAN setup and let the first convolution kernel has 1320 channels. In addition, we finetune the released model from the baseline with the learned generator which is fixed during the finetuning. All models are learned using SGD with momentum, and the batch size is 16. Initial learning rates are set to $10^{-}6$ and $10^{-}7$ for *Cityscapes* and *ADE20K*, and we use the "poly" learning rate policy where current learning rate is the initial one multiplied by $(1-\frac{iter}{max\_iter})^{power}$, and we set power to 0.9. Momentum and weight decay are set to 0.9 and .0005 respectively. Last, we set ϵ in Eq. (3) for label smoothing regularization is set to 0.0001 and 0.1 for *Cityscapes* and *ADE20K*.

4.2 Results on Cityscapes

Table 1 compares models using different sources of synthetic data. During the training of listed models, each batch contains 70% real images and 30% synthetic data. First, we train a segmentation model [33], with using synthetic

Table 1. Comparison results on *Cityscapes* validation set. Masks from training (T) and validation (V) set are used to synthesize data.

Models	Mask	PixelAcc	ClassAcc	mIoU	fwIoU
Baseline		96.34	86.34	79.73	93.15
+Img [37]	T	95.84	82.54	76.55	92.17
+Img [29]	T	96.21	85.61	79.52	93.07
+Img [19]	V	96.33	85.99	79.60	93.11
Ours	T	96.40	87.29	80.30	93.27
Ours	T+V	96.40	87.47	80.33	93.29

Table 2. Comparison results on *ADE20K* validation set. The top block shows the results of single scale prediction and the bottom is multi-scale prediction.

Models	PixelAcc	ClassAcc	mIoU	fwIoU
Baseline	80.04	51.75	41.68	67.46
Ours	80.00	53.83	42.02	68.19
Baseline	80.76	52.27	42.78	68.75
Ours	80.81	54.70	43.35	69.17

images from previous state-of-the-art generation approaches [19,29,37]. For [37], we can see the performance is significantly decreased compared to the baseline model [33]. For [29], we utilize the training set masks to generate images, which reduces performance across all four metrics comparing the baseline model as well. In addition, we test another state-of-the-art image generator [19]. To know if it is possible to improve semantic segmentation, we even utilize validation set masks. Despite providing additional layouts from validation, the performance still decreases at the validation set. Both experiments demonstrate that generated images often do not lead to improved performance. In contrast, applying our synthetic features successfully improves results by applying different semantic masks. Particularly, our feature synthesis has the same learning objective to image synthesis [37], however, the performance is able to be improved clearly with training set only as well as additional validation set. The results demonstrate the effectiveness of utilizing our synthetic features for data augmentation in semantic segmentation, while synthetic images are hard to use.

In addition to quantitative results, we present some visualization plots compared to PSPNet [33] in Fig. 3. First, our model predicts more smooth results. In the second example, our model successfully recognizes the fence and wall along the street and accurately segment the boundary between them. Besides, for some large regions, i.e., the truck, the train, and the walls in the first, third and fourth examples, our approach achieves clear improvement and segment the whole objects, while baseline predicts unsmooth results. Particularly, we notice that our model with training masks only is also able to achieve a clear improvement over baseline and very similar results to our model with additional validation masks, which means our data augmentation is very effective.

4.3 Results on ADE20K

To understand if our approach works well, we also test our approach on *ADE20K* dataset. We present a quantitative comparison in Table 2 with single-scale prediction as well as multi-scale prediction. We emphasize that our model achieves 2.08/2.43 and 0.34/0.57 improvements on ClassAcc and mIou for single scale

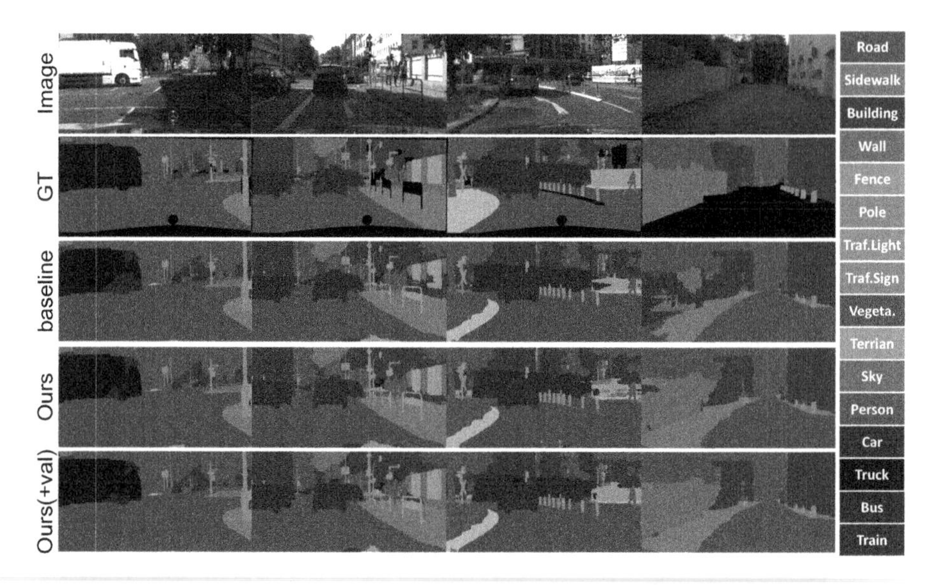

Fig. 3. Qualitative results on the Cityscapes validation set. We show the augmentation results as well as using additional masks from validation set. Best viewed in color. (Color figure online)

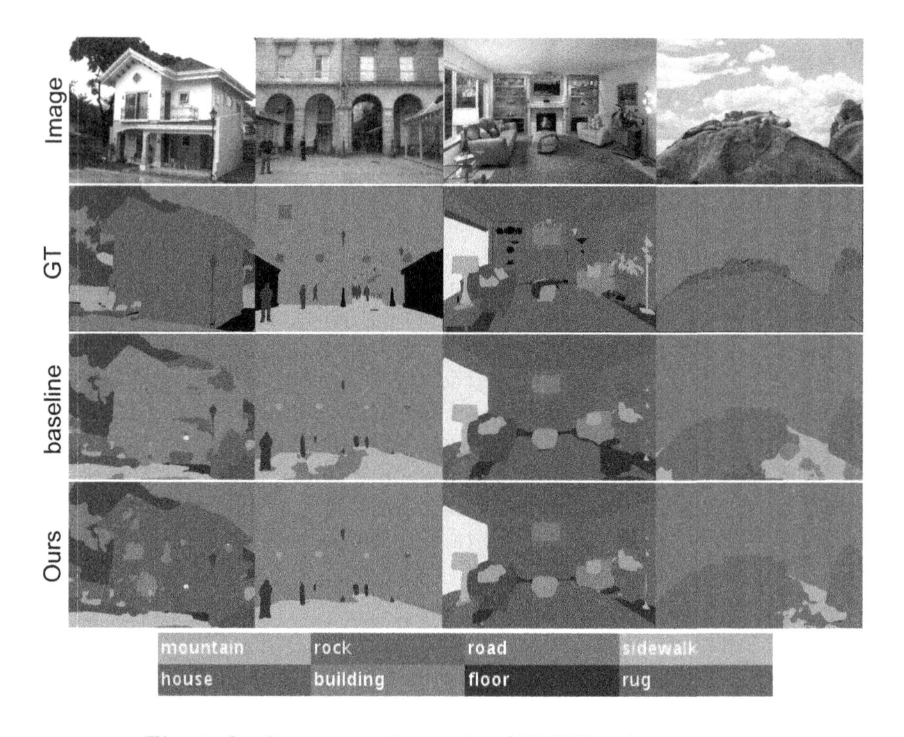

Fig. 4. Qualitative results on the *ADE20K* validation set.

prediction and multiple scale prediction, respectively. Besides, we show several qualitative comparisons in Fig. 4, that our predictions are more smooth and accurate. We are able to distinguish ambiguous classes, such as house/building, mountain/rock. As a result, our model recognizes the entire region for the objects, instead of generating multiple cracked regions.

4.4 Discussion

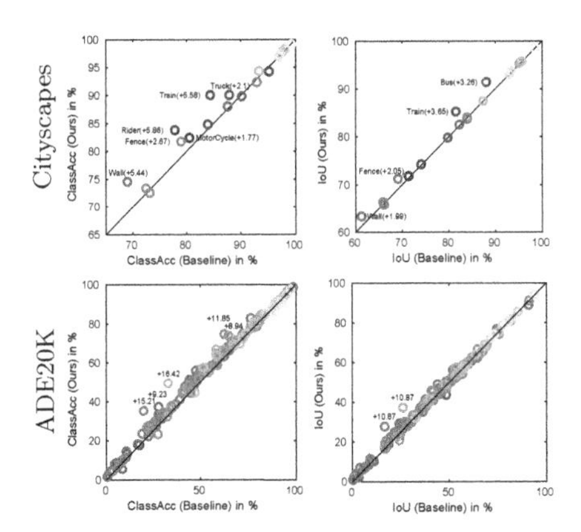

Fig. 5. Per class analysis of ClassAcc and IoU. Yellow indicates more training data, and blue indicates less. (Color figure online)

Per-Class Analysis. To understand how our method boosts a semantic segmentation model, we analyze the per-class accuracy and compare the baseline results and our augmentation results. From Fig. 5, we clearly observe substantial improvements for some classes, especially for those with less training data. Besides, there are only a few points significantly falling below the diagonal, i.e., the negative effects are negligible. In *Cityscapes*, Rider, Train and Wall have more than 5% improvements on ClassAcc; Train and Bus have more than 3% improvements on IoU. In *ADE20K*, we can see some classes achieved more than 10 points improvement on ClassAcc and mIoU.

Overall, we show there are 134 classes obtaining improved ClassAcc and 94 classes have improved IoU in all 150 classes. Particularly, 3 and 2 classes achieve more than 10% improvements on ClassAcc and IoU, respectively. Finally, we also mention that the quantitative comparisons we showed in our main submission and Supplementary Material apparently show the significant improvements.

Synthetic Convolutional Features. To explore the reason for improved results, we analyze the features from different network architectures: (1) baseline architectures in [37], whose output size is same to input; (2) our architectures described in Sect. 3.1.

To begin with, we provide a comparison of PSPNet-scores for those two architectures. We sample 3000 patches from *Cityscapes* training set to compute PixelAcc, ClassAcc, mIoU and fwIoU, as PSPNer-scores. Second, we train a segmentation model with synthetic features from different architectures. A good generator is supposed to have more close number to real features, and achieves larger improvements. Table 3 lists those numbers, showing features from our architecture lead to higher PSPNet-scores in four metrics. For improvements, even baseline has larger ClassAcc, the improvement for PixelAcc, fwIoU is less, and mIoU is quickly decreased. While using features from our architectures

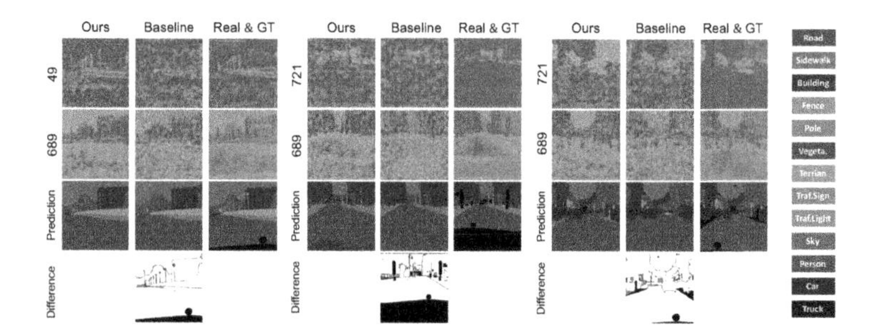

Fig. 6. Visualization of synthetic features conv4_12. In difference maps, cyan color indicates our architecture is better, while red color means baseline is better. Best viewed in color. (Color figure online)

Table 3. Comparison of utilizing different model architectures.

Evaluation	ASPP	PixelAcc	ClassAcc	mIoU	fwIoU
PSPNet-score		93.82	66.85	63.40	88.63
PSPNet-score	✓	96.44	80.05	74.33	93.44
Improvement		+0	+1.44	−0.52	+0.04
Improvement	✓	+0.06	+1.17	+0.55	+0.13

achieves consistent improvements. As a result, we remark that the effectiveness of our proposed architecture is the key to synthesizing successful features.

To understand more about synthetic features from different architectures, we visualize them as well as the extracted features from real images in Fig. 6. Even though both architectures output activations for different channels at similar locations as real features, our architecture with ASPP is able to produce better details. For example, activations in 49-th channel are synthesized for the pole class, which is very small, while the baseline cannot do. In addition, we feed the synthetic features to PSPNet and compare the prediction difference. Observing all the difference maps, we further conclude that our model with high resolution input is stronger in generating features with accurate boundaries and details.

Besides, we perform synthesizing different stages of CNN features. The earliest stage is the image itself, which has shown to be hard to use for data augmentation in Table 1. To understand the synthesis of good features for semantic segmentation, we extract features (before ReLU layers) at different stages and provide statistics for them from PSPNet, as listed in Table 4. Due to the architecture of PSPNet, the latest stage of features we extract is conv4_23, which is the input of the pyramid spatial pooling module.

The statistics are collected from 3000 randomly cropped image patches. First, we compute the entropy per class and channel, which is reported the average number in the table. To eliminate the distribution inconsistency for different stages, we normalize all patches to the same norm ball (L_2 norm 100) for

Table 4. Statistics on different stages of features. The PSPNet-score for the real is 86.15.

	Images	conv1_3	conv2_3	conv3_4	conv4_6	conv4_12	conv4_18	conv4_23
Channels	3	128	256	512	1024	1024	1024	1024
Resolution	1	1/2	1/4	1/8	1/8	1/8	1/8	1/8
Entropy	4.8290	3.0987	3.1584	4.0767	3.3799	3.3659	3.2535	3.4065
mIoU of hist	0.5596	0.5257	0.3171	0.3431	0.4295	0.4087	0.3651	0.2802
PSPNet-score	2.22	9.76	26.97	56.77	70.66	74.33	61.54	83.78
ΔClassAcc	−3.80	+0.02	+0.65	+0.81	+0.99	+1.17	+1.75	+0.08
ΔmIoU	−3.18	−1.09	−0.37	−0.1	+0.11	+0.55	+0.13	+0.05

computing the entropy. A good feature that is easier to generate is supposed to have smaller entropy. As we can see, the statistic is consistent to our requirement, that high-level features usually have smaller entropy.

Second, we compute the similarity of activation distributions across classes and channels. We use the intersection of union between two histograms to measure the similarity. As shown in the 4-th rows in Table 4, similarities for higher level features are less, revealing stronger discrimination and ease to generate. Particularly, color pixels have the largest similarity, which further confirms our motivation and is consistent to the difficulty of synthesizing qualified images to augment semantic segmentation, which is consistent to Table 1.

Third, we directly feed the synthetic features to PSPNet to test if they can be recognized by PSPNet. Good synthetic features should be recognized well. We use mIoU of PSPNet as a measurement, as reported in the 5-th row of the table. As shown, scores are gradually increased from low-level features to high-level features.

Fourth, we applied different stages of synthetic features as training sets, and report the improvements on ClassAcc and mIou on *Cityscapes* validation set. We observe that improvements on two metrics are obtained by using high level features `conv4_6` to `conv4_23`. On the contrary, it is hard to achieve clear improvements with low-level synthetic features `conv1_3` to `conv3_4`. Applying synthetic images leads to the worst performance. We also observe that the largest improvement comes from `conv4_12` instead of `conv4_23`. Even `conv4_23` has the highest PSPNet-score, it can boost less layers of a segmentation model, comparing to `conv4_12`. Finally, we choose to apply synthetic `conv4_12` for *Cityscapes* in the rest of this section.

4.4 Additional Semantic Layout Masks

Since our approach generates a paired data from a semantic mask, we are not only able to augment a training set, but also to leverage new masks to generate more novel examples. To know if more masks are beneficial to boost the performance, we seek for additional masks as listed in the following:

Table 5. Comparison of PSPNet and our approach on Citycapes validation set. The top block is single scale prediction and the bottom is multi-scale prediction.

Models	Additional Mask	PixelAcc	ClassAcc	mIoU	fwIoU
Baseline	–	96.34	86.34	79.73	93.15
Ours	`No`	96.40	87.29	80.30	93.27
Ours	`Val`	96.40	87.47	80.33	93.29
Ours	`Synscapes`	96.39	87.55	80.03	93.27
Ours	`Pseudo GT`	96.40	87.49	80.31	93.28
Baseline	–	96.59	87.23	80.89	93.58
Ours	`No`	96.64	88.16	81.48	93.69
Ours	`Val`	96.64	88.46	81.52	93.71
Ours	`Synscapes`	96.64	88.43	81.34	93.70
Ours	`Pseudo GT`	96.64	88.34	81.51	93.70

- Validation GT masks. It is very easy to acquire high quality masks, which has a similar distribution to the training set. It is used to test if providing more masks with the same distribution is helpful.
- Rendering system. It provides us large amounts of data at very low cost. In our experiments, we apply recent released *Synscapes* dataset [30]. Finally, it provides us 25000 extra semantic masks.
- Pseudo GT masks. We regard prediction from unlabeled images as pseudo ground truth. Although the prediction is not perfect, we generate paired data from a mask, as a result, it still generates features aligned with the input mask. To alleviate the unsmooth prediction, we only save the prediction with posterior larger than 0.7, and complete the holes with nearest neighbor interpolation. To provide novel scenes, we leverage the unlabeled video frames and coarse annotation frames in *Cityscapes*, leading to 29823 extra masks.

Table 5 presents the performance with single scale prediction and multi-scale prediction. To have a fair comparison, we set the ratio between training masks and additional masks for all the additional mask choices. That is $3:1$, in each sampled batch for synthesizing. Besides, the ratio between real images and synthetic data follows previous experiments $7:3$, and online hard negative mining is applied for various versions of our models.

We observe that after adding validation mask, the performance is further improved than pure augmentation (second row). Besides, applying pseudo GT also achieves better performance than training set only. Interestingly, when we apply the semantic masks from *Synscapes* [30], the improvement is less, the reason is the distribution gap of layouts between *Cityscapes* and *Synscapes*. As a result, to leverage synthetic data from the game engine better, not only similar appearance, but also similar semantic layouts are needed, such as shapes, ratios between different objects, etc.

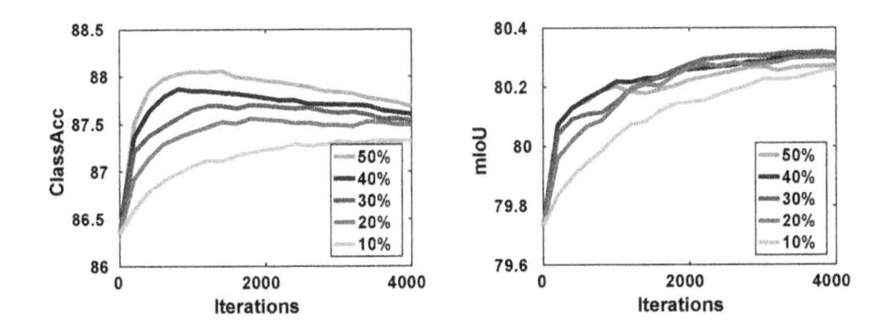

Fig. 7. ClassAcc and mIoU on *Cityscapes* at varying training iterations w.r.t. different percentages of our synthetic features in a training batch.

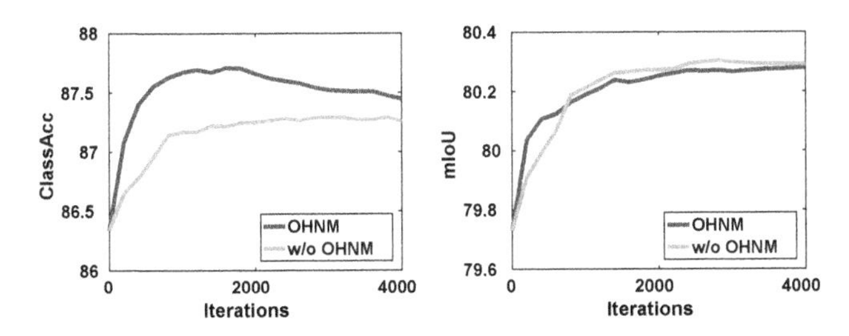

Fig. 8. ClassAcc and mIoU on *Cityscapes* w.r.t. applying online hard negative mining (OHNM).

4.5 Ablation Study

Ratio Between Real and Synthetic Data. In Fig. 7, we provide a study on using different percentages of synthetic features in a batch. First, it clearly shows that incorporating our synthetic features with different percentages brings improvements. Besides, we observe that incorporating more synthetic features will gain more in ClassAcc, which further shows the effectiveness of our approach. On the other hand, better fitting on synthetic data might lead to less improvement on other metrics, as a result, we mix 70% real images and 30% synthetic features in each batch for the rest of the experiments.

Effectiveness of Online Hard Negative Mining. In Fig. 8, we compare the models with randomly sampled masks and mined hard negative during training as described in Sect. 3.3. To begin with, we are able to observe that simply adding synthetic features is already helping to improve the performance, which demonstrates the effectiveness of our augmentation pipeline with synthetic features. Additionally, how to make full use of synthetic features to compensate for the data distribution for boosted performance is also interesting. Even though there may be better designs to incorporate synthetic features, the online hard negative mining is simple and effective, which shows the necessity of feature-level data augmentation.

5 Conclusion

Designing advanced network architectures to incorporate context more suitable or extract more representative features are important in semantic segmentation. We improve semantic segmentation from a different avenue according to utilizing data more effectively. We propose synthesizing convolutional features for improving semantic segmentation. Our pipeline is simple but surprisingly effective to reach stronger segmentation results. A powerful architecture for feature synthesis is presented, enabling us to synthesize realistic features with rich details. Also, several techniques are presented to leverage synthetic features more effectively for the success of improved semantic segmentation.

References

1. Antoniou, A., Storkey, A., Edwards, H.: Data augmentation generative adversarial networks. arXiv preprint arXiv:1711.04340 (2017)
2. Bowles, C., et al.: GAN augmentation: augmenting training data using generative adversarial networks. arXiv preprint arXiv:1810.10863 (2018)
3. Chen, L.C., Papandreou, G., Kokkinos, I., Murphy, K., Yuille, A.L.: DeepLab: semantic image segmentation with deep convolutional nets, atrous convolution, and fully connected CRFs. arXiv preprint arXiv:1606.00915 (2016)
4. Chen, Q., Koltun, V.: Photographic image synthesis with cascaded refinement networks. In: ICCV (2017)
5. Cordts, M., et al.: The cityscapes dataset for semantic urban scene understanding. In: CVPR (2016)
6. Deng, J., Dong, W., Socher, R., Li, L.J., Li, K., Fei-Fei, L.: ImageNet: a large-scale hierarchical image database. In: CVPR (2009)
7. Frid-Adar, M., Klang, E., Amitai, M., Goldberger, J., Greenspan, H.: Synthetic data augmentation using GAN for improved liver lesion classification. arXiv preprint arXiv:1801.02385 (2018)
8. Ghiasi, G., Fowlkes, C.C.: Laplacian pyramid reconstruction and refinement for semantic segmentation. In: Leibe, B., Matas, J., Sebe, N., Welling, M. (eds.) ECCV 2016. LNCS, vol. 9907, pp. 519–534. Springer, Cham (2016). https://doi.org/10.1007/978-3-319-46487-9_32
9. Goodfellow, I., et al.: Generative adversarial nets. In: NIPS (2014)
10. Hoffman, J., Wang, D., Yu, F., Darrell, T.: FCNs in the wild: pixel-level adversarial and constraint-based adaptation. arXiv preprint arXiv:1612.02649 (2016)
11. Hung, W.C., Tsai, Y.H., Liou, Y.T., Lin, Y.Y., Yang, M.H.: Adversarial learning for semi-supervised semantic segmentation. In: BMVC (2018)
12. Isola, P., Zhu, J.Y., Zhou, T., Efros, A.A.: Image-to-image translation with conditional adversarial networks. In: CVPR (2017)
13. Luc, P., Couprie, C., Chintala, S., Verbeek, J.: Semantic segmentation using adversarial networks. In: NIPS Workshop on Adversarial Training (2016)
14. Mirza, M., Osindero, S.: Conditional generative adversarial nets. arXiv preprint arXiv:1411.1784 (2014)
15. Mueller, F., et al.: Ganerated hands for real-time 3d hand tracking from monocular RGB. In: CVPR (2018)
16. Odena, A., Olah, C., Shlens, J.: Conditional image synthesis with auxiliary classifier GANs. In: ICML (2017)

17. Park, T., Liu, M.Y., Wang, T.C., Zhu, J.Y.: Semantic image synthesis with spatially-adaptive normalization. In: Proceedings of the IEEE Conference on Computer Vision and Pattern Recognition (2019)
18. Peng, X., Tang, Z., Yang, F., Feris, R.S., Metaxas, D.: Jointly optimize data augmentation and network training: adversarial data augmentation in human pose estimation. In: CVPR (2018)
19. Qi, X., Chen, Q., Jia, J., Koltun, V.: Semi-parametric image synthesis. In: CVPR (2018)
20. Reed, S., Akata, Z., Yan, X., Logeswaran, L., Schiele, B., Lee, H.: Generative adversarial text to image synthesis. In: ICML (2016)
21. Ronneberger, O., Fischer, P., Brox, T.: U-Net: convolutional networks for biomedical image segmentation. In: Navab, N., Hornegger, J., Wells, W.M., Frangi, A.F. (eds.) MICCAI 2015. LNCS, vol. 9351, pp. 234–241. Springer, Cham (2015). https://doi.org/10.1007/978-3-319-24574-4_28
22. Saleh, F.S., Aliakbarian, M.S., Salzmann, M., Petersson, L., Alvarez, J.M.: Effective use of synthetic data for urban scene semantic segmentation. In: Ferrari, V., Hebert, M., Sminchisescu, C., Weiss, Y. (eds.) ECCV 2018. LNCS, vol. 11206, pp. 86–103. Springer, Cham (2018). https://doi.org/10.1007/978-3-030-01216-8_6
23. Sankaranarayanan, S., Balaji, Y., Jain, A., Lim, S.N., Chellappa, R.: Learning from synthetic data: addressing domain shift for semantic segmentation. In: CVPR (2018)
24. Shrivastava, A., Pfister, T., Tuzel, O., Susskind, J., Wang, W., Webb, R.: Learning from simulated and unsupervised images through adversarial training. In: CVPR (2017)
25. Sixt, L., Wild, B., Landgraf, T.: RenderGAN: generating realistic labeled data. arXiv preprint arXiv:1611.01331 (2016)
26. Souly, N., Spampinato, C., Shah, M.: Semi supervised semantic segmentation using generative adversarial network. In: 2017 IEEE International Conference on Computer Vision (ICCV), pp. 5689–5697. IEEE (2017)
27. Szegedy, C., Vanhoucke, V., Ioffe, S., Shlens, J., Wojna, Z.: Rethinking the inception architecture for computer vision. In: CVPR (2016)
28. Tsai, Y.H., Hung, W.C., Schulter, S., Sohn, K., Yang, M.H., Chandraker, M.: Learning to adapt structured output space for semantic segmentation. arXiv preprint arXiv:1802.10349 (2018)
29. Wang, T.C., Liu, M.Y., Zhu, J.Y., Tao, A., Kautz, J., Catanzaro, B.: High-resolution image synthesis and semantic manipulation with conditional GANs. In: CVPR (2018)
30. Wrenninge, M., Unger, J.: Synscapes: a photorealistic synthetic dataset for street scene parsing. arXiv preprint arXiv:1810.08705 (2018)
31. Wu, Z., et al.: DCAN: dual channel-wise alignment networks for unsupervised scene adaptation. In: Ferrari, V., Hebert, M., Sminchisescu, C., Weiss, Y. (eds.) ECCV 2018. LNCS, vol. 11209, pp. 535–552. Springer, Cham (2018). https://doi.org/10.1007/978-3-030-01228-1_32
32. Xian, Y., Lorenz, T., Schiele, B., Akata, Z.: Feature generating networks for zero-shot learning. In: CVPR (2018)
33. Zhao, H., Shi, J., Qi, X., Wang, X., Jia, J.: Pyramid scene parsing network. In: CVPR (2017)
34. Zheng, Z., Zheng, L., Yang, Y.: Unlabeled samples generated by GAN improve the person re-identification baseline in vitro. In: ICCV (2017)
35. Zhou, B., Zhao, H., Puig, X., Fidler, S., Barriuso, A., Torralba, A.: Scene parsing through ADE20K dataset. In: CVPR (2017)

36. Zhu, J.Y., Park, T., Isola, P., Efros, A.A.: Unpaired image-to-image translation using cycle-consistent adversarial networks. In: ICCV (2017)
37. Zhu, J.Y., et al.: Toward multimodal image-to-image translation. In: NIPS (2017)

Autonomous Car Chasing

Pavel Jahoda[2](✉), Jan Cech[1], and Jiri Matas[1]

[1] Visual Recognition Group, Czech Technical University in Prague, Prague, Czechia
[2] Faculty of Information Technology, Czech Technical University in Prague, Prague, Czechia
pjahoda6@gmail.com

Abstract. We developed an autonomous driving system that can chase another vehicle using only images from a single RGB camera. At the core of the system is a novel dual-task convolutional neural network simultaneously performing object detection as well as coarse semantic segmentation. The system was firstly tested in CARLA simulations. We created a new challenging publicly available CARLA Car Chasing Dataset collected by manually driving the chased car. Using the dataset, we showed that the system that uses the semantic segmentation was able to chase the pursued car on average 16% longer than other versions of the system. Finally, we integrated the system into a sub-scale vehicle platform built on a high-speed RC car and demonstrated its capabilities by autonomously chasing another RC car.

Keywords: Autonomous driving · Chasing · Multi-task convolutional neural network · RC car · Deep learning · CARLA · Simulation

1 Introduction

An outgoing effort is made to enable autonomous systems to drive in a safe manner. To be safe, such a system needs to quickly react to unexpected situations in a dynamic environment. Especially at high speed. A car chase is a scenario, which requires dynamic maneuvers and tests the limits of an autonomous driving system. The key aspects of driving autonomously include an understanding of the surrounding environment, detecting other cars, and reacting to them on the road. In this paper, we present a novel neural architecture that gives information about other cars as well as the surrounding drivable surface. We then test an autonomous driving system that we developed, in the car chasing scenario.

To the extent of our knowledge, a car chasing scenario has not been presented. On the other hand, car following has been studied for more than half a century. Car following models how vehicles should follow each other in a traffic stream. Most of the theoretical models assume having precise continuous information about speed, distance, and acceleration of each car [2]. With the progress

Electronic supplementary material The online version of this chapter (https://doi.org/10.1007/978-3-030-66823-5_20) contains supplementary material, which is available to authorized users.

A. Bartoli and A. Fusiello (Eds.): ECCV 2020 Workshops, LNCS 12538, pp. 337–352, 2020.
https://doi.org/10.1007/978-3-030-66823-5_20

of machine learning and computer vision, more practical car-following models have been developed and tested. In 1991, Kehtarnavaz et al. developed the first autonomous car-following system called BART [11]. The system was able to follow the leading car at 20 km/h speed. However, car-following models have typically not been tested in non-cooperative scenarios. Not to mention scenarios involving high speeds and sudden fast acceleration and deceleration encountered in car chases.

A car chase is a vehicular pursuit that typically involves high speeds and therefore dynamic driving maneuvers are necessary. In the car chasing scenario, a vehicle being pursued is driven by a person, while the vehicle that is chasing it is controlled by an autonomous system. Addressing an autonomous car chasing problem typically requires a complex methodology including modules of computer vision, planning, and control theory. Our vision-based system has an architecture that is similar in spirit to the DARPA Urban Challenge vehicles [4,5,25] and consists of three parts: perception and localization, trajectory planning, and a trajectory controller.

In the perception part, we introduce a novel neural network that detects objects and segments an image at the same time, i.e. in a single pass of the image through the network. It is used to find a 2D bounding box of the pursued object and segments an image into a coarse grid of $S \times S$ cells. In our case, each cell is either drivable or not. Based on the 2D bounding box detection, the angle and the distance between the two cars is estimated. After localizing the chased car, the trajectory is planned. The planning algorithm takes the segmentation of the drivable surface into account. Finally, we calculate the steer and throttle to drive the vehicle. A Pure Pursuit [23] inspired algorithm is developed to control the steering and a PID controller [24] is used to control the throttle.

The contributions of the paper are as follows: (1) A novel dual-task CNN performing object detection and semantic segmentation simultaneously, (2) an algorithm for autonomous car chasing was proposed and both extensively evaluated on a simulated dataset and tested on a real sub-scale platform, and (3) the CARLA Car Chasing dataset together with a test protocol and error statistics.

2 Related Work

To the best of our knowledge, no system for a non-cooperative car chasing scenario has been described in the literature. However, related problems have been extensively researched. Many of the early autonomous vehicles were able to drive by driving along a pre-specified path [19]. In 1985, Carnegie Mellon University demonstrated the first autonomous path-following vehicle [23]. A year later, their Navlab vehicle was able to drive on a road at the top speed of 28 km/h [22].

Many attempts to address the car following problem focused on different parts of an autonomous system independently. In the 1990s, the first attempts of localization of the "lead" chased car were made [17,18,20]. Schwarzinger et al. focused on estimating dimensions of the "lead" car [18] while Gohring et al. focused on estimating the velocity of the chased car [8]. Our solution circumvents these non-trivial problems and works without the need of knowing the velocity of the chased car.

The problem of autonomously chasing an object has been investigated for unmanned aerial vehicles. Target chasing using UAVs is different to the car chasing problem only in two aspects. Firstly, UAVs have very different kinematics and dynamics and move in more directions, which also means that altitude control is necessary. Secondly, frequently there is an assumption that no obstacles would prevent the UAV to move in all directions [13,15,21]. This is in contrast with autonomous chasing using road vehicles, where obstacle avoidance is a crucial part of the problem. At a high level, the architecture of an autonomous UAV is similar to the architectures used in autonomous vehicles [15,21]. As an example of an autonomous UAV target chasing, Telière et al. developed an autonomous quad-rotor UAV that chases a moving RC car [21].

3 Proposed Method

3.1 Perception

Dual-task Neural Network simultaneously detects 2D bounding box of the pursued vehicle and predicts coarse semantic segmentation, i.e., segments an image into a coarse grid of $S \times S$ cells. The network shares the same backbone for both tasks – a 53 layer feature extractor called Darknet-53 [16]. Attached to the feature extractor are two sets of heads – for the object detection and for the image segmentation. The architecture of the neural network is depicted in Fig. 1.

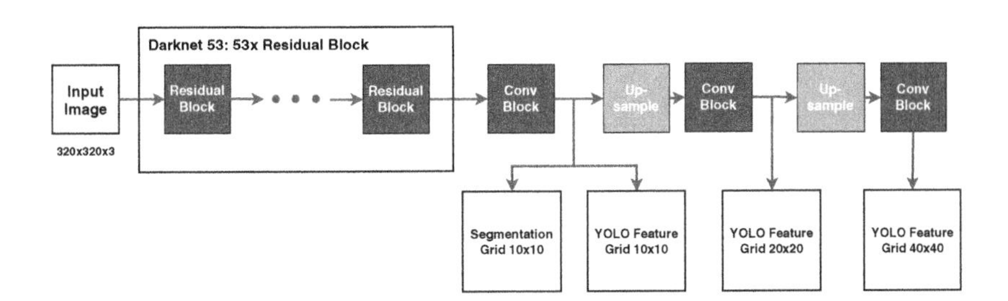

Fig. 1. Architecture of the novel dual-task convolutional neural network. The three YOLO Feature Grids are used to predict bounding boxes at 3 different scales [16]. The Segmentation Grid is the output of the coarse semantic segmentation.

The network was trained by alternating optimization – in every second batch, the network is optimized only for detection, while the segmentation is optimized in the remaining batches. The neural network uses different loss functions depending on the batch.

During inference, an image is passed through the network just once. The network provides an object detection as well as the semantic segmentation outputs. While the training is slightly slower than training a single-task neural network, the extra cost is negligible during inference since the backbone features are shared. Fast execution and a smaller memory footprint of the dual-task network are especially important for real-time tests on embedded hardware with limited resources.

The YOLOv3 backbone [16] was pre-trained on the COCO dataset consisting of more than 330k images with objects from 80 different classes [14]. For the detection problem, the network was fine-tuned on a collected dataset of a few hundred images of the chased car called "Car Chasing Dataset". Only the final layers that follow the Darknet-53 feature extractor were updated.

The segmentation output provides a semantic map of the input image consisting of $S \times S$ cells, see Fig. 2. We set $S = 10$ and two classes: a drivable surface and a background. The logistic function is used to train the network.

Fig. 2. Segmented image from the RC car camera. The image is segmented into 10×10 grid of cells. The green cells represent a drivable surface. The image is taken from the test set of the Car Chasing Dataset. (Color figure online)

For coarse segmentation, we annotated the images from the Car Chasing Dataset. The images were first annotated by pixel-level precision, then 10×10 maps were generated by summing drivable pixels belonging to a cell. If the sum exceeds 50%, the cell was set as drivable, and as the background otherwise.

3.2 Chased Car Localization

Angle and Distance Estimation. A relative angle and distance to the pursued car are sketched in Fig. 3.

To estimate the distance and the angle to the chased car from a camera image, we employ a PnP (Perspective-n-Point) problem [12] solver. Having a 3D model – image correspondences and the intrinsic camera calibration matrix $\mathbf{K}$, a relative camera–model pose $\mathbf{R}, \mathbf{t}$ is estimated.

Known dimensions of the chased car (obtained from an RC manufacturer documentation), namely a tight virtual 3D bounding-box gives 3D model coordinates, and the image bounding box of the detection gives the corresponding image projections. We assume that the bounding box encloses only the corresponding image points, i.e., the back of the RC car. We acknowledge that the bounding box oftentimes encloses more than just the back of the car, but we

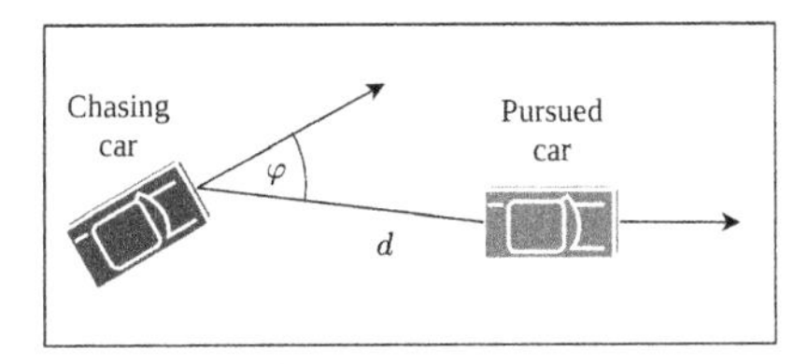

Fig. 3. Angle and distance definition: Chasing car in blue and pursued car in gray. Distance d angle φ are estimated using a camera mounted on the chasing car.

found the resulting inaccuracies negligible. The camera calibration is known. An OpenCV [3] solver is used to estimate the model pose. Then, the distance d is calculated as Euclidean distance between the translation vector $\mathbf{t}$ and the origin. Similarly, we extract angle φ.

Angle and Distance Extrapolation. The image-based detector may fail, especially due to poor lighting, motion blur, or because the pursued vehicle is only partially visible or fully occluded. Therefore, we extrapolate the angle and the distance of the chased car based on the previous estimates. The extrapolation algorithm uses an exponential moving average. Six variables store information about the whole history of estimated angles and distances. The idea is explained by the process of extrapolating distances.

For every frame, we update the exponential moving average variable using the following equation:

$$\hat{e_i} = \begin{cases} \alpha \cdot d_i + (1 - \alpha) \cdot \hat{e}_{i-1}, & \text{if bbox found,} \\ \alpha \cdot d_{ex} + (1 - \alpha) \cdot \hat{e}_{i-1}, & \text{otherwise.} \end{cases} \quad (1)$$

Variable $\hat{e_i}$ is the exponential moving average at frame i, parameter α is in range [0,1] and sets the weight of the last predicted distance. We set $\alpha = 0.5$ in all experiments.

If the chased car is detected at frame i, distance d_i is set to the predicted estimate received from the PnP solver. Then $\hat{e_i}$ is updated using d_i. In case no bounding box was detected, $\hat{e_i}$ is updated by

$$d_{ex} = 2 \cdot d_{i-1} - d_{i-2}. \quad (2)$$

Estimate $\hat{e_i}$ is the output for frame i in case no bounding box was detected. We set $d_i = d_{ex}$ for future computations.

The angle extrapolation is done similarly. Finally, each extrapolated angle is processed by a saturation block that limits the output values. The limits are -175 and $175°$. This method allows to extrapolate the angle and the distance very fast in a $\mathcal{O}(1)$ time and space complexity with almost no overhead.

3.3 Planning and Control

Pure Pursuit [23] **Inspired Algorithm** implements the lateral control of the vehicle. The Pure pursuit is a path tracking algorithm that moves a vehicle from its current position to a look-ahead position on the path. The algorithm was invented in the 1980s as it was used in the first demonstration of an autonomous vehicle capable of following a road using imagery from a black and white camera [23]. The goal of the demonstration was to keep the vehicle centered on the road while it was driving at a constant speed.

We were inspired by the pure pursuit algorithm, however, a chasing car does not move at a constant speed and we are not following a path, but rather chasing a moving target, a car. Therefore, for each frame, the target position is updated with respect to the current position of the car. Unlike the pure pursuit algorithm, we do not calculate the curvature to control the steering. The steering is updated at every frame, such that the steering angle is set as the estimated angle φ described in Sect. 3.2. The steering angle is then mapped to the steering input of a test vehicle, see Sect. 3.4.

The lateral and longitudinal controllers are completely independent. The velocity of the car is not considered for turning, no limit of the steering angle based on current speed is set. In this basic algorithm, we simply assume that the dynamics of the chasing car is better or equal to the pursued car.

Semantic Segmentation Based Planning. The above control algorithm assumes that the entire surface is drivable. If it is not the case, the car may collide with obstacles when pursuing the other car blindly by driving to the latest position of the car. The neural network provides a coarse grid of drivable surface segmentation, so we propose a simple algorithm that resolves the issue.

The idea is illustrated in Fig. 4. First, we construct a line segment from the center of the bottom side of the image to the bounding box of the detected car. If this line segment goes only through cells that are drivable, the trajectory planning phase stays unchanged and we try to follow the car directly as described above. Otherwise, a new target is found. A line segment going to other cells on the same row as the bounding box of the car. If the line segment is all drivable, the endpoint is the new target. If not, the process is repeated until a possible target is found or the algorithm stops if an acceptable target does not exist.

PID Controller. We used a PID controller to maintain a desired distance between the two cars. The controller actuates the throttle of the vehicle. The three constants of the controller were set with respect to the longitudinal dynamics of the vehicle. The setting was found by testing over a set of weights and picked the one that resulted in the lowest absolute average error $\varepsilon = d - d_{desired}$ in simulations and adjusted to fit the real sub-scale vehicle dynamics empirically. We set the controller to be rather slow but highly stable in our experiments.

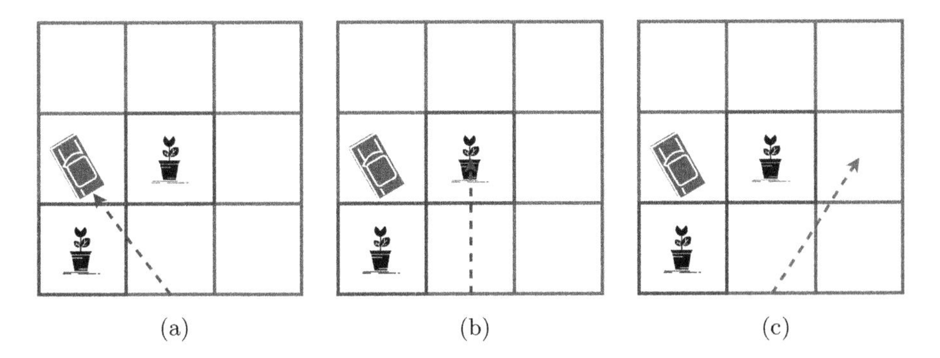

Fig. 4. Illustration of the segmentation based planning. First, the algorithm checks if it is possible to drive directly towards the car **(a)**. This is not possible as the constructed line segment goes through a non-drivable grid cell. Then, the remaining cells on the same row as the detected car are checked **(b,c)**. The algorithm finally proposes a direction represented by a line segment that goes only through drivable surfaces and that is closest to the direction of the chased car **(c)**.

3.4 Implementation Details

We tested the proposed algorithm in CARLA simulations and in real sub-scale platform. The steering angle to the steering input was mapped to respective vehicles. Steering input is in range $[-1, 1]$, which corresponds to steering from maximum left to maximum right. A linear mapping was found for the simulator, while a non-linear mapping $steer \propto (\varphi/180)^2$ was used for controlling the RC car platform. Similarly with the throttle input, a value in range $[0, 1]$ is expected. No torque is 0, while full throttle is 1. PID controller setting was: $w_p = 0.1$, $w_i = 0$, and $w_d = 1$ (proportional, integration and derivative component weights) respectively for CARLA simulator, while $w_p = 0.25$, $w_i = 0$, and $w_d = 0.07$ for the real vehicle. The output of the controller was clipped to $[0, 1]$ range.

4 Simulation Experiments

4.1 CARLA Environment

All the simulation experiments were performed in CARLA – an open-source simulator for autonomous driving research [6]. The experiments were done in version 0.9.8. CARLA has a wide selection of sensors including camera, segmentation camera, LIDAR, etc. It provides multiple maps with an urban layout. Users can control the weather, as well as many different actors (vehicles, pedestrians) at the same time.

4.2 Experimental Setup

We chose Tesla Model 3 as the chased car and Jeep Wrangler Rubicon as the chasing car. In each experiment, the pursued Tesla Model 3 was placed in front of the autonomously driven Jeep, as shown in Fig. 5. At each frame, we simulated a bounding box detection of the chased car. We did so by projecting the ground truth world coordinates of the chased vehicle bounding box to the camera image coordinates. Then, we randomly perturbed the bounding box by adding random exponential noise proportional to the size of the bounding box. Furthermore, we simulated detection failure (false negative) in 10% of the frames. The detection was not provided when the chased car was out of the field of view or occluded.

All the experiments were performed in a synchronous mode in which the simulation waits for the sensor data to be ready before sending the measurements as well as halts each frame until a control message is received [6].

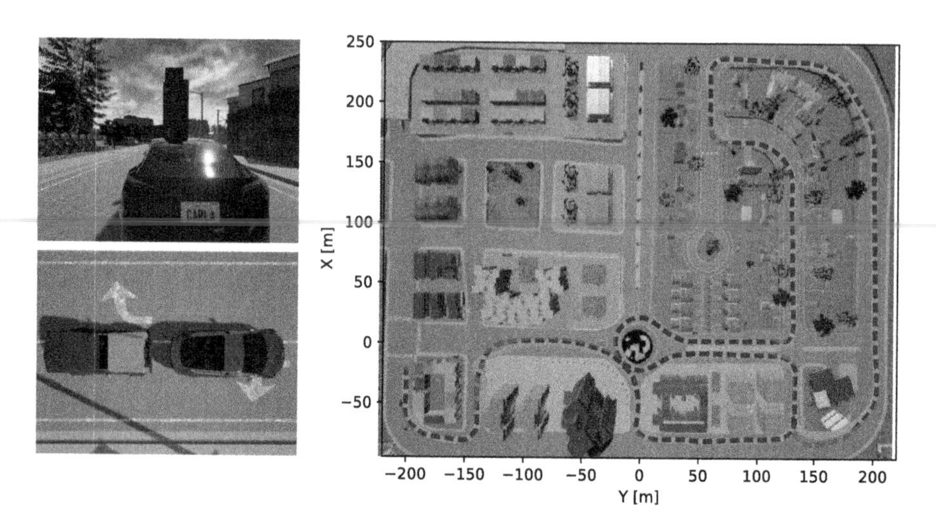

Fig. 5. CARLA experiment. **Top-left image:** view from the chasing car. **Bottom-left image:** chasing and pursued car from above. **Right image:** An example of a difficult drive trajectory. The blue and dashed-black lines are trajectories of the chasing and pursued car respectively. (Color figure online)

CARLA Car Chasing Dataset. To perform the experiments we first collected a driving dataset (available on our GitHub repository [9]). We used the Tesla Model 3 and manually drove it around the city using a connected steering wheel and pedals Hama uRage GripZ to generate the trajectory of the pursued car. The location and orientation were recorded at every frame. This way, we have collected a database of 20 different drives with varying difficulty: 10 easy, and 10 challenging rides. The rides are about 1 min long on average.

The easy subset contains simple drives in which the car was driven slowly and without sudden turning and braking. Any changes such as turns were slow. The average speed in these drives is 35.01 km/h. While the difficult subset aims to test the algorithm to its limits. The drives include sudden braking, fast turns as well as cutting corners. Sometimes even drifts were performed. The average speed in these drives is 49.68 km/h with the fastest drive having an average speed of 63.29 km/h.

We used the database to evaluate the autonomous chasing algorithm. At the beginning of the following experiments, a chasing car was placed half of a meter behind the chased car. Then, we updated every frame the location and orientation of the chased car based on the saved coordinates in the dataset. Meanwhile, the chasing vehicle was driven autonomously and trying to maintain a desired distance from the pursued car. Figure 5 depicts the initial setup as well as an example of a full trajectory of an autonomous chase in CARLA.

4.3 Evaluation for the Autonomous Chasing Algorithm

In the first experiment, we compared different versions of the algorithm. We evaluate the algorithm with its full functionality enabled, and also ablated algorithm without using coarse image segmentation and without the segmentation and distance and angle extrapolation. We compare these versions on both of the easy and difficult sets described in Sect. 4.2.

The evaluation was done using five statistics: (1) An average root-mean-square error (RMSE), and (2) mean absolute error (MAE) of the desired distance and the actual distance between the two cars calculated over frames of all trials. (3) The average number of crashes per drive. A crash can be a collision with the environment (building for example) or with the pursued car. (4) The average drive completion, which is based on how long the autonomous system was chasing the pursued car. It is calculated as follows. The closest point of the pursued car trajectory to the last position of the chasing car is found. The drive completion is the percentage of the length of the trajectory to this point over the entire trajectory length. The statistics are averaged over the test rides. (5) The chase is considered to be successfully finished if the drive was at least 95% completed.

Table 1. Results on CARLA chasing dataset

Easy set	Finished	Avg. completion	Crashes	MAE	RMSE
Full Algorithm	10	97.48%	0.1	9.28	10.91
W/o Segmentation	10	97.41%	0.1	9.26	10.90
W/o Segm. + Ex.	9	97.03%	0.11	9.34	11.23
Difficult set	Finished	Avg. completion	Crashes	MAE	RMSE
Full Algorithm	4	**63.84%**	1.5	14.39	18.30
W/o Segmentation	4	54.76%	1.25	14.30	18.22
W/o Segm. + Ex.	3	53.29%	1	14.49	18.89

The results are summarized in Table 1. As opposed to the difficult driving set, all three versions of the system performed well on the easy test set with even the simplest version finishing most of the drives. On the difficult subset, we can see, that the full algorithm performed significantly better than the remaining two versions. It achieved almost 10 % points higher drive completion on average than the next best-evaluated version.

Qualitative Summary. While performing the experiments, we saw that the use of coarse semantic segmentation allowed the system to stay longer on the road in situations when the pursued car was not detected for a prolonged period. For example, when the pursued car drove behind a building, both versions of the algorithm that use extrapolation correctly predicted the position of the pursued car. However, the algorithm that did not use segmentation drove straight into the building and collided as it did not have any information about the surrounding environment. Only the full version of the algorithm drove around the building and continued the chase. For demonstrations, see [9].

4.4 Impact of the Detector Quality on Chasing Success

We measured the effect of the detector accuracy on the chasing results. We simulated the algorithm with a detector of different miss rates. Failed detection have a uniform distribution over the frames of the sequence. All three versions of the algorithm were tested. Results are shown in Fig. 6.

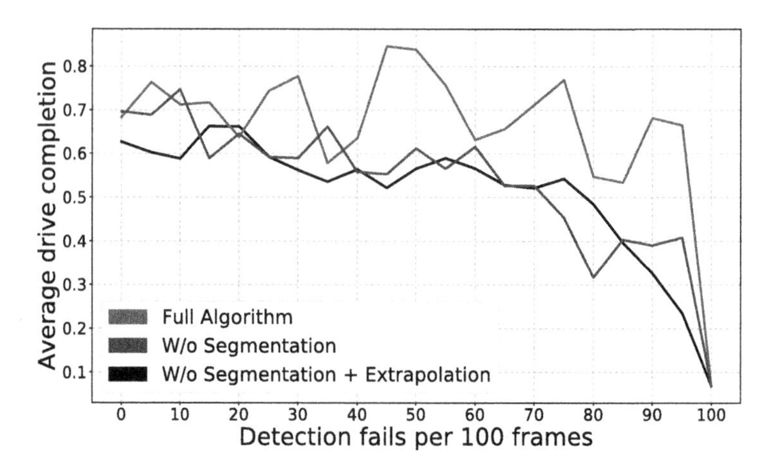

Fig. 6. Average drive completion plotted against detection recall. Only difficult drives were included.

We can see, that for the two versions that do not use semantic segmentation, the average drive completion steadily declines with the detection failure rate. The full version consistently maintains the highest drive completion for all detectors. When the recall is at 60% or lower, the average drive completion of the full version is almost 20% higher than the average drive completion of the remaining two versions.

Note that non-monotonic behaviour is caused by the randomness of the detection failure events. Averaging over several random trials is computationally very expensive since the simulations for all test rides need to be computed. Nevertheless, the trend is apparent.

5 Real Experiments

Chasing RC Car. For the deployment of the autonomous chasing algorithm, a sub-scale vehicle platform called ToMi was used [1]. The platform is built around a large 1:5 scale RC car "Losi Desert Buggy XL-E 4WD". This electrically powered car has 0.9×0.5 m length and width dimensions with reported maximum speed of up to 80 km/h. The platform has a Raspberry Pi generating pulse width modulation (PWM) signals for the throttle and to the steering servomotor. It is equipped with ZED stereo camera (used as a monocular camera) for taking color HD images at 30 FPS and NVIDIA Jetson AGX Xavier with GPU for image processing and neural network inference.

The whole simplified process flow is as follows. First, a single image of the stereopair is taken by the camera, which then goes to the Jetson and the GPU where it is analyzed by the neural network. Then, the planning and control algorithm computes the steer and throttle commands, that are transmitted to the Raspberry Pi which finally generates signals to the motor controller and servomotor that actuates the vehicle.

Chased RC Car. The RC car used as the pursued vehicle is a 1:10 scale "Losi XXX-SCT Brushless RTR". This electrically powered car has 0.55×0.29 m length and width dimensions with reported maximum speed of up to 55 km/h. Both vehicles are shown in Fig. 8.

5.1 Vehicle Detection

Dataset. In order to train and evaluate an object detector capable of predicting a 2D bounding box around the chased RC car in an image, the Car Chasing Dataset was collected. The dataset consists of 460 manually annotated images. All image annotations are in the PASCAL VOC data format introduced in the PASCAL Visual Object Classes Challenge [7]. Each image has its annotation that contains pixel coordinates of the bounding box corners for each object in the image. A single object per image is present – the chased RC car. The data were collected on various surfaces, and under different lighting conditions. The majority of images were collected by the platform camera. Around 5% of the collected images were taken with a smartphone.

Detection Results. The collected dataset was randomly split into three sets: Training (80%), test (10%), and validation (10%) sets. To evaluate the model, we calculated several statistics – recall, precision, XY loss, and WH loss.

Following Pascal VOC, the prediction is true positive (tp) when IoU (Intersection over Union) between the predicted bounding box and the ground truth bounding box is greater than 0.5. XY loss is a mean squared error (MSE) between the center of the predicted bounding box and the center of the ground truth bounding box. WH loss is the difference between width and height of the predicted bounding box and the ground truth bounding box. When calculating both of these statistics, coordinates and image dimensions were scaled to the [0, 1] range.

During training, the model with the smallest sum of the WH and XY losses on the validation subset was selected. The evaluation of the detector on the test set is shown in Table 2. Results indicate that the model detects the chased car accurately. This is achieved with only a few hundred training examples. Examples of the detection, together with angle and distance estimates are shown in Fig. 7.

Table 2. Evaluation of the detector on a test set.

	Precision	Recall	XY loss	WH loss
Test set	99.8%	97.7%	0.066	0.227

Fig. 7. Detection bounding boxes, distance, and angle estimation performed on test images acquired by autonomous car camera.

5.2 Segmentation Results

The segmentation accuracy of the dual-task network was evaluated on the independent test set of 46 images, the same instances as for the detection test. The neural network achieved 94.4% classification accuracy. False positive error (non-drivable cell predicted as drivable) is 3.4% while false negative rate 2.2%. The distribution of the drivable/non-drivable cells was 32/68% respectively. See Fig. 2 for an example of the segmentation result.

5.3 Test Drive Using RC Cars

We performed several live tests under different weather and lighting conditions. The system was tested on an empty roundabout as well as in a residential area as depicted in Fig. 8. The chasing trajectories driven on a road leading to the roundabout included straight and curvy segments and a turn on the intersection. On the roundabout, the trajectories consisted of arc segments and full circles.

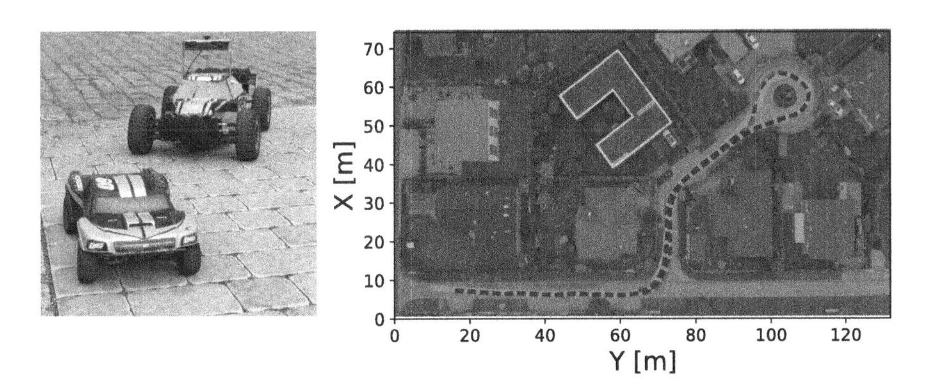

Fig. 8. Left image: RC cars used for testing. The autonomous car with the camera is farther away. **Right image:** an example of a real-world test captured by a drone flying 103 m above the ground. Trajectories of both vehicles are shown: blue of the chasing and dashed black of the pursued car.

The autonomous system followed the other car smoothly without jerky movements. For the most part, it was able to successfully chase the other vehicle. It was maintaining the desired distance when the pursued car was driving in a straight line. If the chased car stopped, so did the autonomous system. A limitation of the system comes from its current reactive nature, which in certain rides affected the ability to make a U-turn on a narrow road.

For better insight, we provide several variables recorded over time for an autonomous chasing instance, see Fig. 9. The trajectory with timestamps of both cars is obtained by tracking their images in stabilized drone recordings. Thus we calculate the distance between the two cars and their velocities. Besides that, we show the throttle and steering commands. Note that the throttle was limited for safety reasons. The small fluctuation of the steering input is probably caused by improperly trimmed mechanics of the steering. The outlier of the steering at around 30 s was caused by a false positive detection.

All autonomous chasing drives were documented and a highlight video is available online [10] for illustration.

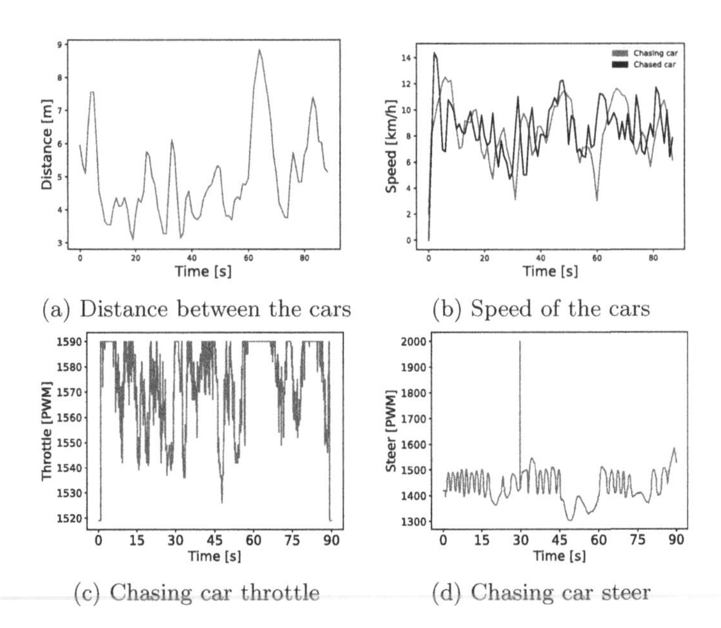

(a) Distance between the cars

(b) Speed of the cars

(c) Chasing car throttle

(d) Chasing car steer

Fig. 9. Trajectory-related and control variables in time for a ride shown in Fig. 8. The sudden spike in the graph **(d)** is caused by a false positive detection in a single frame

6 Conclusion

We have developed a system, capable of autonomously chasing another vehicle, using the novel dual-task network that simultaneously detects objects and predicts coarse semantic segmentation. The proposed system was extensively tested in CARLA simulator and on a real sub-scale vehicle platform [10].

A new challenging publicly available [9] chasing dataset was collected by manually driving the pursued car. We proposed evaluation statistics to measure success of chasing algorithms. Different versions of the system were compared. We have shown that the system benefits from using the coarse drivable surface segmentation and the trajectory extrapolation. Especially when the detector has a low detection recall. We believe that the CARLA Car Chasing dataset will serve as a benchmark for novel algorithms that could compare with ours.

Despite the simplicity of the proposed system, it shows robust chasing capabilities by using only information from a single RGB camera. One of the system limitations is its reactive nature. We believe that the system could improve by using a more sophisticated trajectory planning algorithm that would include predictive modeling of the chased car. In the future, we will evaluate the system more extensively with the RC cars.

Acknowledgement. The research was supported by Toyota Motor Europe.

References

1. Bahnik, M., et al.: Visually assisted anti-lock braking system. In: IEEE Intelligent Vehicles Symposium (2020)
2. Brackstone, M., McDonald, M.: Car-following: a historical review. Transp. Res. Part F Traffic Psychol. Behav. **2**(4), 181–196 (1999). https://doi.org/10.1016/s1369-8478(00)00005-x
3. Bradski, G.: The OpenCV Library. Dr. Dobb's J. Softw. Tools **25**, 120–125 (2000)
4. Broggi, A., et al.: PROUD-public road urban driverless test: architecture and results. In: 2014 IEEE Intelligent Vehicles Symposium Proceedings. IEEE, June 2014. https://doi.org/10.1109/ivs.2014.6856478
5. Buehler, M., Iagnemma, K., Singh, S.: The DARPA Urban Challenge: Autonomous Vehicles in City Traffic. Springer, Heidelberg (2009). https://doi.org/10.1007/978-3-642-03991-1
6. Dosovitskiy, A., Ros, G., Codevilla, F., Lopez, A., Koltun, V.: CARLA: an open urban driving simulator. In: Proceedings of the 1st Annual Conference on Robot Learning, pp. 1–16 (2017)
7. Everingham, M., Gool, L., Williams, C.K., Winn, J., Zisserman, A.: The pascal visual object classes (VOC) challenge. Int. J. Comput. Vis. **88**(2), 303–338 (2010). https://doi.org/10.1007/s11263-009-0275-4
8. Gohring, D., Wang, M., Schnurmacher, M., Ganjineh, T.: Radar/lidar sensor fusion for car-following on highways. In: The 5th International Conference on Automation, Robotics and Applications. IEEE, December 2011. https://doi.org/10.1109/icara.2011.6144918
9. Jahoda, P.: Autonomous car chase (2020). https://github.com/JahodaPaul/autonomous-car-chase
10. Jahoda, P.: Autonomous car chasing—Czech technical university in Prague (2020). https://www.youtube.com/watch?v=SxDJZUTOygA
11. Kehtarnavaz, N., Griswold, N., Lee, J.: Visual control of an autonomous vehicle (BART)-the vehicle-following problem. IEEE Trans. Veh. Technol. **40**(3), 654–662 (1991). https://doi.org/10.1109/25.97520
12. Lepetit, V., Moreno-Noguer, F., Fua, P.: EPnP: an accurate o(n) solution to the PnP problem. Int. J. Comput. Vis. **81**(2), 155–166 (2008). https://doi.org/10.1007/s11263-008-0152-6
13. Lin, F., Dong, X., Chen, B.M., Lum, K.Y., Lee, T.H.: A robust real-time embedded vision system on an unmanned rotorcraft for ground target following. IEEE Trans. Ind. Electron. **59**(2), 1038–1049 (2012). https://doi.org/10.1109/tie.2011.2161248
14. Lin, T.-Y., et al.: Microsoft COCO: common objects in context. In: Fleet, D., Pajdla, T., Schiele, B., Tuytelaars, T. (eds.) ECCV 2014. LNCS, vol. 8693, pp. 740–755. Springer, Cham (2014). https://doi.org/10.1007/978-3-319-10602-1_48
15. Rafi, F., Khan, S., Shafiq, K., Shah, M.: Autonomous target following by unmanned aerial vehicles. In: Gerhart, G.R., Shoemaker, C.M., Gage, D.W. (eds.) Unmanned Systems Technology VIII. SPIE, May 2006. https://doi.org/10.1117/12.667356
16. Redmon, J., Farhadi, A.: YOLOv3: an incremental improvement. ArXiv preprint (2018). http://arxiv.org/abs/1804.02767
17. Schneiderman, H., Nashman, M., Lumia, R.: Model-based vision for car following. In: Schenker, P.S. (ed.) Sensor Fusion VI. SPIE, August 1993. https://doi.org/10.1117/12.150245

18. Schwarzinger, M., Zielke, T., Noll, D., Brauckmann, M., von Seelen, W.: Vision-based car-following: detection, tracking, and identification. In: Proceedings of the Intelligent Vehicles 92 Symposium. IEEE (1992). https://doi.org/10.1109/ivs.1992.252228
19. Snider, J.M.: Automatic steering methods for autonomous automobile path tracking. Tech. Rep. CMU-RI-TR-09-08, Carnegie Mellon University, Pittsburgh, PA, February 2009
20. Sukthankar, R.: RACCOON: a real-time autonomous car chaser operating optimally at night. In: Proceedings of the Intelligent Vehicles 93 Symposium, pp. 37–42. IEEE, August 1993. https://doi.org/10.1109/ivs.1993.697294
21. Teuliere, C., Eck, L., Marchand, E.: Chasing a moving target from a flying UAV. In: 2011 IEEE/RSJ International Conference on Intelligent Robots and Systems. IEEE, September 2011. https://doi.org/10.1109/iros.2011.6094404
22. Thorpe, C., Herbert, M., Kanade, T., Shafer, S.: Toward autonomous driving: the CMU navlab. i. perception. IEEE Expert **6**(4), 31–42 (1991)
23. Wallace, R., Stentz, A., Thorpe, C., Maravec, H., Whittaker, W., Kanade, T.: First results in robot road-following. In: Proceedings of the 9th International Joint Conference on Artificial Intelligence, vol. 2, pp. 1089–1095. IJCAI-85, Morgan Kaufmann Publishers Inc., San Francisco (1985)
24. Ziegler, J.G., Nichols, N.B.: Optimum settings for automatic controllers. J. Dyn. Syst. Meas. Control **115**(2B), 220–222, June 1993. https://doi.org/10.1115/1.2899060
25. Ziegler, J., et al.: Making bertha drive—an autonomous journey on a historic route. IEEE Intell. Transp. Syst. Mag. **6**(2), 8–20 (2014). https://doi.org/10.1109/mits.2014.2306552

SignSynth: Data-Driven Sign Language Video Generation

Stephanie Stoll(✉), Simon Hadfield, and Richard Bowden

Centre for Vision, Speech and Signal Processing, University of Surrey, Guildford, UK
s.m.stoll@surrey.ac.uk

Abstract. We present SignSynth, a fully automatic and holistic approach to generating sign language video. Traditionally, Sign Language Production (SLP) relies on animating 3D avatars using expensively annotated data, but so far this approach has not been able to simultaneously provide a realistic, and scalable solution. We introduce a gloss2pose network architecture that is capable of generating human pose sequences conditioned on glosses. (For sign languages a gloss is a written representation that describes a specific sign.) Combined with a generative adversarial pose2video network, we are able to produce natural-looking, high definition sign language video. For sign pose sequence generation, we outperform the SotA by a factor of 18, with a Mean Square Error of 1.0673 in pixels. For video generation we report superior results on three broadcast quality assessment metrics. To evaluate our full gloss-to-video pipeline we introduce two novel error metrics, to assess the perceptual quality and sign representativeness of generated videos. We present promising results, significantly outperforming the SotA in both metrics. Finally we evaluate our approach qualitatively by analysing example sequences.

Keywords: Sign language · Pose generation · Human motion

1 Introduction

Computational research into sign languages is an important, yet under-researched problem. Whilst there are some applications to translate sign languages into spoken languages [10,32], their success is limited. The inverse process of translating spoken languages to sign languages is widely neglected. However, to provide the Deaf and Hard of Hearing with equal access and opportunities as hearing people, sign languages must become present in all parts of today's society. While sign language transcription is possible using human interpreters, it is simply infeasible to employ interpreters 24/7 at public places such as train stations and post offices, or to record video transcriptions for all web based content. An automatic, scalable solution is needed that can generate naturalistic sign language video from spoken or written language.

Electronic supplementary material The online version of this chapter (https://doi.org/10.1007/978-3-030-66823-5_21) contains supplementary material, which is available to authorized users.

A. Bartoli and A. Fusiello (Eds.): ECCV 2020 Workshops, LNCS 12538, pp. 353–370, 2020.
https://doi.org/10.1007/978-3-030-66823-5_21

Traditionally, research into Sign Language Production (SLP) has focused on animating 3D avatars using sequences of parametrised glosses. However, given the complexity of sign language, the task of manually annotating data requires tremendous effort and expert knowledge. Sign languages are different from country to country and have the same local variations as dialects do in spoken languages. They also rely on much more than hand shape and motion to convey meaning, such as mouth/face gestures, eye gaze, and body pose. These non-manual features need to be annotated correctly and aligned with the gloss they belong to. Most sign avatars largely ignore non-manuals, making them hard to understand, and unnatural looking. Avatars using motion capture data provide a better sign quality, but are limited in their vocabulary. This is due to the cost associated with recording and storing high fidelity motion capture data.

Stoll et al. [34] were the first to present a neural network approach to SLP. They first translate written German into German sign gloss sequences using Neural Machine Translation (NMT), and use a look-up table (LUT) of mean sequences to generate 2D motion from automatically extracted pose information. This data provides the input to a Generative Adversarial Network (GAN) to produce sign language video. Their results for text to gloss translation are impressive and the approach is naturally scalable. However, the quality of the produced videos is lacking, given a low resolution of 128 $\times$ 128 pixels. Furthermore, the use of a LUT severely limits this approach and introduces artefacts and discontinuities in co-articulation between signs.

To further the field of SLP and address the shortcomings of previous approaches we present the following contributions:

1. In order to dramatically increase the quality of synthetic sign video generation, we propose a gloss-to-pose (gloss2pose) network capable of producing sign motion data of high fidelity, conditioned on sign glosses, and trained on weakly labelled data. To our knowledge, we are the first to address the generation of manuals and non-manuals in a holistic, data-driven way.
2. We combine the gloss2pose network with a pose-to-video (pose2video) network and are able to produce high definition sign language video.
3. We introduce two error metrics to assess the quality of automatically generated sign language videos.

We implemented our approach in Pytorch and it is available at SignSynth.

2 Related Work

We provide an overview of recent developments in SLP, before describing the concept of motion graphs and recent approaches relevant to our work. Finally, the field of conditional image generation is presented.

Approaches for Sign Language Production. Automatic SLP is traditionally achieved by animating avatars, using a sequence of parametrised glosses.

Examples of these include VisiCast [2], eSign [40], Tessa [5], dicta-sign [9], and JASigning [17]. All these approaches rely on manually annotated data using a purpose-specific transcription language such as HamNoSys [29] or SigML [18]. Annotating the data requires expert knowledge and is generally carried out by trained linguists. The resulting animations suffer from unnatural, under-articulated motion, that makes the avatars look robotic, hard to understand, and at times uncomfortable to view, due to the uncanny valley effect.[1] Given the tremendous annotation effort required, non-manuals are often neglected. Some work on integrating non-manuals has been carried out in recent years [7,8,20,24], but it remains an unsolved problem. A possible method to circumvent these issues is to animate avatars directly from motion capture data [12]. This results in highly realistic, and expressive animations, including non-manuals. However, these systems are limited to pre-recorded phrases, or need complex re-assembly taking into account the effects of co-articulation. Additionally, the recording and cleaning of high-fidelity motion capture data is costly and time consuming, making this approach not scalable.

To make automatic SLP feasible, Stoll et al. [34] propose generating synthetic sign language video using a LUT and GAN. Whilst this potentially overcomes some of the limitations of avatar technology, the low resolution of the produced videos and the use of a LUT to provide the pose information restricts the approach, particularly in terms of co-articulation between signs. Furthermore, non-manuals such as facial expressions are not addressed. In contrast, our approach learns to generate detailed pose and video sequences. Sequences of varying speed and expressiveness are automatically generated, including non-manuals.

Motion Graphs. A popular concept in computer graphics, they are used to animate characters using a directed graph constructed from motion data, i.e. novel animations are created by re-combining short sequences of recorded motion. They were first introduced independently by Kovar et al. [21], Arikan et al. [1], and Lee et al. [22]. In recent years, deep-learning based approaches have emerged, most relevant to our work being Holden et al. [15], and Zhang, Starke et al. [39]. Holden et al. developed a regression network for generating cyclic motion such as walking and running, by predicting character joint positions, velocities, angles, and the character's global trajectory at $t+1$ given the joint positions, velocities, the character's global trajectory, and a semantic variable describing the type of gait at time t. A Catmull-Rom Spline to calculate the weights of the regression network helps enforce the cyclic nature of the data to be generated. The system is trained on motion capture data. Zhang, Starke et al. [39] build on this approach, and apply it to quadruped characters. They use heavy supervision, such as the character's 3D joint positions, velocities, rotations, as well as a user-defined global character trajectory & velocity, plus action variables describing

[1] The uncanny valley is a concept aimed at explaining the sense of unease people often experience when confronted with simulations that closely resemble humans, but are not quite convincing enough [26].

the type of gait, and footfall pattern. In contrast our system does not require heavy supervision and instead learns to decompose signs into simpler sub-units.

Conditional Image Generation. The field has seen a number of different techniques emerge over the last few years. Convolutional Neural Networks (CNNs) [4,27], as well as Recurrent Neural Networks (RNNs) [14,28] have been explored for generating images. Variational Auto-Encoders (VAEs) [19], and later conditional VAEs [37] have proven a popular choice. Since their initial conception, GANs [13] have provided many approaches to the task of image generation, such as conditional GANs [25,30,31]. VAEs and GANs are often combined to harness the VAE's stability, and the GAN's discriminative nature. Most relevant to our work, VAE - GAN hybrids have been applied to image generation of people [23,33], and more specifically to produce videos of people performing sign language [34]. In image-to-image translation, pix2pix [16], and recently pix2pixHD [35] were able to produce high-definition images from semantic label maps using a multi-stage generator, and a multi-scale discriminator.

In our work we build on the recent success of pix2pixHD, and develop a VAE-GAN-based network that is capable of producing high resolution sign language productions from semantic label maps.

3 Synthetic Sign Video Generation

Our approach to generating sign language video from glosses works in two stages, see Fig. 1. First a gloss is translated into a human pose sequence by our gloss2pose network. The acquired poses are then used to condition a generative network called pose2video.

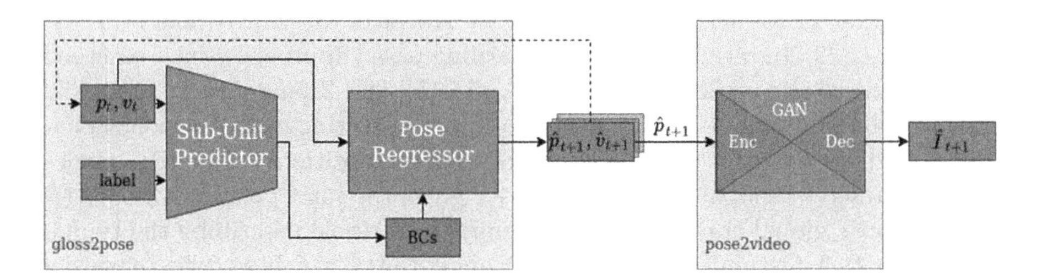

Fig. 1. Overview of our method at runtime. A sub-unit predictor estimates blending coefficients. These are used to generate the weights for the pose regressor. This network predicts poses and velocities for the next time step and is autoregressive. The generated poses are used as input to the video generator which produces sign video frames

3.1 Gloss to Pose

In contrast to previous work [39] our system works on 2D data, and does not require heavy supervision like user-defined global trajectories. We only use 2D skeletal pose and velocity data, as well as a label indicating the desired gloss(es). As the pose data can be automatically extracted using a detector such as [3] the data annotation effort is minimal. At run time the user only has to specify the desired gloss to generate a motion sequence of the target sign(s). The gloss2pose network predicts the state of pose keypoints for a future time step Y, given the current keypoints' state X. X is defined as $X = \{p_t, v_t, l_t\}$ being a vector of the joint positions p_t, velocities v_t, and gloss label l_t at time t. The velocity v_t is defined as $v_t = p_t - p_{t-1}$, and l_t is encoded as a one-hot vector, representing all gloss classes. The output Y is defined as $Y = \{\hat{p}_{t+1}, \hat{v}_{t+1}\}$ where $\hat{p}_{t+1}$ and $\hat{v}_{t+1}$ are the predicted positions and velocities of all keypoints at time $t + 1$.

As shown in Fig. 1, the gloss2pose network consists of two sub-networks, which are trained end-to-end. The main network, called pose regressor, predicts Y given a subset of X, $\hat{X} = \{p_t, v_t\}$:

$$Y = \Phi(\hat{X}|\omega_\Phi), \tag{1}$$

where Φ is the pose regressor network and ω_Φ are its weights. The pose regressor is a three-layer network consisting of two 1D-convolutional residual layers, and one fully connected layer. We found a convolutional architecture to be superior to a fully connected one, for both spatial accuracy in predicting joint poses per frame, as well as learning trajectories given the frame velocities. We achieved this by reshaping the input $\hat{X}$ to the pose regressor from a 1D to a 2D vector, with p_t occupying the first, and v_t the second row. Using a filter of height two we teach the network to learn the relationship between keypoint positions and velocities. To express in theory any number of signs and smoothly blend between them, we want the gloss2pose network to learn their composition in terms of sub-units. We achieve this by learning a set of blending coefficients given X, using a secondary neural network, called the sub-unit predictor. This is a fully connected network, consisting of four fully connected layers. We estimate the set of blending coefficients for X:

$$BC = \iota(\varsigma(X|\omega_\varsigma), l_t|\omega_\iota), \tag{2}$$

where ς is the sub-unit predictor, ω_ς the weights of ς, and ι is the reduction layer, with weights ω_ι. The blending coefficients BC are a vector of length b, with $BC \in \mathbb{R}^b$. BC is used to generate the pose regressor network weights ω_Φ:

$$\omega_\Phi = \sum_{i=1}^{b} BC_i \Omega_i, \tag{3}$$

where Ω is a bank of pose regressor weights ω_Φ of size b. This allows us to dynamically blend between weights depending on the target sign's sub-unit composition over time.

The predicted output Y is compared to the ground truth using the mean-square error. Therefore the loss of the gloss2pose network is defined as:

$$L_P = \frac{1}{N}\sum_{k=1}^{N}(Y_{gt}^k - Y^k)^2, \quad (4)$$

where k is an iterator over all keypoints up to the maximum number of keypoints N, and Y_{gt} is the ground truth position and velocity.

Error-Correcting Data Augmentation. We developed a two-time-step training scheme that allows us to double the amount of training data, and teach the network to correct its own mistakes in pose and velocity predictions. At time t we let the network predict the output Y for $t+1$. After calculating a loss, we use this result as input X for the next time step, to predict Y at $t+2$. After this the next ground truth from the original training data is used and the process repeated. The generated training samples are discarded after their use, to keep the ratio of ground truth and generated training data constant.

In addition to serving as a data augmentation scheme, we argue that this training scheme helps the network to correct itself from mis-predictions. At $t+1$ the predicted result $Y = \hat{p}_{t+1}, \hat{v}_{t+1}$ has a prediction error of ϵ_{t+1}, making the total error for this time step $E = \epsilon_{t+1}$. When using the predicted result as the input for the next time step the total error at $t+2$ now becomes $E = \epsilon_{t+1} + \epsilon_{t+2}$. This is penalised with a higher loss, than just $E = \epsilon_{t+2}$. This means the network learns to correct for drift from error accumulation, and to handle data samples from previously unexplored space.

3.2 Pose to Video

Like pix2pixHD [35], our pose2video network is the combination of a convolutional image encoder and a Generative Adversarial Network (GAN), conditioned on semantic label maps. Inside the GAN a generator G is engaged in a minimax game against a set of multi-scale discriminators D. G is generating new data instances, which are evaluated by D to be either "fake" (as in not belonging to the same data distribution), or "real" (part of the same data distribution). G's aim is to maximise the likelihood of D choosing incorrectly, whereas D tries to maximise its chance of choosing correctly. Trained in conjunction, the networks improve each other, with G creating more and more realistic data samples.

The input to the generator G is the positional information generated by the gloss2pose network $\hat{p}$. The discriminator D evaluates either the generated image $G(\hat{p}_t)$, or the real image I_t. We can therefore define the adversarial loss as

$$L_{GAN}(G, D_k) = \mathop{\mathbb{E}}_{\hat{p}_t}[log(D_k(\hat{p}_t, I_t))] + \mathop{\mathbb{E}}_{\hat{p}_t}[log(1 - D_k(\hat{p}_t, G(\hat{p}_t)))], \quad (5)$$

where k indicates the discriminator scale. To combine the adversarial losses of all D_k, we sum:

$$L_{GAN}(G, D) = \sum_{k=1,2,3} L_{GAN}(G, D_k). \quad (6)$$

Additionally we apply a feature matching loss as presented in [35]:

$$L_f(G, D_k) = \mathop{\mathbb{E}}_{\hat{p_t}} \sum_{i=1}^{T} \frac{1}{N_i} \Big[\sum |D_k^{(i)}(\hat{p_t}, I_t) - D_k^{(i)}(\hat{p_t}, G(\hat{p_t}))| \Big], \tag{7}$$

where T is the total number of layers in D_k, i is the current layer of D_k, and $D_k^{(i)}$ is the ith layer feature extractor of D_k. Again we sum the L_f losses of all D_k to obtain the overall L_f loss:

$$L_f(G, D) = \sum_{k=1,2,3} L_f(G, D_k). \tag{8}$$

The total loss L_{p2v} is a combination of adversarial and L1 loss:

$$L_{p2v} = L_{GAN} + \delta L_f, \tag{9}$$

where δ weighs the influence of L_f.

4 Experiments and Results

We evaluate the gloss2pose and pose2video parts of our approach separately, before quantitatively and qualitatively analysing their combined performance.

We use the SMILE Sign Language Assessment Dataset [6] for training and testing the gloss2pose as well as pose2video networks. It consists of 42 signers performing 105 signs in isolated form, with three repetitions each, in Swiss-German Sign Language (DSGS). The SMILE dataset is multi-view, however we only utilise the Kinect colour stream, which is of 1920×1080 resolution at 30fps. We extract 2D human pose estimations from the video data, using OpenPose [3] for the upper body (14 keypoints), face (70 keypoints) and hands (21 keypoints per hand). For the pose2video network, the extracted keypoints of one chosen signer are used to generate semantic label maps encoding the position and type of joint of each keypoint. At run time the positional information to generate these maps is provided from the output of the gloss2pose network, Y. We split our remaining dataset into train, test and validation sets. We use the training set to train the gloss2pose network, and evaluate it using the test set. The validation set is used to assess the performance of our whole system.

4.1 Gloss to Pose

We compare our gloss2pose network against the approach of Stoll et al. [34], who use a LUT to transform glosses to a dynamically time-warped mean sequence built from all example sequences for the gloss. They populate their LUT with the RWTH-PHOENIX-Weather 2014**T** [11] dataset, which is of continuous German Sign Language (DGS). However as it is of low resolution (227×227), the quality of pose information is poor. Furthermore, as the data is continuous, the glosses were extracted using a forced-alignment approach, meaning boundaries between glosses are not exact. These factors led us to decide it would be unfair to directly compare their results to our results obtained using high-resolution isolated data. We therefore contacted the authors of [34] for their code and populated their LUT using the SMILE data instead. The quantitative comparison in this section will be against this LUT of dynamically time warped mean sequences created out of all examples per gloss.

Quantitative Evaluation. For our first experiment we train with 10, 30, 50, and 80 blending coefficients to find the optimal configuration to encode the 105 gloss classes into sub-units. The data per epoch consists of 859,522 real, and 859,522 synthetic data frames, given the regime described in Sect. 3.1. Batch size is set to 32, the learning rate to 0.0001, and ADAM optimisation is used.

We evaluate by sampling 20 different sequences per gloss and taking the Mean Squared Error (MSE) between sampled positions & velocities, and their ground truth. The MSE per frame is accumulated and divided by the total number of frames across all sequences for all glosses. A sequence is sampled by giving the network a starting input X_t from the test dataset. The network generates a prediction Y_{t+1} which we use as the input for the next time step, and let the network feed back on itself for 150 frames in total. This is a very challenging test environment as there is a huge scope for drift caused by errors propagating as the network feeds back on itself over such a long time. To put our findings into context we compare it to the performance of the network when feeding the next ground truth frame at $t + 1$ instead of a generated frame. We found that after 20 epochs the best and most stable performance was achieved with 50 blending coefficients, see Fig. 2. We did not find that, in general, increasing the number of blending coefficients improves performance, meaning our subunit-predictor is able to dissect sign motions into sub-motions, rather than learning one set of coefficients for each of the 105 signs in the dataset.

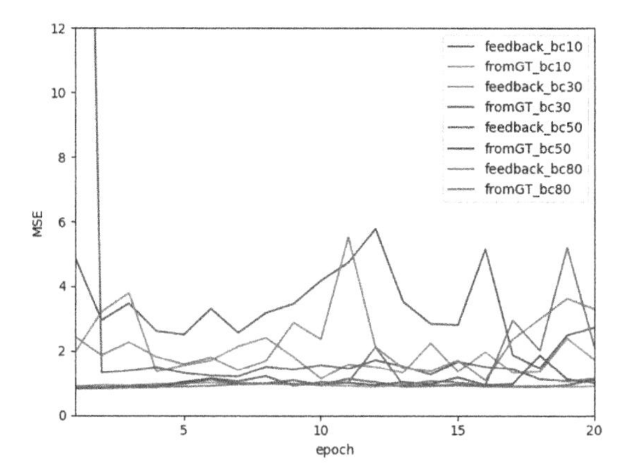

Fig. 2. The performance of our network in terms of MSE over 20 epochs. Different numbers of blending coefficients are explored (10, 30, 50, 80). As a means of comparison we also provide results when the network is fed the next ground truth frame, rather than the sample generated at the previous time step (*fromGT*)

We next compare the MSE of our best performing network against the LUT approach of [34]. The result is presented in Table 1. To test the performance of [34] we compare the dynamically time warped mean sequence for each gloss against all ground truth sequences. As before, the MSE per frame is accumulated and divided by the total number of frames across all sequences. Our network outperforms [34] by a factor of 18. We suspect this has two causes. Firstly, averaging across all signers in the training set to acquire one representative mean sequence per gloss robs the LUT of the capability to express different people's skeletal builds, and a signer's natural variance in expressiveness. However, our network is capable of intelligently managing this spatial variance, as it learns from the data. Secondly, the dynamic time warping needed by [34] removes any variability in speed. However, a sign's duration can vary immensely, depending on repetition, or to convey e.g. excitement. In contrast, our network can express this natural variance in speed.

Table 1. MSE for [34]'s LUT and our gloss2pose (g2p) network

	LUT [34]	g2p (ours)
MSE	19.2886	**1.0673**

Qualitative Evaluation. We analyse two generated pose sequences, see Fig. 3. Each sequence and canonical example is sampled at every 10th frame. We would like to point out that the example video sequences cannot be considered a direct ground truth, but merely a reference for the reader. The sequences depict the signs JAHR (YEAR) and ERZÄHLEN (TELL), respectively. We chose two signs that are close in their overall motion to showcase our network's ability to still

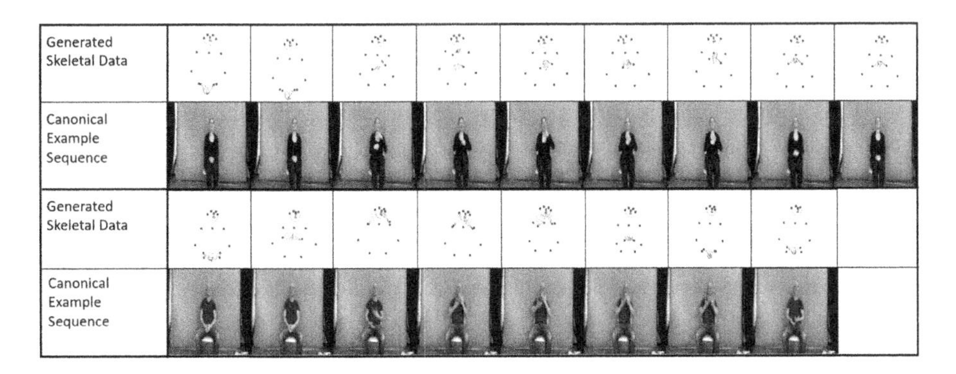

Fig. 3. Two pose sequences generated by the gloss2pose network. Canonical example sequences are provided to give the reader a reference of the signs. The top example is conditioned on the gloss JAHR (YEAR), the bottom example on ERZÄHLEN (TELL). This figure is best viewed in colour and digital format (Color figure online)

generate them distinctively. Both hands for the sign JAHR form fists, whereas for ERZÄHLEN they are flat and open. This is recreated in detail in the pose sequences generated, together with the correct motion for each sign. We also want to point out the variability in speed of the signs produced. The sequence generated for ERZÄHLEN is shorter than the video example provided, whereas for JAHR it is slightly longer. Our network can produce sequences of variable length and expressiveness, something a mean sequence based approach such as [34] is incapable of. Furthermore, we are able to automatically produce aligned non-manuals such as facial expressions, which is not addressed in the SotA [34].

4.2 Pose to Video

We evaluate the performance of pose2video against the Pose-Conditioned Sign Generation Network (PSGN) by Stoll et al. [34]. We trained the pose2video network for 100 epochs, with a training set of 12,500 and a test set of 1,400 image-semantic label map pairs. We used ADAM optimisation with an initial learning rate of 0.0002 that we linearly decay to zero.

Quantitative Evaluation. We evaluate the image generation of each network using Structural Similarity Index Measurement (SSIM) [36], Peak Signal-to-Noise Ratio (PSNR)[38] and MSE. SSIM measures a perceptual degradation of down-sampled or corrupted images compared to their originals. We use this to measure the perceptual degradation of a generated image $\hat{I}_t = G(P_t)$ to its ground truth I_t. PSNR and MSE are also used to assess compressed or corrupted image quality compared to their original. For images the MSE is used to calculate the average squared per-pixel error between $\hat{I}_t$ and I_t. Using the MSE, PSNR measures the peak error in dB. Table 2 compares the SSIM, PSNR, and MSE scores for PSGN [34] and our pose2video network over 1200 frames respectively. PSGN scores marginally higher for SSIM, however as it is a metric focussing on overall image structure and appearance rather than fine details, this is unsurprising. As the original ground truth of 1920 × 1080 is encoded to 128 × 128 pixels by PSGN most fine detail is already lost. In contrast pose2video beats PSGN by a large margin for both PSNR and MSE. This is due to high resolution and detail of the generated images.

Table 2. Mean SSIM, PSNR, and MSE values for PSGN [34] and our pose2video network. SSIM ranges between −1 and +1, with +1 indicating identical images. Lower MSE indicates more likeness, whereas higher PSNR indicates images are more alike

	SSIM	PSNR	MSE
PSGN [34]	**0.9434**	24.7248	226.2474
p2v (ours)	0.9428	**29.0181**	**86.0553**

4.3 Synthetic Sign Video Generation from Gloss

Finally, we evaluate the performance of the gloss2pose and pose2video networks in conjunction (SignSynth). We compare our approach to that of [34] before discussing qualitative results.

Quantitative Evaluation. Evaluating the quality of generated sign language videos in a quantitative fashion is challenging for multiple reasons. For each sign there is a natural variability in terms of size, motion and speed, but also context-specific differences, such as repetitions, or negations. Furthermore, transitions between signs can influence the individual signs' trajectories and positions. A logical option that comes to mind is to use a Sign Language Recognition (SLR) system to assess the quality and accuracy of produced sign language videos. Since the advent of deep learning such systems have certainly improved. However, they are still far from accurate and highly depend on the data they were trained on. To our knowledge there is no publicly available SLR system that is trained on the SMILE dataset. Furthermore, we want to accurately measure the quality of synthetic sign video, rather than diluting the measurements with inevitable errors produced by an SLR network.

We therefore devise two metrics. The first one is a confidence score that assesses the level of detail and human characteristics of the generated videos. The second is a distance measure that assesses how closely generated sign videos resemble the ground truth videos for that gloss. Both metrics make use of the OpenPose [3] human pose detector. To be more specific, we take the generated sign language videos and let OpenPose detect pose, face and hand keypoints. For [34] we do the same on the mean video sequences. For each keypoint k the coordinates x and y as well as a detection confidence c is inferred. For our first metric we utilise c to assess the pose detector's ability to detect human keypoints from the generated videos, the intuition being that the more detailed and "human-like" the generated video, the higher the detector's confidence. We divide our keypoints into regions of interest. This lets us assess confidences on specific body parts, such as the hands and the face. For each region we define the regional confidence as

$$C = \frac{1}{T-1} \sum_{t=0}^{T-1} \left(\frac{1}{I} \sum_{i=1}^{I} \alpha_i c(k_i) \right), \tag{10}$$

where I is the total number of keypoints in the region, T is the total number of frames assessed, and α_i is the importance of a keypoint in the region. We can further obtain the overall confidence by summing over the regions:

$$C_{total} = \frac{1}{R} \sum_{r=1}^{R} \beta_r C_r \tag{11}$$

where β_r is the importance of the region in the total score, and R is the number of regions. In our experiment we define four regions: *pose* (14 keypoints defining the upper body skeleton), *face* (70 facial keypoints), *handl* and *handr* (21 keypoints for the left and right hand respectively). We set α and β to 1.0. Whilst we believe there is scientific value in identifying the importance of specific regions and keypoints for sign language, this is future work and beyond the scope of this manuscript. Table 3 compares confidence scores of [34] and SignSynth. Both methods perform well in the *pose* region. However, the detector fails to detect any facial keypoints in the output of [34] and has very low confidence in the keypoints of both hands. For our method, the detector has a high confidence for the *face* region, and beats the hand confidences of [34] by an order of magnitude. This showcases our approach's superior ability to produce detailed signings with manuals and non-manuals clearly present.

Table 3. Confidence scores for Stoll et al. [34] and SignSynth. Confidences are given for four regions, as well as an overall score. The confidence is measured between 0 and 1

	C_{pose}	C_{face}	C_{handl}	C_{handr}	C_{total}
Stoll et al. [34]	0.499	0.000	0.026	0.025	0.138
SignSynth	**0.791**	**0.766**	**0.120**	**0.266**	**0.485**

For our second experiment we analyse the behaviour of keypoints over time. Rather than just looking for overall smoothness, we want to also relate the trajectory of points to the signs they are meant to represent, whilst taking into account different levels of speed and expression. For this, we first perform hierarchical clustering with average linking of the SMILE validation set. The metric used is based on dynamic time warping (dtw). We define the similarity of two sequences $S1$ and $S2$ as the euclidean distance between $S1$ and $S2$ when the alignment path P is optimal:

$$dtw(S1, S2) = \sqrt{\sum_{(m,n)\in P} ||(S1_m - S2_n)||^2}. \tag{12}$$

We again divide our keypoints into regions and treat each keypoint's trajectory independently:

$$D = \frac{1}{I}\sum_{i=1}^{I} \alpha_i dtw(S1(k_i), S2(k_i)), \tag{13}$$

where D is the regional distance between S1 and S2, α_i again is the importance of a keypoint in the region. To obtain the overall distance between S1 and S2 we sum over the regions as before:

$$D_{total} = \frac{1}{R}\sum_{r=1}^{R} \beta_r D_r, \tag{14}$$

where R is the number of regions and β_r the importance of each region.

After clustering the validation set we use the same metric to measure the distance of each generated sample to each cluster. We then report the mean distance between clusters and samples per sign class. Table 4 shows results for three sign classes. Again, we compare our approach against that of [34]. However, we also provide a reference to put the distances obtained into perspective for the reader. As the reference we take ground truth samples of a signer from the SMILE test set and measure the distance to each cluster. The reference's samples per sign class measure closest to the sign cluster they belong to. [34]'s approach measures more or less the same distances to all clusters per sign class, meaning their sequences are not descriptive of any sign. Overall their distances are significantly larger than that of the reference. Our results lie in the same range as the reference, and their variability per sign class showcases the generated samples' descriptiveness. Two out of three signs are correctly identified, whereas for the third all samples score similarly regardless of which sign class they belong to. When inspecting the samples for the third gloss we saw that our network performs a variable number of repetitions for the circular hand motion in front of the body. While repetition in sign language is common, there are no sequences with repetitions in the data used to form the clusters. We wish to take into account the occurrence of repetitions in future work.

Table 4. D_{total} for test samples of three signs to each sign cluster. ERZÄHLEN is abbreviated to ERZÄ in the table

	SignSynth			Stoll et al. [34]			Reference		
Clusters	ABEND	ABER	ERZÄ	ABEND	ABER	ERZÄ	ABEND	ABER	ERZÄ
ABEND	**12.51**	14.49	16.17	19.82	**16.36**	19.29	**12.28**	14.61	14.69
ABER	15.58	**14.72**	17.02	18.13	**16.29**	19.35	14.86	**13.34**	15.43
ERZÄ	15.51	**15.32**	15.93	20.68	**17.04**	19.78	14.70	15.03	**13.54**

Qualitative Evaluation. We present sequences generated by our SignSynth method, and compare it to results from [34], (see Fig. 4). For each sign, a canonical sequence is provided. For those example sequences and SignSynth results, every 10th frame is shown. For spatial reasons we only show every 20th frame of LUT+PSGN, as the dynamic time warping needed by [34]'s approach results in sequences that are much slower than many real life examples.

We study two signs with similar hand motion, but different hand shape. The first sign ABER (BUT), is shown in the top section of Fig. 4, the second VORGESTERN (DAY-BEFORE-YESTERDAY) in the section below. Our generations for both signs follow the correct trajectory, with slight variations in speed and expressiveness, showcasing our networks' ability to learn natural variations in sign language production. Furthermore, our approach generates significant detail such as an extended index finger in the dominant hand. The hand shape for the sign VORGESTERN (an extended thumb pointing backwards) is also generated. The sequences generated by [34] follow the global trajectories for both signs, but are executed at less than half the speed. Any detail of hands or facial expression is lost completely.

Finally, we show a sequence generated from multiple glosses. Even though our approach is trained on isolated data, it is capable of generating smooth pose and video sequences without artefacts between signs, see Fig. 5. It depicts the generated pose and video data conditioned on the gloss sequence FUSSBALL SPIELEN (PLAY FOOTBALL). As before, every 10th frame is shown. The SignSynth approach generates detailed sequences for pose and video that represent

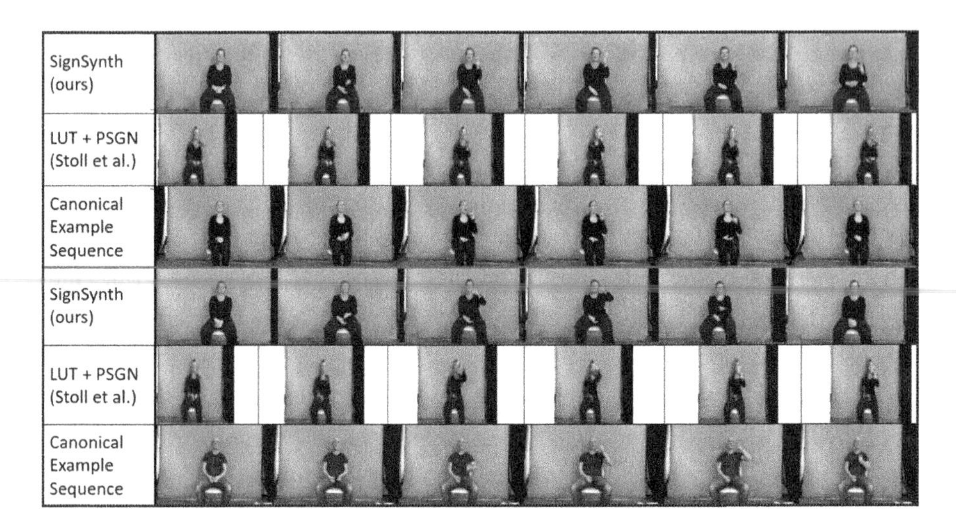

Fig. 4. SignSynth output compared to Stoll et al. [34]. The top half depicts the sign ABER (BUT), the bottom half VORGESTERN (DAY-BEFORE-YESTERDAY). This figure is best viewed in colour and digital format (Color figure online)

Fig. 5. Generated pose and video sequences for the gloss sequence FUSSBALL SPIELEN (PLAY FOOTBALL). This figure is best viewed in colour and digital format (Color figure online)

the gloss input sequence, with smooth transitions between signs. Detail in the hands is well preserved, especially for the sign SPIELEN. For more results we refer to the supplementary material.

5 Conclusion

We presented a novel approach to Sign Language Production (SLP) that only requires minimal user input. Our approach is capable of producing sign language video of high resolution, where sequences contain hand motion with a natural variance in speed, expressiveness, and distinctive hand shape. Non-manuals are also generated and naturally aligned with the rest of the sign, as our approach directly learns from sign language data. Additionally, we are able to smoothly and automatically transition between glosses, making our approach superior to approaches relying on manually enforcing co-articulation. When comparing our method to the current SotA [34], we were able to surpass its performance for pose and video generation, as well as generating videos from gloss information. We evaluated and compared our approach using MSE and popular metrics from broadcast quality assessment. We then developed two new metrics to assess the quality of sign language videos, which we used to compare our approach to [34], and reported highly promising results.

In the future, there are a number of avenues to pursue. The gloss2pose network could be extended to 3D data, and be used to drive an avatar without the drawbacks found with current approaches. We also want to incorporate techniques from NMT to address complete spoken language to sign language translation. Furthermore we want to explore the intricacies of sign language, as mentioned in Sect. 4. Finally, we are excited to continue working with linguists on solving the problem of automatic SLP to further the integration of the Deaf community.

Acknowledgements. This work received funding from the SNSF Sinergia project 'SMILE' (CRSII2 160811), the European Union's Horizon2020 research and innovation programme under grant agreement no. 762021 'Content4All' and the EPSRC project 'ExTOL' (EP/R03298X/1). This work reflects only the authors view and the Commission is not responsible for any use that may be made of the information it contains.

References

1. Arikan, O., Forsyth, D.A.: Interactive motion generation from examples. In: Proceedings of the 29th Annual Conference on Computer Graphics and Interactive Techniques. SIGGRAPH 2002, pp. 483–490. ACM, New York (2002)
2. Bangham, J.A., et al.: Virtual signing: capture, animation, storage and transmission-an overview of the visicast project. In: IEE Seminar on Speech and Language Processing for Disabled and Elderly People (Ref. No. 2000/025), pp. 6/1–6/7 (2000)
3. Cao, Z., Simon, T., Wei, S., Sheikh, Y.: Realtime multi-person 2D pose estimation using part affinity fields. In: 2017 IEEE Conference on Computer Vision and Pattern Recognition (CVPR), vol. 00, pp. 1302–1310, July 2017

4. Chen, Q., Koltun, V.: Photographic image synthesis with cascaded refinement networks. In: ICCV, pp. 1520–1529. IEEE Computer Society (2017)
5. Cox, S., et al.: Tessa, a system to aid communication with deaf people. In: Proceedings of the Fifth International ACM Conference on Assistive Technologies, pp. 205–212. ACM (2002)
6. Ebling, S., et al.: Smile Swiss German sign language dataset. In: 11th Edition of the Language Resources and Evaluation Conference (LREC) (2018)
7. Ebling, S., Glauert, J.: Exploiting the full potential of jasigning to build an avatar signing train announcements, October 2013
8. Ebling, S., Huenerfauth, M.: Bridging the gap between sign language machine translation and sign language animation using sequence classification. In: SLPAT@Interspeech (2015)
9. Efthimiou, E.: The Dicta-Sign Wiki: Enabling Web Communication for the Deaf (2012)
10. Elwazer, M.: Kintrans (2018). http://www.kintrans.com/
11. Forster, J., Schmidt, C., Koller, O., Bellgardt, M., Ney, H.: Extensions of the sign language recognition and translation corpus rwth-phoenix-weather. In: Language Resources and Evaluation, Reykjavik, Island, pp. 1911–1916, May 2014
12. Gibet, S., Lefebvre-Albaret, F., Hamon, L., Brun, R., Turki, A.: Interactive editing in French sign language dedicated to virtual signers: requirements and challenges. Univers. Access Inf. Soc. **15**(4), 525–539 (2016)
13. Goodfellow, I.J., et al.: Generative adversarial nets. In: Advances in Neural Information Processing Systems 27: Annual Conference on Neural Information Processing Systems 2014, 8–13 December 2014, Montreal, Quebec, Canada, pp. 2672–2680 (2014)
14. Gregor, K., Danihelka, I., Graves, A., Rezende, D., Wierstra, D.: Draw: a recurrent neural network for image generation. In: Bach, F., Blei, D. (eds.) Proceedings of the 32nd International Conference on Machine Learning. Proceedings of Machine Learning Research, 07–09 Jul 2015, Lille, France, vol. 37, pp. 1462–1471. PMLR (Jul 2015)
15. Holden, D., Komura, T., Saito, J.: Phase-functioned neural networks for character control. ACM Trans. Graph. **36**(4), 42:1–42:13 (2017). https://doi.org/10.1145/3072959.3073663
16. Isola, P., Zhu, J.Y., Zhou, T., Efros, A.A.: Image-to-image translation with conditional adversarial networks. In: 2017 IEEE Conference on Computer Vision and Pattern Recognition (CVPR), pp. 5967–5976 (2017)
17. JASigning: Virtual humans research for sign language animation (2017). http://vh.cmp.uea.ac.uk/index.php/Main_Page
18. Kennaway, R.: Avatar-independent scripting for real-time gesture animation. CoRR abs/1502.02961 (2013)
19. Kingma, D.P., Welling, M.: Auto-encoding variational Bayes. In: 2nd International Conference on Learning Representations, ICLR 2014, Banff, AB, Canada, 14–16 April 2014, Conference Track Proceedings (2014)
20. Kipp, M., Héloir, A., Nguyen, Q.: Sign language avatars: animation and comprehensibility. In: IVA (2011)
21. Kovar, L., Gleicher, M., Pighin, F.: Motion graphs. In: Proceedings of the 29th Annual Conference on Computer Graphics and Interactive Techniques. SIGGRAPH 2002, pp. 473–482. ACM, New York (2002)

22. Lee, J., Chai, J., Reitsma, P.S.A., Hodgins, J.K., Pollard, N.S.: Interactive control of avatars animated with human motion data. In: Proceedings of the 29th Annual Conference on Computer Graphics and Interactive Techniques. SIGGRAPH 2002, pp. 491–500. ACM, New York (2002)
23. Ma, L., Jia, X., Sun, Q., Schiele, B., Tuytelaars, T., Van Gool, L.: Pose guided person image generation. In: Guyon, I., et al. (eds.) Advances in Neural Information Processing Systems, vol. 30, pp. 406–416. Curran Associates, Inc. (2017)
24. McDonald, J., et al.: An automated technique for real-time production of life-like animations of American sign language. Univ. Access. Inf. Soc. **15**(4), 551–566 (2016)
25. Mirza, M., Osindero, S.: Conditional generative adversarial nets. CoRR abs/1411.1784 (2014). http://arxiv.org/abs/1411.1784
26. Mori, M., MacDorman, K., Kageki, N.: The uncanny valley [from the field], vol. 19, pp. 98–100, June 2012
27. van den Oord, A., Kalchbrenner, N., Espeholt, L., kavukcuoglu, k., Vinyals, O., Graves, A.: Conditional image generation with pixelCNN decoders. In: Lee, D.D., Sugiyama, M., Luxburg, U.V., Guyon, I., Garnett, R. (eds.) Advances in Neural Information Processing Systems, vol. 29, pp. 4790–4798. Curran Associates, Inc. (2016)
28. Oord, A.V., Kalchbrenner, N., Kavukcuoglu, K.: Pixel recurrent neural networks. In: Balcan, M.F., Weinberger, K.Q. (eds.) Proceedings of the 33rd International Conference on Machine Learning. Proceedings of Machine Learning Research, 20–22 Jun 2016, vol. 48, pp. 1747–1756. PMLR, New York (2016)
29. Prillwitz, S.: HamNoSys version 2.0. Hamburg notation system for sign languages: an introductory guide. Intern. Arb. z. Gebärdensprache u. Kommunik. Signum Press (1989)
30. Radford, A., Metz, L., Chintala, S.: Unsupervised representation learning with deep convolutional generative adversarial networks. CoRR abs/1511.06434 (2015). http://arxiv.org/abs/1511.06434
31. Reed, S.E., Akata, Z., Mohan, S., Tenka, S., Schiele, B., Lee, H.: Learning what and where to draw. In: Lee, D.D., Sugiyama, M., Luxburg, U.V., Guyon, I., Garnett, R. (eds.) Advances in Neural Information Processing Systems, vol. 29, pp. 217–225. Curran Associates, Inc. (2016)
32. Robotka, Z.: Signall (2018). http://www.signall.us/
33. Siarohin, A., Sangineto, E., Lathuilière, S., Sebe, N.: Deformable GANs for pose-based human image generation. In: IEEE Conference on Computer Vision and Pattern Recognition, Salt Lake City, United States, pp. 3408–3416, June 2018
34. Stoll, S., Camgoz, N.C., Hadfield, S., Bowden, R.: Sign language production using neural machine translation and generative adversarial networks. In: British Machine Vision Conference (BMVC) (2018)
35. Wang, T.C., Liu, M.Y., Zhu, J.Y., Tao, A., Kautz, J., Catanzaro, B.: High-resolution image synthesis and semantic manipulation with conditional GANs. In: Proceedings of the IEEE Conference on Computer Vision and Pattern Recognition (2018)
36. Wang, Z., Bovik, A.C., Sheikh, H.R., Simoncelli, E.P.: Image quality assessment: from error visibility to structural similarity. IEEE Trans. Image Process. **13**(4), 600–612 (2004)
37. Yan, X., Yang, J., Sohn, K., Lee, H.: Attribute2Image: conditional image generation from visual attributes. In: Leibe, B., Matas, J., Sebe, N., Welling, M. (eds.) ECCV 2016. LNCS, vol. 9908, pp. 776–791. Springer, Cham (2016). https://doi.org/10.1007/978-3-319-46493-0_47

38. Yao, S., Lin, W., Ong, E., Lu, Z.: Contrast signal-to-noise ratio for image quality assessment. In: IEEE International Conference on Image Processing 2005, vol. 1, pp. I–397, September 2005. https://doi.org/10.1109/ICIP.2005.1529771
39. Zhang, H., Starke, S., Komura, T., Saito, J.: Mode-adaptive neural networks for quadruped motion control. ACM Trans. Graph. **37**(4), 145:1–145:11 (2018). https://doi.org/10.1145/3197517.3201366
40. Zwitserlood, I., Verlinden, M., Ros, J., Schoot, S.V.D.: Synthetic signing for the deaf: esign (2004)

ALET (Automated Labeling of Equipment and Tools): A Dataset for Tool Detection and Human Worker Safety Detection

Fatih Can Kurnaz(✉), Burak Hocaoğlu, Mert Kaan Yılmaz, İdil Sülo, and Sinan Kalkan

Department of Computer Engineering, KOVAN Research Lab, Middle East Technical University, Ankara, Turkey
{fatih.kurnaz,burak.hocaoglu,kaan.yilmaz,idil.sulo,skalkan}@metu.edu.tr

Abstract. Robots collaborating with humans in realistic environments need to be able to detect the tools that can be used and manipulated. However, there is no available dataset or study that addresses this challenge in real settings. In this paper, we fill this gap with a dataset for detecting farming, gardening, office, stonemasonry, vehicle, woodworking, and workshop tools. The scenes in our dataset are snapshots of sophisticated environments with or without humans using the tools. The scenes we consider introduce several challenges for object detection, including the small scale of the tools, their articulated nature, occlusion, inter-class invariance, etc. Moreover, we train and compare several state of the art deep object detectors (including Faster R-CNN, Cascade R-CNN, YOLOv3, RetinaNet, RepPoint, and FreeAnchor) on our dataset. We observe that the detectors have difficulty in detecting especially small-scale tools or tools that are visually similar to parts of other tools. In addition, we provide a novel, practical safety use case with a deep network which checks whether the human worker is wearing the safety helmet, mask, glass, and glove tools. With the dataset, the code and the trained models, our work provides a basis for further research into tools and their use in robotics applications.

Keywords: Safety detection · Tool detection

1 Introduction

The near future will see a cohabitation of robots and humans, where they will work together for performing tasks that are especially challenging, tiring or unergonomic for humans. This requires robots to have abilities for perceiving the humans, the task at hand, and the environment. An essential perceptual component for these abilities is the detection of objects, especially the tools.

The robotics community has paid marginal importance to tools that are used by humans. For example, there are studies focusing on affordances of tools, or on

A. Bartoli and A. Fusiello (Eds.): ECCV 2020 Workshops, LNCS 12538, pp. 371–386, 2020.
https://doi.org/10.1007/978-3-030-66823-5_22

Fig. 1. Samples from the ALET dataset, illustrating the wide range of challenging scenes and tools that a robot is expected to recognize in a clutter, possible with human co-workers using the tools. Since annotations are too dense, only a small subset is displayed. [Best viewed in color] (Color figure online)

the detection and transfer of these affordances [21,23,24]. However, these studies considered tools mostly in isolated and limited, toy environments. Moreover, they have considered only a limited set of tools (see Table 1). What is more, the literature has not studied detection of tools, nor is there a dataset available for it.

In this paper, we focus on the detection of tools in realistic, cluttered environments (e.g. like those in Fig. 1) where collaboration between humans and robots is expected. To be more specific, we study detection of tools in real work environments that are composed of many objects (tools) that look alike and that occlude each other. For this end, we first collect an extensive tool detection dataset composed of 49 tool categories. Then, we compare the widely used state-of-the-art object detectors on our dataset, as a baseline. The results suggest that detecting tools is very challenging owing to tools being too small and articulated, and bearing too much inter-class similarity. Finally, we introduce a safety usecase from our dataset and train a novel CNN network for this task.

The Necessity for a Dataset for Tool Detection: Tool detection requires a dataset of its own since it bares novel challenges of its own: (i) Many tools are small objects which elicit a problem to standard object detectors that are tuned for detecting moderately larger objects. (ii) Many tools are articulated, and in addition to viewpoint, scale and illumination changes, object detectors need to address invariance to articulation. (iii) Tools are generally used in highly cluttered environments posing challenges on clutter, occlusion, appearance, and illumination – see Fig. 1 for some samples. (iv) Many tools exhibit low inter-class differences (e.g., between screwdriver, chisel and file or between putty knives and scraper).

Table 1. A comparison of the datasets that include tools. The Epic-Kitchens dataset provides videos, which makes analysis of scenes difficult. Moreover, the figures for the Epic-Kitchens dataset are estimated based on the provided data. *SO denotes the number of scenes that only include a single object.

Dataset	Tool categories	Tool classes	# of images	# Instances per tool	Modality	Dense bounding boxes?
RGB-D Part Aff. [24]	Kitchen, Workshop, Garden	17	3 (SO*: 102)	6.17	RGB-D	No
ToolWeb [1]	Kitchen, Office, Workshop	23	0 (SO*: 116)	5.03	3D model	No
Visual Aff. of Tools [7]	Toy	3	0 (SO*: 5280)	377	RGB stereo (inc. semantic map)	No
Epic-Kitchens [6]	Kitchen	60+	20M (Videos)	200+	RGB stereo (inc. semantic map)	Yes
ALET (Ours)	Farm, Garden, Office, Stonemasonry, Vehicles, Woodwork, Workshop	49	2788 (SO*: 0)	200+	RGB	Yes

1.1 Related Work

Object Detection: Object detection is one of the most studied problems in Computer Vision with many practical applications in many robotics scenarios. Object detection generally follows a two-stage approach: (i) Region selection, which pertains to the selection of image regions that are likely to contain an object. (ii) Object classification, which deals with the classification of a selected region into one of the object categories. With advances in deep learning both stages have seen tremendous boost in performance in many challenging settings, e.g., [12,13,28]. Faster R-CNN [28] is a well-known representative of such two-stage detectors.

It has been shown that the two stages can be combined and objects can be detected in one stage. Models such as [18,20,26] assume a fixed set of localized image regions (anchors) for each object category, and estimate objectness for each category and for each anchor. Among these, RetinaNet [18] processes features and make classification at multiple scales (called a feature pyramid network) and combines the results, yielding the state of the art results among one-stage detectors.

These one-stage and two-stage detectors have a top-down approach. In contrast, bottom-up object detection forms a detection pipeline from points or keypoints that are likely to be on objects and identifies objects by combining such points (examples include CenterNet [9], RepPoints [34]).

The current trend in deep object detectors is to take a pre-trained object classification network as the feature extractor (called the backbone network), then perform two-stage or one-stage object detection from those features. Alternatives for the backbone network include deep classification networks such as VGG [29], ResNet [14] and ResNext [33].

Tools in Robotics: The robotics community has extensively studied how tools can be grasped, manipulated, and how such affordances can be transferred across tools. For example, Kemp and Edsinger [16] focused on detection and the 3D localization of tool tips from optical flow. For the same goal, Mart et al. [21] proposed using a CNN network (AlexNet) to first classify a blob into one of the three tool labels they considered, and then used 3D geometric features to identify tool tips.

Another study [24] addressed the problem of estimating the grasping positions, scoops and supports of tools from RGB-D data. In a similar setting, Mar et al. [22,23] studied prediction of affordances for different categories of tools separately. For this purpose, they first clustered tools using their 3D geometric descriptors, and then estimated the affordances of each cluster separately.

Related Datasets: Comparing our dataset to the ones used in the robotics literature (e.g., [1,7,24] – see also Table 1), we see that they are limited in the number of categories and the instances that they consider. Moreover, since they mainly focus on detection of tool affordances, they are not suitable for training a deep object detector.

For related problems, there are numerous datasets of objects in the robotics literature, e.g. for 3D pose estimation and robot manipulation (e.g., LINEMOD [15], YCB Objects [3], Table-top Objects [30], Object Recognition Challenge [31]). These datasets generally include table-top objects with 3D models and do not include tool categories.

For object detection, there are several datasets such as PASCAL [11], MS-COCO [19] and ImageNet [8], which do include some tool categories (e.g., hammer, scissors); however, these datasets are designed to be for general purpose objects and scenes, unlike the ones we expect to see in tool-used environments (such as the ones in Fig. 1). Therefore, they do not provide sufficient amount of training instances for training a general purpose tool detector with a reasonable performance.

Safety Detection: There are studies on detecting whether or not a human worker is wearing safety helmet [17,32] or a vest [25]. However, there are no studies or datasets that include safety glass, mask, glove and headphone together with helmet.

1.2 Contributions of Our Work

The main contributions of our work can be summarized as follows:

- **A Tool Detection Dataset**: To the best of our knowledge, ours is the first to provide a dataset on detection of tools in the wild.
- **A Baseline for Tool Detection**: On our dataset, we train and analyze many state-of-the-art deep object detectors. Together with the dataset, the code and the trained models, our work can form as a basis for robotics applications that require detection of tools in challenging realistic work environments with humans.
- **A Novel Usecase for Checking the Safety of Human Coworkers:** Our dataset includes humans performing tasks with and without wearing a safety helmet, mask, glass, headphone and gloves. We form a ALET Safety subset of positive and negative instances of these and train a deep CNN network that checks whether a human coworker is wearing these safety tools.

2 ALET: A Tool Detection Dataset

In this section, we present and describe the details of ALET and how the dataset was collected.

2.1 Tools and Their Categories

In ALET, we consider 49 different tools that are used for six broad contexts or purposes: Farming, gardening, office, stonemasonry, vehicle, woodworking, workshop tools[1]. The 20 most frequent tools from our dataset are: Chisel, Clamp, Drill, File, Gloves, Hammer, Mallet, Meter, Pen, Pencil, Plane, Pliers, Safety glass, Safety helmet, Saw, Screwdriver, Spade, Tape, Trowel, Wrench.

We excluded tools used in kitchen since there is already an exclusive dataset for this purpose [6]. Moreover, we limited ourselves to tools that can be ultimately grasped, pushed, or manipulated in an easy manner by a robot. Therefore, we did not consider tools such as ladders, forklifts, and power tools that are bigger than a hand-sized drill.

[1] It is better to call some of these objects as equipment. However, since they provide similar functionalities (being used by a human or a robot while performing a task), we will just use the term tool to refer to all such objects, for the sake of simplicity.

2.2 Dataset Collection

Our dataset is composed of three groups of images:

- **Images collected from the web**: Using keywords and usage descriptions that describe the tools listed in Sect. 2.1, we crawled and collected royalty-free images from the following websites: Creativecommons, Wikicommons, Flickr, Pexels, Unsplash, Shopify, Pixabay, Everystock, Imfree.
- **Images photographed by ourselves**: We captured photos of office and workshop environments from our campus.
- **Synthetic images**: In order to make sure that there are at east 200 instances for each tool, we developed a simulation environment and collected synthetic images (see Fig. 2 for examples). For this, we used the Unity3D platform with 3D models of tools acquired from UnityStore. For each scene to be generated, the following steps are followed:
  - *Scene Background*: We created a room like environment with 4 walls, 10 different random objects (chair, sofa, corner-piece, television) in static positions. At the center of the room, we spawned one of six different tables selected randomly from $Uniform(1, 6)$. To introduce more randomness, we also dropped unrelated objects like mugs, bottles etc. randomly.
  - *Camera*: Each dimension of the camera position (x, y, z) was sampled randomly from $Uniform(-3, 3)$. Camera's viewing direction was set towards the center of the top of the table.
  - *Tools*: In each scene, we spawned $N \sim Uniform(5, 20)$ tools which are selected randomly from $Uniform(1, 49)$. The spawn tools are dropped onto the table from $[x, y, z]$ selected randomly from $Uniform(0, 1)$ above the table. Initial orientation (each dimension) is sampled from $Uniform(0, 360)$.

  The special cases when the sampled camera not seeing the table-top etc. are handled using hand-designed rules. Some examples from this process can be seen in Fig. 2.

Fig. 2. Some examples from the Synthetic Images. [Best viewed in color] (Color figure online)

For annotating the tools in the downloaded and the photographed images, we used the VGG Image Annotation (VIA) tool [10]. Annotation was performed by the authors of the paper.

2.3 Dataset Statistics

In this section, we provide some descriptive statistics about the ALET dataset.

Cardinality and Sizes of BBs: The ALET dataset includes 19,706 bounding boxes (BBs). As displayed in Fig. 3, *for each tool category, there are more than 200 BBs, which is on an order similar to the widely used object detection datasets such as PASCAL* [11]. As shown in Table 2, ALET includes tools that appear small (area $< 32^2$), medium ($32^2 <$ area $< 96^2$) and large ($96^2 <$ area) – following the naming convention from MS-COCO [19]. Although this is expected, as we will see in Sect. 4, deep networks have difficulty especially detecting small tools.

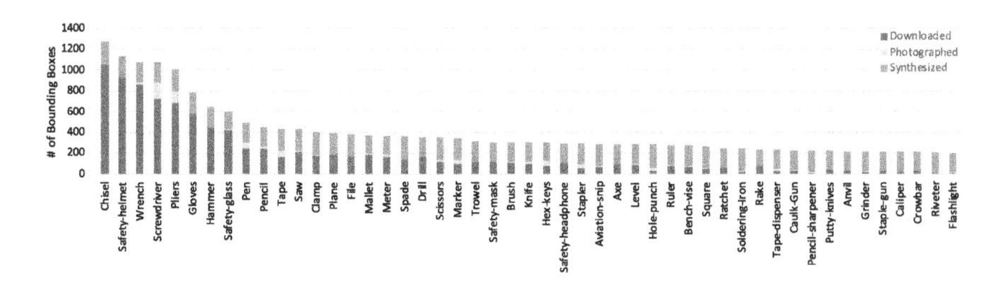

Fig. 3. The distribution of bounding boxes across classes. With the photographed and synthesized photos, each tool category has more than 200 bounding boxes. [Best viewed in color] (Color figure online)

Table 2. The sizes of the bounding boxes (BB) of the annotated tools in ALET. For calculating these statistics, we considered the scaled versions of the images that were fed to the networks, namely, 1333 × 800.

Subset category	Small BBs	Medium BBs	Large BBs	Total
Downloaded	1342	3685	4545	9572
Photographed	23	309	458	790
Synthesized	1784	5314	2246	9344
Total	3149	9308	7249	19706

Cardinality and Sizes of the Images: Our dataset is composed of 2788 images in total, and on average, has size 1138 × 903. See Table 3 for more details. Although the number of images may appear low, the number of bounding boxes (19706) is sufficient since there are more than 200 BBs per tool, and the avg. number of BBs per image is rather large (6.6) compared to PASCAL 2012 (2.3).

Table 3. The cardinality and the resolution of the images in ALET.

Subset	Cardinality	Avg. Resolution
Downloaded	1912	924 × 786
Photographed	89	3663 × 3310
Synthesized	787	1374 × 917
Total/Avg	2788	1138 × 903

2.4 ALET Safety Subset

In this section, we illustrate how the ALET dataset can be used for addressing a critical issue in human-robot collaborative environments; that of checking whether the human worker is conforming to the security guidelines and wearing the required safety tools. ALET dataset contains a good number of safety tool ("helmet, glass, mask, headphone and glove") instances (2159 in total just in real images, and 3104 when combined with synthetic examples), which suffice for training a deep classifier.

For forming the safety subset, first we used OpenPose [4] to both detect the humans in an image and estimate their 2D poses. Then, we compared the positions of safety tools with the positions of the corresponding joints (e.g. for glove, the hand joints are used whereas head joints are used for the helmet etc.). If, for a tool, the Euclidean distance between the positions (normalized wrt. size of the BB of the human) were less than 0.2, the human was considered wearing that safety tool.

3 Methodology

In this section, we briefly describe the deep object detectors that we evaluated as a baseline, and the safety as a straightforward usecase of our dataset.

3.1 Deep Object Detectors

As stated in Sect. 1.1, top-down deep object detectors can be broadly analyzed in two categories: (i) single-stage detectors, and (ii) multi-stage detectors. To form a baseline, we evaluated strong representatives of both single-stage (RetinaNet, YOLOv3) and multi-stage (Faster R-CNN, Cascade R-CNN) detectors. Moreover, we included RepPoints and FreeAnchor, recent bottom-up object detectors.

Faster R-CNN. Faster R-CNN [28] is one of the first networks to use end-to-end learning for object detection. It feeds features extracted from a backbone network to a region proposal network, which estimates an objectness score and the (relative) coordinates of a set of k anchor boxes for each position on a regular grid. For each such box with an objectness score above a threshold, the object classification network (Fast R-CNN) is executed to classify each box into one of the object categories.

For training the network, classification loss and box-regression loss (to penalize the spatial mismatch between the detected box and the ground truth box) are combined. The box-regression loss is weighted with a constant (λ), which we selected as 1.0 as suggested by the paper.

Cascade R-CNN. Cascade R-CNN [2] utilizes a region-proposal network with multiple detection stages with an increasing amount of IoU threshold for each of them. By doing so it eliminates negative samples better in each detection stage while increasing its IoU threshold in each further detection stage.

RetinaNet. RetinaNet [18] is a one-stage detector which forms a multi-scale pyramid from the features obtained from the backbone network and performs classification and bounding box estimation in parallel for each layer (scale) of the pyramid. In order to address the data imbalance problem that affects single-stage detectors owing to background, RetinaNet proposes using *focal loss* that decreases the contribution of the "easy" examples to the overall loss. Compared to other single-stage detectors, RetinaNet considers a denser set of bounding boxes to classify.

YOLOv3. YOLOv3 is the third iteration of YOLO models. Following the YOLOv2 update, which are batch normalization for regularization, 448 × 448 image resolution for classifier compared to 224 × 224, anchor boxes for bounding box prediction, and addition to multi-scale training, Redman and Farhadi [27] used logisctic classifiers for class prediction and a deeper feature extractor named Darknet-53.

RepPoint Detection. Contrary to other approaches, RepPoint [34] is an anchor free, bottom-up detector. It is based on identifying representative points on objects and then combining these points into bounding boxes.

FreeAnchor. Zhang et al. [35] introduced a new system which makes it possible for network to assign bounding boxes for each object from a anchor bag rather than predetermined anhors by hand, using maximum likelihood estimation.

3.2 A Safety Usecase for ALET

We created a CNN architecture consisting of three 2D convolutional layers and two fully connected layers. After each convolutional layer we added a batch normalization layer, and each layer is also followed by ReLu activation. The final layer has five outputs with sigmoid activation. The network performs five-class (one for each safety tool) multi-label classification with binary cross-entropy. The network is trained on ALET Safety Dataset.

An alternative approach could be to combine the results of the tool detector and the pose detector. However, considering that the tool detection networks are having acceptable performance, we adopted an independent network for safety detection. Moreover, a tool detector would be detecting 49 tools in a scene 43 of which are irrelevant for our safety usecase.

4 Experiments

4.1 Training and Implementation Details

We split the ALET dataset (unless otherwise stated, this means real and synthetic images combined together) into 1672 (%60) training, 278 (%10) validation and 838 (%30) testing samples. For training each network, the following libraries and settings are used (in all networks, the output layer is replaced with the tool categories, and the whole network except for the feature extracting backbone is updated during training): For each detector, the pre-trained network from mmdetection [5] is used with backbone ResNet-50-FPN.

4.2 Quantitative Results for Tool Detection

On the testing subset of ALET, we compare the performance of the detectors trained on the training subset of ALET. We use the MS COCO style average precision (AP).

Table 4 lists the AP and mAP values of the baseline networks on our dataset. We notice that the baseline detectors perform well on tools that are very distinctive and different from others; e.g. tape-dispenser, safety-helmet, hole-punch, pencil-sharpener. On such tools, we observe AP of up to 81.1 (Cascade R-CNN for anvil).

However, we see that the detectors have trouble in detecting especially tools that are too narrow, like pen-pencil, knife, file. In fact, YOLOv3 yields an AP of 1.2 on pencil and 2.2 on ruler. There are several reasons for these results: (i) Object detectors have been designed for general object detection and need to extended to consider a wider range of anchors with a wider range of aspect ratios, which would increase the complexities of the networks. (ii) A second reason is that annotated boxes of small tools such as pen-pencil include more pixels of other objects than the annotated small object itself. (iii) Moreover, tools such as screwdriver, chisel, file look very similar to each other. Moreover, these tools appear very similar to parts of other tools from a side view or from far (e.g., the front part of a drill is likely to be classified as a screwdriver, and in many cases, one half of a plier is detected as a chisel).

These suggest that tool detection is indeed a very challenging problem especially owing to small tools and tools having very similar appearances to other tools.

Table 5 evaluates the performance improvement provided by the synthetic images from the simulation environment. We compared our dataset in with and without synthetic images by using RetinaNet and it shows that synthetic images elevate our dataset. The results for other networks are being produced and will be included in the final version.

4.3 Sample Tool Detection Results

In Fig. 4, we display a few detection results on a few of the challenging scenes in Fig. 1. We see that although many tools are detected, many are missed.

Fig. 4. Sample detections from RetinaNet (w 0.4 as confidence threshold) on a few of the challenging scenes from Fig. 1. In the top example, almost all detections are correct. However, detector detected screwdriver and pliers in place of pens, which can be attributed to similarity in size and color. In the second image detector wrongfully annotates a hat as safety-helmet. In addition, it misses some of the crowded wrench examples, which is caused by the complexity of the problem.

Table 4. MS COCO style AP of the baseline networks. Training and testing are both performed with the synthetic + real images.

	Faster R-CNN	Cascade R-CNN	RetinaNET	YOLOv3	RepPoint	FreeAnchor
Anvil	76.1	**81.1**	71.9	78.7	78.6	69.6
Aviation-snip	10.8	**12.4**	2.3	6.1	5.1	10.8
Axe	31.7	51.7	26.5	**66.5**	36.9	39.4
Bench-vise	61.8	69.5	62.6	**72.1**	66.8	65.0
Brush	18.7	**24.2**	4.9	3.1	13.5	4.6
Caliper	24.6	**29.7**	7.6	28.3	23.7	18.3
Caulk-Gun	32.2	**33.7**	6.4	28.6	20.1	21.4
Chisel	7.9	12.4	5.2	**19.4**	7.4	10.4
Clamp	17.8	**19.8**	10.0	16.6	14.8	14.8
Crowbar	28.5	35.8	20.1	**69.3**	22.8	29.2
Drill	49.5	52.7	47.7	**57.3**	50.3	48.8
File	6.4	8.4	1.6	1.2	5.7	**9.8**
Flashlight	51.1	**51.2**	38.7	24.7	41.9	44.1
Gloves	32.6	35.5	33.6	**47.3**	33.3	30.6
Grinder	43.0	**46.8**	37.7	26.3	35.0	43.9
Hammer	17.2	13.6	7.8	**22.6**	12.7	9.6
Hex-keys	38.5	**44.7**	30.3	7.2	30.8	32.6
Hole-punch	61.0	62.2	54.3	25.4	**62.4**	59.0
Knife	19.3	**22.0**	5.0	4.1	15.9	16.1
Level	**34.4**	30.9	21.3	11.1	34.0	26.2
Mallet	22.5	**29.7**	14.0	17.2	22.8	18.0
Marker	17.5	18.7	12.3	11.0	17.2	**18.8**
Meter	21.8	**23.4**	16.1	16.7	15.3	21.7
Pen	11.1	**16.0**	6.3	1.2	14.1	7.8
Pencil	10.2	**13.8**	6.5	1.9	6.3	6.7
Pencil-sharpener	65.5	70.5	63.7	**85.7**	63.8	66.5
Plane	26.7	24.7	12.8	**27.7**	15.0	23.9
Pliers	29.5	**31.6**	25.1	28.0	31.0	31.3
Putty-knives	**14.9**	11.3	4.5	10.0	7.4	5.9
Rake	33.8	50.3	16.2	**80.3**	30.0	33.1
Ratchet	18.6	**19.9**	16.7	10.0	17.3	11.1
Riveter	34.3	**48.0**	19.3	2.4	29.0	28.6
Ruler	**9.4**	8.8	3.5	2.2	4.4	9.2
Safety-glass	20.1	24.3	18.7	**31.3**	16.7	16.8
Safety-headphone	45.9	46.3	45.5	**50.0**	44.0	40.1
Safety-helmet	49.6	54.3	51.1	**84.6**	53.0	51.1
Safety-mask	50.1	50.6	42.3	**59.9**	41.5	46.4
Saw	25.0	**27.8**	7.3	11.9	14.8	20.0
Scissors	16.0	**19.1**	6.9	2.9	10.2	3.8
Screwdriver	23.7	**25.5**	7.4	11.8	17.4	15.1
Soldering-Iron	36.1	**40.7**	13.4	20.9	30.5	24.6
Spade	33.5	40.0	29.4	**62.7**	32.7	29.2
Square	5.6	**6.3**	1.0	1.3	1.8	0.6
Staple-gun	34.5	34.5	18.1	23.0	23.8	**34.6**
Stapler	41.2	**47.3**	35.2	39.6	41.3	36.2
Tape	46.5	46.1	40.5	29.7	43.0	**48.4**
Tape-dispenser	55.4	**60.8**	53.6	29.8	57.8	47.6
Trowel	15.5	**22.4**	12.7	6.6	12.0	8.5
Wrench	22.2	20.9	9.0	**37.0**	19.8	19.1
mAP	30.6	**34.1**	22.7	28.0	27.4	27.1

4.4 Safety Usecase Results

In this usecase, we analyze the method proposed in Sect. 3.2. Our CNN network obtained 0.50 F1 score on the test set, and as illustrated with some examples in Fig. 5 and accuracy results from Table 6, we see that a network and a human detector & pose estimator can be used to easily identify whether a human is wearing a safety equipment or not. However, the network is having difficulty with safety glasses and gloves, because, as illustrated in Fig. 5, the region around eyes may not be visible for a working man and hands tend to be outside the BB in some cases.

Table 5. The gain obtained for RetinaNet using the synthetic examples generated from our simulation environment.

	Training w Real data	Training w Real+Synthetic data
mAp on Real+Syn test images	11.980	20.118

Table 6. Class accuracies for the safety network.

	Helmet	Gloves	Mask	Headphone	Glass
Accuracy	75%	69%	85%	85%	54%

Fig. 5. Sample results for the safety usecase. Symbols: H: Helmet, M: Mask, G: Gloves, L: Glass, D: Headphone. [Best viewed in color] (Color figure online)

5 Conclusion

In this paper, we have introduced an extensive dataset for tool detection in the wild. Moreover, we formed a baseline by training and testing four widely-used state-of-the-art deep object detectors in the literature, namely, Faster R-CNN [28], Cascade R-CNN [2], and RetinaNet [18], YOLOv3 [27], RepPoint Detection [34] and FreeAnchor [35]. We demonstrated that such detectors especially have trouble in finding tools whose appearance is highly affected by viewpoint changes and tools that resemble parts of other tools.

Moreover, we have provided a very practical yet critical usecase for human-robot collaborative scenarios. Combining the detected "helmet, glass, mask, headphone and gloves" categories with the detection results of a human detector & pose estimator, we have demonstrated how our dataset can be used for practical applications other than merely detecting tools in an environment.

Acknowledgment. This work was supported by the Scientific and Technological Research Council of Turkey (TÜBİTAK) through project called "CIRAK: Compliant robot manipulator support for montage workers in factories" (project no 117E002). The numerical calculations reported in this paper were partially performed at TÜBİTAK ULAKBIM, High Performance and Grid Computing Center (TRUBA resources). We would like to thank Erfan Khalaji for his contributions on an earlier version of the work.

References

1. Abelha, P., Guerin, F.: Learning how a tool affords by simulating 3D models from the web. In: IROS (2017)
2. Cai, Z., Vasconcelos, N.: Cascade R-CNN: delving into high quality object detection. arxiv:1712.00726 (2017)
3. Calli, B., Walsman, A., Singh, A., Srinivasa, S., Abbeel, P., Dollar, A.M.: Benchmarking in manipulation research: the YCB object and model set and benchmarking protocols. arXiv:1502.03143 (2015)
4. Cao, Z., Hidalgo, G., Simon, T., Wei, S.E., Sheikh, Y.: OpenPose: realtime multi-person 2D pose estimation using part affinity fields. arXiv:1812.08008 (2018)
5. Chen, K., Wang, J., Pang, J.E.A.: MMDetection: open MMLab detection toolbox and benchmark. arXiv:1906.07155 (2019)
6. Damen, D., et al.: Scaling egocentric vision: the dataset. In: Ferrari, V., Hebert, M., Sminchisescu, C., Weiss, Y. (eds.) ECCV 2018. LNCS, vol. 11208, pp. 753–771. Springer, Cham (2018). https://doi.org/10.1007/978-3-030-01225-0_44
7. Dehban, A., Jamone, L., Kampff, A.R., Santos-Victor, J.: A moderately large size dataset to learn visual affordances of objects and tools using iCub humanoid robot. In: ECCV Workshop on Action and Anticipation for Visual Learning (2016)
8. Deng, J., Dong, W., Socher, R., Li, L.J., Li, K., Fei-Fei, L.: ImageNet: a large-scale hierarchical image database. In: CVPR (2009)
9. Duan, K., Bai, S., Xie, L., Qi, H., Huang, Q., Tian, Q.: CenterNet: keypoint triplets for object detection. In: ICCV (2019)
10. Dutta, A., Gupta, A., Zissermann, A.: VGG image annotator (VIA). http://www.robots.ox.ac.uk/vgg/software/via/ (2016). version: 2.0.5. Accessed 27 Feb 2019

11. Everingham, M., Van Gool, L., Williams, C.K., Winn, J., Zisserman, A.: The pascal visual object classes (VOC) challenge. IJCV **88**(2), 303–338 (2010)
12. Girshick, R.: Fast R-CNN. In: ICCV (2015)
13. Girshick, R., Donahue, J., Darrell, T., Malik, J.: Rich feature hierarchies for accurate object detection and semantic segmentation. In: CVPR (2014)
14. He, K., Zhang, X., Ren, S., Sun, J.: Deep residual learning for image recognition. In: CVPR (2016)
15. Hinterstoisser, S., et al.: Model based training, detection and pose estimation of texture-less 3D objects in heavily cluttered scenes. In: ACCV (2012)
16. Kemp, C.C., Edsinger, A.: Robot manipulation of human tools: autonomous detection and control of task relevant features. In: ICDL (2006)
17. Li, K., Zhao, X., Bian, J., Tan, M.: Automatic safety helmet wearing detection. arXiv:1802.00264 (2018)
18. Lin, T.Y., Goyal, P., Girshick, R., He, K., Dollár, P.: Focal loss for dense object detection. In: ICCV (2017)
19. Lin, T.-Y., et al.: Microsoft COCO: common objects in context. In: Fleet, D., Pajdla, T., Schiele, B., Tuytelaars, T. (eds.) ECCV 2014. LNCS, vol. 8693, pp. 740–755. Springer, Cham (2014). https://doi.org/10.1007/978-3-319-10602-1_48
20. Liu, W., et al.: SSD: single shot multibox detector. In: Leibe, B., Matas, J., Sebe, N., Welling, M. (eds.) ECCV 2016. LNCS, vol. 9905, pp. 21–37. Springer, Cham (2016). https://doi.org/10.1007/978-3-319-46448-0_2
21. Mar, T., Natale, L., Tikhanoff, V.: A framework for fast, autonomous and reliable tool incorporation on iCub. Front. Robot. AI **5**, 98 (2018)
22. Mar, T., Tikhanoff, V., Metta, G., Natale, L.: Multi-model approach based on 3D functional features for tool affordance learning in robotics. In: Humanoids (2015)
23. Mar, T., Tikhanoff, V., Natale, L.: What can i do with this tool? Self-supervised learning of tool affordances from their 3-D geometry. TCDS **10**(3), 595–610 (2018)
24. Myers, A., Teo, C.L., Fermüller, C., Aloimonos, Y.: Affordance detection of tool parts from geometric features. In: ICRA (2015)
25. Nath, N.D., Behzadan, A.H., Paal, S.G.: Deep learning for site safety: real-time detection of personal protective equipment. Autom. Constr. **112**, 103085 (2020)
26. Redmon, J., Divvala, S., Girshick, R., Farhadi, A.: You only look once: unified, real-time object detection. In: CVPR (2016)
27. Redmon, J., Farhadi, A.: Yolov3: an incremental improvement. arXiv (2018)
28. Ren, S., He, K., Girshick, R., Sun, J.: Faster R-CNN: towards real-time object detection with region proposal networks. In: NIPS (2015)
29. Simonyan, K., Zisserman, A.: Very deep convolutional networks for large-scale image recognition. arXiv:1409.1556 (2014)
30. Sun, M., Bradski, G., Xu, B.-X., Savarese, S.: Depth-encoded hough voting for joint object detection and shape recovery. In: Daniilidis, K., Maragos, P., Paragios, N. (eds.) ECCV 2010. LNCS, vol. 6315, pp. 658–671. Springer, Heidelberg (2010). https://doi.org/10.1007/978-3-642-15555-0_48
31. Vaskevicius, N., Pathak, K., Ichim, A., Birk, A.: The jacobs robotics approach to object recognition and localization in the context of the ICRA'11 solutions in perception challenge. In: ICRA (2012)
32. Wu, J., Cai, N., Chen, W., Wang, H., Wang, G.: Automatic detection of hardhats worn by construction personnel: a deep learning approach and benchmark dataset. Autom. Constr. **106**, 102894 (2019)
33. Xie, S., Girshick, R., Dollár, P., Tu, Z., He, K.: Aggregated residual transformations for deep neural networks. In: CVPR (2017)

34. Yang, Z., Liu, S., Hu, H., Wang, L., Lin, S.: Reppoints: point set representation for object detection. In: ICCV (2019)
35. Zhang, X., Wan, F., Liu, C., Ji, R., Ye, Q.: FreeAnchor: learning to match anchors for visual object detection. In: NeurIPS (2019)

DMD: A Large-Scale Multi-modal Driver Monitoring Dataset for Attention and Alertness Analysis

Juan Diego Ortega[1,3](✉), Neslihan Kose[2], Paola Cañas[1], Min-An Chao[2], Alexander Unnervik[2], Marcos Nieto[1], Oihana Otaegui[1], and Luis Salgado[3]

[1] Vicomtech Foundation, Basque Research and Technology Alliance (BRTA), Mendaro, Spain
jdortega@vicomtech.org

[2] Dependability Research Lab, Intel Labs Europe, Intel Deutschland GmbH, Neubiberg, Germany

[3] ETS Ingenieros de Telecomunicación, Universidad Politécnica de Madrid, Madrid, Spain
https://dmd.vicomtech.org/

Abstract. Vision is the richest and most cost-effective technology for Driver Monitoring Systems (DMS), especially after the recent success of Deep Learning (DL) methods. The lack of sufficiently large and comprehensive datasets is currently a bottleneck for the progress of DMS development, crucial for the transition of automated driving from SAE Level-2 to SAE Level-3. In this paper, we introduce the *Driver Monitoring Dataset (DMD)*, an extensive dataset which includes real and simulated driving scenarios: distraction, gaze allocation, drowsiness, hands-wheel interaction and context data, in 41 h of RGB, depth and IR videos from 3 cameras capturing face, body and hands of 37 drivers. A comparison with existing similar datasets is included, which shows the DMD is more extensive, diverse, and multi-purpose. The usage of the DMD is illustrated by extracting a subset of it, the *dBehaviourMD* dataset, containing 13 distraction activities, prepared to be used in DL training processes. Furthermore, we propose a robust and real-time driver behaviour recognition system targeting a real-world application that can run on cost-efficient CPU-only platforms, based on the *dBehaviourMD*. Its performance is evaluated with different types of fusion strategies, which all reach enhanced accuracy still providing real-time response.

Keywords: Driver Monitoring Dataset · Driver actions · Driver behaviour recognition · Driver state analysis · Multi-modal fusion

1 Introduction

Road accidents affect drivers and passengers, but also vulnerable pedestrians, motorcyclists and cyclists. Human factors are a contributing cause in almost

J. D. Ortega, N. Kose, P. Cañas, M.-A. Chao—Equal contribution.

A. Bartoli and A. Fusiello (Eds.): ECCV 2020 Workshops, LNCS 12538, pp. 387–405, 2020.
https://doi.org/10.1007/978-3-030-66823-5_23

90% of those road accidents [14]. Therefore, there has been an active support to the development of fully automated vehicles whose aim is to reduce crashes due to driver factors, eventually achieving the desired EU *Vision Zero objective* road scenario [16]. As automated driving technology advances through SAE-L2 and SAE-L3 [42], there will be a paradigm shift in terms of driver responsibility and function. The driving task will become a shared activity between the human and the machine [19]. For this to be successful, modern driving automation systems shall increasingly include DMS to measure features of the driver inside the cabin and determine the driver's readiness to perform the driving task. Such systems are specially valuable to guarantee a more reliable and safer mode transfer to operate the vehicle [36].

DMS have been developed for the past decade as a measure to increase road safety and also to improve driving comfort [35,46]. The objective of such systems is to understand driver's state according to measurable cues extracted either from direct observation of the driver (i.e. physiological parameters [8], visual appearance of the driver [18]), indirect activity of on-board sensors (i.e. vehicle dynamics, cellphone, smart watches or GPS) [3] or a combination of them [24]. Computer vision methods have acquired important attention particularly due to the demonstrated capabilities of the emerging DL technologies [40]. These methods have a good balance between robustness and applicability, since they could achieve high accuracy rates with non-obtrusive sensors [19]. However, these techniques demand large amounts of visual data to build the necessary models to achieve the desired task (i.e. detect distraction, drowsiness, involvement in secondary tasks, etc.) [2]. There is a significant lack of comprehensive DMS-related datasets open for the scientific community, which is slowing down the progress in this field, while, in comparison, dozens of large datasets exist with sensing data for the exterior of the vehicle [25].

In this paper, we introduce a large-scale multi-modal Driver Monitoring Dataset (DMD) which can be used to train and evaluate DL models to estimate alertness and attention levels of drivers in a wide variety of driving scenarios. The DMD includes rich scene features such as varying illumination conditions, diverse subject characteristics, self-occlusions and 3 views with 3 channels (RGB/Depth/IR) of the same scene. With the DMD, we want to push forward the research on DMS, making it open and available for the scientific community. In addition, we show how to extract a subset of it (the *dBehaviourMD*) and prepare it as a training set to train and validate a DL method for L3 AD by providing a real-time, robust and reliable driver action detection solution.

2 Review of In-Vehicle Monitoring Datasets and Methods

Driver monitoring can be interpreted as observing driver's features related to distraction/inattention, the direction of gaze, head pose, fatigue/drowsiness analysis, disposition of the hands, etc. The state-of-the art shows approaches for these aspects individually and also in a holistic way. Existing literature can be categorised according to the driver's body-part of interest showing visible evidence of different types or aspects of the driver's behaviour.

Hands-Focused Approaches: The position of hands and their actions have shown a good correlation to the driver's ability to drive. The work carried out in this matter includes the creation of datasets focused only on hands, like CVRR-HANDS 3D [37] or the VIVA-Hands Dataset [10], which include the annotation of hands' position in images as bounding boxes. Example methods that use these datasets detect the position of hands using Multiple Scale Region-based Fully Convolutional Networks (MS-RFCN) [29] and detection of interaction with objects such as cellphones to estimate situation which can distract the driver [30].

Face-Focused Approaches: The driver's condition in terms of distraction, fatigue, mental workload and emotions can be derived from observations of the driver's face and head [46]. One aspect widely studied is the direction of eye gaze as in DR(eye)VE [39], DADA [17] and BDD-A [57] datasets. These features plus information about the interior of the vehicle allows identifying which specific areas of the cabin are receiving the attention of the driver, like in DrivFace [13], with the possibility of generating attention heat maps or buffer metrics of distraction such as AttenD [4]. However, many of the available datasets are not obtained in a driving environment, normally they are captured in a laboratory with a simulation setup such as Columbia dataset [47] and MPIIGaze dataset [60]. Moreover, datasets such as DriveAHead [45] and DD-Pose[41] have appeared consisting of images of the driver's head area and are intended to estimate the head pose, containing annotations of yaw, pitch and roll angles.

Fatigue is an important factor that does not allow the driver to safely perform the driving task. Computer vision methods that tackle the detection of fatigued drivers rely on the extraction of parameters from head, mouth and eye activity [38]. In this context, datasets for the identification of drowsy drivers vary in the type of data they offer. They may contain images with face landmarks like DROZY [34] and NTHU-DDD [56]. With these facial points, more complex indicators can be extracted such as head position, 3D head pose and face activity [20], blink duration [5], frequency and PERCLOS [53] with the final goal of estimating the level of drowsiness of the driver [32].

Body-Focused Approaches: A side-view camera can further extend the effective monitoring range to the driver's body action. The first large-scale image dataset for this purpose is the StateFarm's dataset [48], which contains 9 distracted behaviour classes apart from safe driving. However, its use is limited to the purposes of the competition. A similar image-based open dataset is the AUC Distracted Driver (AUC DD) dataset [1]. In [1], driver behaviour monitoring with a side-view camera was approached by image-based models, which cannot capture motion information. By approaching it as video-based action recognition problem, the image-based AUC Dataset [1] is adapted for spatio-temporal models in [28] which proves that just by incorporating the temporal information, significant increase can be achieved in classification accuracy.

The analysis of RGB-D images has become common practice to estimate the position of the body in a 3D space [9]. The Pandora Dataset [6] provides

Table 1. Comparison of public vision-based driver monitoring datasets.

Dataset	Year	Drivers[a]	Views[b]	Size[c]	GT[d]	Streams	Occlusions	Scenarios	Usage
CVRR-Hands [37]	2013	8 (1/7)	1	7k	Hands, Actions	RGB Depth	Yes	Car	Normal driving, Distraction
DrivFace [13]	2016	4 (2/2)	1	0.6k	Face/Head	RGB	No	Car	Normal driving, Head pose
DROZY [34]	2016	14 (11/3)	1	7h	Face/Head Physiological	IR	No	Laboratory	Drowsiness
NTHU-DDD [56]	2017	36 (18/18)	1	210k	Actions	RGB IR	Yes	Simulator	Normal driving, Drowsiness
Pandora [6]	2017	22 (10/12)	1	250k	Face/Head, Body	RGB Depth	Yes	Simulator	Head/Body pose
DriveAHead [45]	2017	20 (4/16)	1	10.5h	Face/Head, Objects	Depth IR	Yes	Car	Normal driving, Head/Body pose
DD-Pose [41]	2019	24 (6/21)	2	6h	Face/Head, Objects	RGB[e] Depth[f] IR[f]	Yes	Car	Normal driving, Head/Body pose
AUC-DD [15]	2019	44 (15/29)	1	144k	Actions	RGB	No	Car	Normal driving, Distraction
Drive&Act [33]	2019	15 (4/11)	6	12h	Hands/Body, Actions, Objects	RGB[e] Depth[e] IR	No	Car	Autonomous driving, Distraction
DMD	2020	37 (10/27)	3	41h	Face/Head, Eyes/Gaze, Hands/Body, Actions, Objects	RGB Depth IR	Yes	Car, Simulator	Normal driving, Distraction, Drowsiness

[a] Number of drivers (female/male)
[b] Simultaneous views of scene
[c] h: hours of video, k: image number
[d] Ground-truth data
[e] Only for side view
[f] Only for face view

high resolution RGB-D images to estimate the driver's head and shoulders' pose. Works like [12], whose objective is to identify driver's readiness, study various perspectives of the body parts (face, hands, body and feet). More recently, the dataset Drive&Act [33] was published containing videos imaging the driver with 5 NIR cameras in different perspectives and 3 channels (RGB, depth, IR) from a side view camera; the material shows participants performing distraction-related activities in an automated driving scenario.

In comparison with our DMD, most of the datasets found in the literature focus either on specific parts of the drivers' body or specific driving actions. With the DMD dataset, we provide a wider range of activities which are directly related to the driver's behaviour and attention state. Table 1 shows a detailed comparison of the studied dataset's features. As can be seen, none of the existing datasets contain DMD's variability, label richness and volume of data.

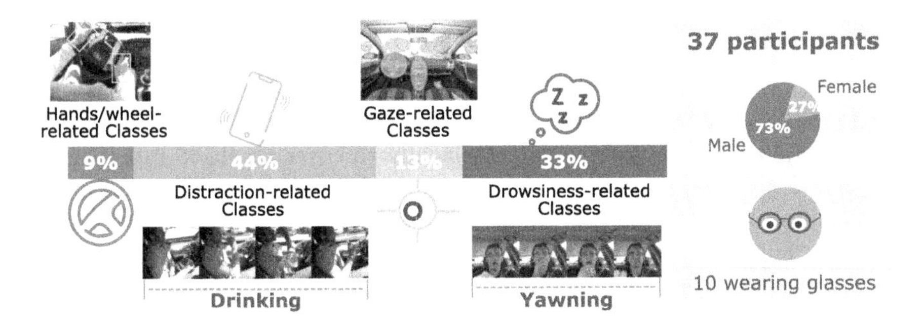

Fig. 1. Class distribution of the complete DMD dataset. Both geometrical and temporal-based annotation are included in the available DMD ground truth data.

The closest dataset to DMD in terms of volume and label richness is the recent Drive&Act [33]. However, there is major difference, as Drive&Act focuses on automated driving actions, where the driver is completely not engaged in the driving task, the participants can perform more diverse actions (e.g. working-on-laptop). In our understanding, this dataset is of interest for future SAE L4-5 automated cars, while DMD focuses on a wider domain of driving behaviours (not only actions) which are still non-solved challenges for the automotive industry for SAE L2-3 AD.

3 The Driver Monitoring Dataset (DMD)

The DMD is a multi-modal video dataset with multimedia material of different driver monitoring scenarios from 3 camera views. In this paper, one example use-case of the dataset is shown for a real-world application. We propose a robust and real-time driver behaviour recognition system using the multi-modal data available in DMD with a cost-efficient CPU-only platform as computing resource.

3.1 Dataset Specifications

The DMD was devised to address different scenarios in which driver monitoring must be essential in the context of automated vehicles SAE L2-3.

The dataset is composed of videos of drivers performing *distraction* actions, drivers in different states of *fatigue and drowsiness*, specific material for *gaze allocation* to interior regions, *head-pose estimation* and different driver's *hands' positions and interaction* with inside objects (i.e. steering wheel, bottle, cellphone, etc.). There was a participation of 37 volunteers for this experiment, the gender proportions are 73% and 27% for men and women, respectively, 10 wearing glasses (see Fig. 1). The age distribution of the participants was homogeneous in the range of 18 to 50 years. The participants were selected to assure novice and expert drivers were included in the recordings. Each participant signed an GDPR informed consent which allows the dataset to be publicly available for research purposes. Moreover, for certain groups of participants, the recording sessions were repeated, one day in the morning and another in the afternoon, to have variation in lighting. Recording sessions were carried out in different days to guarantee variety of weather conditions.

The DMD dataset was designed specifically to cover the lack of data for a fully operational DMS. Different scripted protocols were defined for each of the behaviour domains. However, no further instructions were given about how to perform the specific actions. Therefore, personal variability is introduced by each of the participants, contributing with a realistic component to the available data. In the **distraction recordings**, the drivers performed the most relevant actions which affects attention allocation namely: mobile phone use, use of infotainment, drinking, combing hair, reaching for an object, change gear and talking to the passenger. Regarding **drowsiness recordings**, drivers were asked to perform actions comprised by the most correlated signs of fatigue such as reduce eyelid aperture, microsleeps (i.e. small intervals of closed eyes), yawning and nodding. Moreover, during the **gaze-related recordings** the participants fixated their gaze to 9 predefined regions which covers all the surroundings of the vehicle. Similarly, for the **hand-wheel interaction recordings** the drivers held the steering wheel with different hands positions.

Annotation was done according to the recording type and could include: geometrical features (i.e. landmarks points and bounding boxes), temporal features (i.e. events and actions) or context. Each driver behaviour type has a diverse set of both geometric and temporal classes. Figure 1 depicts the full distribution of classes available in the DMD. This distribution for the different types of recordings was done using a total of 93 classes (temporal, geometric and context). In this paper we focus on a subset of the distraction scenario in which we annotate temporal actions used by the DL algorithm to identify driver actions. The open annotation format VCD[1] was chosen to describe the sequences.

3.2 Video Streams

The dataset was recorded with Intel Realsense D400 Series Depth Cameras which can capture Depth information, RGB and Infrared (IR) images synchronously. Two environments was considered: (i) outdoors in a real car and (ii) indoors in a driving simulator that immerses the person to close-to-real driving conditions. The simulator allows recording the actions or states that are riskier to perform in the car and to obtain the drivers' natural reactions. Such actions were also recorded in the car while being stopped.

The recordings were obtained with 3 cameras placed in the aforementioned environments (see Fig. 2). Two D415 cameras recorded the face area frontally and the hands area from the back; one D435 camera was used to capture the driver's body from the left side. The three cameras capture images at around 30 FPS. After a post-processing step, all camera streams were synchronised to a common frame rate. The resulting material consists of mp4 video files for each

[1] https://vcd.vicomtech.org/.

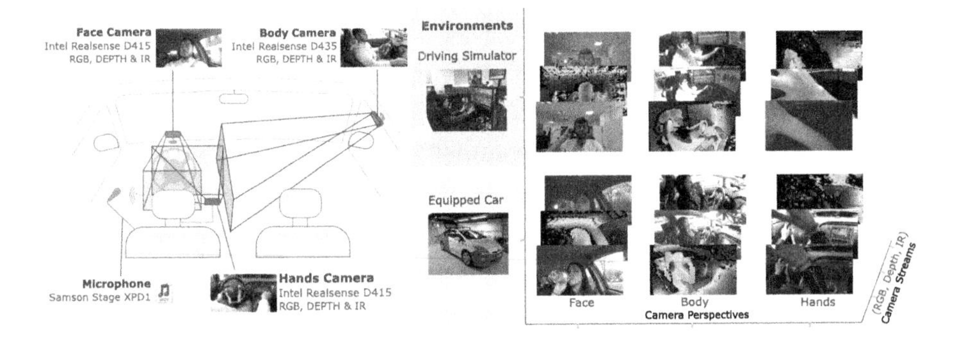

Fig. 2. DMD camera setup for both real car and simulator scenarios.

stream from the 3 cameras (9 mp4 files for each scene). The raw material reaches a volume of 26 TB (source ROS bags), while the mp4 files are slightly compressed with H.264, with target bitrate 15000 kb/s to ease downloading and processing.

In this paper, we also present a driver monitoring function using the data collected with the side-view camera and the sequences of the driving actions scenario only. For this purpose, initially, the side view data is further processed to short video clips, which enables real-time response from the monitoring system, in order to establish a larger scale multi-modal driver behaviour monitoring dataset named as *dBehaviourMD*[2].

3.3 Driver Behaviour Monitoring Dataset (dBehaviourMD)

The side-view recording is done with Intel RealSense D435, which provides RGB frames up to 1920×1080 pixels, and IR and depth frames up to 1280×720 pixels. The FoV of IR and depth are the same, and their pixels are aligned. However, since RGB camera has slightly narrower FoV and different camera intrinsics than IR, pixel-wise alignment of RGB frames with IR and depth frames requires the camera intrinsic parameters after calibration. Although the intrinsic parameters of the cameras are available from the hardware, in this paper, RGB frames are used directly without calibration and rectification for the experiments.

For efficiency purposes, all the frames are resized to 640×360 pixels for the *dBehaviourMD*. Each entry in a depth array d_m is measured in meters, and is converted to pixel value $d_p \in [0, 255]$ by:

$$d_p = \begin{cases} 255 \cdot \min(d_{\min}/d_m, 1) & \text{if } d_m > 0 \\ 0 & \text{otherwise} \end{cases}, \tag{1}$$

where $d_{\min} = 0.5\,\text{m}$, meaning $d_m < d_{\min}$ will be saturated to 255, otherwise the closer the brighter.

For real-time recognition, the machine has to monitor the driver through a sliding window with a fixed time span to recognise the action and give an instant

[2] Both *DMD* and *dBehaviourMD* will be publicly available for the research community.

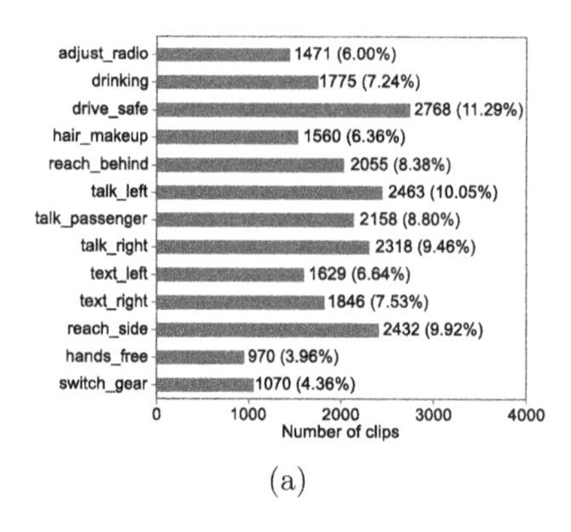

(a)

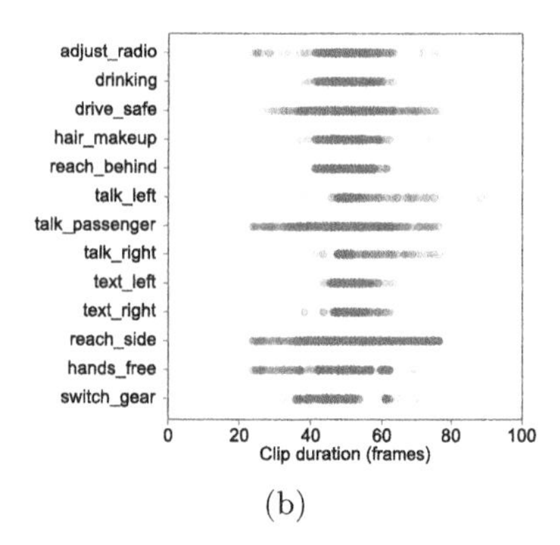

(b)

Fig. 3. *dBehaviourMD* class distribution (a) and clip duration distribution for each class (b) after clip splitting and balance sampling.

Table 2. Specifications of dBehaviourMD and existing driver behaviour datasets.

	AUC DD [1]	Drive&Act [33]	dBehaviourMD
Target actions	Driving behaviour	General behaviour	Driving behaviour
Real driving	No	No	Yes
Video-based	No	Yes	Yes
# Classes	10	34 (Activities)	13
Avg. video clips per class (Min./Max.)	N/A	303 (19/2797)	1886 (970/2768)

response. The time span should be as short as possible, so that the machine can be trained to react faster. In this paper, the time span is set to 50 frames (1.67 s) according to our observations on the actions in our dataset. Shorter clips are preserved as the action can be easily recognised during annotation. Longer clips are split into short clips so that the average duration of clips approaches this target length.

Next, sampling operations[3] have been applied to have a balanced dataset, which has the class distribution and the clip duration distribution for each class as shown in Fig. 3(a) and (b), respectively. It can be noticed that although the average clip duration is similar, there are some classes with relatively various duration than others, meaning those actions are temporally more diverse. In *dBehaviourMD*, the standard deviation of clip duration is kept relatively small, resulting in a dataset that is suitable for deployment with a fixed timing window. Finally, the training, validation, and testing clip sets are split with the ratio roughly 4:1:1, which corresponds to 24:7:6 drivers, respectively.

The comparison of dBehaviourMD with existing datasets for side-view monitoring is shown in Table 2. dBehaviourMD has data collected from more number

[3] Down/upsampling according to the number of samples in each action class.

of drivers in a real-driving scenario compared to existing datasets and it has the 10 classes of the StateFarm [48] and AUC DD [1] datasets, which are shown in Fig. 3, plus 3 classes *reaching side*, *hands free* and *switch gear* which appear quite often during real driving. The most recent multi-modal video dataset is Drive&Act [33], which has overall 83 classes annotated in a multi-hierarchical way, being only 34 of them semantically related to driving behaviour or driver action (initially, 12 high-level instructions are ordered to the drivers. Each of them consists of a fixed routine of several second-level actions. Next, 34 fine-grained activities are defined, including for example sitting still, drinking, etc.).

Being a multi-modal video dataset, which enables to capture temporal information efficiently, and having the most number of average video clips per class, the introduced dBehaviourMD dataset outperforms the existing datasets recorded for behaviour recognition purposes.

4 Real-Time Driver Behaviour Recognition

To develop a real-time system which monitor driver behaviour robustly and reliably, several requirements have to be fulfilled:

- a cost-effective CNN model running on video data, which provides higher accuracy benefiting from both spatial and temporal information;
- a video dataset recorded in a real-driving scenario which not only has RGB modality but also infrared (IR) and depth modalities to benefit from complementary information of different modalities; and
- an architecture which can accommodate more than one modality and fuse them on-the-fly to target a real-time deployment.

Behaviour recognition is achieved via spatio-temporal analysis on video data that could be done in several ways such as applying 2D CNNs on multiple video frames or 3D CNNs on video clips. Temporal segment network (TSN) [54,55] applies 2D CNNs for action recognition with sparse temporal sampling and uses optical flow for the temporal stream. Motion fused frames (MFF) [27] proposes a new data-level fusion strategy and has more capabilities to learn spatial-temporal features with less computational effort in the base model. Optical flow dimensions in MFF contain the frame-wise short-term movement, while long-term movement is kept across the segments. Both TSN and MFF use optical flow which is not very efficient for real-time deployment.

Apart from 2D-CNN models, 3D-CNNs have also been applied for action recognition [7,26,51,52,58]. 2D-CNN models usually do not take all the consecutive frames, and recent studies in 3D-CNN also question the necessity of 3D-CNN [58] and consecutive frames for general action recognition [52]. For a fair comparison, the authors of [52] compare the models with the identical ResNet building blocks which differ only in the temporal dimension, and they observe that the temporal information contributes not as much as the spatial one. [51] examines the accuracy-speed trade-off claiming the 3D model can bring 3% accuracy gain with the double computational effort compared with the 2D

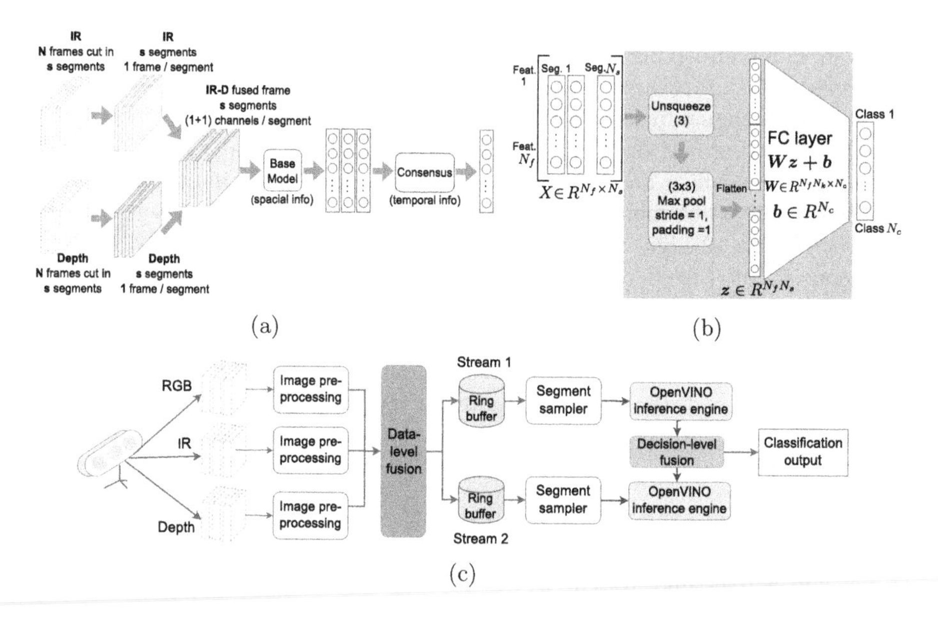

Fig. 4. Proposed approach. (a) Training process flowchart (with fusion of RGB, IR and Depth). (b) Consensus module to capture temporal correlation (c) Real-time inference pipeline using the optimised OpenVINO models as engine.

model on the mini-Kinetics-200 dataset. Although 3D-CNN models could extract more temporal information than 2D-CNN ones, the overhead computational cost is also significant, making it difficult for real-time deployment.

According to [21], the datasets used for transfer learning contribute more than the architectural difference and hence 2D-CNN models pre-trained on ImageNet [11] can still perform better or at least similarly than 3D-CNN models pre-trained on Kinetics [7].

Proposed Architecture for Real-Time Deployment: This work targets a real-world application hence our aim is to find the optimum architecture by analysing the use of different types of architectures for driver monitoring considering all high classification accuracy, fast reaction time, resource efficiency and architectures that can run on cost-efficient CPU-only platforms.

The architecture is built on top of the TSN [54] work, using its partial batch-norm, and data augmentation methods, while disabling horizontal flipping for the dBehaviourMD dataset. The batch size is chosen as 32 or 64 depending on the number of segments selected for the evaluations with 2D-CNN models. The stochastic gradient descent (SGD) is used as the optimiser with initial learning rate of 0.001, momentum 0.9 and weight decay 5×10^{-4}. The dropout rate before the final layer is set as 0.5. The learning rate is reduced by the learning rate decay factor (0.1) at each of the learning rate decay steps [15,30]. Training is finished after a maximum number of 40 epochs.

As shown in Fig. 4, modality fusion can be performed either at data level, decision level or both levels. Inspired by the data-level fusion of RGB and optical flow in MFF [27], in this paper, the fusion of RGB, IR, and depth is evaluated.

We have also explored different consensus modules. We apply max pooling for temporal dimension using the extracted features from each segment, which has the size of $X \in R^{N_f N_s}$ where N_f and N_s represent number of features and number of segments, respectively (Fig. 4(b)). The output is flattened and then passes through a multi-layer perceptron (MLP). Experimental analysis shows that our consensus module outperforms the existing consensus modules such as averaging and MLPs applied in [54] and [27], respectively.

In addition to the deterministic decision-level fusion such as taking average of the confidence scores of each classifier with different modality, this study also explores methods with prior knowledge of how each classifier performs on the validation set, namely the Bayesian weighting and Dempster–Shafer theory (DST) [44,59].

5 Experiments and Results

In this section, we initially evaluate the recognition performances with both 2D-CNN models and 3D-CNNs using different types of feature extractors as base model in order to compare 2D and 3D CNNs and also to understand the impact of base model on recognition performances.

Next, we apply multi-modal fusion in different levels such as data level and decision level. Finally, we provide runtime performances on CPU-only platform with the use of OpenVino for different modalities.

5.1 Comparison of 2D CNN Based Models and 3D CNNs

In order to analyse the impact of different architectures (i.e. base models in Fig. 4) on recognition performances, with each of RGB, IR and depth modalities, we selected one dense and two resource efficient architectures for both 2D and 3D CNN based models, which are *Inception-v3* [49,50] and *3D-ResNeXt* [21] as dense architectures and 2D and 3D versions of *MobileNet-v2* [26,43] and *ShuffleNet-v2* [26,31] as resource efficient architectures for our evaluations in Table 3.

ResNet [22] introduces residual learning and skip paths to achieve deeper architecture. Further improvements push CNN-based models to achieve the accuracy near or beyond human ability such as Inception series with inception modules which are ensembles of kernels of various sizes and bottleneck layers. Apart from targeting high classification accuracy, several works propose cost-effective models which can achieve slightly lower accuracy with greatly reduced computational effort. MobileNet-v2 and ShuffleNet-v2 are the two series of widely applied models in mobile platforms for real-time performance.

In this work, we also evaluated the impact of the dataset used for transfer learning using the same models for both 2D and 3D CNN based analysis (e.g.

Table 3. Behaviour recognition performances with 2D and 3D CNN based models.

Modality	Architecture	Base Model	Segments	Top-1 Acc. (%)
RGB	2D CNN	Inception-v3	4/8	93.2/91.8
		MobileNet-v2 1.0x	4/8	88.7/89.9
		ShuffleNet-v2 1.0x	4/8	81.4/84.8
	3D CNN	3D-ResNeXt	–	91.6
		3D-MobileNet-v2 1.0x	–	90.3
		3D-ShuffleNet-v2 1.0x	–	88.5
IR	2D CNN	Inception-v3	4/8	92/91.6
		MobileNet-v2 1.0x	4/8	90.1/91.1
		ShuffleNet-v2 1.0x	4/8	82.6/85.4
	3D CNN	3D-ResNeXt	–	92.9
		3D-MobileNet-v2 1.0x	–	92.7
		3D-ShuffleNet-v2 1.0x	–	90.6
Depth	2D CNN	Inception-v3	4/8	90.1/91.9
		MobileNet-v2 1.0x	4/8	91.6/91.7
		ShuffleNet-v2 1.0x	4/8	83.4/88.1
	3D CNN	3D-ResNeXt	–	92.8
		3D-MobileNet-v2 1.0x	–	92.4
		3D-ShuffleNet-v2 1.0x	–	89

MobileNet-v2 and 3d-MobileNet-v2 for 2D and 3D based analysis, respectively) for fair comparison. Here, we used ImageNet dataset and Kinetics-600 dataset for the transfer learning of 2D and corresponding 3D CNN models, respectively.

For the training of 3D-CNNs, SGD with standard categorical cross entropy loss is applied and largest fitting batch size is selected for each CNN model. The momentum, dampening and weight decay are set to 0.9, 0.9 and 1×10^{-3}, respectively. For the training of dBehaviourMD, we have used the pre-trained models on Kinetics-600 provided by [26]. For fine-tuning, we start with a learning rate of 0.1 and reduced it two times after 20th and 35th epochs with a factor of 10^{-1}. Optimisation is completed after 10 more epochs. A dropout layer is applied before the final layer of the networks with a ratio of 0.5.

For temporal augmentation, input clips are selected from a random temporal position in the video clip. For the videos containing smaller number of frames than the input size, loop padding is applied. Since the average number of frames in our dataset is around 50 frames and the input to the networks is selected as 16-frame clips, in order to capture the most content of the action in the input clip, we have applied downsampling of 2, which enables to capture temporal information for 32-frame extent. Similar types of spatial augmentation is applied for both 2D and 3D CNN based models. The spatial resolution of the input is selected as 224 and 112 for 2D and 3D CNN networks, respectively.

Table 4. Comparison of the proposed consensus module with averaging and MLP based consensus modules on both validation and test sets.

Base Model	Consensus	N_f	Seg.	Top-1 Acc. (%)							
				Validation Set				Test Set			
				RGB	Depth	IR	IRD	RGB	Depth	IR	IRD
MobileNet-V2	Average	11	4	89.6	92.0	91.1	–	89.4	88.7	89.4	–
	MLP	64	4	90.7	91.4	90.2	–	91.9	90.6	90.6	–
	MaxP-MLP	64	4	**91.5**	**93.5**	**92.6**	**94.3**	**92**	**91.6**	**91.5**	**91.9**
	Average	11	8	90.4	93	91.6	–	88.9	92.3	89.5	–
	MLP	64	8	91.2	93	91.1	–	91.0	**93.2**	91.6	–
	MaxP-MLP	64	8	**92.0**	**94.1**	**92.6**	**95**	**91.6**	92.7	**91.8**	**93.9**

Table 3 shows our results with the approach in Fig. 4 using averaging as consensus module, which simply takes the average of the scores computed for each selected frame of an action video. According to the results in Table 3:

- The overall accuracy can approach around 90%, even with light-weight feature extractor as MobileNet-v2 for RGB modality.
- The benefit of using high-performance base models is more obvious for RGB.
- Depth modality performs better than RGB and IR modality. For the depth and IR modality, light-weight base model MobileneNet-v2 is enough to reach its accuracy limit.
- The number of segments $N_s = 4$ already reaches the accuracy bound for many models compared to $N_s = 8$ case.
- 2D CNN based models can reach almost the similar accuracy of their 3D counterparts for MobileNet-v2 and dense architectures. This result also shows that 2D models pre-trained on a large-scale image dataset (ImageNet dataset) performs as good as 3D models pre-trained on video dataset (Kinetics-600 dataset) as stated in [21] as well.
- IR performs slightly better than RGB. Since the dataset was recorded during the daytime, the major difference between RGB and IR is colour information. This might be the reason for IR models to easier learn the patterns such as shapes and poses compared to RGB models.

5.2 Impact of Different Consensus Modules on Performance

For temporal understanding of action videos, we have used several consensus modules and analysed their impact on recognition performances. Table 4 shows the results we obtained with averaging and MLP based consensus modules in addition to our proposed consensus module (MaxP-MLP in Table 4).[4] With the MobileNet-v2 as base model, the results show that the MaxP-MLP outperforms other consensus modules almost in all cases. The applied pooling operation in MaxP-MLP increases the accuracy considerably compared to other consensus

[4] N_f is set to 11 for averaging-based consensus, excluding *hands free* and *switch gear* actions. N_f is set to 64 for both MLP and MaxP-MLP in all experiments.

Table 5. Fusion of modalities in different levels. Stream1 (IRD) + Stream2 (RGB). Base model is MobileNet-v2. MaxP-MLP is the consensus module.

Fusion Level	Fusion Type	Modalities	Segments	Top-1 Acc (%)
Data level	–	IRD	4	91.9
–	–	RGB	4	92
Decision level	Averaging fusion	RGB + IRD	4	93.7
	Bayesian weighting	RGB + IRD	4	93.7
	DST	RGB + IRD	4	93.7

modules since this process enables to emphasise the significant features in the spatio-temporal features group (X in Fig. 4b) representing an action video.

The evaluations are done on both validation and test sets for both 4 and 8 segment analysis. $N_s = 4$ would already reach its performance upper bound. Similar to the previous outcome, depth modality provides better accuracy compared to RGB and IR. Data level fusion of IR and Depth modalities increases the accuracy on both validation and test sets with both 4 and 8 segment analysis.

5.3 Multi-modal Fusion

Fusion of CNN based classifiers can be done at (i) data-level: two types of data are fused before fed into the CNN model [27]; and at (ii) decision-level: results from two classifiers are fused together.

Table 5 shows the results for the applied data level and decision level fusion. Since IR and depth modalities have pixel-wise correspondence, these modalities are fused at data level which provides an enhanced accuracy compared to their mono-modal analysis. Applying decision level fusion for RGB (Stream 1) and IRD (Stream 2) modalities using the deterministic approach of averaging and the statistics methods based on the prior knowledge of the classifiers such as bayesian weighting and DST, we observed considerable increase in the performances compared to the results of single streams. However, the results also show that statistics methods do not help more for our evaluations.

5.4 Runtime Performance Analysis

The proposed architecture is implemented as a real-time driver behaviour monitoring system using OpenVINO (version 2020.4.255) [23] and selecting MobileNet-v2 as base model for 4-segment analysis. Our results show that MobileNet-v2 provides comparable accuracy with dense architectures with the advantage of having much less number of trainable parameters, achieving real-time deployment, high accuracy, resource efficiency and fast reaction time. The selected models are converted to OpenVINO IR and tested with the inference pipeline on an Intel Core-i9 7940X machine. The single modalities RGB, IR and depth can run up to 8.1 ms, 7.3 ms and 7.3 ms per action video, respectively. The model which is based on data level fusion of IR and Depth modalities can run up to 10.1 ms and two streams (RGB and IRD) together can still reach 18 ms, enabling the whole inference pipeline to operate at more than 30 FPS.

6 Conclusions

In this paper we present the Driver Monitoring Dataset (DMD) as an extensive, varied and comprehensive dataset created for development of Driver Monitoring Systems (DMS), specially designed for training and validating DMS in the context of automated driving SAE L2-3. The dataset has been devised to be multi-purpose, including footage of real driving scenes, with scenes containing driving actions related to driver's distraction, fatigue and behaviour. The dataset is multi-modal as it contains 3 cameras and 3 streams per camera (RGB, Depth, IR), and labels covering a wide range of visual features (e.g. time-lapse actions, body and head-pose, blinking patterns, hands-wheel interactions, etc.).

This paper also reports our work using DMD to extract the *dBehaviourMD* training set, crafted to train a DL model proposed for a real-world application for real-time driver behaviour recognition. The achieved performance of the trained model yields enhanced accuracy for different multi-modal fusion approaches while keeping real-time operation on CPU-only platforms.

Acknowledgements. This work has received funding from the European Union's H2020 research and innovation programme (grant agreement no 690772, project VI-DAS). We would like to thank all the volunteers who participated in the recording process for sharing their time to create the DMD dataset.

References

1. Abouelnaga, Y., Eraqi, H.M., Moustafa, M.N.: Real-time distracted driver posture classification. In: 32nd Conference on Neural Information Processing Systems (NIPS 2018) (2018)
2. Abraham, H., Reimer, B., Mehler, B.: Advanced driver assistance systems (ADAS): a consideration of driver perceptions on training, usage & implementation. In: Proceedings of the Human Factors and Ergonomics Society, pp. 1954–1958 (2017)
3. Aghaei, A.S., et al.: Smart driver monitoring: when signal processing meets human factors: in the driver's seat. IEEE Signal Process. Mag. **33**(6), 35–48 (2016)
4. Ahlstrom, C., Kircher, K., Kircher, A.: A gaze-based driver distraction warning system and its effect on visual behavior. IEEE Trans. Intell. Transp. Syst. **14**(2), 965–973 (2013)
5. Baccour, M.H., Driewer, F., Kasneci, E., Rosenstiel, W.: Camera-based eye blink detection algorithm for assessing driver drowsiness. In: IEEE Intelligent Vehicles Symposium, pp. 866–872 (2019)

6. Borghi, G., Venturelli, M., Vezzani, R., Cucchiara, R.: POSEidon: face-from-depth for driver pose estimation. In: IEEE Conference on Computer Vision and Pattern Recognition (CVPR), pp. 5494–5503 (2017)
7. Carreira, J., Zisserman, A.: Quo vadis, action recognition? A new model and the kinetics dataset. In: IEEE Conference on Computer Vision and Pattern Recognition (CVPR) (2017)
8. Chowdhury, A., Shankaran, R., Kavakli, M., Haque, M.M.: Sensor applications and physiological features in drivers' drowsiness detection: a review. IEEE Sens. J. **18**(8), 3055–3067 (2018)
9. Craye, C., Karray, F.: Driver distraction detection and recognition using RGB-D sensor. CoRR (2015)
10. Das, N., Ohn-Bar, E., Trivedi, M.M.: On performance evaluation of driver hand detection algorithms: challenges, dataset, and metrics. In: IEEE International Conference on Intelligent Transportation Systems, pp. 2953–2958 (2015)
11. Deng, J., Dong, W., Socher, R., Li, L., Li, K., Li, F.F.: ImageNet: a large-scale hierarchical image database. In: IEEE Conference on Computer Vision and Pattern Recognition (CVPR) (2009)
12. Deo, N., Trivedi, M.M.: Looking at the driver/rider in autonomous vehicles to predict take-over readiness. CoRR (2018)
13. Diaz-Chito, K., Hernández-Sabaté, A., López, A.M.: A reduced feature set for driver head pose estimation. Appl. Soft Comput. J. **45**, 98–107 (2016)
14. Dingus, T.A., et al.: Driver crash risk factors and prevalence evaluation using naturalistic driving data. Proc. Natl. Acad. Sci. U.S.A. **113**, 2636–2641 (2016)
15. Eraqi, H.M., Abouelnaga, Y., Saad, M.H., Moustafa, M.N.: Driver distraction identification with an ensemble of convolutional neural networks. J. Adv. Transp. **2019** (2019)
16. European Commission: Roadmap to a single European transport area-towards a competitive and resource efficient transport system. Technical report, European Commission (2011)
17. Fang, J., Yan, D., Qiao, J., Xue, J.: DADA: a large-scale benchmark and model for driver attention prediction in accidental scenarios (2019)
18. Fernández, A., Usamentiaga, R., Carús, J.L., Casado, R.: Driver distraction using visual-based sensors and algorithms. Sensors **16**(11), 1–44 (2016)
19. Fridman, L.: Human-centered autonomous vehicle systems: principles of effective shared autonomy. CoRR (2018)
20. Goenetxea, J., Unzueta, L., Elordi, U., Ortega, J.D., Otaegui, O.: Efficient monocular point-of-gaze estimation on multiple screens and 3D face tracking for driver behaviour analysis. In: 6th International Conference on Driver Distraction and Inattention, pp. 1–8 (2018)
21. Hara, K., Kataoka, H., Satoh, Y.: Can spatiotemporal 3D CNNs retrace the history of 2D CNNs and ImageNet? In: IEEE Conference on Computer Vision and Pattern Recognition (CVPR) (2018)
22. He, K., Zhang, X., Ren, S., Sun, J.: Deep residual learning for image recognition. In: IEEE Conference on Computer Vision and Pattern Recognition (CVPR) (2016)
23. Intel: OpenVINO - develop multiplatform computer vision solutions. https://software.intel.com/en-us/openvino-toolkit
24. Jacobé de Naurois, C., Bourdin, C., Stratulat, A., Diaz, E., Vercher, J.L.: Detection and prediction of driver drowsiness using artificial neural network models. Accid. Anal. Prev. **126**, 95–104 (2019)
25. Janai, J., Güney, F., Behl, A., Geiger, A.: Computer vision for autonomous vehicles: problems, datasets and state-of-the-art. J. Photogram. Remote Sens. (2017)

26. Köpüklü, O., Köse, N., Gunduz, A., Rigoll, G.: Resource efficient 3D convolutional neural networks. CoRR (2019)
27. Köpüklü, O., Köse, N., Rigoll, G.: Motion fused frames: data level fusion strategy for hand gesture recognition. In: IEEE Conference on Computer Vision and Pattern Recognition (CVPR) Workshops (2018)
28. Köse, N., Köpüklü, O., Unnervik, A., Rigoll, G.: Real-time driver state monitoring using a CNN based spatio-temporal approach. In: IEEE Intelligent Transportation Systems Conference (ITSC) (2019)
29. Le, T.H.N., Quach, K.G., Zhu, C., Duong, C.N., Luu, K., Savvides, M.: Robust hand detection and classification in vehicles and in the wild. In: IEEE Conference on Computer Vision and Pattern Recognition (CVPR) Workshops, pp. 1203–1210 (2017)
30. Le, T.H.N., Zheng, Y., Zhu, C., Luu, K., Savvides, M.: Multiple scale faster-RCNN approach to driver's cell-phone usage and hands on steering wheel detection. In: IEEE Conference on Computer Vision and Pattern Recognition (CVPR) Workshops, pp. 46–53 (2016)
31. Ma, N., Zhang, X., Zheng, H.-T., Sun, J.: ShuffleNet V2: practical guidelines for efficient CNN architecture design. In: Ferrari, V., Hebert, M., Sminchisescu, C., Weiss, Y. (eds.) Computer Vision – ECCV 2018. LNCS, vol. 11218, pp. 122–138. Springer, Cham (2018). https://doi.org/10.1007/978-3-030-01264-9_8
32. Mandal, B., Li, L., Wang, G.S., Lin, J.: Towards detection of bus driver fatigue based on robust visual analysis of eye state. IEEE Trans. Intell. Transp. Syst. **18**(3), 545–557 (2017)
33. Martin, M., et al.: Drive&act: a multi-modal dataset for fine-grained driver behavior recognition in autonomous vehicles. In: The IEEE International Conference on Computer Vision (ICCV), pp. 2801–2810 (2019)
34. Massoz, Q., Langohr, T., François, C., Verly, J.G.: The ULg multimodality drowsiness database (called DROZY) and examples of use. In: IEEE Winter Conference on Applications of Computer Vision (WACV), pp. 1–7 (2016)
35. McDonald, A.D., Ferris, T.K., Wiener, T.A.: Classification of driver distraction: a comprehensive analysis of feature generation, machine learning, and input measures. Hum. Factors J. Hum. Factors Ergon. Soc. **62**, 1019–1035 (2019)
36. Mioch, T., Kroon, L., Neerincx, M.A.: Driver readiness model for regulating the transfer from automation to human control. In: International Conference on Intelligent User Interfaces, IUI, pp. 205–213 (2017)
37. Ohn-Bar, E., Trivedi, M.M.: The power is in your hands: 3D analysis of hand gestures in naturalistic video. In: IEEE Computer Society Conference on Computer Vision and Pattern Recognition (CVPR) Workshops, pp. 912–917 (2013)
38. Ortega, J.D., Nieto, M., Salgado, L., Otaegui, O.: User-adaptive eyelid aperture estimation for blink detection in driver monitoring systems. In: Proceedings of the 6th International Conference on Vehicle Technology and Intelligent Transport Systems - Volume 1: VEHITS, pp. 342–352. INSTICC, SciTePress (2020)
39. Palazzi, A., Abati, D., Calderara, S., Solera, F., Cucchiara, R.: Predicting the driver's focus of attention: the DR(eye)VE project. IEEE Trans. Pattern Anal. Mach. Intell. **41**(7), 1720–1733 (2018)
40. Rasouli, A., Tsotsos, J.K.: Autonomous vehicles that interact with pedestrians: a survey of theory and practice. IEEE Trans. Intell. Transp. Syst. **21**, 900–918 (2019)
41. Roth, M., Gavrila, D.M.: DD-pose - a large-scale driver head pose benchmark. In: IEEE Intelligent Vehicles Symposium, pp. 927–934 (2019)
42. SAE International: Taxonomy and definitions for terms related to driving automation systems for on-road motor vehicles. Technical report, SAE International (2018)

43. Sandler, M., Howard, A., Zhu, M., Zhmoginov, A., Chen, L.C.: MobileNetV2: inverted residuals and linear bottlenecks. In: IEEE Conference on Computer Vision and Pattern Recognition (CVPR) (2018)
44. Sarinnapakorn, K., Kubat, M.: Combining subclassifiers in text categorization: a DST-based solution and a case study. IEEE Trans. Knowl. Data Eng. **19**(12), 1638–1651 (2007)
45. Schwarz, A., Haurilet, M., Martinez, M., Stiefelhagen, R.: Driveahead – a large-scale driver head pose dataset. In: IEEE Conference on Computer Vision and Pattern Recognition (CVPR) Workshops, pp. 1165–1174 (2017)
46. Sikander, G., Anwar, S.: Driver fatigue detection systems: a review. IEEE Trans. Intell. Transp. Syst. **20**(6), 2339–2352 (2019)
47. Smith, B., Yin, Q., Feiner, S., Nayar, S.: Gaze locking: passive eye contact detection for human? Object interaction. In: ACM Symposium on User Interface Software and Technology (UIST), pp. 271–280, October 2013
48. StateFarm: State Farm Distracted Driver Detection (2016). https://www.kaggle.com/c/state-farm-distracted-driver-detection. Accessed 04 Mar 2020
49. Szegedy, C., Ioffe, S., Vanhoucke, V., Alemi, A.A.: Inception-v4, Inception-ResNet and the impact of residual connections on learning. In: AAAI Conference on Artificial Intelligence, pp. 3547–3554 (2017)
50. Szegedy, C., Vanhoucke, V., Ioffe, S., Shlens, J., Wojna, Z.: Rethinking the inception architecture for computer vision. In: IEEE Conference on Computer Vision and Pattern Recognition (CVPR) (2016)
51. Tran, D., Bourdev, L., Fergus, R., Torresani, L., Paluri, M.: Learning spatiotemporal features with 3D convolutional networks. In: IEEE International Conference on Computer Vision (ICCV) (2015)
52. Tran, D., Wang, H., Torresani, L., Ray, J., LeCun, Y., Paluri, M.: A closer look at spatiotemporal convolutions for action recognition. In: IEEE Conference on Computer Vision and Pattern Recognition (CVPR) (2018)
53. Trutschel, U., Sirois, B., Sommer, D., Golz, M., Edwards, D.: PERCLOS: an alertness measure of the past. In: 6th International Driving Symposium on Human Factors in Driver Assessment, Training and Vehicle Design, pp. 172–179 (2011)
54. Wang, L., et al.: Temporal segment networks: towards good practices for deep action recognition. In: Leibe, B., Matas, J., Sebe, N., Welling, M. (eds.) ECCV 2016. LNCS, vol. 9912, pp. 20–36. Springer, Cham (2016). https://doi.org/10.1007/978-3-319-46484-8_2
55. Wang, L., et al.: Temporal segment networks for action recognition in videos. IEEE Trans. Pattern Anal. Mach. Intell. **41**(11), 2740–2755 (2019)
56. Weng, C.-H., Lai, Y.-H., Lai, S.-H.: Driver drowsiness detection via a hierarchical temporal deep belief network. In: Chen, C.-S., Lu, J., Ma, K.-K. (eds.) ACCV 2016. LNCS, vol. 10118, pp. 117–133. Springer, Cham (2017). https://doi.org/10.1007/978-3-319-54526-4_9
57. Xia, Y., Zhang, D., Pozdnukhov, A., Nakayama, K., Zipser, K., Whitney, D.: Training a network to attend like human drivers saves it from common but misleading loss functions, pp. 1–14. arXiv: 1711.0 (2017)
58. Xie, S., Sun, C., Huang, J., Tu, Z., Murphy, K.: Rethinking spatiotemporal feature learning: speed-accuracy trade-offs in video classification. In: Ferrari, V., Hebert, M., Sminchisescu, C., Weiss, Y. (eds.) ECCV 2018. LNCS, vol. 11219, pp. 318–335. Springer, Cham (2018). https://doi.org/10.1007/978-3-030-01267-0_19

59. Xu, L., Krzyzak, A., Suen, C.Y.: Methods of combining multiple classifiers and their applications to handwriting recognition. IEEE Trans. Syst. Man Cybern. **22**(3), 418–435 (1992)
60. Zhang, X., Sugano, Y., Fritz, M., Bulling, A.: MPIIGaze: real-world dataset and deep appearance-based gaze estimation. CoRR (2017)

Exploiting Scene-Specific Features for Object Goal Navigation

Tommaso Campari[1,2(✉)], Paolo Eccher[1], Luciano Serafini[2], and Lamberto Ballan[1]

[1] Department of Mathematics "Tullio Levi-Civita", University of Padova, Padua, Italy
tommaso.campari@phd.unipd.it
[2] Fondazione Bruno Kessler, Trento, Italy

Abstract. Can the intrinsic relation between an object and the room in which it is usually located help agents in the Visual Navigation Task? We study this question in the context of Object Navigation, a problem in which an agent has to reach an object of a specific class while moving in a complex domestic environment. In this paper, we introduce a new reduced dataset that speeds up the training of navigation models, a notoriously complex task. Our proposed dataset permits the training of models that do not exploit online-built maps in reasonable times even without the use of huge computational resources. Therefore, this reduced dataset guarantees a significant benchmark and it can be used to identify promising models that could be then tried on bigger and more challenging datasets. Subsequently, we propose the SMTSC model, an attention-based model capable of exploiting the correlation between scenes and objects contained in them, highlighting quantitatively how the idea is correct.

Keywords: Visual Navigation · ObjectGoal Navigation · Reinforcement Learning

1 Introduction

Visual Navigation is a trending topic in the Computer Vision research community. This growth in interest is undoubtedly due to the important practical implications that the development of agent capable of moving in complex environments can have on our society. For example, in an ever closer future we will be able to ask robotic assistants to perform the most disparate tasks in our homes. Before we can ask a robot to take something out of the refrigerator, however, we need to make sure that it is able to find the refrigerator and get to it while avoiding the complex tangle of obstacles that a domestic environment can contain. For this reason, this work focuses on the Object Navigation task, defined in [1] as the search for objects belonging to a specific class by a robotic agent. For humans, this is a very simple task whatever the object to be

A. Bartoli and A. Fusiello (Eds.): ECCV 2020 Workshops, LNCS 12538, pp. 406–421, 2020.
https://doi.org/10.1007/978-3-030-66823-5_24

found is. A human can build a mental link between the object and the room where it is more likely to be found. In this way a human is able to simplify the problem by first searching for the room and then for the required object inside the room. For example, just think of having to look for a sink, unconsciously we will first search for the kitchen or bathroom, and then we will find a sink in them. To implement this intuition we have developed an attention-based [28] policy in which we exploit a joint-representation that integrates inside it visual informations extracted with a scene classifier and encoding of the semantic goal to be searched. This representation allows us to have a significant increase in performance compared to other models taken into consideration.

Several works tried to tackle Navigation through Learning models [6,7]. In particular, [7] leveraged depth images to construct in an online fashion semantic maps of the environment. From these maps they tried to maximize the exploration of the scene. To do this, they placed intermediate subgoals in unexplored areas of the map that the agent was encouraged to reach through planning.

This type of approach inevitably tends to lengthen the agent's paths, at least until the object sought is clearly visible, since wanting to maximize exploration involves a significant amount of moves by the agent. For the simpler PointGoal Navigation [1] task, proposed in the "Habitat Challenge 2019"[1], it was possible to observe how a simple model [29] based on LSTM was able to perform better than competitors based on more complex architectures that exploit maps creation [8,11]. This was made possible by the DD-PPO algorithm [29], a distributed version of the PPO [25] Reinforcement Learning algorithm capable of parallelizing learning in a massive way. In fact, they used 64 NVIDIA V100 GPUs for 3 consecutive days of training. In other words, the model was trained for about 180 days in a single GPU setting. Not all researchers, however, can have access to those massive hardware resources. For this reason, we have generated a reduced version of the dataset produced for the "Habitat Challenge 2020"[2] for the Object Navigation task that would allow the training of Deep Reinforcement Learning models in a few hours. In this way, even using few computational resources, complex models can be trained which on the original dataset would take days.

Furthermore, in [10] it was pointed out how in the Navigation tasks the use of Recurrent Neural Networks is not recommended. These structures usually have the issue of considering as more important the recent past and at the same time gradually forgetting the remote past. On the contrary, by exploiting the principle of attention, described in [28], it is possible to record all the past observations in a memory from which to extract information on every single step that the agent has undertaken in the past, improving that highly penalizing intrinsic aspect in the behavior of the Recurrent Neural Networks.

Summarizing, our contribution is twofold:

- We propose a new reduced dataset for the Object Navigation task extracted from the one proposed in the "Habitat Challenge 2020", on which it is possible

[1] https://aihabitat.org/challenge/2019/.
[2] https://aihabitat.org/challenge/2020/.

to test algorithms that would require a lot of resources for training and that maintain as far as possible the main characteristics that the first had;
- We propose the SMTSC model, an attention-based policy for Object Navigation that is able to exploit, starting from RGB images only, the idea mentioned above, namely that there exist a correlation that binds objects to specific rooms. This intuition improves performance, as demonstrated by the results obtained on a preliminary study performed using the aforementioned dataset.

2 Related Works

Visual Navigation is an increasingly central topic within Computer Vision. However, Visual Navigation involves several different sub-problems and, in this section, we will summarize the most relevant related works to *Simulators and 3D Datasets for Visual Navigation* and to other different research areas connected to *Visual Navigation*.

2.1 Simulators and 3D Datasets for Visual Navigation

In recent years a large number of different simulators have been developed. GibsonEnv [32,33] and AI2Thor [14] both allow to simulate multi-agent situations and to interact with objects, for example lifting them, pushing them, etc. Matterport3DSimulator [2] can provide the agent with photorealistic images extracted from Matterport3d [5] and is mainly used for Room2Room Navigation problem. HabitatAI [23], instead, provides support to 3D datasets such as Gibson, Matterport3D and Replica [27].

2.2 Visual Navigation

Also thanks to the new possibilities offered by these simulators today there are numerous tasks available, as pointed out in [1]. Common Navigation tasks are mainly divided into two categories, namely those that require active exploration of the environment and those that, on the other hand, provide tools that can signal, for example via GPS sensors, the direction to be taken to reach the required goal.

In Classical Navigation, there are numerous approaches that perform path planning on explicit maps [4,13,16].

More recently, however, approaches based on Reinforcement Learning have been presented through policies based on Recurrent Neural Networks [10,15,17,20,22]. Mirowski et al. [17] define an approach that jointly learns the goal-driven Reinforcement Learning problem with auxiliary depth prediction and loop closure classification tasks by exploiting the A3C algorithm [18]. Mousavian et al. [20] propose a Deep Reinforcement Learning framework that uses an LSTM-based policy for Semantic Target Driven Navigation. But LSTMs when they have to analyze very long data sequences tend to focus more on the most recent observations, giving less importance to the first ones that have been seen. On the contrary, Fang et al. [10] propose Scene Memory Transformer, a policy based on attention [28] that is able to exploit even the least recent steps performed by the agent. In this case, the training of the policy is performed through the Deep Q-Learning algorithm [19]. Starting from the work done in [35], Sax et al. [24] show that using Mid-Level Vision results in policies that learn faster and generalize better when compared to learning from scratch. The Mid-Level model achieved high results in the PointGoal Navigation task. In [29], a scalable Reinforcement Learning algorithm on multiple GPUs capable of solving the PointGoal Navigation task almost perfectly has been presented. This solution, in particular, shows how Visual Navigation is a really complex task that requires an impressive amount of resources. In fact, their training was conducted on 64 GPUs for 3 days. Unfortunately, these resources are not within the reach of the whole scientific community and training similar models remain almost prohibitive for most researchers.

For the Target Driven Navigation some works have recently been presented, such as [30,31,34]. Wu et al. [31] construct a probabilistic graphical model over the semantic information to explore structural similarities between the environment. Yang et al. [34], instead, propose a Deep Reinforcement Learning model that exploits the relationships between objects, encoded through a Graph Convolutional Network, to incorporate semantic priors within the policy. Chaplot et al. [7] propose a model for the ObjectGoal Navigation that constructs, during the exploration, a map with the semantic information of the scene extracted through a semantic segmentation model; from the generated map a long-term goal is selected to maximize exploration, through a policy trained with Reinforcement Learning, when the searched object is visible it is set as a new long-term goal. The actions of the agent are selected through the use of the FastMarching algorithm [26].

3 Dataset

Previous works [29] have been able to achieve excellent results on the Point Navigation task [1], a simpler Navigation problem that doesn't require Semantic capabilities, while using an architecture that didn't include complex components such as occupancy maps [9]. However, they leveraged massive parallelism using hardware resources that are inaccessible to most institutions. We investigate on the possibility of solving the same problem in a reduced dataset with a minimal set of computational and time resources. For this reason we decided to concentrate on a subset of the Matterport3D Dataset [5]. We argue that our choice of such subset still offers significant results as its statistical indicators are similar to the one of original Matterport3D.

The extraction of the subset was done restricting the problem to 5 out of 21 objects, choosing `Chair`, `Cushion`, `Table`, `Cabinet`, `Sink`. These objects are among the most frequent in the original set as is shown in Fig. 1. We decided to include the `Sink` object as it is a characterising element of the bathroom. This

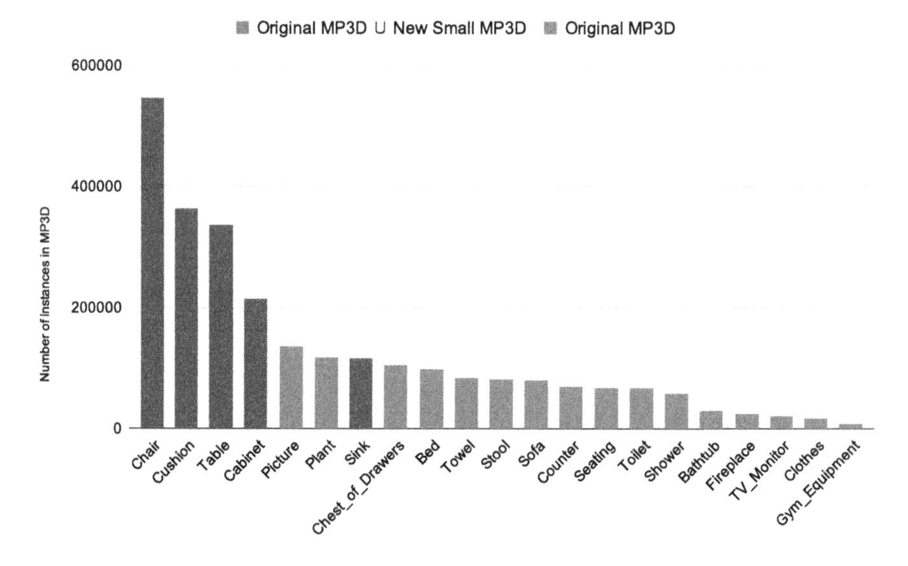

Fig. 1. Distribution of objects in the Matterport3D Dataset.

increase the diversity of the domestic environments represented by our proposed Dataset.

Related to the distributions of objects, one of the main issue in the evaluation of Object Navigation agents using Matterport3D, is the Long Tail Distribution that limits the number of instances of infrequent objects that are seen during training. This, combined with metrics that ignore the precision on single classes, may undermine the development of agents with truly semantic capabilities. Furthermore, we decided to extract our new dataset - from here we will refer to it as *Small MP3D* - using 6 out of 56 scenes of the official Training split. These scenes are `r47D5H71a5s`, `i5noydFURQK`, `ZMojNkEp431`, `jh4fc5c5qoQ`, `HxpKQynjfin` and `GdvgFV5R1Z5`. For each object in each scene we extracted 100 episode for the training split and 20 for both validation and test. In total, Small MP3D possesses 3000 training episodes and 600 episodes for both validation and test. To ensure the ability to generalize to unseen scenes, we decided to generate also an Unseen Test and Validation Set, extracted from the `D7N2EKCX4Sj` and `aayBHfsNo7d` scenes, each with 100 episodes (10 episodes for each object class in each scene).

The characteristics of the new dataset are shown in Table 1. We report the average values of Euclidean and Geodesic Distance as well as the number of Steps required to complete the episodes. Overall, the complexity of the training split of the proposed reduced dataset is lesser than the original data as both Geodesic distance and the required Number of Steps are lower. However, the complexity of the Unseen Test split is significantly higher, with 30% more required steps on average. We think that this additional complexity can guarantee a fair and meaningful benchmark.

Table 1. Statistics of our dataset

	Chair			Cushion			Table		
	Euc	Geo	Steps	Euc	Geo	Steps	Euc	Geo	Steps
Original Train	5.17	6.45	40.05	5.39	7.23	45.56	4.98	6.19	38.82
New Train	**3.50**	**4.00**	**27.71**	**5.79**	**7.23**	**49.14**	**3.12**	**3.85**	**28.42**
Original Val	5.09	7.21	43.32	4.50	6.18	38.77	3.66	5.29	33.83
New Seen Val	**3.43**	**3.89**	**27.01**	**6.09**	**7.58**	**51.37**	**2.77**	**3.42**	**26.42**
New Unseen Val	**4.07**	**5.79**	**37.15**	**9.00**	**12.19**	**68.40**	**4.53**	**5.18**	**32.95**
New Seen Test	**3.64**	**4.15**	**28.76**	**6.25**	**7.61**	**50.17**	**3.15**	**3.79**	**27.82**
New Unseen Test	**4.06**	**5.11**	**32.75**	**10.13**	**12.94**	**70.50**	**4.55**	**5.37**	**34.45**
	Cabinet			Sink			Total Average		
	Euc	Geo	Steps	Euc	Geo	Steps	Euc	Geo	Steps
Original Train	5.24	6.84	42.63	6.33	8.54	51.80	5.27	6.78	42.27
New Train	**4.02**	**4.91**	**34.15**	**4.94**	**5.99**	**39.80**	**4.27**	**5.19**	**35.84**
Original Val	5.40	7.46	46.10	6.52	8.83	53.50	4.73	6.65	40.98
New Seen Val	**4.02**	**4.87**	**34.07**	**4.62**	**5.69**	**38.22**	**4.19**	**5.09**	**35.42**
New Unseen Val	**5.80**	**6.76**	**46.65**	**7.17**	**8.87**	**52.25**	**6.11**	**7.76**	**47.48**
New Seen Test	**4.37**	**5.33**	**36.58**	**4.96**	**6.06**	**40.16**	**4.47**	**5.39**	**36.70**
New Unseen Test	**8.40**	**9.58**	**56.85**	**8.87**	**10.93**	**63.60**	**7.20**	**8.79**	**51.63**

4 Method

In this section we first describe the Problem Setup. Then we introduce our SMTSC model as shown in Fig. 2

4.1 Problem Setup

Our interests fall within the task of Object Navigation. In particular, this task requires finding an occurrence of a certain class starting from a random position in the environment.

This task can be viewed as a Partially Observable Markov Decision Process (POMDP)[12][10] $(\mathcal{S}, \mathcal{A}, \mathcal{O}, R(s,a), T(s'|s,a), P(o|s))$ in which:

- $\mathcal{S}$ is a finite set of states of the world;
- $\mathcal{A}$ is a finite set of actions;
- $\mathcal{O}$ is the observation spaces;
- $R : \mathcal{S} \times \mathcal{A} \to \mathbb{R}$ it is a reward function, which given a state s and an action a to be performed in it, returns a reward for the execution of a in s.
- $T : \mathcal{S} \times \mathcal{A} \to \mathcal{S}$ it is a transition function, which given a state s and an action a to be performed in it, returns the probability of reaching the state s' by executing a in s;

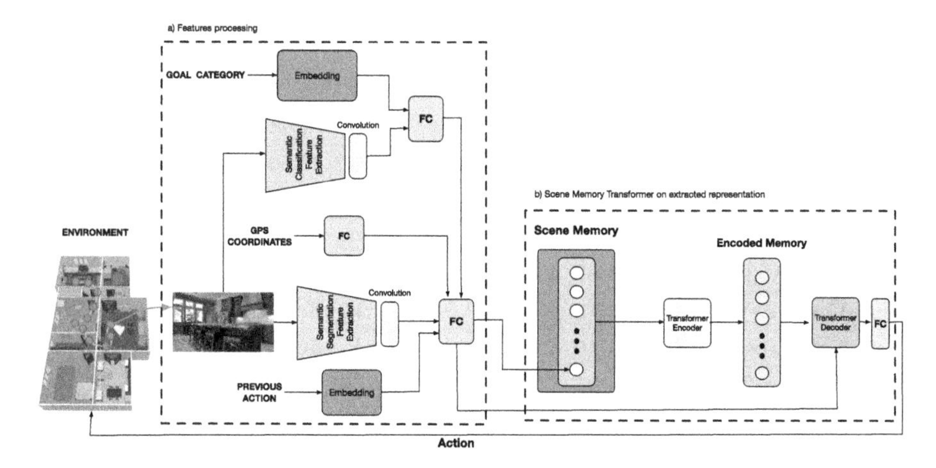

Fig. 2. SMTSC model: a) Features processing: all features are processed and a shared representation is created. This representation is then inserted into a memory (b) Scene Memory Transformer) from which the action that the agent will perform is extracted with a Scene Memory Transformer model.

- $P(o|s)$ is a probability density function that defines the likelihood of observing o in s.

In the setup taken into consideration the set of possible actions is defined as $\mathcal{A} = \{$ **go_forward, turn_left, turn_right, stop** $\}$. The actions are deterministic, that is, apart from possible collisions with objects in the scene, the agent will move in the desired direction without deviations due to noisy dynamics. In particular, with a go_forward action the agent will move forward by 0.25 m, while with the two turn actions, it will rotate 30° in the desired direction. Finally, a stop-action causes the navigation episode to end and, if the agent is less than 0.1 m from an object of the type sought, then the episode will be deemed successfully concluded.

The observation $o = (\text{RGB}, \text{p}, \text{a}_{\text{prev}}, \text{goal}) \in \mathcal{O}$ is the set of features collected by the agent at each step in the environment and passed to the model. RGB is what the agent sees from a given position, it is an RGB image extracted with 640×480 size; p is the agent's position w.r.t. to the starting point, a_{prev} is the action performed in the previous step and finally *goal* is the objective object to be sought.

4.2 Model

The proposed model is visible in Fig. 2, it is composed of two main parts, a first part in which the features are extracted and brought into a joint representation and a second module in which the features of the current observation are added to a memory that keeps track of all past observations and an attention-based policy network extracts a distribution on possible actions.

Features Processing Module. Starting from an observation o = (RGB, p, a_{prev}, goal) $\in \mathcal{O}$ we first define 5 different encoders:

1. $\gamma_{sem\_seg} : \mathbb{R}^{256\times256} \rightarrow \mathbb{R}^{256}$: encodes an RGB observations into a vector of size 256 by using features extracted from a semantic segmentation model.
2. $\gamma_{scene\_class} : \mathbb{R}^{256\times256} \rightarrow \mathbb{R}^{128}$: encodes an RGB observations into a vector of size 128 by using features extracted from a scene classification model.
3. $\gamma_{goal}(goal)$: encodes the goal into a vector of size 32
4. $\gamma_{pos}(p)$: encodes the relative position into a vector of size 32
5. $\gamma_{act}(a_{prev})$: encodes the previous action executed into a vector of size 32

Now we define:

$$\delta(RGB, goal) = FC(\{\gamma_{scene\_class}(RGB), \gamma_{goal}(goal)\}) \tag{1}$$

and:

$$\phi(o) = FC(\{\gamma_{sem\_seg}(RGB), \gamma_{pos}(p), \gamma_{act}(a_{prev}), \delta(RGB, goal)\}) \tag{2}$$

Equation 1 encodes the previously illustrated idea that a goal is intrinsically associated with a specific room. To do this starting from the goal through the γ_{goal} function, a representation of dimension 32 is extracted. In parallel, a scene classifier is used to extract features starting from the RGB image, these two modalities are concatenated and a joint representation is created using a fully connected layer. In this way, we obtain a representation of the goal conditioned step by step from the room in which the agent is located, that is going to add useful information to the agent to understand how to move in the environment.

Finally, Eq. 2 generates a joint representation between all the modalities described above. It, therefore, concatenates the representation obtained from a model for semantic segmentation and the representations of the previous action, position and goal (as defined by Eq. 1). This vector is passed to a fully connected layer which returns a vector of size 256.

Scene Memory Transformer Module. SMT module is based on the one proposed by [10]. Given a new episode, we build an initially empty memory where to save the joint-representations obtained starting from observation o = (RGB, p, a_{prev}, goal) through the application of the $\phi(o)$ function defined in the Eq. 2. This memory is then passed along with $\phi(o)$ to an attention-based encoder-decoder policy [28] which extracts a probability distribution on the actions. The encoder-decoder structure is shown in Fig. 3. The SMT encoder uses a self-attention to encode the M memory, so M is passed to a MultiHeadAttention with 8 heads in which therefore M is both Query and Key and Value as shown in Eq. 3.

$$M' = \text{MultiHeadAttention}(M, M) \tag{3}$$

Subsequently, M is passed together with the joint-representation to the decoding structure. It always uses the attention mechanism to give a

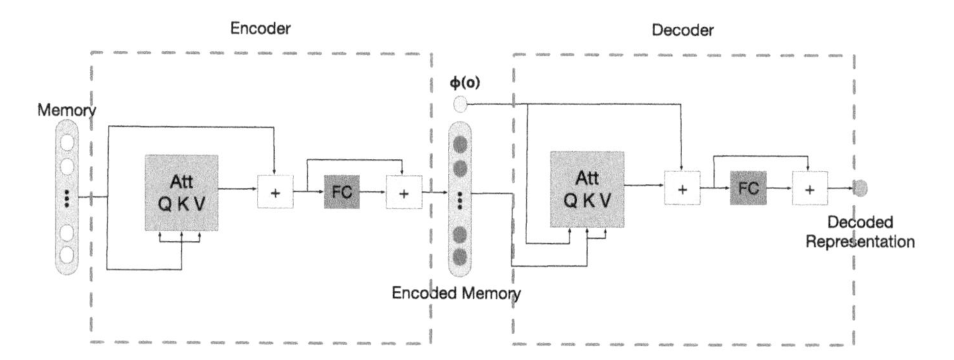

Fig. 3. The structure of the encoder-decoder model based on Multi-Head Attention, as presented in Fang et al. [10].

representation of $\phi(o)$ conditioned by past observations. Again, attention is implemented through an 8-head MultiHeadAttention mechanism as shown in Eq. 4.

$$Q = \text{MultiHeadAttention}(\phi(o), M') \tag{4}$$

Finally, Q is reduced to the dimensionality of the action space $\mathcal{A}$ through a Linear Layer and a Categorical Distribution on the action space is extracted from the latter representation, as defined in Eq. 5.

$$\pi(a|o, M) = \text{Categorical}(\text{Softmax}(\text{FC}(Q))) \tag{5}$$

Implementation and Training Details. We implemented all the models using Python 3.6 and PyTorch [21]. We used the PPO [25] algorithm to train the model using a 32 GB Tesla V100 GPU. We used batch size of 64 and Adam Optimizer with a Learning Rate of 1×10^{-5}. The visual features are extracted from the images using the Taskonomy networks [35] as done in [24]. This allows us to have consistent features as Taskonomy networks have been trained on indoor environments simulated by Gibson [32] and also the number of parameters to be trained on the network drops drastically. All the activation functions within the encoder-decoder structure are ReLU functions and the pose vector is encoded as a quadruple $(x, y, sin(\theta), cos(\theta))$.

5 Experimental Results

In this section an accurate description of the experimental setup will be provided, the results obtained will then be presented.

5.1 Experimental Setup

We used the *Small MP3D* dataset presented in Sect. 3. We decided to test the SMTSC model on both the seen and unseen test sets. Results on the seen set will give as a measure of its capacity to memorize environments seen during training. Conversely, results obtained on the unseen set will quantify the degree of adaptation to unseen scenes that the agent possess. However, as the number of training scenes is only six, we don't expect our agents to develop strong generalization behaviours. The simulator used for all the experiments was Habitat, which provides the agent with 640×480 RGB images. In addition to the images, an odometry system is available that can provide the x and y coordinates and the orientation of the agent with respect to the starting point. The simulated robot has a height of 88 cm from the ground with a radius of 18 cm. The camera of the agent with which he acquires the images is placed 88cm from the ground and allows to capture images with a 79° HFOV. Sliding against objects is not allowed, once a collision has been made the agent must necessarily rotate before being able to proceed again in the environment. The moves that the agent can perform at each step are: 25 cm move forward, 30° turn left, 30° turn right and stop. In particular, when the stop action is called the current episode is declared correct only if the object sought is less than 0.1 m from the agent. For each episode the agent has a maximum of 500 steps to call the stop action, otherwise the episode is considered to be a failure automatically.

The metrics used to evaluate the proposed models are 3: Success Rate, Success weighted by Path Length (SPL) and distance to success (DTS). The Success Rate is simply the ratio between the number of episodes that have been successful and the total number of episodes. The SPL, on the other hand, measures the efficiency in reaching the goal when an episode is successfully completed with a numerical value between 0 and 1. When the episode is not successfully completed 0 is attributed to this metric, otherwise it can be calculated by using Eq. 6, in which N is the number of test episodes, l_i is the shortest-path distance from the agent's starting position to the goal in episode i, p_i the length of the path actually taken by the agent in the episode i and finally S_i is a binary indicator of success.

$$\text{SPL} = \frac{1}{N} \sum_{i=1}^{N} S_i \frac{l_i}{\max(p_i, l_i)} \tag{6}$$

The SPL is considered today as the main metric for the Object Navigation task, but as described in [3] there are numerous problems. In fact, not all the failures are to be considered equal, just think of how an agent arrived at 0.2 m from the object sought is evaluated with the same score obtained by another agent who has rotated on himself for the entire duration of the episode. Finally, DTS is defined as the agent's distance from the threshold boundary around the nearest object. This is mathematically defined as:

$$DTS = max(\text{Geo}(\text{AgentPos}, \text{GoalPos}) - d, 0) \tag{7}$$

In which Geo function measures the geodesic distance between the agent at the end of the episode and the nearest object, d instead is the success distance, in this case 0.1 m.

Table 2. Results on the seen test set

Model	SPL↑	Success↑	DTS↓
Random	0.00	0.00	5.126
Forward-Only	0.009	0.026	5.094
Reactive [24]	0.041	0.126	4.523
LSTM [29]	0.131	0.247	3.199
SMT w/o SC (our)	0.345	0.595	1.562
SMTSC (our)	**0.649**	**0.883**	**0.403**

Table 3. Results on the unseen test set

Model	SPL↑	Success↑	DTS↓
Random	0.00	0.00	7.842
Forward-Only	0.004	0.01	7.922
Reactive [24]	0.001	0.04	7.797
LSTM [29]	0.002	0.04	7.648
SMT w/o SC (our)	0.008	0.04	7.518
SMTSC (our)	**0.039**	**0.080**	**6.817**

Apart from the model presented in Sect. 4, four other different baselines have been tested to assess the performance of the proposed model.

Random Agent: an agent that performs random actions extracted from a uniform distribution.

Forward-Only Agent: an agent who only performs forward actions (with a 1% probability of calling the stop action). These first two baselines were placed to demonstrate the non-triviality of the dataset proposed.

Reactive Agent: a policy that extracts semantic segmentation features through Taskonomy and merges them with position and goal as was done in [24]. The action to be performed is directly extracted from this representation, therefore no type of memory is used.

LSTM Agent: the representation is extracted as for the reactive agent, but in this case, it is passed to an LSTM and the action to be performed is extracted from its output. The model is pretty similar to the RGB-Only presented in [29].

SMT without Scene Classification Features (SMT w/o SC): this is the same model presented in Sect. 4, except for the fact that the goal is coded only through an embedding layer without exploiting the joint representation with the features of the Scene Classifier. The last three models were trained with PPO Reinforcement Learning algorithm.

5.2 Results

In Table 2 are shown the results obtained on the test set in seen environments. The low performance of Random and Forward-Only baselines highlight how the dataset is not trivial, showing that the object to be found is almost never in the immediate proximity of the starting point of the episode. Looking instead at the other baselines based on Reinforcement Learning, we can see how the two models that use the Scene Memory Transformer are able to perform much better in almost all metrics. This big difference is probably attributable to the fact that the SMT can extract crucial information even from actions performed in a fairly remote past, while for example in the case of the LSTM this is very difficult, as more recent information tends to supplant the older ones and this is emphasized especially in very long sequences.

It is also interesting to note that the model that creates a joint representation between the goal coding and the visual features for the scene classification performs much better than the SMT w/o SC model. Even in the SPL, we have an increase of 88%, in the success ratio an increase of 48.4% and finally, the average DTS has fallen by over a meter.

The behavior observed in the seen test set follows the same trend also in the case of the unseen test set whose results are visible in Table 3. In fact, taking as a comparison the average geodesic distance of the unseen test set (7.76 m), we can see how the Forward Only and Random agents tend to conclude their episodes even further away from the starting point. The reactive model instead certifies its performance in line with the average distance reported in Table 1 for the unseen test set. Finally, the two models that exploit the SMT are the ones that perform better, even in an unseen environment. In particular, the proposed model capable of exploiting the joint representation between visual features and goal coding lowers the average distance from the goal by almost a meter, and certifies its successes on 8% of the scenes.

In the next section we report some qualitative example of the SMTSC model on the seen and unseen test sets.

Qualitative Results. In Fig. 4 it is possible to see, in blue, the path taken by the agent to reach an object of the "cushion" class on the seen test set. In green you can see the shortest path to the object. The agent has successfully reached the object sought by stopping less than 0.1m from it. On the contrary, in Fig. 5 is shown an example of failure, still on the seen test set, in which the agent was unable to reach the object sought. The agent, from his initial position was able to recognize an object of the class sought and to head towards it, despite not being the closest chair. At the time of calling the stop action, however, the agent was 7cm beyond the boundary that would have given the episode success. In this case, we can see the strong penalty given by a metric like the SPL, in fact, in our opinion, this episode cannot be considered totally wrong.

In Figs. 6 and 7, on the other hand, it is possible to see two examples extracted from the evaluation of the model on the unseen test set. In the first image, the was able to take a correct path to a "cabinet". However, the cabinet that was found wasn't the nearest so the SPL of this episode was only 0.57. In the second example, on the other hand, a very long route of over 14 m of navigation is presented. The agent here was able to follow the optimal path for about half of its length, then it reached the maximum number of actions allowed. This means that it "wasted" a lot of action to increase its understanding of the scene.

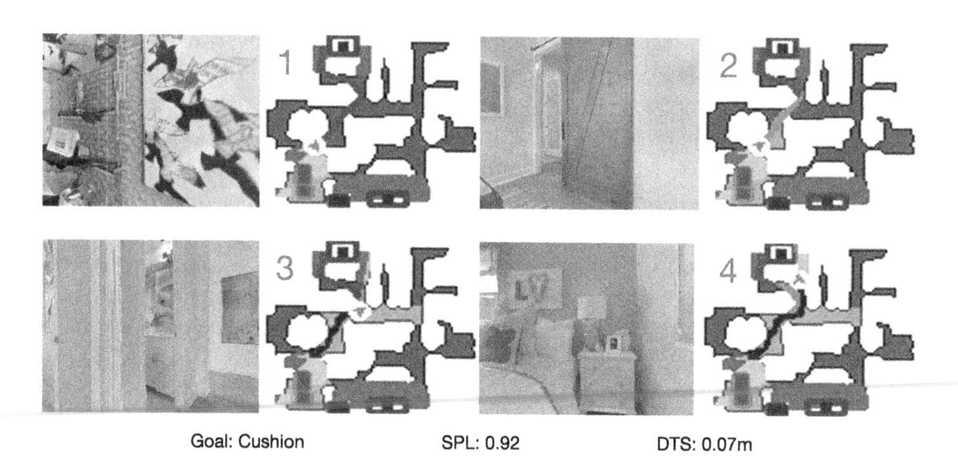

Fig. 4. Successful navigation episode with SMTSC model on seen test set.

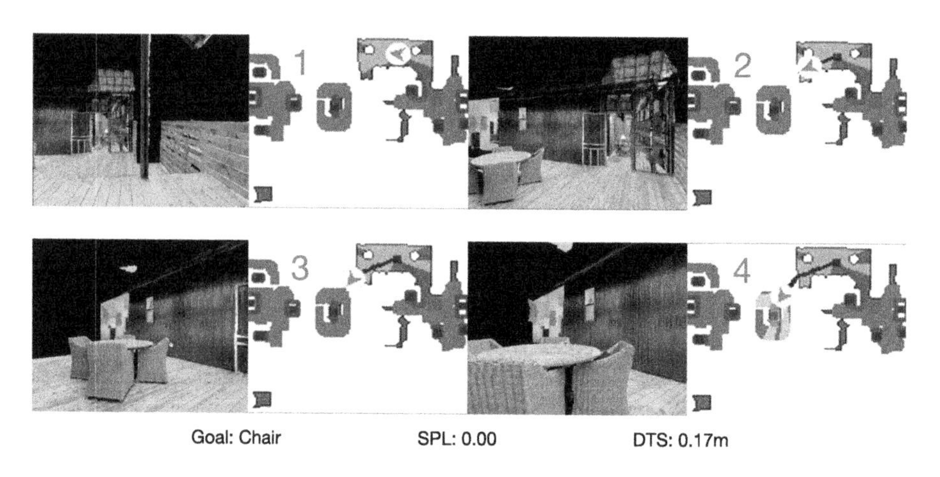

Fig. 5. Unsuccessful navigation episode with SMTSC model on seen test set.

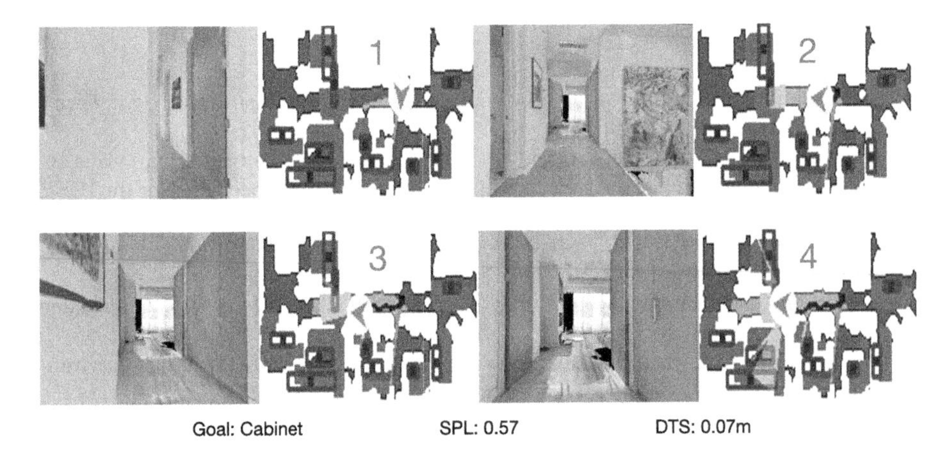

Fig. 6. Successful navigation episode with SMTSC model on unseen test set.

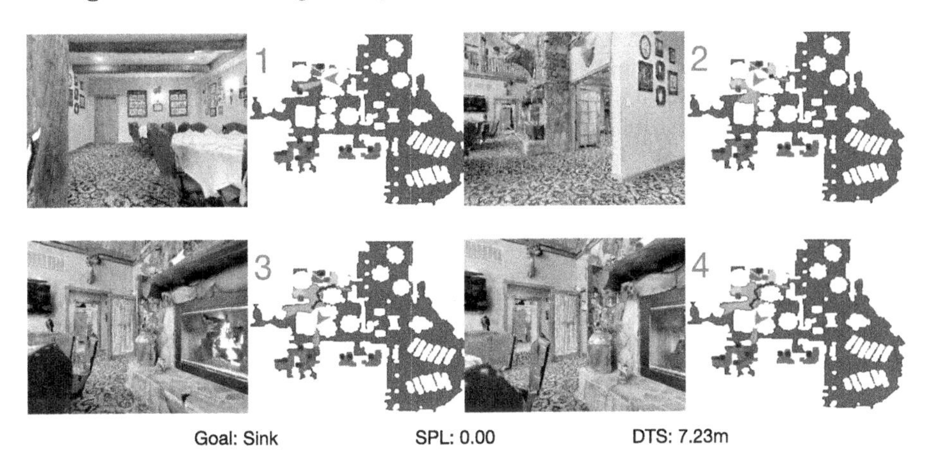

Fig. 7. Unsuccessful navigation episode with SMTSC model on unseen test set.

6 Conclusion

We proposed a subset of the dataset developed for the Habitat 2020 ObjectNav Challenge [23]. This was done to allow the training of models that do not involve the use of planning associated with the construction of maps within them with few computational resources (e.g. a single GPU). Furthermore, we proposed a model capable of exploiting the subtle relationship existing between objects and the rooms in which they are usually located. This intuition, combined with the use of a Scene Memory Transformer showed good results on the proposed dataset. In the future, it would be very interesting to be able to test this model on the complete dataset using the distributed Reinforcement Learning algorithm DD-PPO [29] and a greater number of GPUs in order to perform training in a reasonable time.

References

1. Anderson, P., et al.: On evaluation of embodied navigation agents. arXiv preprint arXiv:1807.06757 (2018)
2. Anderson, P., et al.: Vision-and-language navigation: interpreting visually-grounded navigation instructions in real environments. In: Proceedings of the IEEE Conference on Computer Vision and Pattern Recognition, pp. 3674–3683 (2018)
3. Batra, D., et al.: ObjectNav revisited: on evaluation of embodied agents navigating to objects. arXiv preprint arXiv:2006.13171 (2020)
4. Canny, J.: The Complexity of Robot Motion Planning. MIT Press, Cambridge (1988)
5. Chang, A., et al.: Matterport3D: Learning from RGB-D data in indoor environments. In: International Conference on 3D Vision (3DV) (2017)
6. Chang, M., Gupta, A., Gupta, S.: Semantic visual navigation by watching YouTube videos. arXiv preprint arXiv:2006.10034 (2020)
7. Chaplot, D.S., Gandhi, D., Gupta, A., Salakhutdinov, R.: Object goal navigation using goal-oriented semantic exploration. arXiv preprint arXiv:2007.00643 (2020)
8. Chaplot, D.S., Gandhi, D., Gupta, S., Gupta, A., Salakhutdinov, R.: Learning to explore using active neural slam. arXiv preprint arXiv:2004.05155 (2020)
9. Chen, T., Gupta, S., Gupta, A.: Learning exploration policies for navigation. arXiv preprint arXiv:1903.01959 (2019)
10. Fang, K., Toshev, A., Fei-Fei, L., Savarese, S.: Scene memory transformer for embodied agents in long-horizon tasks. In: Proceedings of the IEEE Conference on Computer Vision and Pattern Recognition, pp. 538–547 (2019)
11. Gupta, S., Davidson, J., Levine, S., Sukthankar, R., Malik, J.: Cognitive mapping and planning for visual navigation. In: Proceedings of the IEEE Conference on Computer Vision and Pattern Recognition, pp. 2616–2625 (2017)
12. Kaelbling, L.P., Littman, M.L., Cassandra, A.R.: Planning and acting in partially observable stochastic domains. Artif. Intell. **101**(1–2), 99–134 (1998)
13. Kavraki, L.E., Svestka, P., Latombe, J.C., Overmars, M.H.: Probabilistic roadmaps for path planning in high-dimensional configuration spaces. IEEE Trans. Robot. Autom. **12**(4), 566–580 (1996)
14. Kolve, E., et al.: AI2-THOR: an interactive 3D environment for visual AI. arXiv preprint arXiv:1712.05474 (2017)
15. Lample, G., Chaplot, D.S.: Playing fps games with deep reinforcement learning. In: Thirty-First AAAI Conference on Artificial Intelligence (2017)
16. LaValle, S.M., Kuffner, J.J.: Rapidly-exploring random trees: progress and prospects. In: Algorithmic and Computational Robotics: New Directions, vol. 5, pp. 293–308 (2001)
17. Mirowski, P., et al.: Learning to navigate in complex environments. arXiv preprint arXiv:1611.03673 (2016)
18. Mnih, V., et al.: Asynchronous methods for deep reinforcement learning. In: International Conference on Machine Learning, pp. 1928–1937 (2016)
19. Mnih, V., et al.: Human-level control through deep reinforcement learning. Nature **518**(7540), 529–533 (2015)
20. Mousavian, A., Toshev, A., Fišer, M., Košecká, J., Wahid, A., Davidson, J.: Visual representations for semantic target driven navigation. In: 2019 International Conference on Robotics and Automation (ICRA), pp. 8846–8852. IEEE (2019)
21. Paszke, A., et al.: Automatic differentiation in PyTorch (2017)

22. Savva, M., Chang, A.X., Dosovitskiy, A., Funkhouser, T., Koltun, V.: Minos: multimodal indoor simulator for navigation in complex environments. arXiv preprint arXiv:1712.03931 (2017)
23. Savva, M., et al.: Habitat: a platform for embodied AI research. In: Proceedings of the IEEE International Conference on Computer Vision, pp. 9339–9347 (2019)
24. Sax, A., et al.: Learning to navigate using mid-level visual priors. arXiv preprint arXiv:1912.11121 (2019)
25. Schulman, J., Wolski, F., Dhariwal, P., Radford, A., Klimov, O.: Proximal policy optimization algorithms. arXiv preprint arXiv:1707.06347 (2017)
26. Sethian, J.A.: Fast-marching level-set methods for three-dimensional photolithography development. In: Optical Microlithography IX, vol. 2726, pp. 262–272. International Society for Optics and Photonics (1996)
27. Straub, J., et al.: The Replica dataset: a digital replica of indoor spaces. arXiv preprint arXiv:1906.05797 (2019)
28. Vaswani, A., et al.: Attention is all you need. In: Advances in Neural Information Processing Systems, pp. 5998–6008 (2017)
29. Wijmans, E., et al.: DD-PPO: Learning near-perfect pointgoal navigators from 2.5 billion frames. arXiv pp. arXiv-1911 (2019)
30. Wortsman, M., Ehsani, K., Rastegari, M., Farhadi, A., Mottaghi, R.: Learning to learn how to learn: self-adaptive visual navigation using meta-learning. In: Proceedings of the IEEE Conference on Computer Vision and Pattern Recognition, pp. 6750–6759 (2019)
31. Wu, Y., Wu, Y., Tamar, A., Russell, S., Gkioxari, G., Tian, Y.: Learning and planning with a semantic model. arXiv preprint arXiv:1809.10842 (2018)
32. Xia, F., Zamir, A.R., He, Z.Y., Sax, A., Malik, J., Savarese, S.: Gibson Env: real-world perception for embodied agents. In: 2018 IEEE Conference on Computer Vision and Pattern Recognition (CVPR). IEEE (2018)
33. Xia, F., et al.: Interactive Gibson benchmark: a benchmark for interactive navigation in cluttered environments. IEEE Robot. Autom. Lett. **5**(2), 713–720 (2020)
34. Yang, W., Wang, X., Farhadi, A., Gupta, A., Mottaghi, R.: Visual semantic navigation using scene priors. arXiv preprint arXiv:1810.06543 (2018)
35. Zamir, A.R., Sax, A., Shen, W., Guibas, L.J., Malik, J., Savarese, S.: Taskonomy: disentangling task transfer learning. In: Proceedings of the IEEE Conference on Computer Vision and Pattern Recognition, pp. 3712–3722 (2018)

Active Crowd Analysis for Pandemic Risk Mitigation for Blind or Visually Impaired Persons

Samridha Shrestha[1,2], Daohan Lu[1,3], Hanlin Tian[1,3], Qiming Cao[1,3], Julie Liu[1,2], John-Ross Rizzo[3], William H. Seiple[4], Maurizio Porfiri[3], and Yi Fang[1,2,3](✉)

[1] NYU Multimedia and Visual Computing Lab, Abu Dhabi and New York, UAE
yfang@nyu.edu
[2] New York University Abu Dhabi, Abu Dhabi 129188, UAE
[3] New York University, New York, NY 10012, USA
[4] Lighthouse Guild, New York, NY 10023, USA

Abstract. During pandemics like COVID-19, social distancing is essential to combat the rise of infections. However, it is challenging for the visually impaired to practice social distancing as their low vision hinders them from maintaining a safe physical distance from other humans. In this paper, we propose a smartphone-based computationally-efficient deep neural network to detect crowds and relay the associated risks to the Blind or Visually Impaired (BVI) user through directional audio alerts. The system first detects humans and estimates their distances from the smartphone's monocular camera feed. Then, the system clusters humans into crowds to generate density and distance maps from the crowd centers. Finally, the system tracks detections in previous frames creating motion maps predicting the motion of crowds to generate an appropriate audio alert. Active Crowd Analysis is designed for real-time smartphone use, utilizing the phone's native hardware to ensure the BVI can safely maintain social distancing.

Keywords: Active crowd analysis · Visually impaired · Human detection · Crowd density · Crowd distance · Crowd motion · Crowd-Risk Alert · Pandemic risk mitigation

1 Introduction

The World Health Organization estimates that there are about 39 million blind and 246 million visually-impaired individuals in the world [5,66]. Numerous reports [28,36,38,63], have stated that even before the start of the current pandemic, low vision already posed significant challenges to the visually impaired

Electronic supplementary material The online version of this chapter (https://doi.org/10.1007/978-3-030-66823-5_25) contains supplementary material, which is available to authorized users.

A. Bartoli and A. Fusiello (Eds.): ECCV 2020 Workshops, LNCS 12538, pp. 422–439, 2020.
https://doi.org/10.1007/978-3-030-66823-5_25

individuals in conducting their day-to-day activities. Recent surveys conducted by the American Foundation for the Blind [10] and the Canadian Council for the Blind [26] found that the COVID-19 outbreak profoundly exacerbated those existing hurdles for the BVI. These challenges included issues in public navigation, transport, and shopping all while avoiding crowds making social-distancing difficult. BVI individuals face these barriers as they have to rely on physical sensation significantly more than the normal-sighted in their everyday lives to locate objects or navigate environments. This reliance on physical touch prevents effective social distancing and puts the sightless at an elevated risk of contracting viruses by being in the vicinity of infected individuals [67]. To ameliorate the reduced perceptive range of the BVI and to mitigate the pandemic health risks, in this paper, we propose Active Crowd Analysis, a system to augment the BVI's environment perception to detect nearby visible crowds and maintain social-distancing in a more intuitive, safe, and independent manner.

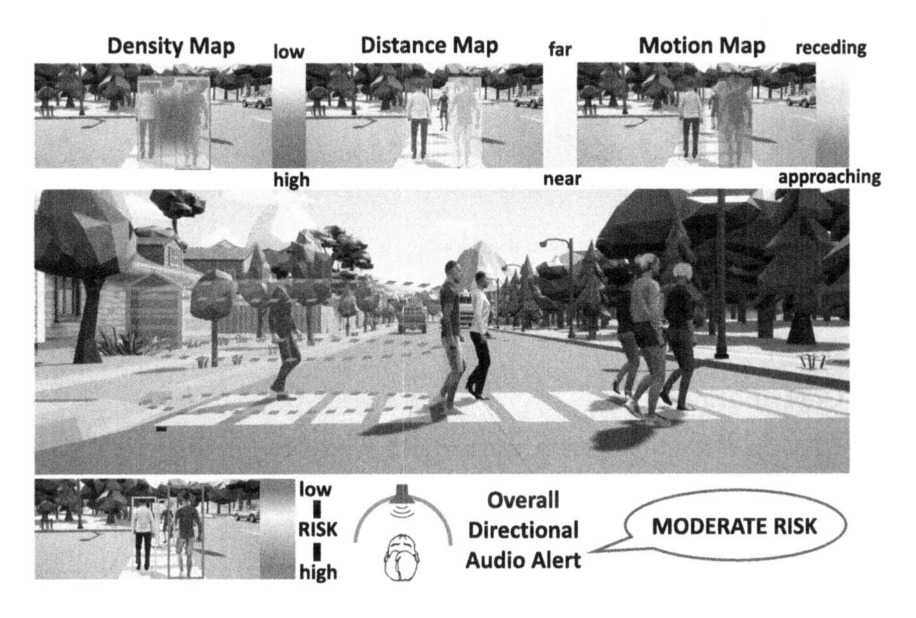

Fig. 1. Active Crowd Analysis System to detect crowds, generate density, distance, and motion maps to finally relay a risk alert to the BVI user

Active Crowd Analysis holistically integrates the density, distance, and the motion of the visible crowd in-front of a BVI person to evaluate the riskiness of crowds and to help the user avoid crowds through audio warnings (Fig. 1). The system is designed for use in a smartphone that is mounted in front of the BVI person with a lanyard. Allowing for wider access, our system requires no specialized hardware except for a standard CPU/GPU and camera-enabled smartphone along with a headphone for audio feedback. We recommend a bone-conduction headphone in specific as such headphones do not obstruct the normal hearing of a visually impaired person while still providing the necessary audio guidance to avoid crowds. The system consists of a backbone feature network to extract features from images from the smartphone camera which is passed to a human detector to detect crowds and create a crowd density map (Fig. 2). The bounding box coordinates of the detected humans are then sent to a distance regressor network to calculate the distances to detected individuals to create a distance map (Fig. 2). Using the detected human and their distances from multiple frames, the system finally generates a motion map (Fig. 2). Using information from the three previous maps, a crowd-risk module then alerts the BVI user of any risk from visible crowds nearby through the bone-conduction headphones as spatialized directional audio (Fig. 6). The system uses computationally-light neural networks that were designed for real-time smartphone use [59] to provide an active and reliable risk mitigation solution for the BVI during a pandemic.

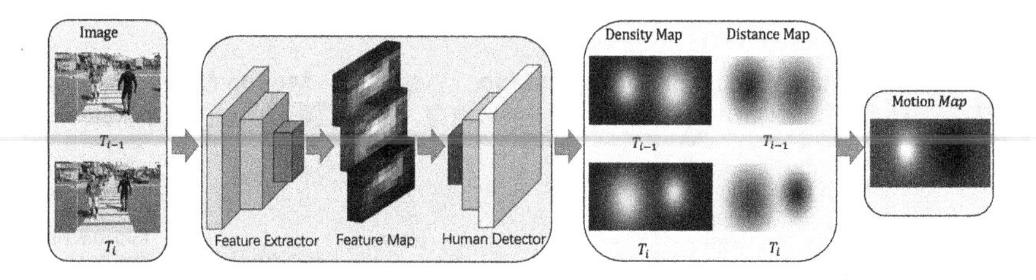

Fig. 2. Crowd-Risk Alert classifies risk-inducing crowds and sends the BVI a concise verbal alert with spatial audio.

1.1 Contributions

In short, our contributions can be summarized as follows

- Crowd Density Analysis: Calculates the density of the visible crowd in-front of the BVI individual using an efficient density-based clustering algorithm for crowd-density evaluation.
- Crowd Distance Analysis: Calculates the distance to each detected human in-front of the BVI person for crowd-distance evaluation.
- Crowd Motion Analysis: Calculates the perceived change in distance to the detected humans to detect motion for crowd-motion evaluation.
- Crowd Risk Analysis: Integrates the risk scores from the density, distance, and motion evaluations to calculate an aggregated risk that is relayed to the BVI user through spatialized 2D audio.

2 Needs of the BVI During a Pandemic

Surveys in [2,10,13,26] stated that one of the paramount needs of the visually impaired during pandemics was maintaining social-distancing while approaching crowded areas where there is a greater risk of contracting the pandemic virus. During a pandemic, the BVI not only have to minimize physical touching for environment exploration but also stay away from crowds to avoid contracting contagions. Therefore, the BVI require real-time information about the surrounding environment such as the existence of crowds to actively avoid them and prevent the transmission of the disease. [11,12,33,49,52] also show that the BVI community is more receptive to lightweight and small wearable solutions with ubiquitous availability (i.e. smartphones) that enable them to perform their daily activities safely, reliably, and independently.

3 Related Work

3.1 Related Assistive Technology for the BVI

Most hardware-based assistive technology solutions for the BVI have had dire adoption rates [23,44,57] owning to significant drawbacks that include high cost, steep learning curves, and heavy and unwieldy hardware. The most widely adopted commercial sensory augmentation devices generally cost upwards of thousands of US dollars due to the high cost of specialized hardware [20,25,30,32,45,65]. Other hardware-based assistive devices often cause discomfort to the user after long use due to carrying additional hardware such as batteries or cameras. Besides, such existing assistive technology solutions were not designed to address the pandemic-specific needs of the BVI. Most assistive technologies focus on limited applications like outdoor navigation [29,37,41,42,57]. Vision substitutes [4,6,15,35] on the other hand do not provide health and safety assistance to the BVI in the context of a pandemic. In contrast, software solutions that run on smartphones are more affordable and accessible for the BVI and include applications like Microsoft Seeing AI [46] and BlindSquare [3]. Unfortunately, these technologies also do not address pandemic risk reduction for the BVI like helping them to maintain social-distance from crowds. These systems also potentially fail due to the lack of onboard visual processing and the dependency on online visual computing platforms (i.e. Microsoft Seeing AI).

3.2 Related Work in Crowd Density, Distance, and Motion Analysis

Human Density Estimation and Motion Tracking: Previous work in crowd density has necessarily involved real-time human detection and motion tracking. [34] use the You Only Look Once V3 (YOLO-V3) object detection algorithm [54] to detect people and implemented background-subtraction with Gaussian Mixture Models (GMM) and contour heat-maps to analyze crowd densities. [62] used the RGB color features of an image and similar background subtraction between frames to filter background noise to detect and track moving objects in video scenes. However, these methods are only applicable to static video surveillance since background subtraction is inapplicable when the camera is moving. Therefore, these methods for people detection and density analysis would fail when a BVI user is moving around. [9] also implement a background subtraction method as a preprocessing step to detect silhouettes of people in still images using a graph-cut segmentation plan but cannot be used in real-time video sequences. [58] proposed an improved algorithm for object motion estimation in videos where they use GMMs for background subtraction and noise cancellation along with an optical flow algorithm to track objects. Background subtraction methods as mentioned previously restrict the BVI user to a stationary position. Besides, background subtraction methods for videos have deteriorated performance if the person(s) detected in the image is not moving since the foreground selection is based on the movement of the target (human) and a static background. Other methods for tracking people used R-CNN [22] and Faster R-CNN [8] object detection networks to detect people and Euclidean distance based object association between subsequent frames to track people. These methods generate accurate object detections. However, R-CNN [22] has a multi-stage region proposal selective search algorithm that generates 2000 regions to be fed into the neural network classifier while Faster R-CNN [55] still uses a region proposal network which drastically reduces the object detection speed. This means these methods are not suitable for real-time crowd detection in smartphones or embedded-devices.

Human Distance Estimation: Crowd-distance estimation is linked to depth estimation and segmentation of objects in images. Depth segmentation methods usually rely on stereo images from two cameras where the distance information is only calculable for the overlapping fields of view between the cameras [51]. For example, [18] used a scene-geometry based method to use depth cues from stereo images to track moving pedestrians from a moving platform. However, stereo images imply a dual-camera requirement which is undesirable as the BVI user would require two smartphone cameras set up infront of them or multiple-camera embedded devices. For distance estimation from monocular images, various works have been documented. [56,64] use an object distance estimation algorithm for monocular images based on inverse-perspective-mapping (IPM) of the camera's 2D image into a bird's eye view coordinate using the camera parameters (focal length, height, etc). Despite working from a single camera, IPM has significant disadvantages as it fails to accurately predict distance for objects on the borders of the image, or curved surfaces. IPM also requires constant calibration (such as the white markings on a road) and a static height for the camera [64] which cannot always be ensured for our specific use with the BVI. [24] proposed another monocular image distance estimation method that used a Support Vector Regressor on the bounding box width and height to predict the distance to the object while in [27], the authors used DistNet to predict the distance from the bounding box features and used a CNN based object detection model (YOLO) to detect the objects themselves. [68] provided an improved and novel method to use features extracted by a neural network such as Resnet or VGG [61] to directly estimate the distance to the detected object. However, these approaches are not suitable for real-time applications on mobile systems such as smartphones due to their excessive memory and computation requirements.

4 Active Crowd Analysis

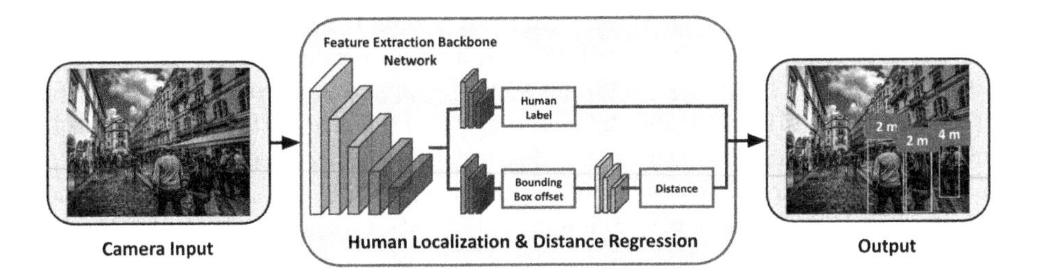

Fig. 3. Human detection and distance estimation network

Active Crowd Analysis is a smartphone-based assistive technology designed for the smartphone's limited memory to achieve real-time human detection, tracking, and distance estimation, traditionally reserved for GPU-based desktop computers. Active-Crowd Analysis identifies and alerts the BVI user of risk-inducing events related to crowds with two sequential modules, Crowd Analysis, and Crowd-Risk Alert. First, the Crowd Analysis module detects all visible people on the camera and clusters people into crowds to generate the crowd density map, crowd distance map, and crowd motion map, as depicted in Fig. 2. The Crowd-Risk Alert module then analyzes the information from the crowd density, distance, and motion analysis to estimate the "riskiness" of crowds per advised health guidelines, such as social distancing [7]. The system finally relays a summary of the crowd's risk analysis (i.e. the level of risk) to the BVI user through a spatialized 2D audio.

For real-time processing on a mobile phone, we develop a fast detection algorithm as shown in Fig. 3 to compute the locations and distances of people on the smartphone camera. It is realized with 1) designing a lightweight feature extractor to simultaneously acquire foundational visual features for different visual tasks in Scene Crowd Analysis 2) designing a shared people-object detector to detect both people and objects simultaneously for Crowd detection, distance, density, and motion map generation. An overview of the people detection algorithm is shown in Fig. 2. The shared feature extractor is modeled based on MobileNet-V2 [59] and further simplified for real-time inference in smartphones. The people-object detector, which classifies the feature maps from the shared feature extractor, is based on SSDLite [60], an object detection algorithm specifically optimized for mobile devices. The detector identifies multiple objects in the image including person and non-person objects, but the Crowd Analysis filters out all the non-human detections.

4.1 Human and Crowd Detection

Detecting humans in the smartphone camera feed is an object detection task. This computer vision task is relatively easy given that many object detection algorithms such as SSD [40] and YOLO [53] can reliably pick up people and objects in an image. However, we require object detection systems that can reliably detect humans in real-time when operated from smartphones. We use a backbone feature extraction network that is based on the MobileNet-V2 network [59] and a bounding box regressor and classification network based on SSD Lite [60]. The outputs from the backbone feature extractor are of sizes 20*20*96, 10*10*1280, 5*5*512, 3*3*256, 2*2*256, and 1*1*64 with attached $4, 6, 6, 6, 4$, and 4 anchor boxes for each output feature map respectively. We also use a non-max suppression threshold of 0.5 to remove multiple detections of the same object. Our loss function is a weighted combination of losses from the object localization (loc) and classification tasks (cls) based on the multi-box detection for multiple classes used in [17,60].

$$L(x, c, l, g) = \frac{1}{N}(L_{cls}(x, c) + \alpha L_{loc}(x, l, g)) \tag{1}$$

where N is the number of matched default box priors. The loss is set to 0 if N is also 0. x is an indicator variable that is set to 1 if the default prior box is matched to the determined ground truth box and 0 otherwise. c is the class confidence score, l and g represent the predicted and ground truth bounding box parameters (center offsets; bounding box width and height) respectively. L_{loc} is the localization loss which is the smooth L1 loss between the predicted (l) and ground-truth box (g) parameters. α is a hyper-parameter that balances the weights of the losses and is determined through cross-validation. L_{cls} is the classification Softmax loss computed over multiple classes. However, after detecting an object, we drop all classes except for the person class in our human detection pipeline.

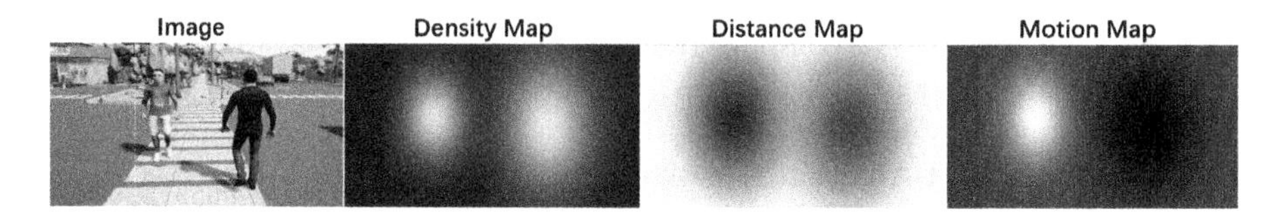

Fig. 4. Illustrations of Crowd Density Map, Crowd Distance Map and Crowd Motion Map.

4.2 Crowd Density Analysis

To create density maps, we use bounding-box parameters ($center : (x, y), width : w, height : h$) generated by the human detection module for each detection. A distance regressor network uses the bounding-box w and h to predict the distance, d to each detected human. We then use the bounding-box centers and distances, (x, y, d) to represent individual persons and form centroid clusters based on 3D euclidean distances between bounding-box centers using a recursive density-based clustering algorithm, DBSCAN [19]. DBSCAN is well suited to our clustering task compared to an algorithm like K-means clustering which is affected by noise and requires the number of clusters apriori. DBSCAN does not require the number of clusters as a hyper-parameter input, instead, it requires the epsilon (*eps*) parameter which is the maximum distance between two points in the same cluster. The most suitable *eps* was experimentally determined to be 18 after re-scaling the dimensions of the input images to $[0, 100]$ for increased generalizability. To generate the density maps, we employ a GMM model for contour heat-map similar to what was used in [34] albeit without using any background subtraction method. The number of components in the GMM model is simply the number of crowd-clusters that was previously computed with DBSCAN.

4.3 Crowd Distance Analysis

For crowd distance map, it is difficult to directly obtain the distances to detected objects because most object detection algorithms do not compute the distance of the detected objects to the camera. To acquire the distances to humans, we use the proportional geometric relationship between the height and the width of the bounding box to the actual size of the person. Such distance estimation of people on a monocular image is intuitively justified as humans also have well-defined shapes when captured by a camera, so their on-camera appearance (size and look) gives a good estimate of their distances. We use a fully-connected deep neural network to regress for the distance to detected human trained on labeled ground-truth data. As shown in Fig. 3, we predict the location and size of the bounding-box and use the width and height of the predicted bounding box to calculate the distance to the detected humans. The Distance regressor network is a deep neural network with five fully connected layers of size 2, 6, 4, 2, 1, experimentally determined to be optimal after hyper-parameter optimization. The network has LeakyRELU activation (slope $= -0.01$) functions between all subsequent layers except for the final layer which has a Softplus activation ensuring all final distance predictions are positive. We train our distance regression network with a supervised Mean Squared Error loss function presented below:

$$MSE(Linear) : \frac{1}{N} \sum_{d \in N} ||d_i - d_i^{gt}||^2 \qquad (2)$$

where d_i is predicted distance and d_i^{gt} is the ground-truth distance to the detected human. The network outputs a distance for every detected human in the input image and assigns the distance to the bounding box center.

4.4 Crowd Motion Analysis

After we detect humans and calculate the distances to each human and cluster center, we can start to track the motion of each detected human or cluster center. In general pedestrian settings, people move relatively slowly as viewed on a camera, so there is a considerable frame-to-frame overlap of their locations. Thus, by comparing the distances of the tracked cluster centers or humans from frame to frame, we can reliably estimate the motion of such crowds or humans as viewed from the BVI user's smartphone camera. By calculating the change in the distance values between consecutive frames, we can calculate the motion of crowds. This velocity will indicate whether detected humans or crowds are moving towards or away from the BVI user and can be used to create the motion map. Motion tracking, however, is essential for estimating motion between frames. To track the detected humans between subsequent frames, we employ a simple bounding box centroid tracking algorithm that links previously-detected bounding boxes with newly detected bounding boxes with the smallest euclidean distance. To suppress noisy detections, we only start tracking humans once they have been visible for a small number of frames and drop missing human detections once they have been absent for several frames (set to 50 in our case). The centroid tracking algorithm is discussed in more detail in Section 1 of the supplementary material. (Sup. Sec. 1)

Fig. 5. Active Crowd Analysis generates a straightforward risk level for detected crowds, where a high-risk level would trigger an audio alert. The closer or denser a crowd is, the more risky it is considered. The crowd on the top left is considered the least risky and the crowd on the top right is considered the risky.

4.5 Crowd Risk Analysis

To evaluate the crowd-risk, we group all people into crowds based on physical proximity. Each person is represented as a point (x, y, d, v) in 4-dimensional space representing their 2D location, distance, and velocity. People are grouped into crowds with the same density-based clustering algorithm, DBSCAN [19], based on their proximity in the 4-dimensional representational space. The resulting crowd is represented by its size (number of people) s, average distance d, and motion v (signed real-number velocity). Then the mathematical formula below is used to evaluate the *riskiness* $r(\mathbf{c})$ of each crowd $\mathbf{c}$ (Detailed in Sup. Eqn. 9):

$$\begin{aligned} \mathbf{c} &\equiv s, d, v \\ r(\mathbf{c}) &= f(s, d, v) \end{aligned} \tag{3}$$

where $r : \mathbb{R}^3 \rightarrow \mathbb{R}^1 \in (0, 1)$ is a function that converts the features of a crowd into a real number representing the risk of the crowd; the crowd feature $\mathbf{c} \equiv s, d, v$ is a three-dimensional vector consisting of its size, distance, and motion. The overall riskiness $R(\mathbf{c})$ is the sum of individual riskiness of crowds:

$$R(\mathbf{c}) = \sum_{\mathbf{c} \in \mathbf{C}} r(\mathbf{c}) \tag{4}$$

where $\mathbf{C}$ is the set of all visible crowds. The individual crowd-risk function r is defined according to social distancing guidelines [7] such as considering 6 feet as the threshold for elevated risk (Sup. Sec. 2). Although other risk evaluation metrics that conform to the official advisory for different pandemics can also be selected. Once the overall crowd-risk is computed, our system sends different levels of audio-risk-alerts to the BVI user as shown in Fig. 5.

4.6 2D Audio Feedback

For every crowd deemed risk-inducing, our system sends a spatialized audio alert with the crowd-risk status (e.g. "Moderate Risk", "High Risk") so that the BVI know the level of risk associated with any nearby crowd and its general direction. The audio spatialization is implemented through open-sourced 3D sound-APIs like OpenSL, for mobile systems like Android [14] or through OpenAL [50] for Apple phones. The efficacy of spatial-directional audio in bone-conduction headphones has been well studied in [43] where participants were able to determine the direction of the audio source with bone-conduction headphones on par with normal headphones. Studies have also shown that blind people are more sensitive to such binaural audio location cues than sighted people [48]. As demonstrated in Fig. 6, after the cluster center of the crowd is computed, the offset from the image center to the cluster center is calculated in Euclidean metrics. To ensure a uniform spread of audio sources in-front of the user regardless of the input image size, we use the image previously

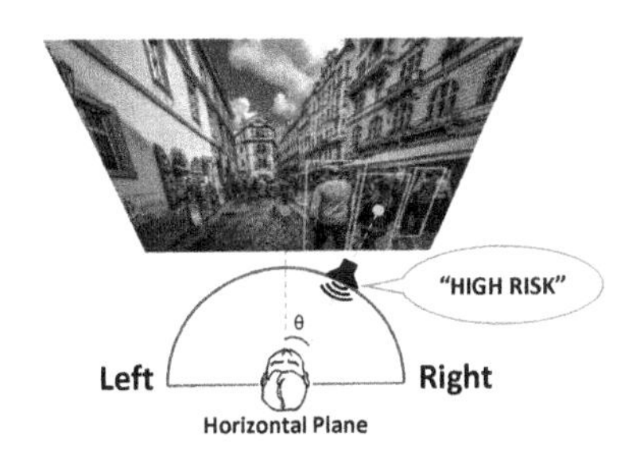

Fig. 6. 2D spatialized audio feedback system (Google Images CC License)

re-scaled to scale of [0, 100]. Since we know the distance to the cluster center, we can calculate the center deviation angle $\theta \equiv \arcsin(\frac{cluster\_to\_img\_center\_offset}{distance\_to\_crowd})$, which is used by the spatial-sound API to generate a 2D directional audio. Finally, the severity of the crowd-risk is also conveyed through the intensity or volume of the audio alerts (i.e. higher risk equates to louder alerts).

5 Experimental Setup and Evaluation

Our tests are run in desktop environments (Sup. Sec. 6), however, our feature extraction module based on MobileNet-V2 [59] has real-time performances in smartphones. [31] experimentally showed that a quantized MobileNet model can run under 25ms at 40 fps with a Snapdragon 845 SoC enabled Qualcomm Hexagon Chipset smartphone. [31] and [39] also provide benchmarked object-detection performances for both Android and iOS smartphones based on other hardware chipsets as well.

5.1 Human Detection Performance

Dataset Setup: We train our human detection network on the PASCAL Visual Object Classes Challenge (VOC) 2007 and VOC 2012 dataset. The trained networks are tested on the PASCAL VOC 2007 test dataset. VOC 2007 and VOC 2012 have 9963 and 11540 images respectively with objects from over 20 different classes. Since we are concerned with human detection, we only report the overall mean average precision performance (mAP) and the results for the person class for object detection and localization task.

Results: Our proposed network, despite having a low mean average precision (mAP) for the people class detection, outperforms all other models in speed as frames per second (fps) and memory usage in both CPU and GPU settings. Our network sacrifices some accuracy for higher speed and lower memory usage, which is paramount if the network is used for detection in smartphones. The MobileNet-v2 backbone in our network has also been demonstrated to achieve a real-time performance 40 fps in a GPU enabled-smartphone[31,59] with model size reduced to 4.3 MB.

5.2 Distance Regression Performance Analysis

Dataset Setup: We train and evaluate our distance regressor networks on the KITTI Vision Benchmark Suite for 2D Object Detection[21], which has 7,481 training and 7,518 test images. Since we are only concerned with human-detection, we drop all classes except for the *Pedestrian* class. The *Cyclist* and *Person Sitting* classes are also dropped as their distributions deviate from that of the *Pedestrian* class when modeled using the bounding box heights and widths (Sup. Sec. 3 Fig. 1, Fig. 2). This leaves us with 1779 images and 4487 human

Table 1. Comparison of different models for human detection on the VOC 2007 test-set. All models were trained on the VOC 2007 and 2012 dataset. The frame rate is calculated for input images of size 500×375 pixels

Model	mAP (Overall)	mAP (Person)	fps (GPU)	fps (CPU)	Model size (MB)
Our network with SSDLite 320×320	0.7814	0.7403	$\mathbf{125 \pm 4}$	$\mathbf{7 \pm 2}$	**25.5**
VGG16 with SSD 300×300	0.9045	0.8751	60 ± 5	3 ± 2	201
Efficient B-Net with SSD 300×300	0.8166	0.7679	79 ± 1	5 ± 2	97.1

Table 2. Distance prediction performance comparison for our validation subset split of the pedestrian class in the KITTI–object-detection dataset

Regressor	Lower is better				
	MSE	RMSE	RMSE_{log}	Squa Rel	Abs Rel
Neural network regressor	6.0088	2.4513	0.1304	0.3911	0.0788
Support vector regressor	7.0425	2.6538	0.1217	0.3215	0.0782

instances. Since the test set is restricted to the KITTI Vision servers, we further divide the 1779 images training set into train and validation subsets with an 80 : 20 split respectively. The images themselves are unnecessary for training the distance regressor as we only require the bounding box coordinates. We train the distance regressor network using the ADAM optimizer with a learning rate of 0.001 for 200 epochs with a batch size of 128. We also use a scheduler to reduce the learning rate to 10% of the previous rate if the validation loss plateaus for five epochs. The Support Vector Regressor (SVR) used for evaluation results in Table 2 is based on the work done by [24]. The SVR is trained and evaluated on the same training and validation subset as the neural network regressor. The SVR is set up to use a radial basis function kernel with the hyper-parameter C set to 1.0 and epsilon set to 0.1.

Evaluation Metrics: Since our aim is to accurately predict the distances to the detected humans, we measure the performance of the distance regression models with metrics used in [16,68], normally reserved for depth estimation. These metrics include the Mean Squared Error (MSE), the root of the mean squared error ($RMSE$), the log root mean squared error ($RMSE_{log}$), the absolute relative difference in distances ($Abs\ Rel$), and the squared relative difference in distances ($Squa\ Rel$) (Sup. Sec. 4).

Results: Fig. 7 and Table 2 demonstrate the effectiveness of our regression neural network compared to the SVR proposed in [24]. The experiments in Fig. 7*b*

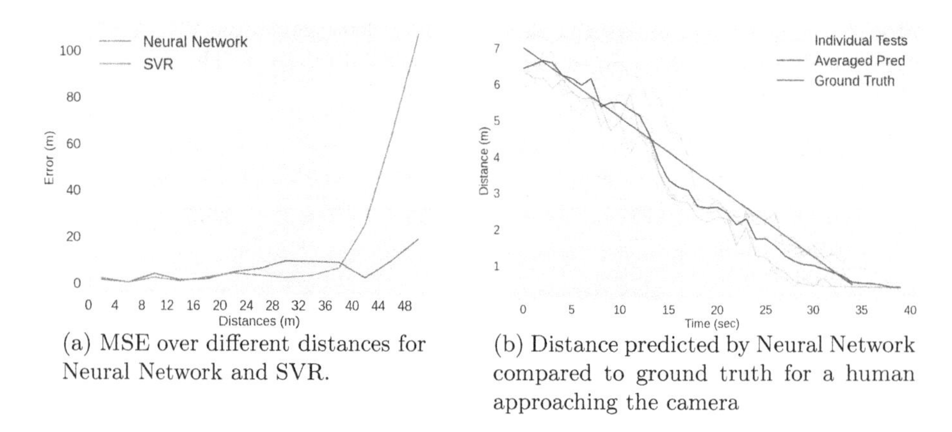

(a) MSE over different distances for Neural Network and SVR.

(b) Distance predicted by Neural Network compared to ground truth for a human approaching the camera

Fig. 7. Distance regressor performance

show the reliability of the distance predictions for a detected human approaching the camera. The SVR was found to be inaccurate for distances closer than 6 m. For the experiment in Fig. 7*b*, the neural network regressor had a RMSE error of 0.4610 while the SVR had a RMSE error of 6.7508. However, Fig. 7*a* shows that both the SVR and the neural network regressor lose distance prediction performance as distances get longer since the detected bounding box shape gets smaller. Deterioration for distance estimation performance was also observed in [64,64,68]. This is not a concern for us since we put more emphasis on how the models perform for short distances within 7 m where social distancing guidelines might come into action.

5.3 Crowd Motion Tracking Performance Analysis

Dataset Setup: To evaluate our crowd-motion analysis setup, we make use of the 2016 Multiple-Object-Tracking Benchmark [47] (*MOT16*) training subset. We use the *MOT16*-02 subset of the *MOT16* data to evaluate different human detection backbone networks as reported in Table 3. We only consider the pedestrian class with a visibility ratio of more than 0.7 in this subset for evaluation and drop all non-pedestrian classes from our test set. We use the *MOT 1602* train subset as the testing subsets are only available for server evaluation on all classes explicitly. This subset contains a video sequence of 600 images of 1920 $\times$ 1080 resolution. The *MOT* benchmarks we report in Table 3 are based on this subset.

Evaluation Metrics: Our primary objective in crowd-motion analysis is to measure the motion (changing distance) of crowds and individuals from the BVI user. To accurately calculate this motion, we need to track the detected humans. Therefore, we use the *CL*assification of *E*vents, *A*ctivities, and *R*elationships (*CLEAR*) metrics [1] for the task of Multiple Object Tracking [47] (*MOT*) to benchmark the human-motion tracking performance. The *CLEAR* metrics in

Table 3. Comparison of the object tracking metrics on the pedestrian class of the *MOT 2016-02 subset* given backbone networks and SSD detectors of different input sizes. (Detection score threshold was set to 0.65 for all models)

Model	Higher is better			Lower is better			
	MOTA(%)	MOTP(%)	MT(%)	ML(%)	FP	FN	IDsW
Our Network with SSDLite 320 × 320	**8.1**	**28.9**	4.65	76.74	**377**	**3378**	**12**
VGG with SSD 300 × 300	4.5	27.8	5.26	73.68	950	3302	29
Efficient Net b3 with SSD 300 × 300	7.3	28.6	5.26	81.57	563	3495	13

Table 1 include standardized metrics such as the Multiple Object Tracking Accuracy (*MOTA*) and the Multiple Object Tracking Precision (*MOTP*) (Sup. Sec. 5). Additional tracking metrics also include the Mostly Tracked (*MT*), Mostly Lost(*ML*), False Positives (*FP*), False Negatives(*FN*), and Identity switches (*IDsW*) (Sup. Sec. 5).

Results: Table 3 shows that our backbone network is the most suitable model for the human tracking task required to create the motion maps in Fig. 4 as our network has the highest *MOTA* and *MOTP*. However, these metric scores should not be alarming since the *MOT16* benchmark contains excessive detectable objects in each frame with frequently occluded paths while even state of the art methods only achieved around 33.7% *MOTA* in the *MOT16* testset [47].

6 Conclusion

Given that previous research and commercial solutions designed to aid the BVI community do not take health risks associated with a pandemic into account, in this paper we have presented and demonstrated the efficacy of an active crowd analysis system to help mitigate these pandemic-related health risks for the BVI. Our smartphone-based system combines crowd density, distance, and motion analysis to detect the risks associated with nearby crowds and relay this risk to the BVI individual through a directional 2D audio. Active Crowd Analysis, in aggregate, enables such sight-impaired persons to maintain a safe physical distance from other humans or crowds meeting the official social distancing guidelines to avoid the spread of contagions during a pandemic.

Acknowledgement. The authors gratefully acknowledge the financial support from the NYUAD Institute (Research Enhancement Fund - RE132).

References

1. Bernardin, K., Stiefelhagen, R.: Evaluating multiple object tracking performance: the clear mot metrics. EURASIP J. Image Video Process. **2008**, 1–10 (2008). https://doi.org/10.1155/2008/246309
2. Bhowmick, A., Hazarika, S.M.: An insight into assistive technology for the visually impaired and blind people: state-of-the-art and future trends. J. Multimodal User Interfaces **11**(2), 149–172 (2017). https://doi.org/10.1007/s12193-016-0235-6
3. BlindSquare: Pioneering accessible navigation - indoors and outdoors, May 2020. https://www.blindsquare.com/
4. Bologna, G., Deville, B., Pun, T., Vinckenbosch, M.: Transforming 3D coloured pixels into musical instrument notes for vision substitution applications. EURASIP J. Image Video Process. **2007**(1), 1–14 (2007). https://doi.org/10.1155/2007/76204
5. Bourne, R.R., et al.: Causes of vision loss worldwide, 1990–2010: a systematic analysis. Lancet Glob. Health **1**(6), e339–e349 (2013)
6. Brainport: Disabilities technology: Brainport technologies: United states. https://www.wicab.com/
7. CDC: Social distancing, quarantine, and isolation, May 2020. https://www.cdc.gov/coronavirus/2019-ncov/prevent-getting-sick/social-distancing.html
8. Chahyati, D., Fanany, M.I., Arymurthy, A.M.: Tracking people by detection using CNN features. Procedia Comput. Sci. **124**, 167–172 (2017). https://doi.org/10.1016/j.procs.2017.12.143
9. Coniglio, C., Meurie, C., Lézoray, O., Berbineau, M.: People silhouette extraction from people detection bounding boxes in images. Pattern Recogn. Lett. **93**, 182–191 (2017). https://doi.org/10.1016/j.patrec.2016.12.014
10. Corp, A.T.: Flattening the inaccessibility curve. https://flatteninaccessibility.com/
11. Coughlan, J.M., Miele, J.: AR4VI: AR as an accessibility tool for people with visual impairments. In: 2017 IEEE International Symposium on Mixed and Augmented Reality (ISMAR-Adjunct), pp. 288–292. IEEE (2017)
12. Dakopoulos, D., Bourbakis, N.G.: Wearable obstacle avoidance electronic travel aids for blind: a survey. IEEE Trans. Syst. Man Cybern. Part C (Appl. Rev.) **40**(1), 25–35 (2009)
13. Dale, Z.: Experiences of deafblind persons during the covid-19 outbreak. International Disability Alliance (2020). http://www.internationaldisabilityalliance.org/content/experiences-deafblind-amid-covid-19-outbreak
14. Deveopers, A.: Android NDK. https://developer.android.com/ndk/guides/audio
15. Dewhurst, D.C.: Audiotactile vision substitution system, 7 August 2012, US Patent 8,239,032
16. Eigen, D., Puhrsch, C., Fergus, R.: Depth map prediction from a single image using a multi-scale deep network. arXiv:1406.2283 [cs], June 2014
17. Erhan, D., Szegedy, C., Toshev, A., Anguelov, D.: Scalable object detection using deep neural networks. In: 2014 IEEE Conference on Computer Vision and Pattern Recognition, pp. 2155–2162 (2014)
18. Ess, A., Leibe, B., Van Gool, L.: Depth and appearance for mobile scene analysis. In: 2007 IEEE 11th International Conference on Computer Vision, pp. 1–8 (2007)
19. Ester, M., Kriegel, H.P., Sander, J., Xu, X.: A density-based algorithm for discovering clusters in large spatial databases with noise. In: Proceedings of the Second International Conference on Knowledge Discovery and Data Mining. KDD 1996, pp. 226–231. AAAI Press (1996)

20. Fox, D., Kar, R., Li, A., Pandey, A.: Augmented reality for visually impaired people. https://www.ischool.berkeley.edu/projects/2019/augmented-reality-visually-impaired-people
21. Geiger, A., Lenz, P., Urtasun, R.: Are we ready for autonomous driving? The KITTI vision benchmark suite. In: Conference on Computer Vision and Pattern Recognition (CVPR) (2012)
22. Girshick, R., Donahue, J., Darrell, T., Malik, J.: Rich feature hierarchies for accurate object detection and semantic segmentation. In: 2014 IEEE Conference on Computer Vision and Pattern Recognition, pp. 580–587. IEEE, June 2014. https://doi.org/10.1109/CVPR.2014.81
23. Giudice, N.A., Legge, G.E.: Blind Navigation and the Role of Technology (chap. 25), pp. 479–500. Wiley (2008). https://doi.org/10.1002/9780470379424.ch25
24. Gökçe, F., Üçoluk, G., Sahin, E., Kalkan, S.: Vision-based detection and distance estimation of micro unmanned aerial vehicles. Sensors (Basel, Switzerland) **15**, 23805–23846 (2015)
25. Google: Google glass. https://www.google.com/glass/start/
26. Gordon, K.D.: Survey: the impact of the COVID-19 pandemic on Canadians who are blind deaf-blind, and partially-sighted (2020). http://ccbnational.net/shaggy/wp-content/uploads/2020/05/COVID-19-Survey-Report-Final-wb.pdf
27. Haseeb, M., Guan, J., Ristić-Durrant, D., Gräser, A.: DisNet: a novel method for distance estimation from monocular camera. In: IEEE/RSJ International Conference on Intelligent Robots and Systems (IROS, Spain 2018). IEEE (2018)
28. Haymes, S.A., Johnston, A.W., Heyes, A.D.: Relationship between vision impairment and ability to perform activities of daily living. Ophthalmic Physiol. Opt. **22**(2), 79–91 (2002)
29. Helal, A., Moore, S.E., Ramachandran, B.: Drishti: an integrated navigation system for visually impaired and disabled. In: Proceedings Fifth International Symposium on Wearable Computers, pp. 149–156. IEEE (2001)
30. HTC: HTC VIVE. https://www.vive.com/us/product/
31. Ignatov, A., Timofte, R., Chou, W., Wang, K., Wu, M., Hartley, T., Van Gool, L.: AI benchmark: running deep neural networks on android smartphones. arXiv:1810.01109 [cs], October 2018
32. IrisVision: Wearable low vision glasses for visually impaired, May 2020. https://irisvision.com/
33. Jackson, A.: The hidden struggles America's disabled are facing during the coronavirus pandemic. CNBC News (2020). https://www.cnbc.com/2020/05/10/the-struggles-americas-disabled-are-facing-during-coronavirus-pandemic.html
34. Kajabad, E.N., Ivanov, S.V.: People detection and finding attractive areas by the use of movement detection analysis and deep learning approach. Procedia Comput. Sci. **156**, 327–337 (2019). https://doi.org/10.1016/j.procs.2019.08.209
35. Kajimoto, H., Kanno, Y., Tachi, S.: Forehead electro-tactile display for vision substitution. In: Proceedings of the EuroHaptics (2006)
36. Kempen, G.I., Ballemans, J., Ranchor, A.V., van Rens, G.H., Zijlstra, G.R.: The impact of low vision on activities of daily living, symptoms of depression, feelings of anxiety and social support in community-living older adults seeking vision rehabilitation services. Qual. Life Res. **21**(8), 1405–1411 (2012). https://doi.org/10.1007/s11136-011-0061-y
37. Kulyukin, V., Gharpure, C., Nicholson, J., Pavithran, S.: RFID in robot-assisted indoor navigation for the visually impaired. In: 2004 IEEE/RSJ International Conference on Intelligent Robots and Systems (IROS) (IEEE Cat. No. 04CH37566), vol. 2, pp. 1979–1984. IEEE (2004)

38. Lamoureux, E.L., Hassell, J.B., Keeffe, J.E.: The determinants of participation in activities of daily living in people with impaired vision. Am. J. Ophthalmol. **137**(2), 265–270 (2004)
39. Lee, J., et al.: On-device neural net inference with mobile GPUs. arXiv:1907.01989 [cs, stat], July 2019
40. Liu, W., et al.: SSD: single shot multibox detector. In: Leibe, B., Matas, J., Sebe, N., Welling, M. (eds.) ECCV 2016. LNCS, vol. 9905, pp. 21–37. Springer, Cham (2016). https://doi.org/10.1007/978-3-319-46448-0_2
41. Loomis, J.M., Golledge, R.G., Klatzky, R.L.: GPS-based navigation systems for the visually impaired. In: Fundamentals of Wearable Computers and Augmented Reality, p. 429, 46 (2001)
42. Loomis, J.M., Golledge, R.G., Klatzky, R.L., Speigle, J.M., Tietz, J.: Personal guidance system for the visually impaired. In: Proceedings of the First Annual ACM Conference on Assistive Technologies, pp. 85–91 (1994)
43. MacDonald, J.A., Henry, P.P., Letowski, T.R.: Spatial audio through a bone conduction interface: Audición espacial a través de una interfase de conducción ósea. Int. J. Audiol. **45**(10), 595–599 (2006). https://doi.org/10.1080/14992020600876519
44. Maidenbaum, S., Abboud, S., Amedi, A.: Sensory substitution: Closing the gap between basic research and widespread practical visual rehabilitation. Neurosci. Biobehav. Rev. **41**, 3–15 (2014)
45. Microsoft: Microsoft Hololens. https://www.microsoft.com/en-us/hololens
46. Microsoft: Seeing AI. https://www.microsoft.com/en-us/ai/seeing-ai
47. Milan, A., Leal-Taixe, L., Reid, I., Roth, S., Schindler, K.: MOT16: a benchmark for multi-object tracking. arXiv:1603.00831 [cs], May 2016
48. Nilsson, M.E., Schenkman, B.N.: Blind people are more sensitive than sighted people to binaural sound-location cues, particularly inter-aural level differences. Hear. Res. **332**, 223–232 (2016).https://doi.org/10.1016/j.heares.2015.09.012, https://linkinghub.elsevier.com/retrieve/pii/S0378595515300174
49. Okonji, P.E., Ogwezzy, D.C.: Awareness and barriers to adoption of assistive technologies among visually impaired people in Nigeria. Assist. Technol. **31**(4), 209–219 (2019)
50. OpenAL: Open audio library (2020). https://openal.org/. Accessed 20 July 2020
51. Praveen, S.: Efficient depth estimation using sparse stereo-vision with other perception techniques (chap. 7). In: Radhakrishnan, S., Sarfraz, M. (eds.) Coding Theory. IntechOpen, Rijeka (2020). https://doi.org/10.5772/intechopen.86303
52. Qiu, S., Han, T., Osawa, H., Rauterberg, M., Hu, J.: HCI design for people with visual disability in social interaction. In: Streitz, N., Konomi, S. (eds.) DAPI 2018. LNCS, vol. 10921, pp. 124–134. Springer, Cham (2018). https://doi.org/10.1007/978-3-319-91125-0_10
53. Redmon, J., Divvala, S., Girshick, R., Farhadi, A.: You only look once: unified, real-time object detection. In: Proceedings of the IEEE Conference on Computer Vision and Pattern Recognition, pp. 779–788 (2016)
54. Redmon, J., Farhadi, A.: YOLOv3: an incremental improvement. arXiv:1804.02767 [cs], April 2018
55. Ren, S., He, K., Girshick, R., Sun, J.: Faster R-CNN: towards real-time object detection with region proposal networks. In: Advances in Neural Information Processing Systems, pp. 91–99 (2015)
56. Rezaei, M., Terauchi, M., Klette, R.: Robust vehicle detection and distance estimation under challenging lighting conditions. IEEE Trans. Intell. Transp. Syst. **16**(5), 2723–2743 (2015)

57. Roentgen, U.R., Gelderblom, G.J., Soede, M., De Witte, L.P.: Inventory of electronic mobility aids for persons with visual impairments: a literature review. J. Vis. Impair. Blindness **102**(11), 702–724 (2008)
58. Kanagamalliga, S., Vasuki, S.: Contour-based object tracking in video scenes through optical flow and Gabor features. Optik **157**, 787–797 (2018). https://doi.org/10.1016/j.ijleo.2017.11.181
59. Sandler, M., Howard, A., Zhu, M., Zhmoginov, A., Chen, L.C.: MobileNetV2: inverted residuals and linear bottlenecks (2018)
60. Sandler, M., Howard, A., Zhu, M., Zhmoginov, A., Chen, L.C.: MobileNetV2: inverted residuals and linear bottlenecks. arXiv:1801.04381 [cs], March 2019
61. Simonyan, K., Zisserman, A.: Very deep convolutional networks for large-scale image recognition. arXiv:1409.1556 [cs], April 2015
62. Singh, P., Deepak, B., Sethi, T., Murthy, M.D.P.: Real-time object detection and tracking using color feature and motion. In: 2015 International Conference on Communications and Signal Processing (ICCSP), pp. 1236–1241. IEEE, April 2015. https://doi.org/10.1109/ICCSP.2015.7322705
63. Stelmack, J.: Quality of life of low-vision patients and outcomes of low-vision rehabilitation. Optom. Vis. Sci. **78**(5), 335–342 (2001)
64. Tuohy, S., O'Cualain, D., Jones, E., Glavin, M.: Distance determination for an automobile environment using inverse perspective mapping in OpenCV. In: IET Irish Signals and Systems Conference (ISSC 2010), pp. 100–105. IET (2010). https://doi.org/10.1049/cp.2010.0495
65. Valve: Valve index. https://store.steampowered.com/valveindex
66. WHO: Global data on visual impairment 2010 (2010). https://www.who.int/blindness/GLOBALDATAFINALforweb.pdf
67. Zhang, R., Li, Y., Zhang, A.L., Wang, Y., Molina, M.J.: Identifying airborne transmission as the dominant route for the spread of COVID-19. Proc. Natl. Acad. Sci. **117**(26), 14857–14863 (2020). https://doi.org/10.1073/pnas.2009637117
68. Zhu, J., Fang, Y.: Learning object-specific distance from a monocular image. In: 2019 IEEE/CVF International Conference on Computer Vision (ICCV), pp. 3838–3847. IEEE, Seoul, Korea, October 2019. https://doi.org/10.1109/ICCV.2019.00394

Enhancing Robot-Assisted WEEE Disassembly Through Optimizing Automated Detection of Small Components

Ioannis Athanasiadis(✉), Athanasios Psaltis, Apostolos Axenopoulos, and Petros Daras

The Visual Computing Lab - Centre for Research and Technology Hellas/Information Technologies Institute, Thessaloniki, Greece
{athaioan,at.psaltis,axenop,daras}@iti.gr

Abstract. Automated detection of small objects poses additional challenges, compared to bigger-sized ones, due to the former's limited resolution for extracting discriminative information. In such cases, even a slight misalignment between a candidate region and its ground truth target has a huge impact on their IoU which significantly increases the amount of noisy information. Given the fact that state of the art two-stage detection algorithms generate predefined shaped and sized candidate regions in pixel-level interval, the aforementioned misalignments are very likely to occur. In this work, a scalable object detection approach is introduced -specifically dedicated to small object parts- incorporating both learnable and handcrafted features. In particular, a set of simplified Gabor waveforms (SGWs) is applied to the raw data, ultimately producing an improved set of anchors for the region proposal network. These Gabor filters are further utilized generating a soft attention mask. Additionally, the interaction of a human with the object is also exploited by taking advantage of affordance-based information for further improvement of detection performance. Experiments have been conducted in a newly introduced device disassembly segmentation dataset, demonstrating the robustness of the method in detection of small device components.

Keywords: Small object detection · WEEE disassembly

1 Introduction

With the advent of industry 4.0, the role of robotics in the industrial environments has evolved. Traditional industrial robots have started being replaced by collaborative robots. The rationale behind this selection is, instead of using high-precision but also dangerous traditional robots in fully automatized processes, to exploit the ability of collaborative robots to coexist with humans in a fence-less way, in order to assist the latter in solving complex cognitive tasks. Some

A. Bartoli and A. Fusiello (Eds.): ECCV 2020 Workshops, LNCS 12538, pp. 440–454, 2020.
https://doi.org/10.1007/978-3-030-66823-5_26

examples include automated parts assembling or disassembling. More specifically, in the context of a Waste Electrical and Electronic Equipment (WEEE) disassembly scenario, within an industrial WEEE recycling environment, the fully-robotised disassembly is not feasible due to the complexity and high variability of devices. Thus, the role of a collaborative robot in assisting the human worker in detecting and removing hazardous components from the electronic devices is much appreciated. In this direction, Computer Vision is necessary to assist the robot's perception of the surrounding environment. Nevertheless, recognition of the small components to be disassembled is a challenging task, since current state of the art computer vision approaches fail to detect objects in very low resolution. In this work, a novel methodology is proposed for effective detection of low-resolution objects, which makes it suitable for automated detection of very small components in robot-assisted WEEE disassembly tasks.

2 Related Work

Object detection is the process of localizing and classifying the objects appeared on an image. Moreover, its numerous applications, ranging from self driving cars to medical image processing along with its importance in providing the machines with the ability to perceive the world, have attracted many researchers to this field. A plethora of approaches have been proposed for the object detection task that can be categorized into two broad groups, namely the two-stage and the one-stage methods, while there is a complementary set of algorithms that aims at enhancing the previous two. One of the first attempts to utilise Convolutional Neural Networks (CNNs) in object detection was the R-CNN [3] in which a number of class-agnostic candidate regions are proposed and fed to a CNN to extract a fixed-length feature descriptor for each region. Thereafter, a set of class-specific Support-Vector Machines (SVMs), classifies these regions based on their extracted descriptors. Built upon R-CNN success, the Fast R-CNN [2] targets the inefficiency of having to pass each of the candidate regions individually through the CNN by forward passing the input image to the network once, generating its feature map and applying ROI pooling for each of the candidate regions to extract their feature representations. Based on the previously mentioned methods, Faster R-CNN [12] introduced a trainable mechanism for the purpose of proposing candidate regions called Regional Proposal Network (RPN). Given a number of fixed shape and size regions, called *anchors*, the RPN distinguishes them between foreground and background before passing the former to the classifier. Mask R-CNN [4] extended the Faster R-CNN by adding an extra head for segmentation and replaced the ROI pooling with ROI align resulting in higher accuracy predictions. Guided Anchoring [16] proposed an approach detaching the hyper-parameterizing needed for the anchoring process. Additional classifiers are added in Cascade R-CNN [1] aiming at progressively increasing the IOU's of the proposed regions with the ground truth objects resulting in improved prediction performance. Prior-knowledge in interpreted in the object detection process by Reasoning R-CNN [17] which consists of two cascade classification levels, in the first one only visual information is considered, while the second one capitalizes on more informative feature descriptors complemented by high-level information as encoded by the reasoning module. In contrary to the R-CNN family methods in which the processes of region proposing and region classification are done by discrete modules, in the one-stage methods the regions are generated and classified in a single pass manner. When it comes to one-stage object detection approaches YOLO [11] and the SSD [9] are the most indicative ones. Although this category of methods offers faster performance compared to the RPN-based one, they are limited in terms of accuracy due to having a high imbalance between positive and negative regions fed to the classifier, where the positive and negative terms refer to the presence and the absence of ground truth object respectively. A novel Focal loss has been proposed in [8] addressing that imbalance by having the ambiguous regions contribute more in the loss calculation, thus valuing the hard examples more than the easily classified ones. The potential of exploiting heuristic information for the purpose of sampling the generated anchors, that are more likely to include objects, are presented in [19] and [18] with promising results. Another method based on handcrafted features is [14], where the High Possible Regions Proposal Network, similarly to how RPN operates, proposes candidate regions given an additional feature map as generated by the application of a set of simplified Gabor wavelets (SGWs) on the input image. The use of context information is adapted in [13] focusing on detecting small objects, that the baseline one and two-stage methods struggle with. In [6] the Perceptual Generative Adversarial Network is introduced targeted at enriching the poor visual representation of small objects by super-resolving them. Although Deep NNs have proved to be quite powerful, given sufficient data, they still rely heavily on large datasets and informative visual representations. In the case of small objects specifically, often none of the previous requirements are met thus their detection frequently fails. The motivation behind this work is to effectively boost the performance of current state of the art object detection methods in detecting small objects by exploiting additional streams of information to make up for the poor visual quality small object posses. The main contributions of our work comprise the effective incorporation of handcrafted features into the object detection process through either an anchoring or a soft attention mechanism guided by SGWs. Finally, we have also tested the introduction of an additional input stream based on *object affordances*, with the objective of further improving the detection accuracy. The aforementioned term of object affordance, refers to all possible ways a specific object can be manipulated during its usage.

3 Our Approach

3.1 Overview

Most of the Computer Vision domains have been greatly benefited from the Deep Learning (DL) era. Regarding the object detection specifically, replacing the initial handcrafted feature extraction process with deep data-driven architectures, has shown considerable potential and displayed remarkable results. Although current state of the art object detection algorithms achieve sufficient object detection accuracy, they perform disproportionately better at detecting big and medium sized objects compared to the smaller ones. The representation generated by the DL-based models, in the case of small objects, are mostly noisy and poor in terms of quality. The former relying heavily on rich feature representation for both object localization and classification combined with the lacking small objects visual information, results in unsatisfactory object detection. Nevertheless, there are applications in which detecting small objects is of high importance. The motive behind this work is to investigate various ways of increasing small object detection performance by exploiting additional information streams. A suitable baseline method is chosen which we progressively enhance through using both handcrafted feature as well as an additional stream of human-to-object interaction information. In the context of this work the detection performance is our main priority, thus we focus on enhancing the two-stage object detection algorithms, nevertheless our proposed method can be mildly modified in order to be applicable to one-stage approaches as well.

3.2 Two-Stage Detection Pipeline

State of the art two-stage detection approaches are consisted of two discrete modules responsible for region proposing and classifying respectively. In the first stage, a set of regions are being validated on their objectness; namely the possibility of containing a ground truth object. The most confident regions, in terms of objectness, are passed to the second stage. In the second stage, a feature representation is extracted for each proposed region which is indicative of the area the latter occupy on the image. Finally each region is classified into one the available categories based on their extracted feature.

First Stage - RPN: Given that target objects may appear anywhere on the image, an anchoring scheme is deployed to generate a number of densely distributed anchors. These anchors are generated on a pixel basis across the feature map and have their size and shape defined by the hyperparameters of scale and ratio respectively. These parameters shall be fine-tuned based on the specific application through carefully considering the input image resolution, the potential shapes of the objects interested in detecting as well as their size relatively to the input. Finally, all the uniformly generated anchors constitute the *candidate regions* and are passed to the RPN. Each candidate region is labeled as foreground region if its intersection over union (IOU) with any ground truth object exceeds a predefined threshold and as background otherwise. Finally the RPN is trained to classify its input regions between foreground and background as well as refines those falling in the former category, in order to better fit their corresponding ground truth targets. The refined foreground regions compose the *proposed regions.*

Second Stage - Region Classification: For each proposed region a feature representation is extracted by pooling onto the feature maps that were previously used for regional proposing. Then these regions are classified into one of the available categories and are further refined by class-aware regressors. The feature maps being shared among the two modules allow for region proposing and classifying modules to be trained simultaneously.

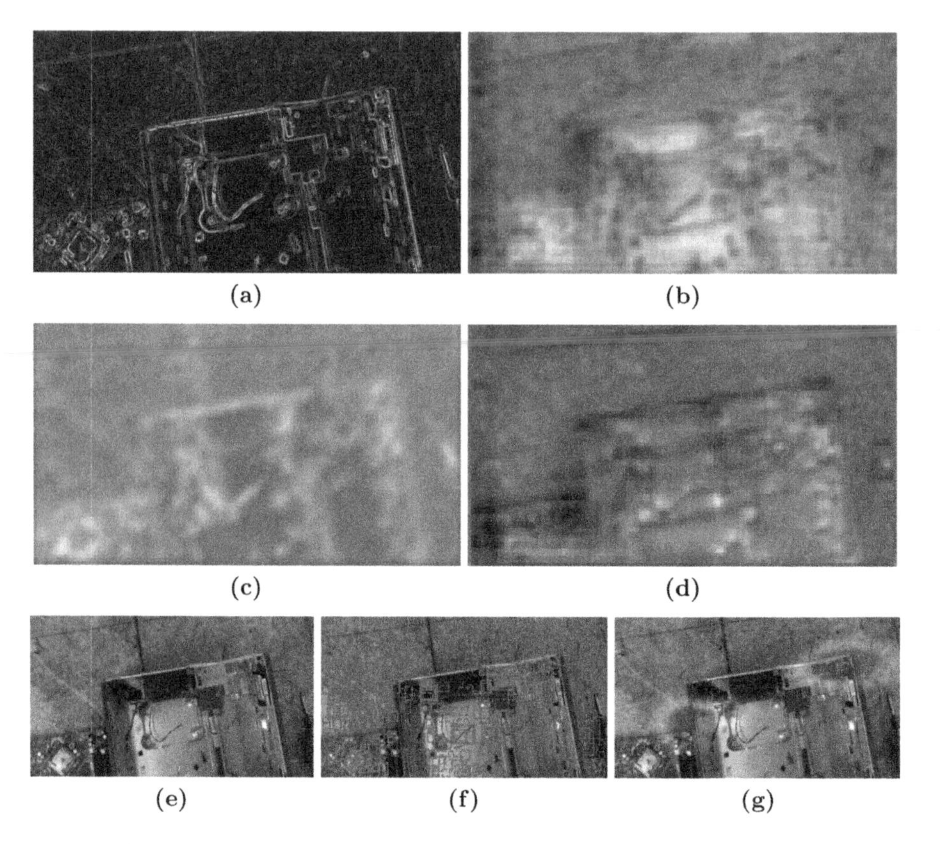

Fig. 1. (**a**) Gabor feature. (**b**) F_s. (**c**) Gabor-driven soft attention. (**d**) Soft attention gain (**e**) Input image. (**f**) Edge anchors. (**g**) Affordance mask.

3.3 Background

Mask R-CNN: In cases where heavy overlapping between the relevant objects occurs, detecting them by their bounding boxes would result in high ambiguity thus instance segmentation is deemed to be more appropriate. For that purpose Mask R-CNN [4] was chosen as a base architecture for both its state of the art performance and its efficiency. Besides that, Mask R-CNN architecture having discrete feature maps for different object sizes through utilizing the Feature

Pyramid Network (FPN) as proposed in [7], renders it even more appealing approach in cases where small sized object detection is required.

Cascade R-CNN: The method as described in [1] is applied on the baseline architecture, with the purpose of training higher *quality* classifiers. The term quality of a classifier refers to the IOU threshold, a proposed region needs to exceed to be considered as ground truth target, which the classifier was trained with. The problem with directly increasing the IOU threshold is that only a handful of candidate region would meet such strict IOU criterion resulting in insufficient training samples. In the context of each candidate being refined to better fit its target, the candidate regions are fed through multiple classifiers of increasing quality to progressively increase their IOU before being passed to the following classifier of higher quality.

3.4 Anchoring by SGWs

EA-CNN$_P$: Although the uniform anchoring has proved to be quite effective in most cases, it is still limited to generating anchors in discrete pixel intervals with fixed shapes and scales resulting in misalignment between the anchors and their respective ground truth targets, which can be crucial in the case of small objects detection. In order to restrict these deviations, additional anchors are generated considering heuristic information which, unlike the densely distributed anchors, are not bounded by any extrinsic hyper parameters; thus tend to better align with the ground truth objects. In Fig. 3, we depict a candidate-target IoU comparison between the edge and the default anchors. In order to maintain efficiency while still improving in detection quality, only the small-sized heuristically generated regions, referred to as *edge anchors* (Fig. 1f), are considered; since the minor misalignments the bigger object exhibit have barely any effect on their

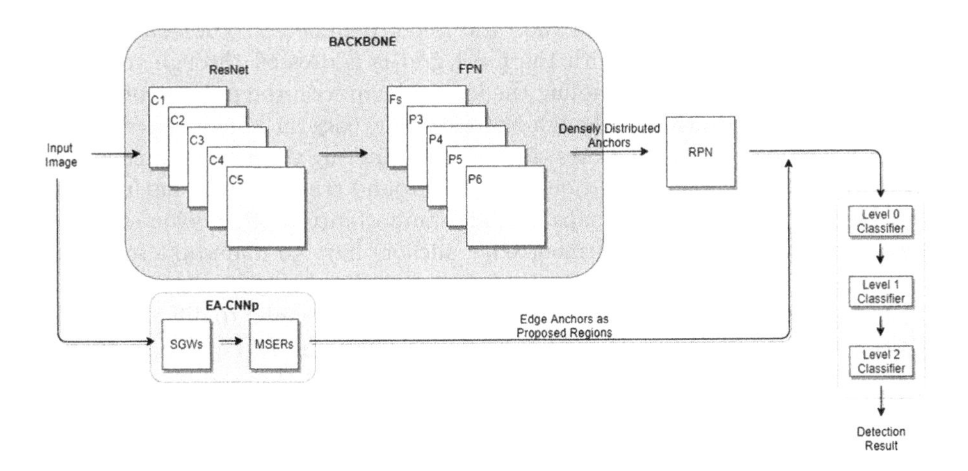

Fig. 2. The architecture combining Cascade with EA-CNN$_P$.

detection. Inspired by [14], the input image is filtered by a set of simplified Gabor wavelets (SGWs) producing an edge-enhanced image, termed as *Gabor feature* (Fig. 1a). The MSER algorithm is then applied on that feature, to extract the edge anchors. Finally in this approach, all edge anchors are merged with the proposed regions and are fed jointly to the classification stage. The architecture described above combined with the cascade R-CNN is shown in Fig. 2.

Edge anchor IoU: **0.74**
Default anchor IoU: **0.54**

Edge anchor IoU: **0.74**
Default anchor IoU: **0.64**

Fig. 3. Displaying the anchors that align the most with the groud truth target. The green, red and blue regions refer to edge anchors, densely distributed (default) anchors and their corresponding ground truth targets respectively. (Color figure online)

EA-CNN$_C$: As a next step, we aim at integrating the edge anchors into the RPN. Due to the former being of varying scale and having continuous center coordinates, some modifications are required so as to be compatible with the RPN training procedure. In order for the FPN feature map to remain scale specific and have the bounding box regressors referring to identical shaped anchors, the edge anchors are refined to match the closest available shape and size configuration, as dictated by the scale and ratio hyper parameters. The issue of edge anchor centers not aligning with the pixel grid is addressed through rounding their centers along with upsampling the feature map with the purpose of lessening the quantization error. Although restricting the edge anchors into predefined shapes and sizes, partially opposes the sense of acquiring the best alignment possible between the candidate regions and the ground truth targets, having scale and shape consistent feature maps is of high importance for regression stability. In order to minimize the refinement edge anchors have to undertake to fit the predefined scale configurations, we introduce additional feature map dedicated to the edge anchors, called *edge* maps. These maps correspond to different scales relevant to small objects and are identical to the feature map of the first FPN level (F_s). After the modifications described above the RPN is able to evaluate regions given both edge and regular anchors as input. Based that on that, MSER is applied to both grayscale input image and its edge-enhanced version resulting in more but less precise edge anchors. The proposed architecture is shown bellow in Fig. 4.

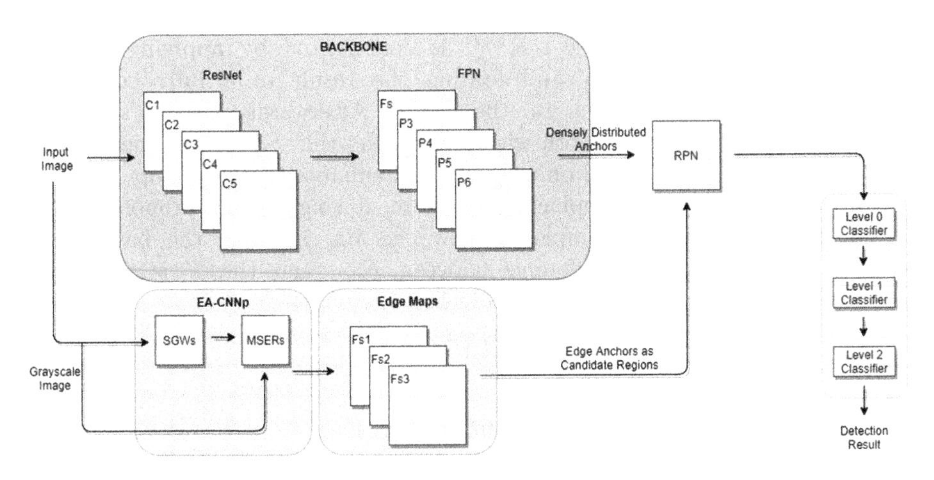

Fig. 4. The architecture combining Cascade with EA-CNN$_C$.

3.5 Attention-Based by SGWs

Attention-SE: In the previously described approach, the Gabor feature (G_f) is used to guide an additional anchoring mechanism targeted at small objects. Although the anchors generated by these methods seem to be reasonable, the pre-processing step required, does not allow for an end-to-end training. Therefore, we propose an architecture in which a soft attention mask $Att_{\text{soft}} \in \mathbb{R}^{\frac{M}{4} \times \frac{N}{4} \times D}$ is generated driven by the G_f, where D is a hyperparameter defining the depth of the FPN feature maps while M and N refer to the input image dimensions. More specifically, at first the $G_f \in \mathbb{R}^{M \times N \times 3}$ is passed through a CNN called *GabA* resulting in $G_A \in \mathbb{R}^{\frac{M}{4} \times \frac{N}{4} \times D}$ as shown in the Eq. 1. Thereafter based on the Squeeze and Excitation [5] approach, a

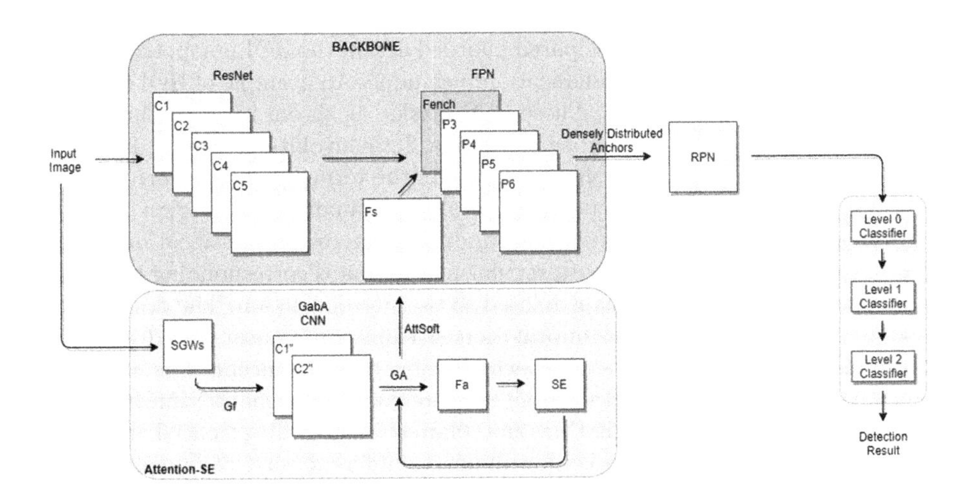

Fig. 5. The architecture combining Cascade with Attention-SE.

depth-wise attention vector $SE \in \mathbb{R}^{1 \times D}$ is constructed by applying average global pooling onto the G_A and feeding the result to a fully connected layer $F_a \in \mathbb{R}^{D \times D}$ as shown in the Eq. 2. Afterwards, the soft attention mask is calculated through depth-wise multiplying the SE with the G_a as shown in Eq. 3. Finally, based on the Eq. 4, an enhanced feature map F_{enh} is generated, which will be replacing F_s both during region proposing and classification stages. An example on how the Att_{soft} alters the base F_s is shown in Fig. 1d, where the difference between F_{enh} and the F_s is visualised. The proposed architecture can be seen in Fig. 5.

$$G_A = ReLU[GabA(G_f)] \tag{1}$$

$$SE = Softmax[F_a(G_A)] \tag{2}$$

$$Att_{\text{soft}} = ReLU[SE \otimes G_A] \tag{3}$$

$$F_{enh} = F_s \otimes Att_{\text{soft}} \tag{4}$$

Attention-SE Enhanced by Human-Object Interaction (HOI) Information: Motivated by the work presented in [15], which achieves increased object recognition performance by exploiting affordance-based knowledge, we incorporate a stream related to that kind of information. More specifically, we enhance the Attention-SE by using the method presented in [10], where potential interaction hotspots are predicted given a static image. A brief description on how their proposed method works is required, in order to better interpret its prediction results. The so-called hotspot prediction is achieved through three discrete stages. At first a typical action recognition model has been trained to predict the various action occurring in video sequences where objects of interest are being manipulated by the human. As a second step, an action anticipation model is trained to map a static image into its corresponding afforded actions. Finally in the third step, given the anticipated afforded action the HOI hotspots are generated by applying a feature visualization technique. An example of HOI hotspot prediction can be seen Fig. 1g. These HOI masks, as shown in the architecture of Fig. 6, are passed through a set of vanilla 2D convolutional layers in order to generate the *affordance feature maps*, where the term *affordance* corresponds to the way the human interacts with the object. Finally, while region proposing in based solely on visual information, during the region classification stage, the ROI-pooling is applied to both regular feature maps corresponding to visual information and the feature maps related to the afforded action. The descriptors generated from these discrete information streams, are concatenated and fed to the classifier. Through this approach classification enhancement is achieved by capitalizing on both visual (sensor) and human-object (motor) information. Finally, the hotspots prediction are not limited in providing spatial information solely, since discrete colorization indicates different ways of human-object manipulations. The described architecture is shown in Fig. 6.

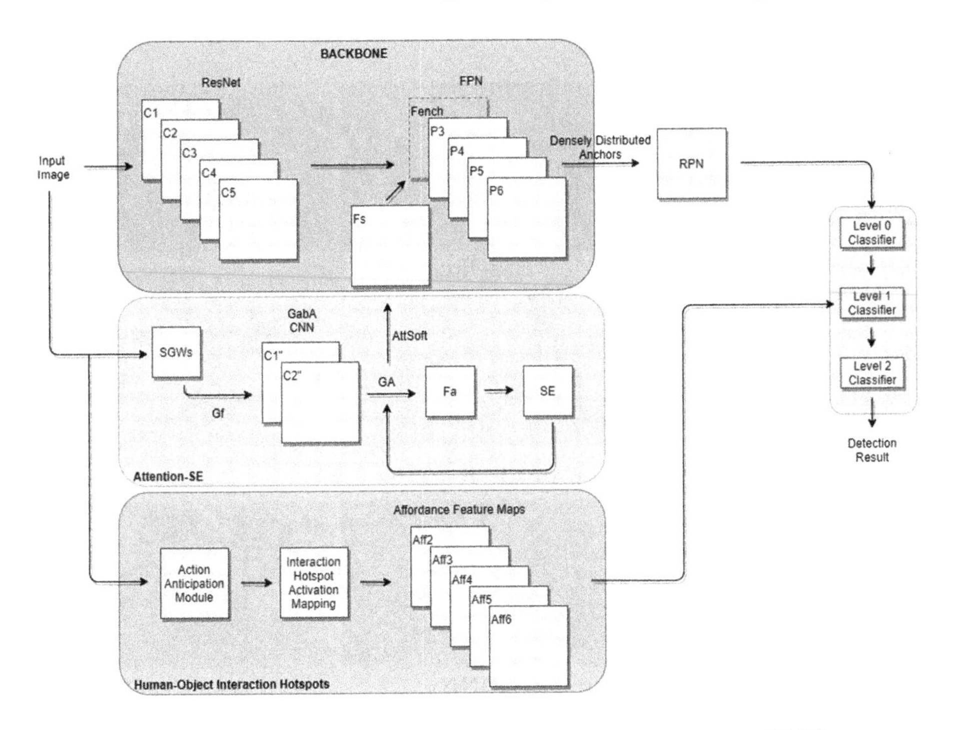

Fig. 6. Attention-SE enhanced by Human-Object Interaction (HOI).

4 Experiments

4.1 Dataset Construction

A set of WEEE disassembly procedures were recorded, in the form of video at the industrial environment. All the recordings were carried out through three cameras, two of which were in fixed position while the other one, being hand-held, resulted in more informative views. Thereafter, a number of frames utilizing all three views, were manually annotated aiming at maximizing the variance in terms of unique annotations. A binary mask was defined for each WEEE component as well as their corresponding WEEE device, distinguishing them from the background environment as well as other instances.

Dataset: Although, our dataset consists of multiple WEEE categories, in this work we focus on PC-Towers explicitly, since it was found to be the most challenging due to the presence of class ambiguities, high occlusions and significant small components. The overall annotated frames referring to the PC-Tower is 395 originated from five unique PC-Tower disassembly procedures. In order to avoid having similar looking frames during training and evaluation, we split the dataset on a procedure basis, that is, the training set is formed by getting the frames only from four disassembly session and leaving the remaining one for evaluation purposes. The dataset split as described previously, results in 325 and 70

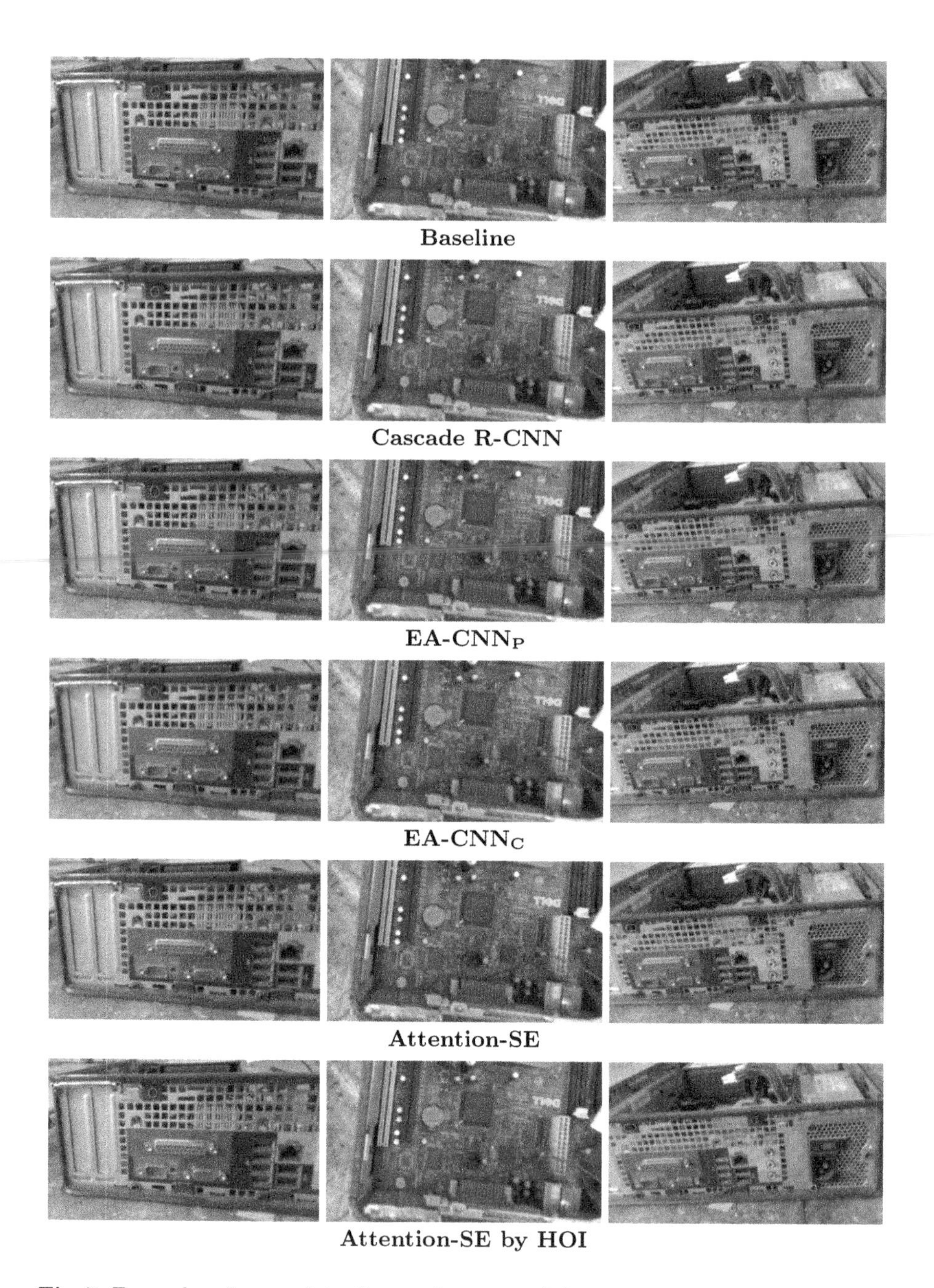

Fig. 7. Examples of screw detection performance of the various approaches. The green and red outlines indicate the successful and the failed screw detection respectively. (Color figure online)

of fully annotated frames in the training and validation sets respectively. Moreover, there are 23 unique components-categories with all of them being present in the training set while only 16 of them in validation set.

4.2 Evaluation Metrics

To evaluate the performance of each of the modalities added, we report the standard COCO mean Average Precision (mAP) $AP_{0.5:0.95}$, $AP_{0.5}$, $AP_{0.75}$, and mean Average Recall (mAR) $AR_{0.5:0.95}$ metrics. Moreover, since we are aiming at small object detection specifically, we also report the metrics of AP_S and AR_S averaged over the $[0.5 : 0.95]$ IoU range thresholds.

4.3 Implementation Details

In all implementations, the ResNet-101 backbone was used for feature extraction combined with the FPN neck. The network's weights were initialized using a model pretrained on MS COCO dataset. The input images were resized such that they biggest dimension is 512 pixel wide, while their aspect ratio is retained. The Stochastic gradient descent (SDG) optimizer was used with a momentum value of 0.9. The model was trained for 400 epochs, using a mini batch size of one image, with an initial learning rate of 0.001 decayed by a factor of 3 at epoch 100 and 250. During the first 100 epochs the backbone layers were kept frozen while the whole network was trained thereafter. Non Maximum Suppression (NMS) was applied both during the proposing and classification stages with thresholds of 0.9 and 0.3 respectively. Regarding the uniform distributed anchors, they are generated using ratios of 0.5, 1, and 2 while their scales were set to $[10, 32, 64, 128, 256]$ targeting objects of various sizes. The RPN was set to propose 2000 candidate regions at training, while during testing 1000 regions are proposed and only the 100 most confident predictions are kept. Finally, data augmentation was applied to address the relatively small dataset, by applying random rotation as well as adding random motion blur noise during training.

Gabor-Based Anchoring: Finally, for the purpose of generating the *edge* anchors, the regions produced by the MSER algorithm occupying an area larger than 15^{1^5} pixels or having their aspect ratio fall outside the $[0.5, 2]$ interval, are filtered out. Moreover, when the edge anchors are treated as candidate regions additional scales of 8,15 and 25 pixels are introduced.

Human-Object Interaction Information: The visual information is encoded into a descriptor of 256 length while the ones referring to based on the Human-Object Interaction information have a length of 64.

Table 1. Object detection results on PC-Tower dataset.

Method	Cascade	$AP_{0.5:0.95}$	$AP_{0.5}$	$AP_{0.75}$	AP_S	$AR_{0.5:0.95}$	AR_S
Baseline	-	40.3	69.8	39.5	23.1	47.3	28.0
	✓	39.5	66.2	40.1	23.6	**48.8**	29.0
EA-CNN$_P$	-	38.7	68.2	38.0	22.9	46.9	28.8
	✓	40.9	**70.5**	42.3	24.6	48.2	30.1
EA-CNN$_C$	-	39.6	69.5	40.1	25.5	47.2	31.4
	✓	40.6	69.5	41.9	25.2	48.4	**32.3**
Attention-SE	-	**41.0**	70.1	41.2	24.0	46.4	31.0
	✓	40.7	69.5	**42.6**	25.0	48.3	28.4
Attention-SE by Affordance	✓	38.9	68.6	38. 9	**25.7**	48.7	31.5

4.4 Results

Based on the results presented in Table 1, the best overall $AP_{0.5}$ is achieved when edge anchors are fed directly to the classification stage without any sort of refinement. On the other hand, when more strict IoU thresholds are required, Attention-SE is the most dominant method in terms of mAP. Comparing the methods utilising Gabor-based anchoring, superior small object detection is accomplished when edge anchors are integrated into the RPN module. Regarding the Attention-based approaches, the one capitalizing on the affordance modality performs better at detecting the small objects. Moreover, it is evident that the best performances have been achieved when the cascade architecture is deployed. Additionally, the qualitative displayed in Fig. 7, referring to screws, are indicative of how each of the proposed method can enhance the baseline in terms of small objects detection quality. Finally, our disassembly dataset consisted of highly variant views, greatly occluded and significantly small components arises challenges that affect the object detection performance. Although, the mAP metrics are relatively low compared to other object detection benchmarks, the detection performance is deemed to be reasonable considering the challenging nature of our dataset.

4.5 Conclusions

In this work we investigated various approaches aiming at boosting the small object detection. At first a set of handcrafted features are generated in order to guide an additional anchoring mechanism targeted specifically at small-sized objects. Applying such anchoring mechanism into the detection pipeline yielded promising results. Additionally, Attention-based approaches were also considered, making use of the previously mentioned features to generate a soft attention mask targeted at small objects. Moreover, build upon the proposed soft attention approach, we further enhance the object detection by taking advantage of an additional stream of information based on Human-Object interaction. Finally our proposed approach has generalization capabilities and is seamlessly applicable to any two-stage object detection approach.

4.6 Future Work

In the future, experiments are to be conducted on the whole disassembly dataset including all four WEEE devices. Moreover, our dataset will be further enriched by annotating additional disassembly sessions on different WEEE disassembly plant premises. Finally, although the authors of [10] state that relatively accurate interaction hotspots can be generated even in cases of unseen object categories, as a future work we consider training their proposed model on the WEEE domain in order to obtain more accurate hotspot interaction prediction and subsequently increase the detection performance.

Acknowledgment. This work was supported by the European Commission under contract H2020-820742 HR-Recycler.

References

1. Cai, Z., Vasconcelos, N.: Cascade R-CNN: high quality object detection and instance segmentation. IEEE Trans. Pattern Anal. Mach. Intell. (2019). https://doi.org/10.1109/tpami.2019.2956516
2. Girshick, R.: Fast R-CNN object detection with caffe. Microsoft Research (2015)
3. Girshick, R., Donahue, J., Darrell, T., Malik, J.: Rich feature hierarchies for accurate object detection and semantic segmentation. In: Proceedings of the IEEE Conference on Computer Vision and Pattern Recognition, pp. 580–587 (2014)
4. He, K., Gkioxari, G., Dollar, P., Girshick, R.: Mask R-CNN. In: The IEEE International Conference on Computer Vision (ICCV), October 2017
5. Hu, J., Shen, L., Sun, G.: Squeeze-and-excitation networks. In: Proceedings of the IEEE Conference on Computer Vision and Pattern Recognition, pp. 7132–7141 (2018)
6. Li, J., Liang, X., Wei, Y., Xu, T., Feng, J., Yan, S.: Perceptual generative adversarial networks for small object detection. In: Proceedings of the IEEE Conference on Computer Vision and Pattern Recognition, pp. 1222–1230 (2017)
7. Lin, T.Y., Dollár, P., Girshick, R., He, K., Hariharan, B., Belongie, S.: Feature pyramid networks for object detection. In: Proceedings of the IEEE Conference on Computer Vision and Pattern Recognition, pp. 2117–2125 (2017)
8. Lin, T.Y., Goyal, P., Girshick, R., He, K., Dollár, P.: Focal loss for dense object detection. In: Proceedings of the IEEE International Conference on Computer Vision, pp. 2980–2988 (2017)
9. Liu, W., et al.: SSD: single shot multibox detector. In: Leibe, B., Matas, J., Sebe, N., Welling, M. (eds.) ECCV 2016. LNCS, vol. 9905, pp. 21–37. Springer, Cham (2016). https://doi.org/10.1007/978-3-319-46448-0_2
10. Nagarajan, T., Feichtenhofer, C., Grauman, K.: Grounded human-object interaction hotspots from video. In: Proceedings of the IEEE International Conference on Computer Vision, pp. 8688–8697 (2019)

11. Redmon, J., Divvala, S., Girshick, R., Farhadi, A.: You only look once: unified, real-time object detection. In: Proceedings of the IEEE Conference on Computer Vision and Pattern Recognition, pp. 779–788 (2016)
12. Ren, S., He, K., Girshick, R., Sun, J.: Faster R-CNN: towards real-time object detection with region proposal networks. In: Advances in Neural Information Processing Systems, pp. 91–99 (2015)
13. Ren, Y., Zhu, C., Xiao, S.: Small object detection in optical remote sensing images via modified faster R-CNN. Appl. Sci. **8**(5), 813 (2018)
14. Shao, F., Wang, X., Meng, F., Zhu, J., Wang, D., Dai, J.: Improved faster R-CNN traffic sign detection based on a second region of interest and highly possible regions proposal network. Sensors **19**(10), 2288 (2019)
15. Thermos, S., Papadopoulos, G.T., Daras, P., Potamianos, G.: Deep affordance-grounded sensorimotor object recognition. In: Proceedings of the IEEE Conference on Computer Vision and Pattern Recognition, pp. 6167–6175 (2017)
16. Wang, J., Chen, K., Yang, S., Loy, C.C., Lin, D.: Region proposal by guided anchoring. In: Proceedings of the IEEE Conference on Computer Vision and Pattern Recognition, pp. 2965–2974 (2019)
17. Xu, H., Jiang, C., Liang, X., Lin, L., Li, Z.: Reasoning-RCNN: unifying adaptive global reasoning into large-scale object detection. In: Proceedings of the IEEE Conference on Computer Vision and Pattern Recognition, pp. 6419–6428 (2019)
18. Zhang, J., Zhang, J., Yu, S.: Hot anchors: a heuristic anchors sampling method in RCNN-based object detection. Sensors **18**(10), 3415 (2018)
19. Zitnick, C.L., Dollár, P.: Edge boxes: locating object proposals from edges. In: Fleet, D., Pajdla, T., Schiele, B., Tuytelaars, T. (eds.) ECCV 2014. LNCS, vol. 8693, pp. 391–405. Springer, Cham (2014). https://doi.org/10.1007/978-3-319-10602-1_26

Structural Plan of Indoor Scenes with Personalized Preferences

Xinhan Di[1(✉)], Pengqian Yu[2], Hong Zhu[1], Lei Cai[1], Qiuyan Sheng[1], Changyu Sun[1], and Lingqiang Ran[3]

[1] Technique Center Ihome Corporation, Nanjing, China
deepearthgo@gmail.com, jszh0825@gmail.com, caileitx1990@gmail.com, shenqiuyan123@gmail.com, sunchangyu@gmail.com
[2] IBM Research, Singapore, Singapore
peng.qian.yu@ibm.com
[3] Shandong University of Finance and Economics, Jinan, China
ranlingqiang@sdufe.edu.cn

Abstract. In this paper, we propose an assistive model that supports professional interior designers to produce industrial interior decoration solutions and to meet the personalized preferences of the property owners. The proposed model is able to automatically produce the layout of objects of a particular indoor scene according to property owners' preferences. In particular, the model consists of the extraction of abstract graph, conditional graph generation, and conditional scene instantiation. We provide an interior layout dataset that contains real-world 11000 designs from professional designers. Our numerical results on the dataset demonstrate the effectiveness of the proposed model compared with the state-of-art methods.

Keywords: Interior layout · Personalised preferences · Conditional graph generation · Conditional scene instantiation

1 Introduction

People spend plenty of time indoors such as bedrooms, kitchen, living rooms, and study rooms. Online virtual interior design tools are available to help designers to produce solutions of indoor-redecoration in tens of minutes. However, the online indoor-redecoration in a shorter time is demanding in recent years. One of the bottlenecks is to automatically process the indoor layout of furniture, and the goal is to come up with an assistive model that helps designers produce indoor-redecoration solutions at the industry-level in a few seconds.

Data-hungry models are trained for computer vision and robotic navigation [1,4], and many approaches have emerged in the line of auto-design work, which includes the object-oriented models that the objects in the indoor-scene and their properties are represented [2]. In PlanIT [10], a model is developed to take the advantages of spaces and objects. However, these assistive models are unsatisfactory since they can only satisfy partial requirements from the customers, and the

A. Bartoli and A. Fusiello (Eds.): ECCV 2020 Workshops, LNCS 12538, pp. 455–468, 2020.
https://doi.org/10.1007/978-3-030-66823-5_27

resulting auto-layouts are towards the same kind. Motivated by the challenge, we propose an assistant model that produces different types of auto-layout solutions of indoor-scene for different groups of customers (see Fig. 1). Specifically, we propose a conditional auto-layout assistive model where the condition is the representation of the preferences/requirements from customers/property owners. Firstly, the proposed model generates a structural representation according to the conditional representation of different requirements. This conditional generation is produced by a trained conditional model. Secondly, a generated graph and the representation of the condition are mixed through a deep module. Finally, a location module is trained to give the prediction of the location of furniture for different requirements.

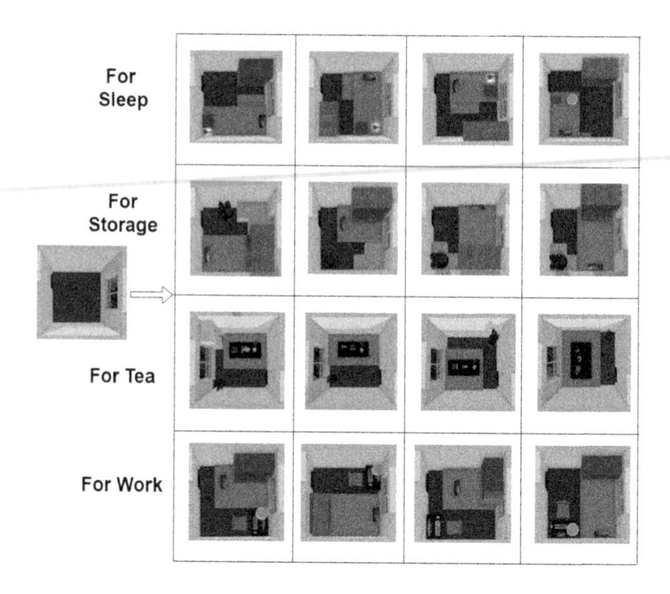

Fig. 1. Hierarchical indoor scenes generation. For example, given an empty tatami room, the layout of objects are produced according to four different requirements from the property owners, including the sleep-functional tatami, the tea-functional tatami, the work-functional tatami and the storage-functional tatami.

In this paper, we first introduce the related work in Sect. 2. The indoor synthesis pipeline which consists of three proposed modules is described in Sect. 3. In Sect. 4, we propose a graph representation procedure to extract the geometric indoor-scene into the abstract graph. We also propose a dataset of the indoor layout which is collected from the designs of professional interior designers. In Sect. 5, a model for a graph generation with conditions is developed where the condition is based on the preferences of the property owners. In Sect. 6, a conditional scene instantiation module is developed to produce the location of furniture for different property owners. In Sect. 7, we evaluate the proposed model for three types of rooms including the tatami, the balcony room, and the kitchen.

Finally, we discuss the advantage and the future work of the proposed model in Sect. 8.

2 Related Work

We discuss the related work of indoor furniture layout including the indoor-scene representation, graph generative models and the location of furniture models.

2.1 Indoor-Scene Representation

The study of indoor-scene representation starts two decades ago. In the early years, rule-based formulation is developed to generate the 3D object layouts for pre-specified sets of objects [11]. Interior design principles are applied through the optimization of the cost function [8]. The object-object relationships are analyzed in the interior scenes [12]. Data-driven models are then developed in this field. The earliest data-driven work is modeled to produce the object co-occurrences through a Bayesian network [2]. The followup work make use of the undirected factor graph learned from RGB-D images [6]. Besides, other interior scene representations are applied to models including RGB-D frames, 2D sketches of the scenes, natural language text, and RGB-D representations. However, these representations of the indoor scenes are only practical to a limited range of indoor scenes.

The learning frameworks for the representation of the indoor scene are developed recently. These learning methods apply a range of techniques including human-centric probabilistic grammars [9], generative adversarial networks on a matrix representation of the interior scene [7], recursive neural networks trained on 3D scene hierarchies [7], and convolutional neural networks trained on top-down image representation of interior scenes [10]. However, these learning methods are not able to obtain a high-level representation of the indoor scene. Besides, they are not practical as the generated in-door scenes cannot meet the personalized preferences of the property owners.

2.2 Graph Generative Models

Another line of recent work on learning the graph-structural representation through deep neural networks is explored. These work aim to learn generative models of graphs. The early work along the line is to develop a graph convolutional neural network (GCN) to perform graph classification [5]. The message passing between nodes in a graph [3] is applied to the graph generation of interior scenes [10]. However, these graph generative models are not able to produce graphs based on different conditions and meet the personalized preferences of property owners.

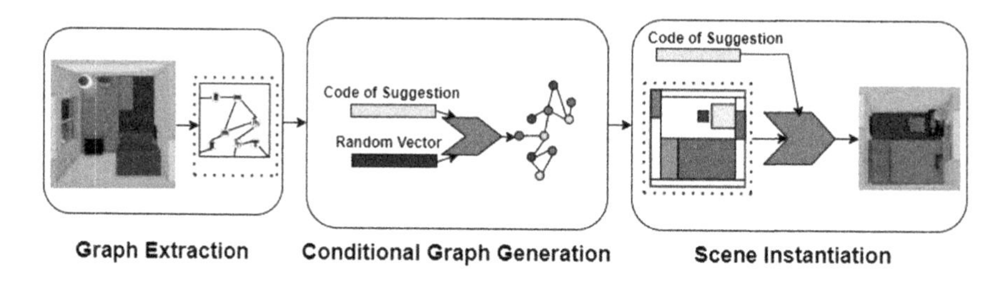

Fig. 2. The proposed indoor synthesis assistive model for personalized preferences. It consists of graph extraction, conditional graph generation and conditional scene instantiation.

3 Indoor Synthesis Pipeline

In this section, we tackle the following indoor-scene layout problem: given the architectural specification of a room (e.g., walls, doors, and windows) of a particular type (e.g., kitchen or tatami), and the personalized preferences of a house owner for the indoor-decoration, we choose and locate a set of objects to decorate that type of a room. We aim to build an assistant model that can support the designer to produce a decoration solution quickly through auto-layout of the furniture for a room, as shown in Fig. 2. In order to synthesize a scene from an empty room according to the owner's suggestion, the assistant model should complete the following three steps. Firstly, partial scenes of the rooms are completed. Secondly, a high-level representation of the layout of a room in the form of a relation graph is extracted. Thirdly, the house owner's requirements for the indoor decoration are encoded and adapted in the model. As a consequence, the proposed system generates a relation graph following the owner's requirements and then produces a solution to the house layout.

This assistant model starts with the extraction of the relation graphs from 3D indoor scenes. This extraction applies the geometric rules and a hierarchy rule to build a structural graph of the indoor scenes, as shown in Fig. 3 and Fig. 4. Next, this corpus of the extracted graphs of the indoor scenes and the customers' requirements on the scenes are encoded. The encoded representation is applied to learn a generative model of graphs, as shown in Fig. 5. The generated graph is therefore corresponding to the representation of an indoor scene given a particular condition. Finally, the abstract conditional graph and the representation of a suggestion are encoded to instantiate into a concrete scene, as shown in Fig. 6.

Formally, given a set of indoor scenes $(x_1, y_1, s_1), \ldots, (x_N, y_N, s_N)$ where N is the number of the scenes, and x_i is an empty indoor scene with basic elements including walls, doors and windows. y_i is the corresponding layout of the furniture for x_i. Each y_i contains the elements p_j, s_j, d_j, l_j: p_j is the position of the jth element; s_j is the size of the jth element; d_j is the direction of jth element; and l_j is the category label of the jth element. Each element represents a furniture in an indoor scene i. Besides, s_i is the requirement from the property

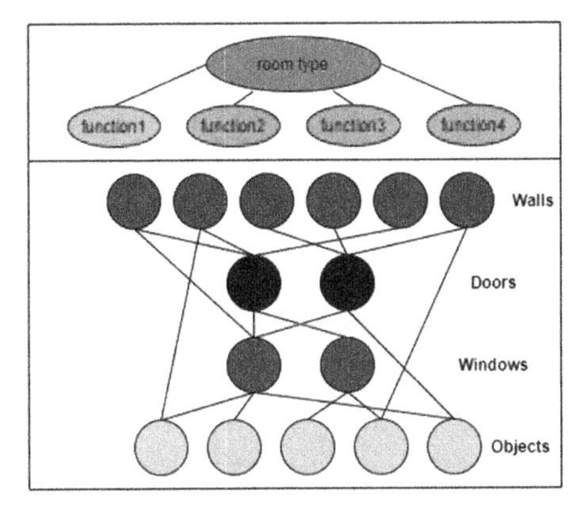

Fig. 3. Graph extraction from the proposed rule of encoder hierarchy. The walls are at the first level, the doors are at the second level, the windows are at the third level and the objects are at the last level.

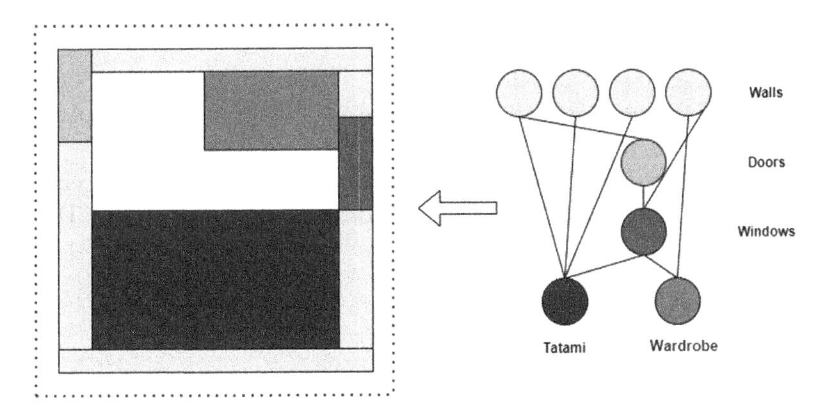

Fig. 4. Scene instantiation hierarchy from an abstract graph.

owner that determines the layout type of the room. Representation of the each indoor scene (x_i, y_i, s_i) is first extracted to structural representation gp_i, en_i. gp_i is the graph representation of y_i, and en_i is the code of s_i. A generative model is then trained to produce the abstract graph $gpout_i$ by applying hidden vectors z and s_i. Finally, given $gpout_i$ and s_i, an instantiation model is learned to convert the abstract representation $gpout_i$ and s_i to the completed scene y_i. An industrial rendering software is then applied to produce an industrial solution for the property owner. The whole assistant system is aimed to help interior designers to produce the layout of the furniture in a shorter time.

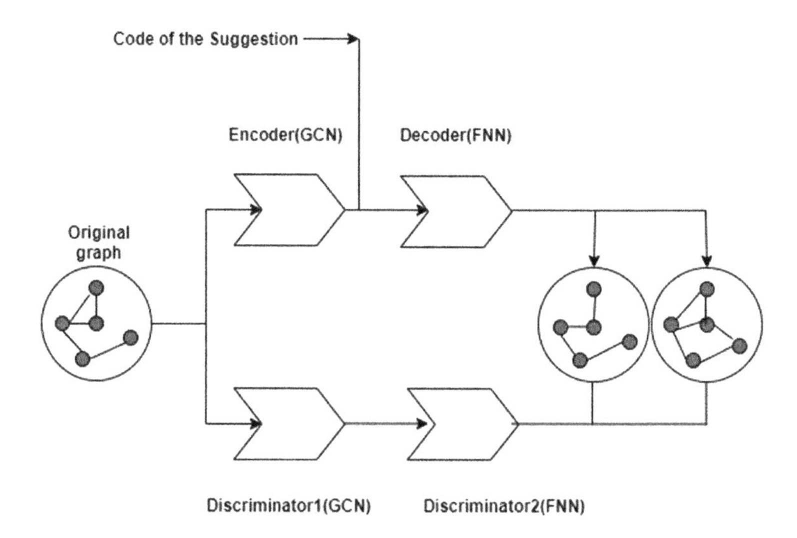

Fig. 5. Conditional graph generation module.

4 Graph Representation of Scenes

In this section, we define the graph representation of the indoor scenes, and introduce a procedure of automatically extracting the graphs from 3D indoor scenes.

4.1 Dataset

We propose a dataset which is a collection of over fifteen thousand indoor-scenes designs from professional interior designers who use an industrial design software in a daily basis. These designs contain three common types of rooms including tatami, kitchen and balcony. Pre-processing on this data is performed to filter out uncompleted layout of the rooms and the mislabeled room types, resulting in 5000 kitchens (with 41 unique object categories), 5000 balconies (with 37 unique categories) and 1000 tatami rooms (with 53 unique categories). Each room has a list of properties including its category, the geometry of each object and the requirements for the layout. There are no hierarchical and semantic relationships among objects in each room.

4.2 Graph Definition

The hierarchical and semantic relationship among objects in a room is encoded in the form of a relation graph. In each graph, each node denotes an object, each edge denotes spatial or semantic relationship between objects. Each node is labeled with a category label of the object. In particular, the walls, doors, windows and different objects have different category labels. The distance between objects contains three labels including near, middle and further. The spatial relationship between objects is represented with this three labels. Similarly, the semantic relationship is represented with three labels as following, walls-doors, walls-windows, walls-objects and objects-objects.

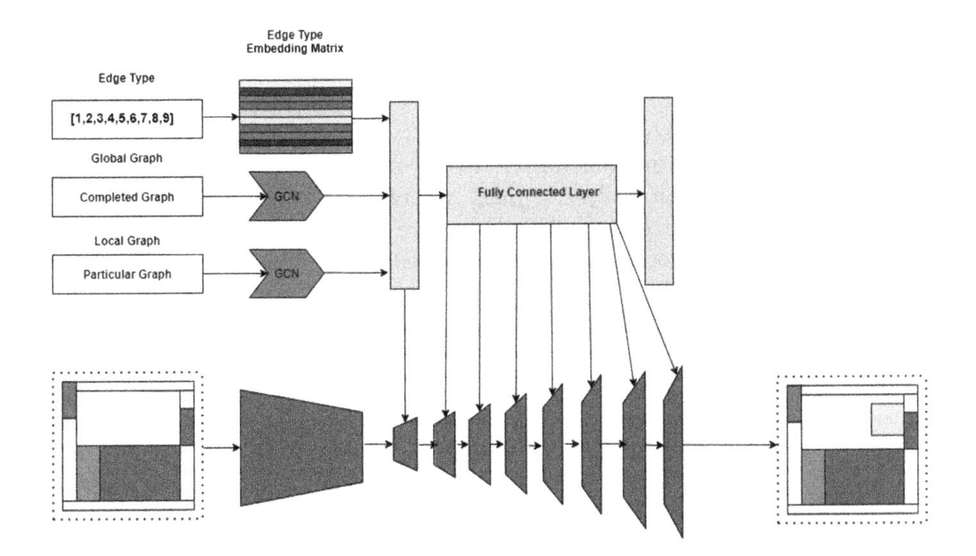

Fig. 6. Conditional scene instantiation module.

4.3 Graph Extraction

To convert an indoor scene into the above graph representation, We first label each node: the node of walls is 1, the node of doors is 2, the node of windows is 3, and the nodes of objects of different categories are $4, 5 \ldots, n-1, n$. The number of edge types is set to 9 following the above definition, and there are 3 types of spatial relationship and 3 types of semantic relationship. For example, the edge type is 2 if the nodes are a wall and a door, and the distance between them is middle. The edge type is 6 if the nodes are a wall and an object and the distance between them is further.

The graph extracted is dense (average of 5 edges node), and it contains many non-semantically meaningful relationships. This density may lead to problems for the learning of a generative graph model. The density is very likely to confuse a neural network as it is hard to recognize the most important structural relationships among many edges. Therefore, less important edges are pruned from the extracted graph. We set a pruning procedure for the extracted graph. Edges between nodes in which both nodes are the wall are deleted. Edges between nodes where the spatial relationship is further deleted. For each node that is labeled as a door or window, we keep the links to the closest wall and the object. Similarly, for each node that is labeled as objects, we keep the links to the closest objects and door/window. Finally, to ensure the graph connectivity, we make use of the same technique to reconnected the nodes as in [10].

5 Deep Graph Generation with Conditions

Once the graph is extracted, the indoor scenes and the corresponding customers' requirements $(x_1, y_1, s_1), \ldots, (x_N, y_N, s_N)$ are converted into graph representation and nodes $(gp1_1, gp2_1, en_1), \ldots, (gp1_N, gp2_N, en_N)$. We focus on the problem of conditional structure generation of the completed indoor scene y.

Given a set of graph $gp2_1, \ldots, gp2_n$, where $gp2_i$ is the extracted graph representation of y_i and $gp2_i = \{V_j, E_j\}$ corresponds to the graph structure described by the set of nodes V_j and the set of edges E_j. The semantic attribute of the graph is denoted as the code en_i. As already introduced above, en_i describes the requirements from the corresponding property owner. It is basically the context of the graph. For example, the codes en_1, en_2, en_3, and en_4 are $1000, 0100, 0010, 0001$ for 4 types of tatami rooms, where these four types are the most four popular choices of customers (see Fig. 1).

A model M is trained on a set of graphs with conditions $\{G_i, E_i, i = 1, ..., n\}$, and this M is used to generate more graphs g mimicking the structure of those in the training set. We apply a conditional graph generation model, which applies three tricks to generates a graph for different conditions. Firstly, latent space conjugation is applied to effectively convert node-level encoding into permutation-invariant graph-level encoding. This conjugation allows the learning on arbitrary numbers of graphs and generation of graphs with variable sizes. Formally, given a graph $G = \{V, E\}$, G is regarded as a plain network with the adjacency matrix A. It generates node features $X = X(A)$ as the standard $k-$-dim spectral embedding based on A. The stochastic latent variable Z is then introduced and $q(Z|X, A) = \prod_{i=1}^{n} q(z_i|X, A), z_i \in Z$ is the regarded as the node embedding of $v_i \in V$. A single distribution $\bar{z}$ is used to model all z_i's by enforcing:

$$q(z_i|X, A) \sim \mathcal{N}(\bar{z}|\bar{\mu}, \mathrm{diag}(\bar{\sigma}^2)) \tag{1}$$

where $\bar{\mu} = \frac{1}{n}\sum_{i=1}^{n} g_\mu(X, A)$ and $\bar{\mu}^2 = \frac{1}{n_2} g_\theta(X, A)_i^2$. $g(X, A) = \bar{A}ReLU(\bar{A}XW_0)W_1$ is a two-layer GCN model, (see encoder(GCN) in Fig. 5). $g_\mu(X, A)$ and $g_\theta(X, A)$ compute the matrices of mean and standard deviation vectors, which share the first-layer parameters W_0. $g(X, A)_i$ is the i-th row of $g(X, A)$. $\bar{A} = D^{-\frac{1}{2}} A D^{-\frac{1}{2}}$ is the symmetrically normalized adjacency matrix of G, where D is its degree matrix with $D_{ii} = \sum_{j=1}^{n} A_{ij}$. After sampling a desirable number of z_i's to improve the capability of the graph decoder, a few fully connected neural network (FNN) layers(see decoder(FNN) in Fig. 5) f are applied to z_i before computing the logistic sigmoid function for link prediction which can be described by

$$p(A|Z) = \prod_{i=1}^{n}\prod_{j=1}^{n} p(A_{ij}|z_i, z_j) \tag{2}$$

where $p(A_{ij} = 1|z_i, z_j) = \theta(f(z_i)^T f(z_j))$, and $\theta(z) = \frac{1}{1+e_{-z}}$. The model is optimized by minimizing the minus variational lower bound:

$$L_{vae} = L_{rec} + L_{prior} = \mathcal{E}_{q(Z|X,A)}[\log p(A|Z)] - D_{KL}(q(Z|X,A)||p(Z)) \quad (3)$$

where L_{rec} is a link reconstruction loss and L_{prior} is a prior loss based on the Kullback-Leibler divergence towards the Gaussian prior $p(Z)_{i=1}^n = \mathcal{N}(Z|0,I)^n$. The model now consists of a GCN-based graph encoder $\xi(A) = \frac{1}{n}\sum_{i=1} ng(X(A),A)_i$, (see encoder(GCN) in Fig. 5) and an FNN-based graph decoder/generator (see encoder(FNN) in Fig. 5) $\Upsilon(Z) = f(z_i)^T f(z_j)$.

GCN is leveraged by devising a permutation-invariant graph discriminator. This discriminator is applied to enforce the intrinsic structural similarity between $A^{'}$ and A under arbitrary node ordering. In particular, a discriminator D of a two-layer GCN is constructed followed by a two-layer FNN. This discriminator is trained together with the above encoder ξ and generator Υ through following the GAN loss of a two-player minimax game:

$$L_{GAN} = \log(D(A)) + \log(1 - D(A^{'})) \quad (4)$$

where $D(A) = f^{'}(g^{'}(X(A),A))$, and X, $g^{'}$ and $f^{'}$ are the spectral embedding for GCN and FNN, respectively.

6 Scene Instantiation

In this section, we describe the procedure of taking a relationship graph and instantiating it into an actual indoor scene. In the extraction of graph, the edge pruning steps is likely to lead to the incomplete graph. Besides, the graph representation of the scene is abstract. It does not contain enough relationship edges to uniquely determine the spatial positions and orientations of all object nodes. The scene instantiation is also required to follow the preferences of the property owner.

Formally, a procedure is required for sampling from the following conditional probability distribution corresponding to the property owner's requirements en:

$$p(S|\Upsilon(V,E,en)) = p(S)en_i \in en\ 1(en_i \in S)v \in V\ 1(v \in S)(u,v,r) \in E(S,u,v) \quad (5)$$

where S is a scene, $\Upsilon(V,E,en)$ is a graph with vertices V, edges E, and requirements en. r is a predicate function indicating whether the relationship implied by the edge (u,v,r) is satisfied in S. en_i is a sample of the requirements en.

6.1 Object Instantiation Order

Before the prediction of the location of objects in the room, the order of the instantiation of objects should be set. The algorithm of determining the order in which to instantiate objects follows a similar logic in [10]. It applies the structure of the graph, along with statistics about typical sizes for nodes of different categories. For example, in the instantiation of the tatami room, the order is set as tatami, work-desk, and cabinet.

6.2 Neural-Guided Object Instantiation

After the order of instantiation is determined and an object of category c is selected to add to the scene, the completed layout prediction of this object is expected to be produced following the requirements of the property owner's suggestion en. These prediction includes the location p, orientation s and physical dimensions d.

Ideally, a generator function $g(p|en, s|en, d|(c, en), S, \Upsilon(V, E, en))$ is expected to produce the outputs values with probability proportional to the true conditional scene category $p(\bar{(S)} = S \cup \{c, p, s, d, en\}|\Upsilon(V, E, en))$. As already explored in the previous work through iterative object insertion [1], the location, orientation and dimensions are sampled iteratively according on the corresponding to en.

Structural Representation of Requirements. The requirements from the property owner are encoded as a vector in the graph generation phase. For example, if the balcony is in two categories leisure balcony and living balcony. The requirements vector is then one-hot vector $(1, 0)$ and $(0, 1)$, respectively. However, this one-hot vector loses much information in the conditional scene instantiation phase. Therefore, we develop a mixture embedding module which applies the embedding of global graph representation, local graph representation, and edge type representation.

We first employ a GCN [5] network to produce the embedding of the global representation of the generated graph gp. Then, we employ the GCN [5] and the setting instantiation order to produce the embedding of the local representation of the generated graph gn. Specifically, we apply the order to get the vector of the adjacency matrix of the selected objects. The embedding of the local graph representation can be obtained. The embedding of the edge-type is obtained in a similar method as in the previous model [10]. Note that this embedding follows a different rule as introduced above. Finally, we apply two-level FCN to combine these three types of embedding.

Conditional Instantiation. After the structural representation of the suggestion is obtained, we train a model that is based on the condition of the requirements. This model receives the inputs of the generated graph gp and code the requirements en, and then predicts the location p, the orientation s, and the size d of the selected object. We train both the embedding module and instantiation module together. Here the instantiation module is similar to the instantiation network in the previous work [10].

7 Evaluation

In this section, we show both qualitative and quantitative results that demonstrate the unity of the proposed assistant model to auto-layout of the indoor scenes with property owners' personalized requirements. The generated examples are shown in Fig. 7.

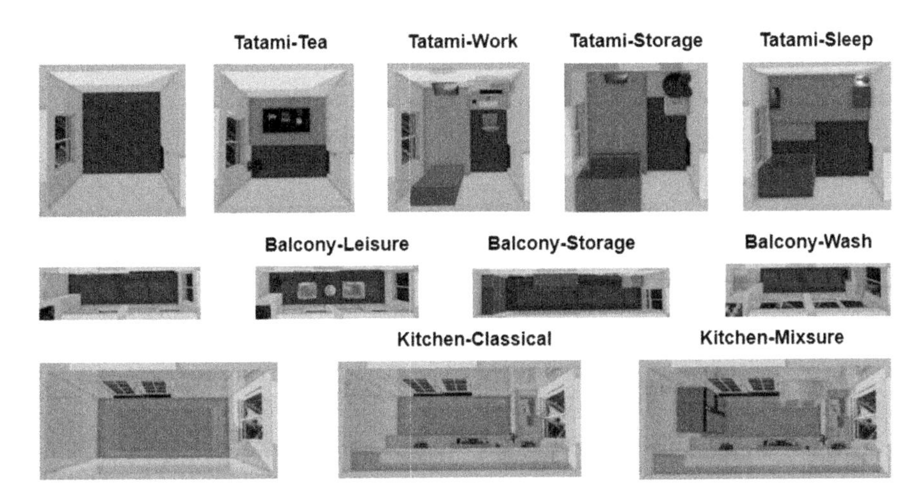

Fig. 7. Generated layout examples with personalized preferences from the property owners.

7.1 Generation Accuracy

The accuracy of the graph generation for each type of rooms is defined by:

$$ACC_G = \frac{1}{N_c} \frac{\sum_{k=1}^{n} \text{label}(g_k) = i}{N_i} \tag{6}$$

where N_i is the number of generated graph for label i and N_c is the number of the category of the graph.

We measure the accuracy of graphs for three types of rooms. For the evaluation of tatami room, we collect four types which are most popular among property owners from our sold interior decoration solutions. These four types are sleeping-functional tatami, storage-functional tatami, tea-functional tatami, and working-functional tatami. Similarly, There are three types of balconies chosen: the leisure-functional balcony, the washing-functional balcony, and the storage-functional balcony. There are two types of kitchen: the classical-functional kitchen and the multi-functional kitchen.

Table 1. Accuracy for tatami rooms.

Averaged accuracy	For sleep	For tea	For storage	For work
0.81	0.82	0.83	0.79	0.78

Table 2. Accuracy for balcony rooms.

Averaged accuracy	For leisure	For wash	For storage
0.84	0.83	0.85	0.84

Table 3. Accuracy for kitchen rooms.

Averaged accuracy	For classical	For mixture
0.86	0.88	0.84

In Table 1, Table 2 and Table 3, we show the averaged accuracy for different types rooms with different functionalities. Note that we measure 10000 generated interior scenes for all functional-type of rooms.

7.2 Perceptual Study

In addition to the quantitative evaluation of the labeled graph generation, a two-alternative forced-choice (2AFC) perceptual study is conducted to compare the images from generated scenes with the corresponding scenes from the sold industrial solutions. The generated scenes are rendered using industrial rendering software. We remark that the rendering process is different from the previous work [10] that only produces solid color for the objects. Participants are shown two top-view scene images side by side and required to pick out the more plausible one. For each comparison and each room type, 10 professional interior designers were recruited as the participants. As shown in Table 4, the generated scenes are similar to the scenes from the sold solutions.

Table 4. Percentage ($\pm$ standard error) of 2AFC perceptual study where the real sold solutions are judged as more plausible than the generated scenes.

Room	Ours	PlanIT	Grains
Tatami	57.12 ± 3.48	71.93 ± 6.42	69.42 ± 5.87
Balcony	58.21 ± 2.95	75.23 ± 2.74	67.51 ± 1.53
Kitchen	54.21 ± 4.51	74.68 ± 4.62	77.94 ± 2.90

Another two-alternative forced-choice (2AFC) perceptual study is conducted to compare images from the generated scenes with the corresponding generative scenes of the state-of-the-art models. 10 professional interior designers were recruited as the participants in this perceptual study. As Table 4 shows, the generated scenes of the proposed system outperforms the state-of-the-art models such as PlanIT [10] and Grains [7]. As shown, the generated images of the proposed models is more similar to the real rendering layout. It is obvious that our model is able to generate indoor scenes for a variety of function-category rooms of a particular room type while the baseline models are not able to produce such layout. We show an example of the generated tatami room with four functionalities in Fig. 1.

8 Discussion

In this paper, we present an assistive method for interior decoration. The assistive method consists of the extraction of the abstract graph, conditional graph generation, and the conditional scene instantiation. The proposed system supports the interior designers in the industrial process to produce the decoration solution more quickly. In particular, this system produces the auto-layout of objects which is plausible according to the choice of the professional designers. Besides, the proposed system is shown to meet the personalized preferences from the property owners.

There are many challenges in automatic interior decoration. The proposed method is only able to support the auto-layout of objects in the interior scenes where the only global auto-layout follows the preferences of the property owners. The type, color, and brand of each object for the personalized preferences are not supported. In addition, the proposed method produces good results when the shape of the room is towards regular while there are a variety of different shapes in the real world. It is worthwhile to explore smarter systems to solve these challenges.

References

1. Dai, A., Ritchie, D., Bokeloh, M., Reed, S., Sturm, J., Nießner, M.: ScanComplete: large-scale scene completion and semantic segmentation for 3D scans. In: The IEEE Conference on Computer Vision and Pattern Recognition (CVPR), June 2018
2. Fisher, M., Ritchie, D., Savva, M., Funkhouser, T., Hanrahan, P.: Example-based synthesis of 3d object arrangements. ACM Trans. Graph. **31**(6) (2012). https://doi.org/10.1145/2366145.2366154
3. Gilmer, J., Schoenholz, S.S., Riley, P.F., Vinyals, O., Dahl, G.E.: Neural message passing for quantum chemistry. CoRR (2017)
4. Gordon, D., Kembhavi, A., Rastegari, M., Redmon, J., Fox, D., Farhadi, A.: IQA: visual question answering in interactive environments. In: The IEEE Conference on Computer Vision and Pattern Recognition (CVPR), June 2018
5. Jiang, B., Zhang, Z., Lin, D., Tang, J., Luo, B.: Semi-supervised learning with graph learning-convolutional networks. In: 2019 IEEE/CVF Conference on Computer Vision and Pattern Recognition (CVPR), pp. 11305–11312 (2019)
6. Kermani, Z., Liao, Z., Tan, P., Zhang, H.: Learning 3D scene synthesis from annotated RGB-D images. Comput. Graph. Forum **35**, 197–206 (2016). https://doi.org/10.1111/cgf.12976
7. Li, M., et al.: Grains: generative recursive autoencoders for indoor scenes. ACM Trans. Graph. **38**(2) (2019). https://doi.org/10.1145/3303766
8. Merrell, P., Schkufza, E., Li, Z., Agrawala, M., Koltun, V.: Interactive furniture layout using interior design guidelines. In: SIGGRAPH 2011, August (2011, to appear). http://graphics.berkeley.edu/papers/Merrell-IFL-2011-08/
9. Qi, S., Zhu, Y., Huang, S., Jiang, C., Zhu, S.C.: Human-centric indoor scene synthesis using stochastic grammar. In: The IEEE Conference on Computer Vision and Pattern Recognition (CVPR), June 2018

10. Wang, K., Lin, Y.A., Weissmann, B., Savva, M., Chang, A.X., Ritchie, D.: Planit: planning and instantiating indoor scenes with relation graph and spatial prior networks. ACM Trans. Graph. **38**(4) (2019). https://doi.org/10.1145/3306346.3322941
11. Xu, K., et al.: Constraint-based automatic placement for scene composition. In: Graphics Interface, pp. 25–34 (2002)
12. Yu, L.F., Yeung, S.K., Tang, C.K., Terzopoulos, D., Chan, T.F., Osher, S.J.: Make it home: automatic optimization of furniture arrangement. ACM Trans. Graph. **30**(4) (2011). https://doi.org/10.1145/2010324.1964981

Behavioural Pattern Discovery from Collections of Egocentric Photo-Streams

Martín Menchón[1(✉)], Estefanía Talavera[2], José Massa[1], and Petia Radeva[3]

[1] INTIA, UNCPBA, CONICET, Tandil, Argentina
{mmenchon,jmassa}@exa.unicen.edu.ar
[2] University of Groningen, Groningen, The Netherlands
e.talavera.martinez@rug.nl
[3] University of Barcelona, Barcelona, Spain
petia.ivanova@ub.edu

Abstract. The automatic discovery of behaviour is of high importance when aiming to assess and improve the quality of life of people. Egocentric images offer a rich and objective description of the daily life of the camera wearer. This work proposes a new method to identify a person's patterns of behaviour from collected egocentric photo-streams. Our model characterizes time-frames based on the context (place, activities and environment objects) that define the images composition. Based on the similarity among the time-frames that describe the collected days for a user, we propose a new unsupervised greedy method to discover the behavioural pattern set based on a novel semantic clustering approach. Moreover, we present a new score metric to evaluate the performance of the proposed algorithm. We validate our method on 104 days and more than 100k images extracted from 7 users. Results show that behavioural patterns can be discovered to characterize the routine of individuals and consequently their lifestyle.

Keywords: Behaviour analysis · Pattern discovery · Egocentric vision · Data mining · Lifelogging

1 Introduction

The automatic discovery of patterns of behaviour is a challenging task due to the wide range of activities that people perform in their daily life and their diversity, when it comes to behaviour. Advances in human behaviour understanding

---

Petia Radeva, Fellow IAPR

Electronic supplementary material The online version of this chapter (https://doi.org/10.1007/978-3-030-66823-5_28) contains supplementary material, which is available to authorized users.

A. Bartoli and A. Fusiello (Eds.): ECCV 2020 Workshops, LNCS 12538, pp. 469–484, 2020.
https://doi.org/10.1007/978-3-030-66823-5_28

were made possible, in a great measure, by the widespread development of wearable devices in the last decade, which can be incorporated into clothing, or used as implants or accessories. These devices provide an immeasurable amount of information, which allows to collect meaningful details about a person's routine. With these sensors it is possible to monitor health [1], locate a person at a certain time [2], the activity carried out at that moment [3], to determine if the person was anxious [4], and to detect group social interaction [5]. However, works that rely on accelerometers or GPS sensors only acquire partial information missing the whole context of the event. The only devices that allow to get visual information about the context and behaviour of the person are wearable cameras. These devices collect egocentric photo-streams that show an objective and holistic view of where the person has been, what activities he/she has been doing, what kind of social interactions he has performed and what objects he has been surrounded, i.e. the camera wearer's behaviour.

Therefore, egocentric photo-streams have shown to be a rich source of information for behaviour analysis [6]. This particular type of images tries to approximate the field of vision of a user. Thanks to the first-person focus of the images, a faithful representation of an individual's day can be obtained. Figure 1 visually shows a collection of egocentric photo-streams by one of the users. A lot of works has been developed to extract semantic information from egocentric photo-streams like: activity and action recognition [7], places detection [8,9], social interaction characterization [10,11], activities prediction [12], routine days discovery [6,13], etc. However, up to our knowledge there is no work on automatic detection of routine patterns in collection of egocentric photo-streams.

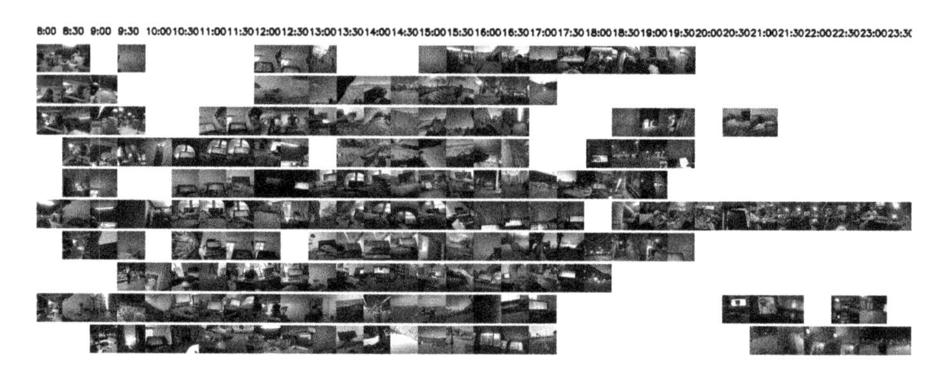

Fig. 1. Visual time-line with collected images by one of the users in the *EgoRoutine* dataset. Rows represent the different collected days and columns indicate time. White spaces correspond to not recorded moments.

In this work, we propose a new method for discovery of sets of patterns describing someone's routine from collections of egocentric photo-streams. Given a collection of images, our model is able to describe the collected days with time-frames characterized by recognized objects, activities, and location. Afterwords, a new greedy clustering algorithm is defined to detect patterns of time-frames with similar context based on computed semantic similarity between time-frames labels. Hence, patterns are presented by time-slots with high semantic similarity, and can be used to describe behavioural habits.

The contributions of this work are three-fold:

- To the best of our knowledge, this is the first work that addresses the automatic discovery of patterns of behaviour from collections of egocentric photo-streams.
- A novel greedy clustering algorithm is proposed to efficiently find patterns in collections of time-lines composed by similar concepts describing time-slots. Special attention is paid to redefine the distance between time-slots mixing categorical and numerical information.
- A new score measure is defined to validate extracted patterns from egocentric images.

The rest of the paper is organized as follows: in Sect. 2 we describe related works. In Sect. 3, we describe our proposed pipeline for behavioural pattern discovery from egocentric photo-streams. In Sect. 4, we describe the performed experiments and obtained results. Finally, we summarize our findings and future research lines in Sect. 5.

2 Related Works

Pattern discovery is a well-studied topic in computer vision, covering from molecular pattern discovery [14], to bird population analysis [15], passing through text mining or speech analysis applications, such as [16] and [17], respectively.

The analysis of behavioural patterns is a relevant topic when studying an ecosystem. However, describing the behaviour of human beings is a difficult task due to the high diversity among individuals, as described in [18] as well as lack of common devices and tools to acquire rich and complete information about individual's behaviour. Human behaviour was studied employing several strategies that use different sources of data, as described in the survey in [19]. However, most of the approaches propose supervised learning, using techniques to build models that learn from predefined activities. In contrast, proposing models to discover patterns in an unsupervised manner from the collected data, without restricting the procedure to a limited and predefined set of activities, still remains an open question.

In [20], the authors describe the discovery of patterns as information extracted "directly from low-level sensor data without any predefined models or assumptions". They proposed a tool for activity classification, where activities are represented as a pattern of values from the collected sensors. They focused on the classification of simple predefined human activities. One of the shortcomings is that there is only a limited and pre-defined set of activities that not always describe the variety of human lifestyle. Moreover, in many situations specially using wearable cameras there is no information about the activity of the person. This fact could be compensated by information about the place and the context i.e. surrounding objects appearing in the egocentric photo-streams.

Thematic pattern discovery in videos was addressed in [21]. The authors proposed a sub-graph mining problem and computed its solution by solving the binary quadratic programming problem. However, this method is not capable of seeking for common patterns in several time-sequences. Moreover, someone's behaviour is not an one-theme topic, but a non-defined task where an efficient algorithm for multiple pattern discovery is needed.

Related to egocentric vision, behaviour analysis from egocentric photo-streams has been previously addressed in [6]. The authors proposed to use topic modelling for the discovery of common occurring topics in the daily life of users of wearable cameras. Their aim was of discovering routine days in individual's diaries. However, their work is not capable of automatically finding patterns as part of the day activities, and what they represent. In this work, instead of aiming to describe what a routine day is, we claim to discover what are the patterns (e.g. breakfast at 8 am, office at 10 am) that compose the lifestyle of the camera wearer, in an unsupervised manner.

3 Behavioural Pattern Discovery

In this section, we describe our method for pattern discovery from egocentric photo-streams.

3.1 Days Characterization by Semantic Features Extraction

Many empirical studies show that most people lead low-entropy lives [22], following certain regular daily routines over long time, although such routines vary from one person to another. Hence, it can be meaningful to develop automatic methods to capture the regular structures in users' daily lives and make use of that information for user classification or behavior prediction. People used to perform similar events in their routine days like working at office in the morning, having lunch at 1pm, etc. Here, we define a pattern as a sequence of elements regarding human behaviour (activities, place, environment represented by its objects around) which descriptors are shared and recurrently occur throughout time. In our case, elements are extracted from the egocentric images that compose photo-streams. In order to obtain a person routine pattern, we propose a novel unsupervised learning approach focusing on the three important factors that are able to describe what a user is doing at any certain moment: the scene, the activity that he/she is carrying out and his/her context expressed by the objects present around him/her:

Activity detection: For characterization of the activities we use the network proposed in [7] to classify a given image as belonging to one of 21 activities of Daily Living.

Object detection: We use the Yolo3 Convolutional Neural Network (CNN) model [23] to extract the appearing objects in the images.

Scene detection For classifying the scene in the single images from the photo-stream, we applied the *VGG16-Places365* [24] CNN, trained with the "Places-365" dataset consisting of 365 classes.

Note that both object detection and scene detection are not fine-tuned to the egocentric photo-streams that probably leads to suboptimal performance on label detection. This fact is compensated by defining a method for patterns discovery tolerant (as it will be made clear in the Validation section) to suboptimal labels extraction.

3.2 Graph Representation of the Day

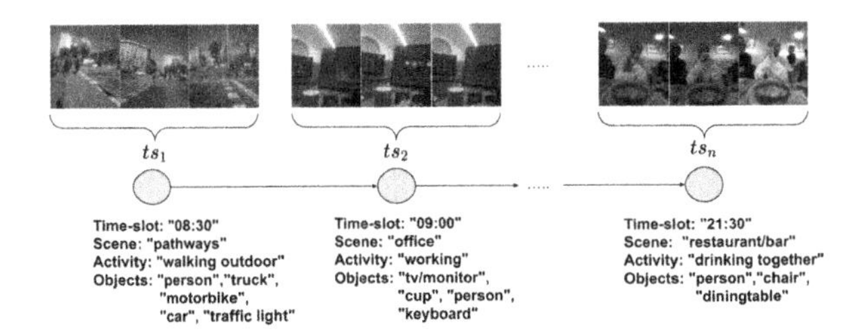

Fig. 2. Illustration of created time-slots for sequences of images corresponding to 30' within the images time series. We indicate the automatic extracted and selected labels to describe the time-slot.

We divide the user's days into half-hour time-slots each one. Each of the time-slots is considered a node in a linear graph (see Fig. 2). We characterize every node by evaluating the detected concepts in the images that compose them. We analyze the images by using pre-trained networks to recognize the scene, activity and objects in the scene that they depict. At a node level, we select the scene and the activity label with highest occurrence as the ones describing the time-slot. We also add as descriptor the objects that appeared in several images (in our case, in more than ten) in order to ensure that they are representative enough of that time-frame. Figure 2 illustrates how time-slots are formed and their descriptors at labels level.

3.3 Detecting Similar Time-Slots

Once linear graphs are obtained, we define a metric to compare them. In particular we compare nodes corresponding to the same time-slot assuming that time should be an inherent feature of patterns (e.g. eating at different time means having breakfast, lunch or dinner). The distance between nodes is represented by:

$$\begin{aligned} d((S_1, A_1, O_{11}, O_{12}, ..O_{1n1}), (S_2, A_2, O_{21}, O_{22}, .., O_{2n2})) \\ = D(S_1 <> S_2) + D(A_1 <> A_2) + DJ(O_1, O_2) \end{aligned} \quad (1)$$

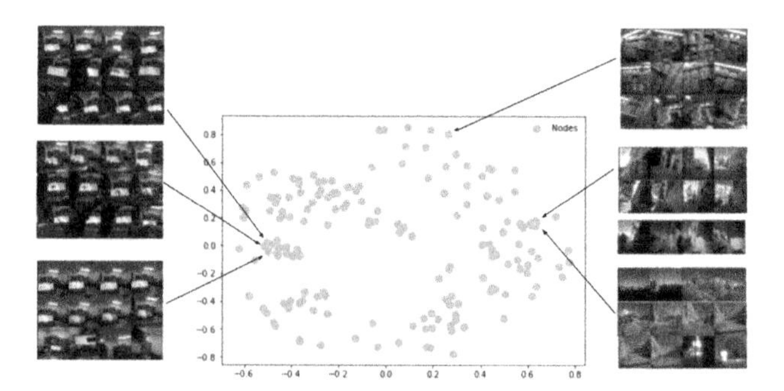

Fig. 3. Output of the multi-dimensional scaling. The input is the distance matrix among nodes computed by the defined distance in Eq. (1). Data samples correspond to nodes, i.e. time-slots in the collection of egocentric photo-sequences.

where $S_i, i = 1, 2$ is the scene, A_i is the activity and $O_{i,j},\ j = 1, ...n_i$ are the objects of the i-th node. $D(S_1 <> S_2)$ is the binary distance between nodes scenes, $D(A_1 <> A_2)$ is the binary distance between nodes activities and $DJ(O_1, O_2)$ is the Jaccard distance [25] between nodes objects. Once obtained the distances between all pair of nodes for the same time-slot, we apply multi-dimensional scaling (MDS) [26] to arrange all the nodes in a two-dimensional space. In this way, nodes that were very different from each other are sent far apart and very similar nodes from each other are grouped together. In Fig. 3, the nodes spatial distribution after the application of the MDS is shown. Note that since nodes are represented by their labels, they do not have a numerical representation to visualize in space. However, applying the distance defined in Eq. (1) and the MDS allow us to observe the spatial relation between the nodes and to visually identify groups of highly similar images.

3.4 Aggregating Similar Time-Slots

To find the patterns, an heuristic aggregating greedy algorithm is proposed. We start with a seed that is settled from the two nodes

$$n_1 = n(i_1, j_1), n_2 = n(i_2, j_1)$$

that are at a minimum distance being from the same time-slot:

$$d(n(i_1, j_1), n(i_2, j_1)) = min_{k,l,m}(d(n(i_k, j_l), n(i_m, j_l))).$$

The idea behind the algorithm is to find similar nodes considering as neighbours contiguous time-slots and/or different days from the same user. By this way, the algorithm looks for a candidate pattern consisting in recurrent similar activities.

A cluster agglomeration of these nodes is carried out where the nodes from the same time-slot or neighbour time-slots are agglomerated using the variance as a criterion to decide which neighbour time-slot to merge. Let us consider that the time-slots aggregated to the pattern P are $P = \{n_1, n_2\}$. And S is the rest of the nodes. From all neighbour nodes n_k, we select the node that added to P contributes causing the smallest variance increase. In order to compute the variance we applied the position of the node provided by the multi-dimensional scaling. See Algorithm 1 for the formalization of our proposed method.

Algorithm 1. Pattern discovery from collections of egocentric photo-streams

initialization: characterization of nodes (n) with the images' detected concepts

- Let $n_0 = n(i_0, j_0)$ and $n_1 = n(i_1, j_0)$ are the closest nodes so that $d(n(i_0, j_0), n(i_1, j_0)) = min_{k,l,m}(d(n(i_k, j_l), n(i_m, j_l))$
- Let $v_1 = d(n_0, n_1)$
- $S_0 = \{n_i\}$ is the set of all nodes
- $P_1 = \{n_0, n_1\}$, $S_1 = S_0 \setminus \{n_0, n_1\}$, $t = 1$
- Until $S_t <> empty$ set or $v_t > K$, do:
- Let $n_t = argmin_{n_l \in neighbours(S_{t-1})} var(P_{t-1} + n_l)$ # now we evaluate all neighbours from the aggregated sets
- $P_t = P_{t-1} \bigcup \{n_t\}$, $S_t = S_{t-1} \setminus \{n_t\}$, $t = t + 1$

The computational complexity of this algorithm is $O(ds * ts * N)$ where ds is the number of days to analyze, ts is the amount of time-slots of the day, and N is the number of nodes.

3.5 Finding Patterns

Let us consider that the nodes aggregated are $(n_1, n_2, ...n_M)$ and the variance evolution is $v = (v_1, v_2, ...v_M)$ adding all neighbour nodes. We are interested in extracting a cluster of semantically similar neighbour nodes (e.g. time-slots). We apply the assumption that similar nodes would have small variance while adding dissimilar nodes would lead abrupt increase in the variance. Note that by the aggregation process the variance evolution is monotonically increasing function. Abrupt change can be detected by looking for the maximum of the first derivative of v and zero-crossing of the second derivative. So to find the cluster with similar nodes, we smooth v through a Gaussian filter and then apply the first and second derivative to detect the separation between clusters. We find n_k so that there is a zero-crossing for the second derivative

$$n_k : d^2(v_k)/dt^2 = 0$$

and the first derivative has a value over a given threshold T

$$d(v_k)/dt > T$$

we consider that $(n_1, n_2, ...n_{k-1})$ form a cluster. In practice, there could be several candidates that fulfill the conditions for the first and second derivatives. In order to choose the optimal one, we define a new criterion for pattern discovery. On one hand, we are interested in selecting patterns that form separated clusters maximizing inter-class distance and minimizing intra-class distance. A good measure for this is given by the silhouette measure of the clustering. The Silhouette score is described by Eq. (2), where $a(i)$ is the average distance between point i and points withing the same cluster, and $b(i)$ is the minimum average distance from i to points of the other clusters:

$$sl = silhouette_{score} = \frac{b(i) - a(i)}{max(a(i), b(i))}. \tag{2}$$

On the other hand, we are interested in obtaining routine patterns that should be recurrent that is being spread on as much as possible days and they should cover significant time. To this purpose, we introduce the scoring function shown in Eq. (3) to evaluate the goodness of the detected pattern:

$$sc(P_{k-1}) = sl(P_{k-1}) + t_{rpr} \tag{3}$$

where t_{rpr} expressed by Eq. (4) is measuring the representativity of the pattern. To this purpose, this term measures the number of days the pattern covers (the higher the better) and the average time of prolongation $\hat{l}$ of the pattern during the days:

$$t_{rpr} = \frac{1}{\#patterns} \sum_{i=1}^{\#patterns} (\#ds_i/\#ds + \#I_i/max(\#I(1hour), (tm(I_{i,n_i}) - tm(I_{i,1})) * frq). \tag{4}$$

where ds stands for days, I stands for image, tm stands for time and frq stands for camera frequency resolution (in our case, $frq = 0.5$ fpm).

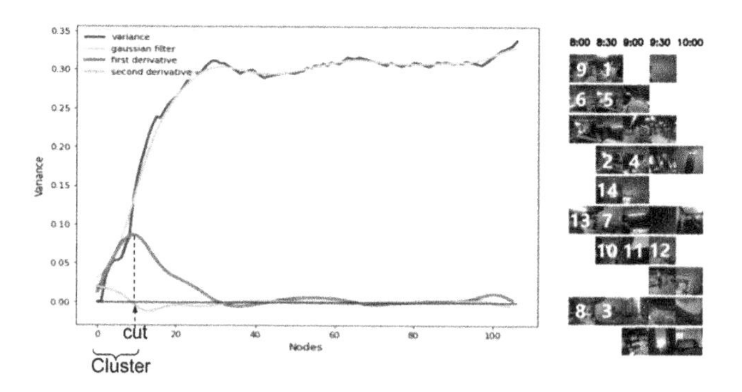

Fig. 4. Example of a fluctuation of the cluster variance as time-slots are added (blue). Filtered variance signal with a Gaussian filter (orange). First and second derivative (green and red, respectively). Cluster cut with the maximum value of the second derivative. On the right, visual representation of the sequentially added time-slots to one identified cluster, i.e. a discovered pattern. (Color figure online)

Once a pattern is detected, those time-slots are discarded and the rest of time-slots are processed using the same procedure in order to find other patterns of behaviour. Figure 4 visually describes the process of finding a pattern.

4 Experiments

In this section, we present the dataset used, the validation metrics, the experimental setup and the final results.

4.1 Dataset

In this work, we employ the *EgoRoutine* dataset proposed in [6]. The *EgoRoutine* dataset consists of unlabelled egocentric data of 115430 images collected on 104 days captured by 7 distinct users. The recorded images capture the daily lives of the users. In Fig. 1 we visually present a subsampled mosaic of images collected by one of the users. Table 1 presents an overview of the dataset distribution.

Table 1. Distribution of the data in the *EgoRoutine* dataset.

	User 1	User 2	User 3	User 4	User 5	User 6	User 7	Total
#Days	14	10	16	21	13	18	13	104
#Images	20521	9583	21606	19152	17046	16592	10957	115430

4.2 Validation Metrics

The proposed model is applied following a personalized approach since patterns are highly individual, i.e. we evaluate the performance at user level by evaluating over collected days by each user. The model applies unsupervised learning thus, no labels nor dataset split are required. Since our method discovers the patterns in an unsupervised manner, a proper performance should be evaluated quantitatively and qualitatively.

We quantify the quality of the found patterns with the Silhouette score [27] and the temporal spread of the patterns. The Silhouette metrics describes the relatedness of each point w.r.t. the cluster group it has been assigned to.

Qualitative measures usually come down to human judgement. To this end, we created a survey that showed to 13 different individuals: (1) the set of images that visualize the collected days of the users, and (2) the found patterns. Individuals were asked the following four questions:

1. Given the following collected egocentric photo-streams, reflecting about the life of the camera wearer, can you see patterns of behaviour? What patterns do you see in this mosaic of images showing the days of user #ID? *Note that "a pattern is defined as a concurrent habit of the camera wearer".*

2. Given these two discovered sets of patterns, which pattern set do you think better represents the user's behaviour?.
3. Given the previous recorded days for user #ID, do you think the found patterns adequately represent the user's behaviour?."
4. Do you consider that the found patterns are predominantly: a) sub-patterns, b) no patterns, c) merged patterns, or d) correct patterns?

4.3 Experimental Setup

In this work, we use several available models for the extraction of descriptors from the images with the aim of characterizing the user's days.

- We rely on Places365 [24] to recognize the place depicted in the images. We use the top-1 label to represent the scene.
- Objects are detected in the frames using the Yolo3 [23] CNN model. We keep only the detected objects with a given class probability > 0.5.
- The network introduced in [7] is used to identify the activity that is described in the frame. We use the top-1 label to represent the scene.

4.4 Results

In this subsection, we evaluate our proposed method for pattern discovery.

In Fig. 5 we show some samples of discovered patterns for User 1. Each pattern is described by the places, activities, and objects recognized in the images

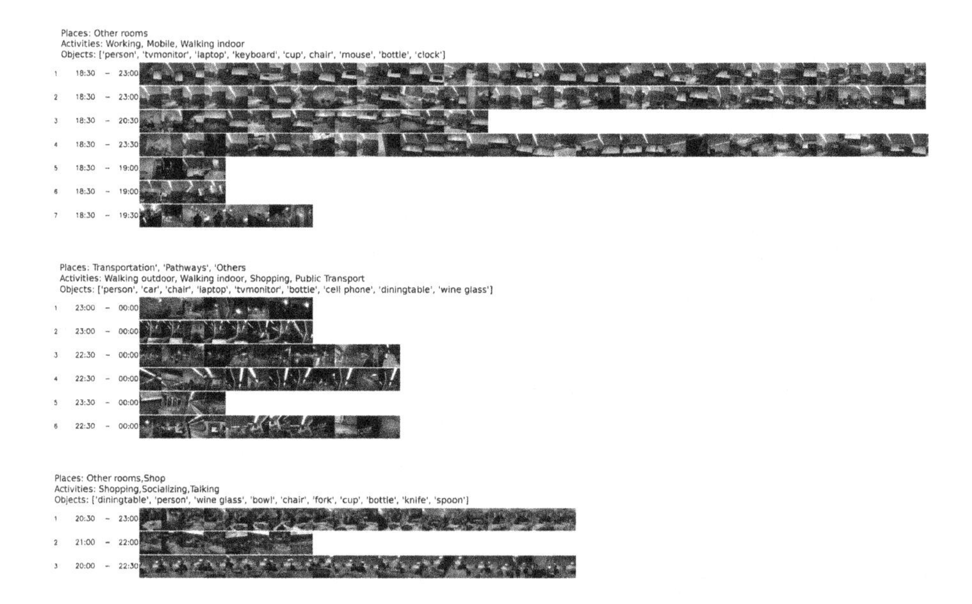

Fig. 5. Some discovered patterns for User 1 and the labels used to analyze them.

that compose it. This allows us to get a better understanding of what the pattern represents. Given a pattern, each row indicated the number of times it appears in the collection of photo-streams. The time-span of such event is indicated on the left side of the pattern. We can observe how the first pattern describes an office-related activity, while the second and third pattern describe a commuting- and social eating-related activities, respectively.

We rely on the Silhouette score metric to evaluate the quality of the found cluster of images that are finally considered as a pattern. In Fig. 6, we present an example of clusters silhouettes. On the right figure we can see the Silhouette coefficient values for the different found clusters. High values of Silhouette score indicate high intra-class and low inter-class similarity among samples within the different clusters. On the right, we have a 2D visualization of the nodes under analysis. The coordinates are obtained from the computed distance matrix with multi-dimensional scaling method. Colours indicate the clusters that are assigned to the nodes. We can observe that even though some elements overlap, clusters can be visually identified.

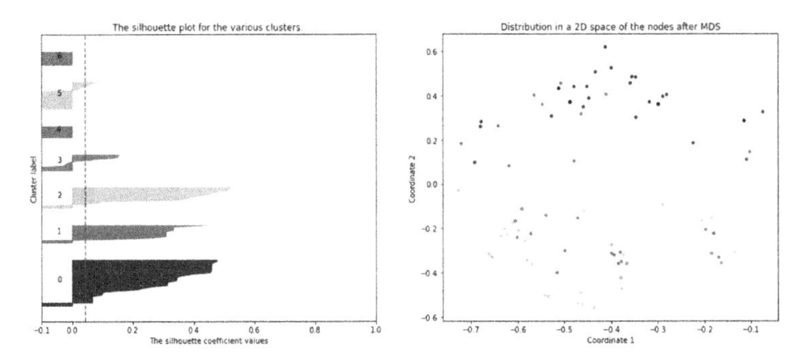

Fig. 6. (Left) Distribution in a 2D space of the nodes after MDS. Colours indicate the cluster label. (Right) Clusters quality measure with the Silhouette Score for the computed clusters for User 2. (Color figure online)

Furthermore, we evaluate the occurrence of patterns in the daily life of the camera wearer. A good estimation is to visually evaluate with a histogram the discovered patterns. Figure 7 indicates the occurrence of patterns for User 2. Samples images have been added to give visual information of what the pattern depicts. We solely indicate the number of days in which the pattern appears. We can observe how the user goes to the supermarket 5 times (pattern 4), bikes 3 days (pattern 0), or works with the computer 10 days (patterns 3 and 6).

In Fig. 8, we can observe the discovered patterns for User 2. We use different colours to represent different patterns. The time-slots that are not associated to any pattern are not included. We can see how the randomly placed seed spread throughout time-slots that are visually alike, i.e. define similar scenes. For instance, the user has the habit of going to the beach at lunch time (blue colour). The user also works in the morning (red colour) with a computer. We

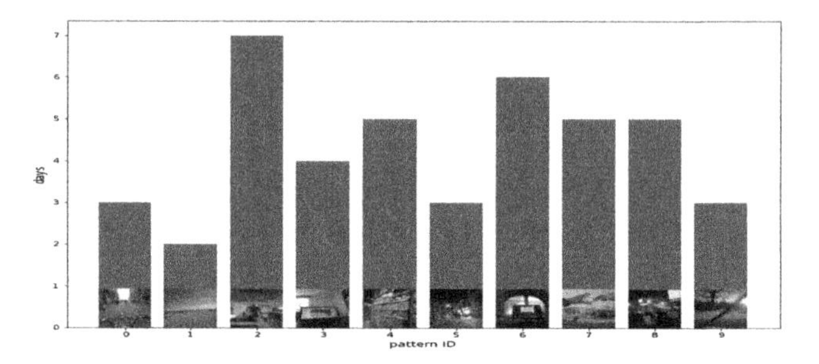

Fig. 7. Histogram of discovered patterns occurrence. X-axis indicate the pattern ID - we add a sample image as an illustration of the scene that the pattern represents. Y-axis indicate the number of days in which the pattern appears.

Fig. 8. Discovered patterns given a user's collection of photo-streams. Rows represent collection of days. Columns represent time-slots of 30'. One sample image is shown per time-slot. Colours indicate different discovered patterns. (Color figure online)

can observe how some visually similar time-slots composed of the same activity belong to different patterns. After visually inspecting these cases, we observed that this is due to the fact that different objects appeared in these scenes. As an example of this, the purple and red patterns found from 10:00 h to 12:00 h, indicate that the user was working in front of the PC while interacting with different objects.

Table 2. Automatically selected thresholds for each user

	User 1	User 2	User 3	User 4	User 5	User 6	User 7
Threshold	0.002	0.006	0.008	0.01	0.006	0.01	0.006

We evaluated the influence of the parameter sigma of the Gaussian applied to the variance evolution curve in order to assess the first and second derivatives. In Fig. 9 two set of patterns can be shown that correspond to sigma 1 and sigma 9

Sigma = 1

10:00	10:30	11:00	11:30	12:00	12:30	13:00	13:30	14:00
			Bedroom	Transportation	Transportation	Office	Pathways	Living room
			Pathways	Restaurant/Bar	Restaurant/Bar	Transportation	Transportation	Sport fields
Pathways	Pathways		Pathways	Transportation	Pathways	Office	Office	Office
Transportation	Office	Office	Office	Office	Office	Office	Office	Office
Office	Office	Office	Other rooms	Other rooms	Other rooms	Office	Restaurant/Bar	Restaurant/Bar
		Pathways	Pathways			Pathways	Transportation	Pathways
Pathways	Transportation	Pathways	Office	Office	Office	Office	Office	Office
	Pathways	Transportation	Pathways	Pathways	Office	Office	Office	Office
	Pathways	Pathways		Transportation	Transportation	Pathways	Kitchen	Office
Pathways			Pathways	Transportation	Pathways	Sport fields	Sport fields	Pathways
		Transportation	Transportation	Office	Office	Pathways	Restaurant/Bar	Restaurant/Bar
Restaurant/Bar	Transportation	Pathways	Education/science	Office	Office	Pathways	Office	Office
Education/science	Education/science	Office	Office	Office	Sport fields	Sport fields		Restaurant/Bar
Restaurant/Bar			Transportation	Transportation	Others	Office	Office	Office

Sigma = 9

10:00	10:30	11:00	11:30	12:00	12:30	13:00	13:30	14:00
			Bedroom	Transportation	Transportation	Office	Pathways	Living room
			Pathways	Restaurant/Bar	Restaurant/Bar	Transportation	Transportation	Sport fields
Pathways	Pathways		Pathways	Transportation	Pathways	Office	Office	Office
Transportation	Office	Office	Office	Office	Office	Office	Office	Office
Office	Office	Office	Other rooms	Other rooms	Other rooms	Office	Restaurant/Bar	Restaurant/Bar
		Pathways	Pathways			Pathways	Transportation	Pathways
Pathways	Transportation	Pathways	Office	Office	Office	Office	Office	Office
	Pathways	Transportation	Pathways	Pathways	Office	Office	Office	Office
	Pathways	Pathways		Transportation	Transportation	Pathways	Kitchen	Office
Pathways			Pathways	Transportation	Pathways	Sport fields	Sport fields	Pathways
		Transportation	Transportation	Office	Office	Pathways	Restaurant/Bar	Restaurant/Bar
Restaurant/Bar	Transportation	Pathways	Education/science	Office	Office	Pathways	Office	Office
Education/science	Education/science	Office	Office	Office	Sport fields	Sport fields		Restaurant/Bar
Restaurant/Bar			Transportation	Transportation	Others	Office	Office	Office

Fig. 9. Found User 1's patterns shown with different colours, applying Gaussian smoothing with Sigma = 1 and Sigma = 9. (Color figure online)

respectively. It can be seen that with sigma 1 a greater number of small patterns are found, while with sigma 9 long-term patterns (in times-lots and days) are observed that are more consistent.

A cluster varies its inner variance when including new elements. We use the score defined in Eq. (3) to assess the quality of the formed clusters with different threshold values T on the first derivative. We tested threshold values within the range [0–0.05]. In Table 2, we present as quantitative results the automatically selected thresholds on the first derivative for every user that yields the maximum score.

Given that there are no previous works addressing the task of behavioural pattern discovery from egocentric photo-streams, we compare the performance of the proposed pipeline against a native version using a CNN. The native approach uses as images descriptor the softmax output values of a CNN (e.g. Places365 [24] model) and the normalized time-stamp. A time-frame is described as the aggregation of the features vectors that compose it. The DBSCAN clustering technique [28] is used to group similar time-frames. To balance the time characteristic with the rest of the 365 characteristics, we apply a weighting procedure to the input feature vector according to [29]. The clusters were grouped into the corresponding time slots and the resulting clusters were considered as patterns. To evaluate quantitatively this patterns and compare them to the ones obtained with our method, we calculated for both approaches, the Silhouette Score for all the users. For our proposed method the Average Silhouette is **0.12**, with a minimum of –0.13 and a maximum of 0.47. For DBSCAN the average is **0.01**, with a minimum of –0.49 and a maximum of 0.44. It is important to highlight that the average of clusters found by DBSCAN was 3.0 and in our method was 5.6, even so in this last one there was less overlap.

Regarding the survey responses, as shown in Figs. 10, 11, 12 and 13 all the individuals answered they found different patterns on the images. Also, regarding if the patterns found by our method represent the user's behaviour, 61.5% of them answered "yes", 36.3% answered "maybe", while 2.2% answered "no".

For the question about which method gets better results, 73.6% answered our method is better, while 26.4% answered DBSCAN get better results. Finally, 68.1% answered our method found "correct patterns", 18.7% answered "Merged Patterns", 9.9% answered "Sub-Patterns", while 3.3% answered "No-Patterns". All these values were averaged over all users.

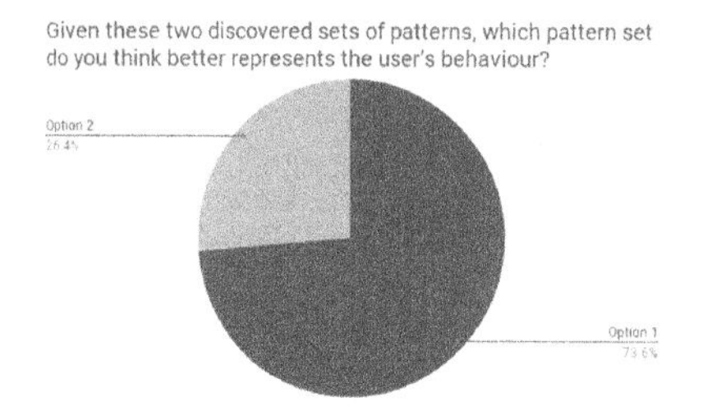

Fig. 10. Answers to Question 2.

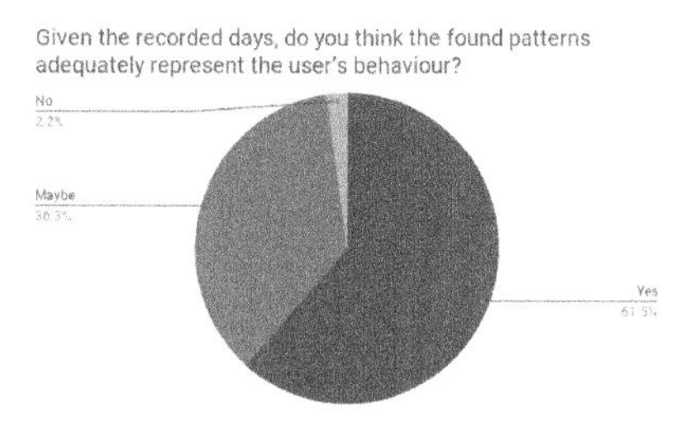

Fig. 11. Answers to Question 3.

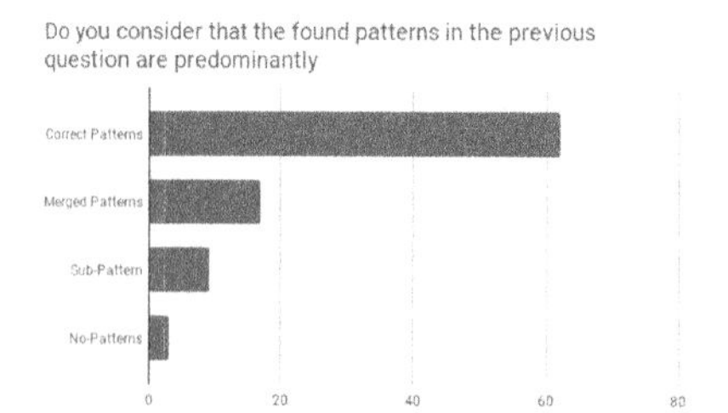

Fig. 12. Answers to Question 4.

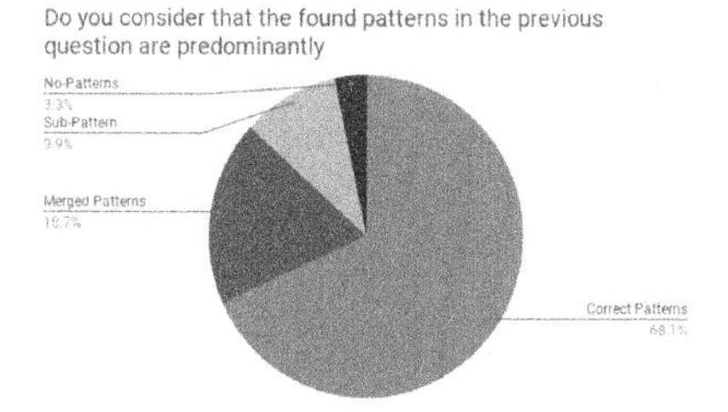

Fig. 13. Answers to Question 4 - Percentages.

5 Conclusions

In this work, we presented a novel pattern discovery model for the automatic detection of behavioural habits from collections of visual logs collected by wearable cameras. The validation of the obtained patterns indicates that our model is valuable and robust, and sets the basis for further research in the field of behaviour understanding from egocentric images. Our proposed model can be applied to other tasks for pattern discovery since it evaluates the similarity among samples of different time-sequences. Further research will explore its applicability to other tasks.

Results show that the chosen descriptors allow us to discover patterns. However, more analysis is required to fully understand the relevance of the different subsets of descriptors ('activity', 'scene', and 'objects'). For instance, objects with a high recognition accuracy are considered but frequency of appearance of these objects within the event is not considered. We could use tools such as Word2Vec [30] to evaluate the semantic similarity of recognized objects. These are questions that we plan to continue evaluating to reach a wider overview about the relevance of the different descriptors. On the other hand, we are interested in discovering more complex patterns composed of combination of events (e.g. the individual used to walk on the street after work). We consider that this work is only the beginning of the research line to discover meaningful patterns of human behaviour from egocentric photo-streams.

Acknowledgment. This work was partially founded by projects RTI2018-095232-B-C2, SGR 1742, CERCA, Nestore Horizon2020 SC1-PM-15-2017 (n° 769643), and Validithi EIT Health Program. The founders had no role in the study design, data collection, analysis, and preparation of the manuscript.

References

1. Pantelopoulos, A., Bourbakis, N.G.: A survey on wearable sensor-based systems for health monitoring and prognosis. IEEE Trans. Syst. Man Cybern. Part C **40**(1), 1–12 (2009)
2. Siła-Nowicka, K., Vandrol, J., Oshan, T., Long, J.A., et al.: Analysis of human mobility patterns from GPS trajectories and contextual information. Int. J. Geogr. Inf. Sci. **30**(5), 881–906 (2016)
3. Lara, O.D., Labrador, M.A.: A survey on human activity recognition using wearable sensors. IEEE Commun. Surv. Tutor. **15**(3), 1192–1209 (2012)
4. Huang, Y., et al.: Assessing social anxiety using GPS trajectories and point-of-interest data. In: ACM International Joint Conference on Pervasive and Ubiquitous Computing, pp. 898–903 (2016)
5. Atzmueller, M., Thiele, L., Stumme, G., Kauffeld, S.: Analyzing group interaction on networks of face-to-face proximity using wearable sensors. In: IEEE International Conference on Future IoT Technologies, pp. 1–10 (2018)
6. Talavera, E., Wuerich, C., Petkov, N., Radeva, P.: Topic modelling for routine discovery from egocentric photo-streams. Pattern Recogn. **104**, 107330 (2020)
7. Cartas, A., Marín, J., Radeva, P., Dimiccoli, M.: Recognizing activities of daily living from egocentric images. In: Alexandre, L.A., Salvador Sánchez, J., Rodrigues, J.M.F. (eds.) IbPRIA 2017. LNCS, vol. 10255, pp. 87–95. Springer, Cham (2017). https://doi.org/10.1007/978-3-319-58838-4_10
8. Furnari, A., Farinella, G.M., Battiato, S.: Recognizing personal locations from egocentric videos. IEEE Trans. Hum.-Mach. Syst. **47**(1), 6–18 (2016)
9. Matei, A., Glavan, A., Talavera, E.: Deep learning for scene recognition from visual data: a survey. In: de la Cal, E.A., Villar Flecha, J.R., Quintián, H., Corchado, E. (eds.) HAIS 2020. LNCS (LNAI), vol. 12344, pp. 763–773. Springer, Cham (2020). https://doi.org/10.1007/978-3-030-61705-9_64
10. Aimar, E.S., Radeva, P, Dimiccoli, M.: Social relation recognition in egocentric photostreams. In: IEEE International Conference on Image Processing, pp. 3227–3231 (2019)
11. Aghaei, M., Dimiccoli, M., Ferrer, C.C., Radeva, P.: Towards social pattern characterization in egocentric photo-streams. Comput. Vis. Image Underst. **171**, 104–117 (2018)

12. Damen, D., et al.: Scaling egocentric vision: the epic-kitchens dataset. In: Ferrari, V., Hebert, M., Sminchisescu, C., Weiss, Y. (eds.) ECCV 2018. LNCS, vol. 11208, pp. 753–771. Springer, Cham (2018). https://doi.org/10.1007/978-3-030-01225-0_44
13. Talavera, E., Petkov, N., Radeva, P.: Unsupervised routine discovery in egocentric photo-streams. In: Vento, M., Percannella, G. (eds.) CAIP 2019. LNCS, vol. 11678, pp. 576–588. Springer, Cham (2019). https://doi.org/10.1007/978-3-030-29888-3_47
14. Brunet, J.-P., Tamayo, P., Golub, T.R., Mesirov, J.P.: Metagenes and molecular pattern discovery using matrix factorization. Proc. Natl. Acad. Sci. **101**(12), 4164–4169 (2004)
15. Hassell, M.P., Lawton, J.H., May, R.: Patterns of dynamical behaviour in single-species populations. J. Anim. Ecol. **45**, 471–486 (1976)
16. Zhong, N., Li, Y., Wu, S.-T.: Effective pattern discovery for text mining. IEEE Trans. Knowl. Data Eng. **24**(1), 30–44 (2010)
17. Park, A.S., Glass, J.R.: Unsupervised pattern discovery in speech. IEEE Trans. Audio Speech Lang. Process. **16**(1), 186–197 (2007)
18. Benchetrit, G.: Breathing pattern in humans: diversity and individuality. Respir. Physiol. **122**(2–3), 123–129 (2000)
19. Borges, P.V.K., Conci, N., Cavallaro, A.: Video-based human behavior understanding: a survey. IEEE Trans. Circuits Syst. Video Technol. **23**(11), 1993–2008 (2013)
20. Kim, E., Helal, S., Cook, D.: Human activity recognition and pattern discovery. IEEE Pervasive Comput. **9**(1), 48–53 (2009)
21. Zhao, G., Yuan, J.: Discovering thematic patterns in videos via cohesive sub-graph mining. In: IEEE 11th International Conference on Data Mining, pp. 1260–1265 (2011)
22. Eagle, N., (Sandy) Pentland, A.: Reality mining: sensing complex social systems. Pers. Ubiquit. Comput. **10**(4), 255–268 (2006). https://doi.org/10.1007/s00779-005-0046-3
23. Redmon, J., Farhadi, A.: YOLOv3: An incremental improvement, arXiv (2018)
24. Zhou, B., Lapedriza, A., Khosla, A., Oliva, A., Torralba, A.: Places: a 10 million image database for scene recognition. IEEE Trans. Pattern Anal. Mach. Intell. **40**, 1452–1464 (2017)
25. Real, R., Vargas, J.M.: The probabilistic basis of Jaccard's index of similarity. Syst. Biol. **45**(3), 380–385 (1996)
26. Borg, I., Groenen, P.J.F.: Modern Multidimensional Scaling: Theory and Applications. SSS. Springer, New York (2005). https://doi.org/10.1007/0-387-28981-X
27. Rousseeuw, P.J.: Silhouettes: a graphical aid to the interpretation and validation of cluster analysis. J. Comput. Appl. Math. **20**, 53–65 (1987)
28. Ester, M., Kriegel, H.-P., Sander, J., Xu, X., et al.: A density-based algorithm for discovering clusters in large spatial databases with noise. In: Kdd, vol. 96, no. 34, pp. 226–231 (1996)
29. Dousthagh, M., Nazari, M., Mosavi, A., Shamshirband, S., Chronopoulos, A.T.: Feature weighting using a clustering approach. Int. J. Model. Optim. **9**(2), 67–71 (2019)
30. Mikolov, T., Chen, K., Corrado, G., Dean, J.: Efficient estimation of word representations in vector space. In: Bengio, Y., LeCun, Y. (eds.) 1st International Conference on Learning Representations, ICLR (2013)

Motion Prediction for First-Person Vision Multi-object Tracking

Ricardo Sanchez-Matilla(✉) and Andrea Cavallaro

Centre for Intelligent Sensing, Queen Mary University of London, London, UK
{ricardo.sanchezmatilla,a.cavallaro}@qmul.ac.uk

Abstract. Tracking multiple independently moving objects with cameras mounted on moving robots is becoming increasingly common. However, most causal trackers rely on linear motion models that may be inaccurate in these scenarios. To overcome this problem, we present a real-time multi-object tracker based on the Early Association Probability Hypothesis Density Particle Filter with a prediction model that disentangles the motion of objects from that of the camera. Moreover, the prediction model allows us to intentionally reduce the video frame rate at which the tracker operates, with only a minor reduction in accuracy. Specifically, the model allows us to halve the processed frames while still outperforming alternative prediction models, including traditional linear motion predictors in moving-camera sequences. Experimental results show that the proposed model improves both accuracy and precision of the tracker.

Keywords: Motion model · Prediction · Mobile cameras · Mobile agents · Tracking

1 Introduction

Robots are being introduced in human-inhabited environments, such as houses, retail spaces and cities. For robots and humans to safely share the space, robots must perceive and understand their surroundings to perform a variety of tasks, such as socially-aware navigation [2,24], assistance of disabled people [17,23,39,44] and assistance in household chores [35]. In the last scenario, robotic arms equipped with a camera on the *moving* end-effector (e.g. a gripper or a hand) may be used to assist *moving* humans by dynamically grasping *moving* objects handed over by the human. Tracking and prediction are essential components of a pipeline that successfully performs these tasks [35].

Unlike offline trackers that can use future object detections to solve the target association problem, causal trackers, which are needed for robotic tasks, can only rely on past object detections. To predict the expected future location of objects, causal trackers use motion models [3,10,19,28]. However, the motion of an object may be masked by the global motion of the camera, thus making the

A. Bartoli and A. Fusiello (Eds.): ECCV 2020 Workshops, LNCS 12538, pp. 485–499, 2020.
https://doi.org/10.1007/978-3-030-66823-5_29

prediction task very challenging. Moreover, miss-detections increase the difficulty of accurately predicting future object locations.

Predictors can describe the motion of an object with a Brownian model [30] or with linear predictions based on its past motion [7,29,37]. Linear predictors may perform accurately with static cameras, where the linearity assumption may hold. However, when objects and camera move independently, linear predictors may cause drifts. Higher frame rates can alleviate this problem within short temporal windows [15], but a higher computation cost is required and, therefore, higher energy budget, undesirable for battery-operated robots.

With static cameras, the prediction can use Kalman Filters [16], linear [29] or Gaussian regression models [41], or deep learning models [2,4,14,31]. With deep learning methods, the location and scale of objects can be predicted with an LSTM [4], by modelling person-to-person interactions [2], person-to-person interactions considering static objects [31], or by learning socially acceptable trajectories with a combination of sequence prediction and generative adversarial networks [14]. The Trajectory Forecasting benchmark (TrajNet) [5] aims to enable fair comparisons among prediction models for static cameras, with a special focus on human-to-human interaction.

With moving cameras, the problem becomes more challenging as the motion of the objects is masked by that of the camera, and predictors generally require additional information, such as camera parameters (i.e. a calibrated camera) [3, 10], additional sensors (e.g. a depth sensor) [10], or strong constraints on the type of scenario, such as a road recording vehicles [8,9], a high-altitude drone recording a top-down view [3], or having all objects sharing a common plane [3, 19,22,28]. A desirable prediction model should instead be agnostic to the type and location of objects as well as the scene type and, ideally, not require any additional hardware. To this end, we recently proposed an accurate prediction model to forecast the position of moving objects by disentangling global and object motion without the need for additional hardware, camera calibration or planarity assumptions [34].

In this paper, we present a causal multi-object tracker with a prediction model that accurately forecasts the motion of objects despite the motion of the camera (see Fig. 1). The model, inspired by the Global Motion (GM) prediction model [34], is informed by the global motion estimation and separates the object motion from that of the camera. We use the Early Association Probability Hypothesis Density Particle Filter (EA-PHD-PF) [36] as base tracker for this prediction model. In order to determine how tracking performance can be improved in moving cameras with an enhanced prediction model while maintaining constrained computation requirements, we do *not* use any appearance information. Experiments are performed on the Multiple Object Tracking Benchmark 2017 (MOTB17) [1]. Results show that the proposed model improves tracking accuracy by 23% and precision by 3% compared with a traditional motion model on moving-camera videos, and achieves comparable overall tracking results to alternative approaches over both static and moving camera scenarios, while only requiring to process half of the frames in moving camera scenarios.

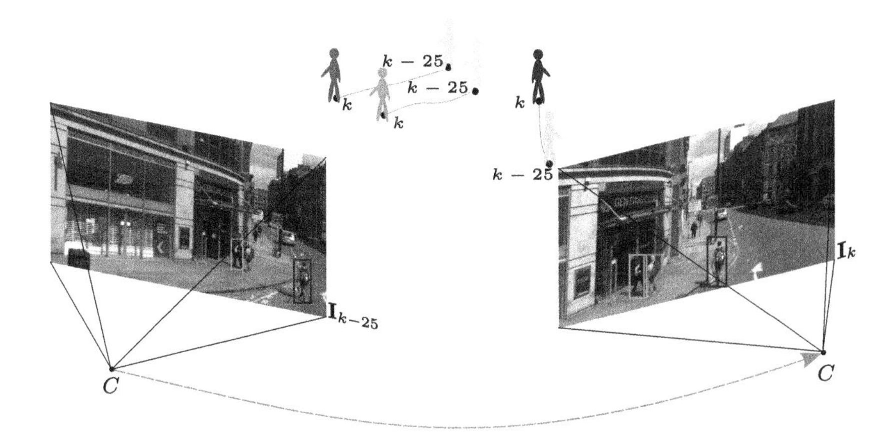

Fig. 1. Multi-object tracking with the proposed prediction model that facilitates maintaining object identities when camera and objects move independently, and without using any appearance information. KEY – $\mathbf{I}_k$: frame at time k; C: camera centre; grey arrow indicates the motion of the camera between time $k - 25$ and time k. The video was recorded at 25 frames per second.

2 Causal Multi-object Tracker with Mobile Cameras

Let $\mathbb{Z}_k = \left\{\mathbf{z}_k^j\right\}_{j=1}^{N_k}$ represent N_k (potentially noisy) detections at time k with $\mathbf{z}_k^j = (u, v, w, h)$, where (u, v) is the centre, and w and h are the width and height of the bounding box representing the object on the image plane. Let $\mathbb{Z}_{1:k}$ represent the detections up to time k.

If λ indicates the object identity, our objective is to estimate $\mathbb{X}_k = \left\{\mathbf{x}_k^\lambda\right\}_{\lambda=1}^{M_k}$, the state (i.e. location, dimensions, and identity) of M_k objects by handling detection errors, which include noisy detections, false positive and false negatives. To perform global motion estimation, let us define the state of an object in homogeneous coordinates as, $\mathbf{x}_k^\lambda = (u, v, 1, \dot{u}, \dot{v}, w, h)$, where $\dot{u}$ and $\dot{v}$ are the horizontal and vertical components of the velocity of the object.

To solve the tracking problem, let us consider a Bayesian framework where the posterior probability distribution, $p(\mathbb{X}_k|\mathbb{Z}_{1:k})$, can be recursively estimated as [27]:

$$\underbrace{p(\mathbb{X}_k|\mathbb{Z}_{1:k})}_{\text{Posterior}} \propto \underbrace{p(\mathbb{Z}_k|\mathbb{X}_k)}_{\text{Observation Model}} \int \underbrace{p(\mathbb{X}_k|\mathbb{X}_{k-1})}_{\text{Prediction Model}} \underbrace{p(\mathbb{X}_{k-1}|\mathbb{Z}_{1:k-1})}_{\text{Previous Posterior}} d\mathbb{X}_{k-1},$$

where the prediction model, $p(\mathbb{X}_k|\mathbb{X}_{k-1})$, defines the temporal evolution of the state and the observation model, $p(\mathbb{Z}_k|\mathbb{X}_k)$, quantifies the likelihood that tracking states generate the detections. In practice, the posterior for the λ-th object can be approximated by a set of weighted particles (i.e. tracking state samples) when using the EA-PHD-PF [36], as an identity is assigned to each particle so that the state of λ-th object can be estimated as

$$\mathbf{x}_k^\lambda = \sum_{i=1}^{P^\lambda} \pi_{k,i}^\lambda \, \mathbf{x}_{k,i}^\lambda, \tag{1}$$

where $\left\{\mathbf{x}_{k,i}^\lambda\right\}_{i=1}^{P^\lambda}$ is the set of P^λ particles estimating the object λ, and $\pi_{k,i}^\lambda$ is the weight of the i-th particle.

The prediction and observation models are discussed next.

2.1 Prediction Model

The proposed prediction model decomposes the observed motion of an object into two independent components, global and object motion, as

$$\mathbf{x}_{k,i}^\lambda = \underbrace{\mathbf{G}_{k|k-1}^\lambda \, \mathbf{x}_{k-1,i}^\lambda}_{\text{Global Motion}} + \underbrace{\mathbf{T}\,\mathbf{x}_{k-1,i}^\lambda}_{\text{Object Motion}} + \mathbf{N}_k^\lambda, \tag{2}$$

where $\mathbf{G}_{k|k-1}^\lambda$ is the global motion matrix, $\mathbf{T}$ is the object motion matrix, and $\mathbf{N}_k^\lambda$ is the noise matrix.

The global motion matrix, $\mathbf{G}_{k|k-1}^\lambda$, describes the temporal evolution of the λ-th object on the image plane when the object is stationary in the 3D world between times $k-1$ and k and it is defined as:

$$\mathbf{G}_{k|k-1}^\lambda = \begin{bmatrix} \frac{1}{\alpha_{k|k-1}^\lambda}\mathbf{H}_{k|k-1} & \mathbf{0}_{3\times 2} & \mathbf{0}_{3\times 2} \\ \mathbf{0}_{2\times 3} & \mathbf{I}_{2\times 2} & \mathbf{0}_{2\times 2} \\ \mathbf{0}_{2\times 3} & \mathbf{0}_{2\times 2} & \mathbf{I}_{2\times 2} \end{bmatrix}, \tag{3}$$

where $\mathbf{H}_{k|k-1} \in \mathbb{R}^{3\times 3}$ is a homography transformation that approximates the global motion between $k-1$ and k, $\alpha_{k|k-1}^\lambda = \langle \mathbf{h}_3, (u, v, 1)\rangle$ is a normalisation factor, $\mathbf{h}_3$ is the third row of $\mathbf{H}_{k|k-1}$ and $\langle\cdot\rangle$ is the dot product, and $\mathbf{0}$ and $\mathbf{I}$ are the zero and identity matrices with dimensions indicated in the subscript, respectively. As the prediction at time k must be calculated only using past information (i.e. the frame at time k is not available for the prediction), then, we approximate the global motion between frames at times $k-1$ and k, by that between the frames $k-2$ and $k-1$, as $\mathbf{H}_{k|k-1} \approx \mathbf{H}_{k-1|k-2}$. We estimate the global motion between two consecutive frames by finding the transformation between the location of a selected set of keypoints that lie on the background of the frames, assuming that, in ground-level first-person vision, the global motion between consecutive frames can be approximated by the camera rotation (i.e. translation effects are negligible) [34]. Specifically, we first detect keypoints following a policy for maintaining a uniform spatial distribution of keypoints across the frame [40]. Next, keypoints lying inside the tracked objects are filtered out as they are potentially lying on objects that move in the 3D world, whose motion is not useful for the calculation of the global motion. Then, keypoints are tracked to the consecutive frame. Finally, a homography is calculated by estimating the transformation that relates the position of the tracked keypoints between these consecutive frames using the random sample consensus (RANSAC) algorithm [38]. We follow the estimation of the global motion estimation (i.e. homography matrix) with the method defined in [34].

Once the global motion has been estimated, the object motion matrix, $\mathbf{T}$, that describes the temporal evolution of the object on the image plane, can assume that the camera is stationary and, therefore it can be simply formulated with a linear prediction as:

$$\mathbf{T} = \begin{bmatrix} \mathbf{I}_{3\times 3} & \mathbf{A} & \mathbf{0}_{3\times 2} \\ \mathbf{0}_{2\times 3} & \mathbf{I}_{2\times 2} & \mathbf{0}_{2\times 2} \\ \mathbf{0}_{2\times 3} & \mathbf{0}_{2\times 2} & \mathbf{I}_{2\times 2} \end{bmatrix}, \tag{4}$$

where

$$\mathbf{A} = \begin{bmatrix} S & 0 \\ 0 & S \\ 0 & 0 \end{bmatrix}, \tag{5}$$

and S is the time interval between two consecutive frames.

We account for unmodeled factors with the noise matrix, $\mathbf{N}_k^\lambda$, defined as:

$$\mathbf{N}_k^\lambda = \left[\mathcal{N}_k^u, \mathcal{N}_k^v, 0, \mathcal{N}_k^{\dot{u}}, \mathcal{N}_k^{\dot{v}}, \mathcal{N}_k^w, \mathcal{N}_k^h\right]^T, \tag{6}$$

where each component is a sample drawn from a Gaussian distribution, $\mathcal{N}_k^* = \mathcal{N}(0; \sigma_p^*)$, with zero mean and standard deviation σ_p^*. These parameters are commonly selected by estimating the uncertainties of each state component in a training dataset but can also be a constant [37], a function of the video frame rate [26], or a function to the object size [36]. We set the standard deviations for the object size, σ_p^w and σ_p^h, as constants, and the standard deviations for the object location and its velocity, σ_p^u, σ_p^v, $\sigma_p^{\dot{u}}$ and $\sigma_p^{\dot{v}}$, as a function of the object size on the image plane to account for the perspective [36] (i.e. an object closer to the camera has higher uncertainty on its location on the image plane compared to another object that is farther from it) and as a function of the video frame rate to account for frame rate variations [26] (i.e. an object seen by a lower frame-rate video has higher uncertainty on its location on the image plane compared to when seen by a higher frame-rate video).

2.2 Observation Model

The observation model estimates the importance of each particle towards the estimation of the state. This importance is encoded in the particle weight, $\pi_{k,i}^\lambda$. As we aim to understand how tracking performance can be improved in moving cameras by an enhanced prediction model while maintaining constrained computation requirements, we use an observation model that only uses the location and dimensions of detections and states, and *no* appearance information.

The particle weight, assuming that the state components are independent, can be updated as [36]:

$$\pi_{k,i}^\lambda = \pi_{k-1,i}^\lambda \, p(\mathbf{z}_k^u|\mathbf{x}_{k,i}^{\lambda,u}) \, p(\mathbf{z}_k^v|\mathbf{x}_{k,i}^{\lambda,v}) \, p(\mathbf{z}_k^w|\mathbf{x}_{k,i}^{\lambda,w}) \, p(\mathbf{z}_k^h|\mathbf{x}_{k,i}^{\lambda,h}), \tag{7}$$

where the likelihood of each particle component is estimated as:

$$p(\mathbf{z}_k^*|\mathbf{x}_{k,i}^{\lambda,*}) \propto exp\left[-\frac{d(\mathbf{z}_k^*, \mathbf{x}_{k,i}^{\lambda,*})^2}{2(\sigma_o^{\lambda,*})^2}\right], \tag{8}$$

where $\mathbf{x}_{k,i}^{\lambda,*}$ indicates the component $*$ of the i-th particle and $d(\cdot,\cdot)$ is the Euclidean distance. Similarly to the prediction model, the standard deviations for the object size, $\sigma_o^{\lambda,w}$ and $\sigma_o^{\lambda,h}$, are constant, whereas the standard deviations for the object location, $\sigma_o^{\lambda,u}$ and $\sigma_o^{\lambda,v}$, are modelled as a function of the detection width, to account for the perspective effects, and of the video frame rate, to account for frame rate variations.

2.3 State Estimation and Resampling

The state of each object is estimated using Eq. 1. Then, we resample the particles so that those with larger weights are maintained and replicated, whereas those with lower weights are removed. Specifically, we perform multi-stage resampling [26], which uses multiple stages depending on the frame where the particles were created. This ensures that particles recently created for newly appeared objects, and likely to have a lower weight than older ones, are not removed.

3 Validation

3.1 Experimental Setup

We compare our tracker against nine state-of-the-art causal trackers: EA-PHD-PF [36], MOTDT [25], HAM-SADF [43], DMAN [45], AM-ADM [21], PHD-GSDL [13], GM-PHD-KCF [18], FPSN [20], and GM-PHD [11]. To examine the contribution of the global motion module, we also make comparisons against the proposed tracker with the global motion matrix disabled, i.e. $\mathbf{H}_{k|k-1} = \mathbf{I}_{3\times 3}$; and the proposed tracker where the prediction is replaced by four prediction models (see Table 1) including two global motion estimators: a homography-based method (SH) [4], the global motion-aware camera motion (GM) [34] (the one we use in our proposed tracker), a Brownian model (BM) [30] and a linear predictor (LP) [37] which do not estimate the global motion generated by the camera motions. SH and GM estimate the global motion in the form of a homography, $\mathbf{H}_{k|k-1}$, and this is provided to the tracker using Eq. 3. We do not consider deep-learning-based prediction models in the evaluation as they are outperformed by the considered predictors when both objects and camera move [34].

We use the MOTB17 [1] dataset which provides three sets of person detections generated by: Deformable Part-based Model (DPM) [12], Fast-RCNN (FRCNN) [32], and Scale Dependent Pooling (SDP) [42]. We evaluate the tracker using each of these detection sets independently. The parameters of the global motion estimators are set as proposed by the authors in [34]. We select the tracking parameters that achieve the best results in the MOTB17 training dataset

Table 1. The prediction models under comparison and their runtime when embedded in the proposed tracker and using DPM as detector on the MOTB17 training dataset. KEY – Ref: reference; OM: considers object motion; CM: considers camera motion; NS: not scene specific; FT: no assumptions on the object location; fps: frames per second.

Ref		Prediction model	OM	CM	NS	FT	Runtime [fps]
[30]	BM	Brownian			✓	✓	**58.0**
[37]	LP	Linear	✓		✓	✓	**58.0**
[4]	SH	Planar homography	✓	✓			*29.7*
[34]	GM	Global motion	✓	✓	✓	✓	22.5

using a grid-search approach and then we fix them for all the experiments on the MOTB17 test dataset. The standard deviations for the prediction model are set to: $\sigma_p^u = \sigma_p^v = Sw/18$, $\sigma_p^{\dot{u}} = \sigma_p^{\dot{v}} = Sw/36$, $\sigma_p^w = \sigma_p^h = 5$; and for the observation model: $\sigma_o^u = \sigma_o^v = Sw/12$ and, $\sigma_o^w = \sigma_o^h = 10$, where w is the width of the object state. The time interval between frames is considered constant and, therefore, we set it as $S = 1$. The number of particles per object is $P^\lambda = 200$.

To evaluate the quality and compare tracking performance, we compare the discrepancy between the tracking estimations and a ground-truth. Specifically, we use Multiple Object Tracking Accuracy (MOTA) and Multiple Object Tracking Precision (MOTP) as performance scores [6]. For a given sequence, MOTA is defined as

$$\text{MOTA} = 1 - \frac{\sum_{k=0}^{K} (\text{FN}_k + \text{FP}_k + \text{IDSW}_k)}{|\tilde{\mathbb{X}}_{1:K}|}, \tag{9}$$

where $|\tilde{\mathbb{X}}_{1:K}|$ is the number of annotations in the sequence, K is the number of frames in the sequence, FN_k, FP_k and IDSW_k are the number of false negatives, false positives, and identity switches at time k. MOTP is defined as

$$\text{MOTP} = \frac{\sum_{\forall\lambda} \sum_{k=0}^{K} \text{IOU}(\mathbf{x}_k^\lambda, \tilde{\mathbf{x}}_k^\lambda)}{\sum_{k=0}^{K} c_k}, \tag{10}$$

where $\text{IOU}(\cdot,\cdot)$ is the intersection over union operator, $\tilde{\mathbf{x}}_k^\lambda$ is the ground-truth annotation for the λ-th object at time k, and c_k is the number of estimations with intersection over union with an annotation higher than 0.5 at time k.

3.2 Confidence Intervals

Generating high-quality annotations is a tedious and time-consuming task. Therefore, datasets often are annotated using semi-automatic techniques, for example by interpolating between manually annotated key-frames [33]. While this approach speeds up the annotation of large datasets (e.g. MOTB17), it introduces inaccuracies in the evaluation that should be taken into account when benchmarking trackers. Specifically, the annotations provided in MOTB17 used linear interpolation [33]. To account for the associated inaccuracies, which are

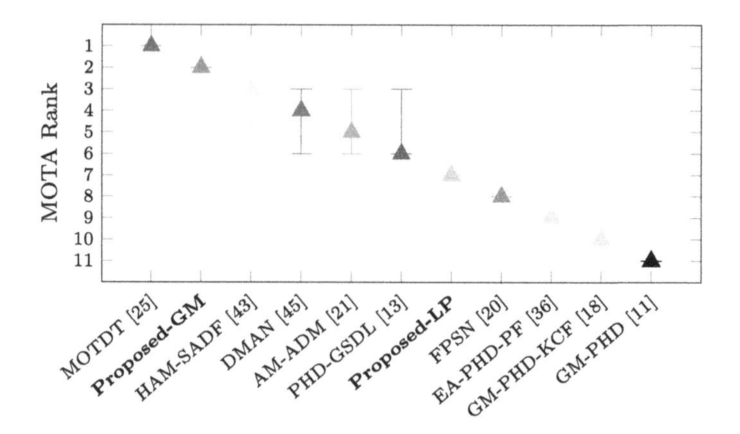

Fig. 2. MOTA ranks and associated confidence intervals [33] for causal trackers on the MOTB17 test dataset.

larger for moving cameras, we calculate a confidence interval for each evaluation measure [33]. The confidence intervals for the MOTB17 test dataset are 0.225 for MOTA and 5.678 for MOTP. The use of the confidence intervals allows us to fairly compare and rank trackers with respect to the chosen tracking performance measure. We can only ensure that a tracker outperforms another one when the value of a performance measure differs more than twice the confidence interval for that measure. Given the probabilistic nature of the proposed tracker, we run all the experiments five times and report the average and standard deviation of the results.

3.3 Tracking Evaluation

Figure 2 shows the MOTA Ranking for the MOTB17 test dataset for all available causal trackers. The results, considering the confidence intervals, indicate that the proposed tracker outperforms all trackers under evaluation except for MOTDT, that uses a deep-learning-based appearance observation model, and slightly outperforms the proposed tracker with a 1.1 higher MOTA, but being 4 frames per second (fps) slower than the proposed tracker. Regarding MOTP, we do not show its ranking as it is not informative in this dataset according to the confidence interval (i.e. the MOTP difference between all the trackers is smaller than the confidence interval). Note that the proposed tracker achieves state-of-the-art MOTP results.

Next, we quantify the contribution of different tracking components on the tracking performance. Table 2 shows the results when modifying the input set of detections (DPM [12], FRCNN [32], or SDP [42]) and the prediction model (BM [30], LP [37], SH [4] and GM [34]) on the training dataset. The detector is the component that has a higher impact on the tracking accuracy. Regarding the prediction model, GM achieves the most accurate and precise tracking results obtaining, on average, an increment in 9.0 and 3.2 MOTA points, and 3.0 and

Table 2. Tracking performance (average and standard deviation) comparison of the proposed tracker with different detectors and prediction models on the MOTB17 training dataset. KEY – PM: prediction model; M: moving-camera sequences only; C: complete dataset; BM: Brownian model [30]; LP: linear prediction [37]; SH: homography-based global motion [22]; GM: global motion estimation [34]. The higher the score, the better the performance. Best and second best performing methods are shown in bold and italic, respectively.

Detector	PM	MOTA				MOTP			
		M		C		M		C	
DPM	BM	30.3	(0.1)	34.8	(0.1)	74.2	(0.1)	*76.3*	*(0.1)*
	LP	*33.1*	*(0.3)*	*36.1*	*(0.2)*	*75.0*	*(0.1)*	**76.5**	**(0.1)**
	SH	31.8	(0.6)	35.5	(0.2)	74.5	(0.1)	*76.3*	*(0.1)*
	GM	**35.5**	**(0.1)**	**37.0**	**(0.1)**	**75.3**	**(0.1)**	**76.5**	**(0.1)**
FRCNN	BM	42.6	(0.1)	46.8	(0.2)	79.2	(0.1)	84.3	(0.2)
	LP	*49.5*	*(0.3)*	*49.6*	*(0.2)*	*83.5*	*(0.1)*	*86.5*	*(0.1)*
	SH	46.5	(0.2)	48.4	(0.1)	82.2	(0.0)	86.0	(0.2)
	GM	**52.9**	**(0.2)**	**50.9**	**(0.1)**	**84.0**	**(0.1)**	**86.6**	**(0.3)**
SDP	BM	48.9	(0.3)	59.9	(0.4)	78.7	(0.1)	83.0	(0.1)
	LP	*56.5*	*(0.4)*	*62.6*	*(0.1)*	*81.0*	*(0.0)*	*83.7*	*(0.0)*
	SH	52.6	(0.6)	61.7	(0.2)	80.4	(0.0)	83.3	(0.1)
	GM	**60.3**	**(0.3)**	**64.7**	**(0.2)**	**81.6**	**(0.1)**	**83.9**	**(0.1)**

0.5 MOTP points, when compared with BM and LP, respectively, in moving cameras. Note that only MOTA improvements are remarkable as they overcome the confidence interval for that measure. Among the alternatives that consider the global motion, SH performs worse than GM due to a more inaccurate global motion estimation, probably because of this model assumes that the whole frame is a planar surface, which might not apply in certain situations in this dataset.

Figure 3 shows tracking results of the proposed tracker[1] with two different motion models: LP (first row) and GM (second row) in a moving-camera sequence (MOTB17-13) of the MOTB17 test set. During this segment of the sequence, the camera heavily jaws clockwise producing a large displacement of the object in the image plane. LP forecasting is not able to account for the global motion and it produces a large number of wrong false trajectory initialisation (note the identities on the figure). Also, identity switches happen, producing tracking errors. GM can accurately predict the motion of the objects in this scenario and maintains the correct object identities with no false-positive initialisations.

[1] Tracking results with the proposed Global Motion prediction are available here, and with Linear Prediction here.

Fig. 3. Sample tracking results in MOTB17-13 sequence while the moving camera jaws clockwise with two prediction models (PM): linear prediction (LP) and the proposed global motion (GM). Colours and numbers in the bounding boxes indicate the object identity. The blue line on the ground indicates the past locations of a static object (the rubbish bin) and it is drawn to help understanding the camera motion. Note that GM successfully maintains the tracker identities, unlike LM. The video source is recorded at 25 frames per second. (Color figure online)

3.4 Robustness to Frame-Rate Reduction

We evaluate the impact on tracking performance when intentionally reducing the video frame rate. In this experiment, we use the proposed tracker where we replace the proposed prediction model, GM, by BM, LP and SH. Figure 4 shows the tracking results when downsampling the video frame rate by keeping one frame and skipping the next γ (e.g. $\gamma = 1$ indicates 50% frame rate reduction). With moving cameras and $\gamma = 3$, tracking with GM consistently obtains the highest tracking accuracy outperforming BM, LP and SH by 26.6, 23.3 and 5.4 MOTA points, respectively. The proposed prediction formulation, when using SH and GM as global motion estimators, allows the tracker to reduce the video frame rate of 75% ($\gamma = 3$) while maintaining 60% and 66% of their original tracking accuracy, whereas BM and LP are only able to maintain 25% and 28% of its original tracking accuracy in moving-camera sequences, respectively. Moreover, GM using only 50% of the frames ($\gamma = 1$) outperforms BM, LP and SH when using all the frames. This is a remarkable result as one could use GM and save half of the computational resources, which is an important aspect in moving agents with limited resources. All predictors perform similarly in static-camera sequences, where the tracking accuracy barely decreases when reducing the video frame rate.

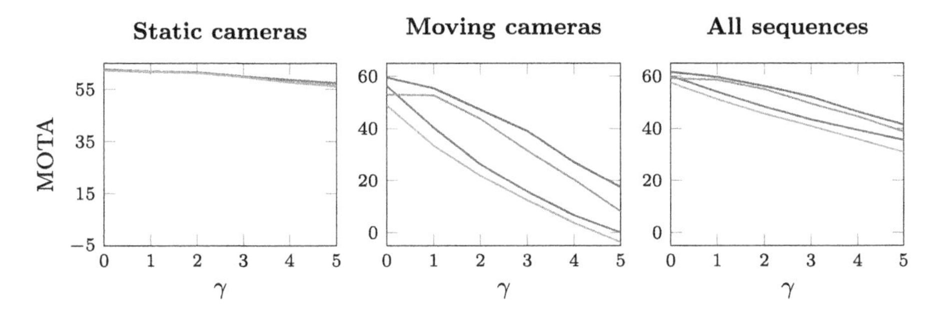

Fig. 4. Tracking performance of the proposed tracker with reduced video frame rate (keeping one frame and skipping the next γ) in static-camera, moving-camera, and all sequences of the MOTB17 training dataset for four prediction models: Brownian model [30] ▬, linear prediction [37] ▬, homography-based global motion [22] ▬ and global motion estimation [34] ▬. Standard deviations are less than one point. (Color figure online)

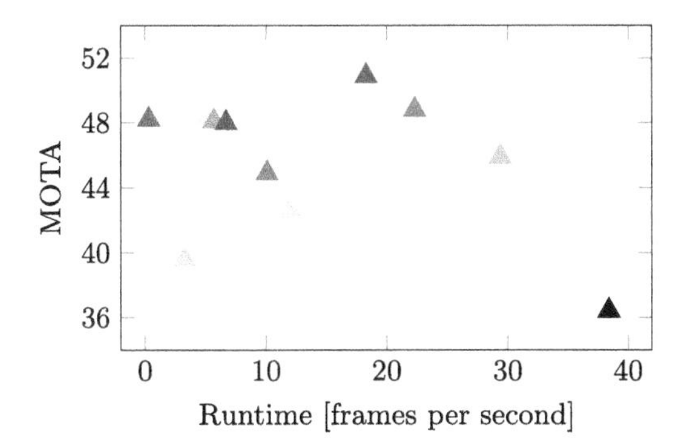

Fig. 5. MOTA vs. runtime of causal trackers in the MOTB17 test dataset. KEY – MOTDT [25] ▲, HAM-SADF [43] , DMAN [45] ▲, AM-ADM [21] ▲, PHD-GSDL [13] ▲, FPSN [20] ▲, EA-PHD-PF [36] , GM-PHD-KCF [18] ▲, GM-PHD [11] ▲, proposed with linear prediction (LP) [37] ▲ and proposed with global motion (GM) [34] ▲. (Color figure online)

3.5 Runtime Analysis

Table 1 shows the runtime of the different prediction models when they are embedded in the proposed tracker, when DPM is used as detector. As expected, the fastest alternatives are the predictors that do not estimate the global motion (BM or LP). The alternatives that use image processing techniques to estimate the global motion achieve runtime suitable for real-time computation for videos under 20 fps.

Figure 5 shows the results when comparing accuracy and runtime (the upper the most accurate and the most to the right the fastest). Considering both measures, the best performing trackers are MOTDT and the proposed one, shown respectively in red and green. The rest of the trackers are slower and less accurate except for GM-PHD (9.2 lower MOTA but 70% faster) and the proposed method with LP as predictor (2.9 lower MOTA but 30% faster).

4 Conclusion

We presented a causal multi-object tracker with a generic prediction model formulation that allows us to cope with camera motion without the need for camera calibration or assumptions on the scene. We validated the proposed tracker on the Multiple Object Tracking Benchmark 2017 dataset and compared against state-of-the-art predictors and trackers. Experimental results show that the use of the proposed model improves tracking accuracy by up to 11.4 MOTA points with respect to using other predictors. The proposed tracker is highly competitive among causal trackers in both accuracy and runtime, even against trackers that use complex (e.g. deep-learning based) appearance models. Moreover, the proposed tracker obtains comparable tracking results to using a linear prediction model but by processing only half of the frames, which is a desirable property for tracking with resource-constrained platforms. Finally, the proposed tracker performs very competitively with respect to published trackers in both accuracy and runtime on the MOTB17 Challenge.

Our future work includes using the prediction model within a framework that addresses the problem of occlusions.

Acknowledgements. This work is supported in part by the CHIST-ERA program through the project CORSMAL, under UK EPSRC grant EP/S031715/1.

References

1. MOT17: A benchmark for multi-object tracking (2016). https://motchallenge.net/data/MOT17/. Accessed 15 July 2020
2. Alahi, A., Goel, K., Ramanathan, V., Robicquet, A., Fei-Fei, L., Savarese, S.: Social LSTM: human trajectory prediction in crowded spaces. In: Proceedings of the IEEE Conference on Computer Vision and Pattern Recognition, Las Vegas, NV, pp. 961–971, June 2016
3. Arrospide, J., Salgado, L., Nieto, M., Mohedano, R.: Homography-based ground plane detection using a single on-board camera. IET Intell. Transp. Syst. **4**(2), 149–160 (2010)
4. Babaee, M., Li, Z., Rigoll, G.: Occlusion handling in tracking multiple people using RNN. In: Proceedings of the IEEE International Conference on Image Processing, Athens, Greece, pp. 2715–2719, October 2018
5. Becker, S., Hug, R., Hübner, W., Arens, M.: An evaluation of trajectory prediction approaches and notes on the trajnet benchmark. arXiv:1805.07663 (2018)
6. Bernardin, K., Stiefelhagen, R.: Evaluating multiple object tracking performance: The CLEAR MOT Metrics. J. Image Video Proc. **2008**(1), 246–309 (2008). https://doi.org/10.1155/2008/246309
7. Breitenstein, M.D., Reichlin, F., Leibe, B., Koller-Meier, E., Gool, L.V.: Online multiperson tracking-by-detection from a single, uncalibrated camera. IEEE Trans. Pattern Anal. Mach. Intell. **33**(9), 1820–1833 (2011)

8. Chandra, R., Bhattacharya, U., Bera, A., Manocha, D.: Traphic: trajectory prediction in dense and heterogeneous traffic using weighted interactions. In: Proceedings of the IEEE Conference on Computer Vision and Pattern Recognition, pp. 8483–8492 (2019)
9. Chandra, R., Bhattacharya, U., Roncal, C., Bera, A., Manocha, D.: RobustTP: end-to-end trajectory prediction for heterogeneous road-agents in dense traffic with noisy sensor inputs. In: ACM Computer Science in Cars Symposium, pp. 1–9 (2019)
10. Choi, W., Pantofaru, C., Savarese, S.: A general framework for tracking multiple people from a moving camera. IEEE Trans. Pattern Anal. Mach. Intell. **35**(7), 1577–1591 (2013)
11. Eiselein, V., Arp, D., Pätzold, M., Sikora, T.: Real-time multi-human tracking using a probability hypothesis density filter and multiple detectors. In: Proceedings of the IEEE Conference on Advanced Video and Signal Based Surveillance, Beijing, China, pp. 325–330, September 2012
12. Felzenszwalb, P.F., Girshick, R., Ramanan, D.: Object detection with discriminatively trained part based models. IEEE Trans. Pattern Anal. Mach. Intell. **32**(9), 1627–1645 (2010)
13. Fu, Z., Feng, P., Angelini, F., Chambers, J., Naqvi, S.M.: Particle PHD filter based multiple human tracking using online group-structured dictionary learning. IEEE Access **6**, 14764–14778 (2018). https://doi.org/10.1109/ACCESS.2018.2816805
14. Gupta, A., Johnson, J., Fei-Fei, L., Savarese, S., Alahi, A.: Social GAN: socially acceptable trajectories with generative adversarial networks. In: Proceedings of the IEEE Conference on Computer Vision and Pattern Recognition, June 2018
15. H. K. Galoogahi, F.H., Fagg, A., Huang, C., Ramanan, D., Lucey, S.: Need for speed: a benchmark for higher frame rate object tracking. In: Proceedings of the IEEE International Conference on Computer Vision, Honolulu, HI, pp. 1125–1134, October 2017
16. Kalman, R.: A new approach to linear filtering and prediction problems. Trans. ASME, J. Basic Eng. **82**, 35–45 (1960)
17. Kutbi, M., Chang, Y., Sun, B., Mordohai, P.: Learning to navigate robotic wheelchairs from demonstration: is training in simulation viable? In: Proceedings of the IEEE International Conference on Computer Vision Workshops, October 2019
18. Kutschbach, T., Bochinski, E., Eiselein, V., Sikora, T.: Sequential sensor fusion combining probability hypothesis density and kernelized correlation filters for multi-object tracking in video data. In: Proceedings of the IEEE International Conference on Advanced Video and Signal Based Surveillance, Lecce, Italy, pp. 1–5, August 2017
19. Lankton, S., Tannenbaum, A.: Improved tracking by decoupling camera and target motion. In: Proceedings SPIE, San Jose, CA, pp. 6811–6819 (2008)
20. Lee, S., Kim, E.: Multiple object tracking via feature pyramid siamese networks. IEEE Access **7**, 8181–8194 (2019)
21. Lee, S., Kim, M., Bae, S.: Learning discriminative appearance models for online multi-object tracking with appearance discriminability measures. IEEE Access **6**, 67316–67328 (2018)
22. Li, S., Yeung, D.: Visual object tracking for unmanned aerial vehicles: a benchmark and new motion models. In: Proceedings of the Association for the Advancement of Artificial Intelligence, San Francisco, CA, pp. 4140–4146, June 2017
23. Lin, Y., Wang, K., Yi, W., Lian, S.: Deep learning based wearable assistive system for visually impaired people. In: Proceedings of the IEEE International Conference on Computer Vision Workshops, October 2019

24. Lisotto, M., Coscia, P., Ballan, L.: Social and scene-aware trajectory prediction in crowded spaces. In: Proceedings of the IEEE International Conference on Computer Vision Workshops, October 2019
25. Long, C., Haizhou, A., Zijie, Z., Chong, S.: Real-time multiple people tracking with deeply learned candidate selection and person re-identification. In: Proceedings of the IEEE International Conference on Multimedia and Expo, San Diego, CA (2018)
26. Maggio, E., Taj, M., Cavallaro, A.: Efficient multitarget visual tracking using random finite sets. IEEE Trans. Circuits Syst. Video Technol. **18**(8), 1016–1027 (2008)
27. Mahler, R.: A theoretical foundation for the Stein-Winter Probability Hypothesis Density (PHD) multitarget tracking approach. In: Proceedings of MSS National Symposium on Sensor and Data Fusion, San Diego, CA, USA, June 2002
28. Máttyus, G., Benedek, C., Szirányi, T.: Multi target tracking on aerial videos. In: Proceedings of the ISPRS Workshop, Istanbul, Turkey (2010)
29. McCullagh, P.: Generalized Linear Models. Routledge, Boca Raton (2018)
30. Montemerlo, M., Thrun, S., Whittaker, W.: Conditional particle filters for simultaneous mobile robot localization and people-tracking. In: Proceedings of the IEEE International Conference on Robotics and Automation, pp. 695–701 (2002)
31. Pfeiffer, M., Paolo, G., Sommer, H., Nieto, J., Siegwart, R., Cadena, C.: A data-driven model for interaction-aware pedestrian motion prediction in object cluttered environments. In: Proceedings of the IEEE International Conference on Robotics and Automation, Brisbane, Australia, pp. 1–8, May 2018
32. Ren, S., He, K., Girshick, R., Sun, J.: Faster R-CNN: towards real-time object detection with region proposal networks. IEEE Trans. Pattern Anal. Mach. Intell. **39**(6), 1137–1149 (2017)
33. Sanchez-Matilla, R., Cavallaro, A.: Confidence intervals for tracking performance scores. In: Proceedings of the IEEE International Conference on Image Processing, Athens, Greece, pp. 246–250, October 2018
34. Sanchez-Matilla, R., Cavallaro, A.: A predictor of moving objects for first-person vision. In: Proceedings of the IEEE International Conference on Image Processing, Taipei, Taiwan, pp. 246–250, September 2019
35. Sanchez-Matilla, R., et al.: Benchmark for human-to-robot handovers of unseen containers with unknown filling. IEEE Robot. Autom. Lett. **5**(2), 1642–1649 (2020)
36. Sanchez-Matilla, R., Poiesi, F., Cavallaro, A.: Online multi-target tracking with strong and weak detections. In: Hua, G., Jégou, H. (eds.) ECCV 2016. LNCS, vol. 9914, pp. 84–99. Springer, Cham (2016). https://doi.org/10.1007/978-3-319-48881-3_7
37. Shafique, K., Lee, M.W., Haering, N.: A rank constrained continuous formulation of multi-frame multi-target tracking problem. In: Proceedings of the IEEE Conference on Computer Vision and Pattern Recognition, Anchorage, AK, USA, pp. 1–8 (2008)
38. Szeliski, R.: Image alignment and stitching: a tutorial. Trans. Found. Trends Comp. Graph. Vis. **2**(1), 1–104 (2006)
39. Tapu, R., Mocanu, B., Zaharia, T.: Dynamic subtitles: a multimodal video accessibility enhancement dedicated to deaf and hearing impaired users. In: Proceedings of the IEEE International Conference on Computer Vision Workshops, October 2019
40. Wang, H., Oneata, D., Verbeek, J., Schmid, C.: A robust and efficient video representation for action recognition. Int. J. Comput. Vis. **119**(3), 219–238 (2016). https://doi.org/10.1007/s11263-015-0846-5

41. Williams, C.K.: Prediction with Gaussian processes: from linear regression to linear prediction and beyond. In: Jordan, M.I. (ed.) Learning in Graphical Models. NATO ASI Series (Series D: Behavioural and Social Sciences), vol. 89, pp. 599–621. Springer, Dordrecht (1998). https://doi.org/10.1007/978-94-011-5014-9_23
42. Yang, F., Choi, W., Lin, Y.: Exploit all the layers: fast and accurate CNN object detector with scale dependent pooling and cascaded rejection classifiers. In: Proceedings of the IEEE Conference on Computer Vision and Pattern Recognition, Las Vegas, NV, pp. 2129–2137, June 2016
43. Young-Chul, Y., Abhijeet, B., Kwangjin, Y., Moongu, J.: Online multi-object tracking with historical appearance matching and scene adaptive detection filtering. CoRR abs/1805.10916 (2018)
44. Yu, S., Lee, H., Kim, J.: Street crossing aid using light-weight CNNs for the visually impaired. In: Proceedings of the IEEE International Conference on Computer Vision Workshops, October 2019
45. Zhu, J., Yang, H., Liu, N., Kim, M., Zhang, W., Yang, M.-H.: Online multi-object tracking with dual matching attention networks. In: Ferrari, V., Hebert, M., Sminchisescu, C., Weiss, Y. (eds.) ECCV 2018. LNCS, vol. 11209, pp. 379–396. Springer, Cham (2018). https://doi.org/10.1007/978-3-030-01228-1_23

i-Walk Intelligent Assessment System: Activity, Mobility, Intention, Communication

Georgia Chalvatzaki, Petros Koutras(✉), Antigoni Tsiami, Costas S. Tzafestas, and Petros Maragos

School of E.C.E., National Technical University of Athens, Athens, Greece
gchal@mail.ntua.gr, {pkoutras,antsiami,ktzaf,maragos}@cs.ntua.gr

Abstract. We present the i-Walk system, a novel framework for intelligent mobility assistance applications. The proposed system is capable of automatically understanding human activity, assessing mobility and rehabilitation progress, recognizing human intentions and communicating with the patients by giving meaningful feedback. To this end, multiple sensors, i.e. cameras, microphones, lasers, provide multimodal data in order to allow for user monitoring, while state-of-the-art and beyond algorithms have been developed and integrated into the system to enable recognition, interaction and assessment. More specifically, i-Walk performs in real-time and consists of four main sub-modules that interact automatically to provide speech understanding, activity recognition, mobility analysis and multimodal communication for seamless HRI. The i-Walk assessment system is evaluated on a database of healthy subjects and patients, who participated in carefully designed experimental scenarios that cover essential needs of rehabilitation. The presented results highlight the efficacy of the proposed framework to endow personal assistants with intelligence.

Keywords: Intelligent assessment system · Human-robot interaction · Activity recognition · 3D pose estimation · Speech understanding · Gait tracking · Gait stability · Multimodal communication

1 Introduction

The rapid increase of people with special needs, such as the elderly population, and the simultaneous reduction of personal care staff, reinforce the need for robotic assistants [29]. When designing an intelligent assistant platform for people with mobility and/or cognitive impairment, special care should be given in developing a system that will monitor and promote rehabilitation in a natural and seamless way.

G. Chalvatzaki, P. Koutras, A. Tsiami—Equal Contribution

Electronic supplementary material The online version of this chapter (https://doi.org/10.1007/978-3-030-66823-5_30) contains supplementary material, which is available to authorized users.

A. Bartoli and A. Fusiello (Eds.): ECCV 2020 Workshops, LNCS 12538, pp. 500–517, 2020.
https://doi.org/10.1007/978-3-030-66823-5_30

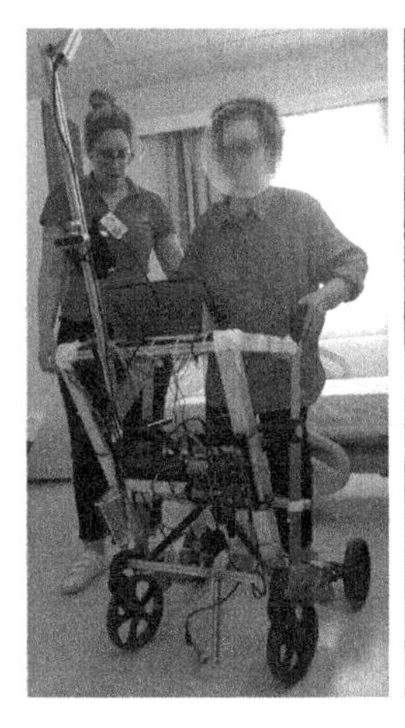

Fig. 1. Left: A patient walking supported by the i-Walk assistant platform. Right: A patient performing rehabilitation exercises while being monitored by i-Walk.

In order for an intelligent robotic assistant to achieve these goals, the development of advanced Human-Robot Interaction (HRI) components and their integration under a unified autonomous system is more than essential. More specifically, the platform should be capable of understanding human activities, intentions and needs, but also analysing multi-sensory signals related to gait and postural stability, so as to provide support and communication.

The i-Walk platform (Fig. 1) has been carefully designed to fill this need, by combining multisensory streams to perform a multitask understanding of human behavior, i.e., speech intention recognition, generalized human activity recognition, and mobility analysis. The multimodal interaction framework of i-Walk aims to provide natural communication and valuable feedback to the user and the medical experts regarding rehabilitation progress, in a way close to that of a personal carer.

Design and development of personal robotic assistants for elderly is prominent in scientific research [17,27]. The role of personal care robots is multiple, covering physical, sensorial and cognitive assistance [22], health and behavior monitoring and companionship [39]. Most intelligent assistive platform designs aim to solve only specific problems, e.g. GUIDO [36] and iWalker [23] provided navigation assistance. Considerable amount of research has focused on analysing anthropometric data from various sensors for assessing human state [19] and eventually control a robotic platform, like CAIROW [13]. The ISR-AIWALKER employed RGB-D data for monitoring the users [31]. MOBOT [29] was equipped with various sensors, attempting to model human activities from multimodal data [20,35] and perform gait and stability analysis [9,12]. In [11] a method was proposed that integrated a human motion intention model, exploiting RGB-D and laser data of the user, into a decision making framework for adapting the platform's motion.

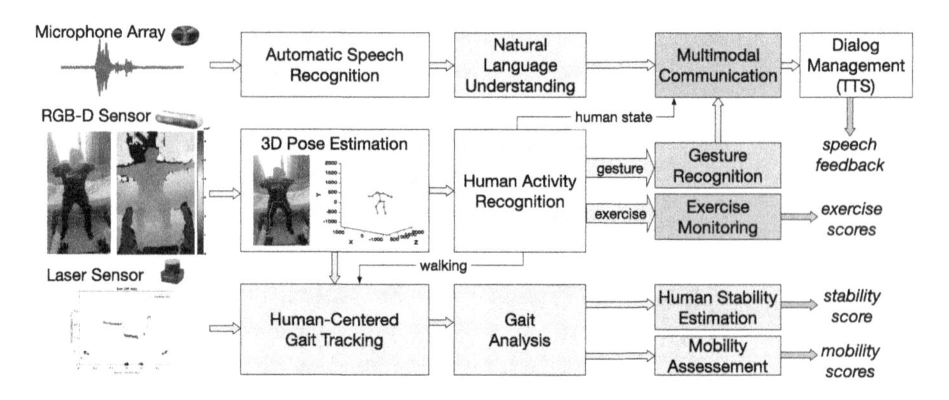

Fig. 2. Overview of the multimodal i-Walk intelligent assessment system.

Several works aim to integrate visual perception into assistive robotic applications [24,41]. Robotic assistive vision is an important topic for HRI systems [18], while also multi-sensory systems integrating visual activity with speech recognition [15] for communicating with robots have recently emerged [43,44]. Human activity recognition is a long-term research problem [47] where human skeleton representations from single images [14] have been extensively used [46,48–50]. Deep Learning progress led to efficient methods for human pose estimation [8] while action recognition benefits from these improved skeleton estimations by employing recurrent methods like Long Short Term Memory (LSTM) networks, which have the ability to model temporal information of action sequences [16,26,38,40,45,51].

However, those methods present difficulties in getting integrated in robotics, since most of them rely on datasets from constrained environments with static cameras, which makes them vulnerable to camera motion effects. Moreover, the most recent implementations incorporating human pose estimation and/or action recognition for robotic applications [33,34,37,52] usually focus on specific tasks rather than provide a holistic approach that demands multimodal human perception and real-time feedback from the system.

In this paper we present the i-Walk intelligent assessment system, a multisensory, multimodal HRI framework destined to provide simple rollators with intelligence, aiming to be of use by patients with mobility and/or cognitive impairment. The system has three important goals: monitoring the patient in terms of his/her activity and mobility, interacting with the patient by allowing him/her to communicate his/her intentions to the platform, engaging in dialog with him/her, giving feedback, etc., and assess the patient's status in the context of a rehabilitation procedure. To this end our main contributions are the design, development, integration and evaluation of the assessment system with its individual interacting sub-modules: the speech understanding, the activity recognition, the mobility analysis and the multimodal communication system. All sub-modules communicate automatically, providing direct feedback. The proposed system has been extensively evaluated using a large corpus of multi-sensory data, both from healthy subjects and real patients from a rehabilitation center, who participated in experimental scenarios that meet the real needs of daily living and mobility rehabilitation.

2 System Overview

The i-Walk Assessment System with its respective sub-modules is presented in Fig. 2. The upper flow (orange blocks) represents the *Speech Understanding* module, the middle flow (light blue blocks) the *Activity Recognition*, while the lower one (light pink blocks) the *Mobility Analysis* module. Each module exerts certain outputs for assessing the activity state and performance, the spoken and gestural intentions and the mobility performance of the user. The *Multimodal Communication* module (red block) is responsible for triggering the dialog management (light green block) providing speech feedback to the users. These modules along with their respective parts are described below.

2.1 Speech Understanding

Speech is the most natural and instant means of communication. Thus, a speech understanding module that will enable the patient to communicate his/her intentions to the robot is a key component of the system and involves two sub-modules, as depicted in Fig. 2: An Automatic Speech Recognition (ASR) module and a Natural Language Understanding (NLU) one. For ASR, a state-of-the-art system has been integrated, where speech recorded through a microphone array serves as input to Google speech-to-text API [2] and is transformed into text. Subsequently, the transcribed text serves as input to the NLU module, in order to be translated into a human intention. The integrated NLU system has been built with RASA [3,4,7]: A set of pre-defined intentions, both general purpose and specific to the current application has been designed. The former category includes 7 general intents, namely greeting, saying my name, saying goodbye, thanking, affirming, denying, asking to repeat, while the latter one includes 7 intents designed for the HRI: standing up, sitting down, walking, stopping, ending interaction, going to the bathroom, doing exercises. Each intention is associated with various phrases to express this particular intention. For example, a user can express his/her will to stand up by saying "I want to stand up", "Can you please help me stand up", or any other variation. A RASA NLU pipeline called tensorflow embeddings [3] is then employed to predict the current intention based on the speech transcription. The predicted intent is the input to the multimodal communication module, that is described later in this section.

2.2 Activity Recognition

Apart from speech, gestures and human activities convey crucial information about a person's intent or state as well. We have designed and implemented a novel sub-module for human activity and gesture recognition that consists of two

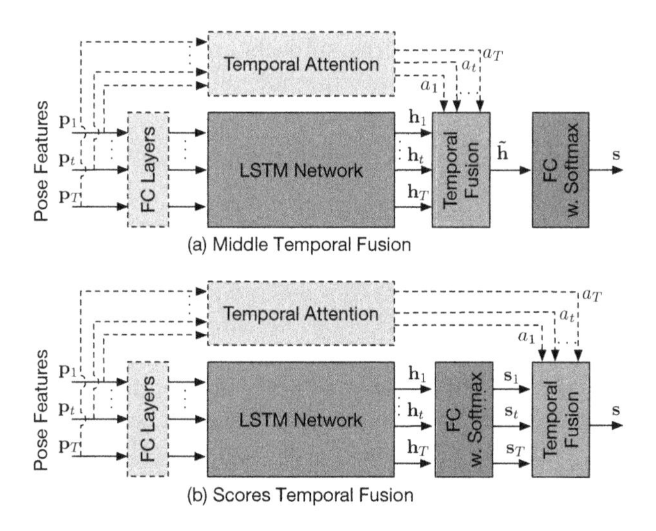

Fig. 3. Neural Network architectures for human activity recognition based on LSTM units. The temporal fusion of the predictions can be applied in two different stages. (Color figure online)

different subsystems: The first one performs 3D human pose estimation (Fig. 2) using the RGB-D sensor, which is mounted on the the robotic rollator (Fig. 1), while the second one recognizes human activity by employing a LSTM-based network architecture (Fig. 3). In case the recognized activity is an exercise, the exercise monitoring module presents (on a screen placed on the rollator) and stores the corresponding recognition scores, while in case a gesture is detected, the gesture recognition module is triggered.

3D Pose Estimation: For the detection of the 2D body keypoints on the image plane we employ Open Pose Library [8] with the accompanied models trained on large annotated datasets [6,25]. The third dimension of the 3D body keypoints is obtained by the corresponding depth maps. Subsequently, given a pair of pixel coordinates for a body joint and the depth value at this pixel, we calculate the corresponding 3D joint's coordinates through the inverse perspective mapping using the calibration matrix of the camera. For the final human skeleton we discard the keypoints of face, hands and feet either because in many cases they are not detected, or the depth values at these points are not reliable.

For activity recognition, either the 2D keypoints on the image plane or the 3D locations of the human joints are used as features. In the former case, since locations are pixel coordinates, we apply a standardization scheme (STD) in order to have features with zero mean and unit variance for each instance sequence. In the latter case, we transform the 3D body joint locations which are provided in the camera coordinate system, to the body coordinate system with the middle-hip joint as origin and normalized by the length between the left- and right-hip joints (BNORM scheme). In addition, we can optionally enhance the pose feature vector with the 3D velocity and acceleration of each joint, computed from the sequence of the normalized 3D joints' positions.

LSTM-based Network for Activity Recognition: In our deep learning based module for human activity recognition we employ a Neural Network architecture based on LSTM units [21]. LSTM constitutes a special kind of recurrent neural networks that can effectively learn long-term dependencies that exist in sequential data, such as human joint trajectories. Our network architecture consists of two LSTM layers stacked on top of each other (Fig. 3 - blue boxes) and a fully connected (FC) layer, followed by softmax activation, to obtain per-class scores (Fig. 3 - red boxes). The sequence of the pose features $\mathbf{p}_t$ in a temporal window of length T, possibly transformed by a sequence of FC layers, is used as input to the above network. The usage of FC layers acts as a static transformation of the initial pose features independently of time dependencies that are modelled by the LSTM units. The network output consists of a sequence of per-class labels, one for each kind of human activity.

Usually, an input sequence of length T is classified by choosing the class with the highest score $\mathbf{s}_T$ that corresponds to the the hidden state $\mathbf{h}_T$ of the last frame. However, in this work we have investigated two different approaches for the temporal fusion of the network's predictions. In the first (middle fusion: Fig. 3.a) we fuse the hidden states $\mathbf{h}_t$ according the aggregation function $\tilde{\mathbf{h}} = \mathscr{F}(\mathbf{h}_t)$, while in the second (scores fusion: Fig. 3.a) we apply temporal pooling of the softmax scores $\mathbf{s}_t$: $\mathbf{s} = \mathscr{F}(\mathbf{s}_t)$. For each one of the above fusion schemes we have experimented with four different types of aggregation functions $\mathscr{F} : \mathbf{g} = \mathscr{F}(\mathbf{f})$, where $\mathbf{f}$ can be either the states $\mathbf{h}_t$ or the scores $\mathbf{s}_t$.

Average Pooling: In this approach, $\mathscr{F}$ represents the frame predictions average, inside a temporal window: $\mathbf{g} = \frac{1}{T}\sum_{t=1}^{T}\mathbf{f}_t$. While average pooling function is able to capture information from the whole temporal video segment, it is sensitive to noisy background.

Max Pooling: In this function we apply max pooling to the elements of $\mathbf{f}$ among the T frames of the temporal window: $\mathbf{g} = \mathscr{F}(\mathbf{f}) = \max_{t\in 1,\cdots,T}\mathbf{f}_t$. With max pooling we emphasize the most discriminative frame of the video segment, which has the strongest activation, and ignore the other parts that may contain noisy information.

Weighted Average: An alternative way of temporal pooling is to use weights γ_t for each frame prediction, learned during the network training: $\mathbf{g} = \sum_{t=1}^{T}\gamma_t \cdot \mathbf{f}_t$. This way, we learn the relative importance of each temporal part but the weights are data independent since they are learned from the whole dataset and are not affected by the content of each video clip.

Attention Weighting: To deal with the above limitations we can apply a weighting scheme where the importance weights a_t for each frame depend on the content of the video: $\mathbf{g} = \sum_{t=1}^{T} a_t(\mathbf{p}_t)\cdot\mathbf{f}_t$. More specifically, the weights are learned using an attention mechanism based on the pose features $\mathbf{p}_t$. In the first phase, the features $\mathbf{p}_t$ of each frame are transformed to attention activation values b_t using a fully connected layer: $b_t = \text{FC}_{\text{att}}(\mathbf{p}_t)$. Then, we compute the weights a_t by applying a temporal softmax normalization:

$$a_t = \frac{\exp(b_t)}{\sum_{\tau=1}^{T} \exp(b_\tau)}. \tag{1}$$

Network Training: We trained from scratch the proposed network using mini-batches of 256 clips for 500 epochs, with initial learning rate 0.1, momentum 0.9 and weight decay 10^{-5}. The learning rate is divided by 10 after 100 epochs. For training we employed the Stochastic Gradient Descend (SGD) optimizer with the a weighted cross-entropy loss while we have augmented the corpus by flipping the original data in the vertical dimension:

$$\mathcal{L} = -\frac{1}{N} \sum_{j}^{N} \mathbf{w}_{c_j} \log \mathbf{s}^{c_j}(\mathbf{p}_{1:t}^{j}, \mathbf{W}), \tag{2}$$

where c_j denotes the class of the j-th sample in the minibatch, $\mathbf{W}$ the trainable parameters of the network and $\mathbf{w}$ is a vector containing the weights for each class, based on the appearance frequencies of each class in the dataset.

2.3 Mobility Analysis

Gait stability and mobility assessment are important for evaluating the rehabilitation progress. The Mobility Analysis module, triggered when activity "Walking" is recognized, consists of the following sub-systems (Fig. 2):

Human-Centered Gait Tracking and Gait Analysis: The tracking module exploits the RGB-D data capturing the upper-body and the laser data detecting the legs. An hierarchical tracking filter based on an Unscented Kalman Filter estimates the positions and velocities of the human Center-of-Mass (CoM), which is computed by the estimated 3D human pose, while an Interacting Multiple Model Particle Filter performs the gait tracking and the recognition of the gait phases at each time frame [9,12]. Considering gait analysis literature [32], the walking periods are segmented into distinct strides given the gait phases recognition and certain gait parameters are computed [10,28].

Gait Stability Assessment: A deep neural network was designed and evaluated in [9], as an encoder-decoder sequence to sequence model based on LSTMs. The input features are the estimated positions of the CoM and the legs along with the respective gait phase at each time frame, while the output predicts the gait stability state considering two classes: stable walking and risk-of-fall state. In particular, the stability score used here is the probability of performing stable walking. If this probability is below a specific threshold (defined by the the experts usually at around 30% for the most patients) the rollator's screen flashes red in order to inform the user to improve his/her gait.

Mobility Assessment: For assessing the patient's mobility status, we compute gait parameters, such as stride length, gait speed, etc., which serve as a feature vector for an SVM classifier [10]. The classes are associated with the Performance Oriented Mobility Assessment (POMA) [42]. POMA scores less than 18 refer to

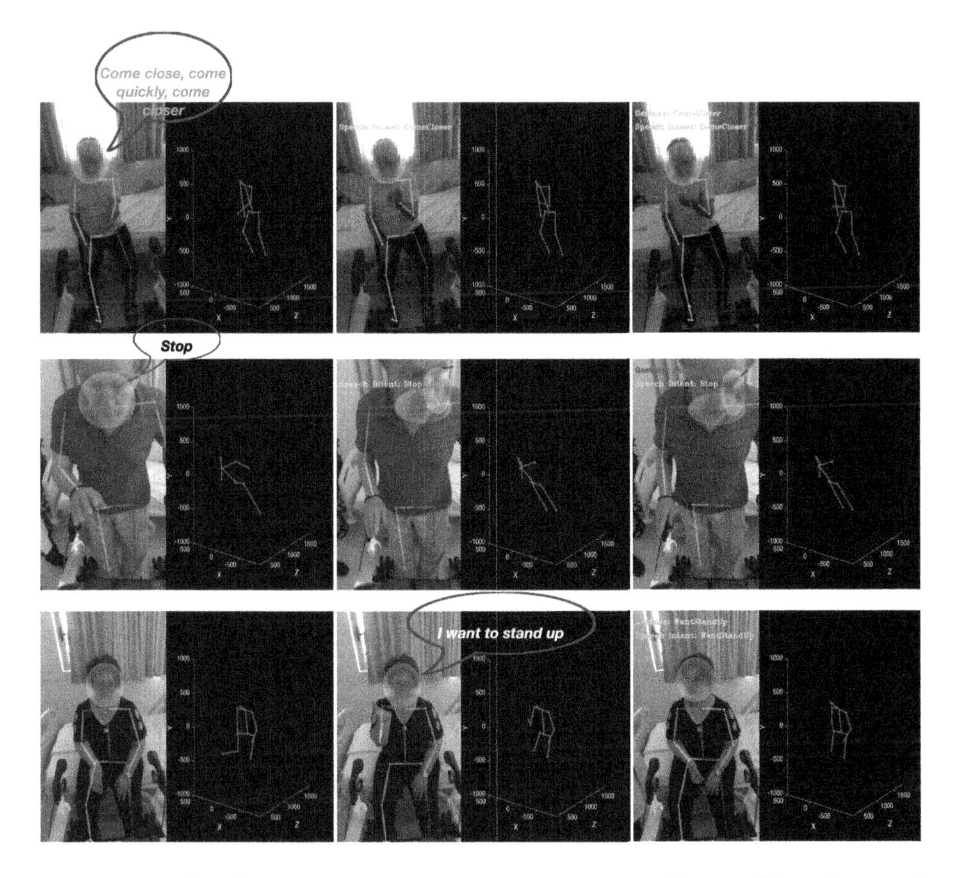

Fig. 4. Examples of the multimodal communication system that combines the speech understanding and the gesture recognition outputs. Note that in most cases the user's intention is successfully recognized in both modalities (green: correctly recognized intention, red: intent not recognized). (Color figure online)

high risk of fall, while a score between 19 and 23 indicates a moderate risk. In this work, the mobility score is taken as the probability for a patient to belong in the high risk mobility class.

The stability and mobility assessment scores, along with several other walking parameters, like gait speed, swing phase, etc., are stored for being reported to the experts responsible for the patients' rehabilitation, so they can acquire a day-by-day quantitative information about the patient's progress.

2.4 Multimodal Communication and Feedback

As depicted in Fig. 2, the multimodal communication module is responsible for gathering and combining the outputs of the speech understanding and the gesture recognition modules and the human state (i.e. standing, sitting, etc.) so as to produce feedback for the user. It should be noted (see Fig. 4) that it can have either both inputs (speech, gesture) at the same time or asynchronous, or a single one [44]. The feedback to the user is given via a TTS (text-to-speech) system [44].

3 Experimental Results

In this section, we present the data collection scenarios and procedure, as well as the performance evaluation of the several components of the system.

3.1 Data Collection and Evaluation Setup

For the training and evaluation of the proposed multimodal assessment system, several sessions of data collection took place, involving both healthy subjects and patients with various mobility and/or cognitive inabilities. The experimental platform depicted in Fig. 1, was equipped with a RealSense camera, a Hokuyo UST-LX10 laser, and an eMeet microphone, which provided the multi-sensory data. The communication of the multiple sensors and sub-modules of the proposed system (Fig. 2) was implemented via ROS [5], while the platform was also equipped with NVIDIA Jetson TX2 modules that allow us to have deep learning processing capability on the rollator. In this work, we present two databases (DB) regarding the collected multimodal data:

1. i-Walk DB: This DB incorporates data both from healthy users and patients. The data collection with patients took part in DIAPLASIS Rehabilitation center [1] in Kalamata, Greece in July 2019. Thirteen patients have participated in the experiments, after approval by the medical staff. The demographic data (age and gender) along with the respective Mini-Mental Mean Score (MMSE) indicating cognitive capacity and the POMA scores regarding mobility efficiency are showcased in Table 6. All patients have signed written consent and are protected under GDPR. For development and comparison purposes, we conducted additional data collection experiments with twenty healthy users (ages 23–32). The data collection scenarios have been carefully designed in collaboration with medical experts, covering real needs regarding ambulation and rehabilitation of the patients.

Scenario 1: Rehabilitation Exercises. The users are seated on a bed/ chair and are asked to perform certain exercises, part of a rehabilitation program. Such exercises include hands raises, torso turns, sit-to-stand transfers, etc. The complete list of the performed human activities is depicted in Table 1. The users were also prompted to perform other meaningful activities, like gestures, and express their respective intentions in free speech (Table 1).

Scenario 2: Transfer to Bathroom. This scenario is essential to patients with mobility problems, as it aims to assist them in a fundamental daily-living need. The users are initially seated on a bed/chair, and express the intend of standing up and going to the bathroom. The walking includes navigating through the hospital room, entering the bathroom, sitting on the toilet, standing up, return to the bed/chair.

Table 1. Description of the activities and gestures classes.

Human Activities		
ID	Codename	Description
1	Sitted	Sitted on chair or bed
2	StandUpPrep	Preparing to stand up using rollator
3	StandUp	Standing up from sitted position
4	SitDown	Sitting down from standing position
5	Walking	Walking using rollator
6	Standing-still	Standing still without make any action
7	HandCross	Place hands crossed on torso while sitted
8	HandCrossTurn	Turning torso left/right with hands crossed while sitted
9	HandOpenTurn	Turning torso left/right with hands opened while sitted
10	HandOpen	Raise hands horizontally while sitted
11	WeightMoves	Body weight transfers from one leg to another while standing supported by rollator
12	StepsHigh	In-place steps with high knees supported by rollator
13	TurnStanding	Turning torso left/right while standing supported by rollator
14	Gesture	Performing gesture towards rollator
Gestures		
ID	Codename	Description
a	ComeCloser	Ask rollator to come closer
b	WantStandUp	Want to stand up
c	WantSitDown	Want to sit up
d	Stop	Stop the procedure
e	End	Procedure completed

Table 2. Confusion matrix for the activity recognition module (middle fusion using MAX) in the i-Walk DB (patients activities).

Target Class \ Predicted Class	1	2	3	4	5	6	7	8	9	10	11	12	13	14
1	89.8%	2.5%	1.6%	1.0%	0.0%	0.0%	0.0%	1.2%	0.0%	0.0%	0.0%	0.0%	0.0%	3.9%
2	7.1%	60.7%	0.0%	0.0%	0.0%	14.3%	0.0%	0.0%	0.0%	0.0%	0.0%	0.0%	0.0%	17.9%
3	0.0%	0.0%	83.7%	0.0%	0.0%	3.5%	0.0%	0.0%	0.0%	0.0%	1.9%	0.0%	0.0%	10.9%
4	1.9%	0.0%	5.1%	76.4%	0.0%	5.6%	0.0%	0.0%	0.0%	0.0%	0.0%	3.8%	0.0%	7.1%
5	0.0%	0.0%	0.0%	0.0%	88.5%	0.0%	0.0%	0.0%	0.0%	0.0%	0.0%	0.0%	0.0%	11.5%
6	0.0%	0.0%	1.7%	3.3%	0.9%	66.0%	0.0%	0.0%	0.0%	0.0%	5.4%	3.2%	0.6%	18.8%
7	0.0%	0.0%	0.0%	0.0%	0.0%	0.0%	81.8%	0.0%	0.0%	0.0%	0.0%	0.0%	0.0%	18.2%
8	0.0%	0.0%	0.0%	0.0%	0.0%	0.0%	0.0%	100.0%	0.0%	0.0%	0.0%	0.0%	0.0%	0.0%
9	3.8%	0.0%	0.0%	0.0%	0.0%	0.0%	0.0%	1.5%	91.3%	0.8%	0.0%	0.0%	0.0%	2.6%
10	4.6%	7.4%	0.0%	0.0%	0.0%	0.0%	0.0%	0.0%	0.0%	87.9%	0.0%	0.0%	0.0%	0.0%
11	0.0%	0.0%	0.0%	1.1%	2.6%	23.8%	0.0%	0.0%	0.0%	0.0%	43.8%	10.0%	3.1%	15.6%
12	0.0%	0.0%	3.2%	15.7%	1.0%	23.4%	0.0%	0.0%	0.0%	0.0%	0.9%	52.1%	0.0%	3.8%
13	0.0%	0.0%	0.0%	1.7%	0.0%	36.7%	0.0%	0.0%	1.7%	1.7%	5.8%	0.4%	32.6%	19.5%
14	1.7%	6.2%	0.7%	2.2%	0.5%	7.3%	0.0%	0.0%	0.0%	0.2%	0.0%	0.0%	0.0%	81.1%

2. MOBOT DB: This DB was collected in the context of the EU project MOBOT [30] in Agaplesion Bethanien Hospital in Heidelberg Germany, in 2014. This DB includes multimodal data from 14 patients who performed a minimal

Table 3. Confusion matrix for the gestures recognition module (middle fusion using MAX) in the i-Walk DB (patients gestures).

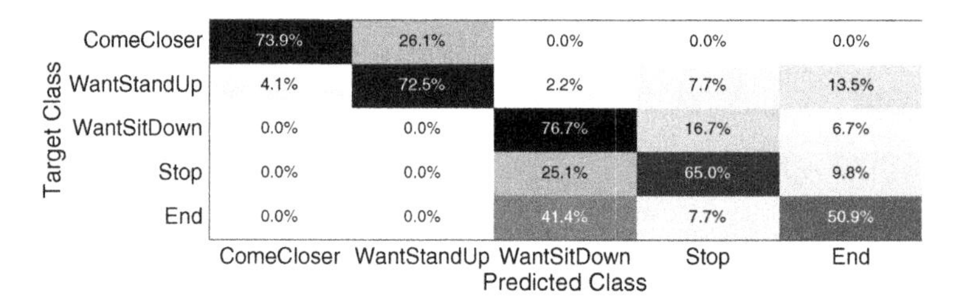

Target Class \ Predicted Class	ComeCloser	WantStandUp	WantSitDown	Stop	End
ComeCloser	73.9%	26.1%	0.0%	0.0%	0.0%
WantStandUp	4.1%	72.5%	2.2%	7.7%	13.5%
WantSitDown	0.0%	0.0%	76.7%	16.7%	6.7%
Stop	0.0%	0.0%	25.1%	65.0%	9.8%
End	0.0%	0.0%	41.4%	7.7%	50.9%

set of activities, namely the activities with ID: 1,3,4,5,6 listed in Table 1. We aim to showcase the transferability of the proposed activity recognition model across different setups and sets of activities.

3.2 Evaluation Results

Speech Intent Recognition: In order to assess the speech understanding module performance, we evaluate the percentage of the correctly recognized intents from speech. For this purpose, we employ the aforementioned i-Walk DB. The collected data contain free speech uttered by both healthy subjects and patients expressing a specific intent. We evaluate the performance of the intention recognition. For the healthy subjects, the DB contains 445 utterances and the accuracy is 94.83%, which is relatively high. For the patients, results are presented in Table 6, in the Speech Intent Recognition column, both for each patient separately and in average. The performance for 173 utterances reaches 74.25%, which is lower than the healthy ones', but this can be attributed to two factors: First, most of the patients did not only have mobility issues, but also cognitive/mental ones, as can be seen in Table 6, consequently their speech was often unintelligible and thus difficult to be transcribed correctly into text. Second, since the patients' age is significantly higher than the healthy ones', their utterances for the same intents are different and have larger variability. Overall, the performance is satisfying, but it could be improved by collecting more data or creating more specific acoustic models.

Activity Recognition: We have conducted a series of experiments to verify our design choices in the proposed system, and investigated several variants in order to achieve the best performance, which is crucial in robotics applications.

Pose Features Selection and Network Architecture: The ablation study (following a leave-one-out cross validation approach) is presented in Table 4, using the MOBOT DB (862 clips in total) that contains simple actions and is thus suitable for running exploratory experiments regarding several parameters. We observe that the 3D features with the body normalization scheme (BNORM) outperform the 2D features, and perform even better when enhanced with the 3D velocities. We also observe that the 2-layer LSTM network without FC layers achieves the

Table 4. Ablation study for the activity recognition system w.r.t. different employed pose features (top) and different network architectures (bottom).

2 LSTM layers (without any FC layers before the LSTMs) with average score fusion						
Pose Features	2D	2D-STD	3D	3D-BNORM	3D-BNORM w. veloc.	3D-BNORM w. veloc. & accel.
MOBOT DB.	50.38	77.62	59.65	86.15	**87.72**	86.80

3D-BNORM pose features and the average score fusion scheme					
Arch.	LSTM	LSTM	FC1+LSTM	FC1+FC2+LSTM	FC1+FC2
FCs sizes	–	–	512	[1024, 512]	[1024, 512]
LSTM Layers	1	2	2	2	0
Hidden sizes	256	[256, 256]	[256, 256]	[256, 256]	–
MOBOT DB	85.47	**86.15**	85.08	83.42	71.92

Table 5. Evaluation results for the different temporal fusion approaches. As pose features we have used 3D-BNORM with velocities while the network architecture consist of 2 LSTM layers without any FC layers before the LSTMs. Bold fonts stand for the best performance while the underlined fonts denote the second best.

Method	Last Hidden	Middle Fusion				Score Fusion			
Aggre.g. Func	–	AVG	MAX	Weighted	Attention	AVG	MAX	Weighted	Attention
MOBOT DB (patients, 5 activities)	85.83	86.21	**90.36**	75.55	86.89	87.72	87.91	80.61	87.56
i-Walk DB (healthy users, 14 activities)	92.74	90.87	94.59	91.86	90.44	90.36	**95.20**	91.60	90, 86
i-Walk DB (healthy users, 5 gestures)	80.07	67.58	88.11	78.96	68.54	70.89	**88, 48**	82.89	71.30
i-Walk DB (patients, 14 activities)	68.90	72.01	73.99	72.96	69.53	68.28	70.76	**75.82**	65.45
i-Walk DB (patients, 5 gestures)	64.42	51.63	**67, 81**	61.72	55.11	56.86	66.21	58.88	51.91

best performance. Note that the network containing only FC layers has significantly lower accuracy since it does not count for the temporal information that is necessary for recognizing dynamic activities.

Temporal Fusion Schemes: In Table 5 we present the evaluation results for the different fusion levels and the employed aggregation function using the features and network architecture that have achieved the best performance in the previous ablation analysis. These experiments have been conducted on the new i-Walk DB, which contains 4 different subsets, in order to validate the generalization of the proposed fusion schemes in more challenging cases. For the healthy users we conduct experiments by employing a leave-one-out cross validation strategy resulting in training and testing sets of 1891 and 107 clips respectively. For the patients we employ the whole corpus of the healthy users as training set (1998 clips) while for the testing we use patients data (1072 clips in total). Results indicate that the best performance overall is achieved for fusion using the MAX function, since this scheme can detect the most discriminative part of the video clip and ignore the other parts, thus recognizing better highly confusing activity classes. We can also observe that "Weighting" and "Attention" schemes perform in many cases quite higher than simple average. Regarding the different fusion levels, middle fusion with max-pooling has the best performance across

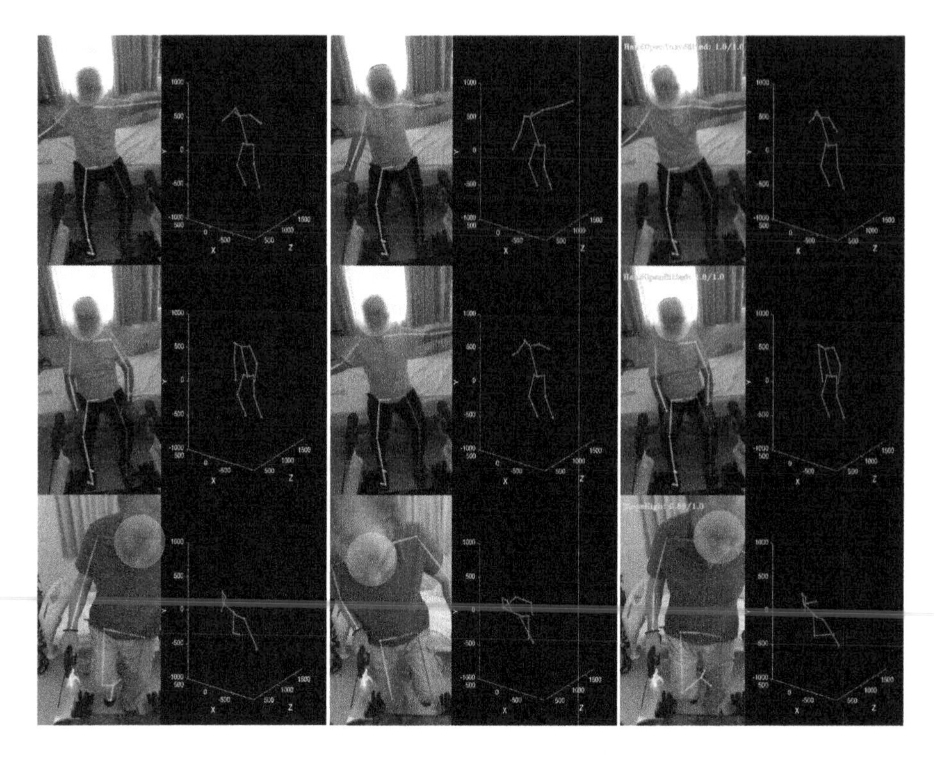

Fig. 5. Examples of the activity recognition system. The system monitors the patients exercises and outputs the scores of their performance (green: correctly performed exercise). (Color figure online)

the different datasets, users and action types, since it has at least the second best performance in all cases.

Evaluation Analysis: Regarding i-Walk DB results, the patients' performance is lower compared to the healthy users', for both activities and gestures subsets. Moreover, gesture recognition accuracy is lower than the activity recognition one, since patients had a large variability in the way they performed gestures compared to healthy users. The confusion matrix presented in Table 2, indicates that activities performed in standing position are often confused with "still-standing" position (due to many patients' difficulty in executing actions while standing). For the same reason, gestures performed in standing position achieve also lower recognition rates (Table 3). Moreover, activities with small duration variation in pose (i.e., "StandUpPrep" or "TurnStanding") are classified sometimes as gestures. For the per patient performance (Table 6) we note that users with high MMSE achieve quite high accuracy for both activities and gestures (see examples in Fig. 5. Combining gesture recognition results with speech intention recognition ones, we can observe that in some cases users with low speech intention performance achieve quite high rates in gesture recognition, a fact that highlights gestures as an alternative way for communication.

Table 6. Demographics and evaluation scores for the thirteen patients of the i-Walk database.

Patients	Age	Gender	MMSE	POMA	Activity performance (%)	Gesture recognition (%)	Speech intent recognition (%)	Gait speed (cm/sec) (mean±std)	Stability score (mean±std) (%) ↑	Mobility Score (mean±std) (%) ↓
1	80	M	29/30	18/28	75.29	90.00	80.00	23.17 ± 8.36	41.93 ± 31.92	62.48 ± 30.35
2	86	M	27/30	18/28	87.01	92.31	77.77	24.83 ± 9.24	46.59 ± 34.86	63.34 ± 27.71
3	25	F	29/30	18/28	83.54	73.33	100.00	25.09 ± 8.16	40.44 ± 32.83	60.68 ± 27.59
4	83	F	23/30	11/28	80.56	75.00	78.57	21.43 ± 8.62	45.57 ± 32.42	64.54 ± 30.49
5	84	F	17/30	11/28	64.67	54.55	62.50	20.09 ± 8.74	42.64 ± 33.78	63.39 ± 27.76
6	50	M	29/30	13/28	79.75	73.33	93.33	15.54 ± 8.37	45.16 ± 34.73	61.79 ± 27.06
7	78	M	27/30	15/28	83.58	40.00	54.54	19.04 ± 9.43	50.77 ± 31.27	57.30 ± 32.31
8	73	M	18/30	12/28	52.78	52.94	86.67	21.98 ± 8.68	47.23 ± 32.37	62.14 ± 30.62
9	72	F	19/30	11/28	59.49	35.71	46.15	18.12 ± 8.62	47.92 ± 36.88	67.47 ± 28.91
10	75	M	25/30	16/28	59.77	75.00	56.25	18.96 ± 9.60	44.79 ± 33.58	65.20 ± 29.02
11	85	F	19/30	14/28	66.23	56.25	84.62	19.67 ± 9.60	47.62 ± 31.72	66.07 ± 27.03
12	55	F	28/30	16/26	90, 79	70.59	100.00	19.98 ± 9.95	31.93 ± 31.72	64.19 ± 28.67
13	75	M	28/30	11/28	91.67	91.91	46.15	26.55 ± 11.71	41.31 ± 33.33	66.24 ± 29.02
Average	–	–	–	–	75.01	67.69	74.25	21.24 ± 9.97	45.10 ± 33.86	63.80 ± 28.95

Mobility Analysis: The individual components of the mobility analysis module are adopted from works in [9, 11, 12] proving tracking robustness [9, 11] and high performance scores in stability and mobility status recognition [9, 10], hence are suitable for the multimodal setting of i-Walk framework. Building on this mobility analysis system, we fine-tuned the models with walking data of some trials of users in the i-Walk DB (from a different walking scenario not included in this work), in order to provide the necessary mobility assessment scores for monitoring rehabilitation. In particular, we evaluate the stability performance, the mobility classification, and present the gait speed parameter for each patient w.r.t. their categorization by the medical experts.

The average stability scores of each patient along with the respective standard deviations (std) are depicted in Fig. 7. The solid red line represents the average stability score of the healthy subjects and the dotted lines the upper and lower confidence levels of the healthy stability measure. It is evident that all patients present low stability while walking, and only some of them can achieve instances of stability close to the lower bound of the healthy performance. This can also be affirmed by the results in Table 6 (Stability score), where the average stability score across all patients is 45.1%, while the healthy score is 81.46%. The upward arrow means that higher scores correspond to more stable performance.

Table 6 also presents the Mobility scores and the mean and std for gait speed. All patients have been classified to the high risk-of-fall class, which is also confirmed by the POMA scores. Here, the downward arrow denotes that lower scores refer to better performance. It is interesting that patients with higher POMA perform slightly better, e.g. patient #3 (POMA 18) had average gait speed 25 cm/sec w.r.t #6 (POMA 13) with average speed 15.5 cm/sec. Although, this is not the norm, as more gait parameters are important for mobility classification [10], it is an indication for examining interclass categorization, for which a larger database from patients, presenting higher variation in terms of mobility, should be collected. Figure 6 presents snapshots from patients #4 and #7 with a depiction of the center of mass and legs' state estimation, that feed the stability and mobility analysis classifiers. In the current setting patient #4 is performing

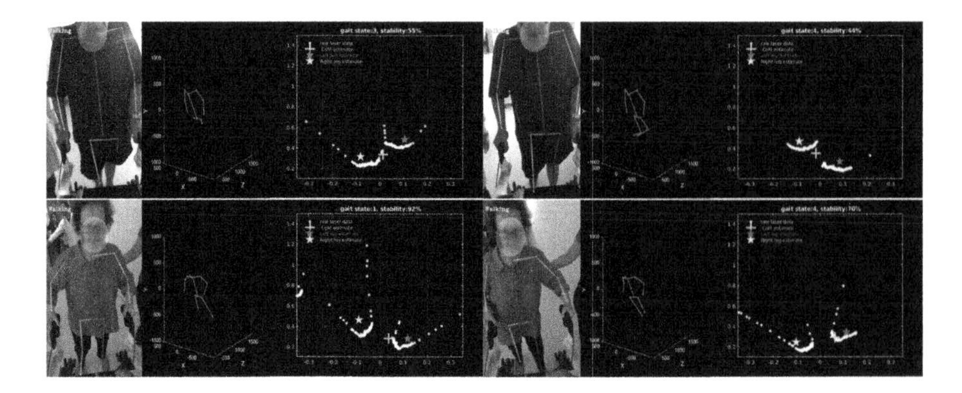

Fig. 6. Mobility analysis examples. The center of mass from the detected pose is fused with the legs' state estimation for an accurate patient tracking. These observations feed the gait stability network and the mobility assessment classifier. Upper: Snapshots from patient #7 with current estimate of stability at 44%. Lower: Snapshots of patient #4 with current estimate of stability at 70% (higher score is better).

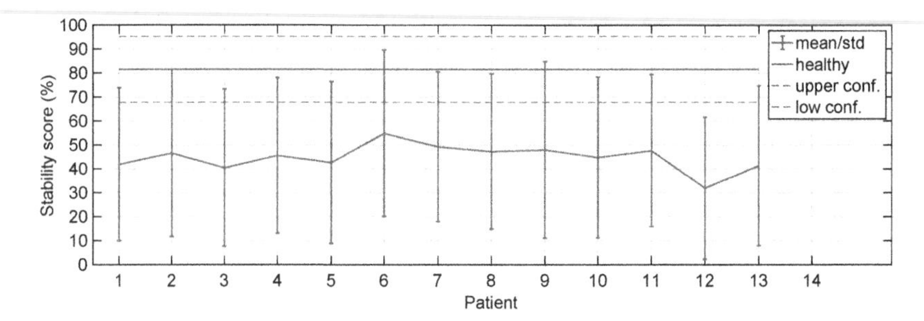

Fig. 7. Average per patient stability performance w.r.t. the average stability of the healthy subjects and their confidence levels.

less stable walking than patient #7 (scores 44% against 70% of stability). Such a detected instability shall trigger a red alarm on the rollator's screen. In general, the mobility analysis results follow the respective POMA categorization of the patients highlighting the ability of this module to successfully assess measures essential for measuring rehabilitation progress.

4 Conclusions

This paper presents a multi-sensory multimodal framework, the i-Walk assessment system, that endows assistant platforms with the ability to successfully recognize human activities, understand audio-gestural intentions, monitor user's stability and mobility and assess rehabilitation progress giving meaningful feedback to the user. The i-Walk assessment system is extensively evaluated on a database of healthy subjects and patients where the presented results show quite high performance of the developed system components as well as the efficacy of the proposed framework not only to provide passive rollators with "intelligence", but also to be integrated into a general decision making strategy for natural and user-adaptive HRI of robotic assistant platforms.

Acknowledgment. This research has been co-financed by the European Union and Greek national funds (project code:T1EDK-01248, acronym: i-Walk).

References

1. DIAPLASIS Rehabilitation Center. https://www.diaplasis.eu/
2. Google cloud speech-to-text. https://cloud.google.com/speech-to-text/
3. RASA. http://rasa.com
4. RASA. http://github.com/RasaHQ
5. Robot Operating System (ROS). http://www.ros.org/about-ros/
6. Andriluka, M., Pishchulin, L., Gehler, P., Schiele, B.: 2D human pose estimation: new benchmark and state of the art analysis. In: CVPR (2014)
7. Bocklisch, T., Faulkner, J., Pawlowski, N., Nichol, A.: Rasa: open source language understanding and dialogue management. In: NIPS (2017)
8. Cao, Z., Simon, T., Wei, S., Sheikh, Y.: Realtime multi-person 2D pose estimation using part affinity fields. In: CVPR (2017)
9. Chalvatzaki, G., Koutras, P., Hadfield, J., Papageorgiou, X.S., Tzafestas, C.S., Maragos, P.: LSTM-based network for human gait stability prediction in an intelligent robotic rollator. In: 2019 International Conference on Robotics and Automation (ICRA), pp. 4225–4232. IEEE (2019)
10. Chalvatzaki, G., Papageorgiou, X.S., Maragos, P., Tzafestas, C.S.: User-adaptive human-robot formation control for an intelligent robotic walker using augmented human state estimation and pathological gait characterization. In: 2018 IEEE/RSJ International Conference on Intelligent Robots and Systems (IROS), pp. 6016–6022. IEEE (2018)
11. Chalvatzaki, G., Papageorgiou, X.S., Maragos, P., Tzafestas, C.S.: Learn to adapt to human walking: a model-based reinforcement learning approach for a robotic assistant rollator. IEEE Rob. Autom. Lett. **4**(4), 3774–3781 (2019)
12. Chalvatzaki, G., Papageorgiou, X.S., Tzafestas, C.S., Maragos, P.: Augmented human state estimation using interacting multiple model particle filters with probabilistic data association. IEEE Rob. Autom. Lett. **3**(3), 1872–1879 (2018)
13. Chang, M., Mou, W., Liao, C., Fu, L.: Design and implementation of an active robotic walker for Parkinson's patients. In: Proceedings of SICE, pp. 2068–2073 (2012)
14. Chen, Y., Tian, Y., He, M.: Monocular human pose estimation: a survey of deep learning-based methods. Comput. Vis. Image Underst. **192**, 102897 (2020)
15. Deriu, J., et al.: Survey on evaluation methods for dialogue systems. Artif. Intell. Rev. 1–56 (2020). https://doi.org/10.1007/s10462-020-09866-x
16. Du, Y., Wang, W., Wang, L.: Hierarchical recurrent neural network for skeleton based action recognition. In: CVPR (2015)
17. Dubowsky, S., et al.: PAMM-A robotic aid to the elderly for mobility assistance and monitoring: a" helping-hand" for the elderly. In: ICRA (2000)
18. Efthymiou, N., Koutras, P., Filntisis, P.P., Potamianos, G., Maragos, P.: Multi-view fusion for action recognition in child-robot interaction. In: 2018 25th IEEE International Conference on Image Processing (ICIP), pp. 455–459. IEEE (2018)

19. Frizera-Neto, A., Ceres, R., Rocon, E., Pons, J.: Empowering and assisting natural human mobility: the Simbiosis walker. Int. J. Adv. Rob. Syst. **8**(3), 29 (2011)
20. Guler, A., et al.: Human joint angle estimation and gesture recognition for assistive robotic vision. In: ECCV (2016)
21. Hochreiter, S., Schmidhuber, J.: Long short-term memory. Neural Comput. **9**(8), 1735–1780 (1997)
22. Jenkins, S., Draper, H.: Care, monitoring, and companionship: views on care robots from older people and their carers. Int. J. Soc. Rob. **7**(5), 673–683 (2015)
23. Kulyukin, V., Kutiyanawala, A., LoPresti, E., Matthews, J., Simpson, R.: iWalker: toward a rollator-mounted wayfinding system for the elderly. In: IEEE International Conference on RFID (2008)
24. Leo, M., Furnari, A., Medioni, G.G., Trivedi, M., Farinella, G.M.: Deep learning for assistive computer vision. In: ECCV (2018)
25. Lin, T., et al.: Microsoft COCO: common objects in context. In: ECCV (2014)
26. Liu, J., Wang, G., Hu, P., Duan, L., Kot, A.: Global context-aware attention LSTM networks for 3D action recognition. In: CVPR (2017)
27. Morris, A., et al.: A robotic walker that provides guidance. In: ICRA (2003)
28. Muro-De-La-Herran, A., Garcia-Zapirain, B., Mendez-Zorrilla, A.: Gait analysis methods: an overview of wearable and non-wearable systems, highlighting clinical applications. Sensors **14**(2), 3362–3394 (2014)
29. Papageorgiou, X.S., Chalvatzaki, G., Dometios, A.C., Tzafestas, C.S., Maragos, P.: Intelligent assistive robotic systems for the elderly: two real-life use cases. In: Proceedings of the 10th International Conference on PErvasive Technologies Related to Assistive Environments, pp. 360–365 (2017)
30. Papageorgiou, X.S., et al.: Advances in intelligent mobility assistance robot integrating multimodal sensory processing. In: Stephanidis, C., Antona, M. (eds.) UAHCI 2014. LNCS, vol. 8515, pp. 692–703. Springer, Cham (2014). https://doi.org/10.1007/978-3-319-07446-7_66
31. Paulo, J., Peixoto, P., Nunes, U.: ISR-AIWALKER: robotic walker for intuitive and safe mobility assistance and gait analysis. IEEE Trans. Hum. Mach. Syst. **47**(6), 1110–1122 (2017)
32. Perry, J.: Gait Analysis: Normal and Pathological Function. Slack Incorporated (1992)
33. Piyathilaka, L., Kodagoda, S.: Human activity recognition for domestic robots. In: Field and Service Robotics, pp. 395–408 (2015)
34. Rezazadegan, F., Shirazi, S., Upcrofit, B., Milford, M.: Action recognition: from static datasets to moving robots. In: ICRA (2017)
35. Rodomagoulakis, I., et al.: Multimodal human action recognition in assistive human-robot interaction. In: ICASSP (2016)
36. Rodriguez-Losada, D., Matia, F., Jimenez, A., Galan, R., Lacey, G.: Implementing map based navigation in guido, the robotic SmartWalker. In: ICRA (2005)
37. Roitberg, A., Perzylo, A., Somani, N., Giuliani, M., Rickert, M., Knoll, A.: Human activity recognition in the context of industrial human-robot interaction. In: Signal and Information Processing Association Annual Summit and Conference (2014)
38. Shahroudy, A., Liu, J., Ng, T., Wang, G.: NTU RGB+ D: a large scale dataset for 3D human activity analysis. In: CVPR (2016)
39. Sharkey, A., Sharkey, N.: Children, the elderly, and interactive robots. IEEE Robot. Autom. Mag. **18**(1), 32–38 (2011)
40. Song, S., Lan, C., Xing, J., Zeng, W., Liu, J.: An end-to-end spatio-temporal attention model for human action recognition from skeleton data. In: AAAI Conference on Artificial Intelligence (2017)

41. Stavropoulos, G., Giakoumis, D., Moustakas, K., Tzovaras, D.: Automatic action recognition for assistive robots to support mci patients at home. In: PETRA (2017)
42. Tinetti, M., Williams, T., Mayewski, R.: Fall risk index for elderly patients based on number of chronic disabilities. Am. J. Med. **80**(3), 429–434 (1986)
43. Tsiami, A., Filntisis, P.P., Efthymiou, N., Koutras, P., Potamianos, G., Maragos, P.: Far-field audio-visual scene perception of multi-party human-robot interaction for children and adults. In: 2018 IEEE International Conference on Acoustics, Speech and Signal Processing (ICASSP), pp. 6568–6572. IEEE (2018)
44. Tsiami, A., Koutras, P., Efthymiou, N., Filntisis, P.P., Potamianos, G., Maragos, P.: Multi3: multi-sensory perception system for multi-modal child interaction with multiple robots. In: 2018 IEEE International Conference on Robotics and Automation (ICRA), pp. 1–8. IEEE (2018)
45. Veeriah, V., Zhuang, N., Qi, G.: Differential recurrent neural networks for action recognition. In: ICCV (2015)
46. Vemulapalli, R., Arrate, F., Chellappa, R.: Human action recognition by representing 3D skeletons as points in a lie group. In: CVPR (2014)
47. Vrigkas, M., Nikou, C., Kakadiaris, I.A.: A review of human activity recognition methods. Front. Rob. AI **2**, 28 (2015)
48. Wang, J., Liu, Z., Wu, Y., Yuan, J.: Learning actionlet ensemble for 3D human action recognition. PAMI **36**(5), 914–927 (2013)
49. Wang, P., Yuan, C., Hu, W., Li, B., Zhang, Y.: Graph based skeleton motion representation and similarity measurement for action recognition. In: ECCV (2016)
50. Zanfir, M., Leordeanu, M., Sminchisescu, C.: The moving pose: an efficient 3D kinematics descriptor for low-latency action recognition and detection. In: ICCV (2013)
51. Zhu, W., et al.: Co-occurrence feature learning for skeleton based action recognition using regularized deep LSTM networks. In: AAAI Conference on Artificial Intelligence (2016)
52. Zimmermann, C., Welschehold, T., Dornhege, C., Burgard, W., Brox, T.: 3D human pose estimation in RGBD images for robotic task learning. In: ICRA (2018)

Adaptive Virtual Reality Exergame for Individualized Rehabilitation for Persons with Spinal Cord Injury

Shanmugam Muruga Palaniappan[1], Shruthi Suresh[1], Jeffrey M. Haddad[2], and Bradley S. Duerstock[1,3](✉)

[1] Weldon School of Biomedical Engineering, Purdue University, West Lafayette, USA
[2] Department of Health and Kinesiology, Purdue University, West Lafayette, USA
[3] School of Industrial Engineering, Purdue University, West Lafayette, USA
bsd@purdue.edu

Abstract. Typical exergames used for rehabilitative therapy can be either too difficult to play or monotonous leading to a lack of adherence. Adapting exergames by tuning various gameplay parameters based on the individual's physiological ability maintains a constant challenge to improve a participant's level of engagement and to encourage the physical performance of the user to achieve rehabilitation goals. In this paper we developed a pilot exergame using a commercially available virtual reality (VR) system with varied and customizable gameplay parameters and accessible interface. A baseline task VR tool was previously developed to determine an individual player's initial 3-D spatial range of motion and areas of comfort. We observed the effects of adjusting gameplay parameters on a participant's physiological performance by measuring velocity of motions and frequency and effort of targeted movements. We calculated joint torques through inverse kinematics to serve as an analysis tool to quantitatively gauge muscular effort. The system can provide an improved rehabilitation experience for persons with tetraplegia in home settings while allowing oversight by clinical therapists through tracking of physiological performance metrics and movement analysis from mixed reality videos.

Keywords: Virtual reality · Rehabilitation · Spinal cord injury · Shoulder torque · Exergame

1 Introduction

Inpatient rehabilitation after sustaining a traumatic spinal cord injury (SCI) plays a critical role in recovering functional capacity, maintaining bone density, endurance, muscle strength, and improving psychological well-being [13,20,24]. Rehabilitative therapy of the upper limbs is also crucial for providing functional independence after a high-level SCI [8]. The inpatient rehabilitation process involves physical and occupational therapy to determine the primary goals and

A. Bartoli and A. Fusiello (Eds.): ECCV 2020 Workshops, LNCS 12538, pp. 518–535, 2020.
https://doi.org/10.1007/978-3-030-66823-5_31

interventions that are important to maximize recovery of motor function and the ability to perform activities of daily living (ADL). Therapists periodically re-evaluates the patient to assess progress and determine if new therapy protocols are required. However, over the past 40 years the length of stays for rehabilitation for patients with SCIs have decreased by two-thirds from a median stay of 98 days in 1973–1979 to 36 days in 2010–2014 [33]. This substantial decrease of supervised therapy for newly spinal cord injured persons make it imperative to find solutions to continue regular therapy after they are discharged from the rehabilitation hospital [29].

Despite the advantages of regular home-based therapy, there are several barriers to individuals with SCI performing the prescribed therapy regularly and correctly [48]. These include physical barriers such as equipment, availability of resources as well as psychological or social barriers such as perceptions or attitudes towards disability, motivation, and fear of injury. Therefore, the desire to exercise does not match behavior and various studies have found that lack of motivation was a ubiquitous factor in reduced exercise [25,42]. In order to combat the monotony of traditional exercise regimens, exercise gaming "exergaming" was introduced for individuals with sedentary lifestyles or other ailments such as obesity and cardiovascular disease [57].

Exergames have also increasingly become popular to enable persons with disabilities to participate in appropriate exercises to achieve the required physical intensity but are also engaging to play [32]. Motivation through serious exergaming has been shown to raise patients' interest and improve their adherence to rehabilitation at home[27,35,53]. Variety of research and commercial game platforms, such as the Nintendo Wii^{TM} (Nintendo, Kyoto, Japan) and Leap Motion (Leap Motion Inc, CA) [1,10,14,19], have been developed for exergaming by tracking players' movements [23,55]. Several studies involve individuals standing using a balance board, dancing, or stepping in place as the actions being performed [11]. A relatively new foray into the exergaming space has been with virtual reality (VR) tools.

VR is emerging as a useful tool to facilitate exergaming therapy with the potential to support home-based exercise programs [41]. VR-based exergames have emerged as a tool for rehabilitation for diseases such as stroke, traumatic brain injury, SCI, cerebral palsy, Parkinson's disease (PD) and other developmental issues [28]. The use of VR in rehabilitation is particularly attributed to its ability to provide immersive experiential learning experiences in an engaging, realistic but safe environment [44,49]. The VR system encourages the repetition of active movement, making it ideally suited as a tool for motor rehabilitation. VR's ability to automatically deliver stimulus at known timepoints allows clinicians and therapists to focus on the patients' performance and observe whether they are using effective strategies [43,52]. Clinicians can also use VR to allow patients to achieve a variety of objectives through the varying of task complexity as well as type and amount of feedback [56].

However, past VR exergaming systems often encompass extremely large hardware setups that are generally not portable in nature. Their expensive nature also makes them impossible for use as part of a home-based rehabilitation system. Additionally, the existing setups are designed for individuals who do not have limitations with manual dexterity since they often require substantial finger strength [26]. This lack of hand function common in tetraplegics make it difficult to hold and depress various buttons on a typical VR game controller. Design customization to meet the need and limitations of tetraplegics needs to be a significant consideration in gameplay development. Most games available in the market require the use of buttons, making it impossible to use in individuals with limited hand function, and difficult to be modified for use in individuals with impaired dexterity [39,52]. Thus, there is a critical need for VR exergames to be developed that are customized to the unique limitations of tetraplegics, which are fun to use and provide therapeutic benefit.

Additionally, when considering development of exergames, it is critical to track and quantify the progress of movements made by individuals performing at-home therapies [15,17,18]. Current techniques to determine functional ability include the functional independence measure (FIM), manual muscle testing (MMT), Range of Motion Scale (ROMS), and the Modified Ashworth Scale [22,34,46]. These subjective tests require trained clinical therapists to obtain a good inter-rater reliability [15]. Moreover, these methods are time consuming, labor and resource intensive and are often dependent on patient compliance [28,47]. Developing objective tests that can track patient progress and functionality at home would alleviate many of the aforementioned issues and nicely fits into the more recent migration of many health care providers to offer more telemedicine home-based assessments. In order to fill this gap, we developed a baseline tool [37] and an exergame that uses a commercial off-the-shelf head mounted VR gaming system, the HTC Vive©. The Vive is a relatively low-cost, portable system which can be easily setup in the home environment. Since shoulder torques are positively correlated with perceived muscular effort [9], the tool we developed also incorporates a method to calculate static joint forces and torques at the shoulder, as a measure of muscular effort. Moreover, we also explored the role of altering various gameplay parameters to identify the impact on perceived user effort as well as the impact on muscular effort. Surveys were conducted at the end of the games to understand the perceived fatigue levels and feedback about gameplay mechanics. This work could quantify perceived effort that the users likely felt during gameplay enabling a better/iterative rehabilitation experience for the individuals with SCI and their therapists.

2 Methods

2.1 Recruitment of Participants

We recruited six participants (one female and five males) from the Rehabilitation Hospital of Indiana, and the mean age of the participants was 37.5 ± 9.9. All participants had a cervical (C) SCI ranging from C4- C7 level injuries. At the time of their participation, participants had been injured, on average, for 15 ±

11.2 years. All study protocols were approved by an Institutional Review Board. Prior to the study, informed consent was obtained from all the participants.

2.2 HTC Vive and Mixed Reality Setup

The HTC Vive© is a commercially available VR system which uses a head mounted display to immerse the individual into a virtual world where virtual distances match real world distance (Fig. 1a). Therefore, the gestures and tasks performed virtually are directly translatable to the physical reach of the individual during rehabilitation. Previous rehabilitation exergaming systems have employed 2D flatscreens, which do not permit any depth perception. Completion of this visual feedback loop is vital for effective reaching tasks during rehabilitation [40]. The Vive comprises of two base stations, called lighthouses, which have spinning IR lasers that flash and sweep a beam of light alternatingly. The head mounted display (HMD) and trackers have a constellation of IR receivers that use flashes and beams of IR light to determine the position and orientation. The position data is calculated at a 90 Hz refresh rate [2]. In order to make the controllers more accessible, we used Vive trackers that were mounted to game objects to be tracked by the base stations. A 3D printed shim was designed to allow the tracker to be secured firmly to a participant's end effector with a Velcro strap (Fig. 1b). VR tools and games were developed using the Unity3D™ (Unity Technologies, San Francisco, CA) game engine to work with the Vive trackers. The games involved designing 3D models to be rendered during gameplay written in C# to detect and handle collision events.

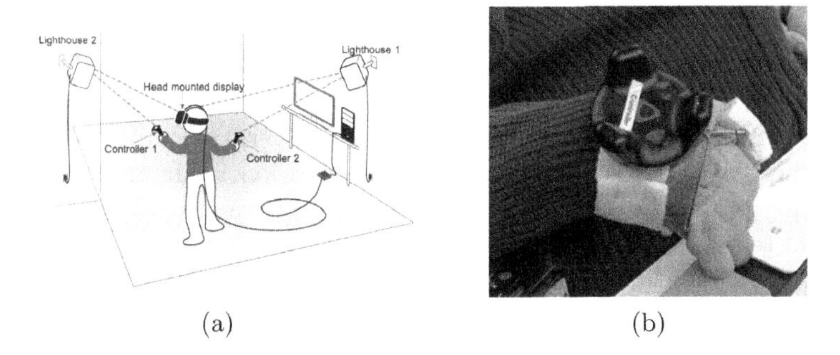

Fig. 1. a) Mixed reality setup with HTC Vive and Green Screen b) 3D printed mount to fix camera and tracker geometry (Color figure online)

Mixed Reality System. A mixed reality system was developed to allow clinicians to observe participant interaction within the virtual environment. It provided a combined view of the gameplay that incorporated both a real video with the virtual environment and virtual objects overlaid in the correct position and orientation. This mixed reality system utilizes a position tracked camera using an HTC Vive© tracker, a green screen and the traditional HTC Vive© VR setup Fig. 2a). The mixed reality videos were generated through the Liv software (LIV Inc, San Francisco CA) in real time, and saved for analysis. A virtual camera was setup in Unity3D in the exact same position the camera in the real-world. A one-time calibration was required to set up the mixed reality system. This was done through the Liv software wherein the physical distance of the camera relative to the participant and the Vive tracker are mapped to the virtual space. The camera's field of view and orientation are also calibrated through the software to match the representation in the virtual environment.

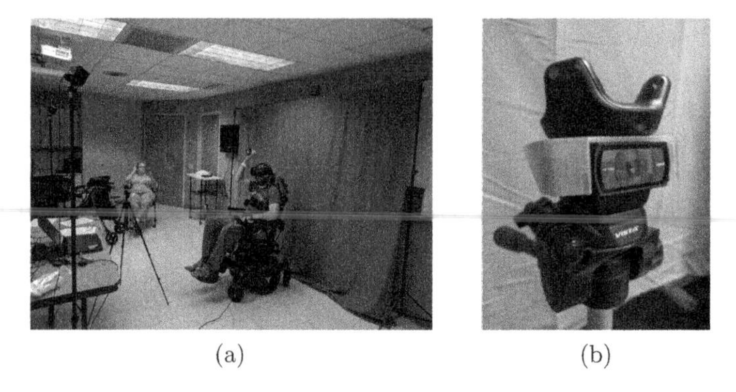

(a) (b)

Fig. 2. a) Mixed reality setup with HTC Vive and Green Screen b) 3D printed mount to fix camera and tracker geometry (Color figure online)

In order to tackle the challenge of calibrating the camera lens' optical parameters such as focus and field of view we accounted for a translational and rotational offset between the camera and the tracker. Several attempts were required to align the real and virtual videos and re-calibration required large overhead times during participant studies. To prevent this, a tracker camera mount was 3D printed to keep the translational and rotational offsets fixed (Fig. 2b). This greatly reduced the time needed for re-calibration during each set up. The video feed from the real and virtual camera were combined using the Chroma Key composting technique (Unity Mixed Reality Capture).

2.3 Exergame Design

An exergame was developed based on physical therapists' recommendations to physically challenge the participant but not overly so as to discourage continued gameplay [5,31]. The exergame developed involved targeting virtual balloons that were spawned randomly around the participant (Fig. 3). A virtual model of a light saber was attached to the participant's tracker that was used to target virtual balloons. These multicolored balloons were designed to pop when the lightsaber targeted them for a specific duration. The color of the balloon would change to a fluorescent pink when it was successfully targeted, i.e. the light saber was inside the balloon. In order to avoid inadvertent pops resulting from flailing motion or other unplanned motion, a small delay was added before a balloon would pop. The baseline delay was chosen to be 100ms. At the end of a successful pop, a popping animation and a loud realistic balloon pop sound was played as visual and auditory notifications to participants.

We modified two different parameters to measure the resultant change in participant biomechanical responses. The first gameplay parameter was the scale (size) of the balloons and the second was the delay required for the balloon to pop after targeting. Two different trials were designed to investigate how participants interacted with the game and how each of these parameters affected the participants' interaction/performance. During both trials, the coordinates of the participant's HMD, end effector tracker, number of popped balloons, and the location of the tip of the light saber were all logged throughout gameplay.

The scale trial comprised a 2-minute-long gameplay with balloons were either at full scale (1) or at half scale (0.5) which were presented/spawned at an equal probability. All the balloons had the baseline pop delay threshold of 100ms. The participants were asked to play the game and try to pop as many balloons as they could during the duration of gameplay. We measured the 1) number of balloons that were popped, 2) average velocity while approaching the balloon [36], 3) fatigue levels reported on a five-point Likert scale, and 4) static torques. The delay trial comprised a 2-minute-long gameplay where spawned balloons were all at full scale. However, half of the spawned balloons would pop at the baseline pop delay threshold (100 ms); whilst the other half of spawned balloons would only pop at the increased pop threshold (300 ms). Participants were asked to pop as many balloons as possible. They were not provided information regarding which balloons had the longer delay. In addition to the aforementioned measurement metrics, we also measured the time to failure, which was defined as giving up on popping the balloon.

2.4 Joint Muscular Force Calculation

Joint reaction forces and torques were calculated to understand the level of muscular exertion [21]. Four important assumptions were used to perform inverse dynamics to calculate joint forces: 1) anthropometric data, including the participants' arm segment's center of mass and weight of arm segment in proportion to total body weight, 2) link segment model of the human body, which is a simplified representation of the complex limb joint as simple revolute joints and arm segments with masses and moment of inertias located at the center of mass of the segments [54], 3) kinematic data obtained using HTC Vive trackers, and 4) external force measurement performed through inverse kinematics. In order to calculate joint forces for each gesture performed by participants during gameplay, the human arm was modelled kinematically as a serial-link robot following the Denavit-Hartemberg (D-H) notation [38,51] and implemented through the

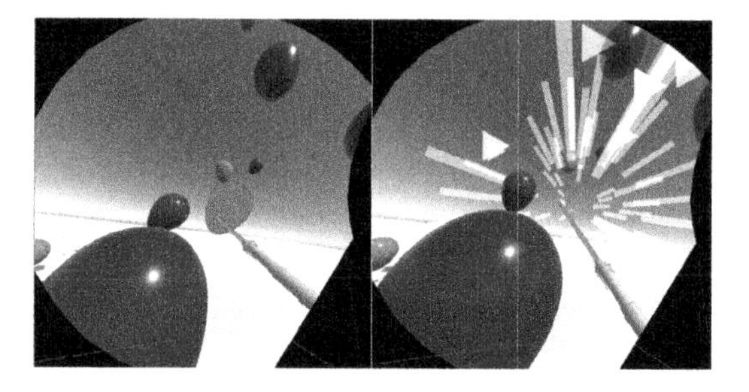

Fig. 3. Left: balloon turning fluorescent pink indicating it was targeted. Right: Balloon popping animation indicates that the light saber targeted the balloon for a sufficient amount of time (> 100 ms).

MATLAB Robotic Toolbox [6]. The D-H notation consist of five parameters [7] that are used to describe each link to the previous link in the series (Fig. 4).

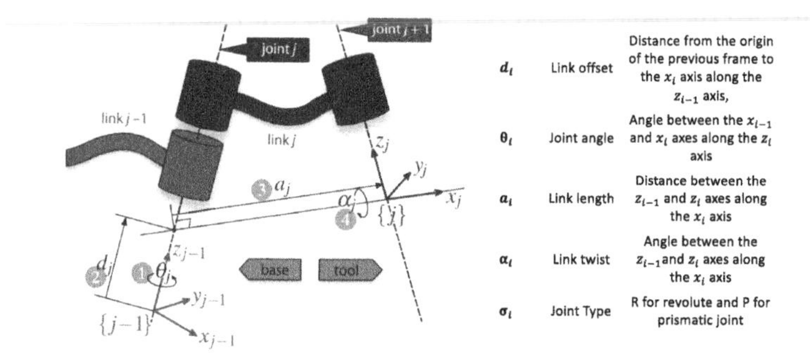

Fig. 4. Geometry of D-H parameters. [7]

In addition to these default D-H parameters, absolute joint constraints [3] were added to each joint to prevent orientations that are unachievable biomechanically. Figure 5a illustrates that the first three joints are at the exact same point in 3D space but offset by 90°, these joints correspond to the degrees of freedom at the shoulder joint. The fourth joint is at the elbow offset from the first three joints by the length of the humerus or upper arm. The fourth joint has a length of the ulna or forearm. The wrist joint was not modelled, as the gameplay did not involve wrist motion.

The length of individual participant's upper and forearm were derived from participants' video recordings taken during gameplay. Figure 5b demonstrates how anthropometric measurements were made using the open source physics video tracking tool, Physlets Tracker, from the video recording of the participant. The known value of the HMD's width (shown by the blue in Fig. 5b) was used as reference for calibration. This measurement was performed three times for each participant and the average length was used for data analysis. A new coordinate system was defined with the HMD coordinates as the origin. Therefore, the tracker coordinates and all the virtual objects were referenced to the HMD position.

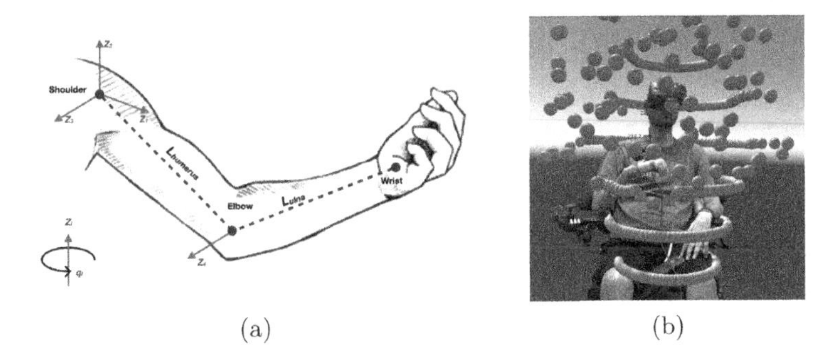

Fig. 5. a) Kinematic model of the human arm with reference frames associated to the various degrees of freedom b) Anthropometric measurements of the participant (shown in red) from recorded mixed reality videos. (Color figure online)

Inverse kinematics was performed using the kinematic arm model from the tracked and translated coordinates relative to the HMD coordinates of the end effector. Inverse kinematics returned the joint angles necessary to achieve the specific pose and yielded several possible solutions for the position of the elbow as there is no unique solution. Locations of the elbow which were not biomechanically viable were discarded and in order to obtain a conservative estimate, a natural and comfortable "elbow down" [6] start pose was determined empirically for each individual (Fig. 6). The inverse kinematics tool accepts a start pose as an input argument to use as a starting point, which determines the final orientation of the calculated pose. Inverse kinematic algorithms have been developed in the past with maximizing human comfort by minimizing joint torques [58]. An inertial model of the arm was also created by modelling the two arm segments as cylinders.

The mass and center of mass of the arm segments were calculated as a percentage of the body weight obtained from standardized anthropometric data. The upper arm and forearm masses were 2.66% and 1.82% of the entire body weight for men and 2.6% and 1.82% for women respectively. The distance of center of mass from the proximal joints were 48.5% and 44% respectively [4,12]. The inertial parameters for the principle axes for the arm segments modelled as solid cylinders were then calculated using the equations below.

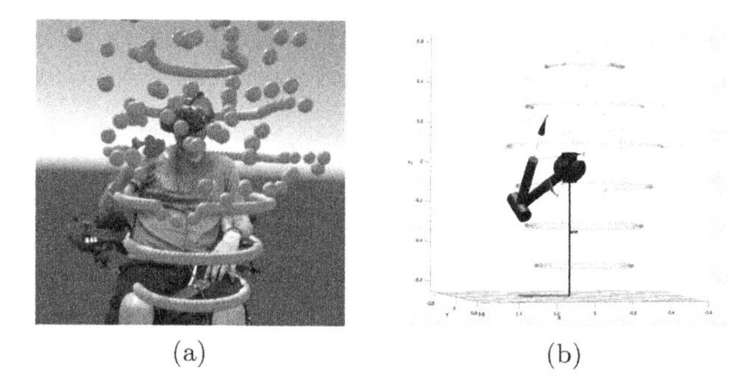

(a) (b)

Fig. 6. Initial location of the elbow based on comfort shown in a) mixed reality view with "elbow down" orientation and simulated with b) a kinematic model of the arm with "elbow down" orientation

$$I_{xx} = \frac{1}{2}mr^2 \tag{1}$$

$$I_{yy} = I_{yy} = \frac{1}{2}m(3r^2 + h^2) \tag{2}$$

3 Results

Scale Trial Metrics. Participants targeted 40% more larger balloons than smaller balloons (Fig. 7), however this was shown to be non-significant ($p>0.05$) through ANOVA testing. Non-parametric permutation testing showed a significant difference ($p<0.05$) in the velocities of arm movement when popping the smaller balloons compared to the larger balloons. The average velocity of the arm was 25.4% higher for the larger balloons compared to the smaller ones. Representative velocity profiles (Fig. 8) show that for the larger balloons, the change of velocity of the arm was sharper at or immediately prior to the balloon popping event (indicated by red dashed lines). On average the participants reported a fatigue score of 3.0 ± 1.26 after playing the balloon popping game with different sized balloons.

We identified a positive correlation between the total torque at the shoulder and the distance of the balloons from the shoulder ($r^2 = 0.73$). The further the distance of the balloon- the more muscular effort was required to reach the balloon. Figure 9a shows the performance of an individual with greater upper limb motor function compared to an individual with lesser motor function (Fig. 9b). The individual with greater motor function did not show preference in popping either large or small -scaled balloons and exerted almost equal effort. However, the individual with lesser motor function popped larger balloons more frequently (Fig. 9b).

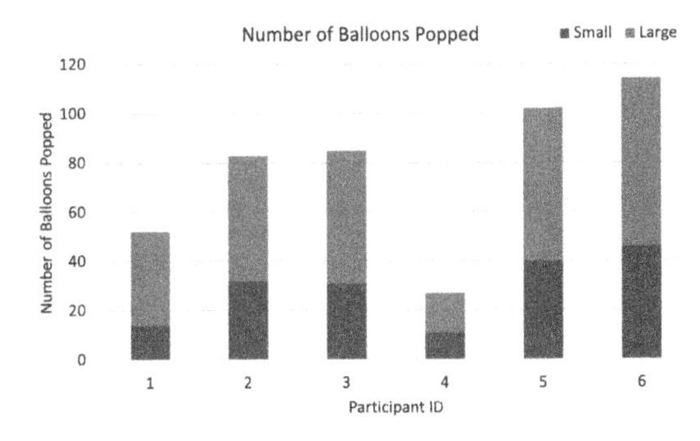

Fig. 7. Preference of balloon based on number of balloons popped

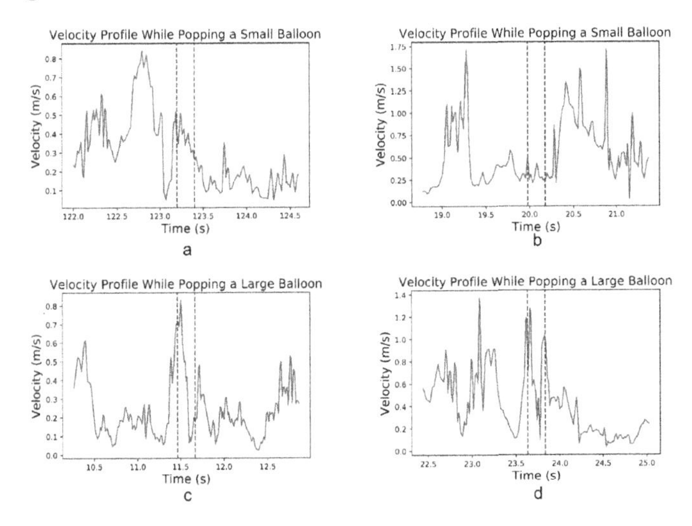

Fig. 8. Representative velocity profiles while popping balloons of different sizes. Red dashed line indicates popping event. (a and b) Small Balloons for two different participants. (c and d) Large Balloons for two different participants. (Color figure online)

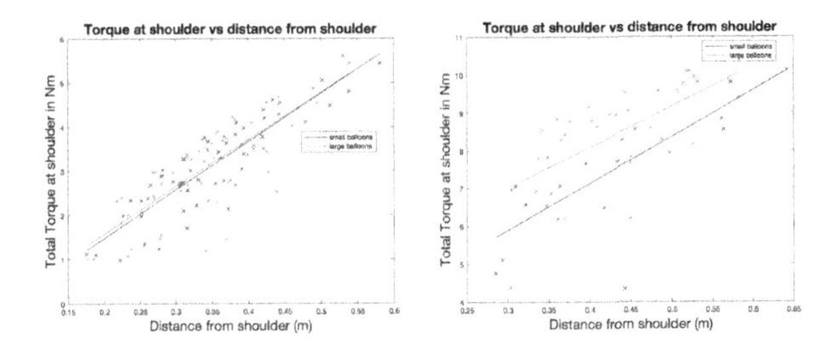

Fig. 9. Static Torque calculations at the shoulder for different sized balloons versus distance from shoulder for two different subjects – a) participant with greater upper limb motor function and b) participant with lesser motor function. The red markers represent large balloons whilst the blue markers represent small balloons. (Color figure online)

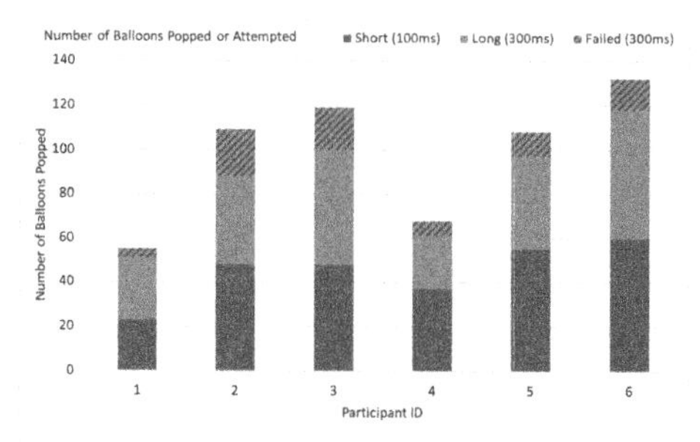

Fig. 10. Number of balloons attempted/successfully popped with different pop delays for each participant.

Delay Trial Metrics. Participants successfully popped 10% more short delay (100 ms) balloons than long delay balloons. On average participants had a failure rate of 23.8% of all the long delay (300 ms) balloons attempted (Fig. 10). The overall average duration a participant spent inside a long delay balloon before failing was 190.36 ms. The non-parametric permutation test showed a significant difference in the velocities between popping balloons with short pop delays (100 ms) and those with long pop delays (300 ms) ($p < 0.05$). The average velocity was 33.3% higher for the balloons with short pop delays. The velocity profiles of the different pop delays (Fig. 11) showed a trend wherein the velocity of the hand changes at a lower rate when popping the balloons with longer pop delays.

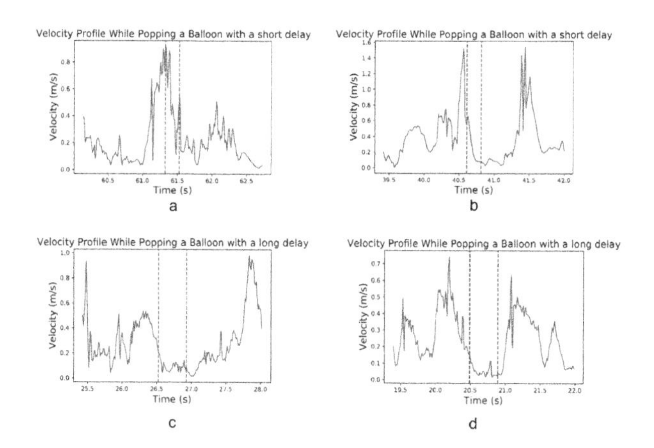

Fig. 11. Representative velocity profiles while popping balloons of different pop delays. Red dashed line indicates popping event. (a and b) Profiles for short delay to pop. (c and d) Profiles for long delay to pop.

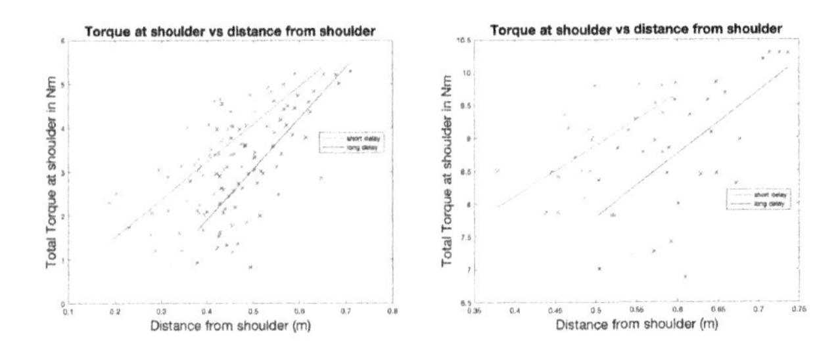

Fig. 12. Static Torque calculations when popping balloons with different delays related to distance from shoulder for two different subjects – a) participant with greater upper limb motor function and b) participant with lesser motor function.

On average the participants reported a higher fatigue score of 3.83 ± 0.94 after playing the balloon popping with different pop delays. There was a positive correlation between the total torque at the shoulder and the distance of the balloon from the shoulder ($r^2 = 0.62$). As anticipated overall participants popped longer delay balloons further away from the shoulder (Fig. 12). Individuals with limited motor function had a larger difference in targeting short delay balloons over long delay balloons (Fig. 12b).

4 Discussion

Through adaptive exergaming approaches such as modifying the gameplay parameters explored in this paper, it is possible to manipulate the performance and required exertion levels of players. Games are perceived to be boring if they are determined to be too easy [5,50] and frustrating if they are found to be too difficult [31,59]. Thus, it is imperative to operate within the individuals' physiological limits including functional reach, spatial areas of comfort [37], and within the peak static torques to achieve maximal engagement and thereby, the therapeutic potential of the exergame.

Impact of Changing Exergame Parameters on Gameplay Performance. Changing the two parameters of the exergame resulted in changes in performance of the participant, which has clear benefits toward providing individualized rehabilitative therapy. Visible cues, such as using different-sized balloons, in the exergame example that we developed resulted in a clearly defined user preferences, which could be manipulated to encourage progress of rehabilitation metrics of individuals with upper limb motor impairments.

Rehabilitative metrics that could be automatically calculated during gameplay included changes in the velocity profile. For smaller balloons, the change in gesture velocity occurred over a longer period of time, suggesting the need for more precise, deliberate but slower movements when targeting small balloons. Wherein the participants' change in velocity was sharper at or immediately prior to balloon popping event for larger balloons. This is in agreement with Fitts' law, which states that the time required to rapidly move to a target area is affected by the width of the target and the distance to the target [60]. Fitts' law has been widely used to describe reaching motions and has been applied to a variety of different upper-extremity exercises. With this in mind, the size of the balloon in the exergame was manipulated to encourage participants to perform more deliberate motor skills through the targeting of smaller balloons.

Participants had significantly preferred to pop larger balloons over smaller ones targeting almost 40% more large balloons than small balloons. This could be accounted to the higher visibility of the larger balloons as well as providing an easier target for participants. The participants also described targeting the smaller balloons as "requiring more finesse". Thus for rehabilitative purposes balloon sizes could be manipulated during gameplay to either improve the success rate of individuals with profound motor impairments to prevent frustration or making play more challenging. Likewise, the distribution of large to small balloons can be modified according to therapeutic needs. During this study, there was an equal proportion of large and small balloons being populated during gameplay [31,59]. Additionally, the participants were presented with an equal number of balloons with short or long pop delays. From the data we observed that on average participants successfully popped 10% more short delay than long delay balloons, though they were visually indistinguishable, unlike the balloon scale trial. On average participants had a failure rate of 23.8% for all the long delay balloons attempted. The lack of visual differences coupled with the failure rate suggests increased difficulty lies in holding the hand in position while tracking the balloon for 300ms to pop. This is in agreement with literature which shows that static holding tasks are harder to perform than a dynamic task [16,30]. The gameplay results showed that participants spent 190ms inside a long delay balloon before ultimately failing suggesting that if the pop delay was lowered to this number, we would observe a greater success rate. The results also showed a significant difference in the velocities between popping balloons with short pop delays and those with long pop delays as expected. While popping a balloon with a long delay, the individual needed to slow down and use fine motor control to maintain the hand position inside the balloon before popping. By altering the delay parameter can change rehabilitation outcomes, such as rewarding the user for performing holding tasks as opposed to a flinging motion often used in short delay popping events.

Joint Force Calculation to Measure Physical Effort. We developed a method of calculating static joint forces based on a number of assumptions that were characteristic of participants playing our exergame. In previous studies, joint forces at the knee and hip were calculated using strain gauges to measure the pedal forces [54] with kinematic data recorded using video cameras and retroreflective markers. However, we were able to estimate static joint forces from the data collected when using our VR-based exergaming system, including kinematic data obtained from the HTC Vive trackers and participants' anthropometric data, and calculated using traditional inverse dynamics methods.

As expected, a high positive correlation was shown between the total torque at the shoulder and the distance of the balloon from the shoulder. With increasing distance, the center of mass of the arm is farther away from the shoulder. The change in position of the center of mass leads to a mechanical disadvantage due to the increased moment arm. Therefore, the torque required at the shoulder needs to be higher to support the arm at this extended pose. When evaluating static torques for balloons with different delays, we observed a trend wherein balloons with long delays were generally popped at a farther distance from the individual's shoulder. This could be due to balloons being popped in positions where skeletal loading could be high, thus reducing the torque on the shoulder. This could also be due to balloons slowly drifting upwards causing participants to track the balloons upwards and further away from the shoulder. Through the calculation of torque at the shoulder it is possible to estimate the level of exertion [9]. This would help clinicians develop a better understanding of how challenging a specific movement might be. Moreover, it would be possible to adapt the gameplay parameters to maintain an individual at a constant level of exertion in order to attain therapeutic goals by the clinician and prevent overexerting muscles causing injury.

5 Conclusions

We developed an adaptive exergame using commercial VR systems that incorporated a variety of quantifiable, physiological-based measurements including gesture velocity and muscular exertion through joint force calculation. The VR exergaming system we developed in this paper was focused on providing an improved rehabilitation experience for persons with tetraplegia in home settings while allowing oversight by clinical therapists. Current therapies are not engaging or not adapted to the upper limb motor functions of individual players. Through mixed reality videos, therapists can receive patients' physiological results and observe their movements in real-time during gameplay for evaluation. Observation of gameplay allows clinicians to become aware of any compensatory movements, such as overusing the shoulder muscles to assist in raising the arm, which could improve function in the short term but might be detrimental in the long term [45]. This would allow patients to get regular feedback from their rehabilitation therapists on their progress. Such telehealth applications with clinical oversight are becoming more and more critical as rehabilitation stays have decreased due to limitations in insurance.

Acknowledgement. This project was supported through the Indiana State Department of Health. We thank our colleagues, Dr. Eric Nauman, George Takahashi and Dr. Juan Wachs for their insight. We thank Becky Runkel, Emily Hursey, Dr. Dawn Neumann, Dr. T. George Hornby, Larissa Swan and Reann Ray from Rehabilitation Hospital of Indiana for their help with subject recruitment.

References

1. Agmon, M., Perry, C.K., Phelan, E., Demiris, G., Nguyen, H.Q.: A pilot study of Wii Fit exergames to improve balance in older adults. J. Geriatr. Phys. Ther. **34**(4), 161–7 (2011). https://doi.org/10.1519/JPT.0b013e3182191d98
2. Borrego, A., Latorre, J., Alcañiz, M., Llorens, R.: Comparison of oculus rift and HTC vive: feasibility for virtual reality-based exploration, navigation, exergaming, and rehabilitation. Games Health J. (2018). https://doi.org/10.1089/g4h.2017.0114
3. Cabrera, M.E., Wachs, J.P.: Biomechanical-based approach to data augmentation for one-shot gesture recognition. In: 2018 13th IEEE International Conference on Automatic Face & Gesture Recognition (FG 2018), pp. 38–44. IEEE (5 2018). https://doi.org/10.1109/FG.2018.00016
4. Clauser, C.E., McConville, J.T., Young, J.W.: Weight, volume, and center of mass of segments of the human body (1969)
5. Colombo, R., et al.: Design strategies to improve patient motivation during robot-aided rehabilitation. J. NeuroEng. Rehabilitation **4**(1), 3 (2007). https://doi.org/10.1186/1743-0003-4-3, https://jneuroengrehab.biomedcentral.com/articles/10.1186/1743-0003-4-3
6. Corke, P.: A robotics toolbox for MATLAB. IEEE Robot. Autom. Mag. **3**(1), 24–32 (1996). https://doi.org/10.1109/100.486658
7. Corke, P.: A simple and systematic approach to assigning denavit - hartenberg parameters. IEEE Trans. Robot. **23**(3), 590–594 (2007). https://doi.org/10.1109/TRO.2007.896765
8. Cortes, M., et al.: Improved motor performance in chronic spinal cord injury following upper-limb robotic training. NeuroRehabilitation 33(1), 57–65 (2013). https://doi.org/10.3233/NRE-130928
9. Dickerson, C.R., Martin, B.J., Chaffin, D.B.: The relationship between shoulder torques and the perception of muscular effort in loaded reaches. Ergonomics **49**(11), 1036–1051 (2006). https://doi.org/10.1080/00140130600730960
10. van Diest, M., Stegenga, J., Wörtche, H.J., Postema, K., Verkerke, G.J., Lamoth, C.J.: Suitability of Kinect for measuring whole body movement patterns during exergaming. J. Biomech. **47**(12), 2925–2932 (2014). https://doi.org/10.1016/J.JBIOMECH.2014.07.017
11. Dockx, K., et al.: Virtual reality for rehabilitation in Parkinson's disease (2016). https://doi.org/10.1002/14651858.CD010760.pub2
12. Drillis, R., Contini, R., Bluestein, M.: Body segment parameters; a survey of measurement techniques. Artif. Limbs **25**, 44–66 (1964)
13. Durán, F.S., Lugo, L., Ramírez, L., Lic, E.E.: Effects of an exercise program on the rehabilitation of patients with spinal cord injury. Arch. Phys. Med. Rehabil. **82**(10), 1349–1354 (2001). https://doi.org/10.1053/APMR.2001.26066
14. Esculier, J., Vaudrin, J., Bériault, P., Gagnon, K., Tremblay, L.E.: Home-based balance training program using the Wii and the Wii Fit for Parkinson's disease (2011)
15. Fan, E., Ciesla, N.D., Truong, A.D., Bhoopathi, V., Zeger, S.L., Needham, D.M.: Inter-rater reliability of manual muscle strength testing in ICU survivors and simulated patients. Intensive Care Med. **36**(6), 1038–1043 (2010). https://doi.org/10.1007/s00134-010-1796-6

16. Frey Law, L.A., Lee, J.E., McMullen, T.R., Xia, T.: Relationships between maximum holding time and ratings of pain and exertion differ for static and dynamic tasks. Appl. Ergon. **42**(1), 9–15 (2010). https://doi.org/10.1016/J.APERGO.2010.03.007
17. Gitkind, A.I., Olson, T.R., Downie, S.A.: Vertebral artery anatomical variations as they relate to cervical transforaminal epidural steroid injections. Pain Med. (United States) **15**(7), 1109–1114 (2014). https://doi.org/10.1111/pme.12266
18. Herbison, G.J., Isaac, Z., Cohen, M.E., Ditunno, J.F.: Strength post-spinal cord injury: myometer vs manual muscle test. Spinal Cord **34**(9), 543–8 (1996)
19. Herz, N.B., Mehta, S.H., Sethi, K.D., Jackson, P., Hall, P., Morgan, J.C.: Nintendo Wii rehabilitation ("Wii-hab") provides benefits in Parkinson's disease. Parkinsonism Relat. Disord. **19**(11), 1039–1042 (2013). https://doi.org/10.1016/J.PARKRELDIS.2013.07.014, https://www.sciencedirect.com/science/article/pii/S135380201300268X
20. Hicks, A.L., et al.: Long-term exercise training in persons with spinal cord injury: effects on strength, arm ergometry performance and psychological well-being. Spinal Cord **41**(1), 34–43 (2003). https://doi.org/10.1038/sj.sc.3101389
21. Hull, M., Jorge, M.: A method for biomechanical analysis of bicycle pedalling. J. Biomech. **18**(9), 631–644 (1985). https://doi.org/10.1016/0021-9290(85)90019-3
22. (IMACS), I.C.S.G.: Key to Muscle Grading. Muscle Grading and Testing Procedures (1993)
23. Janarthanan, V.: Serious video games: games for education and health. In: Proceedings of the 9th International Conference on Information Technology, ITNG 2012, pp. 875–878 (2012). https://doi.org/10.1109/ITNG.2012.79
24. Jones, L.M., Legge, M., Goulding, A.: Intensive exercise may preserve bone mass of the upper limbs in spinal cord injured males but does not retard demineralisation of the lower body. Spinal Cord **40**(5), 230–235 (2002). https://doi.org/10.1038/sj.sc.3101286
25. Kinne, S., Patrick, D.L., Maher, E.J.: Correlates of exercise maintenance among people with mobility impairments. Disabil. Rehabil. **21**(1), 15–22 (1999). https://doi.org/10.1080/096382899298052
26. Lange, B., Flynn, S., Rizzo, A.: Initial usability assessment of off-the-shelf video game consoles for clinical game-based motor rehabilitation. Phys. Therapy Rev. **14**(5), 355–363 (2009). https://doi.org/10.1179/108331909x12488667117258
27. Laut, J., Cappa, F., Nov, O., Porfiri, M.: Increasing patient engagement in rehabilitation exercises using computer-based citizen science. PLoS ONE **10**(3), e0117013 (2015). https://doi.org/10.1371/journal.pone.0117013
28. Laver, K.E., George, S., Thomas, S., Deutsch, J.E., Crotty, M.: Virtual reality for stroke rehabilitation (2015). https://doi.org/10.1002/14651858.CD008349.pub3
29. Li, B., Maxwell, M., Leightley, D., Lindsay, A., Johnson, W., Ruck, A.: Development of Exergame-based Virtual Trainer for Physical Therapy using Kinect. In: Games for Health 2014, pp. 79–88. Springer Fachmedien Wiesbaden, Wiesbaden (2014). https://doi.org/10.1007/978-3-658-07141-7_11
30. Lin, C.L., Wang, M.J.J., Drury, C.G., Chen, Y.S.: Evaluation of perceived discomfort in repetitive arm reaching and holding tasks. Int. J. Ind. Ergonomics **40**(1), 90–96 (2010). https://doi.org/10.1016/J.ERGON.2009.08.009
31. Lopes, R., Bidarra, R.: Adaptivity challenges in games and simulations: a survey. IEEE Trans. Comput. Intell. AI Games **3**(2), 85–99 (2011). https://doi.org/10.1109/TCIAIG.2011.2152841. http://ieeexplore.ieee.org/document/5765665/

32. Mat Rosly, M., Mat Rosly, H., Davis OAM, G.M., Husain, R., Hasnan, N.: Exergaming for individuals with neurological disability: a systematic review, April 2017. https://doi.org/10.3109/09638288.2016.1161086, https://www.tandfonline.com/doi/full/10.3109/09638288.2016.1161086
33. NSCSIC: Annual Report Complete Public Version. Technical report (2018). https://www.nscisc.uab.edu/PublicDocuments/reports/pdf/2014NSCISCAnnualStatisticalReportCompletePublicVersion.pdf
34. Oosterwijk, A.M., Nieuwenhuis, M.K., Schouten, H.J., van der Schans, C.P., Mouton, L.J.: Rating scales for shoulder and elbow range of motion impairment: call for a functional approach. PLOS ONE **13**(8), e0200710 (2018). https://doi.org/10.1371/journal.pone.0200710. https://dx.plos.org/10.1371/journal.pone.0200710
35. Palaniappan, S.M., Duerstock, B.S.: Developing rehabilitation practices using virtual reality exergaming. In: 2018 IEEE International Symposium on Signal Processing and Information Technology (ISSPIT), pp. 090–094. IEEE, December 2018. https://doi.org/10.1109/ISSPIT.2018.8642784, https://ieeexplore.ieee.org/document/8642784/
36. Palaniappan, S.M., Suresh, S., Duerstock, B.S.: Gesture segmentation of continuous movements during virtual reality gameplay. Virtual Reality (Under Review). https://doi.org/10.1089/g4h.2017.0114
37. Palaniappan, S.M., Zhang, T., Duerstock, B.S.: Identifying comfort areas in 3D space for persons with upper extremity mobility impairments using virtual reality. In: ASSETS 2019–21st International ACM SIGACCESS Conference on Computers and Accessibility, pp. 495–499. Association for Computing Machinery, Inc., October 2019. https://doi.org/10.1145/3308561.3353810
38. Paul, R.P.: Mathematics, Programming and Control. The Computer Control of Robot Manipulators. MIT Press (1983)
39. Pietrzak, E., Pullman, S., McGuire, A.: Using virtual reality and videogames for traumatic brain injury rehabilitation: a structured literature review. Games Health J. **3**(4), 202–214 (2014). https://doi.org/10.1089/g4h.2014.0013, http://www.liebertpub.com/doi/10.1089/g4h.2014.0013
40. Piron, L., Cenni, F., Tonin, P., Dam, M.: Virtual reality as an assessment tool for arm motor deficits after brain lesions. In: Studies in Health Technology and Informatics (2001). https://doi.org/10.3233/978-1-60750-925-7-386
41. Powell, V., Powell, W.: Therapy-led design of home-based virtual rehabilitation. In: 2015 IEEE 1st Workshop on Everyday Virtual Reality, WEVR 2015, pp. 11–14. IEEE, March 2015. https://doi.org/10.1109/WEVR.2015.7151688
42. Rimmer, J.H., Rubin, S.S., Braddock, D.: Barriers to exercise in African American women with physical disabilities. Arch. Phys. Med. Rehabil. **81**(2), 182–188 (2000). https://doi.org/10.1053/apmr.2000.0810182
43. Rizzo, A., Kim, G.J.: A SWOT analysis of the field of virtual reality rehabilitation and therapy (2005). https://doi.org/10.1162/1054746053967094
44. Rizzo, A.A., Schultheis, M., Kerns, K.A., Mateer, C.: Analysis of assets for virtual reality applications in neuropsychology (2004). https://doi.org/10.1080/09602010343000183
45. Roby-Brami, A., Feydy, A., Combeaud, M., Biryukova, E.V., Bussel, B., Levin, M.F.: Motor compensation and recovery for reaching in stroke patients. Acta Neurologica Scandinavica **107**(5), 369–381 (2003). https://doi.org/10.1034/j.1600-0404.2003.00021.x
46. van Rooijen, D.E.: Reliability and validity of the range of motion scale (ROMS) in patients with abnormal postures. Pain Med. (United States) **16**(3), 488–493 (2015). https://doi.org/10.1111/pme.12541

47. Saposnik, G., et al.: Effectiveness of virtual reality using Wii gaming technology in stroke rehabilitation: a pilot randomized clinical trial and proof of principle. Stroke (2010). https://doi.org/10.1161/STROKEAHA.110.584979
48. Scelza, W.M., Kalpakjian, C.Z., Zemper, E.D., Tate, D.G.: Perceived barriers to exercise in people with spinal cord injury (2005). https://doi.org/10.1097/01.phm.0000171172.96290.67
49. Schultheis, M.T., Himelstein, J., Rizzo, A.A.: Virtual reality and neuropsychology: Upgrading the current tools. Journal of Head Trauma Rehabilitation (2002). https://doi.org/10.1097/00001199-200210000-00002
50. Shapi'i, A., Arshad, H., Baharuddin, M.S., Mohd Sarim, H.: Serious games for post-stroke rehabilitation using microsoft kinect. Int. J. Adv. Sci. Eng. Inf. Technol. **8**(4–2), 1654 (2018). https://doi.org/10.18517/ijaseit.8.4-2.6823
51. Spong, M.W., Hutchinson, S., Vidyasagar, M.: Robot modeling and control (2006). https://doi.org/10.1109/MCS.2006.252815https://www.wiley.com/en-us/Robot+Modeling+and+Control-p-9780471649908
52. Sveistrup, H.: Motor rehabilitation using virtual reality (2004). https://doi.org/10.1186/1743-0003-1-10
53. Sveistrup, H., McComas, J., Thornton, M., Marshall, S., Finestone, H., McCormick, A., Babulic, K., Mayhew, A.: Experimental studies of virtual reality-delivered compared to conventional exercise programs for rehabilitation. CyberPsychol. Behav. **6**(3), 245–249 (2003). https://doi.org/10.1089/109493103322011524
54. Wangerin, M., Schmitt, S., Stapelfeldt, B., Gollhofer, A.: Inverse Dynamics in Cycling Performance. In: Advances in Medical Engineering, pp. 329–334. Springer, Heidelberg (2007). https://doi.org/10.1007/978-3-540-68764-1_55
55. Wattanasoontorn, V., Boada, I., García, R., Sbert, M.: Serious games for health. Entertainment Comput. **4**(4), 231–247 (2013). https://doi.org/10.1016/j.entcom.2013.09.002
56. Weiss, P.L., Sveistrup, H., Rand, D., Kizony, R.: Video capture virtual reality: A decade of rehabilitation assessment and intervention. Phys. Ther. Rev. **14**(5), 307–321 (2009). https://www.tandfonline.com/action/journalInformation?journalCode=yptr20
57. Whitehead, A., Johnston, H., Nixon, N., Welch, J.: Exergame effectiveness: What the numbers can tell us. In: Proceedings - Sandbox 2010: 5th ACM SIGGRAPH Symposium on Video Games pp. 55–61. ACM Press, New York (2010). https://doi.org/10.1145/1836135.1836144, http://portal.acm.org/citation.cfm?doid=1836135.1836144
58. Yang, F., Ding, L., Yang, C., Yuan, X.: An algorithm for simulating human arm movement considering the comfort level. Simul. Modell. Pract. Theory **13**(5), 437–449 (2005). https://doi.org/10.1016/J.SIMPAT.2004.12.004
59. Yannakakis, G., Hallam, J.: Real-time game adaptation for optimizing player satisfaction. IEEE Trans. Comput. Intell. AI Games **1**(2), 121–133 (2009). https://doi.org/10.1109/TCIAIG.2009.2024533, http://ieeexplore.ieee.org/document/5067382/
60. Zimmerli, L., Krewer, C., Gassert, R., Müller, F., Riener, R., Lünenburger, L.: Validation of a mechanism to balance exercise difficulty in robot-assisted upper-extremity rehabilitation after stroke. J. NeuroEng. Rehabil. **9**(1), 6 (2012). https://doi.org/10.1186/1743-0003-9-6. http://jneuroengrehab.biomedcentral.com/articles/10.1186/1743-0003-9-6

W26 - Computer Vision for UAVs Workshop and Challenge

W26 - Computer Vision for UAVs Workshop and Challenge

With great pleasure we present the proceedings of the 1st Computer Vision for UAVs Workshop and Challenge (UAVision/VisDrone2020). This online workshop was held on the 28th of August 2020 in conjunction with the 16th European Conference on Computer Vision (ECCV), originally planned to take place in Glasgow, Scotland. Note that this workshop originates from two merged workshops: the UAVision workshop (International Workshop on Computer Vision for UAVs) and the VisDrone workshop (Vision Meets Drones: A Challenge). The focus of this merged workshop was twofold. First, there was a regular paper submission track on state-of-the-art real-time image processing on-board Unmanned Aerial Vehicles (UAVs), making efficient use of specific embedded hardware and highly optimizing implementations. Second, the workshop originally consisted of 5 different challenges, ranging from real-time person detection on specific hardware to object detection, tracking and crowd counting. For the regular paper submissions track we received 15 submissions, which were all double-blind reviewed. For this, we assembled an expert panel consisting of 34 reviewers, which in total produced 65 reviews. All papers had a minimum of 4 reviews each (with the exception of one paper, which had 3 reviewers). Based on these reviews the program committee selected eight papers as full papers. These authors each submitted a full recorded presentation and a 5-minute spotlight presentation during one of two live workshop sessions (organised by the UAVision committee). The other live session (organised by the VisDrone committee) featured more information on the different challenges, and a talk from the respective winners of each challenge. Concerning the challenges, all competitors together submitted around 70 different algorithms. The sessions had one keynote speaker each, being Prof. Davide Scaramuzza and Prof. Jiri Matas. In total we estimated that our workshop gathered around 80 attendees. Apart from all accepted papers as mentioned above, these workshop proceedings also include three discussion papers written by the workshop organizers (Van Beeck et al., Dawei Du et al. and Heng Fan et al., which were not peer-reviewed). These papers aim to summarize the results of the challenges, and present work and highlight a number of common challenges and diverse proposed solutions for UAV vision applications that were identified by multiple authors. We would like to thank all members of the program committee and all contributing authors for the work they invested in assuring that our UAVision/VisDrone2020 workshop was a great success and achieved a high standard of quality.

August 2020

Dawei Du
Heng Fan
Toon Goedemé
Qinghua Hu
Haibin Ling
Davide Scaramuzza
Mubarak Shah
Tinne Tuytelaars
Kristof Van Beeck
Longin Wen
Pengfei Zhu

ATG-PVD: Ticketing Parking Violations on a Drone

Hengli Wang[1], Yuxuan Liu[1], Huaiyang Huang[1], Yuheng Pan[2], Wenbin Yu[2], Jialin Jiang[2], Dianbin Lyu[2], Mohammud J. Bocus[3], Ming Liu[1], Ioannis Pitas[4], and Rui Fan[2,5](✉)

[1] HKUST Robotics Institute, Hong Kong, China
{hwangdf,yliuhb,hhuangat,eelium}@ust.hk
[2] ATG Robotics, Hangzhou, China
{panyuheng,yuwenbin,jiangjialin,lvdianbin}@atg-itech.com
[3] University of Bristol, Bristol, UK
junaid.bocus@bristol.ac.uk
[4] Aristotle University of Thessaloniki, Thessaloniki, Greece
pitas@csd.auth.gr
[5] UC San Diego, La Jolla, USA
rui.fan@ieee.org

Abstract. In this paper, we introduce a novel suspect-and-investigate framework, which can be easily embedded in a drone for automated parking violation detection (PVD). Our proposed framework consists of: 1) SwiftFlow, an efficient and accurate convolutional neural network (CNN) for unsupervised optical flow estimation; 2) Flow-RCNN, a flow-guided CNN for car detection and classification; and 3) an illegally parked car (IPC) candidate investigation module developed based on visual SLAM. The proposed framework was successfully embedded in a drone from ATG Robotics. The experimental results demonstrate that, firstly, our proposed SwiftFlow outperforms all other state-of-the-art unsupervised optical flow estimation approaches in terms of both speed and accuracy; secondly, IPC candidates can be effectively and efficiently detected by our proposed Flow-RCNN, with a better performance than our baseline network, Faster-RCNN; finally, the actual IPCs can be successfully verified by our investigation module after drone re-localization.

Dataset and Demo Video:
sites.google.com/view/atg-pvd

1 Introduction

We are currently experiencing an unprecedented crisis due to the ongoing Coronavirus Disease 2019 (COVID-19) pandemic. Its worldwide escalation has taken us by surprise, causing major disruptions to global health, economic and social

H. Wang, Y. Liu, H. Huang, Y. Pan—Equal contributions.

A. Bartoli and A. Fusiello (Eds.): ECCV 2020 Workshops, LNCS 12538, pp. 541–557, 2020.
https://doi.org/10.1007/978-3-030-66823-5_32

systems. Indeed, our lives have changed overnight – businesses and schools are closed, most employees are working from home, and many have found themselves without a job. Millions of people across the globe are confined to their homes, while healthcare workers are at the frontline of the COVID-19 response [1]. With the increase in COVID-19 cases, public transport use has plummeted, as commuters shun buses, trams, and trains in favor of private cars and taxis. For instance, USA Today reported that the transit ridership demand in April 2020 was down by about 75% nationwide, compared to normal, with figures of 85% in San Francisco, 67% in Detroit and 60% in Philadelphia [2].

With the increasing number of vehicles on the roads, parking spaces have become scarce and many vehicles are parked just by the roadside, which in turn results in a significant increase in parking violations. In late March 2020, the Department of Transportation in Los Angeles [3] announced relaxed parking enforcement regulations as part of the emergency response to COVID-19, so that their citizens could practice safe social distancing without being concerned about a ticket. As the Return-to-Work Plan progresses, the relaxed parking enforcement regulations are no longer in force, consequently increasing the workload of the local traffic law enforcement officers. The demand for automated and intelligent parking violation detection (PVD) systems has thus become greater than ever.

The existing automated PVD systems typically recognize illegally parked cars (IPCs) by analyzing the videos acquired by closed-circuit televisions (CCTVs) through 2D/3D object detection algorithms [4] or video surveillance analysis algorithms [5]. However, the efficiency of such methods relies on CCTV camera positions, as IPCs cannot always be detected, especially if they are at a distant location. Deploying more CCTVs can definitely minimize misdetections, but this will also incur a high cost, and/or may not be practical. Therefore, many researchers have turned their focus towards mobile PVD systems, which can be mounted on any vehicle type. For example, the Birmingham City Council in England utilizes surveillance cars to detect IPCs and record their plate numbers [6]. However, such surveillance cars are expensive and typically require drivers. Therefore, autonomous machines, especially drones, have emerged as more efficient and cheaper alternatives.

The cars in the street can be grouped into three categories: 1) moving cars (MCs), 2) legally parked cars (LPCs) and 3) IPCs. MCs can be distinguished from LPCs and IPCs using dynamic object detection techniques, such as optical flow analysis, while IPCs can be distinguished from LPCs using object detection networks, such as Faster-RCNN [7], with the assistance of parking spot information. In this paper, we introduce a novel *suspect-and-investigate* PVD system (see Fig. 1) embedded in a drone. In the suspicion phase, we first employ a novel unsupervised optical flow estimation network, referred to as *SwiftFlow*, to estimate the optical flow F_t between I_t and I_{t+1}. F_t is then incorporated into a novel object detection and classification network, referred to as *Flow-RCNN*, to detect cars and classify them into MCs, LPCs and IPC candidates. A visual simultaneous localization and mapping (VSLAM) module then builds a localizable map

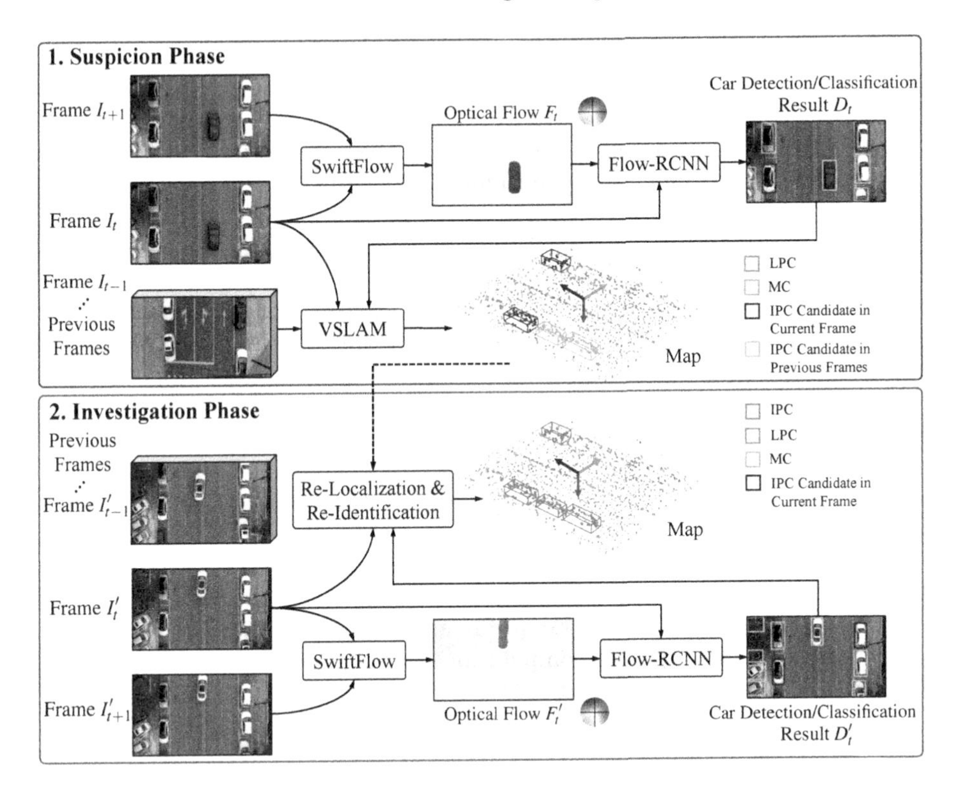

Fig. 1. The framework of our proposed suspect-and-investigate PVD system: the first phase identifies suspected IPC candidates, and the second phase investigates the suspected IPC candidates and issues tickets to the actual IPCs. The frame I_t in the suspicion phase corresponds to the frame I'_t in the investigation phase.

containing the suspected IPC candidates. After a parking grace period (which is typically five minutes) has elapsed, the drone flies back to the same location. The VSLAM module in the investigation phase subsequently detects loop closure and re-localizes the drone in the pre-built map. Finally, the suspected IPC candidates are re-identified, and the actual IPCs are marked in the map. Our main contributions are summarized as follows:

- A novel suspect-and-investigate PVD framework;
- SwiftFlow, a novel unsupervised optical flow estimation network;
- Flow-RCNN, a novel car detection and classification network;
- A large-scale PVD dataset, published for research purposes.

2 Related Work

2.1 Optical Flow Estimation

Traditional approaches generally formulate optical flow estimation as a global energy minimization problem [8–11]. Recently, convolutional neural networks

(CNNs) have achieved impressive performance in optical flow estimation. FlowNet [12] was the pioneering work in end-to-end deep optical flow estimation. Its key component is a so-called correlation layer, which can provide explicit matching capabilities. Later methods, PWC-Net [13] and LiteFlowNet [14] introduced the popular coarse-to-fine architecture, which provides a good trade-off between optical flow accuracy and computation efficiency. Meanwhile, IRR-PWCNet [15] demonstrates that occlusion prediction integrated into optical flow estimation can effectively enhance the optical flow estimation accuracy.

Although the aforementioned supervised optical flow estimation methods perform impressively, they generally require a large amount of optical flow ground truth to learn the best solution. Acquiring such ground truth, especially for real-world datasets, is extremely time-consuming and labor-intensive, making these supervised approaches difficult to apply in real-world applications. For these reasons, unsupervised learning has recently become the preferred technique for such applications. For instance, DSTFlow [16] employs a photometric loss and a smooth loss in CNN training, which are similar to the global energy used in traditional methods. Additionally, some methods, such as UnFlow [17], DDFlow [18] and SelFlow [19] integrate occlusion reasoning into unsupervised optical flow estimation frameworks to further improve their accuracy. However, such approaches are typically computationally intensive, and they are difficult to embed in a drone.

2.2 Object Detection

Discovering objects and their locations in images is still a challenging problem in computer vision. Due to their promising results, CNNs have emerged as a powerful tool for object detection. The modern deep object detection algorithms can be grouped into two main types: a) anchor-based and b) anchor-free.

Anchor-based methods predict bounding boxes based on initial guesses. According to the pipelines and primary proposal sources, they can be further categorized as either one-stage or two-stage methods. The former make predictions directly from hand-crafted anchors. For example, RetinaNet [20] employs a feature pyramid network (FPN) to produce dense predictions at multiple scales. On the other hand, the two-stage methods make predictions using the proposals produced by a one-stage detector. For instance, Fast-RCNN [21] and Faster-RCNN [7] perform cropping and resizing on images or feature maps, according to the bounding box proposals. The RCNN branch in Faster-RCNN utilizes a field of view (FOV), that is larger than the bounding box proposals, so as to extract regions of interest (RoIs) directly from the feature maps.

Anchor-free methods usually do not rely on human-designed region proposals to bootstrap the detection process. For example, CornerNet [22] translates the object detection problem into a keypoint detection and matching problem, where specially-designed pooling layers construct biased receptive fields for corner point detection. CenterNet [23], which is based on CornerNet [22], utilizes two customized modules: a) cascade corner pooling and b) center pooling, to enrich information collected by both the top-left and bottom-right corners. It detects each object as a triplet, rather than a pair, of keypoints.

In recent years, incorporating additional visual information, such as semantic predictions, into object classification is becoming an increasingly ubiquitous part of object detection. Since MCs can be easily distinguished from optical flow images, we incorporate the latter into our framework to improve IPC candidate detection.

2.3 VSLAM

Traditional VSLAM approaches leverage visual features and the geometric relations between multiple views of a 3D scene (typically known as multi-view geometry) to estimate camera poses and construct/update a map of the 3D scene. The state-of-the-art VSLAM approaches are classified as either indirect [24–26] or direct [27–29]. Both types extract visual features from images and associate them with descriptors. However, the indirect methods sample corners and associate them with higher dimensional descriptors, while the direct methods typically sample pixels with a relatively large local intensity gradient and associate them with a patch of pixels surrounding their sampled location. Furthermore, these two types of methods typically minimize different objective functions: the indirect methods resort to geometric residuals, whereas the direct methods resort to photometric residuals.

In order to combine the advantages of these two types of methods, Froster *et al.* [30] proposed semi-direct visual odometry (SVO), which tracks camera poses via sparse image alignment and utilizes hierarchical bundle adjustment (BA) as the back-end to optimize the geometry structure and camera motion. Furthermore, many researchers have integrated other computer vision tasks, such as 2D object detection [31–33], instance segmentation [34,35] and flow/depth prediction [36,37], into their SLAM systems, so as to address the problem of the existence of dynamic objects, by exploiting high-level semantic information. For example, Huang *et al.* [32] proposed ClusterVO, which uses a multi-level probabilistic association scheme to both track low-level visual features and realize high-level object detection. Moreover, Yang *et al.* [31] introduced CubeSLAM, which performs single image 3D cuboid object detection, together with multi-view object SLAM.

3 ATG-PVD Framework

3.1 SwiftFlow

Since our proposed SwiftFlow network is based on the pipeline of PWC-Net [13], we first provide readers with some preliminaries about the latter. In PWC-Net [13], feature maps are first extracted from video frames using a Siamese pyramid network. Then, the feature map x_{t+1}^l of the $(t+1)$-th video frame at level l is aligned with the feature map x_t^l of the t-th video frame at level l

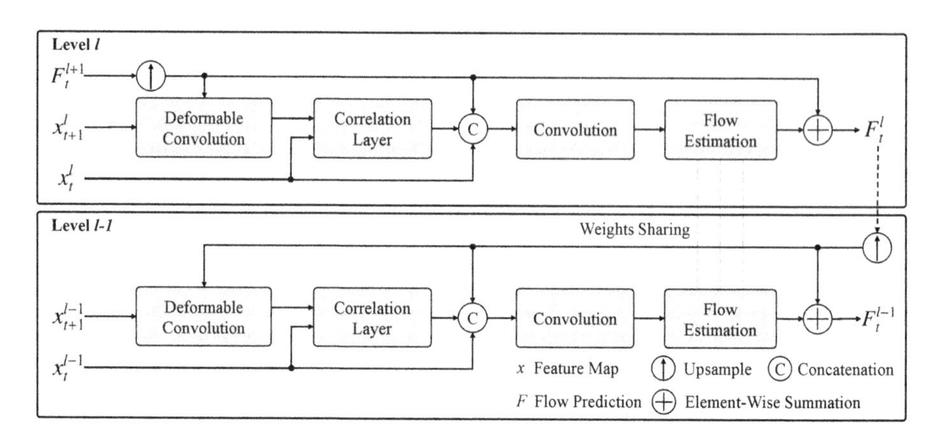

Fig. 2. The decoder architecture of our proposed SwiftFlow. The pipeline of two adjacent levels in the decoder are displayed for simplicity.

via a warping operation based on the upsampled flow prediction F_t^{l+1} at level $l+1$. A correlation layer is then employed to compute the cost volume, which is subsequently concatenated with x_t^l as well as the upsampled flow prediction F_t^{l+1} at level $l+1$. Finally, the flow residual, predicted by the flow estimation module, is combined with the upsampled flow prediction F_t^{l+1} at level $l+1$ using an element-wise summation to generate the flow prediction F_t^l at level l. We iterate this process and obtain the flow predictions at different scales.

SwiftFlow improves on PWC-Net [13] in terms of computational efficiency, so that it can perform in real time on a drone. The decoder in PWC-Net [13] has too many parameters, so we make three major modifications to the decoder architecture (see Fig. 2) to minimize the model size and improve accuracy. As the decoder in PWC-Net [13] employs a dense connection scheme in each pyramid level, making the network computationally intensive, SwiftFlow establishes connections only between two adjacent levels, which can reduce the number of network parameters by 50%. Furthermore, the optical flow estimation modules at different pyramid levels of PWC-Net [13] have different learnable weights to estimate optical flow residuals. Considering that the optical flow estimation modules at different levels have the same functionality and the optical flow residuals at different levels have similar value ranges, we believe sharing the weights of optical flow estimation modules at all pyramid levels can be a more effective and efficient strategy. We also add an additional convolutional layer before the optical flow estimation module at each level for feature map alignment. Moreover, we notice that the warping operation can induce ambiguity to occluded areas, which breaks correlation layer symmetry. We propose to add an asymmetric layer before the correlation layer to alleviate this problem and improve optical flow estimation accuracy. Therefore, we replace the warping operation with a deformable convolutional layer [38], as shown in Fig. 2.

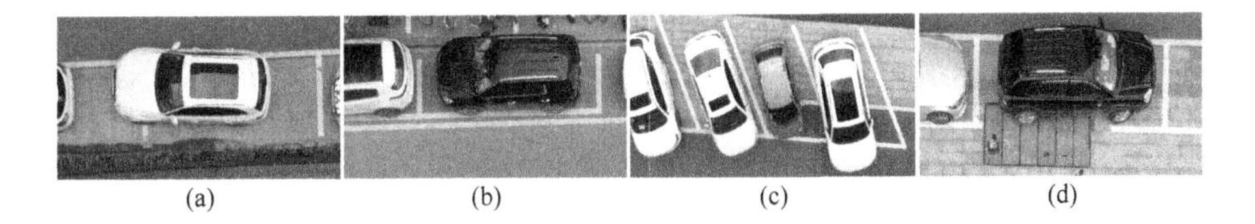

(a) (b) (c) (d)

Fig. 3. Challenging cases for parked car detection and classification.

Referring to the commonly applied unsupervised training strategy, we train SwiftFlow by minimizing the following weighted sum of losses:

$$L = \lambda_{\text{photo}} \cdot L_{\text{photo}} + \lambda_{\text{smooth}} \cdot L_{\text{smooth}} + \lambda_{\text{self}} \cdot L_{\text{self}}, \tag{1}$$

where L_{photo} is the photometric loss that considers an occlusion-aware mask [39], L_{smooth} is the smoothness regularization [40], and L_{self} is the self-supervision Charbonnier loss [18]. Following the instructions in [41], we set $\lambda_{\text{photo}} = 1$ and $\lambda_{\text{smooth}} = 2$ in our experiments. Moreover, we use $\lambda_{\text{self}} = 0$ for the first 50% of training steps, and increase it to 0.3 linearly for the next 10% of training steps, after which it stays at a constant value.

3.2 Flow-RCNN

Given an RGB video frame and its corresponding estimated optical flow, the proposed Flow-RCNN detects cars in the video frame and classifies them into MCs, LPCs, and IPC candidates.

Judging whether a car is legally parked is very challenging. Intuitively, we can resort to the parking spot delimitation lines, which are typically painted in white. However, in real-world environments, methods that rely solely on the parking spot information may fail. For instance, in Fig. 3(a), the white car is not parked entirely within the designated parking spot; in Fig. 3(b), only parts of the white car and parking spot appear; and in Fig. 3(c) and Fig. 3(d), the parking spots are not enclosed. Moreover, parking spots are not always bounded by rectangular line markings, as illustrated in Fig. 3(c). It is challenging to design a rule-guided method to solve for these cases, even with perfectly labeled cars and parking spots. Furthermore, various tall objects, such as light poles and trees, often present salient optical flow estimations. In this case, the methods that rely entirely on optical flow information can wrongly characterize an IPC/LPC as an MC. Therefore, an end-to-end, optical flow-guided, and detect-and-classify architecture for IPC candidate detection provides a better alternative.

The architecture of our proposed Flow-RCNN is illustrated in Fig. 4. It incorporates the optical flow information, obtained by SwiftFlow in Sect. 3.1, into the conventional Faster-RCNN [7] architecture for IPC candidate detection, and it outputs the position and category (MC, LPC or IPC candidate) of each car in the video frame in an end-to-end manner. The RGB video frame is first passed through a backbone CNN to produce multi-scale feature maps y^i. The features

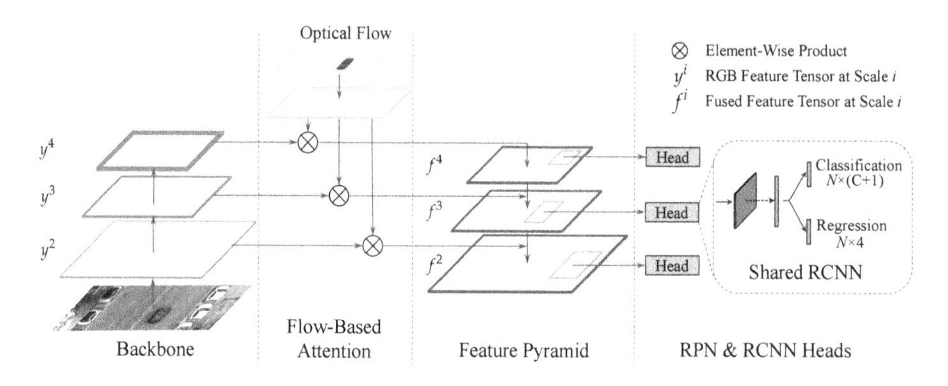

Fig. 4. Flow-RCNN architecture. The optical flow image, obtained from SwiftFlow, is fed into multiple convolutional layers. The optical flow features then dynamically weigh each element in the multi-scale feature maps extracted from the RGB image.

extracted from the optical flow image then dynamically weigh the activation of each element in the multi-scale feature maps y^i, which enables the detector to focus more on MCs. We then fuse the multi-scale feature maps to produce a feature pyramid for the subsequent region proposal network (RPN) and RCNN heads [7]. Since our dataset is highly imbalanced (see Fig. 7), *i.e.*, most vehicles are regarded as IPC candidates or IPCs, we apply focal loss [20] to mitigate the class imbalance problem in the classification stage.

3.3 Mapping, Re-localization and Re-identification

Given RGB images and the corresponding detected IPC candidates, our next target is to build a 3D map, investigate each IPC candidate and mark it in the map. To this end, we develop a mapping, re-localization and re-identification module, as illustrated in Fig. 5, on top of ORB-SLAM2 [42].

Our proposed system applies a suspect-and-investigate scheme to mark IPCs in 3D. In the suspicion phase, we leverage ORB-SLAM2 [42] to build a 3D localizable map and mark the detected IPC candidates in the map. Given an RGB image containing detected IPC candidates, the system first extracts ORB [43] features $\{\mathbf{u}_0, \ldots, \mathbf{u}_t\}$ and associates them with 2D bounding boxes $\{\mathcal{B}_0^{2D}, \ldots, \mathcal{B}_h^{2D}\}$. We explicitly exclude the ORB features extracted from MCs in the subsequent procedures, *i.e.*, tracking and mapping. The rest of the features are then matched with the 3D keypoints $\{\mathbf{x}_0, \ldots, \mathbf{x}_m\}$ in the map. With these 3D-2D correspondences $\mathcal{K} \doteq \{(i_k, j_k)\}_{k=1:N}$, the current camera pose $\mathbf{T} = [\mathbf{R}, \mathbf{t}]$ is estimated in a perspective-n-point (PnP) scheme by minimizing the reprojection error as follows [42]:

$$\mathbf{R}^*, \mathbf{t}^* = \arg\min_{\mathbf{R},\mathbf{t}} \sum_{(i,j)\in\mathcal{K}} \|\mathbf{u}_i - \pi(\mathbf{R}\mathbf{x}_j - \mathbf{t})\|, \quad (2)$$

where $\pi(\cdot)$ is the camera projection function. After solving the camera pose, the inlier correspondences $\mathcal{K}^* \doteq \{(i_k, j_k)\}_{k=1:N'}$ can be determined via their

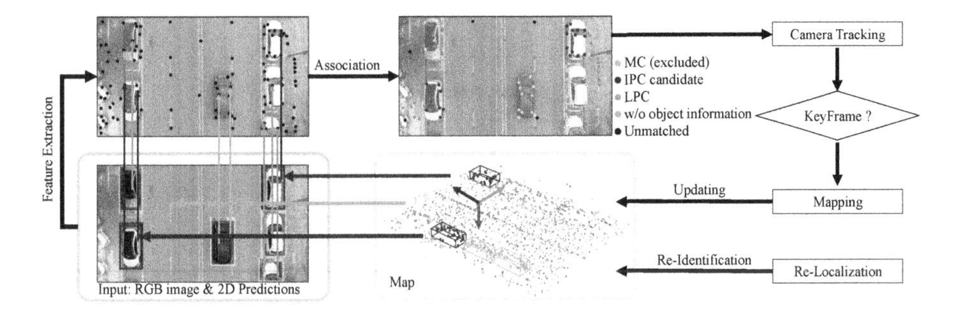

Fig. 5. The pipeline of our mapping, re-localization and re-identification module.

reprojection errors. Then we attempt to associate 2D bounding boxes in the current frame with candidates in the map. A pair of 3D and 2D bounding boxes $(\mathcal{B}_i^{3D}, \mathcal{B}_j^{2D})$ is associated if $|\mathcal{K}_{ij}| > \delta_{\text{obj}}$, where $\mathcal{K}_{ij}$ is a subset of $\mathcal{K}$, $(\mathbf{u}_i, \mathbf{x}_j)$ with $(i, j) \in \mathcal{K}^*$ is a pair of 2D/3D keypoints belonging to a pair of 2D/3D bounding boxes respectively and δ_{obj} is the threshold. In the mapping module, the system triangulates 2D feature correspondences into 3D keypoints, which are assigned with their corresponding 3D bounding box information. Then, it jointly optimizes the camera poses of keyframes $\{\mathbf{T}_0, \ldots, \mathbf{T}_n\}$ and the 3D keypoint positions $\{\mathbf{x}_0, \ldots, \mathbf{x}_m\}$. We consider the 3D bounding boxes in the suspicion phase as IPC candidates and mark them in the map. In the investigation phase, the system detects loop closure to re-localize the drone in the pre-built map. After the drone is successfully re-localized, we further verify existing IPC candidates. In the re-localization stage, if sufficient semantic keypoints belonging to a candidate $\mathcal{B}_i^{3D}$ are associated with a detected vehicle $\mathcal{B}_j^{2D}$ in the current frame, we re-identify the candidate as an IPC and mark it in the map. The proposed solution does not take into account that the local traffic law enforcement officers already have 2D street maps with labeled parking spots, but the drone map can be registered with such 2D street maps to greatly improve IPC detection.

4 Experiments

4.1 Experimental Setup

Our proposed PVD system is embedded in an ATG-R680 drone[1] (see Fig. 6), controlled by a Pixhawk 4[2] advanced autopilot. The maximum take-off weight of the drone is 5.6 kg. We utilize an Argus zoom pot[3] microminiature tri-axis gimbal camera to capture images with a resolution of 2160 × 3840 pixels at 25 fps. The captured images are then processed by an NVIDIA Jetson TX2

[1] atg-itech.com.
[2] docs.px4.io/v1.9.0/en/flight_controller/pixhawk4.html.
[3] topxgun.com/en/product-argus.html.

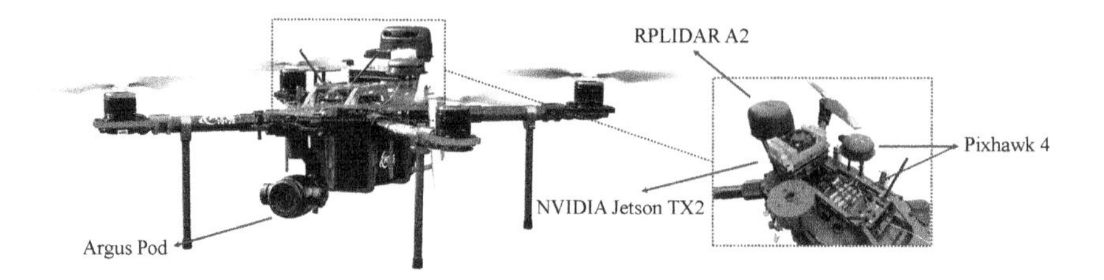

Fig. 6. Experimental setup.

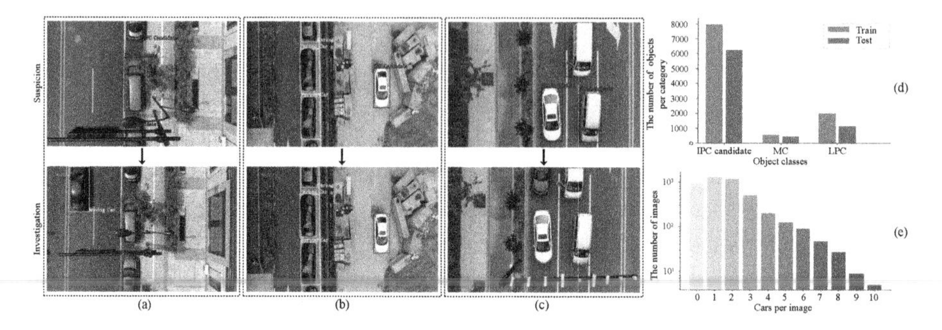

Fig. 7. Our created ATG-PVD dataset: (a)–(c) the images on the first row are used in the suspicion phase, while the images on the second row are used in the investigation phase; (d) and (e) the statistical analysis of the dataset.

GPU[4], which has an 8 GB LPDDR4 memory and 256 CUDA cores, for IPC detection. Furthermore, we also equip our drone with an RPLIDAR A2[5], which can perform 360° omnidirectional laser range scanning.

4.2 ATG-PVD Dataset

Using the aforementioned experimental setup, we created a large-scale real-world dataset, named the ATG-PVD dataset, for parking violation detection. Our dataset is publicly available at sites.google.com/view/atg-pvd for research purposes. The ATG-PVD dataset contains seven sequences (resolution: 2160×3840 pixels) and the corresponding 2D bounding box annotations for car detection and classification. The ground truth used in the suspicion phase has three classes: a) IPC candidates, b) MCs and c) LPCs, while in the investigation phase, the IPC ground truth is also provided. Examples of the images used in the suspicion and investigation phases are shown in Fig. 7(a)–(c).

In our experiments, we divide our ATG-PVD dataset into a training set and a testing set, which respectively contains 4924 and 4398 images. The statistics for these two sets are shown in Fig. 7(d) and (e), where it can be observed that there are more IPC candidates or IPCs than MCs and LPCs. Additionally, most

[4] developer.nvidia.com/embedded/jetson-tx2.
[5] slamtec.com/en/Lidar/A2.

Table 1. Ablation study of our SwiftFlow on the KITTI flow 2015 [44] training dataset. Best results are shown in bold font.

Backbone	Reduce Dense	Shared Weights	Deformable Convolution	F1-all (%)	# Params (M)
PWC-Net	–	–	–	8.37	8.75
	✓	–	–	7.22	5.26
	✓	✓	–	6.95	**2.18**
	✓	✓	✓	**6.51**	2.51

Table 2. The evaluation results on the KITTI flow benchmarks, where DDFlow [18], UnFlow [17], Flow2Stereo [45] and SelFlow [19] are the state-of-the-art self-supervised approaches. Best results are shown in bold font.

Approach	KITTI 2012		KITTI 2015		Runtime (s)
	Out-Noc (%)	Rank	F1-all (%)	Rank	
DDFlow [18]	4.57	60	14.29	91	0.06
UnFlow [17]	4.28	53	11.11	66	0.12
Flow2Stereo [45]	4.02	48	11.10	65	0.05
SelFlow [19]	3.32	34	8.42	51	0.09
SwiftFlow (Ours)	**2.64**	**24**	**7.23**	**35**	**0.03**

images contain fewer than five cars. Furthermore, our experiments are conducted on downsampled images with a resolution of 540×960 pixels. Sections 4.3, 4.4, and 4.5 respectively discuss the performances of SwiftFlow, Flow-RCNN and our PVD system in terms of both qualitative and quantitative experimental results.

4.3 Evaluation of SwiftFlow

Ablation Study. We conduct an ablation study to validate the effectiveness of SwiftFLow. The experimental results are presented in Table 1. We can see that, by removing dense connections between different levels, our approach can reduce many parameters, but still retain a similar optical flow estimation performance, compared with the PWC-Net [13] baseline. Moreover, sharing weights of flow estimation modules can yield a performance improvement with fewer parameters. Furthermore, thanks to deformable convolution, our proposed SwiftFlow achieves the best performance with only a few additional parameters.

Evaluation. Since our ATG-PVD dataset does not contain optical flow ground truth, we evaluate our proposed SwiftFlow on the KITTI flow 2012 [46] and 2015 [44] benchmarks. According to the online leaderboard of the KITTI flow benchmarks, as shown in Table 2, our SwiftFlow ranks 24th on the KITTI flow 2012 benchmark[6] and 35th on the KITTI flow 2015 benchmark[7], outperforming

[6] cvlibs.net/datasets/kitti/eval_stereo_flow.php?benchmark=flow.

[7] cvlibs.net/datasets/kitti/eval_scene_flow.php?benchmark=flow.

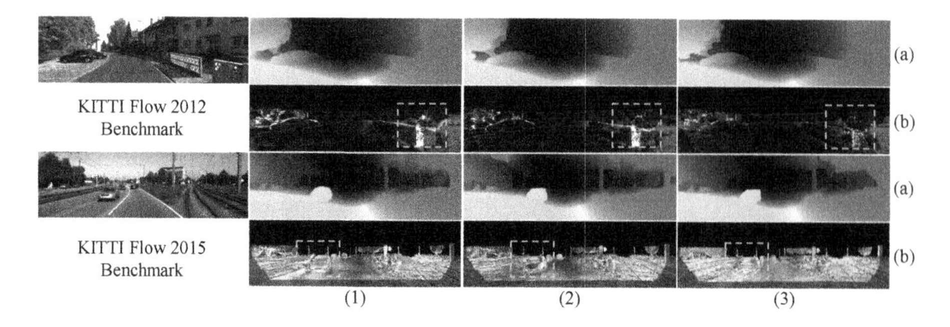

Fig. 8. Examples from the KITTI flow benchmarks, where rows (a) and (b) on columns (1)–(3) show the optical flow estimations and the corresponding error maps of (1) UnFlow [17], (2) SelFlow [19] and (3) our SwiftFlow, respectively. Significantly improved regions are highlighted with green dashed boxes. (Color figure online)

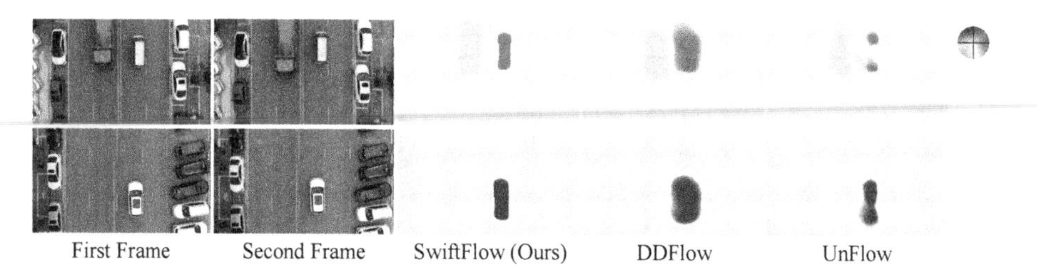

Fig. 9. Examples of the optical flow estimation results on our ATG-PVD dataset. Our proposed SwiftFlow is compared with DDFlow [18] and UnFlow [17].

all other state-of-the-art unsupervised optical flow estimation approaches, with a faster running speed (in real time) achieved in the mean time. Figure 8 presents examples from the KITTI flow benchmarks, where we can see that SwiftFlow yields more robust results than others. Furthermore, Fig. 9 shows optical flow estimation results on our ATG-PVD dataset, indicating that our proposed SwiftFlow performs much more accurately than both DDFlow [18] and UnFlow [17], another two well-known unsupervised optical flow estimation approaches, especially on the boundary of the MCs.

4.4 Evaluation of Flow-RCNN

In our experiments, we compute the mean average precision (mAP) over ten IoU thresholds between 0.50 and 0.95 (refer to [47] for more details) to quantitatively evaluate the performance of our proposed Flow-RCNN. It should be noted that IPCs are regarded as IPC candidates in both training and testing experiments, due to the fact that IPCs are re-identified as IPC candidates.

We compute mAP for all three categories (IPC candidate, MC and LPC) so as to comprehensively evaluate the performance of our proposed Flow-RCNN. The quantitative results are provided in Table 3, where it can be observed that

Table 3. Car detection mAP, where the best results are shown in bold font.

Method	Total	IPC candidate	MC	LPC
Faster-RCNN [7]	0.770	0.844	0.672	0.789
Flow-RCNN (ours)	**0.789**	**0.845**	**0.733**	**0.796**

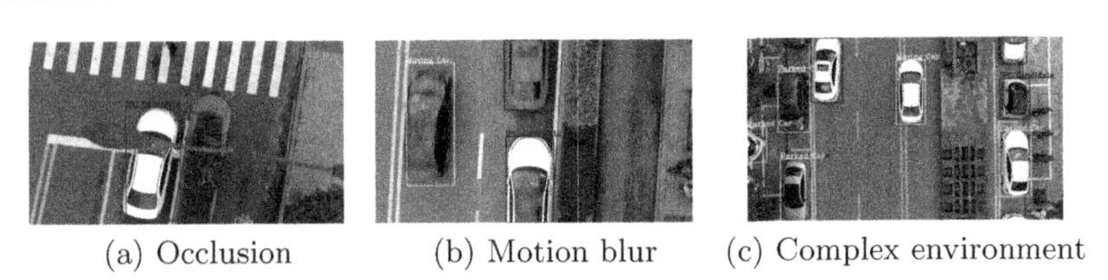

(a) Occlusion (b) Motion blur (c) Complex environment

Fig. 10. Examples of our Flow-RCNN results.

Flow-RCNN outperforms the baseline network Faster-RCNN [7] (especially for MC detection) in terms of both car detection and classification. It is rather astonishing that Faster-RCNN can still successfully detect many MCs from only RGB images, even without using optical flow information. We speculate that the baseline network might also consider the road textures around a car when inferring its category. For instance, an MC is typically at the center of a lane, and the road textures around it are similar, which can weaken the influence caused by motion blur problem.

Experimental results of our Flow-RCNN are given in Fig. 10, showing the robustness of our proposed approach. For example, in Fig. 10(a), the light pole, that occludes part of an IPC candidate, can produce a similar optical flow estimation to an MC. Fortunately, our Flow-RCNN which fuses both RGB and flow information can still detect the IPC candidate correctly. Furthermore, although it is hard to extract features from a blurred car image, it can be seen in Fig. 10(b) that our proposed approach can avoid such misdetections by leveraging additional optical flow information. Moreover, in complex environments, such as the case shown in Fig. 10(c), car with different categories can still be successfully detected and classified.

4.5 Evaluation of Parking Violation Detection

We also comprehensively evaluate the performance of the entire system for parking violation detection using our ATG-PVD dataset, and a precision of 91.7%, a recall of 94.9% and an F1-Score of 93.3% are achieved. An example of the detected IPCs in the map is illustrated in Fig. 11, where readers can observe that our proposed suspect-and-investigate system can detect parking violations effectively and efficiently.

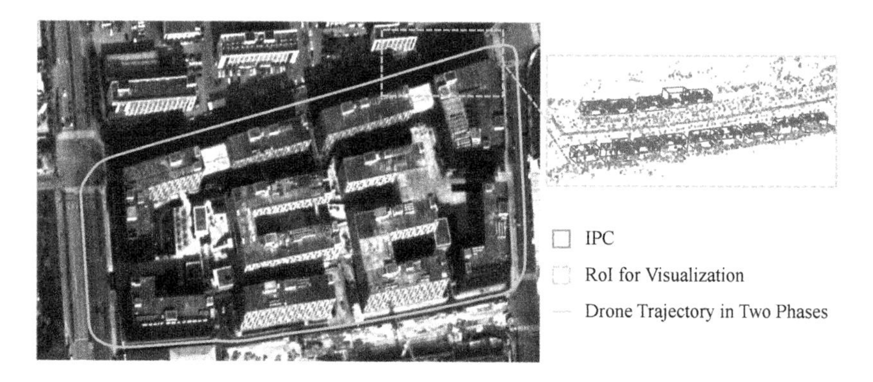

Fig. 11. An example of the detected IPCs in the map.

5 Conclusion

In this paper, we proposed a novel, robust and cost-effective parking violation detection system embedded in an ATG-R680 drone equipped with a TX2 GPU. Our system utilizes a so-called suspect-and-investigate framework, which consists of: 1) an unsupervised optical flow estimation network named SwiftFlow, 2) a novel flow-guided object detection network named Flow-RCNN, and 3) a drone re-localization and IPC re-identification module based on VSLAM. On the KITTI flow 2012 and 2015 benchmarks, our proposed SwiftFlow outperforms all other state-of-the-art unsupervised optical flow estimation approaches in terms of both speed (real-time performance was achieved) and accuracy. By incorporating the inferred optical flow information into our object detection framework, IPC candidates, MCs and LPCs can be effectively detected and classified, even in many challenging cases. In the investigation phase, our VSLAM module detects loop closure to re-localize the drone in the pre-built map. After the drone is successfully re-localized, we further re-identify whether an existing IPC candidate is an actual IPC. The experimental results both qualitatively and quantitatively demonstrate the effectiveness and robustness of our proposed parking violation detection system.

Acknowledgements. This work was supported by the National Natural Science Foundation of China, under grant No. U1713211, Collaborative Research Fund by the Research Grants Council Hong Kong, under Project No. C4063-18G, and the Research Grant Council of Hong Kong SAR Government, China, under Project No. 11210017, awarded to Prof. Ming Liu.

References

1. McKee, M., Stuckler, D.: If the world fails to protect the economy, covid-19 will damage health not just now but also in the future. Nat. Med. **26**(5), 640–642 (2020)
2. Hughes, T.: Poor, essential and on the bus: coronavirus is putting public transportation riders at risk, April 2020

3. Garcetti, E.: Mayor garcetti relaxes parking enforcement, March 2020
4. Zhou, Y., Liu, L., Shao, L., Mellor, M.: DAVE: a unified framework for fast vehicle detection and annotation. In: Leibe, B., Matas, J., Sebe, N., Welling, M. (eds.) ECCV 2016. LNCS, vol. 9906, pp. 278–293. Springer, Cham (2016). https://doi.org/10.1007/978-3-319-46475-6_18
5. Regazzoni, C.S., Cavallaro, A., Wu, Y., Konrad, J., Hampapur, A.: Video analytics for surveillance: theory and practice [from the guest editors]. IEEE Signal Process. Mag. **27**(5), 16–17 (2010)
6. Council, B.C.: Codes of practice for operation of CCTV Enforcement Cameras. Birmingham City Council (2013)
7. Ren, S., He, K., Girshick, R., Sun, J.: Faster R-CNN: towards real-time object detection with region proposal networks. IEEE Trans. Pattern Anal. Mach. Intell. **39**(6), 1137–1149 (2017)
8. Horn, B.K., Schunck, B.G.: Determining optical flow. In: Techniques and Applications of Image Understanding, vol. 281, pp. 319–331. International Society for Optics and Photonics (1981)
9. Mémin, E., Pérez, P.: Dense estimation and object-based segmentation of the optical flow with robust techniques. IEEE Trans. Image Process. **7**(5), 703–719 (1998)
10. Brox, T., Bruhn, A., Papenberg, N., Weickert, J.: High accuracy optical flow estimation based on a theory for warping. In: Pajdla, T., Matas, J. (eds.) ECCV 2004. LNCS, vol. 3024, pp. 25–36. Springer, Heidelberg (2004). https://doi.org/10.1007/978-3-540-24673-2_3
11. Zach, C., Pock, T., Bischof, H.: A duality based approach for realtime TV-L^1 optical flow. In: Hamprecht, F.A., Schnörr, C., Jähne, B. (eds.) DAGM 2007. LNCS, vol. 4713, pp. 214–223. Springer, Heidelberg (2007). https://doi.org/10.1007/978-3-540-74936-3_22
12. Dosovitskiy, A., et al.: Flownet: learning optical flow with convolutional networks. In: Proceedings of the IEEE International Conference on Computer Vision, pp. 2758–2766 (2015)
13. Sun, D., Yang, X., Liu, M.Y., Kautz, J.: PWC-NET: Cnns for optical flow using pyramid, warping, and cost volume. In: Proceedings of the IEEE Conference on Computer Vision and Pattern Recognition, pp. 8934–8943 (2018)
14. Hui, T.W., Tang, X., Change Loy, C.: Liteflownet: a lightweight convolutional neural network for optical flow estimation. In: Proceedings of the IEEE Conference on Computer Vision and Pattern Recognition, pp. 8981–8989 (2018)
15. Hur, J., Roth, S.: Iterative residual refinement for joint optical flow and occlusion estimation. In: Proceedings of the IEEE Conference on Computer Vision and Pattern Recognition, pp. 5754–5763 (2019)
16. Ren, Z., Yan, J., Ni, B., Liu, B., Yang, X., Zha, H.: Unsupervised deep learning for optical flow estimation. In: Thirty-First AAAI Conference on Artificial Intelligence (2017)
17. Meister, S., Hur, J., Roth, S.: Unflow: unsupervised learning of optical flow with a bidirectional census loss. In: Thirty-Second AAAI Conference on Artificial Intelligence (2018)
18. Liu, P., King, I., Lyu, M.R., Xu, J.: DDFlow: learning optical flow with unlabeled data distillation. In: Proceedings of the AAAI Conference on Artificial Intelligence, vol. 33, pp. 8770–8777 (2019)
19. Liu, P., Lyu, M., King, I., Xu, J.: SelFlow: self-supervised learning of optical flow. In: Proceedings of the IEEE Conference on Computer Vision and Pattern Recognition, pp. 4571–4580 (2019)

20. Lin, T., Goyal, P., Girshick, R., He, K., Dollár, P.: Focal loss for dense object detection. IEEE Trans. Pattern Anal. Mach. Intell. **42**(2), 318–327 (2020)
21. Girshick, R.: Fast R-CNN. In: International Conference on Computer Vision (ICCV) (2015)
22. Law, H., Deng, J.: Cornernet: detecting objects as paired keypoints. In: The European Conference on Computer Vision (ECCV), September 2018
23. Zhou, X., Wang, D., Krähenbühl, P.: Objects as points. In: arXiv preprint arXiv:1904.07850. (2019)
24. Klein, G., Murray, D.: Parallel tracking and mapping for small ar engel2018dso. In: Proceedings of the 2007 6th IEEE and ACM International Symposium on Mixed and Augmented Reality, pp. 1–10. IEEE Computer Society (2007)
25. Strasdat, H., Davison, A.J., Montiel, J.M., Konolige, K.: Double window optimisation for constant time visual slam. In: 2011 International Conference on Computer Vision, IEEE (2011) 2352–2359
26. Mur-Artal, R., Montiel, J.M.M., Tardos, J.D.: ORB-SLAM: a versatile and accurate monocular slam system. IEEE Trans. Rob. **31**(5), 1147–1163 (2015)
27. Newcombe, R.A., Lovegrove, S.J., Davison, A.J.: DTAM: dense tracking and mapping in real-time. In: IEEE 2011 International Conference on Computer Vision, pp. 2320–2327 (2011)
28. Engel, J., Schöps, T., Cremers, D.: LSD-SLAM: large-scale direct monocular SLAM. In: Fleet, D., Pajdla, T., Schiele, B., Tuytelaars, T. (eds.) ECCV 2014. LNCS, vol. 8690, pp. 834–849. Springer, Cham (2014). https://doi.org/10.1007/978-3-319-10605-2_54
29. Engel, J., Koltun, V., Cremers, D.: Direct sparse odometry. IEEE Trans. Pattern Anal. Mach. Intell. **40**(3), 611–625 (2018)
30. Forster, C., Pizzoli, M., Scaramuzza, D.: SVO: fast semi-direct monocular visual odometry. In: IEEE International Conference on Robotics and Automation (ICRA), pp. 15–22. IEEE (2014)
31. Yang, S., Scherer, S.: CubeSLAM: monocular 3-D object SLAM. IEEE Trans. Rob. **35**(4), 925–938 (2019)
32. Huang, J., Yang, S., Mu, T.J., Hu, S.M.: ClusterVO: clustering moving instances and estimating visual odometry for self and surroundings. In: Proceedings of the IEEE/CVF Conference on Computer Vision and Pattern Recognition, pp. 2168–2177 (2020)
33. Nicholson, L., Milford, M., Sünderhauf, N.: QuadricSLAM: dual quadrics from object detections as landmarks in object-oriented SLAM. IEEE Robot. Autom. Lett. **4**(1), 1–8 (2018)
34. Runz, M., Buffier, M., Agapito, L.: MaskFusion: real-time recognition, tracking and reconstruction of multiple moving objects. In: IEEE International Symposium on Mixed and Augmented Reality (ISMAR), IEEE 2018, pp. 10–20 (2018)
35. McCormac, J., Clark, R., Bloesch, M., Davison, A., Leutenegger, S.: Fusion++: Volumetric object-level slam. In, : international conference on 3D vision (3DV). IEEE **2018**, 32–41 (2018)
36. Zhang, T., Zhang, H., Li, Y., Nakamura, Y., Zhang, L.: FlowFusion: dynamic dense RGB-D slam based on optical flow. arXiv preprint arXiv:2003.05102 (2020)
37. Tateno, K., Tombari, F., Laina, I., Navab, N.: CNN-SLAM: real-time dense monocular slam with learned depth prediction. In: Proceedings of the IEEE Conference on Computer Vision and Pattern Recognition, pp. 6243–6252 (2017)
38. Dai, J., et al.: Deformable convolutional networks. In: Proceedings of the IEEE International Conference on Computer Vision, pp. 764–773 (2017)

39. Wang, Y., Yang, Y., Yang, Z., Zhao, L., Wang, P., Xu, W.: Occlusion aware unsupervised learning of optical flow. In: Proceedings of the IEEE Conference on Computer Vision and Pattern Recognition, pp. 4884–4893 (2018)
40. Tomasi, C., Manduchi, R.: Bilateral filtering for gray and color images. In: Sixth International Conference on Computer Vision (IEEE Cat. No. 98CH36271), pp. 839–846. IEEE (1998)
41. Jonschkowski, R., Stone, A., Barron, J.T., Gordon, A., Konolige, K., Angelova, A.: What matters in unsupervised optical flow. arXiv preprint arXiv:2006.04902 (2020)
42. Mur-Artal, R., Tardós, J.D.: ORB-SLAM2: an open-source slam system for monocular, stereo, and RGB-D cameras. IEEE Trans. Rob. **33**(5), 1255–1262 (2017)
43. Rublee, E., Rabaud, V., Konolige, K., Bradski, G.: Orb: an efficient alternative to sift or surf. In: International Conference on Computer Vision. IEEE **2011**, 2564–2571 (2011)
44. Menze, M., Heipke, C., Geiger, A.: Joint 3d estimation of vehicles and scene flow. In: ISPRS Workshop on Image Sequence Analysis (ISA) (2015)
45. Liu, P., King, I., Lyu, M.R., Xu, J.: Flow2Stereo: effective self-supervised learning of optical flow and stereo matching. In: Proceedings of the IEEE/CVF Conference on Computer Vision and Pattern Recognition, pp. 6648–6657 (2020)
46. Geiger, A., Lenz, P., Urtasun, R.: Are we ready for autonomous driving? the kitti vision benchmark suite. In: Conference on Computer Vision and Pattern Recognition (CVPR) (2012)
47. Lin, T.: Microsoft COCO: common objects in context. CoRR abs/1405.0312 (2014)

Next-Best View Policy for 3D Reconstruction

Daryl Peralta(✉), Joel Casimiro, Aldrin Michael Nilles, Justine Aletta Aguilar, Rowel Atienza, and Rhandley Cajote

Electrical and Electronics Engineering Institute, University of the Philippines, Quezon City, Philippines
{daryl.peralta,joel.casimiro,aldrin.michael.nilles,justine.aguilar, rowel,rhandley.cajote}@eee.upd.edu.ph

Abstract. Manually selecting viewpoints or using commonly available flight planners like circular path for large-scale 3D reconstruction using drones often results in incomplete 3D models. Recent works have relied on hand-engineered heuristics such as information gain to select the Next-Best Views. In this work, we present a learning-based algorithm called **Scan-RL** to learn a Next-Best View (NBV) Policy. To train and evaluate the agent, we created **Houses3K**, a dataset of 3D house models. Our experiments show that using **Scan-RL**, the agent can scan houses with fewer number of steps and a shorter distance compared to our baseline circular path. Experimental results also demonstrate that a single NBV policy can be used to scan multiple houses including those that were not seen during training. The link to **Scan-RL**'s code is available at https://github.com/darylperalta/ScanRL and **Houses3K** dataset can be found at https://github.com/darylperalta/Houses3K.

Keywords: 3d reconstruction · View planning · Reinforcement learning · 3d model dataset

1 Introduction

In recent years, there is an increased demand in 3D model applications including autonomous navigation, virtual and augmented reality, and 3D printing. Of particular interest in this work are 3D models of large infrastructure such as buildings and houses which can be used for construction monitoring, disaster risk management, and cultural heritage conservation.

Common methods in creating large 3D scenes include the use of color or depth images. Image-based methods use Structure from Motion (SfM) algorithms [20,23] that simultaneously estimate the camera poses and 3D structure from images. Depth-based methods such as [5,13] fuse depth images from different

Electronic supplementary material The online version of this chapter (https://doi.org/10.1007/978-3-030-66823-5_33) contains supplementary material, which is available to authorized users.

A. Bartoli and A. Fusiello (Eds.): ECCV 2020 Workshops, LNCS 12538, pp. 558–573, 2020.
https://doi.org/10.1007/978-3-030-66823-5_33

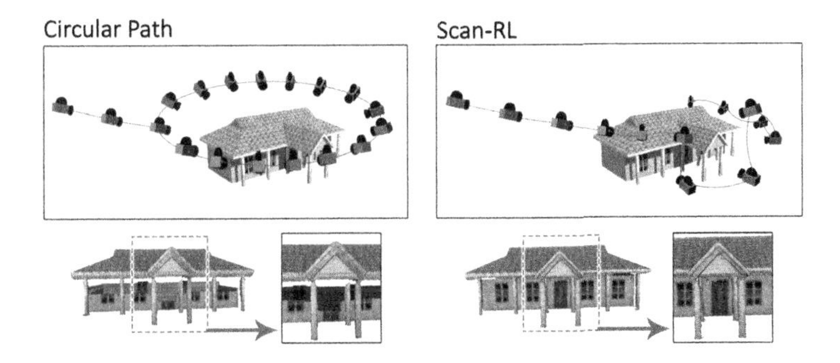

Fig. 1. *Upper left*: Viewpoints using a traditional circular path. *Lower left*: Output 3D model using circular path with 87% Surface Coverage. *Upper right*: Viewpoints using *Scan-RL* which was able to capture occluded regions under the roof. *Lower right*: Output 3D model using *Scan-RL* with 97% Surface Coverage.

sensor positions to create the 3D model. These algorithms rely heavily on the quality of the viewpoints that were used. Lack of data in some parts of an object creates holes in the 3D reconstruction. A solution to this is to add more images from different viewpoints. However, using more images becomes computationally expensive and results to longer processing time [9].

In doing large-scale 3D reconstruction with drones, a pilot normally sets way-points manually or uses commonly available planners like circular path. However, these methods often result in incomplete or low quality 3D models due to lack of data in occluded areas. Multiple flight missions are needed to complete a 3D model. This iterative process is time consuming and prohibitive due to the limited flight time of drones.

The problem of minimizing the views required to cover an object for 3D reconstruction is known as the View Planning Problem [14]. This problem is also addressed by selecting the Next-Best View (NBV) which is the next sensor position that maximizes the information gain. These problems are widely studied because of their importance not only for 3D reconstruction but also for inspection tasks, surveillance, and mapping.

In this paper, an algorithm is proposed to answer the question: can an agent learn how to scan a house efficiently by determining the NBV from monocular images? Humans can do this task by looking at the house since we can identify occluded parts such as under the roof. We explore this idea and propose *Scan-RL*, a learning-based algorithm to learn an NBV policy for 3D reconstruction. We cast the problem of NBV planning to a reinforcement learning setting. Unlike other methods, that rely on manually crafted criteria to select NBV, *Scan-RL* trains a policy to choose NBV. Furthermore, *Scan-RL* only needs images in making its decisions and does not need to track the current 3D model. Depth images and the reconstructed 3D model are only used during training.

Training *Scan-RL* requires a dataset of textured 3D models of houses or buildings. To the best of our knowledge, there is no sufficiently large dataset of watertight 3D models of big structures suitable for our experiments. Most datasets focus on the interior like House3D [29]. Thus, we created *Houses3K*, a dataset made of 3,000 watertight and textured 3D house models. We present a modular approach to creating such dataset and a texture quality control process to ensure the quality of the models. While *Scan-RL* was the motivation behind *Houses3K*, it could also be used for other applications where 3D house models may be needed such as training geometric deep learning algorithms and creating realistic synthetic scenes for autonomous navigation of drones.

Experiments were done in a synthetic environment to evaluate the algorithm. Results show that using *Scan-RL*, an NBV policy can be trained to scan a house resulting in an optimized path. We show that the path using the NBV Policy is shorter than the commonly used circular path. A comparison of the circular path and the resulting path from *Scan-RL* is shown in Fig. 1. Further experiments also show that a single policy can learn to scan multiple houses and be applied to houses not seen during training.

To summarize, our paper's main contributions are:

1) *Scan-RL*, a learning-based algorithm to learn an NBV policy for 3D reconstruction for scanning in an optimal path.
2) *Houses3K*, a dataset of 3D house models that can be used for future works including view planning, geometric deep learning, and aerial robotics.

2 Related Work

The challenge of view planning and active vision is widely studied in the field of robotics and computer vision. A survey of early approaches for sensor planning was done by Tarabanis et al. [26]. Scott et al. [21] presented a survey of more recent works in view planning. Our work is also related to active vision which deals with actively positioning the sensors or cameras to improve the quality of perception [1,2].

Approaches on planning views for data acquisition may be divided into two groups. The first group are those that tackle the view planning by reducing the views required to cover an object from a set of candidate views. These include [10,14,22]. The second group are those from the robotics community which aim to select the NBV in terms of information gain. These works include [6,12,15,18,24,27]. Our work is closer to NBV algorithms. *Scan-RL* tries to train a policy that commands the drone where to position next given its current state to maximize improvement in the 3D reconstruction.

For view planning, Smith et al. [22] proposed heuristics to quantify the quality of candidate viewpoints. They also created a dataset and benchmark tool for path planning. Our method aims to learn a policy that will select the NBV for each step instead of simultaneous optimization of camera positions.

In NBV algorithms, a way to quantify information gain from each candidate viewpoint is needed. Delmerico et al. [7] presented a comparative study of existing volumetric information gain metrics. Isler et al. [12] proposed a way to quantify information gain from a candidate view using entropy contained in 3D voxels. Their algorithm chooses the candidate view with maximum information gain. Daudelin et al. [6] also quantifies entropy in voxels and proposed a new way of generating candidate poses based on the current object information. However, these works are limited to the resolution of voxels.

Surface-based methods were also proposed [15,24]. Kriegel et al. [15] proposed to use the mesh representation of the 3D model together with a probabilistic voxelspace. Mesh holes and boundaries were detected for the NBV algorithm. Song et al. [24] used truncated Signed Distance Field (TSDF), a surface representation. TSDF provides the information in improving the quality of the model based on confidence. Both algorithms track a volumetric representation for exploration and find boundaries between explored and unexplored voxels.

Wu et al. [28] used points to estimate the NBV. The Poisson field from point cloud scans was used to identify the low quality areas and compute the next best view to improve the 3D reconstruction. This approach aims for completeness and quality of the 3D reconstruction. Huang et al. [11] extended the work by introducing a fast MVS algorithm and applying the algorithm to drones.

All works presented previously relied on hand-engineered algorithms to compute the NBV. We aim to train a policy that will learn to select the NBV instead of manually proposing criteria for view quality.

Yang et al. [30] used deep learning to predict the 3D model given the images coupled with reinforcement learning for a view planner. Their main contribution was the view planner which selects views to improve the 3D model predictions. However, the network only predicts 3D models of small objects with simple shapes and low resolution. Also, both the training of the 3D prediction network and the reward for the view planner rely on ground truth models.

Choudhury et al. [3] proposed a learning-based approach to motion planning but not for 3D reconstruction. Kaba et al. [14] formulated a reinforcement learning approach to solve the view planning problem. However, their algorithm assumes a 3D model is available to minimize the number of views. Han et al. [8] used a deep reinforcement learning algorithm for view planning as part of scene completion task. However, the policy was trained to maximize the accuracy of the depth inpainting task and was not trained to optimize the path. They also used the 3D point cloud as state. *Scan-RL* aims to train a policy that optimizes the path using only monocular images as state.

3 Scan-RL: Learning a Next-Best View Policy

A block diagram of *Scan-RL* is shown in Fig. 2. The system is composed of two main components: the NBV policy π and the 3D reconstruction module. The NBV policy π selects the next pose to scan the target house based on images generated by the Unreal Engine[1] environment. The 3D reconstruction module reconstructs the 3D point cloud model of the target structure using the depth images collected. Rewards are then extracted from the output point cloud.

[1] https://www.unrealengine.com/en-US/feed.

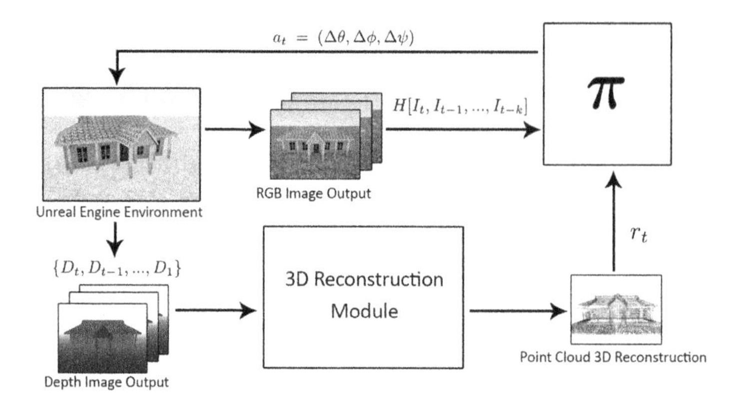

Fig. 2. Block diagram of *Scan-RL*. From the Unreal Engine environment, RGB images are rendered and used by the policy to generate the Next-Best View. Depth is rendered in the next view for the 3D Reconstruction module to create a point cloud 3D model. During training, rewards are extracted from the 3D model for the policy to learn scanning houses.

Scan-RL is modular. The NBV policy π can be any policy trained using any reinforcement learning algorithm. The 3D reconstruction module can be any 3D reconstruction algorithm that may utilize the monocular images, depth images, and camera poses from the synthetic scene. To make training feasible, a fast 3D reconstruction algorithm is needed. We utilized depth fusion algorithm based on Truncated Signed Distance Function (TSDF) representation introduced by Curless et al. [4]. TSDF is being used in recent depth-based reconstruction algorithms [5,13]. For the reinforcement learning algorithm, we used Deep Q-Network (DQN) [17] and Deep Deterministic Policy Gradient (DDPG) [16] which had success in high dimensional state space.

Unreal Engine, a game engine, was used to create realistic synthetic scenes for our experiments. Unreal Engine allows controlling the camera positions and rendering color and depth images. We created our own synthetic game scenes where 3D house models from *Houses3K* were used as shown in Fig. 2. To make the scene more realistic and feature-rich, we applied a grass texture for the ground. We built on top of Gym-unrealcv [31] and UnrealCV [19] to implement the reinforcement learning algorithm and control the agent position in our synthetic environment.

We cast the NBV selection process as a Markov Decision Process (MDP). MDP is made of (S, A, T, R, γ) where S is the state, A is the action, T is the transition probability, R is the reward and γ is the discount factor. We approximate an MDP process by making the state be the image captured by the camera at a viewpoint and action be the change in pose. We then applied reinforcement learning to find an optimal policy which we call NBV policy to select the next views.

3.1 State Space

In reinforcement learning, the state contains the necessary information for the agent to make its decision. Normally, robots and drones have multiple sensors like GPS, IMU, depth sensors, and cameras. To make our algorithm not dependent on the robot platform, we only used monocular images as state since cameras are commonly found in most drones. We refrained from using the 3D model as part of the state because it will require more computational resources.

We define a preprocessing function H shown in Eq. 1. H generates the state vector by concatenating the current image I_t with previous k frames, converting the color images to grayscale, and resizing them to a smaller resolution.

$$s_t = H[I_t, I_{t-1}, ..., I_{t-k}] \tag{1}$$

3.2 Action Space

Change in relative camera pose was used as action because we want our method to be independent of the dynamics of the robot platform. This also separates the low level control making it applicable to any robot which can measure its relative camera pose.

Camera pose was parameterized to (θ, ϕ, ψ) [25]. θ is the azimuth angle, ϕ is the elevation angle and ψ is the distance from the object. This assumes that the object is at the origin. Using this parameterization, the camera pose is reduced to three degrees of freedom.

We then defined the discrete action space to be the increase or decrease in the camera poses resulting in six actions namely: increase θ, decrease θ, increase ϕ, decrease ϕ, increase ψ or decrease ψ. The resolution for the change in camera pose is $(\pm 45°, \pm 35°, \pm 25$ units). For the continuous action space, we used the same relative camera pose (θ, ϕ, ψ) but the change in angles is continuous resulting in $a_t = (\Delta\theta, \Delta\phi, \Delta\psi)$. Distance ψ is in Unreal Engine distance units.

3.3 Reward Function

The task to learn is selecting the NBV to completely scan an object. With this task, the reward should lead to a complete 3D model. To quantify completeness of the 3D model, we used surface coverage C_s from [12]. Surface coverage C_s given by Eq. 2 is the percentage of observed surface points N_{obs} over the total number of surface points N_{gt} in the ground truth model. N_{obs} are points in the ground truth model with a corresponding point in the output reconstruction whose distance is less than some threshold.

$$C_s = \frac{N_{obs}}{N_{gt}} \tag{2}$$

Reward r_t for each step is defined by Eq. 3. During steps where the terminal surface coverage is not achieved, reward is the change in surface coverage $k_c * \Delta C_s$ minus a negative penalty equal to -2 per step and a penalty proportional to distance ΔX. The scaled change in surface coverage $k_c * \Delta C_s$ is introduced for the agent to maximize the surface coverage while the negative penalties are added to minimize the number of steps and distance. A reward of 100 is given when terminal surface coverage is achieved to emphasize the completion of the scanning task. The constants k_c and k_x were set to 1 and 0.02 respectively.

$$r_t = \begin{cases} k_c * \Delta C_s - k_x * \Delta X - 2 & C_s \leq C_{s,terminal} \\ 100 & C_s > C_{s,terminal} \end{cases} \tag{3}$$

To efficiently compute the reward r_t and make training feasible, we needed a fast algorithm for the 3D reconstruction module. We implemented depth fusion algorithm [4] that generates a point cloud from all depth images $\{D_t, D_{t-1}, ..., D_1\}$ and camera poses. Depth-based algorithms are faster than image-based 3D reconstruction algorithms since depth information is already available.

3.4 Training

Deep Q-Network (DQN) [17] was implemented for the deep reinforcement learning algorithm in the discrete action space. In DQN, a deep neural network is trained to maximize the optimal action-value function Q^* expressed in Eq. 4. The optimal action-value function Q^* is the maximum sum of rewards with discount factor γ given state s_t and action a_t following a policy π.

$$Q^* = \max_{\pi} \mathbb{E}(r_t + \gamma r_{t+1} + \gamma^2 r_{t+2} + ...|s_t = s, a_t = a, \pi) \tag{4}$$

For the continuous action space, we implemented an actor-critic algorithm Deep Deterministic Policy Gradient (DDPG) in [16]. DDPG is composed of two neural networks namely actor and critic networks. Similar to DQN, the critic network in DDPG with parameters θ_w^Q is also trained to maximize Eq. 4. The actor network with parameters θ_w^μ predicts the action that maximizes the expected reward. It is trained by applying the sampled policy gradient $\nabla_{\theta^\mu} J$ for a minibatch of N transitions (s_j, a_j, r_j, s_{j+1}) expressed in Eq. 5.

$$\nabla_{\theta^\mu} J \approx \frac{1}{N} \sum_j \nabla_a Q(s, a|\theta_w^Q)|_{s=s_j, a=\mu(s_j)} \nabla_{\theta_w^\mu} \mu(s|\theta_w^\mu)|s_j \tag{5}$$

From the preprocessing function H, color images are converted to grayscale and are resized to 84×84. The preprocessing function H then concatenates the previous k frames which we set to 5 frames. This will result in a state vector of $84 \times 84 \times 6$.

At every step, depth and pose are extracted from the synthetic scene. These are used in the 3D reconstruction module. Surface coverage is then computed from the output point and the reward r_t is computed based on the reward function in Eq. 3. To compute the surface coverage, we generated the ground truth 3D point cloud by sampling 10,000 points from the ground truth mesh of *Houses3K*.

A house model is placed in the scene in each training episode. The episode terminates when the terminal surface coverage is achieved. When multiple models are being used during training, a model is drawn randomly from the dataset at the start of each episode. This can be done for the agent to learn features from different types of houses to perform the scanning task. Additional details about the training are presented in the supplementary material.

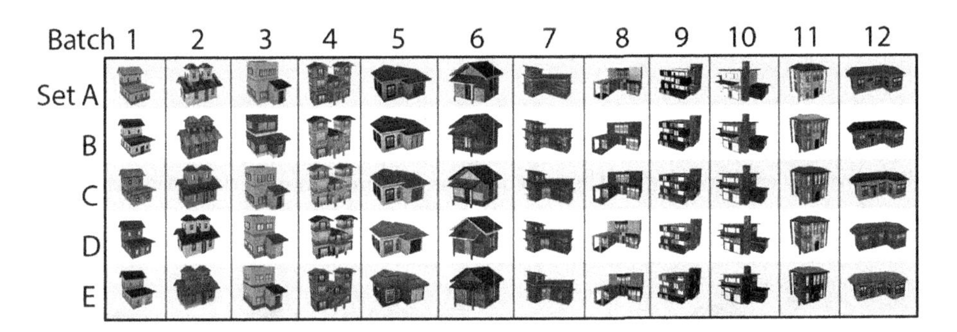

Fig. 3. Sample houses for each set of *Houses3K*. A dataset of 3000 3D house models grouped into 12 batches and 5 sets in each batch.

3.5 Houses3K

To train the policy, a dataset of textured 3D models of houses is needed. However, there is no suitable dataset that fits our need of 3D mesh focusing on the exterior of the house. Thus we created *Houses3K*, a dataset consisting of 3,000 watertight and textured 3D house models.

The dataset is divided into twelve batches, each containing 50 unique house geometries. For each batch, there are five different textures applied on the structures, multiplying the house count to 250 unique houses. Various architectural styles were adopted, from single-storey bungalow house type, up to a more contemporary and modern style. Figure 3 shows sample houses from *Houses3K*.

We present a modular approach to creating the dataset illustrated in Fig. 4. We created our own *style vocabulary* which consisted of different house structure types, roof shapes, windows, and door styles, and even paints/textures for the roof and walls. Based on the *style vocabulary*, modular 3D assets such as different styles of roofs, windows, doors, wall structures, and surface textures were created. These modular assets were then assembled in various combinations to make the different houses that comprise the dataset.

To ensure the quality of the textures, we implemented a texture quality control process illustrated in Fig. 5. Two representative models were sampled from each set and 108 images in a circular path with varying elevations were rendered at different viewpoints from these 3D models using Unreal Engine. These images were then used as input to an image-based 3D reconstruction

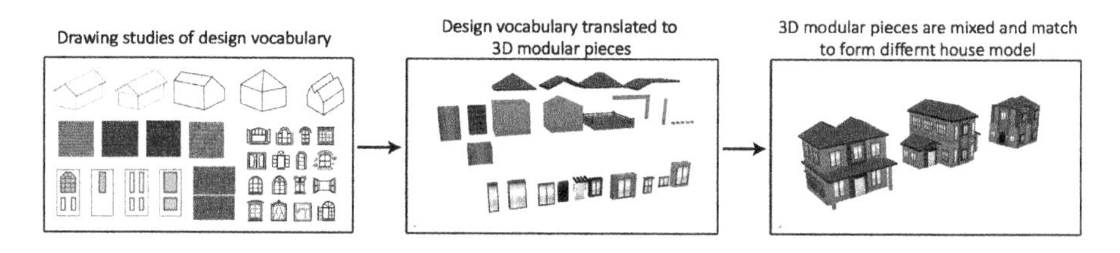

Fig. 4. Modular approach to creating the *Houses3K* dataset using 3D modular pieces.

system to generate a 3D model. If the model was not reconstructed properly as shown in the left side of Fig. 5, we redesigned the texture to make it feature rich. This process was done until the texture was detailed enough to reconstruct the model.

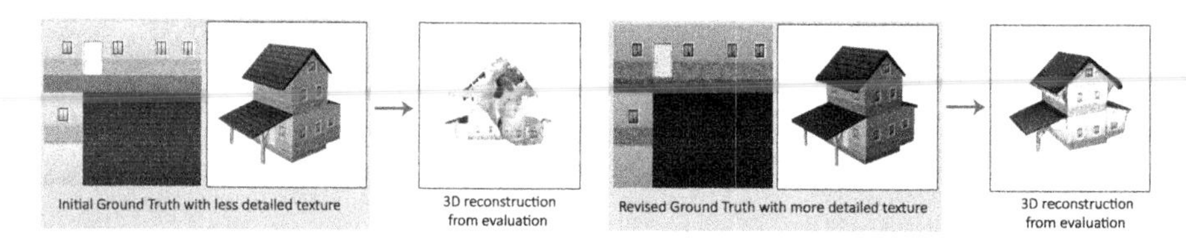

Fig. 5. Texture quality control process. Rendered images from *Houses3K* models were used as input to a multiple view image-based 3D reconstruction system to check if the models will be reconstructed.

4 Experiments

Three experiments were performed to evaluate *Scan-RL*. The first experiment was the *single house policy experiment* which aims to test if an agent can learn an NBV policy to scan a house efficiently using *Scan-RL*. The second experiment was the *multiple houses single policy transfer experiment* which aims to test if a single policy can learn to scan multiple houses and be transferred to houses not seen during training. The last experiment was on Stanford Bunny[2]. This was conducted to test if *Scan-RL* can be used for non-house objects and to compare our work with Isler et al. [12].

4.1 Single House Policy Experiment

We used the house model from batch 6 of *Houses3K* shown in Fig. 1 for this experiment. The house has self-occluded parts that will challenge the algorithm and will allow us to observe if the NBV policy can be used to efficiently scan

[2] Available from Stanford University Computer Graphics Lab.

houses. A terminal surface coverage of 96% was set during training. We implemented both discrete and continuous action space versions of *Scan-RL*.

For the discrete action space, the number of distance ψ levels were varied to two and three levels and azimuth θ was varied to 45.0° and 22.5° resolutions. Elevation ϕ was fixed to three levels. In the continuous action space, the maximum allowed distance from the origin was varied. The number of steps and the distance moved by the agent for the scanning task were measured and compared for all setups.

Fig. 6. *Left*: Output 3D model using circular path with 87% surface coverage from 17 steps. *Right*: Output 3D model using NBV Policy with 97% surface coverage from 13 steps.

We used the images and depth from the circular path and using the trained NBV policy from *Scan-RL* to create the 3D models for each. The viewpoints selected for both methods are illustrated in Fig. 1. Output 3D models are shown in Fig. 6. Notice in left of Fig. 6 that the output 3D model has a large hole under the roof resulting in only 87% surface coverage with 17 steps. In the right of Fig. 6, the 3D model shows that parts under the roof were reconstructed resulting in 97% surface coverage using only 13 steps. This occurred since the policy was able to focus on the occluded regions under the roof. This show that the NBV policy learned to maximize the surface coverage to create a more complete 3D reconstruction while minimizing the number of steps.

To further evaluate *Scan-RL*, three types of circular paths with varying elevations were implemented for the discrete action space. The first type named *Circ 1* path starts with the agent at the farthest distance level and at 45.0° elevation. The agent moves up to the highest elevation level, moves closer to the object which is 100 units from the origin, circulates around the object, and then moves to the middle and bottom elevation levels. *Circ 2* path starts with the same position, the agent then moves straight to the nearest allowed distance which is 100 units from the origin and circulates around the object as shown in Fig. 1. It then moves to the bottom then to the top elevation. *Circ 3* path is similar to *Circ 1* but starts with the bottom elevation. For the continuous action space setup, the range of action a_t is ($\pm 45.0°, \pm 35.0°, \pm 25$ units). A similar circular path baseline that chooses the maximum value in the action range was implemented for the continuous action space.

Table 1 shows results for the discrete action space with varying azimuth resolution and distance levels. In all setups, using *Scan-RL*'s NBV policy resulted to fewer steps and shorter distance compared to the baseline circular path. Table 2 shows results for the continuous action space for two different maximum distance allowed from the origin. In the continuous setups, *Scan-RL*'s NBV policy was able to optimize the steps but not the distance. This occurred because both are being optimized and it is possible to have fewer steps but longer distance. In Fig. 7, the cumulative reward of *Scan-RL* during training is shown. It can be observed that for all setups, the policy learned to maximize the reward throughout the training.

Table 1. Single House Policy Experiment in discrete action space. Number of steps and distance covered for 96% terminal coverage were compared with different baselines.

Distance	Azimuth	Steps				Distance covered (units)			
Levels	Resolution	Circ 1	Circ 2	Circ 3	Scan-RL	Circ 1	Circ 2	Circ 3	Scan-RL
2	45.0°	23	12	8	**7**	1087.30	662.87	510.62	**410.13**
2	22.5°	44	23	13	**12**	1162.93	702.12	492.60	**416.81**
3	45.0°	24	13	9	**8**	1097.54	687.87	520.85	**435.13**
3	22.5°	43	24	18	**15**	1082.13	727.12	641.38	**533.34**

Table 2. Single House Policy Experiment in continuous action space. Number of steps and distance covered for 96% terminal coverage were compared with different baselines.

Max Allowed	Steps				Distance covered (units)			
Distance (units)	Circ 1	Circ 2	Circ 3	Scan-RL	Circ 1	Circ 2	Circ 3	Scan-RL
125	23	12	8	**7**	1087.30	662.87	**510.62**	520.08
150	24	13	9	**7**	1097.54	687.87	**520.85**	550.80

These results show that the agent was able to learn an NBV policy to scan the house and achieve the terminal surface coverage while minimizing steps and distance travelled by the agent. The agent learned to move forward, move around the house, and go down for the occluded parts as shown in Fig. 1 resulting in an optimized path.

These results also show that it is indeed possible to scan a house efficiently using *Scan-RL* without manually crafting criteria for NBV. The trained NBV policy uses only images as state vector. It means that the policy learned all the necessary information including the concept of relative position from the images.

Being able to train an NBV policy that optimizes the path is important for large-scale 3D reconstruction using drones with limited flight. It also has potential applications to other tasks like inspection and exploration.

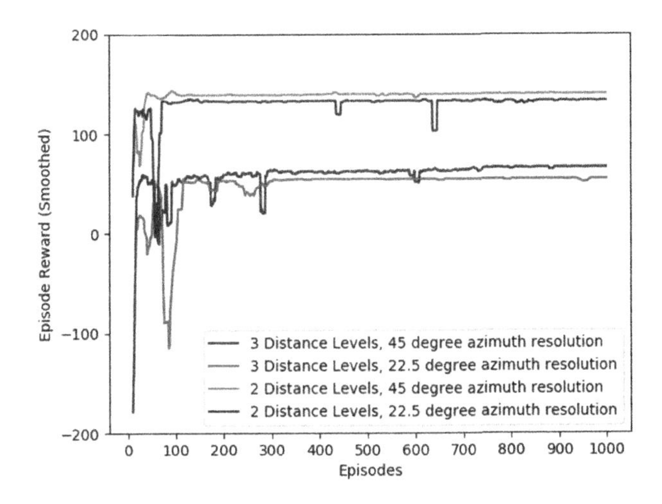

Fig. 7. *Scan-RL* performance during training for the different action space setups.

4.2 Multiple Houses Single Policy Experiment

In this experiment, we implemented the discrete action space version of *Scan-RL*. *Houses3K* models were split into train and test sets per batch. House models in the train set were used for training a single policy to scan multiple houses. Models in the test set were used to test if the trained NBV policy can scan the unseen test models. We implemented two kinds of split namely *random* and *geometry* split. Random split randomly splits the houses to 90% for training and 10% per batch. *Geometry* split also splits with the same ratio but makes sure that the geometry in the train set is not in the test set.

For each batch, a single policy was trained using *Scan-RL* discrete space. Hyperparameters were held constant for all batches. Solved house models in *Houses3K* are defined to be models where the terminal coverage was achieved within 50 steps. A solved model means that the agent was not stuck during the process. For each batch, we also ran the circular path baseline which we ran for 27 steps since this is the number of steps required to cover all the possible viewpoints at the closest distance level.

For Random Split, the terminal surface coverage for all batches were set to 96% except for batches 8 and 10 which were set to 87% due to the models' complex geometry. For Geometry Split, the terminal surface coverages for all batches were also 96% except for batches 7, 8 and 10 which were set to 87%. Table 3 presents the ratio of the number of solved houses using *Scan-RL* over the total number of houses used for each batch. In this experiment, we are interested in the number of houses that can be solved so the number of steps were not included. It is also expected that the circular path will have a higher score since it will cover all the views in the closest distance level.

Results using the train set show that *Scan-RL* can train a single NBV policy to scan multiple houses in *Houses3K* as seen in the high number of houses solved except for the more complex batches 7 and 9. Results from the test set show that for most batches, the NBV policy was able to scan the unseen houses. Failure cases were mainly caused by the complex geometries which may require more degrees of freedom in the camera pose.

Table 3. Ratio of solved houses using *Scan-RL* over the total number of houses used for each *Houses3K* batch.

	Random Split				Geometry Split			
	Train Set		Test Set		Train Set		Test Set	
Batch	Scan-RL	Circ	Scan-RL	Circ	Scan-RL	Circ	Scan-RL	Circ
1	217/225	225/225	23/25	24/25	218/225	224/225	25/25	25/25
2	224/225	225/225	25/25	25/25	224/225	225/225	24/25	25/25
3	182/225	186/225	18/25	22/25	203/225	190/225	16/25	20/25
4	204/225	206/225	24/25	24/25	193/225	210/225	17/25	20/25
5	222/225	225/225	25/25	25/25	214/225	225/225	25/25	25/25
6	213/225	215/225	25/25	25/25	204/225	220/225	20/25	20/25
7	76/210	137/210	10/20	13/20	197/210	190/210	17/20	20/20
8	183/225	189/225	18/25	21/25	131/225	185/225	16/25	25/25
9	155/225	191/225	13/25	19/25	145/225	190/225	11/25	20/25
10	194/225	214/225	14/25	20/25	214/225	211/225	20/25	25/25
11	191/225	225/225	18/25	25/25	220/225	225/225	25/25	25/25
12	195/225	189/225	18/25	18/25	184/225	190/225	19/25	25/25

4.3 Non-house Target Model Experiment

We replicated the setup of Isler et al. [12] in our synthetic environment. The main differences in the setups are that we used the depth rendered from Unreal Engine for the 3D reconstruction module and that we generated our own ground truth point cloud with 10,000 points from the ground truth mesh. To have the same set of viewpoints to the one used by [12], we used the 22.5° azimuth resolution of our discrete action space setup. We also set the depth sensor range such that we get the maximum surface coverage of 90%. Training was similar with the single house policy experiment including the hyperparameters.

The plot of surface coverage per reconstruction step on Stanford Bunny is shown in Fig. 8. Data from Isler et al. [12] were compared with the performance of *Scan-RL*'s trained NBV policy. Both methods were able to increase the surface coverage per step and converge to a high surface coverage. This experiment shows that *Scan-RL* is not only limited to houses.

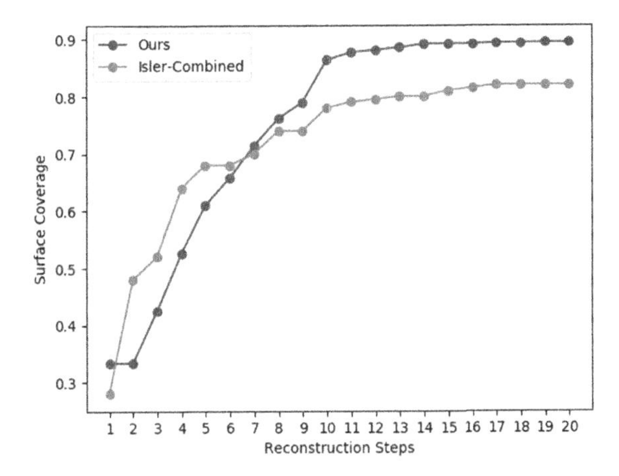

Fig. 8. Reconstruction of Stanford Bunny. Surface Coverage per step of our method is compared with Isler [12].

5 Conclusion

We presented *Scan-RL*, a learning-based algorithm to train an NBV policy for 3D reconstruction inspired by how humans scan an object. To train and evaluate *Scan-RL*, we created *Houses3K*, a dataset of 3,000 watertight and textured 3D house models which was built modularly and with a texture control quality process. Results in the single house policy experiment show that *Scan-RL* was able to achieve 96% terminal surface coverage in fewer steps and shorter distance than the baseline circular path. Results in the multiple houses single policy evaluation show that the trained NBV policy for each batch in *Houses3K*, can learn to scan multiple houses one at a time and can also be transferred to scan houses not seen during training. Future works can look into deploying the NBV policy trained in the synthetic environment to a drone for real world 3D reconstruction.

Acknowledgements. This work was funded by CHED-PCARI Project IIID-2016-005.

References

1. Aloimonos, J., Weiss, I., Bandyopadhyay, A.: Active vision. Int. J. Comput. Vis. **1**(4), 333–356 (1988)
2. Chen, S., Li, Y., Kwok, N.M.: Active vision in robotic systems: a survey of recent developments. Int. J. Robot. Res. **30**(11), 1343–1377 (2011)
3. Choudhury, S., et al.: Data-driven planning via imitation learning. Int. J. Robot. Res. **37**(13–14), 1632–1672 (2018)
4. Curless, B., Levoy, M.: A volumetric method for building complex models from range images (1996)

5. Dai, A., Nießner, M., Zollhöfer, M., Izadi, S., Theobalt, C.: Bundlefusion: real-time globally consistent 3d reconstruction using on-the-fly surface reintegration. ACM Trans. Graph. (ToG) **36**(4), 76a (2017)
6. Daudelin, J., Campbell, M.: An adaptable, probabilistic, next-best view algorithm for reconstruction of unknown 3-d objects. IEEE Robot. Autom. Lett. **2**(3), 1540–1547 (2017)
7. Delmerico, J., Isler, S., Sabzevari, R., Scaramuzza, D.: A comparison of volumetric information gain metrics for active 3d object reconstruction. Auton. Robot. **42**(2), 197–208 (2018)
8. Han, X., et al.: Deep reinforcement learning of volume-guided progressive view inpainting for 3d point scene completion from a single depth image. In: Proceedings of the IEEE Conference on Computer Vision and Pattern Recognition, pp. 234–243 (2019)
9. Heinly, J., Schonberger, J.L., Dunn, E., Frahm, J.M.: Reconstructing the world* in six days*(as captured by the yahoo 100 million image dataset). In: Proceedings of the IEEE Conference on Computer Vision and Pattern Recognition, pp. 3287–3295 (2015)
10. Hepp, B., Nießner, M., Hilliges, O.: Plan3d: viewpoint and trajectory optimization for aerial multi-view stereo reconstruction. ACM Trans. Graph. (TOG) **38**(1), 4 (2018)
11. Huang, R., Zou, D., Vaughan, R., Tan, P.: Active image-based modeling with a toy drone. In: 2018 IEEE International Conference on Robotics and Automation (ICRA), pp. 1–8. IEEE (2018)
12. Isler, S., Sabzevari, R., Delmerico, J., Scaramuzza, D.: An information gain formulation for active volumetric 3d reconstruction. In: Robotics and Automation (ICRA), 2016 IEEE International Conference on, pp. 3477–3484. IEEE (2016)
13. Izadi, S., et al.: Kinectfusion: real-time 3d reconstruction and interaction using a moving depth camera. In: Proceedings of the 24th Annual ACM Symposium on User Interface Software and Technology, pp. 559–568. ACM (2011)
14. Kaba, M.D., Uzunbas, M.G., Lim, S.N.: A reinforcement learning approach to the view planning problem. In: CVPR, pp. 5094–5102 (2017)
15. Kriegel, S., Rink, C., Bodenmüller, T., Suppa, M.: Efficient next-best-scan planning for autonomous 3d surface reconstruction of unknown objects. J. Real-Time Image Process. **10**(4), 611–631 (2015)
16. Lillicrap, T.P., et al.: Continuous control with deep reinforcement learning. arXiv preprint arXiv:1509.02971 (2015)
17. Mnih, V., et al.: Human-level control through deep reinforcement learning. Nature **518**(7540), 529 (2015)
18. Potthast, C., Sukhatme, G.S.: A probabilistic framework for next best view estimation in a cluttered environment. J. Vis. Commun. Image Representation **25**(1), 148–164 (2014)
19. Qiu, W., Yuille, A.: UnrealCV: connecting computer vision to unreal engine. In: Hua, G., Jégou, H. (eds.) ECCV 2016. LNCS, vol. 9915, pp. 909–916. Springer, Cham (2016). https://doi.org/10.1007/978-3-319-49409-8_75
20. Schonberger, J.L., Frahm, J.M.: Structure-from-motion revisited. In: Proceedings of the IEEE Conference on Computer Vision and Pattern Recognition, pp. 4104–4113 (2016)
21. Scott, W.R., Roth, G., Rivest, J.F.: View planning for automated three-dimensional object reconstruction and inspection. ACM Comput. Surv. (CSUR) **35**(1), 64–96 (2003)

22. Smith, N., Moehrle, N., Goesele, M., Heidrich, W.: Aerial path planning for urban scene reconstruction: a continuous optimization method and benchmark. In: SIGGRAPH Asia 2018 Technical Papers, p. 183. ACM (2018)
23. Snavely, N., Seitz, S.M., Szeliski, R.: Photo tourism: exploring photo collections in 3d. In: ACM Transactions on Graphics (TOG), vol. 25, pp. 835–846. ACM (2006)
24. Song, S., Jo, S.: Surface-based exploration for autonomous 3d modeling. In: 2018 IEEE International Conference on Robotics and Automation (ICRA), pp. 1–8. IEEE (2018)
25. Su, H., Qi, C.R., Li, Y., Guibas, L.J.: Render for cnn: viewpoint estimation in images using cnns trained with rendered 3d model views. In: Proceedings of the IEEE International Conference on Computer Vision, pp. 2686–2694 (2015)
26. Tarabanis, K.A., Allen, P.K., Tsai, R.Y.: A survey of sensor planning in computer vision. IEEE Trans. Robot. Autom. **11**(1), 86–104 (1995)
27. Wenhardt, S., Deutsch, B., Angelopoulou, E., Niemann, H.: Active visual object reconstruction using d-, e-, and t-optimal next best views. In: 2007 IEEE Conference on Computer Vision and Pattern Recognition, pp. 1–7. IEEE (2007)
28. Wu, S., et al.: Quality-driven poisson-guided autoscanning. ACM Trans. Graph. **33**(6) (2014)
29. Wu, Y., Wu, Y., Gkioxari, G., Tian, Y.: Building generalizable agents with a realistic and rich 3d environment. arXiv preprint arXiv:1801.02209 (2018)
30. Yang, X., et al.: Active object reconstruction using a guided view planner. arXiv preprint arXiv:1805.03081 (2018)
31. Zhong, F., Qiu, W., Yan, T., Alan, Y., Wang, Y.: Gym-unrealcv: realistic virtual worlds for visual reinforcement learning (2017)

A Flow Base Bi-path Network for Cross-Scene Video Crowd Understanding in Aerial View

Zhiyuan Zhao, Tao Han, Junyu Gao, Qi Wang(✉), and Xuelong Li

School of Computer Science and Center for OPTical IMagery Analysis and Learning (OPTIMAL), Northwestern Polytechnical University, Xi'an 710072, Shaanxi, People's Republic of China
tuzixini@163.com, hantao10200@mail.nwpu.edu.cn, gjy3035@gmail.com, crabwq@gmail.com, li@nwpu.edu.cn

Abstract. Drones shooting can be applied in dynamic traffic monitoring, object detecting and tracking, and other vision tasks. The variability of the shooting location adds some intractable challenges to these missions, such as varying scale, unstable exposure, and scene migration. In this paper, we strive to tackle the above challenges and automatically understand the crowd from the visual data collected from drones. First, to alleviate the background noise generated in cross-scene testing, a double-stream crowd counting model is proposed, which extracts optical flow and frame difference information as an additional branch. Besides, to improve the model's generalization ability at different scales and time, we randomly combine a variety of data transformation methods to simulate some unseen environments. To tackle the crowd density estimation problem under extreme dark environments, we introduce synthetic data generated by game Grand Theft Auto V(GTAV). Experiment results show the effectiveness of the virtual data. Our method wins the challenge with a mean absolute error (MAE) of 12.70[1]. Moreover, a comprehensive ablation study is conducted to explore each component's contribution.

Keywords: Crowd counting · Optical flow · Data augmentation · Synthetic data

1 Introduction

Video analysis [2,46] has become an increasingly important part of computer vision, which involves a wide range of fields, such as object detection and tracking [1,21], crowd segmentation [18], density estimation and localization [15,37], and group detection [31]. Among them, a new challenge that understands the crowd

Z. Zhao and T. Han—Equal Contribution.

[1] Finally, reach the MAE of 12.36, ranked the second.

A. Bartoli and A. Fusiello (Eds.): ECCV 2020 Workshops, LNCS 12538, pp. 574–587, 2020.
https://doi.org/10.1007/978-3-030-66823-5_34

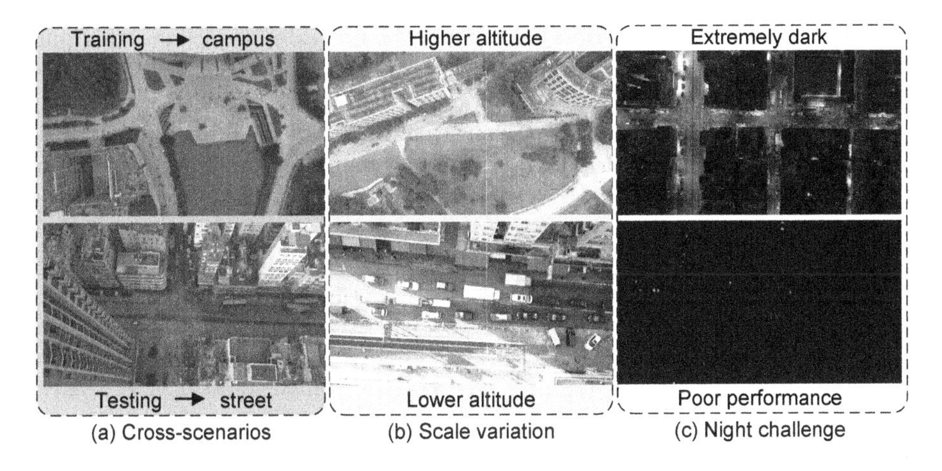

Fig. 1. Three challenges in the DroneCrowd Dataset.

from a drone perspective has recently attracted lots of attention. Compared with the density estimation in the ground scenes, unmanned aerial vehicles (UAVs) have a broader surveillance scope. It can play a more crucial role in public safety and urban management.

The scale of the datasets limited the development of crowd counting tasks over the past few years. However, it has been booming driven by the release of a series of large-scale datasets recently, such as UCF-QNRF [15], JHU-Crowd [29], NWPU-Crowd [32]. Unlike the counting task from the ground perspective, there are extra challenges in the UAV scenarios. In this paper, we work on density and crowd estimation for the DroneCrowd dataset [37]. As illustrated in Fig. 1, we found that the difficulties of UAV's task of counting people lie in the following three aspects:

- *Cross-scenarios.* The training set and validation set of the DroneCrowd dataset are divided according to the scenario, which means that it is a cross-scene crowd counting task. The test set scenarios are almost unknown to the model, which increases the difficulty of the task.
- *Scale variation.* The flying height of drones is constantly changing during data collection, which significantly affects the scales of objects. The diversity of scale increases the difficulty of model fitting.
- *Night challenge.* There are some extremely dark scenes in the test set, but there are no scenes with similar illumination conditions in the training set. These night scenarios have a significant impact on the test results.

The biggest problem of cross-scene attributes is that it has serious background noise in testing unknown scenes, which is impossible to be tackled directly by existing counting models. The effective measure is to reduce background noise by enhancing the model's ability to distinguish between background and moving objects. Note that the DroneCrowd dataset is composed of video sequences.

We can extract the optical flow information of the video to achieve this purpose. Therefore, in our counting framework, an additional branch is designed to input optical flow information to enhance the feature extraction of moving objects. Besides, to improve the generalization ability of the model for the scale variables, multi-scale images are transferred to simulate the scenes at different heights. Finally, to alleviate the large test deviation caused by low illumination scenes, a variety of image brightness adjustment algorithms and synthetic data are used to simulate different illumination conditions in the training sequences. All three ideas contribute to the competitive performance, which we will explore in detail in Sect. 4.

As a summary, we propose a bi-path framework based on RGB image and optical flow information to improve the model's ability to weaken the background noise for unknown scenes and exploit multi-scale transformation and multi-luminance adjustment algorithms to make data augmentation for reducing the adverse effects of scale and illumination. Besides, we generate synthetic data under dark scenes, which expands training data using low-cost methods and improves the performance of the module in extreme scenarios.

2 Related Work

Scale-aware Crowd Counting. With the development of crowd counting, it is found that scale change has the greatest negative impact among the influencing factors (occlusion, scale, and viewpoint, etc.). Many crowd counting algorithms focus on scale variability in recent years. Zhang *et al.* [47] propose a three-columns network with different kernels for scale perception in 2016. Onoro-Rubio *et al.* [25] introduce a Hydra CNN with three-columns, where each column is fed by a patch from the same image with a different scale. Wu *et al.* [38] develop a powerful multi-column scale-aware CNN with an adaptation module to fuse the sparse and congested columns. In the same year, Adaptively Fusing Predictions (AFP) [17] generates a density map by fusing the attention map and intermediary density map in each column. ic-CNN [27] generates a high-resolution density map via passing the feature and predict map from the low-resolution CNN to the high-resolution CNN. SDA-MCNN [40], a scale-distribution-aware neural network that resolves scale change by processing a crowd image with multiple Convolutional Neural Network columns and minimizing the per-scale loss. Hossain *et al.* [13] employs a scale-aware attention network, where each column is weighted with the output of a global scale attention network and local scale attention network. ACM-CNN [50], Adaptive Capacity Multi-scale convolutional neural networks assign different capacities to different portions of the input based on three modules: a coarse network, a fine network, and a smooth network. Except for multi-column scale-aware architecture, some works focus on the single-column scale-aware CNN, such as SANet [3], SaCNN [45]. To combine the multi-column and single-column scale-aware CNN, CSRNet [20], CAN [23] and FPNCC [4] develop a model containing multiple paths only in several part of the networks. In 2020, SCAN [41], SRN+PS [7], SRF-Net [5], and ASNet [16] also explore the scale-aware to improve the counting performance.

Cross-scene/domain Crowd Counting. Due to the laborious data annotation required of crowd datasets, cross-scene, and cross-domain crowd counting attract researchers' attention in recent years. In this task, the model is trained on a labeled dataset and then adapted to an unseen scene. Authors in [43] establish the earliest cross-scene dataset, which includes 103 scenes for training and the remaining five scenes for testing, a total of 3980 images. In [22], they propose a Fully Convolutional Neural Network(FCN) and a weighted adaptive human Gaussian model for person detection and then apply it to the new scene with few labeled data. DA-ELM [39], a counting model based on domain adaptation-extreme learning machine, counts the people in a new scene with only a half of the training samples compared with counting without domain adaptation. In [14], they propose a one-shot learning approach for learning how to adapt to a target scene using one labeled example. In [28], they apply the MAML [8] to learn scene adaptive crowd counting with few-shot learning. Inspired by the synthetic data can automatically label as the source domain, Wang *et al.* [33] launch a large-scale synthetic dataset to pre-train a model and adapt it to real-world datasets by a fine-tune operation. Except fine-tuning, they also complete counting without any real-world labeled information by using the Cycle GAN [48] and SE Cycle GAN [33] to generate realistic images. Recently, several efforts have been made to follow them. DACC [9], a method for domain adaptation based on image translation and Gaussian-prior reconstruction, achieves a better performance on several mainstream datasets. At the same time, some works [10,12,24,35] extract domain invariant features based on adversarial learning. In [34], the authors propose a Neuron Linear Transformation (NLT) method that models domain differences and uses few labeled target data to train the domain shift parameters. Experimental results show that cross-domain crowd counting can alleviate the problem of data annotation to some extent.

3 Proposed Method

A perfect crowd counter should have stable and excellent performance in practical applications. However, as mentioned above, the actual application scenarios are complex and changeable, and the training data is limited. Therefore, in our proposed method, we promote the model's performance from the following two aspects:

- Improve the feature extraction and representation ability of the crowd count network through better network structure design. Therefore, it boosts the cross-scene stability and accuracy of the model.
- Try to mine the information contained in the limited training data. This includes using different kinds of data transform methods to manually create unseen verity scenario from existing data, and make use of the joint information between adjacent frames while reading continuous sequential data.

The remaining paragraphs follow these two aspects to introduce the work done by the proposed method.

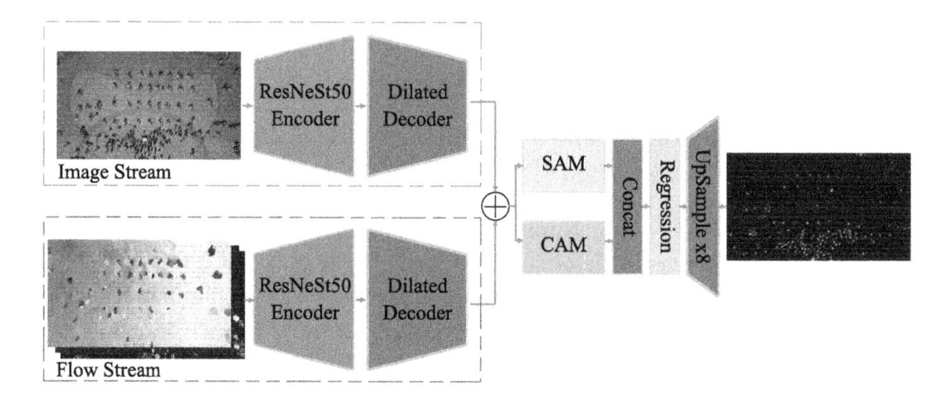

Fig. 2. Network structure of the proposed bi-path crowd counter.

3.1 Network Design

As shown in Fig. 2, we design a bi-path flow-based crowd counting network to make use of inter-frame information together with the input RGB image. Each pathway consists of one powerful feature extractor and several convolutional layers as the decoder. The encoder is the first three layers of the newly proposed ResNeSt [44] with weights pre-trained on the ImageNet [6] dataset. The crowd under the aerial view is small, and the building area is large. To better analyze the relationship between humans and background, we follow the idea of CSRNet [20], the decoder contains six dilated convolutions, which reduces the number of channels while parsing the semantic information of the input features. Atrous convolution layers enlarge the convolution kernel's reception field, which enriches spatial contextual information inside the features. The decoder outputs feature vectors of 1/8 size of the original input with 64 channels. The image branch and the flow branch are combined immediately after the decoders. Besides the dilated convolutions, we introduce spatial-wise attention module (SAM) and channel-wise attention module(CAM) from [11] to enhance large-range dependencies on the spatial and channel dimensions.

The detailed network structure is given in Table 1. "ResNeSt50_Conv1_c64" means this layer has the same structure as the Conv1 layer of ResNeSt50, and its output is a 64 channel tensor. "k(3,3)-c512-s1-d2-R" represents convolutional layer with kernel size 3×3, output channel 512, stride 1, dilation rate 2, and followed by a ReLU activation layer.

The image stream accepts the three-channel RGB color image as input, while the flow stream also takes a three-channel input with the same size as the original image.

Generally speaking, the optical flow is a two-channel vector, and each channel stores the optical flow information of the image along the horizontal axis and vertical axis separately. We name this form of optical flow f_x and f_y. It can also be transformed into polar coordinates. Under the polar system, two transformed channels represent the polar radius and polar angle, respectively. Researchers

Table 1. Network structure of the proposed method.

Encoder	
ResNeSt50_Conv1_c64	
ResNeSt50_Layer1_c256	
ResNeSt50_Layer2_c512	
ResNeSt50_Layer3_c1024	
Decoder	
k(3,3)-c512-s1-d2-R	
k(3,3)-c512-s1-d2-R	
k(3,3)-c512-s1-d2-R	
k(3,3)-c256-s1-d2-R	
k(3,3)-c128-s1-d2-R	
k(3,3)-c64-s1-d2-R	
Attention	
SAM	CAM
concat	
Regression	
k(3,3)-c512-s1-d2-R	
UpSample ×8	

usually take these two channels as the first two channels in the HSV color space, and then convert them back to the RGB space for visualization, so we name them f_h and f_s here. The flow stream takes a three-channel vector as input, here we fill the third dimension of the input data with the frame difference vector of the two adjacent frames. Which means take frame t subtracts frame t+1 and get f_{sub}. Different forms of optical flow contain different information, leading to distant results to the network's training. We have tried several combinations; the specific experimental results and analysis are given in Sect. 4.4.

Figure 3 visualizes different forms of optical flow data. During the visualization and analysis process, we find that there may be a slight jitter in the process of data acquisition, which introduces a large area of noise to the optical flow information. Therefore, we add a simple threshold filtering in the actual use, which effectively suppresses part of the noise.

3.2 Data Argumentation

Apart from the network structure design, it is also significant to effectively use existing data.

The training data only contains limited scenes. By observing, analyzing, and summarizing the status of the existing data, we speculate the scenarios that may be encountered during the actual use. We manually augment the existing data of

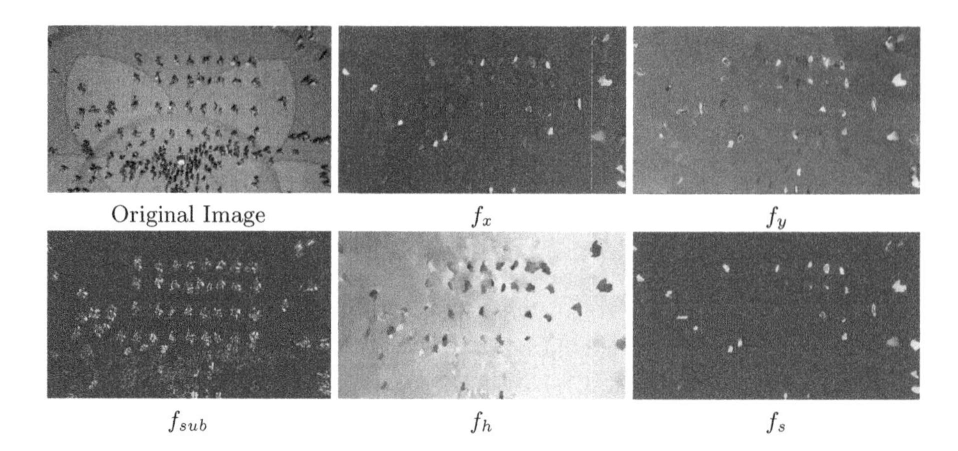

Fig. 3. Visualization of various optical flow forms.

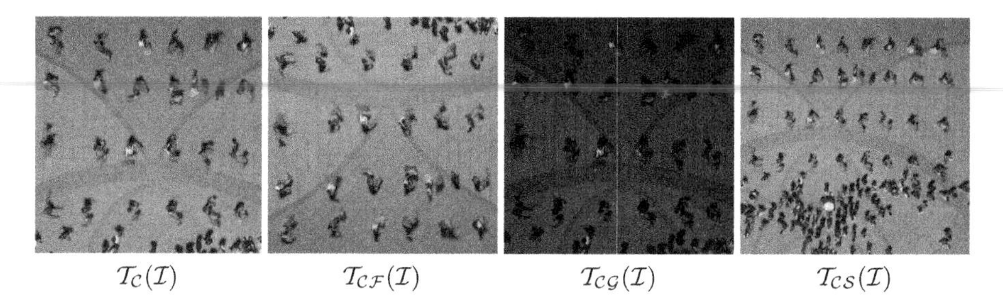

Fig. 4. Visualization of various data argumentation operations. The original image is the same as the one in Fig. 3. Two subscripts write together indicates the usage of two kinds of transformation together.

the scene according to the actual situation. During training, numerous transform methods were used to generate new data, improving the model's generalization ability. Instead of original data group $(\mathcal{I}, F, D)$ (Image, Flow, DotMap), new training data group $(\mathcal{I}_{\mathcal{N}}, F_{\mathcal{N}}, D_{\mathcal{N}})$ is generated by random transform method $\mathcal{T}(\cdot)$:

$$(\mathcal{I}_{\mathcal{N}}, F_{\mathcal{N}}, D_{\mathcal{N}}) = \mathcal{T}_{\{\mathcal{C},\mathcal{F},\mathcal{G},\mathcal{S}\}}(\mathcal{I}, F, D), \tag{1}$$

where subscript $\{\mathcal{C}, \mathcal{F}, \mathcal{G}, \mathcal{S}\}$ represents crop, flip, gamma correction, scale change respectively. The following part of this subsection introduces all data transformation methods we use according to the characteristics of the data. Figure 4 gives examples of several transformations.

Scene Change: Depending on the drone's shooting location, different scenes may vary considerably, whether it is the street or the park. Their style and direction will change. To cope with the scenario's variability, we make random transforms to each group of input data during the training process, which includes random crop into fixed size (576 × 576 here), random flip (up-down and left-right). Of

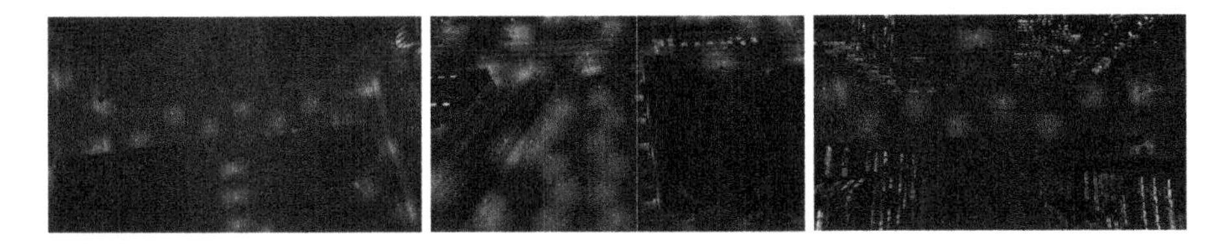

Fig. 5. Some examples of the generated synthetic data.

course, those transformations perform the same processing within each group of data $(\mathcal{I}, F, D)$ used for training. We name those transformations $\mathcal{T}_{\{\mathcal{C},\mathcal{F}\}}(\cdot)$.

Extreme Light Variation: All the training data we have are taken under good light conditions, but the data used in the test and the real world may not always have such good light conditions. There will be overexposure or underexposure. So before we send the data into the model, we carry out random gamma correction on the RGB image with a certain probability (here is 0.5). The range of Gamma values is from 0.4 to 2 to simulate excessive brightness and insufficient brightness. This transformation only applies on the RGB images, and we call it $\mathcal{T}_{\mathcal{G}}(\cdot)$.

Scale Diversity: When the UAV collects data in different states, it may fly under a low or high altitude, seriously affecting the size of the crowd in the picture and ultimately affecting the model's analysis results. To further enhance the model's generalization ability, we force it to deal with images collected at various heights. We change the scale of the images before cropping, which means randomly enlarge or shrink the image, and then crop the new one for training. This transformation works on all three input data together. We name it $\mathcal{T}_{\mathcal{S}}(\cdot)$.

3.3 Synthetic Data

The cost of collecting and labeling real scene data is very high. Moreover, in general, the real scene is limited, and the diversity is weak. The standard data enhancement method can only alleviate this problem to a certain extent. To better increase the diversity of training data, we refer to [33]'s method and generate some synthetic data to assist training. Figure 5 demonstrates some of the synthetic data that we generated.

4 Experiments

4.1 Dataset

VisDrone crowd count dataset (DroneCrowd) [37] is a remote crowd counting dataset with manual labels and 1080p RGB images collected by drones. It consists of 112 sequences, of which 82 contains publicly reached point labels, and the rest of the sequences can only be tested by uploading the results to the official challenge [49] server. Therefore, we randomly divide the labeled 82 sequences

Table 2. Results on the test set of several methods (See footnote 1).

Method	MAE	User	Institution	Comment
MCNN [47] [a]	34.7	-	-	-
MSCNN [42] [a]	58	-	-	-
CSRNet [20] [a]	19.8	-	-	-
'sa_tta' [b]	13.81	L xl		Ranked Third
'CSRNet' [b]	13.80	Shinan liu	Beijing jiaotong univerisity	Ranked Second
Ours	**12.70**	T Xini	NWPU	**Ranked First**

[a] Results from official paper [37].
[b] MAE from official test server [36]. Data collected on 02:41:20, July 15, 2020, GMT.

into 75 for training and 7 for validation. Our model achieves MAE 12.7 in the test server and wins the challenge (See footnote 1). Because there is a limit on the number of test times, part of our following experimental results are based on the validation set.

4.2 Implementation Details

All experiments are build based on one NVIDIA TITAN RTX GPU and Intel(R) Xeon(R) Silver 4110 CPU @ 2.10GHz under Ubuntu 18.04 LTS operating system and PyTorch [26] framework. The ResNeSt50 encoder is initialized with the weights pre-trained on the ImageNet dataset, all other convolutional layers are initialized with normally distributed weights and zero bias. The learning rate is set to 10^{-5}.

4.3 Performance Measures

Researchers use MAE and mean squared error (MSE) to measure the differences between the predicted density map and the ground truth. The definition of MAE and MSE are given by following equations:

$$\begin{aligned} \text{MAE} &= \tfrac{1}{H\times W}\textstyle\sum_{i=1}^{H}\sum_{j=1}^{W}\left|z_{i,j}-\hat{z}_{i,j}\right|, \\ \text{MSE} &= \sqrt{\tfrac{1}{H\times W}\textstyle\sum_{i=1}^{H}\sum_{j=1}^{W}\left|z_{i,j}-\hat{z}_{i,j}\right|^{2}}, \end{aligned} \tag{2}$$

where H and W are the height and width of the test density map, $z_{i,j}$ is the ground truth pixel value of location (i, j), $\hat{z}_{i,j}$ is the corresponding one of the predicted density map.

The official challenge test server only provides MAE values. As for our experimental analysis, both MAE and MSE are calculated on the validation set. Table 2 shows MAE values on the test set of several methods from the challenge [49] and the official paper [37]. Our method outperforms all those methods and wins the challenge with MAE 12.70 (See footnote 1).

Table 3. Results on the validation set of different types and combinations of the flow stream.

Stream	Flow type	Flow combination	MAE (val)	MSE (val)
Image	-	-	28.73	42.16
Image+Flow	PWC	$[f_x, f_y, f_{sub}]$	26.36	36.10
Image+Flow	DIS	$[f_x, f_y, f_{sub}]$	26.31	38.26
Image+Flow	PWC	$[f_h, f_s, f_{sub}]$	23.14	31.09
Image+Flow	DIS	$[f_h, f_s, f_{sub}]$	**22.66**	**34.46**

Table 4. Results on the validation set of different light condition enhancement sets.

Methods	MAE (val)	MSE(val)
Without gamma correction	28.37	39.42
With gamma correction	**22.66**	**34.46**

4.4 Ablation Studies

Influence of Flow Stream: To verify whether the flow stream helps improving the network, we set up experiments to train the network without the flow stream. We use different methods to generate the optical flow; the first one is based on deep learning [30], while the second one is based on dense inverse search [19]. We abbreviated them as PWC and DIS, respectively. We also take different combinations of $[f_x, f_y, f_h, f_s, f_{sub}]$ into consideration. Experiment results are shown in Table 3. Due to time limitations, we only train 30 epochs under each setting and take the best result.

It can be seen that the introduction of the flow stream improves the generalization ability of the model, and its type and composition will also affect the final result. Optical flow generated by DIS, with combination method $[f_s, f_v, f_{sub}]$ achieves the best.

Influence of Gamma Correction: There are some extremely dark scenes in the test set, but there is no scene with such low light conditions in the training set. These night scenarios have a significant impact on the test results. The validation set can not verify the effectiveness of gamma correction. We submit results with/without gamma correction to the official test server. The test results are given in Table 4.

It can be seen that random brightness transformation is performed on the input image during training to increase the diversity of the lighting conditions in the training data, which effectively improves the model's performance in the test set with low light environments.

Influence of Scale Variation: To simulate different cruising altitudes of the data collected by drones, we set experiments under the original image, scale input image between [0.7, 1.2], and [0.6, 1.8] randomly. Experiment results are shown

Table 5. MAE on the validation/test set under different training scale range and data.

Scale range	MAE (val)	MSE (val)	MAE (test)
None	36.55	47.65	24.61
[0.7, 1.2]	29.19	41.58	-
[0.6, 1.8]	**22.66**	**34.46**	17.04
[0.6, 1.8] With Synthetic	-	-	**12.70**

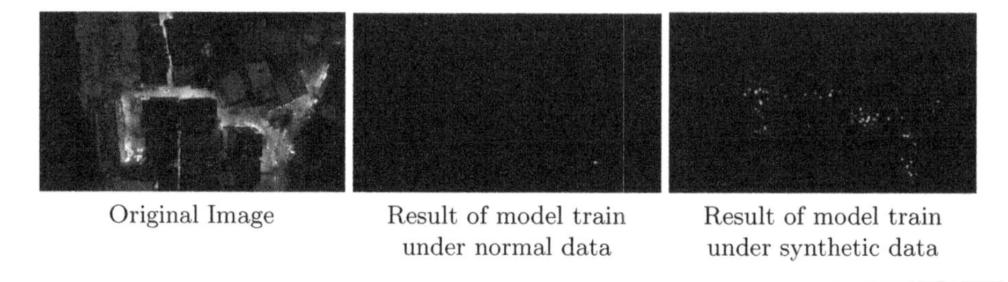

Original Image | Result of model train under normal data | Result of model train under synthetic data

Fig. 6. Visualization of density map generate by different models.

in Table 5, the scale variation during the training process improved the model performance. According to the validation results, a broader range of scale change brings better results. However, if we apply scale change larger than [0.6, 1.8], the changed image size cannot meet the size requirements of the crop operation, so no more experiments are performed here.

Influence of Synthetic Data: We use the synthetic data alone to train a new model to predict the night scene, and then use the original model to predict the daytime scene. Table 5 also shows the results of mixing the two models. It can be seen that the synthetic data has brought a significant improvement to the final result. Figure 6 shows the visualization of the predict density map under normal training data or synthetic training data. It is clear that the model trained under ordinary data can not make good predictions in extremely dark scenes. Nevertheless, the model trained under synthetic data can give good results.

5 Conclusion

This paper proposes a bi-path optical flow-based crowd counter network, together with several data argumentation methods specially designed for remote crowd count tasks. To be specific, two separate ResNeSt50 based encoder-decoder stream extract feature vectors from RGB images and 3-channel optical flow tensors. During the training phase, data argumentation operations like crop, random flip, gamma correction, scale variation are applied to create nonexistent scenes from the train set. Synthetic data are generated for better performance under extreme dark scenarios. With the combination of network design, data expansion and synthetic data, we finally win the official competition (See footnote 1). In addition, we designed sufficient experiments to verify and analyze the effectiveness of network structure design and different data argumentation methods. In the future, we will explore effective network design and data argumentation methods for remote crowd count task.

References

1. Bertinetto, L., Valmadre, J., Henriques, J.F., Vedaldi, A., Torr, P.H.S.: Fully-convolutional siamese networks for object tracking. In: Hua, G., Jégou, H. (eds.) ECCV 2016. LNCS, vol. 9914, pp. 850–865. Springer, Cham (2016). https://doi.org/10.1007/978-3-319-48881-3_56
2. Borges, P.V.K., Conci, N., Cavallaro, A.: Video-based human behavior understanding: a survey. IEEE Trans. Circ. Syst. Video Technol. **23**(11), 1993–2008 (2013)
3. Cao, X., Wang, Z., Zhao, Y., Su, F.: Scale aggregation network for accurate and efficient crowd counting. In: Proceedings of the European Conference on Computer Vision (ECCV), pp. 734–750 (2018)
4. Cenggoro, T.W., Aslamiah, A.H., Yunanto, A.: Feature pyramid networks for crowd counting. Procedia Comput. Sci. **157**, 175–182 (2019)
5. Chen, Y., Gao, C., Su, Z., He, X., Liu, N.: Scale-aware rolling fusion network for crowd counting. In: 2020 IEEE International Conference on Multimedia and Expo (ICME), pp. 1–6. IEEE (2020)
6. Deng, J., Dong, W., Socher, R., Li, L.J., Li, K., Fei-Fei, L.: Imagenet: a large-scale hierarchical image database. In: 2009 IEEE Conference on Computer Vision and Pattern Recognition, pp. 248–255. IEEE (2009)
7. Dong, Z., Zhang, R., Shao, X., Li, Y.: Scale-recursive network with point supervision for crowd scene analysis. Neurocomputing **384**, 314–324 (2020)
8. Finn, C., Abbeel, P., Levine, S.: Model-agnostic meta-learning for fast adaptation of deep networks. In: Proceedings of the 34th International Conference on Machine Learning, vol. 70, pp. 1126–1135. JMLR. org (2017)
9. Gao, J., Han, T., Wang, Q., Yuan, Y.: Domain-adaptive crowd counting via inter-domain features segregation and gaussian-prior reconstruction. arXiv preprint arXiv:1912.03677 (2019)
10. Gao, J., Wang, Q., Yuan, Y.: Feature-aware adaptation and structured density alignment for crowd counting in video surveillance. arXiv preprint arXiv:1912.03672 (2019)
11. Gao, J., Wang, Q., Yuan, Y.: Scar: spatial-/channel-wise attention regression networks for crowd counting. Neurocomputing **363**, 1–8 (2019)
12. Han, T., Gao, J., Yuan, Y., Wang, Q.: Focus on semantic consistency for cross-domain crowd understanding. In: ICASSP 2020–2020 IEEE International Conference on Acoustics, Speech and Signal Processing (ICASSP), pp. 1848–1852. IEEE (2020)
13. Hossain, M., Hosseinzadeh, M., Chanda, O., Wang, Y.: Crowd counting using scale-aware attention networks. In: 2019 IEEE Winter Conference on Applications of Computer Vision (WACV), pp. 1280–1288. IEEE (2019)
14. Hossain, M.A., Kumar, M., Hosseinzadeh, M., Chanda, O., Wang, Y.: One-shot scene-specific crowd counting

15. Idrees, H., et al.: Composition loss for counting, density map estimation and localization in dense crowds, pp. 532–546 (2018)
16. Jiang, X., et al.: Attention scaling for crowd counting. In: Proceedings of the IEEE/CVF Conference on Computer Vision and Pattern Recognition, pp. 4706–4715 (2020)
17. Kang, D., Chan, A.: Crowd counting by adaptively fusing predictions from an image pyramid. In: British Machine Vision Conference 2018, BMVC 2018, Newcastle, UK, September 3–6, 2018, p. 89. BMVA Press (2018)
18. Kang, K., Wang, X.: Fully convolutional neural networks for crowd segmentation. arXiv preprint arXiv:1411.4464 (2014)
19. Kroeger, T., Timofte, R., Dai, D., Van Gool, L.: Fast optical flow using dense inverse search. In: Leibe, B., Matas, J., Sebe, N., Welling, M. (eds.) ECCV 2016. LNCS, vol. 9908, pp. 471–488. Springer, Cham (2016). https://doi.org/10.1007/978-3-319-46493-0_29
20. Li, Y., Zhang, X., Chen, D.: Csrnet: dilated convolutional neural networks for understanding the highly congested scenes. In: Proceedings of the IEEE Conference on Computer Vision and Pattern Recognition, pp. 1091–1100 (2018)
21. Lin, T., et al.: Microsoft COCO: common objects in context. In: Fleet, D., Pajdla, T., Schiele, B., Tuytelaars, T. (eds.) ECCV 2014. LNCS, vol. 8693, pp. 740–755. Springer, Cham (2014). https://doi.org/10.1007/978-3-319-10602-1_48
22. Liu, H., Li, Y., Zhou, Z., Wu, W.: Cross-scene crowd counting via fcn and gaussian model. In: 2016 International Conference on Virtual Reality and Visualization (ICVRV), pp. 148–153. IEEE (2016)
23. Liu, W., Salzmann, M., Fua, P.: Context-aware crowd counting. In: Proceedings of the IEEE Conference on Computer Vision and Pattern Recognition, pp. 5099–5108 (2019)
24. Nie, F., Wang, Z., Wang, R., Wang, Z., Li, X.: Towards robust discriminative projections learning via non-greedy $l_{2,1}$-norm minmax. IEEE Transactions on Pattern Analysis and Machine Intelligence (2019)
25. Oñoro-Rubio, D., López-Sastre, R.J.: Towards perspective-free object counting with deep learning. In: Leibe, B., Matas, J., Sebe, N., Welling, M. (eds.) ECCV 2016. LNCS, vol. 9911, pp. 615–629. Springer, Cham (2016). https://doi.org/10.1007/978-3-319-46478-7_38
26. Paszke, A., et al.: Automatic differentiation in pytorch (2017)
27. Ranjan, V., Le, H., Hoai, M.: Iterative crowd counting. In: Proceedings of the European Conference on Computer Vision (ECCV), pp. 270–285 (2018)
28. Reddy, M.K.K., Hossain, M., Rochan, M., Wang, Y.: Few-shot scene adaptive crowd counting using meta-learning. In: The IEEE Winter Conference on Applications of Computer Vision, pp. 2814–2823 (2020)
29. Sindagi, V.A., Yasarla, R., Patel, V.M.: Jhu-crowd++: large-scale crowd counting dataset and a benchmark method. arXiv preprint arXiv:2004.03597 (2020)
30. Sun, D., Yang, X., Liu, M.Y., Kautz, J.: Pwc-net: Cnns for optical flow using pyramid, warping, and cost volume. In: Proceedings of the IEEE Conference on Computer Vision and Pattern Recognition, pp. 8934–8943 (2018)
31. Wang, Q., Chen, M., Nie, F., Li, X.: Detecting coherent groups in crowd scenes by multiview clustering. IEEE Trans. Pattern Anal. Mach. Intell. **42**(1), 46–58 (2018)
32. Wang, Q., Gao, J., Lin, W., Li, X.: Nwpu-crowd: a large-scale benchmark for crowd counting. IEEE Transactions on Pattern Analysis and Machine Intelligence (2020)
33. Wang, Q., Gao, J., Lin, W., Yuan, Y.: Learning from synthetic data for crowd counting in the wild. In: Proceedings of IEEE Conference on Computer Vision and Pattern Recognition, pp. 8198–8207 (2019)

34. Wang, Q., Han, T., Gao, J., Yuan, Y.: Neuron linear transformation: modeling the domain shift for crowd counting. arXiv preprint arXiv:2004.02133 (2020)
35. Wang, Z., Nie, F., Tian, L., Wang, R., Li, X.: Discriminative feature selection via a structured sparse subspace learning module
36. Wen, L., et al.: Visdrone challenge leaderboard. http://aiskyeye.com/leaderboard/
37. Wen, L., et al.: Drone-based joint density map estimation, localization and tracking with space-time multi-scale attention network. arXiv preprint arXiv:1912.01811 (2019)
38. Wu, X., Zheng, Y., Ye, H., Hu, W., Yang, J., He, L.: Adaptive scenario discovery for crowd counting. In: ICASSP 2019–2019 IEEE International Conference on Acoustics, Speech and Signal Processing (ICASSP), pp. 2382–2386. IEEE (2019)
39. Yang, B., Cao, J.M., Wang, N., Zhang, Y.Y., Cui, G.Z.: Cross-scene counting based on domain adaptation-extreme learning machine. IEEE Access **6**, 17029–17038 (2018)
40. Yang, B., Zhan, W., Wang, N., Liu, X., Lv, J.: Counting crowds using a scale-distribution-aware network and adaptive human-shaped kernel. Neurocomputing **390**, 207–216 (2019)
41. Yuan, L., et al.: Crowd counting via scale-communicative aggregation networks. Neurocomputing **409**, 420–430 (2020)
42. Zeng, L., Xu, X., Cai, B., Qiu, S., Zhang, T.: Multi-scale convolutional neural networks for crowd counting. In: 2017 IEEE International Conference on Image Processing (ICIP), pp. 465–469. IEEE (2017)
43. Zhang, C., Kang, K., Li, H., Wang, X., Xie, R., Yang, X.: Data-driven crowd understanding: a baseline for a large-scale crowd dataset. IEEE Trans. Multimedia **18**(6), 1048–1061 (2016)
44. Zhang, H., et al.: Resnest: split-attention networks. arXiv preprint arXiv:2004.08955 (2020)
45. Zhang, L., Shi, M., Chen, Q.: Crowd counting via scale-adaptive convolutional neural network. In: 2018 IEEE Winter Conference on Applications of Computer Vision (WACV), pp. 1113–1121. IEEE (2018)
46. Zhang, X., Yu, Q., Yu, H.: Physics inspired methods for crowd video surveillance and analysis: a survey. IEEE Access **6**, 66816–66830 (2018)
47. Zhang, Y., Zhou, D., Chen, S., Gao, S., Ma, Y.: Single-image crowd counting via multi-column convolutional neural network. In: Proceedings of the IEEE Conference on Computer Vision and Pattern Recognition, pp. 589–597 (2016)
48. Zhu, J.Y., Park, T., Isola, P., Efros, A.A.: Unpaired image-to-image translation using cycle-consistent adversarial networks. In: Proceedings of the IEEE International Conference on Computer Vision, pp. 2223–2232 (2017)
49. Zhu, P., Wen, L., Du, D., Bian, X., Hu, Q., Ling, H.: Vision meets drones: past, present and future. arXiv preprint arXiv:2001.06303 (2020)
50. Zou, Z., Cheng, Y., Qu, X., Ji, S., Guo, X., Zhou, P.: Attend to count: crowd counting with adaptive capacity multi-scale cnns. Neurocomputing **367**, 75–83 (2019)

Multi-view Convolutional Network for Crowd Counting in Drone-Captured Images

Giovanna Castellano, Ciro Castiello, Marco Cianciotta, Corrado Mencar, and Gennaro Vessio(✉)

Department of Computer Science, University of Bari "Aldo Moro", Bari, Italy
{giovanna.castellano,ciro.castiello,marco.cianciotta,corrado.mencar, gennaro.vessio}@uniba.it

Abstract. This paper proposes a novel lightweight and fast convolutional neural network to learn a regression model for crowd counting in images captured from drones. The learning system is initially based on a multi-input model trained on two different views of the same input for the task at hand: (*i*) real-world images; and (*ii*) corresponding synthetically created "crowd heatmaps". The synthetic input is intended to help the network focus on the most important parts of the images. The network is trained from scratch on a subset of the VisDrone dataset. During inference, the synthetic path of the network is disregarded resulting in a traditional single-view model optimized for resource-constrained devices. The derived model achieves promising results on the test images, outperforming models developed by state-of-the-art lightweight architectures that can be used for crowd counting and detection.

Keywords: Unmanned aerial vehicles · Crowd counting · Computer vision · Convolutional neural networks · Multi-view learning

1 Introduction

Crowd analysis is the underpinning of several cutting-edge applications, including security surveillance, logistic control in critical situations, traffic management, analysis of people behaviour in simulated environments, and so on [4,10,16]. Crowd analysis encompasses several tasks, including crowd counting and crowd density estimation. While the outcome of a crowd counting task is an estimated number of people in a scene, crowd density estimation yields a map representing the concentration of people in the scene [22]. Usually, these tasks are at the basis of more complex analysis processes, such as people tracking.

Both crowd analysis and density estimation are challenging computer vision tasks posing a number of difficulties related to sparseness of people in the scene, variable lighting conditions, occlusion of subjects, and so on. Recently, besides the standard approaches based on hand-crafted features [7,13,26], some other

A. Bartoli and A. Fusiello (Eds.): ECCV 2020 Workshops, LNCS 12538, pp. 588–603, 2020.
https://doi.org/10.1007/978-3-030-66823-5_35

methods have been populating the literature panorama, which are based on Convolutional Neural Network (CNN) models. These tools are able to approximate complex nonlinear relationships by self-learning meaningful features from the low-level pixel representation of images. The success of CNNs in crowd counting is largely acknowledged when common images are investigated [1,21,28].

In this work, we are interested in analyzing a particular category of images, that are those caught from flying drones. These aircrafts, often referred to as unmanned aerial vehicles (UAVs), are getting more and more attention, and their employment in scenarios of crowd detection and crowd investigation is becoming a hot research topic [25]. Using drones for image capturing is quite a simple task: high-resolution cameras can be mounted on them and even some cheap equipment proves its usefulness to produce real-time image acquisition. Also, the availability of embedded GPUs contributes to turn a simple UAV into a flying image processing device.

Although those perspectives are fascinating, there is a downside to be taken into account. Pictures shot from drones often present a scene where several complications are involved, which may hamper plain image analysis. Such difficulties are related to some technical issues ranging from the distance of the framed subjects, to the different viewpoints which may be set during the flight (depending, for instance, on the aircraft inclination). This, in turn, should call for the application of complex and time-consuming techniques which may stand in contrast to the processing capabilities of UAVs, characterized by limited computing power, real-time response, reduced autonomy, and so on. This makes any computer vision task applied to drone-captured images (particularly, the crowd counting estimation) a real challenge [31].

We rise to this challenge by proposing a particular scheme of deep neural learning consisting in a fully-convolutional neural network (FCN). This can prove to be a light model suitable for tackling the crowd counting task applied to aerial scenes shot from drones. Given an aerial image, the idea is to use the FCN as a regressor to evaluate the number of persons involved in the scene. To improve its generalization power, the network is derived from a multi-input model in which two twin FCNs, i.e. sharing the same topology, are simultaneously trained for the same regression task, starting from two different "views" of the same input. For each analyzed image, the couple of input views is represented by: (*i*) the real-world image that has been acquired from the drone; and (*ii*) a corresponding synthetically created "crowd heatmap". This synthetic image resembles a kind of density map, where high pixel values are associated to the original image regions where people are in the scene; low pixel values are related to areas where people are absent. While multi-input models are typically conceived for solving distinct tasks simultaneously, the proposed method exploits the synthetic input as a means to inject additional information (semantically related to the original one), so that the network can focus on the most important regions of the images, thus helping in the extraction of more meaningful patterns. The synthetic path is used to guide the network during training and is disregarded at inference time, resulting in a lighter network that can be used in embedded devices.

Another problem connected with the analysis of drone-captured images is the lack of suitable data for conducting experiments. That is the reason why in our work we resorted to the VisDrone dataset which includes a huge number of aerial images shot from real UAVs in flight [31]. VisDrone can be intended as a major benchmark at this moment, and its images are heterogeneous enough to cover different scenarios in terms of depicted objects, scenes, contexts, people concentrations, as well as weather and lighting conditions. In this way, we mitigate the well-known problems arising from the reduced amount of variance characterizing the data commonly employed in this kind of research [22]. We compared our approach with state-of-the-art lightweight models and obtained encouraging results from both an inference and computational cost perspective.

The rest of the paper is structured as follows. Section 2 discusses the related work. Section 3 presents the proposed method. Section 4 describes the data used for the present study and reports the obtained results. Finally, Sect. 5 concludes the work.

2 Related Work

In order to address the problem of crowd counting, the CNN proposed in [28] is trained on two learning objectives, namely crowd density and crowd count. More precisely, a switchable learning approach is put in action to obtain better local optima for both tasks. The authors are able to perform cross-scene crowd counting without resorting to re-training and extra annotation: a mapping from images to crowd count is firstly learned, and then the CNN model is fine-tuned for each target scene to overcome the domain gap between different scenes. The proposed method does not rely on foreground segmentation (as a number of other approaches do) since only appearance information is exploited.

CrowdNet is a deep learning-based approach designed to analyze highly dense scenarios, i.e. involving thousands of people [1]. This particular context must be investigated through the analysis of images which obviously present additional challenges. Among them, occlusion of the depicted subjects and perspective problems must be faced. Also, the presence of dense people gatherings makes the realization of annotated crowd datasets almost infeasible, thus reducing the availability of data for this kind of experiments. CrowdNet performs people detection working both at a highly semantic level (where faces and bodies are visible) and at lower level (where people are away from the camera, thus resulting as head blob patterns). This is done through a combination of a deep and shallow convolutional neural network: the first one captures the high-level details from people gatherings; the latter produces the low-level head blob detection. The aim is to estimate crowd density and total crowd count. For ensuring robustness to scale variations, the authors perform extensive data augmentation by sampling patches from the multi-scale image representation.

The problem of analyzing dense crowds has been tackled also in [21]. Once again, the authors' model is oriented to take into account a variety of density levels: the adopted strategy consists in incorporating a high-level prior into the network. In other words, images are analyzed to perform a classification of crowds in terms of their count. Count labels are exploited to estimate coarse count of persons in the image, irrespective of scale variations. The high-level prior is jointly learned along with density map estimation using a cascade of CNNs.

It must be underlined that the previously mentioned papers do not focus on drone-captured images. Also, the proposed solutions are quite demanding from a computational point of view since they do not take into account the reduced processing capabilities characterizing the machinery mounted on a UAV.

When we turn to consider methods addressing the analysis of aerial images, the literature panorama is not so vast. The approach illustrated in [6] moves from the assumption that counting methods relying on regression models often prove to be unsuitable to generate precise object positions. Since the authors are interested in performing their counting process in some specific contexts (i.e., drone-captured images of parked cars), they take advantage of certain identifiable layout patterns to improve accuracy. They employ a novel Layout Proposal Network to count objects, introducing a spatially regularized loss to drive its learning. Obviously, such spatial information can be hardly exploited in a different context involving chaotic people gatherings.

The work presented in [15] is related to crowd images. Once again, the starting point is to perform crowd counting by deep neural learning. However, since viewpoints are continuously modified in sequences acquired by drones, the authors' concern relates to perspective distortion and scale change. To produce a kind of real-world density map (instead of an image-based map), the adopted CNN is fed both with the original image and with another image (of the same size) containing the local scale of the scene, in the way it is determined from the camera settings. Also, the authors' attempt is to model the motion of people making use of physically justified constraints, thus imposing the densities in various images to correspond to those which are physically admissible.

The approach followed in [12] to perform crowd detection and counting in drone-captured images is totally different. The authors propose their SARC-CODI method which is based on image segmentation techniques and neighbourhood filtering. The obtained results are declared to be more precise with respect to other CNN-based state-of-the-art approaches. However, the method has been verified on a reduced number of sample images, thus limiting the appreciation of its generalization power (this is a common issue for several of the proposals found in literature).

In such works as [23] and [24] the focus is still on aerial images, but it is shifted toward crowd detection only, to be achieved by lightweight architectures implementing deep neural learning. To this end, a fully-convolutional network is adopted. By resorting to this kind of model, which discards the fully-connected layer characterizing the classical CNNs, a twofold result can be achieved. On the one hand, the convolutional layers of the architecture preserve the spatial information, thus producing feature maps with enriched possibilities to fit computer vision tasks. On the other hand, a fully-connected architecture decreases the amount of parameters to be adjusted, so that computational effort is reduced.

To test their approach, the authors constructed their own *Crowd-Drone* dataset: they retrieved from Youtube drone videos related to crowded events as well as non-crowded videos. Then, they manually annotated them, labelling the crowded regions from the extracted frames. The model is able to effectively distinguish between crowded and non-crowded scenes, providing also semantic heatmaps of the estimated probability of crowd presence inside the scenes.

Following this research line, in a previous paper of ours we presented an approach to detect people gatherings in aerial images exploiting an FCN architecture [2]. The neural learning is based on a two-loss model coupling a classification task (devoted to distinguish between crowded and non-crowded regions) and a regression task (oriented to produce crowd counting). In a following work [3], we improved our proposal by resorting to the construction of a spatial graph to accomplish the regression task. In the present paper, we delve into our previous work by following a different perspective. Instead of training an FCN for a binary classification, we directly train the model to perform people count. Although the initial model is double-sized, half of it is discarded at test time, thus limiting the computational cost.

3 Proposed Method

Multi-view data are very common in real-world applications where information is collected from diverse sources and exhibits heterogeneous properties, leading to different views of the same phenomenon. Multiple views for a particular problem can take different forms. For example, a document can be described by the words contained in the text, or the metadata describing the document (e.g., title, author, and journal), or the co-citation network graph produced for scientific document management. Likewise, the visual content of an image can be represented by diverse views, including color or texture descriptors, local binary patterns, local shape descriptors, spatio-temporal context information captured by multiple cameras, and so on. A common machine learning approach to handle multi-view data is to concatenate all the multiple views into one single view, in order to derive a predictive model. However, when a small-size training set is used, this concatenation causes overfitting. Also it may not be physically meaningful because each view has a specific statistical property [27]. In contrast to single-view learning, the paradigm of multi-view learning [29] introduces one function to model each particular view and jointly optimizes all the functions to exploit the redundant views of the same input data and improve the generalization performance. The underlying principle is that multiple views may also have very different statistical properties but, when processed simultaneously, they can contribute to gain in predictive accuracy.

Following this research direction, we conceived our idea of coupling multi-view data and multi-functional deep learning for efficient crowd counting in aerial images. Specifically, in our proposal we exploit the real-world RGB image and the corresponding crowd heatmap as multiple views of the same scene containing a crowd. A deep neural network is trained on each view, with weights jointly

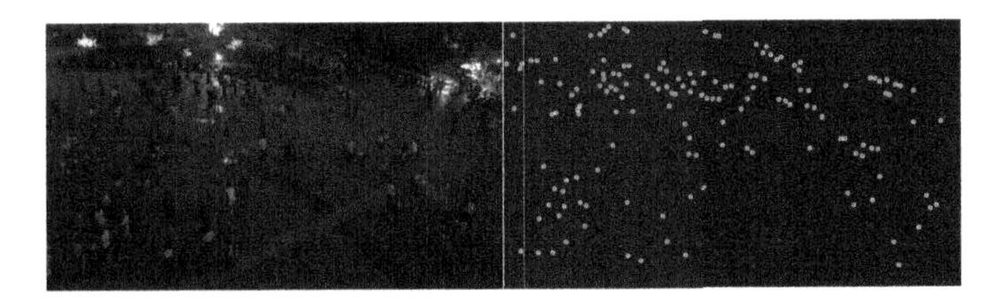

Fig. 1. A crowd scene shot by a drone (on the left), and the corresponding synthetically created crowd heatmap (on the right).

updated at the same time. Finally, only the network that processes the real-world images is retained as final model for crowd counting. This mechanism produces a lightweight model that is suitable to meet the limited computational resources of a UAV. The following subsections describe the proposed method in detail.

3.1 Generation of Crowd Heatmaps

The creation of a crowd heatmap provides an additional view of the real-world scene shot by the drone. In particular, it is meant to help the model better focus on the image's regions where people are actually depicted, thus locating the crowd. We assume the availability of an image dataset which has been suitably annotated to provide information about the presence of people in each scene. For this purpose, it is possible to employ the bounding boxes commonly used for pedestrian detection. Given an annotated dataset, for each image we consider the middle point of the bounding box framing an individual. Then, we draw the ellipsoid of a bivariate zero-mean Gaussian distribution centered on that point:

$$G(x, y) = \frac{1}{2\pi\sigma^2} \exp\left(-\frac{x^2 + y^2}{2\sigma^2}\right),$$

where x and y are the pixel coordinates (with respect to the entire image) of the middle point of an individual. The value of σ, in our case, is a free parameter. A bi-dimensional representation of the distribution can be obtained, with higher pixel values in correspondence with the peak area.

The choice of σ controls the width of the "bell" and is extremely application dependent. After some preliminary trials, we set σ equal to 4: by doing so, we were able to produce (almost everywhere) non-overlapping crowd heatmaps, thus providing precise information about the individuals' location in the scene (see Fig. 1). Such information proves to be very relevant to drive the subsequent learning process of the multi-view network model we are going to describe.

3.2 Multi-view Network Model

Since we are interested in analyzing aerial images taking into account the reduced processing capabilities provided by the machinery mounted on a UAV, the definition of a lightweight model was a first concern of ours. The FCN architecture appears to be an excellent candidate for a number of reasons. First, FCN

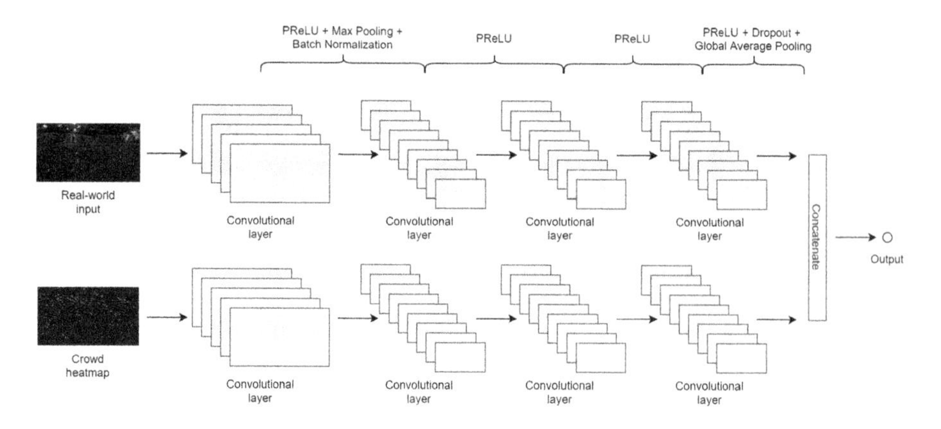

Fig. 2. Scheme of the proposed method.

models may help reduce the computational effort burdening the image processing system. In fact, the idea behind the definition of an FCN is to organize a neural model on the basis of convolutional layers only, thus discarding the fully-connected layers typically introduced at the end of a CNN to produce classification. By doing so, it is possible to decrease the number of parameters to be adjusted in the model for the benefit of the overall reduction of computational complexity. Also, getting rid of the fully-connected layers determines some other advantages related to the specific activity of image processing. On the one hand, images can be presented to the network with any size of representation (the fixed dimension constraint for input images is no longer imposed). On the other hand, the spatial information collected by the network is preserved till the end of the computation, since it is not missed anymore while going through the fully-connected layers (as it is the case in common CNN models).

Moving from the assumptions described above, we defined the FCN architecture illustrated in Fig. 2. The model includes two twin networks (sharing the same topology) whose inputs are represented by the three-channel real-world images and the corresponding crowd heatmaps, respectively. More precisely, both inputs are manipulated beforehand for the sake of processing efficiency, so that they are scaled to 128×128 pixels with values normalized in $[0, 1]$. To prevent the input information from being lost, the convolutional layer is configured to have 32 filters (kernel size is 5×5, and stride is equal to 1). Then, a Parametric ReLU (PReLU) non-linearity follows. A max pooling layer is adopted to downsample the PReLU output (spatial dimensions are divided by a factor of 2). The output is further propagated to reach a batch normalization layer (momentum equal to 0.99 and ϵ equal to 0.001) which is supposed to facilitate generalization [8]. The following three convolutional layers include 64 filters (kernel size is 3×3). It can be observed the increased number of filters with respect to the first convolutional layer: initially, a few low-level features are to be considered in an image; then they may combine in several ways producing a greater number of high-level features. A PReLU activation follows each convolutional layer; there is no pooling layer among them to avoid a further reduction of the feature map resolution (that would be detrimental during the analysis of images presenting some relevant scale variations). A dropout layer (dropout rate set to 50%) and a global average pooling layer complete the scheme of each single network: the first is applied to reduce overfitting, the latter ensures that the final regression process can be performed on the basis of a reduced number of features. The intuition about the design choices for such an FCN architecture is based on previous works of ours [2,3].

The global architecture is finalized by merging the contributes of the twin networks, i.e. the couple of independent branches, encoding the real-world and the synthetic input as globally averaged pooled feature vectors, are concatenated. An output layer, which performs the final regression (thus with no activation function), is added on top of the concatenated representations.

To perform the crowd counting task, the FCN is addressed to the minimization of a loss function. Let N be the number of training samples, and let y_i, y_i' be the actual and the predicted values of crowd count, respectively. The loss function $\mathcal{L}(\theta)$ is represented by the commonly employed mean absolute error:

$$\mathcal{L}(\theta) = \frac{1}{N} \sum_{i=1}^{N} |y_i - y_i'|,$$

which has been preferred among other possible functions due to its simplicity and adherence to the physical meaning.

The overall multi-view network is trained until a termination criterion is met. Then, the network devoted to the analysis of crowd heatmaps is disregarded, while retaining the network trained on real-world images only. Since the concatenating layer has been removed, a new set of weights connecting the global average pooling layer of this final model to the output layer must be randomly initialized. This set of weights is finally fine-tuned on the same data, so that a final prediction can ultimately be obtained. In this way, despite its initial complexity, the ultimate model is still extremely light, thus meeting the strict computational requirements imposed by the UAV. In other words, the initial multi-view architecture is meant to provide pre-trained weights, benefiting from the additional information provided by the crowd heatmaps. It is worth noting that this synthetic information is not available at test time, as the on-board camera realistically captures only RGB images; hence, the network trained on crowd heatmaps can be safely discarded. Obtaining analogous input heatmaps also for the test instances may be feasible with purposely realized models. However, this process would cause a dramatic drop in efficiency, thus making the overall method unpractical for the application in the context of drone-captured image analysis.

3.3 Post-hoc Attention Maps

The proposed model proves its usefulness also to produce a kind of "attention maps" which can be exploited to improve the semantic contents of the flight maps that must be followed by the drone controllers. Such attention maps can spring from a convolutional layer (the last layer or one of the intermediate ones) included in our model. To achieve this aim, we followed a strategy which is similar to the one proposed in [20]. More precisely, a regression activation map (RAM) is proposed, consisting in a weighted sum of the feature maps in a given convolutional layer.

Let $A^k \in \mathbb{R}^{u \times v}$ be the k-th feature map deriving from a convolutional layer, where u and v stand as the values of its height and width, respectively. These maps embed some pieces of information which are useful to identify inside an image the most discriminative regions in regard to the value y' obtained at the end of the regression process. We define a RAM L_{RAM} by applying a ReLU to the linear combination of feature maps:

$$L_{RAM} = ReLU\left(\sum_k \alpha_k A^k\right).$$

According to [20], given the k-th feature map, the corresponding coefficient α_k can be expressed by considering the gradient of y' that is average-pooled over the width and height dimensions:

$$\alpha_k = \frac{1}{uv}\sum_{i=1}^{u}\sum_{j=1}^{v}\frac{\partial y'}{\partial A^k_{i,j}}.$$

In this way, given a feature map, we can measure the impact of each of its pixel on the prediction y'. The RAM can be subsequently upsampled to the input image resolution, so as to spot the regions which are discriminative for the model prediction. This result can be intended as a post-hoc *crowd attention map* drawn from the analysis of the image, without conducting a real learning process to acquire it.

4 Experiment

To test our proposal we derive a crowd dataset from a larger collection of data represented by VisDrone (details on the dataset preparation are provided in the next section).

Moreover, we referred to some baseline methods for the sake of comparison. In a first experiment on crowd counting, we employed MobileNetV2 [19] and TinyYOLOv3 [18]. Concerning the first baseline, it has been pre-trained on ImageNet [11] and subsequently fine-tuned on our crowd dataset. Such an attuning process has been accomplished by discarding the top level classifier and applying a purposely trained layer, as it is commonly the case. MobileNet is a light architecture which is well suited to mobile and embedded computer vision applications [5]. Concerning TinyYOLOv3, it belongs to the "You Only Look Once" (YOLO) family of models for object detection [17] and we employed it to perform crowd counting through a people recognition process (we neglected the identification of other categories of objects). TinyYOLOv3 represents a reduced and faster version of YOLOv3. It has been pre-trained on the MS COCO object detection dataset [14] and fine-tuned to VisDrone.

On the other hand, we performed a second experiment aimed at evaluating our model if employed for crowd detection (i.e., for performing the binary classification crowd vs. non-crowd). To this end, a comparison was made with a couple of baselines (specifically conceived for drone-captured images) we previously described: the FCN proposed by Tzelepi and Tefas [24] and our prior two-loss crowd detector [3].

4.1 Dataset Preparation

The data involved in our experimental sessions come from the VisDrone dataset, that is a benchmark realized by the AISKYEYE team[1] at the Tianjin University, China. VisDrone includes 400 videos that make it the largest reference for drone-captured images among those currently available. The scenes refer to various cities in China and are very different from each other in terms of subjects, scenarios, environmental conditions, and presence of people gatherings. Images in VisDrone are provided with manual annotations (for training and validation sets only), resulting in more than 2.6 million targets (reported as bounding boxes). Among those targets, several categories are considered, ranging from persons to objects and animals. Pedestrians/persons is the specific category of interest in our tests.

We crafted our crowd dataset picking up from VisDrone 2019: taking advantage of the provided annotations (with particular reference to the count of people present in the images), we collected two groups of images including 30 669 and 3394 items, respectively adopted as training and test set. Table 1 shows the distribution of data based on three classes of crowd density (low, medium, high). Low density scenes were much more numerous, therefore, to overcome this imbalance, we artificially augmented the medium and high classes through a number of data augmentation techniques carefully chosen to preserve the image contents and physical meaningfulness (we applied image blurring, horizontal flipping, and contrast adjustment). Since each image is coupled with a crowd heatmap (created by applying the procedure described in Sect. 3.1), the same transformation techniques have been executed also on those synthetic images, to preserve the coherence of the network inputs.

Finally, it is worth noting that the current VisDrone 2020 challenge provides an additional small dataset, specifically conceived for crowd counting. These images concern more overcrowded scenes in bird's-eye view. However, the new data are characterized by only 2420 frames in total. Thus, to evaluate how the proposed method is effectively able to count people in these different frames, we used these data as a separate set and, based on them, we further fine-tuned the model already trained on the bigger purposely designed crowd dataset.

[1] http://aiskyeye.com.

Table 1. Characteristics of our crowd drone dataset derived from VisDrone.

Crowd density	Training set (size)	Test set (size)
Low (less than 10 people)	15 588	1 634
Medium (from 10 to 29 people)	11 882	782
High (30 people and more)	3 199	978
Total	30 669	3 394

4.2 Implementation Details

In our experiments, we used TensorFlow 2.0 and the Keras API. Training was performed offline on Google Colaboratory, which provides a platform for writing and executing Python code and free access to a NVIDIA Tesla K80. To estimate the real-time capabilities of the proposed model, tests were performed on a computational platform which is similar to those usually mounted on drones, that is a NVIDIA Jetson TX2, running Ubuntu 18.04 Operating System on 8GB RAM.

Training of the proposed multi-view FCN model was performed by using the Adam optimizer [9] with randomly sampled mini-batches of 128 images, and learning rate of 2×10^{-4}. Once trained the global model, we discarded the network trained on crowd heatmaps and retained the network fed with real-world images. The latter was fine-tuned with a lower learning rate (set to 5×10^{-5}) to prevent the previously learned weights from being destroyed.

Comparison with the state-of-the-art has been conducted with MobileNetV2, TinyYOLOv3 and the FCNs for crowd detection proposed in [24] and [3]. Since MobileNetV2 had been already trained on ImageNet, the subsequent phase of fine-tuning has been performed on the crowd dataset adopting a low value of learning rate (10^{-4}), so that the previous learned weights could be preserved. To address the specific capacity of this network, we selected a larger size of input image (224×224), and each input channel was re-scaled to the range $[-1, 1]$ (in accordance with the network prerequisites). The input provided to TinyYOLOv3 are images characterized by a larger size (416×416) with pixel values normalized in $[0, 1]$. Concerning the threshold values which are commonly set when object detection process are carried out, we referred to some typical values employed in this context of application. Particularly, the confidence score (related to the capability of a bounding box to provide an accurate object description) and the overlapping threshold (related to the amount of overlapping which regulates the bounding box suppression) have been both set to 0.50. Concerning the model presented in [24], we realized an implementation for it that we tested setting the same ensemble of parameters mentioned in the original paper. Images have been resized to 128×128 (batch size set to 64); the values of learning rate and momentum have been set to 10^{-5} and 0.9, respectively. Finally, training of our previous two-loss model was performed by applying stochastic gradient descent on the basis of randomly sampled mini-batches of 64 images: again, they have been resized to 128×128 and normalized in $[0, 1]$. Learning rate was set at 0.01.

Table 2. Crowd counting results.

Model	MAE	RMSE	Size (MB)	Speed (fps)
MobileNetV2	9.79	14.68	16.74	~35
TinyYOLOv3	11.80	18.41	33.89	~7
Proposed single-view FCN	9.00	14.82	10.56	~125
Proposed multi-view FCN	8.12	13.90	10.56	~125

All the involved models have been trained using an early stopping strategy with patience amounting to 1. This technique was applied by monitoring the loss value on a validation set randomly held out as a fraction of 10% of the training set. Concerning the training time, each model has been trained for few tens of epochs before reaching convergence, which required a few hours of time. Since the data from VisDrone exhibit a high degree of variance (which determines quick overfitting), such results are in line with expectations.

4.3 Experimental Results

Crowd Counting. Evaluation of experimental results for crowd counting has been performed by referring to a couple of standard metrics which are commonly employed to assess the effectiveness of crowd counting methods, i.e. mean absolute error (MAE) and root mean squared error (RMSE). As additional means for assessment, we refer also to some measures of size and speed which evaluate the models during their operation. Size is expressed in HDF5 format (measured in terms of Megabytes); speed is expressed in terms of frames per second (fps).

All the metrics values reported in Table 2 have been derived during the tests conducted on each model applied to our crowd test set. The table includes the results pertaining to two versions of the proposed method. The first one is a version of the proposed FCN designed as a single-input model, without the additional view provided by the crowd heatmaps. The second one is the network related to the proposed multi-view FCN. The baseline results are also reported in table for the sake of comparison.

TinyYOLOv3 exhibits the worst results among all the tested models. Being VisDrone a challenging dataset, an object detection technique (such any of those included in the YOLO family) was expected to meet some difficulties while tackling it [30]. To equalize the assessment with other models, the y_i' value predicted by TinyYOLOv3 for each scene has been calculated by considering both the true positives and the false positives among its predictions. Also, the low speed value can be justified in terms of the intrinsic behaviour of TinyYOLOv3 which, in addition to predicting the number of people, also takes care of specifying their position inside the scene. MobileNetV2 exhibits a reduced size and an increased speed, thus confirming the suitability of regression methods for facing the crowd counting task. The best results have been achieved by the proposed FCN. Particularly, the multi-view network model exhibits some slightly better results both

Table 3. Crowd detection results.

Model	Accuracy	Precision	Size (MB)	Speed (fps)
Replica of [24]	0.79	0.83	17.30	~12
Our previous work [3]	0.88	0.88	2.00	~13
Proposed multi-view FCN	0.85	0.87	10.56	~125

in terms of MAE and RMSE values. The accuracy supremacy with respect to MobileNetV2 can be justified if we consider that our proposal is characterized by a reduced capacity, thus being less susceptible to overfitting. Also, pre-training of MobileNetV2 was performed on data which are quite different from those included in VisDrone.

As previously mentioned, a smaller dataset for crowd counting has been recently released. These data are characterized by a higher altitude, if compared to the other images of VisDrone, making the task particularly challenging. Thus, we used these data as a separate dataset to further validate the effectiveness of the proposed method. Indeed, our model has been submitted for the crowd counting task of the currently running VisDrone challenge. To evaluate the prediction performance on the test set, we further fine-tuned the FCN model derived from the multi-view network on the training data provided, stopping backpropagation again by monitoring the loss value on a held-out validation set. We finally obtained an MAE of 51.63, which improves the baseline results made publicly available but still calls for further developments.

Crowd Detection. As an additional experimentation, it is worth to note that the test images can be divided into categories and then the model can be evaluated by reformulating the task as a classic classification problem. In particular, test images can be divided into two classes: sparse scenes, with less than 10 people in the scene; and crowded scenes, with 10 people and more in the scene. Accordingly, we adjusted the actual and predicted counts as category labels. In this way, we were able to measure standard classification metrics such as classification accuracy and average precision.

As it can be seen in Table 3, which reports the results of the proposed FCN model in discriminating between sparse (less then 10 people) and crowded (10 people and more) scenes, the model was, on average, pretty good in correctly detecting the presence or absence of a crowd in the test images. Evaluation shows that the approach has a competitive performance at higher frame rate compared to state-of-the art crowd detectors from drones.

It is worth noting that the crowd classification task has not been tested for the small crowd counting data, as they do not provide sparse scenes.

Post-hoc Attention Maps. As a final remark, we underline how the proposed method is useful to produce post-hoc crowd attention maps which can be exploited to enrich the semantic contents of the flight maps. In Fig. 3 are reported

Fig. 3. Top row: four test images; bottom row: the corresponding attention maps provided by the method (for a better visualization, they have been re-scaled to their original sizes).

four sample images together with their corresponding heatmaps. The figure lets us appreciate the capability of the model to distinguish the crowded areas in the scene from unoccupied zones. This discriminating capability is highly desirable for UAVs, since it can be exploited to accomplish a new set of tasks (among the most important, autonomous landing in emergency conditions).

5 Conclusion

Unmanned aerial vehicles are increasingly populating our skies finding application in a variety of contexts. Recent research advancements demonstrate that it is relatively easy to provide these vehicles with a minimum set of hardware equipment so that they can be translated into mobile devices for image processing. In this paper we addressed the specific problem of crowd counting which has a relevant impact in several application contexts, such as security surveillance, emergency intervention, attendance checking during public rallies, and so on. We proposed a multi-view lightweight model represented by a fully-convolutional network. The proposed method was able to outperform some well-known models employed as baseline. In addition to performing a regression task, our model is also capable of generating heatmaps which visually highlight the presence of people inside a scene, thus enriching the semantic contents of the flight maps followed by the UAVs.

The extreme lightness and speed make the method a suitable candidate for further investigations. In fact, the idea of training the model with multiple views, of which only one is retained for final use, can be explored in several ways for the sake of model lightness. As a future work, we want also to conceive some methods to manage the video streams shot by a camera mounted on a drone (instead of referring to them in terms of sequences of images). In this way, it could be possible to take into account even the people flow.

Acknowledgement. This work was supported by the Italian Ministry of Education, University and Research within the "RPASInAir" Project under Grant PON ARS01_00820.

References

1. Boominathan, L., Kruthiventi, S.S., Babu, R.V.: Crowdnet: a deep convolutional network for dense crowd counting. In: Proceedings of the 24th ACM International Conference on Multimedia, pp. 640–644. ACM (2016)
2. Castellano, G., Castiello, C., Mencar, C., Vessio, G.: Crowd detection for drone safe landing through fully-convolutional neural networks. In: Chatzigeorgiou, A., et al. (eds.) SOFSEM 2020. LNCS, vol. 12011, pp. 301–312. Springer, Cham (2020). https://doi.org/10.1007/978-3-030-38919-2_25
3. Castellano, G., Castiello, C., Mencar, C., Vessio, G.: Crowd detection in aerial images using spatial graphs and fully-convolutional neural networks. IEEE Access **8**, 64534–64544 (2020)
4. Chaker, R., Al Aghbari, Z., Junejo, I.N.: Social network model for crowd anomaly detection and localization. Pattern Recogn. **61**, 266–281 (2017)
5. Howard, A.G., et al.: Mobilenets: Efficient convolutional neural networks for mobile vision applications. arXiv preprint arXiv:1704.04861 (2017)
6. Hsieh, M.R., Lin, Y.L., Hsu, W.H.: Drone-based object counting by spatially regularized regional proposal network. In: Proceedings of the IEEE International Conference on Computer Vision, pp. 4145–4153 (2017)
7. Idrees, H., Saleemi, I., Seibert, C., Shah, M.: Multi-source multi-scale counting in extremely dense crowd images. In: Proceedings of the IEEE Conference on Computer Vision and Pattern Recognition, pp. 2547–2554 (2013)
8. Ioffe, S., Szegedy, C.: Batch normalization: Accelerating deep network training by reducing internal covariate shift. arXiv preprint arXiv:1502.03167 (2015)
9. Kingma, D.P., Ba, J.: Adam: A method for stochastic optimization. arXiv preprint arXiv:1412.6980 (2014)
10. Kok, V.J., Lim, M.K., Chan, C.S.: Crowd behavior analysis: a review where physics meets biology. Neurocomputing **177**, 342–362 (2016)
11. Krizhevsky, A., Sutskever, I., Hinton, G.E.: Imagenet classification with deep convolutional neural networks. In: Advances in Neural Information Processing Systems, pp. 1097–1105 (2012)
12. Küchhold, M., Simon, M., Eiselein, V., Sikora, T.: Scale-adaptive real-time crowd detection and counting for drone images. In: 2018 25th IEEE International Conference on Image Processing (ICIP), pp. 943–947. IEEE (2018)
13. Liang, R., Zhu, Y., Wang, H.: Counting crowd flow based on feature points. Neurocomputing **133**, 377–384 (2014)
14. Lin, T.Y., et al.: Microsoft COCO: common objects in context. In: Fleet, D., Pajdla, T., Schiele, B., Tuytelaars, T. (eds.) ECCV 2014. LNCS, vol. 8693, pp. 740–755. Springer, Cham (2014). https://doi.org/10.1007/978-3-319-10602-1_48
15. Liu, W., Lis, K.M., Salzmann, M., Fua, P.: Geometric and physical constraints for drone-based head plane crowd density estimation. In: IEEE/RSJ International Conference on Intelligent Robots and Systems (IROS). No. CONF, IEEE/RSJ (2019)
16. Lloyd, K., Rosin, P.L., Marshall, D., Moore, S.C.: Detecting violent and abnormal crowd activity using temporal analysis of grey level co-occurrence matrix (GLCM)-based texture measures. Mach. Vis. Appl. **28**, 361–371 (2017). https://doi.org/10.1007/s00138-017-0830-x
17. Redmon, J., Divvala, S., Girshick, R., Farhadi, A.: You only look once: unified, real-time object detection. In: Proceedings of the IEEE Conference on Computer Vision and Pattern Recognition, pp. 779–788 (2016)

18. Redmon, J., Farhadi, A.: Yolov3: An incremental improvement. arXiv preprint arXiv:1804.02767 (2018)
19. Sandler, M., Howard, A., Zhu, M., Zhmoginov, A., Chen, L.C.: Mobilenetv 2: inverted residuals and linear bottlenecks. In: Proceedings of the IEEE Conference on Computer Vision and Pattern Recognition, pp. 4510–4520 (2018)
20. Selvaraju, R.R., Cogswell, M., Das, A., Vedantam, R., Parikh, D., Batra, D.: Grad-CAM: visual explanations from deep networks via gradient-based localization. In: Proceedings of the IEEE International Conference on Computer Vision, pp. 618–626 (2017)
21. Sindagi, V.A., Patel, V.M.: CNN-based cascaded multi-task learning of high-level prior and density estimation for crowd counting. In: 2017 14th IEEE International Conference on Advanced Video and Signal Based Surveillance (AVSS), pp. 1–6. IEEE (2017)
22. Sindagi, V.A., Patel, V.M.: A survey of recent advances in CNN-based single image crowd counting and density estimation. Pattern Recogn. Lett. **107**, 3–16 (2018)
23. Tzelepi, M., Tefas, A.: Human crowd detection for drone flight safety using convolutional neural networks. In: 2017 25th European Signal Processing Conference (EUSIPCO), pp. 743–747. IEEE (2017)
24. Tzelepi, M., Tefas, A.: Graph embedded convolutional neural networks in human crowd detection for drone flight safety. IEEE Trans. Emerg. Topics Comput. Intell. (2019)
25. Valavanis, K.P., Vachtsevanos, G.J.: Handbook of Unmanned Aerial Vehicles. Springer, Heidelberg (2015). https://doi.org/10.1007/978-90-481-9707-1
26. Xing, J., Ai, H., Liu, L., Lao, S.: Robust crowd counting using detection flow. In: 2011 18th IEEE International Conference on Image Processing, pp. 2061–2064. IEEE (2011)
27. Yu, J., Wang, M., Tao, D.: Semisupervised multiview distance metric learning for cartoon synthesis. IEEE Trans. Image Process. **21**(11), 4636–4648 (2012)
28. Zhang, C., Li, H., Wang, X., Yang, X.: Cross-scene crowd counting via deep convolutional neural networks. In: Proceedings of the IEEE Conference on Computer Vision and Pattern Recognition, pp. 833–841 (2015)
29. Zhao, J., Xijiong, X., Xu, X., Sun, S.: Multi-view learning overview: recent progress and new challenges. Inf. Fus. **38**, 43–54 (2017). https://doi.org/10.1016/j.inffus.2017.02.007
30. Zhu, P., et al.: VisDrone-VID2019: the vision meets drone object detection in video challenge results. In: Proceedings of the IEEE International Conference on Computer Vision Workshops (2019)
31. Zhu, P., Wen, L., Bian, X., Ling, H., Hu, Q.: Vision meets drones: A challenge. arXiv preprint arXiv:1804.07437 (2018)

PAS Tracker: Position-, Appearance- and Size-Aware Multi-object Tracking in Drone Videos

Daniel Stadler[1,2(✉)], Lars Wilko Sommer[2,3], and Jürgen Beyerer[1,2]

[1] Vision and Fusion Lab, Karlsruhe Institute of Technology, Haid-und-Neu-Str. 7, 76131 Karlsruhe, Germany
{daniel.stadler,juergen.beyerer}@iosb.fraunhofer.de
[2] Fraunhofer IOSB, Fraunhoferstr. 1, 76131 Karlsruhe, Germany
lars.sommer@iosb.fraunhofer.de
[3] Fraunhofer Center for Machine Learning, Munich, Germany

Abstract. While most multi-object tracking methods based on tracking-by-detection use either spatial or appearance cues for associating detections or apply one cue after another, our proposed PAS tracker employs a novel similarity measure that combines position, appearance and size information jointly to make full use of object representations. We further extend the PAS tracker by introducing a filtering technique to remove false positive detections, particularly in crowded scenarios, and apply a camera motion compensation model to align track positions across frames. To provide high quality detections as input for the proposed tracker, the performance of eight state-of-the-art object detectors is compared on the VisDrone MOT dataset, on which the PAS tracker achieves state-of-the-art performance and ranks 3rd in the VisDrone2020 MOT challenge. In an ablation study, we demonstrate the effectiveness of the introduced tracking components and the impact of the employed detections on the tracking performance is analyzed in detail.

Keywords: Multi-object tracking · Object detection · Drone imagery

1 Introduction

Multi-object tracking (MOT) in drone-based imagery has several applications in traffic analysis, sports recording or surveillance. The task demands the localization and identification of a defined set of objects in a video sequence so that each individual object gets a unique identity number and keeps it in every frame. While the tracking of multiple objects is challenging because of occlusion, fast moving objects or poor image quality, additional difficulties emerging from drone-based imagery are small object sizes, sparse object distributions and huge appearance variations due to different camera angles and altitudes as can be seen in Fig. 1.

In recent years, the release of large drone-based image and video datasets [15,17,52,55,56] has led to an increased development and evaluation of new

A. Bartoli and A. Fusiello (Eds.): ECCV 2020 Workshops, LNCS 12538, pp. 604–620, 2020.
https://doi.org/10.1007/978-3-030-66823-5_36

Fig. 1. Example images of the VisDrone MOT dataset [56]

methodologies in this growing research field. While some recent approaches try to solve the MOT problem in one step [1,41,50], most of the commonly applied methods follow the *tracking-by-detection* paradigm by first applying an object detector on each frame of the video and then associating the detections across frames [5,18,27,40,46,48,51]. This association is often formulated as a graph problem, where each node corresponds to a detection and each edge represents a cost or likelihood of two detections belonging to the same object instance, whereby the goal is to minimize the overall costs. The existing approaches can be distinguished into batch methods that build a graph over the whole video and solve the MOT problem offline and methods that use only past information and run in an online fashion.

For the association task, typically either spatial or appearance cues are used separately or applied in a sequential manner. We argue, that all available information should be considered at the same time in order to maximize the benefit from each property. Thus, we propose a novel similarity measure that jointly combines **p**osition, **a**ppearance and **s**ize information of objects and, hence, term our approach PAS tracker. To predict accurate object locations for each frame, we extend the constant velocity assumption (CVA) by a camera motion compensation (CMC) model. For visual information, we embed the appearance of objects with a model adopted from person re-identification. Additionally, we propose a filtering technique to remove false positive detections in crowded scenarios. The proposed PAS tracker achieves state-of-the-art results on the VisDrone MOT dataset and ranks 3rd in the VisDrone2020 MOT challenge[1].

In the ablation study, we show the benefits of our tracking modules and the tracking performance is analyzed w.r.t. class-specific detection score thresholds, which we tune carefully. Since no public detections are provided, we train eight state-of-the-art object detection models. Furthermore, we improve the detection performance by leveraging test-time strategies, e.g. multi-scale testing and horizontal flipping. To summarize, the main contributions of this work are as follows:

- Following the tracking-by-detection paradigm, we develop a novel similarity measure that jointly combines position, appearance and size information of objects and propose a filtering technique that further improves the tracking performance by removing false positive detections.

[1] http://www.aiskyeye.com.

- To provide high quality detections as input for our tracker, we train and compare the performance of eight state-of-the-art object detectors on the VisDrone MOT dataset and apply several test-time strategies to improve the detection performance.
- In the ablative experiments, we examine the dependency of a tracking-by-detection based MOT approach on the employed detection score thresholds and analyze the influence of the modules of our PAS tracker in detail.
- The proposed PAS tracker achieves state-of-the-art performance on the VisDrone MOT dataset and ranks 3rd in the VisDrone2020 MOT challenge.

2 Related Work

Most of the MOT methods first run a detector on each frame of the input sequence and then link the detections across frames using spatial or appearance cues. The advantage of this tracking-by-detection procedure is that the ongoing improvements in the field of object detection can be fully utilized. In the following, an overview of recent deep learning based object detection methods and related MOT approaches is provided.

2.1 Object Detection

Nowadays, deep learning based object detection methods outperform conventional detection methods by a large margin in different domains including object detection in drone-based imagery. These methods can be categorized into one-stage and two-stage approaches. While the latter ones generate object candidates in the first stage and classify them in the second stage, one-stage methods perform both localization and classification at once yielding a higher processing speed but at the cost of generally lower detection accuracy.

Besides SSD [25], YOLO [35] and its variants [36,37], RetinaNet [23] belongs to the most popular architectures of one-stage detectors. It introduces a new loss function that addresses the class imbalance problem in anchor based detectors by focusing more on hard examples during training. FreeAnchor [53] makes the matching of objects and anchors more flexible than the predominant matching based on Intersection over Union (IoU) by formulating detector training as a maximum likelihood estimation procedure. FCOS [44] and FoveaBox [19] do not rely on anchors at all by regressing object locations and predicting class scores directly. CornerNet [21] and CenterNet [10] also do not require the computation of anchor boxes, since they represent objects as points.

The most prominent two-stage detector is Faster R-CNN [38] that is comprised of an initial Region Proposal Network (RPN) and a subsequent classification stage, which share the convolutional layers of the backbone network. In the classification stage, the region proposals generated by the RPN are projected onto the output of the last convolutional layer termed feature map. For each region proposal, the corresponding features are aggregated via Region of Interest (RoI) pooling and used as input for the classification and regression head that further improves its localization. Multiple extensions have been proposed to improve the detection accuracy of Faster R-CNN. While Feature Pyramid Network (FPN) [22] introduces a top-down path to combine features from different layers, Cascade R-CNN [7] improves the proposal generation step by training several stages with increasing IoU thresholds making the final proposals more accurate. Grid R-CNN [26] adapts the regression branch so that an object is represented as a grid of points instead of only two points of a bounding box. Libra R-CNN [33] introduces IoU-balanced sampling, balanced feature pyramid and balanced L1 loss in order to reduce the class imbalance problem of anchor based detectors. HRNet [42] maintains high resolution feature maps in the backbone network to enrich the feature representation capabilities.

For drone-based imagery, in particular on the VisDrone MOT dataset, the applicability of most of the aforementioned detectors has been shown with CenterNet and Cascade R-CNN being the best performing one-stage and two-stage detector, respectively [9].

2.2 Multi-Object Tracking

One of the simplest tracking-by-detection approaches is the IOU tracker [4] that associates the frame-wise detections only based on IoU. To circumvent fragmented tracks caused by missing detections, the advanced V-IOU tracker [5] makes use of single object trackers like KCF [14] or Medianflow [16]. Another extension of the IoU Tracker is the so-called KF-IoU Tracker - the winner of the IWT4S Challenge on Advanced Traffic Monitoring [29] - that employs a Kalman Filter (KF) to prevent fragmented tracks and to reduce the number of ID switches. SORT [3] also uses a KF for state prediction and bases the association on IoU, whereas its improved version Deep SORT [48] leverages the KF only for inhibiting impossible matches and associates detections based on visual appearance computed from a person re-identification model. Joint Detection and Embedding [46] integrates an appearance model into a one-stage detector and learns classification, localization and appearance of objects jointly. Similarly, in [45], a Mask R-CNN [13] is extended by an association head to learn object detection, segmentation and appearance within one shared network. Furthermore, 3D convolutions are used to extract spatio-temporal features of a small sequence of images instead of treating each frame at once. All the aforementioned methods apply either position or appearance cues in the association step or apply one cue after another. In contrast, our PAS tracker combines position, appearance and size information of objects into a joint similarity measure in order to use as much information as possible when associating detections to tracks.

Some recent approaches aim at integrating detection and tracking more tightly or performing both tasks simultaneously. Zhang et al. [54] condition the detector in each frame on tracklets computed in the previous frame. In [12], correlation features between two consecutive frames are utilized for track regression and a multi-task loss for simultaneous detection and tracking is applied. Tracktor [1] utilizes the regression head in the second step of a two-stage detector to track objects over time and no extra association step is needed. While the enhanced Tracktor++ [1] also uses a CMC model and a re-identification model for resuming stopped tracks, our PAS tracker integrates the appearance information directly into the association. Xu et al. [50] approximate the Hungarian algorithm [20] by a Deep Hungarian Net and, additionally, propose differentiable approximations of the MOTA and MOTP metrics [2] allowing an end-to-end training of the whole tracking framework. Brasó et al. [6] learn the association step directly based on Message Passing Networks by formulating the MOT task as a network flow formulation problem. TubeTK [32] regresses bounding boxes of objects in multiple frames in one step, after extracting spatio-temporal features with a fully 3D convolutional neural network.

In case of drone-based imagery, MOT methods that follow the tracking-by-detection strategy are commonly applied. Due to the large variations in image content, e.g., object appearance, large effort is rather put in designing a strong object detection model than in developing a sophisticated tracking pipeline. For example, the best reported tracker on the VisDrone MOT dataset is based on the simple SORT algorithm [30]. Similar findings are made analyzing the top performing trackers on the UA-DETRAC [47] benchmark, a dataset of videos captured from real-world traffic scenes. Thus, integrating more information from objects into the association step is a promising research direction for MOT in drone-based imagery.

3 Position-, Appearance- and Size-Aware Tracking

The proposed PAS tracker combines position, appearance and size information for associating detections to tracks following the tracking-by-detection paradigm. Before we delve into our tracking approach in more detail, we give a brief summary of the development of the applied object detection model. How the detector is integrated into the overall tracking pipeline can be seen in Fig. 2.

3.1 Object Detector

For the VisDrone MOT dataset, no public detections are available. To generate high quality detections for our tracker, we train several state-of-the-art object detectors on the train split of the VisDrone MOT dataset and compare their performances on the test-dev split, which is the default test data for general evaluation. Both two-stage detectors, i.e., Faster R-CNN [38], Libra R-CNN [33], Grid R-CNN [26], Cascade R-CNN [7], and one-stage detectors, namely RetinaNet [23], FoveaBox [19], FreeAnchor [53], are used. ResNeXt-101 (32×4d) [49] is utilized for strong feature extraction, while HRNetV2p-W32 [42] is also taken as backbone of the Faster R-CNN framework. In order to make the detector more robust w.r.t. different object scales, we use a FPN [22] as neck in every model. To enhance the feature representation learning during the training process, we follow the SSD data augmentation pipeline [25]. At inference time, we increase the number of region proposals to account for the high object densities

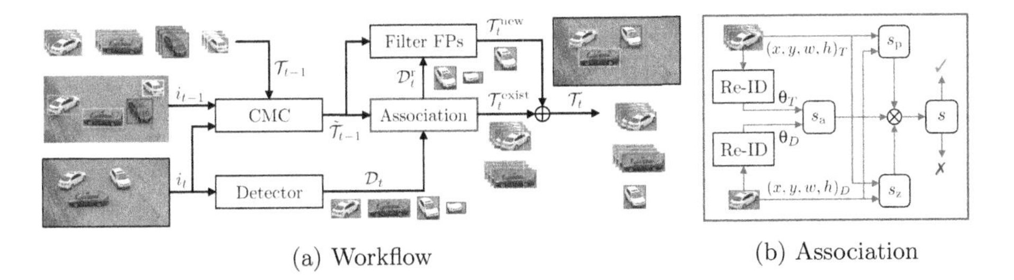

Fig. 2. Overview of the proposed PAS tracker. (a) For each frame, a detector generates a set of detections and the tracks from the previous frame are aligned with a CMC model. Then, detections are associated to tracks based on a similarity measure. Before new tracks are started, remaining false positive detections are filtered. (b) The association is based on a joint similarity measure of position, appearance and size information.

and perform multi-scale testing as well as horizontal flipping to further boost the detection performance.

3.2 PAS Tracker

Our tacker builds upon the tracking-by-detection paradigm and works in an online fashion. After the detection step, position, appearance and size information of objects is considered in the association step. In each frame of a video sequence, similarity measures between all detections and tracks are computed and collected in a cost matrix representing the likelihood of all possible matches. The assignment problem then is solved with the Hungarian method [20]. The entries of the cost matrix measure the similarity between a track and a detection in terms of position, appearance and size. In the following, the components and the workflow of the proposed PAS tracker are described in more detail. For a schematic overview, refer to Fig. 2, whereas the single steps of the pipeline can be found in Algorithm 1.

Position Similarity. Let $\mathcal{D}_t = \{D_t^1, D_t^2, \dots, D_t^m\}$ denote the set of detections generated by the object detector at time t and $\mathcal{T}_{t-1} = \{T_1, T_2, \dots, T_n\}$ the set of tracks at time $t-1$. A detection $D_t = (x, y, w, h, s, q)$, generated by the object detection model on frame i_t of a video sequence $I = [i_1, i_2, \dots, i_L]$, is characterized by a bounding box with center position $\mathbf{p}_D = (x, y)$ and size $\mathbf{z}_D = (w, h)$, a detection score $s_D \in [0, 1]$ and a predicted class membership q_D. When objects of different categories are treated, detections and tracks are processed for each category separately. A track $T = [D_{t_1}, D_{t_2}, \dots, D_{t_N}]$ is represented as an ordered list of detections from the same object, where the time information for each detection is also stored. For estimating the position of a track in the current frame t, we follow a CVA and calculate the velocity of a track $\mathbf{v}_T$ as the displacement of the center of the last two detections of two consecutive frames:

Algorithm 1: Workflow of the Proposed PAS Tracker

```
Data: Ordered sequence of images I = [i_1, i_2, ..., i_L]
Result: Set of tracks T = {T_1, T_2, ..., T_n} with T_n = [D^n_{t_1}, D^n_{t_2}, ..., D^n_{t_N}] as
        ordered list of detections
1  T^a, T^i, T^f ← ∅     // init. sets of active, inactive and finished tracks
2  for i_t ∈ I do
3      D_t = {D^1_t, D^2_t, ..., D^m_t} ← detector(i_t)          // detections at time t
4      T_{t-1} = T^a ∪ T^i                                        // tracks at time t − 1
5      T̃_{t-1} ← camera_motion_compensation(T_{t-1}, i_t, i_{t-1})
6      C = [c_{k,l}] ← compute_cost_matrix(T̃_{t-1}, D_t)          // Equation (5)
7      {(k,l)} ← min_cost_matching(C)                              // Hungarian method
8      for (k,l) ∈ {(k,l)} do                                      // |{(k,l)}| = |T_{t-1}|
9          if c_{k,l} < c_max then
10             T_k ← T_k + D^l_t        // extend track with associated detection
11             if T_k ∈ T^i then
12                 T^a ← T^a ∪ {T_k}, T^i ← T^i\{T_k}              // re-activate track
13             D_t = D_t\{D^l_t}                                   // remove detection
14         else if T_k ∈ T^a then
15             T^a ← T^a\{T_k}, T^i ← T^i ∪ {T_k}                  // inactivate track
16
17     end
18     D^r_t ← filter_false_positive_detections(D_t, T̃_{t-1})
19     for D_t ∈ D^r_t do          // start tracks with remaining detections
20         T^a ← T^a ∪ {[D_t]}
21     end
22     for T ∈ T^i do                // finish tracks inactive for too long
23         if t − t_N(T) > t^i_max then
24             T^f ← T^f ∪ {T}, T^i ← T^i\{T}
25
26     end
27 end
28 T ← T^f ∪ T^a ∪ T^i
```

$\mathbf{v}_T = \mathbf{p}(D_{t_N}) - \mathbf{p}(D_{t_{N-1}})$. New tracks are initialized with $\mathbf{v}_T = 0$. With the velocity and the CVA, the estimated position of a track $\hat{\mathbf{p}}_T$ at time t can be calculated: $\hat{\mathbf{p}}_T = \mathbf{p}(D_{t_N}) + \mathbf{v}_T \cdot (t - t_N)$. For each track-detection pair $(T_k, D^l_t)_{k=1...n, l=1...m}$, the position similarity s_p can be computed as

$$s_\text{p} = \max(1 - \lambda_\text{p} ||(\hat{\mathbf{p}}_T - \mathbf{p}_D) \oslash \mathbf{z}_D||, 0) \tag{1}$$

with $\oslash$ denoting the element-wise division, $||\cdot||$ representing the Euclidean norm and λ_p is a hyper-parameter that tunes the size of the neighbourhood of the detection, where the position similarity is not zero. We set λ_p empirically to 0.3.

Appearance Similarity. For a visual comparison of detections and tracks, we use a re-identification model from [28]. It takes as input image crops of the detection and track bounding boxes, respectively, and outputs 2048-dimensional vectors $\boldsymbol{\theta}$ that represent the appearances of the underlying objects. For a track, the appearance vector is obtained by building the mean vector over at most the last 10 detections. The appearance similarity s_{a} between a track T and a detection D is given by the cosine similarity of their appearance vectors $\boldsymbol{\theta}_T$ and $\boldsymbol{\theta}_D$:

$$s_{\mathrm{a}} = \frac{\boldsymbol{\theta}_T \cdot \boldsymbol{\theta}_D^{\mathrm{T}}}{||\boldsymbol{\theta}_T|| \cdot ||\boldsymbol{\theta}_D||}. \tag{2}$$

While other metrics could be applied to measure the similarity between the two appearance vectors, the cosine similarity has the advantage of being normalized, what makes it straightforward to combine with the position similarity s_{p}.

Size Similarity. As a third metric, we compare the sizes of detections and tracks. Since the sizes of objects in the image frames can change over time due to relative movement w.r.t. the camera, the size of a track is set as the size of the last detection belonging to the track. Again, aiming at a normalized measure for easy combination with the previously defined similarities, we take the following formula for calculating the size similarity s_{z}, which is adopted from [43]:

$$s_{\mathrm{z}} = 1 - ||(\mathbf{z}_T - \mathbf{z}_D) \oslash (\mathbf{z}_T + \mathbf{z}_D)|| \tag{3}$$

Camera Motion Compensation. The sequences of the VisDrone MOT dataset are captured with a high frame rate making the CVA a valid motion model in a lot of situations. But there also exist scenarios with a moving or rotating camera, where the CVA is no good approximation so that the position similarity calculation is harmed. For this reason, we apply a CMC model that uses the Enhanced Correlation Coefficient Maximization from [11]. Given two consecutive frames from the video sequence I, i.e., i_t and i_{t-1}, an Euclidean transformation is computed that is used to align the tracks $\mathcal{T}_{t-1}$ from frame i_{t-1} with the current image i_t yielding the transformed tracks $\tilde{\mathcal{T}}_{t-1}$ and, therefore, the quality of the position similarity measure is improved.

Association and Track Management. The final similarity s that is used for associating detections to tracks is the product of the three previously defined similarities:

$$s = s_{\mathrm{p}} \cdot s_{\mathrm{a}} \cdot s_{\mathrm{z}} \tag{4}$$

In each frame i_t of the video sequence I, the similarities $s_{k,l} \in [0,1]$ of all track-detection pairs $(T_k, D_t^l)_{k=1\ldots n, l=1\ldots m}$ are calculated and subtracted from 1 yielding the cost matrix C:

$$C = [c_{k,l}] = [1 - s_{k,l}] \tag{5}$$

The assignment problem is solved using the Hungarian method that outputs a set of associated detections and tracks $\{(k,l)\}$. If the cost of the association is below a threshold, i.e., $c_{k,l} < c_{\max}$, the track T_k is extended by the detection D_t^l. We empirically set $c_{\max}$ to 1. This corresponds to prohibiting associations if the position similarity is zero. In order to make the tracker robust against missing detections, we do not stop tracks immediately when no detection is assigned but allow a track to be inactive for at most $t_{\max}^{\mathrm{i}} = 10$ frames before declaring it as finished. Both the active tracks $\mathcal{T}^{\mathrm{a}}$ and inactive tracks $\mathcal{T}^{\mathrm{i}}$ are considered in the association step. Detections that are not associated with an existing track ($\mathcal{T}^{\mathrm{exist}}$) start a new one ($\mathcal{T}^{\mathrm{new}}$).

Filtering of False Positives. As the tracking performance depends on the quality of the available detections, we propose a simple approach for filtering false positive detections that appear most notably in crowded scenarios. We hypothesize, that a detection, which has not been associated by the Hungarian method and has a high overlap with existing tracks, is probably a false positive and remove it from the remaining set of detections $\mathcal{D}_t^{\mathrm{r}}$ before starting new tracks. More precisely, for each remaining detection after the association step, the overlapping area o with existing tracks is computed. If $o/(wh) > o_{\max}$, the detection is discarded, where we empirically set $o_{\max} = 0.8$. In addition, all tracks with a length shorter than 10 are removed in a post-processing step presuming that most of them arise from false positive detections.

4 Experiments

Before we compare our PAS tracker with state-of-the-art approaches, we evaluate the performance of different object detection models.

4.1 Object Detector

For training the detectors, the MMDetection toolbox [8] is used and available weights originating from trainings on the COCO [24] dataset are taken for initialization of our models. As the resolution of the VisDrone images is quite large, crops of size 608×608 pixels are randomly sampled from the training images. We train each detector for 12 epochs and batches of 16 with SGD and a learning rate of 0.02, momentum of 0.9 and weight decay of 0.0001. We lower the learning rate by a factor of 10 after epochs 8 and 11. A warm-up strategy and gradient clipping are used. Although we train on all 10 classes of the dataset to learn more diverse feature representations, we only evaluate on the 5 classes *pedestrian*, *car*, *van*, *truck* and *bus* because these are the default classes for the MOT task [30]. To compare the performance of the different detectors, we compute the COCO metrics [24] on the test-dev split. Whereas training is performed on image crops, we process the original-sized images in the inference stage. The results are depicted in Table 1. Note that all detectors except the Faster R-CNN with HRNetV2p-W32 apply a ResNeXt-101 (32×4d) as backbone for feature extraction and a FPN [22] neck is integrated in every model to improve the performance for different object scales. For the calculation of the average precision

Table 1. Performance of different detectors on the test-dev split of the VisDrone MOT dataset. The AP is given in %.

Detector	AP	AP@0.5	AP@0.75	AP_s	AP_m	AP_l
Faster R-CNN	33.5	60.7	32.4	12.3	37.0	53.2
Cascade R-CNN	**35.4**	**62.5**	**35.2**	12.3	**38.9**	54.7
Libra R-CNN	33.4	60.2	32.5	12.2	36.5	53.1
Grid R-CNN	34.7	60.3	35.0	12.0	38.5	53.4
HRNetV2p-W32	34.1	60.5	34.3	**14.2**	36.2	**56.5**
RetinaNet	33.6	60.0	32.7	9.1	37.6	54.0
FreeAnchor	33.3	58.7	32.7	9.8	36.5	55.5
FoveaBox	33.8	61.8	32.7	9.7	37.0	54.7

(AP), the 200 most confident detections are considered. Amongst all detectors, Cascade R-CNN achieves with AP = 35.4% the overall best performance. Therefore, we use it as our object detection model.

To further improve the quality of the detections, we utilize several test strategies: First, we increase the number of region proposals from 1000 to 4000 to take into account the high number of objects in the drone images. Second, multi-scale testing is applied by re-scaling the images so that the shorter size of an image becomes 1000 and 2000, respectively. Lastly, with the additional horizontal flipping, four sets of detections are combined for each image. The results of all test-time improvements are listed in Table 2. Leveraging all test-time strategies, the Cascade R-CNN improves by 3.8 points in AP, achieving AP = 39.2% for the five classes *pedestrian*, *car*, *van*, *truck* and *bus* on the test-dev split of the VisDrone MOT dataset. The detections generated from this configuration serve as input to our PAS tracker.

4.2 PAS Tracker

We first compare the performance of our PAS tracker with three popular tracking approaches from the literature, namely Tracktor++ [1], Deep SORT [48] and V-IOU [5], on the test-dev split of the VisDrone MOT dataset, since this

Table 2. Test setting improvements for Cascade R-CNN on the test-dev split of the VisDrone MOT dataset. The AP is given in %.

Cascade R-CNN	AP	AP@0.5	AP@0.75	AP_s	AP_m	AP_l
Baseline	35.4	62.5	35.2	12.3	38.9	54.7
+ more proposals	35.8	63.6	35.4	12.5	39.2	54.8
+ multi-scale testing	38.4	67.7	37.6	16.7	42.4	55.7
+ horizontal flipping	**39.2**	**68.5**	**38.4**	**17.8**	**43.2**	**56.6**

is the default test data for general evaluation. After that, we show the results of our approach on the test-challenge split participating in the VisDrone2020 MOT challenge. While the V-IOU tracker can handle objects of different categories, Tracktor++ and Deep SORT originally are developed for the *MOT benchmark* [31], which only contains pedestrians. For Deep SORT, we simply extend the framework to a multi-class setting by applying the association step for each category independently. For Tracktor++, we make the following adaptations. First, a track of a class is stopped if the maximum of the score vector produced by the bounding box regression corresponds to a different class. Second, the re-identification is done for each class independently. Third, the CMC model and the re-identification model are the same as in the PAS tracker for a fair comparison. Fourth, the Faster R-CNN is replaced with the Cascade R-CNN detector. In addition to the CMC, a CVA is used for predicting the bounding box positions of tracks in the following frame.

As will be shown in Sect. 5, the detection thresholds, i.e., which detections are used in the tracking process, have a huge influence on the tracking performance. Due to the non-homogeneous distribution of object entities per class in the training set, the detection model performs differently well amongst the classes. For this reason, we tune the detection thresholds for each class and for each tracking method separately. All methods utilize the same detections generated by the Cascade R-CNN detector but apply different detection thresholds. The results, when setting the best detection thresholds for each method, are depicted in Table 3. The tracking AP is calculated following the evaluation protocol in [34]. Note that the AP for multi-object tracking differs from the AP for object detection. Our PAS tracker outperforms the other trackers on all metrics. In contrast to the other approaches, our method makes use of multiple cues at the same time, namely position, appearance and size, whereas V-IOU only uses appearance information for filling gaps caused by missing detections and Deep SORT uses position information only for inhibiting impossible tracks. Tracktor++ handles the association task implicitly by bounding box regression. We find that this causes problems in crowded scenarios, where the bounding box regression is sensitive to drifting onto nearby objects.

We further evaluate the performance of our PAS tracker on the test-challenge split, where no annotations are publicly available, by participating in the Vis-

Table 3. Comparison of our PAS tracker with other tracking approaches from the literature on the test-dev split of the VisDrone MOT dataset. The AP is given in %. Note that all methods use the same object detector.

Tracker	AP	AP@0.25	AP@0.5	AP@0.75	AP_{car}	AP_{bus}	AP_{trk}	AP_{ped}	AP_{van}
V-IOU	26.4	34.5	29.2	15.6	40.9	36.4	22.7	7.8	24.4
Deep SORT	33.2	51.0	35.0	13.6	31.9	58.3	30.0	21.3	24.5
Tracktor++	34.3	48.6	35.5	18.8	50.6	40.0	32.8	20.2	27.8
PAS tracker	**50.8**	**66.1**	**52.5**	**33.8**	**62.7**	**81.2**	**43.9**	**30.3**	**35.9**

Table 4. Best performing entries of the VisDrone2020 MOT challenge and the VisDrone2019 MOT challenge, respectively. The names of the methods are taken from the official leaderboard[a] and from [30]. The AP is given in %.

Method (2020)	AP	Method (2019)	AP
COFE	**61.9**	DBAI-Tracker	43.9
SOMOT	57.7	TrackKITSY	39.2
PAS tracker (ours)	50.8	Flow-Tracker	30.9
Deepsort	42.1	HMTT	28.7
YOLO-TRAC	42.1	TNT_DRONE	27.3

[a] http://www.aiskyeye.com/leaderboard/

Drone2020 MOT challenge. For this, we submit our tracking results to the evaluation server that returns the overall average precision. With an AP of 50.8%, the PAS tracker achieves the 3rd highest tracking performance out of 19 different methods. Table 4 lists the five best performing entries from the 2020 challenge and the 2019 challenge, respectively. Note that no public detections are available for the VisDrone MOT dataset and a fair comparison of the tracking approaches is hardly possible since the quality of the object detector has a large influence on the tracking performance as we show in the ablation study. While we only trained the object detector on the VisDrone MOT dataset, the usage of external data is not prohibited in the VisDrone challenge. We hypothesize that a large performance improvement of our method could be achieved by using more datasets for training the detector.

5 Ablation Study

In this chapter, ablation experiments are performed to examine the impact of applied detection score thresholds on the tracking performance and to analyze the influence of each tracking component in detail.

Influence of Detection Score Tresholds on Tracking Performance. As in all tracking-by-detection based approaches, the performance of our PAS tracker depends on the provided detections and, hence, on the detection score thresholds. To underline this, we analyze the performance of our PAS tracker with different thresholds applied for each class separately. The resulting average precision is visualized for each class in Fig. 3. The AP averaged over the five classes is depicted in solid black. It reaches its maximum at 48.2% for a detection score threshold of 0.1. Using a non-optimal detection score threshold results in clearly worse AP values, which highlights the importance of a well-set detection score threshold. For instance, the AP shrinks to 42.9% in case of using a detection score threshold of 0.5. Furthermore, the figure shows that the best threshold

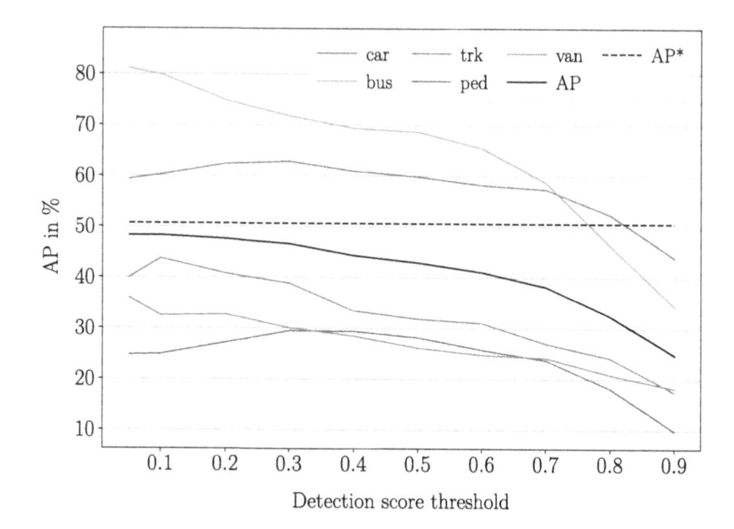

Fig. 3. Influence of the detection score thresholds on the class-wise performance of the PAS tracker. The AP averaged over the five classes is shown in solid black. Setting the best detection score threshold for each class separately yields the best performance marked as dashed line (AP*).

varies between the classes. Hence, using the best threshold for each class separately (*car*: 0.3, *bus*: 0.05, *truck*: 0.1, *pedestrian*: 0.4, *van*: 0.05), results in an increased AP of 50.8%, drawn as dashed line in Fig. 3 (AP*).

Impact of the Components of the PAS Tracker. To get more insights into our tracking approach, we examine the influence of each component by integrating one after another and evaluate, besides AP, also the number of identity switches (IDSW) and false positives (FPs). The results of this ablation study are presented in Table 5. It can be seen that, even if only the position similarity is used for association, the AP is still better than of Tracktor++, Deep SORT and V-IOU (see Table 3). When high quality detections are available and the frame rate is large enough, the CVA is a good approximation of the true motion of objects making the position similarity s_p a suitable association metric. The CMC further improves the position similarity when the camera is moving leading to 8% less IDSW. Adding the appearance similarity s_a increases the AP by 1.9 points and further reduces the IDSW by 20%. Note that the re-identification model from [28] has been trained on DukeMTMC [39], which is a dataset for person re-identification. Thus, the appearance vectors for the appearance similarity s_a are not optimal for the VisDrone MOT dataset. Integrating the size similarity s_z boosts the performance by 5.8 points in AP. It ensures that false positive detections, which only cover parts of objects and are not filtered by the non-maximum suppression, are not associated. This also improves identity preservation lowering the number of IDSW by additional 12%.

Table 5. Influence of the individual components of the PAS tracker on the tracking performance

	Tracking components					
Position similarity s_p	✓	✓	✓	✓	✓	✓
Camera motion compensation (CMC)		✓	✓	✓	✓	✓
Appearance similarity s_a			✓	✓	✓	✓
Size similarity s_z				✓	✓	✓
Filtering of false positive detections					✓	✓
Removal of short tracks (<10 frames)						✓
False positives (↓)	97459	97466	97482	97456	94096	**82805**
Identity switches (↓)	4346	3977	3163	2791	2422	**2159**
Average precision in % (↑)	39.7	42.4	44.3	50.1	50.6	**50.8**

Since unmatched detections start new tracks, it is desirable to remove as many false positive detections as possible. While the previously analyzed tracking components have negligible influence on the number of FPs, our filtering approach is capable of removing wrong detections that appear in crowded scenarios, where the object detector often can not separate the individual objects precisely, reducing the number of FPs by 3% and the number of IDSW by 13%, respectively. Finally, the removal of short tracks with less than 10 frames yields further improvements of 12% in FPs and 11% in IDSW. With all components, our PAS tracker reaches an AP of 50.8% on the test-dev split of the VisDrone MOT dataset and achieves state-of-the-art-performance.

6 Conclusion

In this paper, we propose a tracking-by-detection approach for MOT in drone videos, named PAS tracker, which comprises a novel association step based on a joint similarity measure of position, appearance and size information. To benefit from high quality detections, we compare eight state-of-the-art object detectors on the VisDrone MOT dataset and further improve the detection performance by leveraging test-time strategies. Adapting the detection score threshold for each class separately further boosts the tracking accuracy of the particular class. Furthermore, we propose a filtering technique to remove false positive detections in crowded scenarios. Our PAS tracker achieves state-of-the-art results on the VisDrone MOT dataset and ranks 3rd in the VisDrone2020 MOT challenge. In future works, the appearance similarity could be improved applying re-identification models specialized for vehicles and training directly on image patches of the VisDrone MOT dataset. In addition, the tracking performance is highly correlated with the detection quality as shown in the ablation study. Thus, integrating the tracking and the detection task more tightly is a promising research direction.

References

1. Bergmann, P., Meinhardt, T., Leal-Taixé, L.: Tracking without bells and whistles. In: 2019 IEEE/CVF International Conference on Computer Vision (ICCV) (2019)
2. Bernardin, K., Stiefelhagen, R.: Evaluating multiple object tracking performance: the clear mot metrics. J. Image Video Process. **2008**, 1–10 (2008)
3. Bewley, A., Ge, Z., Ott, L., Ramos, F., Upcroft, B.: Simple online and realtime tracking. In: 2016 IEEE International Conference on Image Processing (ICIP) (2016)
4. Bochinski, E., Eiselein, V., Sikora, T.: High-speed tracking-by-detection without using image information. In: 2017 14th IEEE International Conference on Advanced Video and Signal Based Surveillance (AVSS) (2017)
5. Bochinski, E., Senst, T., Sikora, T.: Extending IOU based multi-object tracking by visual information. In: 2018 15th IEEE International Conference on Advanced Video and Signal Based Surveillance (AVSS) (2018)
6. Brasó, G., Leal-Taixé, L.: Learning a neural solver for multiple object tracking. In: Proceedings of the IEEE/CVF Conference on Computer Vision and Pattern Recognition (2020)
7. Cai, Z., Vasconcelos, N.: Cascade r-cnn: delving into high quality object detection. In: 2018 IEEE/CVF Conference on Computer Vision and Pattern Recognition (2018)
8. Chen, K., et al.: MMDetection: Open mmlab detection toolbox and benchmark. arXiv preprint arXiv:1906.07155 (2019)
9. Du, D., et al.: Visdrone-det2019: the vision meets drone object detection in image challenge results. In: 2019 IEEE/CVF International Conference on Computer Vision Workshop (ICCVW) (2019)
10. Duan, K., Bai, S., Xie, L., Qi, H., Huang, Q., Tian, Q.: Centernet: keypoint triplets for object detection. In: 2019 IEEE/CVF International Conference on Computer Vision (ICCV) (2019)
11. Evangelidis, G.D., Psarakis, E.Z.: Parametric image alignment using enhanced correlation coefficient maximization. IEEE Trans. Pattern Anal. Mach. Intell. **30**(10), 1858–1865 (2008)
12. Feichtenhofer, C., Pinz, A., Zisserman, A.: Detect to track and track to detect. In: 2017 IEEE International Conference on Computer Vision (ICCV) (2017)
13. He, K., Gkioxari, G., Dollár, P., Girshick, R.: Mask r-cnn. In: 2017 IEEE International Conference on Computer Vision (ICCV) (2017)
14. Henriques, J.F., Caseiro, R., Martins, P., Batista, J.: High-speed tracking with kernelized correlation filters. IEEE Trans. Pattern Anal. Mach. Intell. **37**(3), 583–596 (2015)
15. Hsieh, M., Lin, Y., Hsu, W.H.: Drone-based object counting by spatially regularized regional proposal network. In: 2017 IEEE International Conference on Computer Vision (ICCV) (2017)
16. Kalal, Z., Mikolajczyk, K., Matas, J.: Forward-backward error: automatic detection of tracking failures. In: 2010 20th International Conference on Pattern Recognition (2010)
17. Kalra, I., Singh, M., Nagpal, S., Singh, R., Vatsa, M., Sujit, P.B.: Dronesurf: benchmark dataset for drone-based face recognition. In: 2019 14th IEEE International Conference on Automatic Face Gesture Recognition (FG 2019) (2019)
18. Karthik, S., Prabhu, A., Gandhi, V.: Simple unsupervised multi-object tracking. arXiv preprint arXiv:2006.02609 (2020)

19. Kong, T., Sun, F., Liu, H., Jiang, Y., Shi, J.: Foveabox: Beyond anchor-based object detector. arXiv preprint arXiv:1904.03797 (2019)
20. Kuhn, H.W., Yaw, B.: The Hungarian method for the assignment problem. Naval Res. Logist. Q. **21**, 83–97 (1955)
21. Law, H., Deng, J.: Cornernet: detecting objects as paired keypoints. In: Proceedings of the Computer Vision—ECCV 2018–15th European Conference (2018)
22. Lin, T., Dollár, P., Girshick, R., He, K., Hariharan, B., Belongie, S.: Feature pyramid networks for object detection. In: 2017 IEEE Conference on Computer Vision and Pattern Recognition (CVPR) (2017)
23. Lin, T.Y., Goyal, P., Girshick, R., He, K., Dollár, P.: Focal loss for dense object detection. In: 2017 IEEE International Conference on Computer Vision (ICCV) (2017)
24. Lin, T.Y., et al.: Microsoft COCO: common objects in context. In: Fleet, D., Pajdla, T., Schiele, B., Tuytelaars, T. (eds.) ECCV 2014. LNCS, vol. 8693, pp. 740–755. Springer, Cham (2014). https://doi.org/10.1007/978-3-319-10602-1_48
25. Liu, W., et al.: SSD: single shot multibox detector. In: Leibe, B., Matas, J., Sebe, N., Welling, M. (eds.) ECCV 2016. LNCS, vol. 9905, pp. 21–37. Springer, Cham (2016). https://doi.org/10.1007/978-3-319-46448-0_2
26. Lu, X., Li, B., Yue, Y., Li, Q., Yan, J.: Grid r-cnn. In: 2019 IEEE/CVF Conference on Computer Vision and Pattern Recognition (CVPR) (2019)
27. Lu, Z., Rathod, V., Votel, R., Huang, J.: Retinatrack: online single stage joint detection and tracking. arXiv preprint arXiv:2003.13870 (2020)
28. Luo, H., Gu, Y., Liao, X., Lai, S., Jiang, W.: Bag of tricks and a strong baseline for deep person re-identification. In: 2019 IEEE/CVF Conference on Computer Vision and Pattern Recognition Workshops (CVPRW) (2019)
29. Lyu, S., et al.: Ua-detrac 2018: report of avss2018 iwt4s challenge on advanced traffic monitoring. In: 2018 15th IEEE International Conference on Advanced Video and Signal Based Surveillance (AVSS) (2018)
30. Lyu, S., et al.: Visdrone-mot2019: the vision meets drone multiple object tracking challenge results. In: 2019 IEEE/CVF International Conference on Computer Vision Workshop (ICCVW) (2019)
31. Milan, A., Leal-Taixe, L., Reid, I., Roth, S., Schindler, K.: Mot16: a benchmark for multi-object tracking. arXiv preprint arXiv:1603.00831 (2016)
32. Pang, B., Li, Y., Zhang, Y., Li, M., Lu, C.: Tubetk: adopting tubes to track multi-object in a one-step training model. In: Proceedings of the IEEE/CVF Conference on Computer Vision and Pattern Recognition (2020)
33. Pang, J., Chen, K., Shi, J., Feng, H., Ouyang, W., Lin, D.: Libra r-cnn: towards balanced learning for object detection. In: 2019 IEEE/CVF Conference on Computer Vision and Pattern Recognition (CVPR) (2019)
34. Park, E., Liu, W., Russakovsky, O., Deng, J., Li, F.F., Berg, A.: Large scale visual recognition challenge **2017** (2017). http://image-net.org/challenges/LSVRC/2017
35. Redmon, J., Divvala, S., Girshick, R., Farhadi, A.: You only look once: unified, real-time object detection. In: 2016 IEEE Conference on Computer Vision and Pattern Recognition (CVPR) (2016)
36. Redmon, J., Farhadi, A.: Yolo9000: better, faster, stronger. In: 2017 IEEE Conference on Computer Vision and Pattern Recognition (CVPR) (2017)
37. Redmon, J., Farhadi, A.: Yolov3: an incremental improvement. arXiv preprint arXiv:1804.02767 (2018)
38. Ren, S., He, K., Girshick, R., Sun, J.: Faster r-cnn: towards real-time object detection with region proposal networks. IEEE Trans. Pattern Anal. Mach. Intell. **39**(6), (2017)

39. Ristani, E., Solera, F., Zou, R.S., Cucchiara, R., Tomasi, C.: Performance measures and a data set for multi-target, multi-camera tracking. arXiv preprint arXiv:1609.01775 (2016)
40. Sadeghian, A., Alahi, A., Savarese, S.: Tracking the untrackable: learning to track multiple cues with long-term dependencies. In: 2017 IEEE International Conference on Computer Vision (ICCV) (2017)
41. Schulter, S., Vernaza, P., Choi, W., Chandraker, M.: Deep network flow for multi-object tracking. In: 2017 IEEE Conference on Computer Vision and Pattern Recognition (CVPR) (2017)
42. Sun, K., Xiao, B., Liu, D., Wang, J.: Deep high-resolution representation learning for human pose estimation. In: 2019 IEEE/CVF Conference on Computer Vision and Pattern Recognition (CVPR) (2019)
43. Tian, W., Lauer, M., Chen, L.: Online multi-object tracking using joint domain information in traffic scenarios. IEEE Trans. Intell. Transp. Syst. **21**(1), 374–384 (2020)
44. Tian, Z., Shen, C., Chen, H., He, T.: Fcos: fully convolutional one-stage object detection. In: 2019 IEEE/CVF International Conference on Computer Vision (ICCV) (2019)
45. Voigtlaender, P., et al.: Mots: multi-object tracking and segmentation. In: 2019 IEEE/CVF Conference on Computer Vision and Pattern Recognition (CVPR) (2019)
46. Wang, Z., Zheng, L., Liu, Y., Wang, S.: Towards real-time multi-object tracking. arXiv preprint arXiv:1909.12605 (2019)
47. Wen, L., et al.: UA-DETRAC: a new benchmark and protocol for multi-object detection and tracking. Comput. Vis. Image Understanding **193**, 102907 (2015)
48. Wojke, N., Bewley, A., Paulus, D.: Simple online and realtime tracking with a deep association metric. In: 2017 IEEE International Conference on Image Processing (ICIP) (2017)
49. Xie, S., Girshick, R., Dollár, P., Tu, Z., He, K.: Aggregated residual transformations for deep neural networks. In: 2017 IEEE Conference on Computer Vision and Pattern Recognition (CVPR) (2017)
50. Xu, Y., Osep, A., Ban, Y., Horaud, R., Leal-Taixé, L., Alameda-Pineda, X.: How to train your deep multi-object tracker. In: Proceedings of the IEEE/CVF Conference on Computer Vision and Pattern Recognition (2020)
51. Yu, F., Li, W., Li, Q., Liu, Yu., Shi, X., Yan, J.: POI: multiple object tracking with high performance detection and appearance feature. In: Hua, G., Jégou, H. (eds.) ECCV 2016. LNCS, vol. 9914, pp. 36–42. Springer, Cham (2016). https://doi.org/10.1007/978-3-319-48881-3_3
52. Yu, H., Li, G., Zhang, W., Du, D., Tian, Q., Sebe, N.: The unmanned aerial vehicle benchmark: object detection, tracking and baseline. Int. J. Comput. Vis. **128**, 1141–1159 (2019)
53. Zhang, X., Wan, F., Liu, C., Ji, R., Ye, Q.: Freeanchor: learning to match anchors for visual object detection. In: Advances in Neural Information Processing Systems, vol. 32 (2019)
54. Zhang, Z., Cheng, D., Zhu, X., Lin, S., Dai, J.: Integrated object detection and tracking with tracklet-conditioned detection. arXiv preprint arXiv:1811.11167 (2018)
55. Zhu, P., Sun, Y., Wen, L., Feng, Y., Hu, Q.: Drone based rgbt vehicle detection and counting: A challenge. arXiv preprint arXiv:2003.02437 (2020)
56. Zhu, P., Wen, L., Du, D., Bian, X., Hu, Q., Ling, H.: Vision meets drones: Past, present and future. arXiv preprint arXiv:2001.06303 (2020)

Insights on Evaluation of Camera Re-localization Using Relative Pose Regression

Amir Shalev[1,2](✉), Omer Achrack[2](✉), Brian Fulkerson[3](✉), and Ben-Zion Bobrovsky[1](✉)

[1] Tel-Aviv-University, Tel Aviv, Israel
Bobi@eng.tau.ac.il
[2] Intel, Santa Clara, USA
{Amir.Shalev,Omer.Achrack}@intel.com
[3] Intel, London, UK
Bfulkerson@gmail.com

Abstract. We consider the problem of relative pose regression in visual relocalization. Recently, several promising approaches have emerged in this area. We claim that even though they demonstrate on the same datasets using the same split to train and test, a faithful comparison between them was not available since on currently used evaluation metric, some approaches might perform favorably, while in reality performing worse. We reveal a tradeoff between accuracy and the 3D volume of the regressed subspace. We believe that unlike other relocalization approaches, in the case of relative pose regression, the regressed subspace 3D volume is less dependent on the scene and more affect by the method used to score the overlap, which determined how closely sampled viewpoints are. We propose three new metrics to remedy the issue mentioned above. The proposed metrics incorporate statistics about the regression subspace volume. We also propose a new pose regression network that serves as a new baseline for this task. We compare the performance of our trained model on Microsoft 7-Scenes and Cambridge Landmarks datasets both with the standard metrics and the newly proposed metrics and adjust the overlap score to reveal the tradeoff between the subspace and performance. The results show that the proposed metrics are more robust to different overlap threshold than the conventional approaches. Finally, we show that our network generalizes well, specifically, training on a single scene leads to little loss of performance on the other scenes.

Keywords: Re-localization · Relative-pose-regression · Frustum-overlap

---

Electronic supplementary material The online version of this chapter (https://doi.org/10.1007/978-3-030-66823-5_37) contains supplementary material, which is available to authorized users.

A. Bartoli and A. Fusiello (Eds.): ECCV 2020 Workshops, LNCS 12538, pp. 621–637, 2020.
https://doi.org/10.1007/978-3-030-66823-5_37

1 Introduction

Visual Simultaneous Localization and Mapping (V-SLAM) is a widely used method in vision-based applications, including mobile robots, virtual reality (VR), augmented reality (AR) and navigation (from domestic environments to rockets and satellites), as well as in many more applications in multifarious fields. Although GPS is a robust and readily available solution for localization, for many applications it is not affordable. Due to its fundamental importance as a core technology for a wide range of applications, visual localization has been extensively studied with major progress evidenced over the past few years [3,9,10,25,31]. However, accurate and robust localization is still a challenging problem, alongside the growing demands for constructing better maps. If an agent needs to locate itself when a map is given in addition to visual clues, the task is called re-localization [1]. A few examples of re-localization are: loop closure for autonomous navigation, map loading for virtual and augmented reality, and the kidnap robot problem. Re-localization can be viewed as regressing the pose of the camera when prior knowledge is given as a map. In recent years, researchers have used the deep learning approach, and trained a deep neural network to regress the camera pose on a query (test) image, given a set of training images. One of the most effective known techniques is called *relative-pose-regression with image retrieval*; it is computationally efficient and able to generalize to new **unseen** scenes, as we will demonstrate herein. Other published works [2,8,20,29] present different implementations and techniques to improve re-localization performance. Detailed evaluations of these and other works were done independently by [32,34,39]. However, to the best of our knowledge, our work is the first to take into account the elusive tradeoff that has been overlooked in the comparison criteria published to date. It transpires that choosing different parameters on the same model affects the accuracy of the model, when using the current metrics. In this work, we will demonstrate this tradeoff, both qualitatively and empirically. This paper introduces new metrics that take into account the subspace size. To evaluate our new metrics we used a relative-pose-regression framework for camera re-localization as a baseline for future works. Our Siamese network consists of two equivalent backbones with shared weights. Each learns to encode geometric information from an image into a feature vector. In the test stage, only one of these networks is used to estimate the camera pose. This method is able to generalize well, as we will show Sect. 5; we train it on one scene and show that the learned model is able to work well on the other scenes without retraining. We also present a new technique for estimating the amount of correlated information, given a pair of images, based on their location and some prior knowledge on the environment.

To summarize, we offer the following contributions:

- We raise a concern regarding erroneous conclusions drawn in published relative-pose-regression camera methods. We will show that due to neglect of the tradeoff we revealed, some approaches perform favorably in terms of current evaluation metrics, while, in reality, they perform badly.

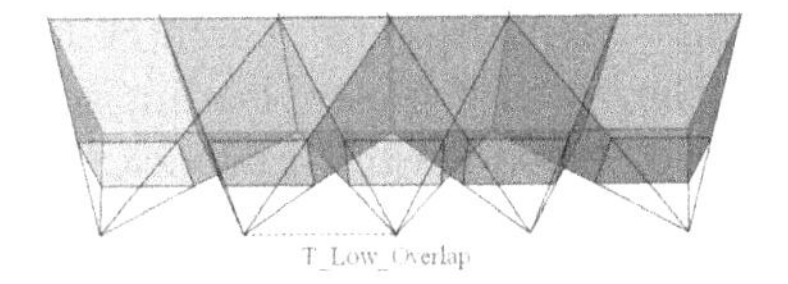

Low overlap spacing

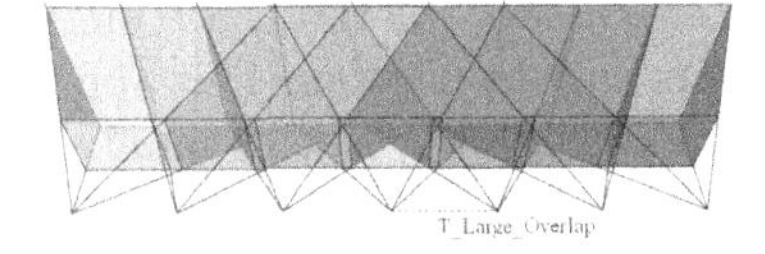

High overlap spacing

Fig. 1. The figure illustrates a 3D space occupied by low (a) and high (b) Overlap spacing. Low overlap creates sparse sampling, which leads to a 'hard' relative-pose-regression problem. High overlap creates dense sampling, which leads to an 'easy' relative-pose-regression problem. The complexity of a regression problem is affected by the regression subspace volume. In terms of absolute-pose-regression, both are equivalent since we use the same space.

- We propose to revise the metrics used for evaluating relative-pose-regression camera methods. We propose three metrics, which incorporate some statistics regarding the scene volume. Our new metrics depend on the 3D volume occupied and on how closely viewpoints are sampled.
- We introduce a new way to estimate the amount of correlated information between pairs of images by 3D frustum-overlap. This approach makes the perception of the neglected tradeoff more intuitive.
- We present a better way to use the overlap score, and we demonstrate the robustness and benefits of the relative-pose-regression method that produces competitive results on untrained environments.

The rest of the paper is organized as follows: Sect. 2 briefly reviews the related work in visual re-localization. The details of our approach, network structure, and overlap score function are detailed in Sect. 3. Sect. 4 is devoted to reviewing the comparison criteria. Evaluation results are provided in Sect. 5. In Sect. 6, we summarize the results and conclusions, and offer some suggestions for future work. Our source code will be publicly available soon.

2 Related Work

2.1 Re-localization Approaches

The task of camera re-localization has a long history of research in various V-SLAM systems. Traditional approaches, as well as some leading algorithms [3,12,30], are built on multi-view geometry theory. Other methods, like appearance-based similarity, the Hough Transform [38], and random-forest based methods [14,23,35], have been investigated and shown good results on some benchmarks. Recently, deep learning approaches have become a popular end-to-end solution for re-localization problems [13,18,28,35], instead of only being a replacement for parts of the re-localization pipeline [5]. The benefits of this approach are reduction of the inference time and of memory consumption, which is crucial for low computation and memory devices like drones, AR/VR and mobile devices. The major methods are now briefly summarized.

Features-Based. Methods are methods that use multi-view geometry features extractors and descriptors. This approach is the base for several leading solutions [3,4,19,30,33]. However, this family of methods has a major drawback: they are limited to a small working area since the computational costs grow significantly with working area size, even after optimization, such as is found in the case presented in [10] and others.

Fiducial Markers. One of the most popular approaches is based on the use of binary square fiducial markers. The main benefit of these markers is that a single marker provides enough correspondences to obtain the camera pose. Every marker is associated with a coordinate system, and poses are given relative to that coordinate system's origin [11,27,40].

Absolute Pose-Regression. The method outlined in [17] suggested learning the re-localization pipeline in an end-to-end supervised manner. The idea of regressing the absolute pose by using machine learning offered several appealing advantages compared to traditional feature-based methods. Deep learning based on absolute pose-regression does not require the design of hand-crafted features. The trained model has a low memory footprint and constant runtime. However, any solution for absolute pose-regression suffers from over-fitting on the training data and will not generalize well - which is a desired property in practical cases. Recently, many learning-based algorithms have been developed [7,21,24,36].

Relative-Pose-Regression with Image Retrieval. The key idea of relative-pose-regression is that once an anchor image has been determined, one can directly regress the relative camera pose between the anchor and test images, and thus obtain the absolute pose. The relative-pose-regression approach requires a definition of similarity. This definition has been proposed and studied recently in several works. The method enjoys many of the advantages of the previously described methods, *e.g.* attractive computational costs stemming from its maintaining a relatively small database, decoupling of the pose-regression process from the coordinate frame of the scene. In addition, it does not require scene-specific training, as we will demonstrate. [8,20,22,23,29,31,37].

Combination of Methods. The authors in [39] presented a multi-task training approach, leveraging relative-pose information during the training, and demonstrated impressive results. Alternatively, a variety of combinations are outlined in [6], where, for sequence localization they combine absolute pose-regression from the current frame to relative pose from the previous frame. In addition, a combination of the new approaches with traditional algorithms has improved the results, such as where the combination was done using visual odometry algorithms, which take as input IMU and GPS sensors. In the test stage, predictions were further refined with pose graph optimization.

2.2 Images Intersection Score

In relative-pose-regression a metric is used to estimate the correlated information between pairs of images, i.e. to determine whether a given pair of images contains enough overlapped information such that the algorithm can retrieve the relative pose between them. A few metrics have evolved during the last few years: the authors in [20] measured the overlap as the percentage of pixels projected onto a candidate image plane using the respective ground truth pose and depth maps; the method in [2] projects depth from one pose, and un-projects it from another, and then counts how many points are on the image plane; the authors in [8] suggested using ORB similarity for outdoor scenes with 2D image data-sets. Both overlap calculation methods use the geometric as well as the visual content of the images. In this work, we propose using pose information from the ground-truth, and do not assume depth information is available. Although each method has its advantages, we think that none provides adequate intuitive understanding of the sub-space that is spread by the span of all the relative pose. To this end we now introduce a new, and more intuitive, image intersection score we call frustum-overlap.

3 Methodology

The main novelty of this work is to revise the problem of re-localization assessment expressed with standard comparison criteria, and to propose a new comparison protocol. We therefore used the popular models and methods that are known to work on the problem, such as camera re-localization.

3.1 Frustum-Overlap - An Intuitive Images Intersection Score

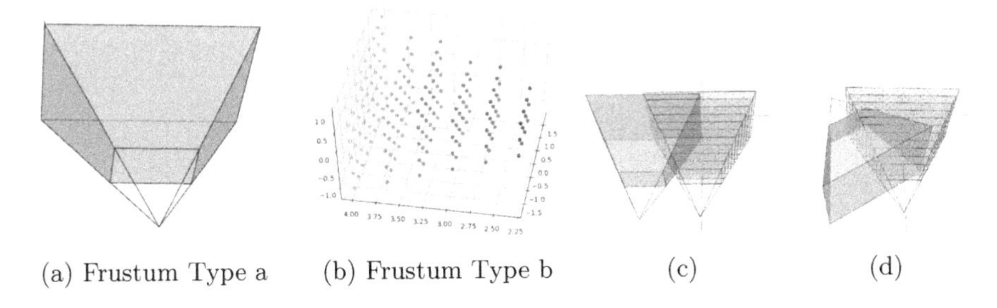

(a) Frustum Type a (b) Frustum Type b (c) (d)

Fig. 2. Frustum types

We introduce a novel definition for image overlap that does not rely on the content of the images. We borrowed the concept of the camera view frustum from computer graphics. A frustum is a truncated pyramid that starts at the focal plane of the camera, and extends up to the maximum viewable distance. We use two different ways to define a frustum:

- **Type a** is presented in Fig. 2a - A set of 6 3D planes, 2 parallel planes and 4 planes delimiting them to a pyramid.
- **Type b** is presented in Fig. 2b - A discrete set of 3D points on a grid, constructing a frustum shape.

To get an overlap score given a pair of two poses, we need to place a frustum of *Type a* in the first pose and a frustum of *Type b* in the other pose, and then extract the percentage of 3D points of the second frustum that are inside the first frustum. This is illustrated in Fig. 2c. In Eq. 1 we describe the counting formula we used, to give some informal intuition behind the calculation.

$$
\begin{aligned}
overlap\_score &= \sum_{p_i \in frustum\_a} \frac{a_i}{N_b} \\
a_i &= \begin{cases} 1, & \text{if } p_i \text{ is in frustum\_b} \\ 0, & \text{otherwise} \end{cases}
\end{aligned}
\tag{1}
$$

Our overlap scoring method. N_b is the total number of points in frustum b.

This approach has a major drawback - it is possible to get a high overlap score for two images without any common visual content (see Fig. 2d for an illustration.) We overcome this drawback by limiting the relative rotation between two captures. It is important to emphasize that, in contrast to other methods, our method scores overlap in 3D space, and therefore has a tight relationship with the span subspace. A formal definition of the calculation of the score is given in the supplementary material.

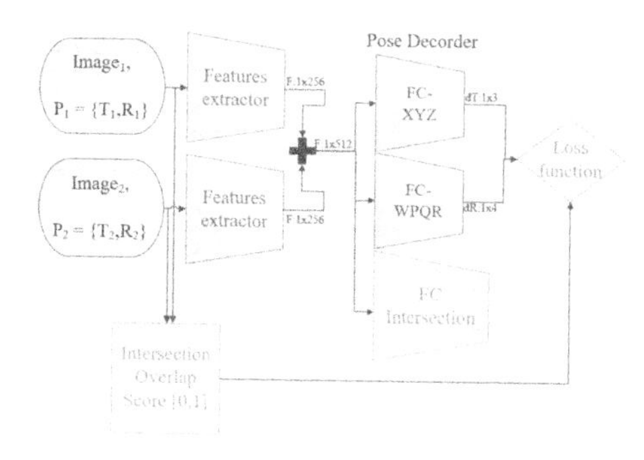

Fig. 3. Diagram of the proposed training architecture: each Siamese branch uses a shared features extractor to get a representation of the image. The pose decoder predicts rotation, represented by a quaternion, and translation, using x, y, z coordinates.

3.2 Training Procedure

Our training procedure is done in a straightforward way. Given the pre-defined pairs of images induced by the selection of the overlap computing method and overlap threshold, as well as their corresponding relative pose as a label, we leverage Siamese architecture to get the estimated relative pose for the network. A schematic of our training procedure is given in Fig. 3.

3.3 Loss Function

Loss definition for supervised pose-regression tasks is challenging because it involves learning two distinct quantities - rotation and translation. Each has its own different units and different scales. The authors in [16] found that outdoor and indoor scenes are characterized by different weights, which motivates to dynamically learn the weights of the loss for each task instead of determining them in a hard-coded manner. We follow this notion and use learned α and β to balance the losses and combine them, as described formally in [16]:

$$L_q(\hat{q_{rel}}, q_{rel}) = \left\| q_{rel} - \frac{\hat{q_{rel}}}{\|\hat{q_{rel}}\|_2} \right\|_2 \tag{2a}$$

$$L_t(\hat{t_{rel}}, t_{rel}) = \left\| t_{rel} - \hat{t_{rel}} \right\|_2 \tag{2b}$$

$$L(\hat{p_{rel}}, p_{rel}) = \alpha^2 + \beta^2 + e^{-\alpha^2} L_t(\hat{t_{rel}}, t_{rel}) + e^{-\beta^2} L_q(\hat{q_{rel}}, q_{rel}) \tag{2c}$$

where t_{rel} and q_{rel} are the relative translation and rotation of the $\hat{}$ superscript that denotes predicted values. This form of loss function makes it possible for us to learn the weighting between the translation and rotation objectives and to find optimal weighting for a specific dataset.

4 Comparison Criteria

4.1 Standard Comparison Criteria

To evaluate the performance of the proposed algorithm, we require a set of images and their corresponding ground truth poses. We estimate the image poses predicted by the algorithm and calculate the pose error for translation and rotation. The popular localization metrics are:

$$t_{err}(\hat{t}_i, t_i) = \|t_i - \hat{t}_i\|_2 \tag{3a}$$

$$q_{err}(\hat{q}_i, q_i) = \frac{180}{\pi} 2\cos^{-1}(< q_i, \hat{q}_i >) \tag{3b}$$

After obtaining a set of translation and rotation errors, a statistical measure to convert them to a single scalar score is used. Common choices [2,6,17] are the mean and the median:

$$T_{err} = mean/median(\{t_{err}(\hat{t}_i, t_i)\}_{i \in N}) \tag{4a}$$

$$Q_{err} = mean/median(\{q_{err}(\hat{q}_i, q_i)\}_{i \in N}) \tag{4b}$$

4.2 The Limitations of Standard Comparison Criteria

In absolute-pose-regression, when one calculates the error using Eq. 4 the scale (regression subspace volume) is implicitly involved in the computation. The error is limited by the scene dimensions, and can be manipulated by normalizing the scene. Moreover, the train and test sets are exactly the same, because there is no **pre-defined overlap based selection**. In relative-pose-regression, this is not the case; the selection of the overlap method and threshold implicitly determines the complexity of the problem, errors are limited by this selection and scene normalization will not solve it. Changing either the overlap calculation method or the threshold might significantly change the train and test data. An illustration of this key insight is depicted in Fig. 1. Qualitatively, if one chooses 'high' overlap, the regression subspace volume is smaller than if one choses 'low' overlap. This, in turn, will lead at first sight to the conclusion that the algorithm that trained and tested on 'high' overlap is better, but we argue that this is not necessarily so. We further argue that there is a strong correlation between the spanned 3D volume induced by the overlap selection and the performance of the selected algorithm, as we will show in Sect. 5. We demonstrate this claim visually in Fig. 4

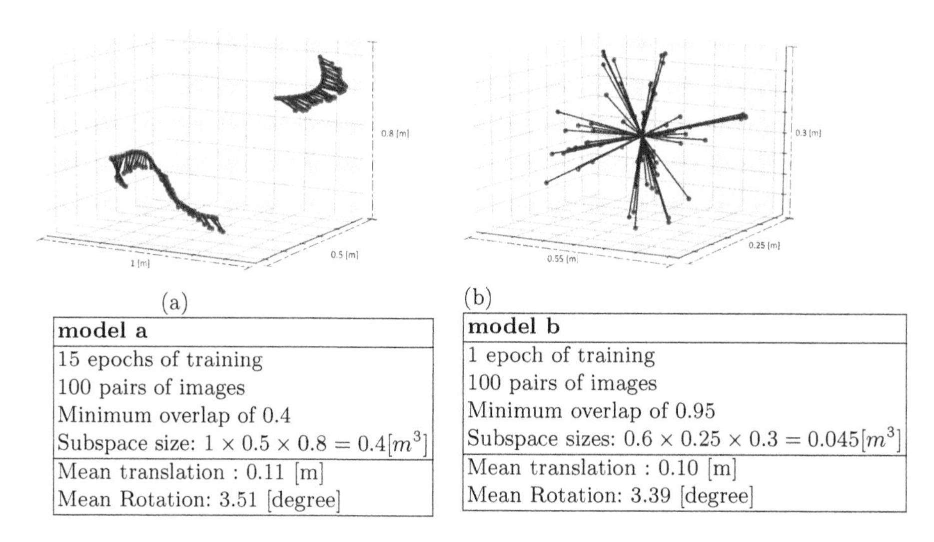

model a
15 epochs of training 100 pairs of images Minimum overlap of 0.4 Subspace size: $1 \times 0.5 \times 0.8 = 0.4[m^3]$
Mean translation : 0.11 [m] Mean Rotation: 3.51 [degree]

model b
1 epoch of training 100 pairs of images Minimum overlap of 0.95 Subspace sizes: $0.6 \times 0.25 \times 0.3 = 0.045[m^3]$
Mean translation : 0.10 [m] Mean Rotation: 3.39 [degree]

Fig. 4. The figures present the translations error of two models. Red and green points are ground truths and estimates, respectively. Considering only the numerical results the naive model (b) is better. However, after visual examination, it is clear that model (b) is worthless, while model (a) learned to predict the pose. The mistake stems from not taking the subspace volume into account. (Color figure online)

Table 1. The table demonstrates the inadequacy of the existing criterion. We adjusted the overlap parameter to get results that seem better than those of the other methods. However, we emphasize that this is not sufficient to determine which method is preferable.

	RelocNet [2]	NNnet [20]	Anchornet [29]	CamNet [8]	Ours
Scene	RPR	RPR	RPR	RPR	RPR
Chess	0.12 m, 4.14°	0.13 m, 6.46°	0.08 m, 4.12°	0.04 m, 1.73°	0.03 m, 1.36°
Fire	0.26 m, 10.4°	0.26 m, 12.72°	0.16 m, 11.1°	0.03 m, 1.74°	0.03 m, 1.92°
Heads	0.14 m, 10.5°	0.14 m, 12.34°	0.09 m, 11.2°	0.05 m, 1.98°	0.03 m, 3.18°
Office	0.18 m, 5.32°	0.21 m, 7.35°	0.11 m, 5.38°	0.04 m, 1.62°	0.03 m, 1.75°
Pumpkin	0.26 m, 4.17°	0.24 m, 6.35°	0.14 m, 3.55°	0.04 m, 1.64°	0.03 m, 0.98°
Kitchen	0.23 m, 5.08°	0.24 m, 8.03°	0.13 m, 5.29°	0.04 m, 1.63°	0.03 m, 1.11°
Stairs	0.28 m, 7.53°	0.27 m, 11.80°	0.21 m, 11.9°	0.04 m, 1.51°	0.03 m, 1.27°

Table 2. Misleading comparison to other methods (APR: absolute-pose-regression, RPR: relative-pose-regression, 3D-classical) using the standard median metric on 7-scenes data-set.

	Active search	DSAC++	Posenet	Posenet Geometric	MapNet	CamNet	Ours
	3D	3D	APR	APR	APR	RPR	RPR
Chess	0.04 m,1.96°	0.02 m, 0.5°	0.32 m, 6.60°	0.13 m, 4.48°	0.08 m, 3.25°	0.04 m, 1.73°	0.03 m, 1.36°
Fire	0.03 m,1.53°	0.02 m, 0.9°	0.47 m, 14.0°	0.27 m, 11.3°	0.27 m, 11.69°	0.03 m, 1.74°	0.03 m, 1.92°
Heads	0.02 m,1.45°	0.01 m, 0.8°	0.30 m, 12.2°	0.17 m, 13.0°	0.18 m, 13.25°	0.05 m, 1.98°	0.03 m, 3.18°
Office	0.09 m,3.61°	0.03 m, 0.7°	0.48 m, 7.24°	0.19 m, 5.55°	0.17 m, 5.15°	0.04 m, 1.62°	0.03 m, 1.75°
Pumpkin	0.08 m,3.10°	0.04 m, 1.1°	0.49 m, 8.12°	0.26 m, 4.75°	0.22 m, 4.02°	0.04 m, 1.64°	0.03 m, 0.98°
Kitchen	0.07 m,3.37°	0.04 m, 1.1°	0.58 m, 8.34°	0.23 m, 5.35°	0.23 m, 4.93°	0.04 m, 1.63°	0.03 m, 1.11°
Stairs	0.03 m,2.22°	0.09 m, 2.6°	0.48 m, 13.1°	0.35 m, 12.4°	0.30 m, 12.08°	0.04 m, 1.51°	0.03 m, 1.27°

4.3 Proposed Metrics

As a direct sequitur to the previous sections, the results obtained using a naive implementation of Eq. 4 is not enough to make a fair comparisons. As mentioned earlier, we claim that in regression methods comparison with Eq. 4 is reliable only if their volume is similar. However, in the case where we do not know the subspace volume we need other information in order to get a reliable comparison. Hence, our first suggestion is to use a sub-space agnostic metric. As a minimal requirement for fair comparison, one should include the subspace volume or a statistical representation of it.

Statistical Criteria Inspired from Financial Analysis.

- **The mean absolute percentage error (MAPE)**, defined as:

$$T_{MAPE} = \frac{1}{N}\sum_{i\in N}(\frac{|t_i - \hat{t_i}|_1}{|t_i|_1}) \quad (5)$$

  The intuition behind this metric is: penalty decreases proportionally with the distance from the origin. Hence, it implies a larger penalty on errors in 'smaller' scenes [26].

- **The mean absolute scaled error (MASE)** is another relative measure of the error. It is defined as the mean absolute error of the model divided by the mean absolute error of a naive random-walk-without-drift model (*i.e.*the mean absolute value of the first difference of the series) [15]. Thus, it measures the relative error compared to a naive model:[1]

$$T_{MASE\_error} = \frac{T_{MAE}}{T_{Naive}} = \frac{\sum_{i\in N}(|t_i - \hat{t_i}|)}{\sum_{i\in N}|t_i - mean(\{|t|\}_{i\in N}|} \quad (6)$$

- **The mean absolute percentage scaled error (MAPSE)** is a combination of the two measures mentioned above. The benefits of using this measure are better normalization and scale-less errors:

$$T_{MAPSE\_error} = \frac{T_{MAPE}}{T_{Naive}} = \frac{\sum_{i\in N}(\frac{|t_i - \hat{t_i}|}{|t_i|})}{\sum_{i\in N}|t_i - mean(\{|t|\}_{i\in N}|} \quad (7)$$

We visually illustrate these measures in Fig. 5.

The claims are valid for both rotation and translation. Subspace volume is not intuitive in quaternions representation, but if we convert to Euler angles we can use the same formulas. The subspace of absolute orientation in most scenes is the full sphere, but for the rotation (relative orientation) it is also much smaller. The "MAPE" formula is given by:

$$R_{MAPE}(EulerAngles) = \frac{1}{N}\sum_{i\in N}(\frac{|r_i - \hat{r_i}|}{|r_i|}) \quad (8)$$

Area Under Curve for All Overlap Ranges. The accuracy of a model depends on how well it estimates the relative pose given a test pair of images. As mentioned earlier, the selection of the overlap method and threshold significantly determines the complexity of the problem. In order to achieve a non-overlap dependence on assessment, one way is to compute the area under the curve (AUC) across the overlaps. The model is better as the area decreases. Quantification of the difference between the two models could be obtained from the difference between their AUCs. This is illustrated in Fig. 6.

[1] By naive model we mean that the model always returns the mean value of the training data.

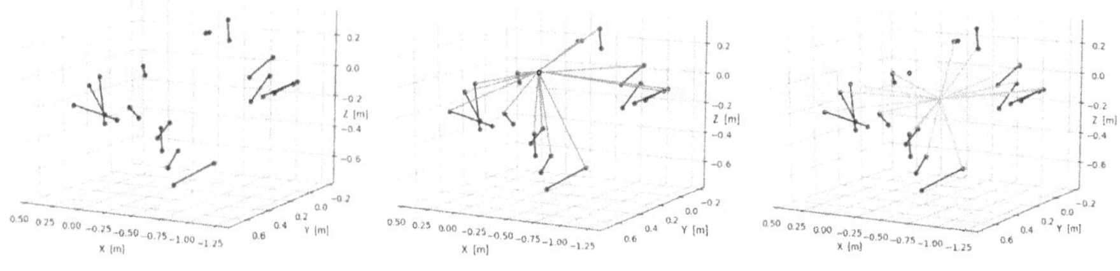

(a) Translation error vector (b) Mean Absolute Percentage Error (MAPE) (c) Mean Absolute Scale Error (MASE)

Fig. 5. An Illustration of the proposed metrics. Ground truth points are in green, estimation points in red, the blue line connected them is the translation error. Green and yellow lines represent the normalization factor of the error in MAPE and MASE, respectively. These types of normalization are necessary when subspace volume is taken into account.

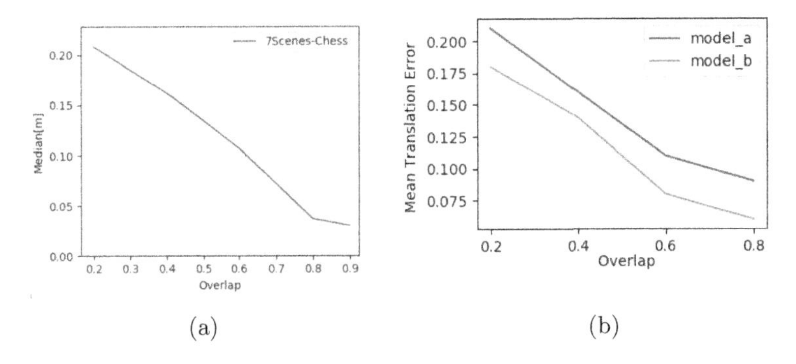

(a) (b)

Fig. 6. (a) Accuracy across different overlaps on 7-scenes data-set. We can see clearly the correlation, when overlap increase accuracy decrease. (b) Two trained models tested on the same data. We can see that Model b performs better than Model a since it valid for any overlap score. To evaluate and compare in cases there is no single model that is always better, we turned to the area under the curve (AUC) approach.

Minimal Requirement for Reliable Comparison. As a relaxation of the previous method, it might be enough to consider the sub-space diameter[2]. Intuitively, the subspace diameter quantifies the spanned 3D space induced by the relative pose targets (labels). The sub-space diameter can be defined as:

$$subspace\_diameter = mean(\{|t|\}_{i \in N}) + 2std(\{|t|\}_{i \in N}) \tag{9}$$

5 Experiments and Results

[2] In [20] a hard-coded metric was used. This provided correct predictions by limiting the translation and rotation errors up to a pre-defined threshold. This metric can be used in addition to considering the sub-space diameter.

Table 3. Relative poses regression subspace diameter in meters as a function of overlap. The subspace created by the span of all relative-poses with enough overlap. The diameters are correlated to overlap threshold, while the scene volume almost not matter.

	Overlap Threshold				
	0.2	0.4	0.6	0.8	0.9
Chess	1.95	1.91	1.54	0.71	0.57
Fire	1.65	1.16	0.80	0.77	0.68
Heads	2.13	1.93	1.13	0.57	0.45
Office	2.79	2.65	2.28	0.92	0.42
Pumpkin	1.24	1.34	1.42	0.80	0.26
Kitchen	3.49	2.47	1.76	0.88	0.38
Stairs	0.91	0.75	0.60	0.50	0.37

(a) Demonstrated on 7-scenes[35] data-set.

	Overlap Threshold				
	0.2	0.4	0.6	0.8	0.9
Kings Collage	6.82	4.87	3.51	2.50	1.26
Great Court	6.54	5.00	3.59	2.39	1.58
Old Hospital	8.08	6.71	5.16	2.73	0.89
St M. Church	8.43	7.21	5.61	2.40	0.69
Streets	6.48	4.78	3.56	2.45	1.14
Shop Facade	6.08	4.28	3.79	2.42	1.43

(b) Demonstrated on Cambridge[17] data-set.

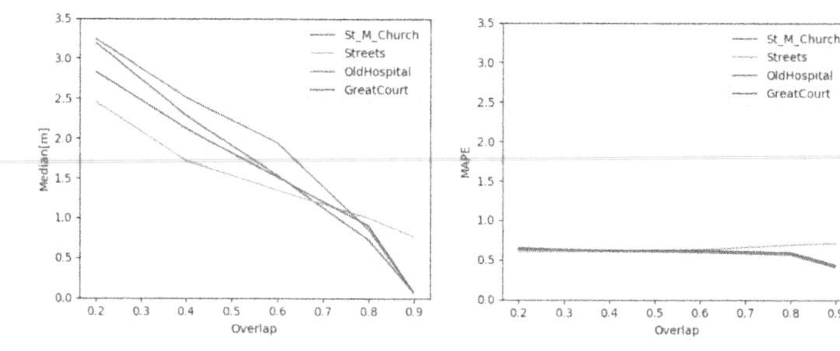

(a) Median accuracy across different overlaps on Cambridge data-set.

(b) MAPE across different overlaps on Cambridge data-set.

Fig. 7. Translation error using median (a) and using MAPE (b) across different overlaps.

Comparison to Other Methods: To demonstrate our key insights we conducted several experiments. We chose the 7-Scenes data-set [35] and trained our architecture (see Sect. 3). As its name implies, this data-set consists of 7 different scenes taken using a Microsoft Kinnect camera. For each of these scenes, we compare our trained network using an appropriate overlap threshold to achieve satisfactory results. These experiments are summarized in Table 2. It is worth noting that our comparison is not complete. The train and test data are **not** the same, due to pairs selection induced by the overlap method. At first sight, our method yields the best results compared to other methods. However, one should note that we chose the overlap accordingly in order to achieve better results using the common criteria metrics. This complies with what was argued in Sect. 4.2 (Fig. 7).

Comparison to Relative-Pose Methods: We compare our method using the standard median to other relative-pose methods. Note that these comparisons are made without relating to the sub-space diameter, and are summarized in Table 1.

Correlation Between Sub-space Diameter and Accuracy: To analyze the correlation with sub-space diameter as defined in Sect. 4, we measure the sub-space diameter for each scene over a range of different overlaps. We introduce here another data-set we used, called Cambridge [17]. Results are summarized in Table 3b. To prove empirically our claim that there is a strong correlation between overlap and accuracy, we trained the Chess scene from the 7-scenes data-set with an overlap of 0.6 and tested it under various overlap ranges. Results are shown in Fig. 6a.

Volume Agnostic Methods: We offer sub-space diameter agnostic metrics. By agnostic, we mean that the proposed measure should be *stable* under different overlap and threshold selection methods. We used our trained model on the Chess scene from the 7-scenes data-set. We evaluated the methods proposed in Sect. 4.3 and showed that the metrics result is stable around the same values, even when we modified the overlap threshold. These results are summarized in Fig. 8. We also changed the common definition of MAPE and MASE, and used $L2$ as a norm to examine the robustness of our method to norm selection. Results are summarized in Fig. 9.

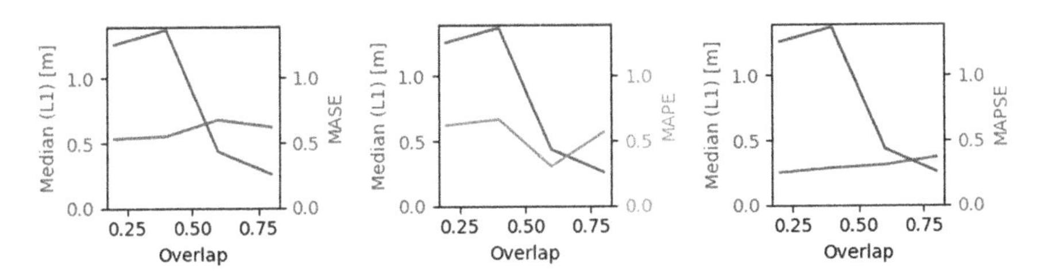

Fig. 8. An illustration of using volume-agnostic metrics vs median metric under L1 norm. The blue line at each plot is the median, when the other line in each plot is one of the 3 proposed metrics. It can be clearly seen that the proposed method is robust to different overlaps.

Comparison Given Sub-space Diameter: As we mentioned earlier, a minimal requirement for fair comparison is referred to as the subspace diameter, as reported in [29]. Hence, we compare our method to theirs, as summarized in Table 4.

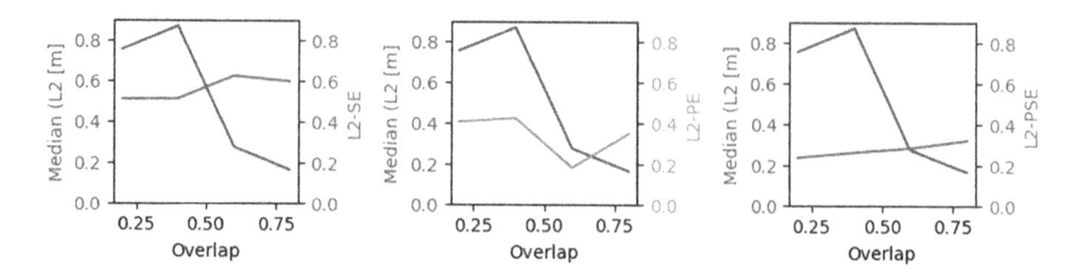

Fig. 9. Proposed metrics are agnostic to norm selection. Unlike Fig. 8 here we use $L2$. It can be seen that the proposed metrics are robust to **norm selection**.

Table 4. Comparison to [29] given the subspace diameter. Note that in the third double-column we train only on one scene and yet are able to generalize to other scenes as well.

	AnchorNet		Ours best		Ours chess model	
	Accuracy	Subspace diameter	Accuracy	Subspace diameter	Accuracy	Subspace diameter
Chess	0.08 m, 4.12°	0.19	0.07 m, 1.63°	0.19	0.07 m, 1.63°	0.19
Fire	0.16 m, 11.1°	0.14	0.02 m, 1.45°	0.14	0.16 m, 6.25°	0.14
Heads	0.09 m, 11.2°	0.05	0.02 m, 1.73°	0.05	0.07 m, 3.43°	0.05
Office	0.11 m, 5.38°	0.18	0.02 m, 1.53°	0.18	0.12 m, 4.35°	0.18
Pumpkin	0.14 m, 3.55°	0.11	0.02 m, 1.09°	0.11	0.09 m, 3.96°	0.11
Red Kitchen	0.13 m, 5.29°	0.22	0.02 m, 1.14°	0.22	0.10 m, 4.06°	0.22
Stairs	0.21 m, 11.9°	0.17	0.01 m, 0.74°	0.17	0.10 m, 2.54°	0.17

6 Conclusions

In this paper, we introduce a novel comparison criterion for relative-pose-regression. We demonstrate the insufficiency of the current standard metrics in terms of their capability to assess relative-pose-regression algorithms. We also show that our proposed metrics overcome many of the drawbacks of existing metrics, and believe that the proposed metrics should be adopted in the field. We believe that relative-pose-regression is crucial for solving the problem of camera re-localization. Generalization (training on one scene and achieving good results on another) is a key to the widely-adopted full solution of the problem; however, we leave its extensive investigation for future work.

In our opinion, due to the lack of generalization capabilities of the absolute pose-regression approach, relative-pose-regression is now the most promising research direction in data-driven visual re-localization.

References

1. Argamon, S., McDermott, D.: Error correction in mobile robot map learning. In: Proceedings 1992 IEEE International Conference on Robotics and Automation, vol. 3, pp. 2555–2560 (1992)

2. Balntas, V., Li, S., Prisacariu, V.: RelocNet: continuous metric learning relocalisation using neural nets. In: Ferrari, V., Hebert, M., Sminchisescu, C., Weiss, Y. (eds.) Computer Vision – ECCV 2018. LNCS, vol. 11218, pp. 782–799. Springer, Cham (2018). https://doi.org/10.1007/978-3-030-01264-9_46
3. Brachmann, E., et al.: DSAC – differentiable RANSAC for camera localization. In: 2017 IEEE Conference on Computer Vision and Pattern Recognition (CVPR), pp. 2492–2500 (2016)
4. Brachmann, E., Rother, C.: Learning less is more - 6D camera localization via 3D surface regression. In: 2018 IEEE/CVF Conference on Computer Vision and Pattern Recognition, pp. 4654–4662 (2017)
5. Brachmann, E., Rother, C.: Expert sample consensus applied to camera re-localization. In: 2019 IEEE/CVF International Conference on Computer Vision (ICCV), pp. 7524–7533 (2019)
6. Brahmbhatt, S., Gu, J., Kim, K., Hays, J., Kautz, J.: Geometry-aware learning of maps for camera localization. In: 2018 IEEE/CVF Conference on Computer Vision and Pattern Recognition, pp. 2616–2625 (2017)
7. Cavallari, T., Golodetz, S., Lord, N.A., Valentin, J.P.C., di Stefano, L., Torr, P.H.S.: On-the-fly adaptation of regression forests for online camera relocalisation. In: 2017 IEEE Conference on Computer Vision and Pattern Recognition (CVPR), pp. 218–227 (2017)
8. Ding, M., Wang, Z., Sun, J., Shi, J., Luo, P.: CamNet: coarse-to-fine retrieval for camera re-localization. In: The IEEE International Conference on Computer Vision (ICCV), October 2019
9. Engel, J., Koltun, V., Cremers, D.: Direct sparse odometry. IEEE Trans. Pattern Anal. Mach. Intell. **40**, 611–625 (2016)
10. Engel, J., Schöps, T., Cremers, D.: LSD-SLAM: large-scale direct monocular SLAM. In: Fleet, D., Pajdla, T., Schiele, B., Tuytelaars, T. (eds.) ECCV 2014. LNCS, vol. 8690, pp. 834–849. Springer, Cham (2014). https://doi.org/10.1007/978-3-319-10605-2_54
11. Fiala, M.: ARTag, a fiducial marker system using digital techniques. In: 2005 IEEE Computer Society Conference on Computer Vision and Pattern Recognition (CVPR 2005), vol. 2, pp. 590–596 (2005)
12. Gálvez-López, D., Tardós, J.D.: Bags of binary words for fast place recognition in image sequences. IEEE Trans. Rob. **28**(5), 1188–1197 (2012). https://doi.org/10.1109/TRO.2012.2197158
13. Gordo, A., Almazán, J., Revaud, J., Larlus, D.: Deep image retrieval: learning global representations for image search. In: Leibe, B., Matas, J., Sebe, N., Welling, M. (eds.) ECCV 2016. LNCS, vol. 9910, pp. 241–257. Springer, Cham (2016). https://doi.org/10.1007/978-3-319-46466-4_15
14. Guzmán-Rivera, A., et al.: Multi-output learning for camera relocalization. In: 2014 IEEE Conference on Computer Vision and Pattern Recognition, pp. 1114–1121 (2014)
15. Hyndman, R.J., Koehler, A.B.: Another look at measures of forecast accuracy. In: FORESIGHT (2006)
16. Kendall, A., Cipolla, R.: Geometric loss functions for camera pose regression with deep learning. In: 2017 IEEE Conference on Computer Vision and Pattern Recognition (CVPR), pp. 6555–6564 (2017)
17. Kendall, A., Grimes, M.K., Cipolla, R.: PoseNet: a convolutional network for real-time 6-DOF camera relocalization. In: 2015 IEEE International Conference on Computer Vision (ICCV), pp. 2938–2946 (2015)

18. Larsson, M., Stenborg, E., Toft, C., Hammarstrand, L., Sattler, T., Kahl, F.: Fine-grained segmentation networks: Self-supervised segmentation for improved long-term visual localization. In: The IEEE International Conference on Computer Vision (ICCV), October 2019
19. Laskar, Z., Huttunen, S., DanielHerrera, C., Rahtu, E., Kannala, J.: Robust loop closures for scene reconstruction by combining odometry and visual correspondences. In: 2016 IEEE International Conference on Image Processing (ICIP), pp. 2603–2607 (2016)
20. Laskar, Z., Melekhov, I., Kalia, S., Kannala, J.: Camera relocalization by computing pairwise relative poses using convolutional neural network. In: 2017 IEEE International Conference on Computer Vision Workshops (ICCVW), pp. 920–929 (2017)
21. Massiceti, D., Krull, A., Brachmann, E., Rother, C., Torr, P.H.S.: Random forests versus neural networks – what's best for camera localization? In: 2017 IEEE International Conference on Robotics and Automation (ICRA), pp. 5118–5125 (2016)
22. Melekhov, I., Ylioinas, J., Kannala, J., Rahtu, E.: Relative camera pose estimation using convolutional neural networks. In: ACIVS (2017)
23. Meng, L., Chen, J., Tung, F., Little, J.J., Valentin, J., de Silva, C.W.: Backtracking regression forests for accurate camera relocalization. In: 2017 IEEE/RSJ International Conference on Intelligent Robots and Systems (IROS), pp. 6886–6893 (2017)
24. Meng, L., Tung, F., Little, J.J., Valentin, J., de Silva, C.W.: Exploiting points and lines in regression forests for RGB-D camera relocalization. In: 2018 IEEE/RSJ International Conference on Intelligent Robots and Systems (IROS), pp. 6827–6834 (2017)
25. Mur-Artal, R., Montiel, J.M.M., Tard'os, J.: ORB-SLAM: a versatile and accurate monocular SLAM system. IEEE Trans. Rob. **31**(5), 1147–1163 (2015)
26. Myttenaere, A.D., Golden, B., Grand, B.L., Rossi, F.: Mean absolute percentage error for regression models. Neurocomputing **192**, 38–48 (2016)
27. Pfrommer, B., Daniilidis, K.: TagSLAM: Robust SLAM with fiducial markers. arXiv abs/1910.00679 (2019)
28. Radenović, F., Tolias, G., Chum, O.: CNN image retrieval learns from BoW: unsupervised fine-tuning with hard examples. In: Leibe, B., Matas, J., Sebe, N., Welling, M. (eds.) ECCV 2016. LNCS, vol. 9905, pp. 3–20. Springer, Cham (2016). https://doi.org/10.1007/978-3-319-46448-0_1
29. Saha, S., Varma, G., Jawahar, C.V.: Improved visual relocalization by discovering anchor points. In: BMVC (2018)
30. Sattler, T., Leibe, B., Kobbelt, L.: Improving image-based localization by active correspondence search. In: Fitzgibbon, A., Lazebnik, S., Perona, P., Sato, Y., Schmid, C. (eds.) ECCV 2012. LNCS, vol. 7572, pp. 752–765. Springer, Heidelberg (2012). https://doi.org/10.1007/978-3-642-33718-5_54
31. Sattler, T., Leibe, B., Kobbelt, L.: Efficient & effective prioritized matching for large-scale image-based localization. IEEE Trans. Pattern Anal. Mach. Intell. **39**, 1744–1756 (2017)
32. Sattler, T., Zhou, Q., Pollefeys, M., Leal-Taixé, L.: Understanding the limitations of CNN-based absolute camera pose regression. arXiv abs/1903.07504 (2019)
33. Schönberger, J.L., Pollefeys, M., Geiger, A., Sattler, T.: Semantic visual localization. In: 2018 IEEE/CVF Conference on Computer Vision and Pattern Recognition, pp. 6896–6906 (2017)
34. Shavit, Y., Ferens, R.: Introduction to camera pose estimation with deep learning (2019)

35. Shotton, J., Glocker, B., Zach, C., Izadi, S., Criminisi, A., Fitzgibbon, A.W.: Scene coordinate regression forests for camera relocalization in RGB-D images. In: 2013 IEEE Conference on Computer Vision and Pattern Recognition, pp. 2930–2937 (2013)
36. Svärm, L., Enqvist, O., Kahl, F., Oskarsson, M.: City-scale localization for cameras with known vertical direction. IEEE Trans. Pattern Anal. Mach. Intell. **39**, 1455–1461 (2017)
37. Taira, H., et al.: InLoc: indoor visual localization with dense matching and view synthesis. In: 2018 IEEE/CVF Conference on Computer Vision and Pattern Recognition, pp. 7199–7209 (2018)
38. Tejani, A., Tang, D., Kouskouridas, R., Kim, T.-K.: Latent-class Hough forests for 3D object detection and pose estimation. In: Fleet, D., Pajdla, T., Schiele, B., Tuytelaars, T. (eds.) ECCV 2014. LNCS, vol. 8694, pp. 462–477. Springer, Cham (2014). https://doi.org/10.1007/978-3-319-10599-4_30
39. Valada, A., Radwan, N., Burgard, W.: Deep auxiliary learning for visual localization and odometry. In: 2018 IEEE International Conference on Robotics and Automation (ICRA), pp. 6939–6946. IEEE (2018)
40. Wang, J., Olson, E.: AprilTag 2: efficient and robust fiducial detection. In: 2016 IEEE/RSJ International Conference on Intelligent Robots and Systems (IROS), pp. 4193–4198 (2016)

A Deep Learning Filter for Visual Drone Single Object Tracking

Xin Zhang, Licheng Jiao(✉), Xu Liu, Xiaotong Li, Wenhua Zhang, Hao Zhu, and Jie Zhang

Key Laboratory of Intelligent Perception and Image Understanding of Ministry of Education, School of Artificial Intelligence, Xidian University, Xi'an, China
{xinZhang1,lixiaotong}@stu.xidian.edu.cn, lchjiao@mail.xidian.edu.cn, xuliu361@163.com, zhangwenhua_nuc@163.com, haozhu@xidian.edu.cn, 1437614843@qq.com

Abstract. Object tracking is one of the most important topics in computer vision. In visual drone tracking, it is an extremely challenging due to various factors, such as camera motion, partial occlusion, and full occlusion. In this paper, we propose a deep learning filter method to relieve the above problems, which is to obtain a priori position of the object at the subsequent frame and predict its trajectory to follow up the object during occlusion. Our tracker adopts the geometric transformation of the surrounding of the object to prevent the bounding box of the object lost, and it uses context information to integrate its motion trend thereby tracking the object successfully when it reappears. Experiments on the VisDrone-SOT2018 test dataset and the VisDrone-SOT2020 val dataset illustrate the effectiveness of the proposed approach.

Keywords: Geometric transformation · Context information · Visual drone tracking

1 Introduction

Visual object tracking has been evolved so fast over the last decades. Given an object bounding box of the first frame from a video, the goal is to learn a classifier about predicting the object bounding box from the next frame. The challenge is that the object is class-agnostic and defined solely by its initial location and scale. With the development of visual tracking, some researchers use the discriminative or generative method [4,9,11] to learn a correlation filter for the object appearance online, while other researchers attempt to train a deep architecture offline, such as the siamese network [2], to select the most similar candidate patch from the object template.

The discriminative and generative methods are mainly based on the correlation filter (CF) trackers. Since MOSSE [4] has shown outstanding results on object tracking benchmarks, the CF based trackers model developed so fast.

A. Bartoli and A. Fusiello (Eds.): ECCV 2020 Workshops, LNCS 12538, pp. 638–650, 2020.
https://doi.org/10.1007/978-3-030-66823-5_38

Numerous methods have been proposed to improve the CF trackers by equipping it with the kernel trick, e.g. KCF [11] and DSST [8]. In general, the CF trackers aim to find the most similar region online among the sampled candidate regions to the given object. With the rapid development of the deep networks, pre-trained deep CNN models are utilized as the feature maps to correlate. For instance, C-COT [9] produces a periodic extension of the feature map and employs the Conjugate Gradient method to solve the loss function. ECO [6] employs a factorized convolution operator to reduce the computational and memory complexity based on C-COT. In recent years, the siamese network based on trackers has received great attention. CFNet [19] back-propagate gradients through the Correlation Filter based on [2]. Some tracking-by-detection methods [3,7,10,13,14,18,20] emerge in an endless stream. Meanwhile, the end-to-end tracking networks [12,17] are trained in an end-to-end manner during offline training.

Nevertheless, compared to the natural object tracking, it has some discrepancy challenges in visual drone tracking, such as camera motion, partial occlusion, and full occlusion. Most methods have poor sensitivity when the search area changes drastically in the consecutive frames. While the object is being occluded, it makes the storage of the CF trackers inaccurate because of the long-term slow change of the object appearance. To address these issues above, our paper proposes a general intensity mapping operator and an energy function. The contributions of this work can be summarized as follows:

(1) We perform an intensity deformation of the surrounding of the object to process the geometric transformation between consecutive samples, and it effectively improves the accuracy of the deep learning filter tracker in case of camera motion.
(2) Context information is exploited to predict the motion trend of the object by minimizing the energy function, which to ensure that the object re-tracked successfully after occlusion.

The remainder of this paper is organized as follows. In Sect. 2, we discuss the related work. The details of our proposed method are presented in Sect. 3. Section 4 reports the implementation details and the experimental results. Finally, Sect. 5 ends this work with some concluding remarks.

2 Related Work

In this section, we briefly survey the relevant literature from the following three aspects: the CF trackers, a similarity measure between images and the motion estimation of the object.

The CF trackers, which can robustly build an object appearance model, are to learn a discriminative filter for the next frame from the current frame and historical information. And in many different tracking scenarios, the discriminative CF trackers have been demonstrated a good capability of locating the object accurately. Among the CF trackers, Continuous Convolution Operator Tracker

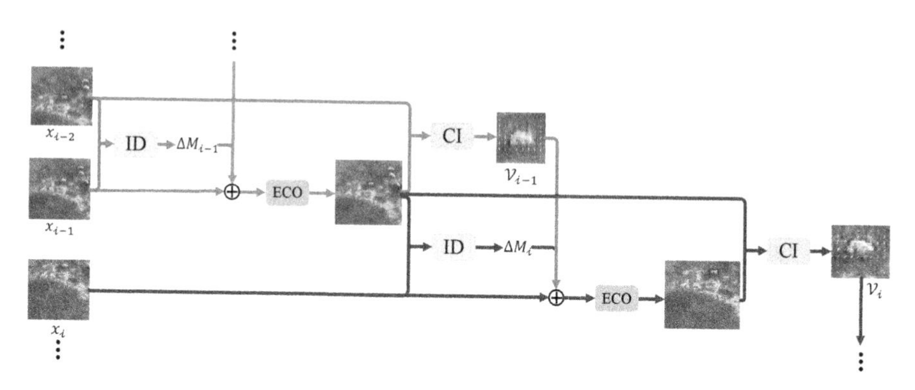

Fig. 1. The structure of our deep learning filter method, which contains the intensity deformation module (ID), the basic tracker Efficient Convolution Operator (ECO), and the context information module (CI).

(C-COT) [9] and Efficient Convolution Operator (ECO) [6], which are achieved the feature maps of the object by performing convolutions in the continuous domain, significantly increase the tracking accuracy compared to the original CF tracker. However, since the learned object-regularized CF model relies only on the appearance but ignores the motion trends of the object, they cannot work well under the complex environments. For instance, they are prone to lose the object when severely occluded. Besides, these CF trackers rarely premeditate dynamic changes caused by external effects, such as camera motion of vision drone tracking.

It's easy to distinguish from the natural video tracking such as camera motion due to the special property of the visual drone tracking. We can easily know that the adjacent frames almost have the same scene since the instantaneous movement of the object is short. The gradient-based measure method, such as the scale-invariant feature transform (SIFT) [15], has been considered a photometric-invariant similarity measures for the images under different viewpoint. It aims to find the best correspondence maximizing specific similarity measures between two consecutive images and helps to overcome issues such as image rotation and scale that are common when overlaying images. But it fails in case of non-monotonic mapping due to the contraction of the image data.

Note that the occlusion problem exists in most object tracking videos, and some successful algorithms rely on the motion estimation of the object. And the trajectory of the object can be predicted well by calculating the motion trend. Optic flow [16] estimation, which is caused by the relative motion between foreground and background, is a motion estimation to the object in a visual scene based on the assumptions of brightness constancy and spatial smoothness. And it usually deal the occlusion problem with predicting the motion trend of the object through more than two consecutive frames in a video tracking sequence.

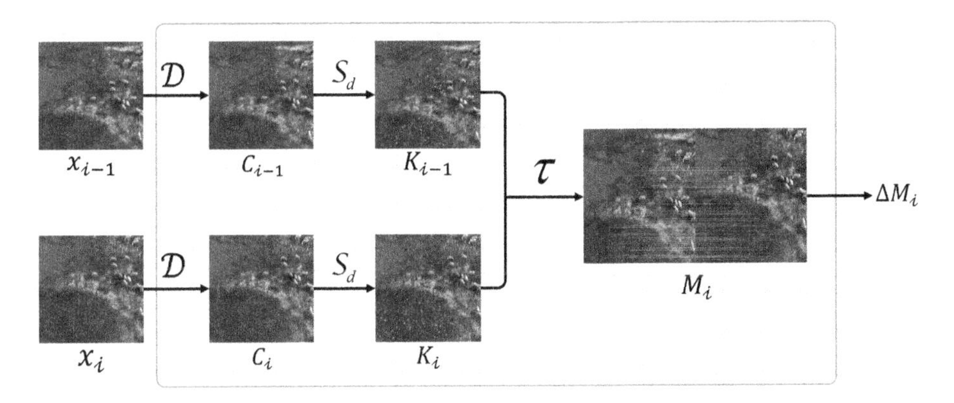

Fig. 2. Main steps of the Intensity Deformation module (ID). It reduced the image offset caused by camera motion.

3 Proposed Method

Figure 1 shows the whole architecture of our tracker. In this section, the intensity deformation method is performed to decrease the difference in similarity measures from adjacent frames. Then, we introduce the basic deep continuous convolution operator tracker, which employs Gauss-Newton and the Conjugate Gradient method to optimize the quadratic subproblems of the loss function. Finally, an energy function is proposed to integrate the motion trend of the object.

3.1 Intensity Deformation

In this subsection, we explore a highly accurate and computationally efficient methodology to solve the correspondence similarity measure estimation in a visual drone tracking video.

We propose a mapping of an image x from the intensity domain to the local area domain, and set $q \in \mathcal{N}_p$ is a neighboring pixel of the current pixel p at the image x, so the local area transform can then be expressed as follows,

$$C(p) = \sum_{q \in \mathcal{N}_p} \phi(x(p), x(q)) \tag{1}$$

where $C(p)$ denotes the local area statistical information. Hence, the logistic function $\phi(m, n)$ records the value 1 when x equals to y, and 0 otherwise, which having two properties, one is $\phi(km, kn) = \phi(m, n)$ where $k \in \mathbb{N}$ and $k \neq 0$, the other is $\phi(m_1, n_1) = 1$ and $\phi(m_2, n_2) = 1 \Rightarrow \phi(m_1 m_2, n_1 n_2) = 1$.

As the particularity of the vision drone tracking, we attempt to a general nonlinear intensity deformation, which defined as $\mathcal{D}\{\tilde{x}_i(p)\} = \delta(x_i(p))x_i(p)$, which $\delta(\cdot)$ represents an intensity mapping operator, and $\tilde{\cdot}$ denotes intensity domain information corresponding to local area domain. Therefore, the relationship between $\tilde{C}_i$ and C_i can be computed by,

$$\begin{aligned}\tilde{C}_i(p) &= \sum_{q\in\mathcal{N}_p} \phi(\tilde{x}_i(p), \tilde{x}_i(q)) \\ &= \sum_{q\in\mathcal{N}_p} \phi(\delta(x_i(p))x_i(p), \delta(x_i(q))x_i(q)) \\ &= \sum_{q\in\mathcal{N}_p} \phi(\alpha x_i(p), \beta x_i(q)) \\ &= \sum_{q\in\mathcal{N}_p} \phi(x_i(p), x_i(q)) = C_i(p)\end{aligned} \tag{2}$$

where α and β are constant values according to $x_i(p)$ and $x_i(q)$. We presume the deformable function as an one-to-one mapping, so it follows that $\tilde{C}_i(p) = C_i(p)$.

As a result, the sample $C_i(p)$ can be transformed into the sample $\tilde{C}_i(p)$ through the non-linear intensity deformation δ in local area domain. And then the SIFT descriptor $\mathcal{S}_d$ is used to detect the keypoint matrices K_{i-1} and K_i from C_{i-1} and C_i respectively. We connect the two keypoint matrices accordingly through the matching threshold $\tau_i = \sum \tau(K_i)\tau^T(K_{i-1})$, which is to find the keypoints with high similarity from the two keypoint matrices, to get the matching matrix M_i. The prior position of the object is based the bounding box of the previous frame on the total offset $\varDelta M_i$ which is averaged the matching matrix M_i. The intensity mapping operator makes the keypoints more accurate when calculating the prior position of the object. The advantage of the local area statistical is that its invariant to exposure deformation and illuminations corresponding to radiometric or photometric deformations, and performs reasonably well on natural images.

3.2 Base Tracker

Our basic tracker, which introduces a set of basis filters f with dimension C and a learned coefficient matrix P, employs a factorized convolution approach to solve the corresponding loss in the Fourier domain. We exploit the below function to minimize the original loss,

$$L(f,P) = \|\hat{z}^T P\hat{f} - \hat{y}\|^2_{\ell^2} + \sum_{c=1}^{C} \|\hat{\omega} * \hat{f}\|^2_{\ell^2} + \lambda\|P\|^2_F \tag{3}$$

Hence, $z = J\{x\}$, which $\{x_j\}_1^M \in \mathcal{X}$ denotes M training samples, encoded the object features (applying the first and the last convolutional layers of the VGG-Net [5]) with the interpolation operator $J : \mathbf{R}^{\mathrm{N}} \rightarrow L^2$. We use $\hat{\cdot}$ to denote its Fourier series, $*$ denotes correlation operators. As the bilinear term $\hat{z}^T P\hat{f}$, we substitute the first-order approximation into (3), and then the loss function can be formulated as,

$$\hat{L}(\hat{f}_{i,\Delta}, \Delta P) = \|\hat{z}^T P_i \hat{f}_{i,\Delta} + (\hat{f}_i \otimes \hat{z})^T \text{vec}(\Delta P) - \hat{y}\|_{\ell^2}^2 + \sum_{c=1}^{C} \|\hat{\omega} * \hat{f}_{i,\Delta}^c\|_{\ell^2}^2 + \mu \|P_i + \Delta P\|_F^2 \tag{4}$$

Δ denotes the size of matrix step. Here we assume the object is centered on the image region. And the Gaussian Mixed Model is employed to separate different Gaussian components with the number of ℓ to reduce the risk of over-fitting from the most recent frames. Besides, we introduce a probabilistic generative model in which the number of samples has decreased.

So the model is learned by minimizing the following objective,

$$L(f) = \mathbb{L}\{\|S_f\{x\} - y\|_{\ell^2}^2\} + \sum_{d=1}^{D} \|\omega f^d\|_{\ell^2}^2 \tag{5}$$

Since the value of the response map corresponds to the confidence of the object, the purpose is predicted to be centered at the maximum value location. As the objective is to find an accurate bounding box of the object from the next frame in a visual drone tracking video. Usually, it is still insufficient to localize the accurate object while the camera motion severely results in a significant deviation of the object in adjacent frames.

3.3 Context Information

For this purpose, there is a need to determine the motion vectors of the object located in consecutive frames. The context information method between successive frames, which belongs to the general matching problem, is often used for motion-based object detection and tracking method. The matching problem is formulated in terms of energy minimization with the energy function and calculated as,

$$L_{CI}(I) = \sum_{p \in box_i} G_p(v_p) + \sum_{N \in \mathcal{N}_p} B_N(v_N) \tag{6}$$

where $p \in box_i$ denotes the pixel value of local search area x_i at current frame and $N \in \mathcal{N}_p$ corresponds to edges of a pixel p neighborhood of the frame graph $\mathcal{G}_i = (box_i, \mathcal{N})$, we set v_p to the label of pixel p which belongs to some discrete set of 2D motion vectors that represents to correspondence search region.

G_p, which is a unary potential, denotes to the normal penalty cost function. And B_N, which corresponds to a binary potential, defines edge interaction between pixel interval. Since our framework can be seen as an optic flow optimization subproblem, $\mathcal{V}_i = \{v_1, \cdots, v_p\}$, which minimizing the energy functional (6),

$$v_p = \arg\min_{v_p} L_{CI}(v_p) \tag{7}$$

To calculate the cost volume in our pipeline, we use the features extracted from an interpolate layer. The cost of the unary potential G_p is the vector dot-product of two matched pixel features,

$$G(x_i, x_j) = 1 - J\{x_i\}J\{x_j\} \tag{8}$$

Energy minimization methods have lately attracted much attention in computer vision, especially in the optical flow estimation. When the occlusion region in the frame i has no corresponded pixels relative to the current frame $i-1$, but the occluded area of the object is visible in the frame $i-2$. That's to say, those pixels which turn invisible in the subsequent frame are usually visible in the previous frame. It is an apparent visual motion that we experience moving through the scene. As mentioned above, more accurate results are obtained with this approach, while the computation time is slower compared to the other method.

4 Experiments

In this section, two aspects will be introduced: the implementation details and the extensive experimental results on tracking datasets. We use the publicly available trained networks of these methods and run them as implemented by their authors on the test and val datasets. All the tracking results are used the reported results to ensure a fair comparison.

4.1 Implementation Details

We implement our tracker in MATLAB on Intel Xeon(R) CPU at 2.20 GHz and the needed libraries are developed with the MatConvNet toolbox. The filter $\hat{f}$ of the first iteration is set to zero. We apply the same feature representation as ECO, along with HOG and Color Names. Note that all parameters setting are kept fixed for all videos in a dataset.

4.2 Comparison with State-of-the-Art Trackers

Here, we compare our approach with state-of-the-art trackers on the Vision Meets Drone Single-Object Tracking (VisDrone-SOT)2018 and the VisDrone-SOT2020. The former dataset involves 132 video sequences, which are further divided into three non-overlapping sets, train dataset (86 sequences with 69,941 frames), val dataset (11 sequences with 7,046 frames) and test dataset (35 sequences with 29,367 frames) sets. The success of the proposed method is achieved by overcoming the following challenges.

We evaluate the proposed method with comparisons to the trackers, including ATOM, ECO, SRDCF, RPN++, SiamRPN++, SiamMask, SiamRPN and DCF. Note that the overall performance for all the trackers is summarized by the precision and success plots as shown in Fig. 3 on the VisDrone-SOT2020 val dataset, where the trackers are initialized at the first frame and start tracking without re-initialization. The proposed LZZ-ECO tracker ranks first among all the trackers, as shown in the left picture of Fig. 3. Among the siamese networks (such as ATOM, RPN++, SiamRPN++, SiamMask and SiamRPN), ATOM, which integrates the object estimation with the discriminative method, achieves the best precision score. Besides, in the CF trackers (such as ECO, SRDCF and DCF), ECO has the highest performance in terms of precision due to the

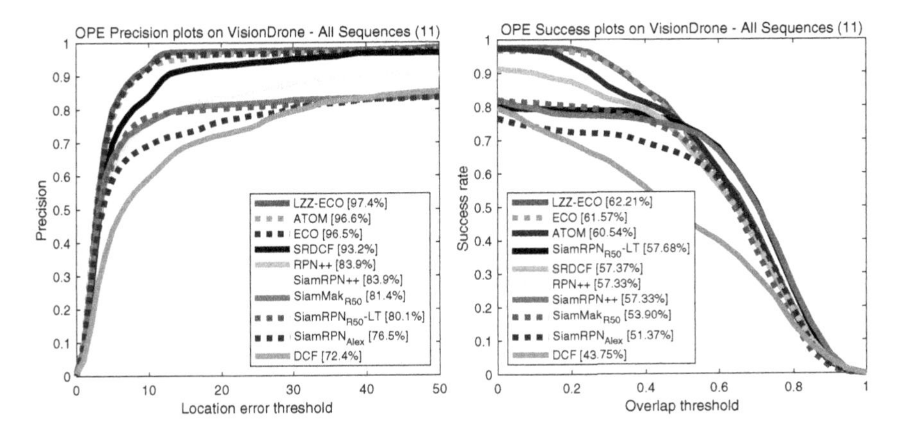

Fig. 3. Precision plots (left) and success plots (right) both the proposed method and state-of-the-art methods on the VisDrone-SOT2020 val dataset. The legend contains the average distance precision score at 20 pixels and the AUC of the success plot of each method. Our trackers show outstanding performance among state-of-the-art trackers.

interpolation at the continuous spatial domain. And our proposed method (LZZ-ECO) achieves the best performance because of the intensity deformation and the context information from continuous frames. We can see that the performance of the first three methods (LZZ-ECO, ATOM and ECO) reaches a higher peak when the Euclidean distance between the center coordinates of the ground-truth box and the center coordinates of the estimated bounding box is about 10 pixels, and then increases very slowly.

Figure 4 presents the results on the VisDrone-SOT2020 val dataset. We can find that our method performs the best in most instances. In particular, our method, which predicting the motion trend of the object before occlusion occurs, improves 3.95% by comparing with the second-placed method ECO in the attribute evaluation of full occlusion. But the performance is slightly lower than the SiamRPN, which is based on classification and regression branches when the object ratio and appearance change.

To access the proposed method, we compare it with twelve recent representative trackers, including MDNet, ECO, DCFNet, SRDCF, SECFNet, CFWCRKF, OST, TRACA+, Staple, KCF, CKCF and IMT3. Figure 5 [1] shows the success plots and the precision plots on the VisDrone-SOT2018 test dataset. ECO tracker obtains AUC scores of 49.0%, and our tracker achieves the best results, outperforming the second-best method by 5.2%.

Figure 6 [1] provides the overlap success plots and precision plots of the competing trackers on the VisDrone-SOT2018 test dataset under three attributes: Camera motion, full occlusion and partial occlusion. It can be seen that our method (LZZ-ECO) performs favorably against the state-of-the-art trackers on these attributes. Figure 7 shows the qualitative evaluation of 4 challenging sequences of the VisDrone-SOT2018 test dataset. Our method can track successfully when occlusion occurs.

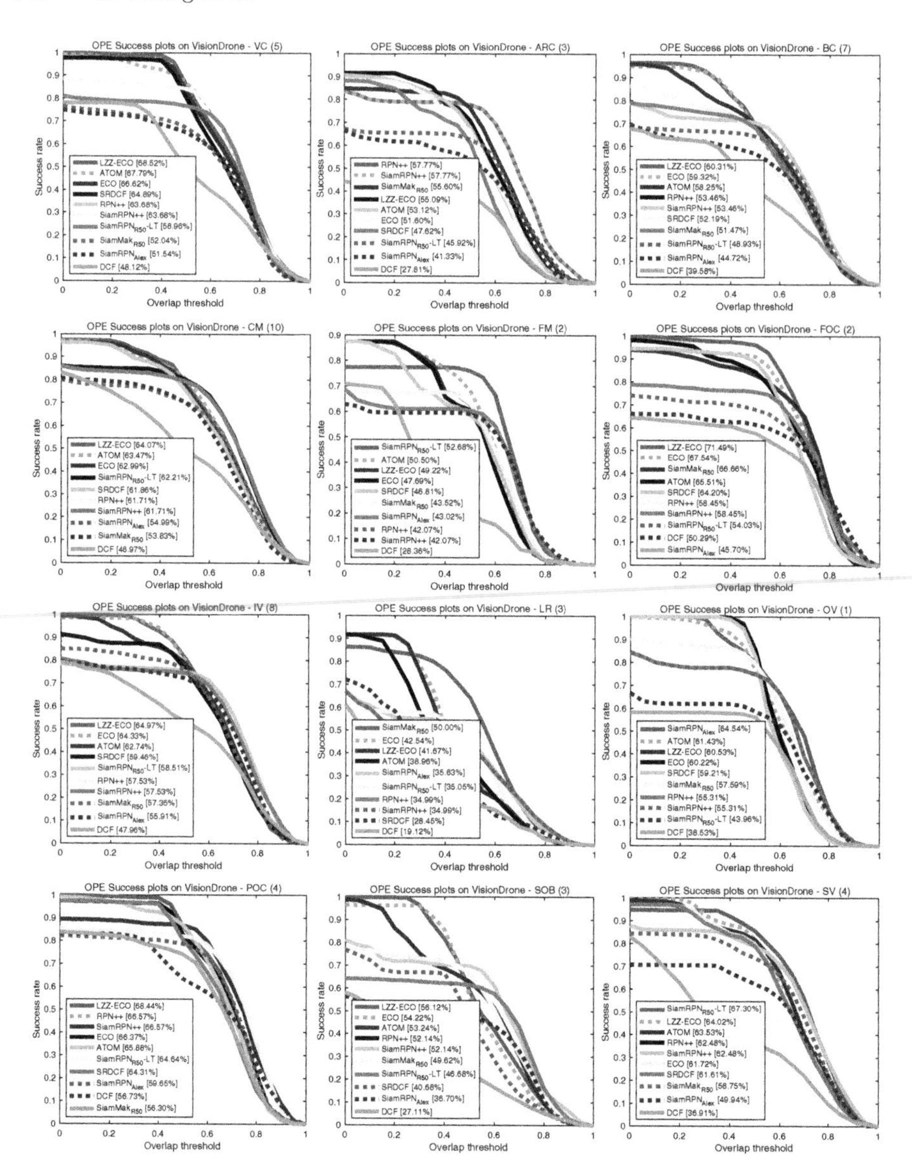

Fig. 4. Overlap success plots of eleven trackers on the VisDrone-SOT2020 val dataset. The challenges are composed of VC (Viewpoint Change), ARC (Aspect Ratio Change), BC (Background Clutter), CM (Camera Motion), FM (Fast Motion), FOC (Full Occlusion), IV (Illumination Variation), LR (Low Resolution), OV (Out-of-View), POC (Partial Occlusion), SOB (Similar Object), and SV (Scale Variation). Our tracker perform favorably against the state-of-the-art methods.

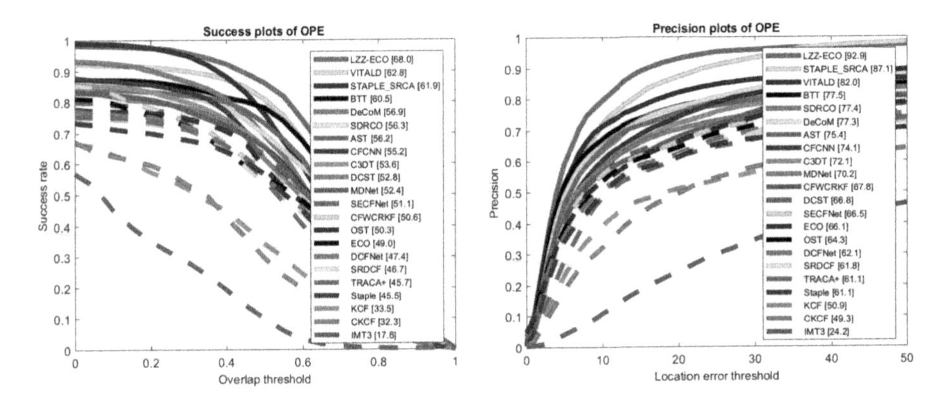

Fig. 5. The overall success plots and precision plots of the VisDrone-SOT2018 test dataset. Our method (LZZ-ECO) achieves the best performance with two metrics.

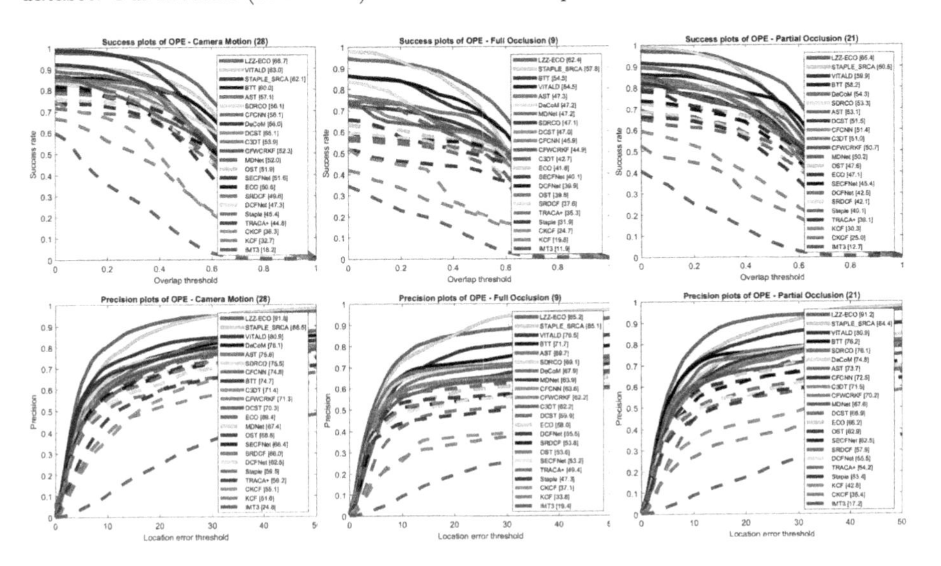

Fig. 6. Some attribute analysis on the VisDrone-SOT2018 test dataset. Our method obtains the best performance on these attributes.

Fig. 7. The tracking results of ours (red line) and baseline (green line) on five VisDrone-SOT2018 test dataset sequences (from top to down, the sequences are uav0000077_00000_s, uav0000083_00783_s, uav0000094_00000_s, uav0000244_00479_s, uav0000286_00001_s, respectively). These results show that our method can effectively use the context information of the object to solve the occlusion problem. (Color figure online)

Fig. 8. Failure cases of the proposed method. The videos are uav0000073_00038_s, uav0000087_00290_s, uav0000094_02070_s, uav0000328_04137_s, uav0000367_00001_s. Our deep learning filter method struggles when a drastic appearance change or a similar object occlusion occurs.

5 Conclusion

The paper proposes a deep learning filter method, which makes it more robust in the problem of visual drone tracking. The intensity deformation makes the similarity measure between two samples more accurate, and the context information predicts the motion of the object well when the occlusion occurs. The results have demonstrated that the method can achieve outstanding performance. But it cannot recover from occlusion when the occlusion and the object have extensive similarities as Fig. 8 shows. The future work is to explore the potentials of our method to more recent tracking schemes such as the Siamese network.

Acknowledgements. This work was supported in part by the State Key Program of National Natural Science of China (No.61836009), in part by the National Natural Science Foundation of China (No.U1701267), in part by the Major Research Plan of the National Natural Science Foundation of China (No.91438201), in part by the Program of Cheung Kong Scholars and Innovative Research Team in University (No.IRT_15R53), and in part by the Fundamental Research Funds for the Central Universities (JBF201905).

References

1. VisDrone-SOT2018: The vision meets drone single-object tracking challenge results. In: ECCV (2018)
2. Bertinetto, L., Valmadre, J., Henriques, J.F., Vedaldi, A., Torr, P.H.S.: Fully-convolutional Siamese networks for object tracking. In: ECCV (2016)
3. Bhat, Goutam, D.M.V.G.L., Timofte, R.: Learning discriminative model prediction for tracking. In: ICCV (2019)
4. Bolme, D.S., Beveridge, J.R., Draper, B.A., Lui, Y.M.: Visual object tracking using adaptive correlation filters. In: CVPR (2010)
5. Chatfield, K., Simonyan, K., Vedaldi, A., Zisserman, A.: Return of the devil in the details: delving deep into convolutional nets. In: BMVC (2014)
6. Danelljan, M., Bhat, G., Khan, F.S., Felsberg, M.: ECO: efficient convolution operators for tracking. In: CVPR (2017)
7. Danelljan, M., Bhat, G., Khan, F.S., Felsberg, M.: ATOM: accurate tracking by overlap maximization. In: CVPR (2019)
8. Danelljan, M., Haumlger, G., Shahbaz, K.F., Felsberg, M.: Accurate scale estimation for robust visual tracking. In: BMVC (2014)
9. Danelljan, M., Robinson, A., Khan, F.S., Felsberg, M.: Beyond correlation filters: learning continuous convolution operators for visual tracking. In: ECCV (2016)
10. Danelljan, M., Van Gool, L., Timofte, R.: Probabilistic regression for visual tracking. In: CVPR (2020)
11. Henriques, J.F., Caseiro, R., Martins, P., Batista, J.: High-speed tracking with Kernelized correlation filters. IEEE Trans. Pattern Ana. Mach. Intell. **37**(3), 583–596 (2015)
12. Jung, I., Son, J., Baek, M., Han, B.: Real-time MDNet. In: ECCV (2018)
13. Li, B., Wu, W., Wang, Q., Zhang, F., Xing, J., Yan, J.: SiamRPN++: evolution of Siamese visual tracking with very deep networks. In: CVPR (2019)
14. Li, B., Yan, J., Wu, W., Zhu, Z., Hu, X.: High performance visual tracking with siamese region proposal network. In: CVPR (2018)
15. Lowe, D.: Distinctive image features from scale-invariant keypoints. In: IJCV (2004)
16. Lucas, B.D., Kanade, T.: An iterative image registration technique with an application to stereo vision. In: IJCAI (1981)
17. Nam, H., Han, B.: Learning multi-domain convolutional neural networks for visual tracking. In: CVPR (2016)
18. Sauer, A., Aljalbout, E., Haddadin, S.: Tracking holistic object representations. In: BMVC (2019)
19. Valmadre, J., Bertinetto, L., Henriques, J.F., Vedaldi, A., Torr, P.H.S.: End-to-End representation learning for correlation filter based tracking. In: CVPR (2017)
20. Wenhua, Z., Haoran, W., Zhongjian, H., Yuxuan, L., Jinliu, Z., Licheng, J.: Accuracy and long-term tracking via overlap maximization integrated with motion continuity. In: ICCVW (2019)

Object Detection Using Clustering Algorithm Adaptive Searching Regions in Aerial Images

Yi Wang, Youlong Yang(✉), and Xi Zhao

School of Mathematics and Statistics, Xidian University, Xi'an, China
wangyi0102@stu.xidian.edu.cn, ylyang@mail.xidian.edu.cn

Abstract. Aerial images are increasingly used for critical tasks, such as traffic monitoring, pedestrian tracking, and infrastructure inspection. However, aerial images have the following main challenges: 1) small objects with non-uniform distribution; 2) the large difference in object size. In this paper, we propose a new network architecture, Cluster Region Estimation Network (CRENet), to solve these challenges. CRENet uses a clustering algorithm to search cluster regions containing dense objects, which makes the detector focus on these regions to reduce background interference and improve detection efficiency. However, not every cluster region can bring precision gain, so each cluster region difficulty score is calculated to mine the difficult region and eliminate the simple cluster region, which can speed up the detection. Then, a Gaussian scaling function(GSF) is used to scale the difficult cluster region to reduce the difference of object size. Our experiments show that CRENet achieves better performance than previous approaches on the VisDrone dataset. Our best model achieved 4.3% improvement on the VisDrone dataset.

1 Introduction

Equipped with cameras and embedded systems, Unmanned Aerial Vehicles (UAV) are endowed with computer vision ability and widely used for traffic monitoring, pedestrian tracking, and infrastructure inspection. With the rapid development of deep neural networks, the object detection framework based on deep neural networks has gradually become the mainstream technology of object detection. Although correlation detectors (such as R-CNN family [9–11,29], YOLO family [1,26–28], SSD family [7,19], etc.) have achieved good performance in natural images, they cannot achieve satisfactory results in aerial images.

Compared to natural images such as COCO [18], ImageNet [4] and Pascal VOC [6], aerial images have the following features:

(1) Small objects with the non-uniform distribution. Generally, a small object refer to object with the area of less than 32×32 in an image. The main problems of small objects are low resolution and small amount of information, which lead

A. Bartoli and A. Fusiello (Eds.): ECCV 2020 Workshops, LNCS 12538, pp. 651–664, 2020.
https://doi.org/10.1007/978-3-030-66823-5_39

to weak feature expression. The traditional method of processing small objects is to enlarge the image, which will increase the processing time and the memory needed to store large feature maps. Another common method is uniform cropping an image into several regions [8,20] and then detect in each region, which solves the problem of storing a large feature map. However, the uniform cropping ignores the sparsity of objects, and some regions may have few or no objects, which will waste a lot of computing resources. As can be seen from Fig. 1, the object distribution in the aerial image is uneven and the object is highly clustered in a certain region. Therefore, one method to improve detection efficiency is to focus the detector on these regions with a large number of objects.

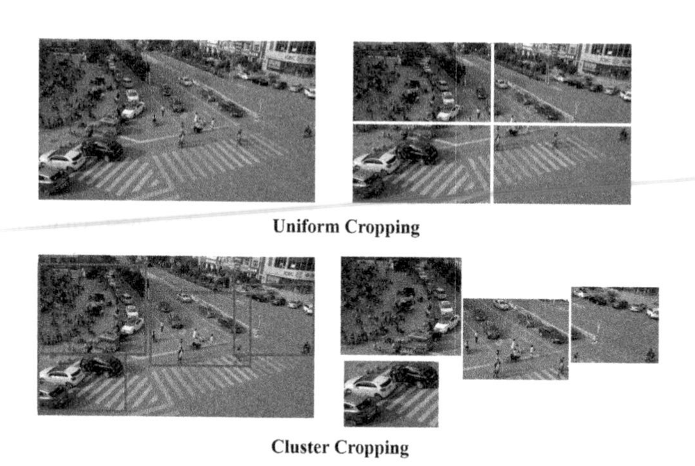

Fig. 1. Visualization of uniform cropping vs cluster cropping. The first row is an example of uniform cropping. The second row is an example of cluster cropping. Compared with uniform cropping, cluster cropping has the following advantages: (1) concentrate on computing resources in cluster regions with a large number of objects, (2) have no background interference.

(2) Diversity of object size. When collecting UAV images, the shooting height varies from tens of meters to hundreds of meters, which leads to huge difference in the size of the same category of objects. It is a big problem for the anchor-based detector to set the size of the anchor. For anchor-free detector, it is difficult to directly regress the width and height of the object. Therefore, it is necessary to reduce the difference in object size among images as much as possible.

To solve the above problems, this paper proposes a coarse-to-fine detection framework CRENet. As shown in Fig. 2, CRENet is composed of three parts: a coarse detection network (CNet), a cropping module, and a fine detection network (FNet). The aerial image is first sent to the coarse detection network CNet to get the initial detection results, which will get a rough distribution of the object. Then the initial detection results are sent to the crop module. The first we mentioned, the cluster region is obtained through a clustering algorithm.

The second step is to calculate the difficult score of each cluster region, it is believed that the cluster region with higher difficult score can bring greater accurate gain to the detector, and the cluster region with a small score can be deleted to improve the detection efficiency of the model. In the third step, we plug the remaining difficult cluster region into a Gaussian scaling function(GSF), which calculates the scaling factor for each of the difficult cluster regions. In particular, we refer to the difficult cluster region after scaling as ROI[1] (region of interest). Finally, ROI is sent to the fine detection network to obtain fine detection results, and the fine detection results are fused with the coarse detection results.

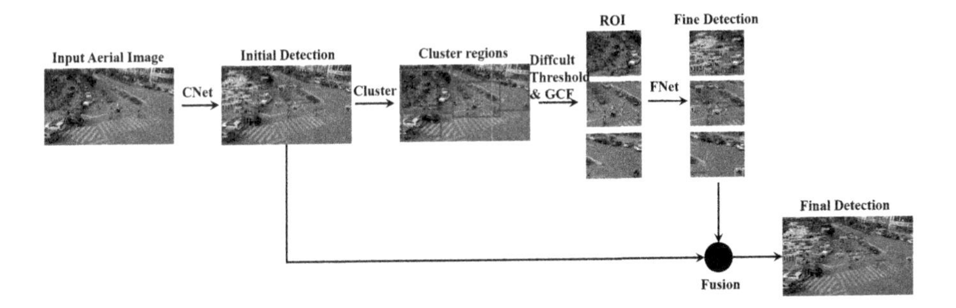

Fig. 2. Overview of CRENet framework. Firstly, CRENet sends the aerial image to the coarse detector CNet to get the initial prediction. Then a clustering algorithm is used to generate cluster regions for the initial prediction. And we mine difficult cluster regions, and then use the Gaussian scaling function(GSF) to scale difficult cluster regions. The difficult cluster regions are sent to the fine detector. Finally, the detection result of the global image is fused with the detection result of ROI to generate the final detection result. See Sect. 3 for more details.

Compared with previous detectors, the proposed CRENet has the following advantages: (1) The computational resources are concentrated on a dense region with a large number of objects, which reduces the computational cost and improves the detection efficiency; (2) Because the cluster region has different size, clustering algorithms are directly used to replace the network to predict the cluster region, which avoids the problem of anchor setting and the cluster region overlapping processing; (3) Calculating the difficult score of each cluster region and eliminating the cluster region that can hardly bring the accuracy gain can speed up the calculation; (4) Using Gaussian scaling function(GSF) can reduce the difference of object size among different images.

To sum up, the contributions of this paper are as follows:

1) A new CRENet detector is proposed that it can adaptively search and sacle regions with dense object for fine detection.

[1] The ROI here is different from Faster RCNN [29]. The ROI of this paper contains not just one object but multiple objects of interest, and it is used to represent the region with dense objects.

2) A Gaussian scaling function(GSF) is proposed to solve the problem of the large size difference of objects in aerial images and improve the detection accuracy.
3) We achieve more advanced performance on representative aerial image dataset VisDrone [45] with fewer images.

The rest of this paper is organized as follows. Section 2 briefly reviews relevant work. In Sect. 3, the proposed approach is described in detail. Section 4 for experimental results and Sect. 5 for the conclusion.

2 Related Work

In this section, we will review the benchmark of anchor-based detectors and anchor-free detectors for natural images and some recent efforts in aerial images. Finally, we focus on searching the region of interest for fine detection.

2.1 Generic Object Detection

At present, the mainstream object detection algorithms are mainly based on deep convolutional neural network, which can be divided into two types: anchor-based detectors and anchor-free detectors. The anchor-based detectors can be further divided into two categories: two-stage detector and single-stage detector. The two-stage detector consists of two steps: proposal region extraction and region classification. The first stage produces proposal regions, containing approximately location information of the object. In the second stage, the proposal regions are classified, and the positions are adjusted. Representatives of two-stage detectors include R-CNN family [9,10,29] and Mask RCNN [11]. The single-stage detector does not need the stage of producing proposal region , but directly generates the classification confidence and position of objects in only one stage. Representatives of single-stage detectors include SSD family [7,19], YOLO family [1,26–28], RetinaNet [17], etc. In general, the two-stage detector has higher accuracy, and the single-stage detector has higher detection speed. However, the anchor-based detector depends on the good prior anchor, and it is difficult to estimate a suitable prior anchor for the large variation in object size. In addition, in order to improve the recall rate, dense anchors are set, and most anchors are negative. This leads to an imbalance between negative anchors and positive anchors, which seriously affects the final detection performance.

The anchor-free detector is a method of object detection based on point estimate. CornerNet [14] uses a convolutional neural network to predict the upper left and lower right corner of an object and predicts embedding Vector of each diagonal corner to determine whether the upper left or lower right corner belong to the same object. CenterNet [45] directly predicts the center of the object and regresses its length and width. ExtremeNet [44] detects vertex, left point, bottom point, rightmost point, and the center point of the object.

It is difficult to find the suitable anchor size for all objects because of the large difference of object size caused by the change of UAV flying height. Therefore, in this paper, the anchor-free detector CenterNet [45] is used as the detection framework to solve this problem. Experiments also show that the anchor-free detector has better performance on the aerial image datasets.

2.2 Aerial Image Detection

Compared with natural image object detection, aerial image object detection faces more challenges: small objects, objects with uneven distribution, objects with various perspectives and objects vary in size. According to the characteristics of aerial images, people have proposed various solutions. Because the focus of this works is deep learning, this paper only reviews the related work of aerial image detection using deep neural networks. In [24], the tile method was used in the training stage and testing stage to improve the detection ability of small objects. In [35], the free metadata recorded by drones are used to learn Nuisance Disentangled Feature Transform (NDFT) to eliminate the interference of the detector caused by flying altitude change, adverse weather conditions, and other nuisances. Objects in the aerial image can be in any direction and any position. [5,21,38] uses rotate anchors to detect objects in any direction. In [16], the shape mask is allowed to flexibly detect objects in any direction without any pre-defined rotate anchors. In [34], the researcher studied the scale variation of aerial image object detection and proposed a Receptive Field Expansion Block (RFEB) to increase the receptive field size for high-level semantic features and a Spatial-Refinement Module (SRM) to repair the spatial details. In [25], a multi-task object detection and segmentation model is proposed. The segmentation map is used as the weight of self-attention mechanism to weight the feature map of object detection, which reduces the signal of non-correlated regions.

2.3 Region Search in Detection

In object detection, searching the region of interest for fine detection is usually used to detect small objects. The work of [20] proposes an adaptive detection strategy, which can continuously subdivide the regions that may contain small objects and spend computing resources in the regions that contain sparse small objects. The method in [31,37], the clustering algorithm was used to get ROI's ground truth on the original datasets, then a special CNN was used to predict ROI, and finally, ROI was sent to the fine detector. [42], using sliding window method on the feature map, then calculating the difficulty score for each window, and send the difficulty region to the fine detector. [8,32,33] solved the problem of small object detection in large images, and used a reinforcement learning method to find ROI for fine detection. [15] proposed an aerial image object detection network based on a density map. According to the density map, we can get a rough distribution of objects to search for the ROI.

Among the methods reviewed above, some use the network to predict the ROI, some use fixed windows to slide on the feature map to search the ROI, and some directly uniform crop the original image to get the ROI. Due to the different shapes and sizes of ROI, it is difficult to set the size of an anchor or regress the width and height of ROI by the network. The size of ROI obtained by using the fixed window sliding method and the uniform crop of the original image is fixed, which is difficult to adapt to the real ROI. Therefore, this paper sends the images to the coarse detection network to get the approximate distribution of objects, and then uses the clustering algorithm to adaptively get the ROI. Through the clustering algorithm, the ROI of various sizes can be obtained, which is more in line with the actual situation.

3 Methodology

As shown in Fig. 2, detection of an aerial image can be divided into three stages: the difficult cluster region extraction, fine detection of the ROI, and fusion of the detection results. In the first stage, aerial images are first sent into CNet to obtain an initial prediction. Then the cluster region is obtained by mean shift [39] for initial detection. Besides, the difficulty score of each cluster region is calculated, and the region with a higher score is regarded as a difficult cluster region. In the second stage, firstly, we use the Gaussian scaling function(GSF) to scale the difficult cluster region, so as to reduce the scale difference of objects. The ROI, scaled difficult cluster region, is then finely detected using the FNet. Finally, the third stage fuses the detection results of each ROI and global image with soft-NMS [2].

3.1 Difficult Cluster Region Extraction

Difficult cluster region extraction consists of three steps: Firstly, aerial images are fed into the trained CNet to obtain coarse detection results of objects. Then the results are used to obtain a cluster region. Finally, the difficulty score for each cluster region is calculated, and the non-difficult region is removed to speed up the detection.

In previous work, [37] proposed to use the clustering algorithm to generate the ground truth of the cluster region for each image and then trained a detector to predict the cluster region. However, there is a large overlap between the cluster regions predicted by the network. It is necessary to use the Iterative Cluster Merging (ICM) for the cluster region, and the number of the cluster region obtained is fixed. Especially, in aerial images, due to different camera angles, shooting time, and other reasons, the number of cluster regions may be different. Therefore, a fixed number of cluster regions is not suitable for all cases. Another problem is that the cluster regions vary in shape. It is difficult to manually set the anchor size in Faster-RCNN [29], and it is also difficult to regress anchor. In this paper, We use the clustering algorithm instead of the network to search the clustering region and avoid the problem of anchor setting. Specifically, aerial images are sent into the CNet, so as to obtain the initial prediction results of the object. The cluster region is obtained by using mean shift [39] from the initial prediction. Because an object can only belong to one region, the overlap between cluster regions is very small. Unlike [37], our algorithm is an unsupervised algorithm, whereas [37] is a supervised algorithm.

The aerial image is acquired at high altitude, so the background is complex and the objects is small. As can be seen from Fig. 1, the objects are usually gathered together. We can use a clustering algorithm to get the cluster region, and then crop and enlarge it for fine detection, which can not only solve the problem of small object detection but also reduce the interference of background. The mean shift [39] is a dense-based clustering algorithm, which assumes that the data of different clusters belong to different probability density distributions. By inputting the initial detection into the mean shift [39] algorithm, the cluster region of the image can be obtained adaptively.

It is worth noting that not every cluster region can get accurate gain. In order to improve the detection efficiency of the detector, it is necessary to eliminate the regions which cannot bring accurate gain or small accurate gain. This paper assumes that the denser the objects are in the cluster region, the lower the average confidence score is. The cluster region with denser objects or low average confidence score can obtain greater the accuracy gain from the fine detection in this region. According to this assumption, similar to [42], the initial prediction results of aerial images are used to calculate a score for each cluster region, and the regions whose score is greater than the difficulty threshold are retained.

$$M = \frac{\sum_{i=1}^{N} score_i}{N} \tag{1}$$

$$S = \frac{N^2}{A \times M} \tag{2}$$

Using Eqs. (1) and (2) to calculate the difficulty score for region p, where N is the number of the predicted boxes in p, M is the average of the confidence scores of all the prediction boxes by the coarse detector for region p. It is believed that the smaller the value of M is, the greater the accuracy gain will be. Therefore, we place it in the denominator. Where A is the area of p, S is the final score of p. $\frac{N^2}{A}$ represents the density of region p. It is believed that the denser the region is, the more accuracy gain it can bring. Because in places where objects are dense, they are usually accompanied by occlusion between objects. When the occlusion is serious, the detector will miss detection. Therefore, for dense regions, enlarging it can effectively reduce the missed detection.

3.2 Fine Detection on Region of Interest

After obtaining the difficult cluster region, a special detector FNet is utilized to perform fine detection on these regions. But the difficult cluster region has different shapes and different sizes of objects, it will bring a problem that it is difficult to regress the width and height of objects. Different from the existing approaches, [8,13,20] that directly send these regions to fine detection, inspired by [41], this paper proposes a Gaussian scaling function(GSF) to reduce the size difference of objects. [41] uses the transformation function Scale Match to scale the extra dataset, so that there is little difference in object size between the targeted dataset and the extra dataset. In [41], MS COCO [18] is used as an extra dataset to pre-train the detector and improve its performance on the targeted dataset. Unlike [41], we do not scale the extra dataset, but the targeted dataset. We use Gaussian scale functions(GSF) to scale difficult cluster regions. First, select a mean value that is suitable for the receptive field of the backbone. Then, we select the standard deviation based on the three Sigma rule of thumb. The Gaussian scale functions(GSF) is made up of the mean value and standard deviation. The Gaussian scale function (GSF) can be used to shrink large objects and enlarge small objects. The implementation process is shown in Algorithm 1.

Algorithm 1: Gaussian Scaling Function

Input: Initial difficult cluster regions $\mathcal{R} = \{\mathcal{R}_i\}_{i=1}^{N_\mathcal{R}}$
Output: ROI $\mathcal{R}' = \{\mathcal{R}_i\}_{i=1}^{N_\mathcal{R}}$
Note:R_i is i-th region in $\mathcal{R}$, B_i represents all predicted boxes set in R_i. *Scaleregion* is a function to scale region and predicted boxes with a given scale factor. *Uniform Crop* is a function that uniform crop an region into two regions; *Padding* is a function that uses the scaling factor to fill in the region from the original image.

```
R' ⟵ ∅
for (R_i, B_i) in R do
    sample k ~ gauss(μ, σ);
    s ⟵ Mean(B_i);
    c ⟵ s/k;
    R_i, B_i ⟵ ScaleRegion(R_i, B_i, c);
    if the size of R_i is to large then
        R' ⟵ R' ∪ UniformCrop(R_i);
        else if the size of R_i is to small then
            R' ⟵ R' ∪ Padding(R_i, c);
            else
                R' ⟵ R' ∪ R_i;
```

$$AS\left(G_{ij}\right) = \sqrt{w_{ij} \times h_{ij}} \tag{3}$$

In Algorithm 1, Eq. (3) is used to calculate the absolute size of each object for each difficult cluster region. The ratio of the value randomly sampled from a Gaussian function to the average absolute size of all objects in the region is used as the scale factor for scaling the difficult cluster region. If the size of the difficult cluster region after scaling is less than a certain range, the padding function is used to pad the region proportionally. Otherwise, the uniform crop function is used to divide it into two equal regions.

After scaling the difficult cluster region, we get the ROI. Then, the detection network (FNet) performs fine object detection. The architecture of the FNet can be any state-of-the-art detectors. The backbone of the detector can be any standard backbone networks, e.g., VGG [30], ResNet [12], Hourglass-104 [23].

3.3 Final Detection with Local-Global Fusion

NMS [22] is a post-processing step commonly used in object detection, which is used to remove the duplicate detection box to reduce false detection. When there are multiple prediction boxes on the same object, NMS will eliminate the remaining prediction boxes whose IOU is greater than the threshold value with the prediction box with the maximum confidence score. It can be seen that NMS is too strict and soft-NMS [2] replaces the original score with a slightly lower score instead of zero. The final detection of an aerial image is obtained by fusing the detection results of ROI and global detection results of the whole image with the soft-NMS [2] post-processing.

4 Experiments

4.1 Datasets and Evaluation Metrics

Datasets. The VisDrone [45] dataset was collected by using drones from 14 different cities in China under different weather and different lighting. The objects in this dataset are mostly small, and the objects are often clustered together. It contains a total of 10,209 images, including 6,471 training images, 548 validation images, 1,610 test-dev images, and 1,580 test-challenge images. Except for the test challenges, all other annotations are publicly available. The dataset contains a total of 10 categories, and its image resolution is about 2000*1500 pixels. In order to make a fair comparison with existing works [15,37], we evaluate the detection performance in the validation set.

Evaluation Metric. We use the same evaluation protocol as proposed in MS COCO [18] to evaluate our method. Six evaluation metrics of AP, AP_{50}, AP_{75}, AP_{small}, AP_{medium}, and AP_{large} are reported. The AP is the average precision of all categories on 10 IOU thresholds, ranging from 0.50 to 0.95 with a step size of 0.05. AP_{50} is the average precision of ten categories when the IOU threshold is set to 0.5, and the IOU threshold of AP_{75} is set to 0.75. The AP_{small} means that the AP for objects with area less than 32×32. The AP_{medium} means that the AP for objects with area less than 96×96. The AP_{large} means that the AP for objects with area greater than 96×96. The number of ROI will affect inference time. In the following experiments, we use $\#img$ to record the total number of images, send to the detector, including both original images and cropped ROI.

4.2 Implementation Details

We implemented the proposed CRENet on pytorch 1.4.0. Using an RTX 2080Ti GPU to train and test the model. In common with training many deep CNNs, we use data augmnetation. Specifically we use horizontal flipping. In this article, CNet and FNet use the same detector, which is CenterNet [43] with the backbone network Hourglass-104 [23], and different detectors may be selected. After obtaining the detection result of the aerial image through CNet, use mean shift [39] to get the preliminary cluster region. The region with an area of fewer than 10000 pixels, the aspect ratio of more than 4 or less than 0.25, and the number of objects less than 3 are excluded. The cluster regions with difficulty scores less than 0.01 will be eliminated. The image input resolution for CNet and FNet both are 1024*1024. We train the baseline detector for 140 epoch with Adam, and the initial learning rate was 2.5×10^{-4}.

4.3 Quantitative Result

CenterNet [43] with Hourglass-104 [23] is chosen as the baseline model. Table 1 shows that our approach with baselines, UC, RC, ClusDet [37], and DMNet [15]. We achieve the best performance using fewer images than other methods, and even the small backbone network DLA-34 [40] with deformable convolution layers [3], modified by CenterNet [43], achieves better performance than ClusDet [37] and DMNet [15] both use Faster R-CNN [29] with ResNeXt [36]. We find that the AP value of RC was lower than the baseline, possibly because RC truncates the object when cropping the image. Experiments show that AP_{small} and AP_{medium} have more improvements, indicating that the method, the clustering algorithm adaptive cropping regions we proposed, is of great help to the detection of small and medium objects.

Table 1. The quantitative results on the validation set of VisDrone. $\#img$ is the number of images that send to the detector. UC refers to the uniform cropping of the aerial image into four parts, while RC refers to the random cropping of four 1024*1024 regions from the image each time.

Method	Backbone	$\#img$	AP	AP_{50}	AP_{75}	AP_{small}	AP_{medium}	AP_{large}	s/img
Baseline	Hourglass-104	548	30.6	53.1	29.9	21.6	44.0	57.6	0.194
ClusDet [37]	ResNeXt 101	2716	28.4	53.2	26.4	19.1	40.8	54.4	0.773
DMNet [15]	ResNeXt 101	2736	29.4	49.3	30.6	21.6	41.0	56.7	–
Baseline+RC(4)	Hourglass-104	2740	30.2	52.3	29.5	21.6	42.7	56.6	1.046
Baseline+UC (2×2)	Hourglass-104	2740	32.3	**55.6**	32.8	24.3	44.1	55.2	1.040
CRENet	DLA-34	2413	30.3	53.7	29.2	21.6	41.9	50.6	0.561
CRENet	Hourglass-104	2337	**33.7**	54.3	**33.5**	**25.6**	**45.3**	**58.7**	0.901

4.4 Ablation Study

In this experiment, we show how the three components of the framework, clustering algorithm, difficulty threshold, and Gaussian scaling function(GSF), affect the final performance. We consider five cases: (a) Baseline: we use CenterNet [43] with hourglass-104 [23] as the baseline model; (b) CRENet w/o difficult threshold and GSF: a clustering algorithm is added to the baseline model to search cluster regions, but the difficulty threshold and the Gaussian scaling function(GSF) are not used to process regions; (c) CRENet w/o difficult threshold: clustering algorithm produces regions that are not filtered using difficulty threshold; (d) CRENet w/o GSF: difficult regions are directly sent to the fine detector without Gaussian scaling function(GSF); (e) CRENet: The complete implementation of our method. As can be seen from Table 2, the performance improvement of searching regions using only the clustering algorithm is limited. Therefore, it is necessary to use a Gaussian scaling function(GSF) to scale regions. Using the difficulty threshold to filter the cluster regions, 767 regions were eliminated without lowering the AP. It shows that the difficulty threshold can effectively eliminate the region which can hardly bring the precision gain, thus speeding up the detection speed. The above experiments show that the two components of our proposed, clustering algorithm, and Gaussian scaling function(GSF), are very important for the full improvement of detection performance. And the component of difficulty threshold is crucial to achieve a high inference speed.

Table 2. Ablation study of detection result on validation set of VisDrone.

Method	$\#img$	AP	AP_{50}	AP_{75}	AP_{small}	AP_{medium}	AP_{large}
Baseline	548	30.6	53.1	29.9	21.6	44.0	57.6
CRENet w/o difficult threshold and GSF	2807	32.0	54.8	31.8	23.8	43.8	57.8
CRENet w/o difficult threshold	2836	33.7	**57.6**	33.3	25.5	**45.5**	**60.0**
CRENet w/o GSF	2040	31.9	54.7	31.8	23.7	43.7	57.2
CRENet	2337	**33.7**	54.3	**33.5**	**25.6**	45.3	58.7

5 Conclusions

In this paper, we propose a new method CRENet for object detection in aerial images. CRENet using the clustering algorithm can adaptively obtain cluster regions. Then, the difficulty threshold can be used to eliminate the cluster region that can not bring precision gain, and speed up detection. We also propose that the Gaussian scaling function(GSF) can scale the difficult cluster region to reduce the scale difference between objects. Experiments show that CRENet performs well for small and medium objects in dense scenarios. A large number of experiments have demonstrated that CRENet achieves better performance over the VisDrone [45] dataset.

Acknowledgements. This work was supported by National Natural Science Foundation of China grant 61573266.

References

1. Bochkovskiy, A., Wang, C.Y., Liao, H.Y.M.: YOLOv4: optimal speed and accuracy of object detection (2020)
2. Bodla, N., Singh, B., Chellappa, R., Davis, L.S.: Soft-NMS - improving object detection with one line of code. In: Proceedings of the IEEE International Conference on Computer Vision (ICCV), October 2017
3. Dai, J., et al.: Deformable convolutional networks. In: Proceedings of the IEEE International Conference on Computer Vision (ICCV), October 2017
4. Deng, J., Dong, W., Socher, R., Li, L., Kai Li, Li Fei-Fei: ImageNet: a large-scale hierarchical image database. In: 2009 IEEE Conference on Computer Vision and Pattern Recognition, pp. 248–255 (2009)
5. Ding, J., Xue, N., Long, Y., Xia, G.S., Lu, Q.: Learning RoI transformer for detecting oriented objects in aerial images (2018)
6. Everingham, M., Eslami, S.M.A., Van Gool, L., Williams, C.K.I., Winn, J., Zisserman, A.: The Pascal visual object classes challenge: a retrospective. Int. J. Comput. Vis. **111**(1), 98–136 (2015)
7. Fu, C.Y., Liu, W., Ranga, A., Tyagi, A., Berg, A.C.: DSSD : deconvolutional single shot detector (2017)
8. Gao, M., Yu, R., Li, A., Morariu, V.I., Davis, L.S.: Dynamic zoom-in network for fast object detection in large images. In: Proceedings of the IEEE Conference on Computer Vision and Pattern Recognition (CVPR), June 2018
9. Girshick, R.: Fast R-CNN. In: Proceedings of the IEEE International Conference on Computer Vision (ICCV), December 2015
10. Girshick, R., Donahue, J., Darrell, T., Malik, J.: Rich feature hierarchies for accurate object detection and semantic segmentation. In: Proceedings of the IEEE Conference on Computer Vision and Pattern Recognition (CVPR), June 2014
11. He, K., Gkioxari, G., Dollar, P., Girshick, R.: Mask R-CNN. In: Proceedings of the IEEE International Conference on Computer Vision (ICCV), October 2017
12. He, K., Zhang, X., Ren, S., Sun, J.: Deep residual learning for image recognition. In: Proceedings of the IEEE Conference on Computer Vision and Pattern Recognition (CVPR), June 2016
13. LaLonde, R., Zhang, D., Shah, M.: ClusterNet: detecting small objects in large scenes by exploiting spatio-temporal information. In: Proceedings of the IEEE Conference on Computer Vision and Pattern Recognition (CVPR), June 2018
14. Law, H., Deng, J.: CornerNet: detecting objects as paired keypoints. In: The European Conference on Computer Vision (ECCV), September 2018
15. Li, C., Yang, T., Zhu, S., Chen, C., Guan, S.: Density map guided object detection in aerial images. In: Proceedings of the IEEE/CVF Conference on Computer Vision and Pattern Recognition (CVPR) Workshops, June 2020

16. Li, Y., Huang, Q., Pei, X., Jiao, L., Shang, R.: RADet: refine feature pyramid network and multi-layer attention network for arbitrary-oriented object detection of remote sensing images. Remote Sens. **12**(3) (2020). https://doi.org/10.3390/rs12030389
17. Lin, T.Y., Goyal, P., Girshick, R., He, K., Dollar, P.: Focal loss for dense object detection. In: Proceedings of the IEEE International Conference on Computer Vision (ICCV), October 2017
18. Lin, T.-Y., et al.: Microsoft COCO: common objects in context. In: Fleet, D., Pajdla, T., Schiele, B., Tuytelaars, T. (eds.) ECCV 2014. LNCS, vol. 8693, pp. 740–755. Springer, Cham (2014). https://doi.org/10.1007/978-3-319-10602-1_48
19. Liu, W., et al.: SSD: single shot MultiBox detector. In: Leibe, B., Matas, J., Sebe, N., Welling, M. (eds.) ECCV 2016. LNCS, vol. 9905, pp. 21–37. Springer, Cham (2016). https://doi.org/10.1007/978-3-319-46448-0_2
20. Lu, Y., Javidi, T., Lazebnik, S.: Adaptive object detection using adjacency and zoom prediction. In: Proceedings of the IEEE Conference on Computer Vision and Pattern Recognition (CVPR), June 2016
21. Ma, J., Shao, W., Ye, H., Wang, L., Wang, H., Zheng, Y., Xue, X.: Arbitrary-oriented scene text detection via rotation proposals. IEEE Trans. Multimed. **20**(11), 3111–3122 (2018)
22. Neubeck, A., Van Gool, L.: Efficient non-maximum suppression. In: 18th International Conference on Pattern Recognition (ICPR 2006), vol. 3, pp. 850–855 (2006)
23. Newell, A., Yang, K., Deng, J.: Stacked Hourglass Networks for Human Pose Estimation. In: Leibe, B., Matas, J., Sebe, N., Welling, M. (eds.) ECCV 2016. LNCS, vol. 9912, pp. 483–499. Springer, Cham (2016). https://doi.org/10.1007/978-3-319-46484-8_29
24. Unel, F.O., Ozkalayci, B.O., Cigla, C.: The power of tiling for small object detection. In: Proceedings of the IEEE/CVF Conference on Computer Vision and Pattern Recognition (CVPR) Workshops, June 2019
25. Perreault, H., Bilodeau, G., Saunier, N., Héritier, M.: SpotNet: self-attention multi-task network for object detection. In: 2020 17th Conference on Computer and Robot Vision (CRV), pp. 230–237 (2020)
26. Redmon, J., Divvala, S., Girshick, R., Farhadi, A.: You only look once: unified, real-time object detection. In: Proceedings of the IEEE Conference on Computer Vision and Pattern Recognition (CVPR), June 2016
27. Redmon, J., Farhadi, A.: YOLO9000: better, faster, stronger. In: Proceedings of the IEEE Conference on Computer Vision and Pattern Recognition (CVPR), July 2017
28. Redmon, J., Farhadi, A.: YOLOV3: an incremental improvement (2018)
29. Ren, S., He, K., Girshick, R., Sun, J.: Faster R-CNN: towards real-time object detection with region proposal networks. In: Cortes, C., Lawrence, N.D., Lee, D.D., Sugiyama, M., Garnett, R. (eds.) Advances in Neural Information Processing Systems 28, pp. 91–99. Curran Associates, Inc. (2015). http://papers.nips.cc/paper/5638-faster-r-cnn-towards-real-time-object-detection-with-region-proposal-networks.pdf
30. Simonyan, K., Zisserman, A.: Very deep convolutional networks for large-scale image recognition (2014)
31. Tang, Z., Liu, X., Shen, G., Yang, B.: PENet: object detection using points estimation in aerial images (2020)
32. Uzkent, B., Ermon, S.: Learning when and where to zoom with deep reinforcement learning. In: Proceedings of the IEEE/CVF Conference on Computer Vision and Pattern Recognition (CVPR), June 2020

33. Uzkent, B., Yeh, C., Ermon, S.: Efficient object detection in large images using deep reinforcement learning. In: Proceedings of the IEEE/CVF Winter Conference on Applications of Computer Vision (WACV), March 2020
34. Wang, H., et al.: Spatial attention for multi-scale feature refinement for object detection. In: Proceedings of the IEEE/CVF International Conference on Computer Vision (ICCV) Workshops, October 2019
35. Wu, Z., Suresh, K., Narayanan, P., Xu, H., Kwon, H., Wang, Z.: Delving into robust object detection from unmanned aerial vehicles: a deep nuisance disentanglement approach. In: Proceedings of the IEEE/CVF International Conference on Computer Vision (ICCV), October 2019
36. Xie, S., Girshick, R., Dollar, P., Tu, Z., He, K.: Aggregated residual transformations for deep neural networks. In: Proceedings of the IEEE Conference on Computer Vision and Pattern Recognition (CVPR), July 2017
37. Yang, F., Fan, H., Chu, P., Blasch, E., Ling, H.: Clustered object detection in aerial images. In: Proceedings of the IEEE/CVF International Conference on Computer Vision (ICCV), October 2019
38. Yang, X., Liu, Q., Yan, J., Li, A., Zhang, Z., Yu, G.: R3Det: refined single-stage detector with feature refinement for rotating object (2019)
39. Cheng, Y.: Mean shift, mode seeking, and clustering. IEEE Trans. Pattern Anal. Mach. Intell. **17**(8), 790–799 (1995)
40. Yu, F., Wang, D., Shelhamer, E., Darrell, T.: Deep layer aggregation. In: Proceedings of the IEEE Conference on Computer Vision and Pattern Recognition (CVPR), June 2018
41. Yu, X., Gong, Y., Jiang, N., Ye, Q., Han, Z.: Scale match for tiny person detection. In: Proceedings of the IEEE/CVF Winter Conference on Applications of Computer Vision (WACV), March 2020
42. Zhang, J., Huang, J., Chen, X., Zhang, D.: How to fully exploit the abilities of aerial image detectors. In: Proceedings of the IEEE/CVF International Conference on Computer Vision (ICCV) Workshops, October 2019
43. Zhou, X., Wang, D., Krähenbühl, P.: Objects as points (2019)
44. Zhou, X., Zhuo, J., Krahenbuhl, P.: Bottom-up object detection by grouping extreme and center points. In: Proceedings of the IEEE/CVF Conference on Computer Vision and Pattern Recognition (CVPR), June 2019
45. Zhu, P., Wen, L., Bian, X., Ling, H., Hu, Q.: Vision meets drones: a challenge (2018)

Real-Time Embedded Computer Vision on UAVs:
UAVision2020 Workshop Summary

Kristof Van Beeck[1(✉)], Tanguy Ophoff[1], Maarten Vandersteegen[1], Tinne Tuytelaars[2], Davide Scaramuzza[3], and Toon Goedemé[1]

[1] EAVISE-PSI, KU Leuven, Sint-Katelijne-Waver, Belgium
kristof.vanbeeck@kuleuven.be
[2] PSI, KU Leuven, Leuven, Belgium
[3] Robotics and Perception Group, ETH Zürich, Zürich, Switzerland

Abstract. In this paper we present an overview of the contributed work presented at the UAVision2020 (International workshop on Computer Vision for UAVs) ECCV workshop. Note that during ECCV2020 this workshop was merged with the VisDrone2020 workshop. This paper only summarizes the results of the regular paper track and the ERTI challenge. The workshop focused on real-time image processing on-board of Unmanned Aerial Vehicles (UAVs). For such applications the computational complexity of state-of-the-art computer vision algorithms often conflicts with the need for real-time operation and the extreme resource limitations of the hardware. Apart from a summary of the accepted workshop papers and an overview of the challenge, this work also aims to identify common challenges and concerns which were addressed by multiple authors during the workshop, and their proposed solutions.

Keywords: Computer vision · Real-time · UAVs · Embedded hardware · Deep learning · GPUs · Hardware optimizations

1 Introduction

This paper contains a summary of the material presented at the International Workshop on Computer Vision for UAVs (UAVision2020). This workshop took place as a merged workshop together with VisDrone2020, in conjunction with ECCV2020, Glasgow, Scotland on Friday the 28th of September 2020. Note that this paper only contains a summary of the regular paper track, and the ERTI challenge (both organised by the UAVision2020 organizing committee). Apart from a brief summarization of each paper, we also identified a number of common concerns, challenges and possible proposed solutions that several authors addressed during the workshop.

This workshop focused on state-of-the-art real-time image processing on-board of Unmanned Aerial Vehicles. Indeed, cameras make ideal sensors for drones as they are lightweight, power-efficient and an enormously rich source

A. Bartoli and A. Fusiello (Eds.): ECCV 2020 Workshops, LNCS 12538, pp. 665–674, 2020.
https://doi.org/10.1007/978-3-030-66823-5_40

of information about the environment in numerous applications. Although lots of information can be derived from camera images using the newest computer vision algorithms, the use of them on-board of UAVs poses unique challenges. Their computational complexity often conflicts with the need for real-time operation and the extreme resource limitations of the platform. Of course, developers have the choice to run their image processing on-board or on a remote processing device, although the latter requires a wireless link with high bandwidth, minimal latency and ultra-reliable connection. Indeed, truly autonomous drones should not have to rely on a wireless datalink, thus on-board real-time processing is a necessity. However, because of the limitations of UAVs (lightweight processing devices, limited on-board computational power, limited electrical power on-board), extreme algorithmic optimization and deployment on state-of-the-art embedded hardware (such as embedded GPUs) is the only solution. In this workshop we focused on enabling embedded processing in drones, making efficient use of specific embedded hardware and highly optimizing computer vision algorithms towards real-time applications.

The remainder of this paper is structured as follows. Section 2 gives an overview and short summary of each presented paper at our workshop. In Sect. 3 we discuss the ERTI challenge that was organised apart from the regular paper track. Section 4 discusses the challenges that were identified by multiple authors during the workshop and their proposed solutions. Finally, we conclude this work in Sect. 5.

2 Contributed Papers

In total eight papers were accepted for publication at the UAVision2020 workshop. Each paper submitted a 15 min pre-recorded video, and gave a 5 min spotlight presentation during the live online sessions of the workshop. Below we list and summarize each paper using the paper abstracts.

2.1 ATG-PVD: Ticketing Parking Violations on a Drone [6]

In this paper [6], we introduce a novel suspect-and-investigate framework, which can be easily embedded in a drone for automated parking violation detection (PVD). Our proposed framework consists of: 1) SwiftFlow, an efficient and accurate convolutional neural network (CNN) for unsupervised optical flow estimation; 2) Flow-RCNN, a flow-guided CNN for car detection and classification; and 3) an illegally parked car (IPC) candidate investigation module developed based on visual SLAM. The proposed framework was successfully embedded in a drone from ATG Robotics. The experimental results demonstrate that, firstly, our proposed SwiftFlow outperforms all other state-of-the-art unsupervised optical flow estimation approaches in terms of both speed and accuracy; secondly, IPC candidates can be effectively and efficiently detected by our proposed Flow-RCNN, with a better performance than our baseline network, Faster-RCNN; finally, the actual IPCs can be successfully verified by our investigation module after drone re-localization.

2.2 Next-Best View Policy for 3D Reconstruction [2]

Manually selecting viewpoints or using commonly available flight planners like circular path for large-scale 3D reconstruction using drones often results in incomplete 3D models. Recent works have relied on hand-engineered heuristics such as information gain to select the Next-Best Views. In this work [2], we present a learning-based algorithm called **Scan-RL** to learn a Next-Best View (NBV) Policy. To train and evaluate the agent, we created **Houses3K**, a dataset of 3D house models. Our experiments show that using **Scan-RL**, the agent can scan houses with fewer number of steps and a shorter distance compared to our baseline circular path. Experimental results also demonstrate that a single NBV policy can be used to scan multiple houses including those that were not seen during training. The link to **Scan-RL**'s code is available at https://github.com/darylperalta/ScanRL and **Houses3K** dataset can be found at https://github.com/darylperalta/Houses3K.

2.3 A Flow Base Bi-Path Network for Cross-Scene Video Crowd Understanding in Aerial View [9]

Drones shooting can be applied in dynamic traffic monitoring, object detecting and tracking, and other vision tasks. The variability of the shooting location adds some intractable challenges to these missions, such as varying scale, unstable exposure, and scene migration. In this paper [9], we strive to tackle the above challenges and automatically understand the crowd from the visual data collected from drones. First, to alleviate the background noise generated in cross-scene testing, a double-stream crowd counting model is proposed, which extracts optical flow and frame difference information as an additional branch. Besides, to improve the model's generalization ability at different scales and time, we randomly combine a variety of data transformation methods to simulate some unseen environments. To tackle the crowd density estimation problem under extreme dark environments, we introduce synthetic data generated by game Grand Theft Auto V(GTAV). Experiment results show the effectiveness of the virtual data. Our method wins the challenge with a mean absolute error (MAE) of 12.70. Moreover, a comprehensive ablation study is conducted to explore each component's contribution.

2.4 Multi-view CNN for Crowd Counting from Drones [1]

This paper [1] proposes a novel lightweight and fast convolutional neural network to learn a regression model for crowd counting in images captured from drones. The learning system is initially based on a multi-input model trained on two different views of the same input for the task at hand: (*i*) real-world images; and (*ii*) corresponding synthetically created "crowd heatmaps". The synthetic input is intended to help the network focus on the most important parts of the images. The network is trained from scratch on a subset of the VisDrone dataset. During inference, the synthetic path of the network is disregarded resulting in a traditional single-view model optimized for resource-constrained devices. The derived model achieves promising results on the test images, outperforming models developed by state-of-the-art lightweight architectures that can be used for crowd counting and detection.

2.5 PAS Tracker: Position-, Appearance- and Size-Aware Multi-Object Tracking in Drone Videos [4]

While most multi-object tracking methods based on tracking-by-detection use either spatial or appearance cues for associating detections or apply one cue after another, our proposed PAS tracker [4] employs a novel similarity measure that combines position, appearance and size information jointly to make full use of object representations. We further extend the PAS tracker by introducing a filtering technique to remove false positive detections, particularly in crowded scenarios, and apply a camera motion compensation model to align track positions across frames. To provide high quality detections as input for the proposed tracker, the performance of eight state-of-the-art object detectors is compared on the VisDrone MOT dataset, on which the PAS tracker achieves state-of-the-art performance and ranks 3rd in the VisDrone2020 MOT challenge. In an ablation study, we demonstrate the effectiveness of the introduced tracking components and the impact of the employed detections on the tracking performance is analyzed in detail.

2.6 Insights on Evaluation of Camera Re-Localization Using Relative Pose Regression [3]

We consider the problem of relative pose regression in visual relocalization [3]. Recently, several promising approaches have emerged in this area. We claim that even though they demonstrate on the same datasets using the same split to train and test, a faithful comparison between them was not available since on currently used evaluation metric, some approaches might perform favorably, while in reality performing worse. We reveal a tradeoff between accuracy and the 3D volume of the regressed subspace. We believe that unlike other relocalization approaches, in the case of relative pose regression, the regressed subspace 3D volume is less dependent on the scene and more affect by the method used to score the overlap, which determined how closely sampled viewpoints are. We propose three new metrics to remedy the issue mentioned above. The proposed metrics incorporate statistics about the regression subspace volume. We also propose a new pose regression network that serves as a new baseline for this task. We compare the performance of our trained model on Microsoft 7-Scenes and Cambridge Landmarks datasets both with the standard metrics and the newly proposed metrics and adjust the overlap score to reveal the tradeoff between the subspace and performance. The results show that the proposed metrics are more robust to different overlap threshold than the conventional approaches. Finally, we show that our network generalizes well, specifically, training on a single scene leads to little loss of performance on the other scenes.

2.7 A Deep Learning Filter for Visual Drone Single Object Tracking [8]

Object tracking is one of the most important topics in computer vision. In visual drone tracking, it is an extremely challenging due to various factors, such as camera motion, partial occlusion, and full occlusion. In this paper [8], we propose a deep learning filter method to relieve the above problems, which is to obtain a priori position of the object at the subsequent frame and predict its trajectory to follow up the object during occlusion. Our tracker adopts the geometric transformation of the surrounding of the object to prevent the bounding box of the object lost, and it uses context information to integrate its motion trend thereby tracking the object successfully when it reappears. Experiments on the VisDrone-SOT2018 test dataset and the VisDrone-SOT2020 val dataset illustrate the effectiveness of the proposed approach.

2.8 Object Detection Using Clustering Algorithm Adaptive Searching Regions in Aerial Images [7]

Aerial images are increasingly used for critical tasks, such as traffic monitoring, pedestrian tracking, and infrastructure inspection. However, aerial images have the following main challenges:1) small objects with non-uniform distribution; 2) the large difference in object size. In this paper [7], we propose a new network architecture, Cluster Region Estimation Network (CRENet), to solve these challenges. CRENet uses a clustering algorithm to search cluster regions containing dense objects, which makes the detector focus on these regions to reduce background interference and improve detection efficiency. However, not every cluster region can bring precision gain, so each cluster region difficulty score is calculated to mine the difficult region and eliminate the simple cluster region, which can speed up the detection. Then, a Gaussian scaling function(GSF) is used to scale the difficult cluster region to reduce the difference of object size. Our experiments show that CRENet achieves better performance than previous approaches on the VisDrone dataset. Our best model achieved 4.3% improvement on the VisDrone dataset.

3 ERTI Challenge

Apart from the regular paper track, the UAVision2020 workshop also featured the Embedded Real-Time Inference challenge (ERTI). In this challenge, competitors need to develop a pedestrian detection framework for UAV videos which runs in real-time on an NVIDIA Jetson TX2. As real-time constraint we enforce a minimum processing speed of 8 FPS. We did not distribute training data for the challenge. We recorded and annotated a private dataset on which we performed speed and accuracy measures ourselves (more information below). For qualitative evaluation we distributed a single video of around 2 min of this footage. Figure 1 displays an example frame (with annotations) form our private ERTI dataset.

Fig. 1. An annotated example frame of our ERTI challenge dataset.

As seen, the dataset is quit challenge due to the moving camera viewpoint, small resolution of the pedestrians and high level of occlusions. During the accuracy evaluation, the videos are downsampled to 8 FPS. The accuracy and framerate were measured for each contestant, and the framework with the highest accuracy (and a processing speed of at least 8 FPS) won the competition. As accuracy measure, we used the average precision (AP) with an IOU of 0.5 as main metric.

The private dataset was recorded with a DJI MAVIC Pro, flying with an altitude around 15 m at a university campus environment. The resolution of these images is 1920×1080 pixels. In total, 17 videos were recorded and annotated, consisting of 35.696 frames, 433 unique pedestrian tracks and 390.509 bounding boxes. Of all bounding boxes, 64.608 were occluded more than 50% (about 17% of the total bounding boxes). The annotations were performed using the VATIC toolbox[1].

Unfortunately, for this challenge only two teams competed, which in total submitted 6 different frameworks (each team was able to submit up to three unique implementations). These two teams were the following:

- Real-time Pedestrian Detection Framework on Embedded Platform (submissions are coined *Orange Control*).
- RetinaNet for Fast Pedestrian Detection on Jetson TX2 (submissions are coined *FPD*)

The abstracts of both submissions are not published in the ECCV2020 workshop proceedings, but can be found on the workshop website[2]. Figure 2 displays

[1] https://github.com/cvondrick/vatic.
[2] https://sites.google.com/site/uavisionvisdrone2020/.

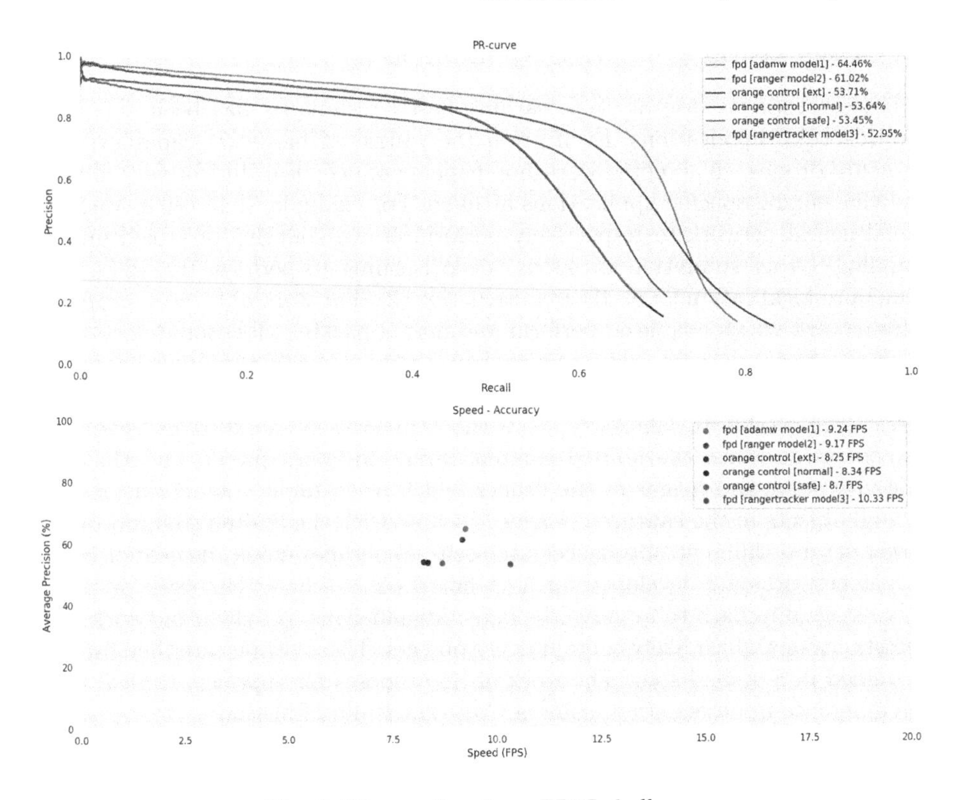

Fig. 2. The results of our ERTI challenge

the results of the ERTI challenge. The top figure displays the precision-recall curve for all implementations, whereas the bottom image displays the average precision versus the processing speed of all algorithms.

All submissions meet the speed requirement of at least 8 FPS. The three PR curves for Orange Control are almost exactly the same, although the highest recall differs slightly. The fastest model is FPD (rangertracker model 3), with an impressive speed of 10.33 FPS (at the cost of the lowest AP - 52.95%). The winner of this year's ERTI challenge, with a frame rate of 9.24 FPS, and an accuracy of 64.46% is the FPD (Adamw Model 1) implementation.

4 Discussion: Trends and Solutions to Common Challenges

Throughout the workshop we identified a number of common trends for UAV vision applications that multiple authors identified and proposed solutions for. Below we give an overview.

4.1 Deep Learning for UAV Applications

It is clear that the use of deep learning based techniques has been extended towards UAV applications. Even given the restricted hardware capabilities of on-board drones, all accepted workshop papers use deep learning for very diverse and challenging specific applications at hand. For example, [1,9] use a convolutional neural network to respectively perform crowd counting or crowd understanding. Other submitted works use deep learning to perform tracking [4,8], camera re-localization [3] or 3D reconstruction [2]. The work of [6] even proposes a framework which is able to perform automated parking violation detection.

However, we do note that many authors propose specifically designed neural network architectures for their drone-inspired applications. Existing architectures seem unable to efficiently solve the challenges at hand. This is mainly due to two reasons: the limited computational on-board power, and the specific viewpoint and height of the drones which often imply low resolutions for ground objects in the captured images. To cope with the limited computational power several different approaches are seen. Sometimes small, dedicated lightweight networks are developed for a specific task [1,6,8]. Other work proposes a two-stage approach to improve the detection efficiency [7]: the framework first detects regions which likely contain dense objects. These regions are then further examined in a second step. The work of [7] proposes an approach to deal with the difference in object sized, using a Gaussian scaling function.

4.2 Using the Time Dimension

We observe that many authors start to take the time dimension into account, e.g. using optical flow or a tracker. Even from an agile moving camera viewpoint, this produces important information, yielding better results than still images. Furthermore, the use of such approaches might enable real-time processing on limited hardware. For example, computationally expensive object detection networks can be evaluated on a limited number of frames only, whereas the missing information in between is predicted using tracking approaches. Examples of submitted works which use such tracking approaches are [6], which proposes Swift-Flow and Flow-RCNN, and [9] which extracts optical flow and frame difference information as an additional information source.

4.3 Collecting Training Data for UAV Applications

The debate continues. As mentioned in the summary paper of a previous edition of our workshop (UAVision2018 - [5]), a difficulty many drone vision researchers struggle with is how to gather enough visual training material to train these neural networks. Although some authors are able to reach satisfying results by training their networks from scratch on real drone data only, many others have to rely on simulated data or transfer learning. Indeed, it is very difficult and expensive to acquire real UAV image data that has enough variance. To cope with these challenges, several authors use synthetic data. For example, [1] uses both real-world images and synthetically created crowd heatmaps during training. Another example is the work of [9] which use synthetic data from the video game Grand Theft Auto to extend their dataset to extreme dark environments.

5 Conclusion

This paper summarized the contributed work which was presented at the UAVision2020 ECCV workshop (together organised with VisDrone), the results of the ERTI challenge, tried to identify common concerns and challenges that were recognized by multiple authors, and their proposed solutions. Three significant trends were discovered. First, all submitted papers use deep learning for embedded UAV applications, despite the limited on board computational power. Secondly, the use of time information seems to be an important aspect to extract additional useful information while simultaneously reducing the computation complexity. Finally, due to the expensive nature of UAV imagery recordings, the use of synthetically generated images remains indispensable to efficiently train specific applications.

Acknowledgements. This work is supported by the agency Flanders Innovation & Entrepreneurship (VLAIO, Tetra project Start To Deep Learn) and FWO (SBO Project Omnidrone).

References

1. Castellano, G., Castiello, C., Cianciotta, M., Mencar, C., Vessio, G.: Multi-view CNN for crowd counting from drones. In: Proceedings of the European Conference on Computer Vision Workshops (ECCVW), UAVision/VisDrone2020 Workshop (2020)
2. Peralta, D., Casimiro, J., Nilles, A.M., Aletta, J., Atienza, R., Cajote, R.: Next-best view policy for 3D reconstruction. In: Proceedings of the European Conference on Computer Vision Workshops (ECCVW), UAVision/VisDrone2020 Workshop (2020)
3. Shalev, A., Achrack, O., Fulkerson, B., Bobrovsky, B.Z.: Insights on evaluation of camera re-localization using relative pose regression. In: Proceedings of the European Conference on Computer Vision Workshops (ECCVW), UAVision/VisDrone2020 Workshop (2020)
4. Stadler, D., Sommer, L.W., Beyerer, J.: PAS tracker: position-, appearance- and size-aware multi-object tracking in drone videos. In: Proceedings of the European Conference on Computer Vision Workshops (ECCVW), UAVision/VisDrone2020 Workshop (2020)
5. Van Beeck, K., Tuytelaars, T., Scarramuza, D., Goedemé, T.: Real-Time embedded computer vision on UAVs. In: Leal-Taixé, L., Roth, S. (eds.) ECCV 2018. LNCS, vol. 11130, pp. 3–10. Springer, Cham (2019). https://doi.org/10.1007/978-3-030-11012-3_1
6. Wang, H., et al.: ATG-PVD: ticketing parking violations on a drone. In: Proceedings of the European Conference on Computer Vision Workshops (ECCVW), UAVision/VisDrone2020 Workshop (2020)

7. Wang, Y., Yang, Y., Zhao, X.: Object detection using clustering algorithm adaptive searching regions in aerial images. In: Proceedings of the European Conference on Computer Vision Workshops (ECCVW), UAVision/VisDrone2020 Workshop (2020)
8. Zhang, X., et al.: A deep learning filter for visual drone tracking. In: Proceedings of the European Conference on Computer Vision Workshops (ECCVW), UAVision/VisDrone2020 Workshop (2020)
9. Zhao, Z., Han, T., Gao, J., Wang, Q., Li, X.: A flow base bi-path network for crowd understanding. In: Proceedings of the European Conference on Computer Vision Workshops (ECCVW), UAVision/VisDrone2020 Workshop (2020)

VisDrone-CC2020: The Vision Meets Drone Crowd Counting Challenge Results

Dawei Du[1], Longyin Wen[2], Pengfei Zhu[3(✉)], Heng Fan[4], Qinghua Hu[3], Haibin Ling[4], Mubarak Shah[5], Junwen Pan[3], Ali Al-Ali[21], Amr Mohamed[12], Bakour Imene[14], Bin Dong[11], Binyu Zhang[13], Bouchali Hadia Nesma[14], Chenfeng Xu[6], Chenzhen Duan[20], Ciro Castiello[16], Corrado Mencar[16], Dingkang Liang[6], Florian Krüger[9], Gennaro Vessio[16], Giovanna Castellano[16], Jieru Wang[8], Junyu Gao[7], Khalid Abualsaud[12], Laihui Ding[15], Lei Zhao[8], Marco Cianciotta[16], Muhammad Saqib[18], Noor Almaadeed[12], Omar Elharrouss[12], Pei Lyu[8], Qi Wang[7], Shidong Liu[13], Shuang Qiu[17], Siyang Pan[13], Somaya Al-Maadeed[12], Sultan Daud Khan[19], Tamer Khattab[12], Tao Han[7], Thomas Golda[9,10], Wei Xu[13], Xiang Bai[6], Xiaoqing Xu[20], Xuelong Li[7], Yanyun Zhao[13], Ye Tian[20], Yingnan Lin[8], Yongchao Xu[6], Yuehan Yao[11], Zhenyu Xu[11], Zhijian Zhao[17], Zhipeng Luo[11], Zhiwei Wei[20], and Zhiyuan Zhao[7]

[1] Kitware, Inc., Clifton Park, NY, USA
[2] JD Finance America Corporation, Mountain View, CA, USA
[3] Tianjin University, Tianjin, China
zhupengfei@tju.edu.cn
[4] Stony Brook University, New York, NY, USA
[5] University of Central Florida, Orlando, FL, USA
[6] Huazhong University of Science and Technology, Wuhan, China
[7] Northwestern Polytechnical University, Xi'an, China
[8] Unisoc, Inc., Shanghai, China
[9] Karlsruhe Institute of Technology, Karlsruhe, Germany
[10] Fraunhofer IOSB, Karlsruhe, Germany
[11] DeepBlue Technology(Shanghai) Co., Ltd., Shanghai, China
[12] Qatar University, Doha, Qatar
[13] Beijing University of Posts and Telecommunications, Beijing, China
[14] University of Science and Technology Houari Boumediene, Bab Ezzouar, Algeria
[15] Ocean university of China, Qingdao, China
[16] University of Bari, Bari, Italy
[17] Supremind, Shanghai, China
[18] University of Technology Sydney, Ultimo, Australia
[19] National University of Technology, Islamabad, Pakistan
[20] Harbin Institute of Technology (Shenzhen), ShenZhen, China
[21] Supreme Committee for Delivery and Legacy, Doha, Qatar

Abstract. Crowd counting on the drone platform is an interesting topic in computer vision, which brings new challenges such as small object inference, background clutter and wide viewpoint. However, there are few algorithms focusing on crowd counting on the drone-captured data due to the lack of comprehensive datasets. To this end, we collect a large-scale

A. Bartoli and A. Fusiello (Eds.): ECCV 2020 Workshops, LNCS 12538, pp. 675–691, 2020.
https://doi.org/10.1007/978-3-030-66823-5_41

dataset and organize the Vision Meets Drone Crowd Counting Challenge (VisDrone-CC2020) in conjunction with the 16th European Conference on Computer Vision (ECCV 2020) to promote the developments in the related fields. The collected dataset is formed by 3, 360 images, including 2, 460 images for training, and 900 images for testing. Specifically, we manually annotate persons with points in each video frame. There are 14 algorithms from 15 institutes submitted to the VisDrone-CC2020 Challenge. We provide a detailed analysis of the evaluation results and conclude the challenge. More information can be found at the website: http://www.aiskyeye.com/.

Keywords: VisDrone · Crowd counting · Challenge · Benchmark

1 Introduction

Crowd counting aims to estimate the number of objects, *e.g.*, pedestrians [43], vehicles [14], commodity [3], animals [1], and cells [20], in images precisely. It has wide applications in video surveillance, crowd analysis, and traffic monitoring, to name a few.

With the developments of deep learning techniques in recent years, many researchers formulate the crowd counting problem as the regression of density maps using deep neural networks. For example, Zhang *et al.* [43] design a multi-column network architecture with three branches to deal with different scales of objects. After that, various algorithms [18,21,22,40] achieve significant advances on several challenging datasets captured on surveillance scenes, such as UCF_CC_50 [16], WorldExpo [41], ShanghaiTech [43], and UCF-QNRF [17].

In contrast to images captured on surveillance scenarios, drone-captured video sequences involve difference challenges, including wide viewpoint variations, small objects and clutter background, which puts forward higher requirements of crowd counting algorithms. However, there are few datasets focus on such scenarios in the community. To advance the developments in crowd counting, we organize the Vision Meets Drone Crowd Counting (VisDrone-CC2020) challenge, which is one track of the "Vision Meets Drone: A Challenge" held on August 28, 2020, in conjunction with the 16th European Conference on Computer Vision (ECCV 2020). In particular, we provide a dataset, which is recorded by various drone-mounted cameras in 70 different scenarios across 4 different cities in China (*i.e.*, Tianjin, Guangzhou, Daqing, and Hong Kong). The objects of interest are pedestrian. We invite researchers to submit the results of counting algorithms and share their latest research in the workshop. There are 14 algorithms from 16 institutes considered in the VisDrone-CC2020 Challenge. The detailed evaluation results can be found on the challenge website: http://www.aiskyeye.com/.

2 Related Work

In this section, we review the related crowd counting datasets and algorithms briefly. More details can be found in the survey [12].

2.1 Crowd Counting Datasets

Recently, numerous datasets have been proposed to deal with the challenges in crowd counting, such as scale variations, background clutter, and illumination variation in the wild. The most frequently used crowd counting datasets include UCF_CC_50 [16], WorldExpo [41], ShanghaiTech [43], and UCF-QNRF [17]. UCF_CC_50 [16] contains only 50 images from different scenes with various densities and different perspective distortions. WorldExpo [41] is collected from videos of Shanghai 2010 WorldExpo, and include $3,920$ frames in total. ShanghaiTech [43] may be the most popular dataset (Part A and Part B) and composed of $1,198$ images with $330,165$ annotations. UCF-QNRF [17] is a large-scale high-resolution dataset that contains $1,535$ images and about 1.25 million people heads.

However, the above-mentioned datasets are of relatively small size, which limits the power of deep learning. To avoid overfitting, recent proposed datasets collect more large-scale data in both the number of images and the number of persons, *e.g.*, GCC [33] and Crowd Surveillance [39]. The GCC dataset [33] is a diverse synthetic dataset collected from 400 scenes in Grand Theft Auto V (GTA5). It includes 15,212 images and 7,625,843 persons, with resolution of 1080×1920. The Crowd Surveillance [39] dataset contains 13,945 high-resolution images and 386,513 annotated people. Instead of surveillance scenes, in this work, we focus on crowd counting on the drone-captured scenes. Specifically, our proposed VisDrone-CC2020 dataset is formed by 3,360 images, including 2,460 images for training, and 900 images for testing, which contains more than $400k$ annotated people heads. In Table 1, we summarize the statistical comparison between our dataset and the previous datasets.

2.2 Crowd Counting Methods

The majority of early crowd counting methods [19,32,36] rely on sliding-window detectors to scan still images or video frames to detect the pedestrians based on the hand-crafted features. However, these methods are easily affected by heavy occlusion, scale and viewpoint variations on crowded scenarios. Benefited from the great success of deep learning, many modern methods [21,43] formulate crowd counting problem as regression of density maps by the networks. Zhang *et al.* [43] develop a multi-column architecture with three branches to deal with different scales of objects. To handle small objects, Li *et al.* [21] employ dilated convolution layers to expand the receptive field while maintaining the resolution as backend network. In [22], a multi-scale deformable network is proposed to generate high-quality density maps, which is more effective to capture the crowd features and more resistant to various noises. To capture interdependence of pixels in density maps, Zhang *et al.* [40] propose a Relational Attention Network with a self-attention mechanism. Jiang *et al.* [18] develop the trellis encoder-decoder network including a multi-scale encoder and a multi-path decoder to generate high-quality density estimation maps.

Table 1. Comparison of existing crowd counting datasets. We summarize the maximal, minimal, average and total count in the datasets.

Dataset	Type	Frames	Max	Min	Ave	Total	Year
UCSD [5]	Surveillance	2,000	46	11	24.9	49,885	2008
UCF_CC_50 [16]	Surveillance	50	4,543	94	1,279.5	63,974	2013
Mall [25]	Surveillance	2,000	53	13	31.2	62,316	2013
WorldExpo [41]	Surveillance	3,980	253	1	50.2	199,923	2015
Shanghaitech A [43]	Surveillance	482	3,139	33	501.4	241,677	2016
Shanghaitech B [43]	Surveillance	716	578	9	123.6	88,488	2016
AHU-Crowd [15]	Surveillance	107	2,201	58	420.6	45,000	2016
CARPK [14]	Aerial	1,448	188	1	62.0	89,777	2017
Smart-City [42]	Surveillance	50	14	1	7.4	369	2018
UCF-QNRF [17]	Surveillance	1,535	12,865	49	815.4	1,251,642	2018
FDST [10]	Surveillance	15,000	57	9	26.7	394,081	2019
GCC [33]	Synthetic	15,212	501	0	3995.0	7,625,843	2019
Crowd Surveillance [39]	Surveillance	13,945	-	-	35.0	386,513	2019
VisDrone-CC2020	Aerial	3,360	421	25	144.7	486,155	2020

To achieve better performance, some recent methods exploit unlabeled or synthetic data for training. Based on the generated synthetic data [33], the supervised learning and domain adaptation strategies are proposed to improve the counting accuracy significantly. In [24], the ranked image sets are generated from unlabeled data for counting applications suffering from a shortage of labeled data. Sam *et al.* [28] propose the Grid Winner-Take-All autoencoder to learn features from unlabeled images such that weight update of neurons in convolutional output maps is restricted to the maximally activated neuron in a fixed spatial cell.

3 The VisDrone-CC2020 Challenge

The VisDrone-CC2020 Challenge aims to count pedestrian heads from video frames taken from drones. Participants are required to submit their algorithm and evaluate on the released VisDrone-CC2020 dataset. They are allowed to use external training data to improve the model. However, it is forbidden to submit different variants of the same algorithm. Meanwhile, the submission with detailed algorithm description obtains the authorship in the ECCV 2020 workshop proceeding.

3.1 Dataset

The VisDrone-CC2020 dataset is formed by 3,360 images with the resolution of 1920×1080. As shown in Fig. 1, the data is captured by various drone-mounted cameras to keep diversity, for 70 different scenarios across 4 different cities in China (*i.e.*, Tianjin, Guangzhou, Daqing, and Hong Kong). Moreover, we divide the dataset into the training and testing subsets, with 2,460 and 900 images, respectively. To avoid overfiting to particular scenes, we collect the images in the training and testing subsets at different but similar locations. To analyze the performance of algorithms thoroughly, we define 3 visual attributes, described as follows.

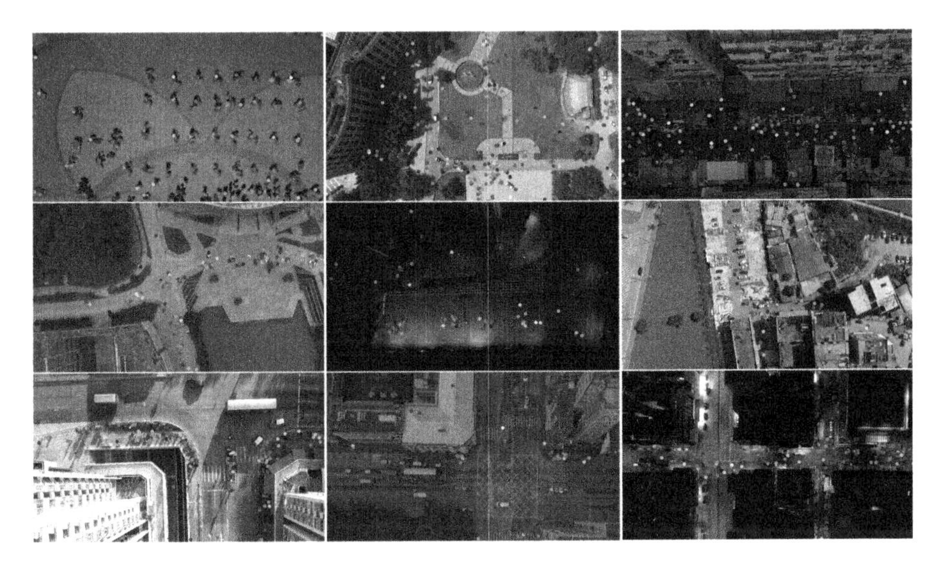

Fig. 1. Annotation exemplars in the VisDrone-CC2020 dataset. Different color indicates different person head. (Color figure online)

- **Scale** indicates the size of objects. We define 2 categories of scales including *Large* (the diameter of objects $>$ 15 pixels) and *Small* (the diameter of objects $\leq$ 15 pixels).
- **Illumination** has significant influence on the appearance of objects. We define 3 kinds of illumination conditions, *i.e.*, *Cloudy*, *Sunny*, and *Night*.
- **Density** is the number of objects in each frame. According to the average number of objects in each frame, we divide the dataset into 2 density levels. *Crowded* density indicates that the number of objects in each frame is larger than 150, and *Sparse* density indicates that the number of objects in each frame is less than 150.

Comparison to Previous Datasets. Compared with the previous datasets focusing on crowd counting on surveillance scenes, our proposed dataset brings new challenges on drone-captured scenes as follows.

- Compared to objects in video sequences recorded on surveillance scenes, the scales of objects in our dataset are extremely small (even less than 30 pixels) because of high shooting altitude by drones. It is difficult for the model to extract sufficient appearance information to describe the objects.
- In our dataset, the crowds are scattered on video frames (see Fig. 1). Each crowd contains a few to dozens of people.
- Since the crowds are dynamically scattered on video frames, each crowd is surrounded by various backgrounds. The clutter background is another challenge in the proposed dataset.

3.2 Evaluation Protocol

Following the previous works [41,43], each algorithm is evaluated through computing the number of people heads, mean absolute error (MAE) and mean squared error (MSE) between the predicted number of people heads and ground-truth in evaluation, which are defined as follows.

$$\begin{aligned} \text{MAE} &= \frac{1}{\sum_{i=1}^{M} N_i} \sum_{i=1}^{M} \sum_{j=1}^{N_i} |z_{i,j} - \hat{z}_{i,j}|, \\ \text{MSE} &= \sqrt{\frac{1}{\sum_{i=1}^{M} N_i} \sum_{i=1}^{M} \sum_{j=1}^{N_i} |z_{i,j} - \hat{z}_{i,j}|^2}, \end{aligned} \tag{1}$$

where M is the number of video clips, N_i is the number of frames in the i-th video. $z_{i,j}$ and $\hat{z}_{i,j}$ are the ground-truth and estimated number of people in the j-th frame of the i-th video clip, respectively. MAE and MSE describe the accuracy and robustness of the estimation, where MAE is the primary metric to rank the counting algorithms.

4 Results and Analysis

In this section, we evaluate the crowd counting methods submitted in the VisDrone-CC2020 Challenge and discuss the results thoroughly in terms of different attributes. Then, we point out several potential research direction in this field.

4.1 Submitted Methods

We have received 37 entries in the VisDrone-CC2020 Challenge, 14 of which submitted the results with correct format and complete algorithm description. In the following we brief overview the submitted algorithms included in the crowd counting task of VisDrone2020 Challenge and provide the corresponding descriptions in the Appendix A.

The majority of the submitted algorithms are improved from state-of-the-art methods such as AutoScale [37], CSRNet [21] and SANet [4]. FPNCC (A.1) is based on AutoScale [37]. BVCC (A.2) is a double-stream network that extracts optical flow and frame difference information. 6 algorithms are variants of CSR-Net [21], including PDCNN (A.4), CSRNet+ (A.6), SCNet (A.8), CSR-SSOF (A.9) and Soft-CSRNET (A.10). To extract multi-scale features of the target object and incorporate larger context, M-SFANet (A.7) improves SFANet [44] by adding two modules called ASSP and CAN. MILLENNIUM (A.12) uses multi-view data (*i.e.*, real-world RGB image and the corresponding crowd heatmap) to construct two deep neural networks for crowd counting. DevaNetv2 (A.5) employs attentional mechanism and feature pyramids to deal with different scales of people heads. SANet (A.13) is a new encoder-decoder based Scale Aggregation Network [4] to extract multi-scale features with scale aggregation modules and generate high-resolution density maps by using a set of transposed convolutions. Besides, two submissions are state-of-the-art methods trained on the VisDrone-CC2020 dataset, *i.e.*, CFF (A.3) and CANet (A.11). CFF (A.3) proposes supervised focus from segmentation to focus on areas of interest and from global density to learn a matching global density. CANet (A.11) combines features obtained using multiple receptive field sizes and learns the importance of each such feature at each image location [23]. ResNet-FPN101 (A.14) is a baseline method by using ResNet-101 backbone to regress the density maps.

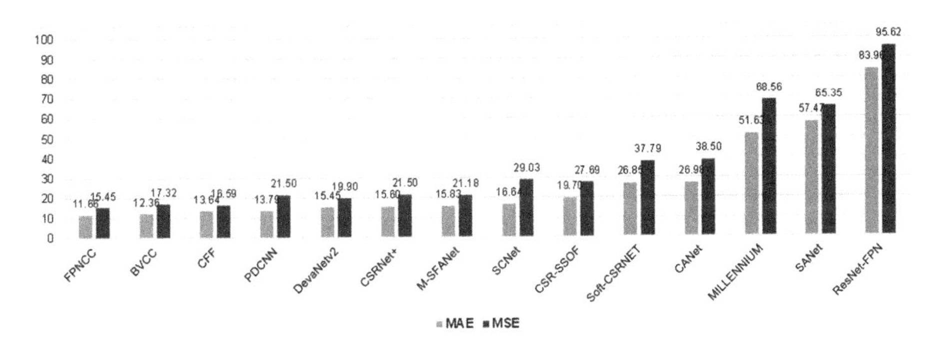

Fig. 2. Comparison of submissions in the VisDrone-CC2020 Challenge.

4.2 Overall Results

The overall results of all submissions are shown in Fig. 2. FPNCC (A.1) obtains the best overall MAE score of 11.66 and MSE score of 15.45. This is attributed to the proposed Learning to Scale (L2S) module to rescale the dense regions into similar density levels, which mitigates imbalance of density values in the dataset. BVCC (A.2) ranks the second place by using a double-stream network, which introduces the external synthetic data generated by GTA5 [33]. After that, CFF

Table 2. Results on the VisDrone-CC2020 dataset.

Method	Large		Small		Cloudy		Sunny		Night		Crowded		Sparse	
	MAE	MSE	MAE	MSE	MAE	MSE	MAE	MSE	MAE	MSE	MAE	MSE	MAE	MSE
FPNCC(A.1)	13.74	18.37	**10.27**	**13.15**	**10.83**	15.70	**11.04**	**13.16**	15.39	18.63	15.40	19.77	**9.49**	**12.28**
BVCC(A.2)	13.48	18.03	11.61	13.76	11.78	**15.55**	12.32	15.29	14.19	16.37	**13.66**	**18.23**	11.61	13.87
CFF(A.3)	13.73	18.38	13.58	16.58	13.28	17.73	13.35	15.99	15.32	18.58	18.69	21.97	10.72	13.93
PDCNN(A.4)	13.01	**15.66**	14.32	17.18	14.84	17.62	13.43	16.56	**11.36**	**13.10**	15.79	18.80	12.63	15.16
DevaNetv2(A.5)	**12.84**	15.84	17.18	22.19	13.59	17.97	15.54	18.66	20.82	26.65	14.29	17.35	16.12	21.23
CSRNet+(A.6)	14.52	20.38	16.31	22.21	12.06	17.55	13.59	16.91	30.23	35.75	17.82	22.70	14.31	20.78
M-SFANet(A.7)	20.59	27.08	12.65	16.08	15.56	19.02	11.22	14.81	25.88	34.16	17.24	21.40	15.01	21.04
SCNet(A.8)	19.64	37.02	14.64	22.15	13.77	22.49	11.53	14.78	35.47	55.69	18.40	26.61	15.62	30.34
CSR-SSOF(A.9)	21.73	30.23	18.34	25.86	12.43	18.53	15.94	19.66	49.02	52.90	17.47	24.23	20.98	29.51
Soft-CSRNET(A.10)	29.31	44.10	25.21	32.92	15.37	20.12	19.73	23.34	75.53	79.15	21.86	26.02	29.74	43.16
CANet(A.11)	30.92	38.33	24.36	38.61	11.97	16.95	23.40	27.22	79.18	80.92	24.06	29.25	28.67	42.95
MILLENNIUM(A.12)	47.39	53.70	54.46	76.89	65.24	85.26	44.71	51.83	24.65	32.00	90.33	103.25	29.22	35.36
SANet(A.13)	56.81	66.56	57.91	64.52	44.77	52.55	69.91	77.17	70.67	73.68	70.14	78.74	50.13	56.16
ResNet-FPN(A.14)	98.64	109.53	74.18	85.09	80.94	94.27	84.51	96.65	91.95	97.54	131.11	133.18	56.67	64.56

(A.3) benefits from two kinds of point annotations (*i.e.*, segmentation, global density) as supervision for density-based counting, achieving MAE score of 13.65 and MSE score of 17.32. PDCNN (A.4) focuses on performing accurate counting under different illumination conditions, which obtains similar performance as CFF (A.3). After that, the following two methods focus on enhancing the training data. DevaNetv2 (A.5) (rank 5) splits each image into 16 sub-images, where each sub-image is processed by several data argumentation steps such as random rotation, random flip, random color process (include brightness, saturation, contrast), and normalization. Finally, 4 sub-images are randomly chosen to merge into one new image. In contrast, CSRNet+ (A.6) (rank 6) upsamples the training sequences to equalize the illumination distribution, which is pretrained on the DLR-Aerial Crowd Dataset [2].

4.3 Attribute-Based Results

For thorough evaluation, we report the results in terms of different attributes in Table 2. It can be seen that the best five performers achieve the best performance on various subsets. Specifically, improved from state-of-the-art L2S method [37], FPNCC (A.1) obtains the best MAE score in 4 attribute subsets, *i.e.*, *Small*, *Cloudy*, *Sunny*, and *Sparse*. With the help of the synthetic dataset [33] to simulate diverse environments, BVCC (A.2) achieves the best MSE score in 2 attribute subsets including *Cloudy* and *Crowded*. PDCNN (A.4) achieves the best MSE score in terms of *Large* and *Night* attributes. This is because PDCNN (A.4) uses different networks to handle different scenarios in day and night illumination. DevaNetv2 (A.5) obtains the best MAE score of 12.84 in the *Large* attribute, which show the effectiveness of the data argumentation strategy.

4.4 Discussion

As presented in Table 2, it can be seen that the best method FPNCC (A.1) achieves 11.66 MAE score. That is, there are 8% errors in average to count the people heads. It is still not satisfying in real applications. We summarize some topics worth to explore in crowd counting on drone-captured scenes as follows.

- **Groundtruth Density Map.** The majority of existing methods convert point based groundtruth to density map using a Gaussian kernel model training. Although the geometry-adaptive kernel-based density map generation [43] is widely used on surveillance scenes, it may fail on the drone-captured scenarios where the people crowd is relative sparse.
- **Unsupervised Learning.** Since point-level annotation is expensive to collect, some methods leverage unlabeled data or synthetic external data to improve the crowd counting performance. To narrow the gap between different datasets, we believe the domain adaptation technique will attract much interest in the field.
- **Head Localization.** Besides crowd counting, head localization is also an important task in safety control application. However, the submitted algorithms focus on estimating the number of people heads in a frame rather than accurate location. Previous works [17,26,35] usually output the localization map and post-process the map by finding the local maximums based on a threshold. The two sub-tasks should be complementary and support each other.

5 Conclusions

In this paper, we summarize the results of all submitted crowd counting algorithms in the VisDrone-CC2020 Challenge. To evaluate the performance of algorithms, we provide a dataset formed by 3, 360 images, *i.e.*, 2,460 images for training, and 900 images for testing. We provide annotated coordinates for people. Specifically, 14 crowd counting algorithms from 15 institutes are submitted to the VisDrone-CC2020 Challenge. The top three performers are FPNCC (A.1), BVCC (A.2), and CFF (A.3), achieving 11.66, 12.36 and 13.65 MAE scores, respectively. However, there still remains much room for improvement such as the localization accuracy of head. For future work, we plan to extend the dataset with more attributes and annotations to advance the state-of-the-art. We hope our work can largely boost the development of crowd counting on the drone-captured scenes.

Acknowledgements. This work was supported in part by the National Natural Science Foundation of China under Grant 61876127 and Grant 61732011, in part by Natural Science Foundation of Tianjin under Grant 17JCZDJC30800.

A Submitted Crowd Counting Algorithms

In this appendix, we provide a short summary of all crowd counting algorithms that were considered in the VisDrone-CC2020 Challenge.

A.1 Feature Pyramid Network for Crowd Counting (FPNCC)

Dingkang Liang, Chenfeng Xu, Yongchao Xu, Xiang Bai
dkliang@hust.edu.cn

FPNCC is based on AutoScale [37] (an extension of L2S [38]) with the VGG16-based FPN backbone, which automatically scales dense regions into similar and appropriate density levels (see Fig. 3). Meanwhile, we separate the overlapped blobs and decompose the original accumulated density values in density map. To preserve sufficient spatial information for accurate counting, we discard the last pooling layer and all following fully connected layers, as well as the pooling layer between stage4 and stage5. Note that we exploit the pre-trained model based on ImageNet Database. Besides using the traditional MSE Loss, we also use SSIM loss to improve the high-quality density map to aid the final counting performance.

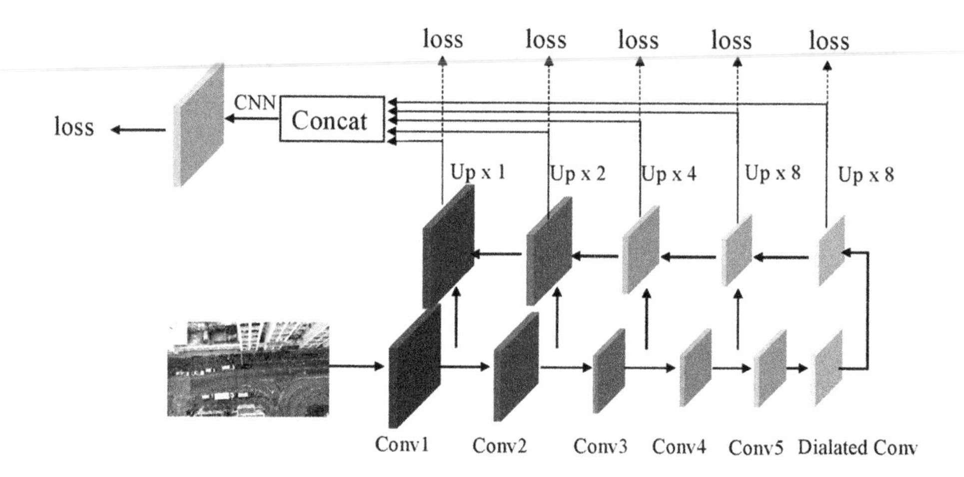

Fig. 3. The framework of FPNCC.

A.2 Bi-path Video Crowd Counting (BVCC)

Zhiyuan Zhao, Tao Han, Junyu Gao, Qi Wang, Xuelong Li
tuzixini@163.com, hantao10200@mail.nwpu.edu.cn, {gjy3035, crabwq}@gmail.com, li@nwpu.edu.cn

To deal with the challenges such as varying scale, unstable exposure, and scene migration, BVCC is proposed to automatically understand the crowd from the visual data collected from drones. First, to alleviate the background noise generated in cross-scene testing, a double-stream crowd counting model is proposed, which extracts optical flow and frame difference information as an additional branch. Besides, to improve the generalization ability of the model at different scales and time, we randomly combine a variety of data transformation methods to simulate some unseen environments. To tackle the crowd density estimation problem under extreme dark environments, we introduce synthetic data generated by GTAV [33].

A.3 Counting with Focus for Free (CFF)

Wei Xu
wxu.bupt@gmail.com

CFF [29] proposes two kinds of free supervision including segmentation maps and global density. Besides, an improved kernel size estimator is proposed to facilitate density estimation and the focus from segmentation. During training, we augment the images by randomly cropping 128 × 128 patches. Code is available at https://github.com/shizenglin/Counting-with-Focus-for-Free.

A.4 Parallel Dilated Convolution Neural Network (PDCNN)

Pei Lyu, Lei Zhao, Jieru Wang, Yingnan Lin
{pei.lv, michael.zhao, jieru.wang, lynn.lin}@unisoc.com

PDCNN is the illumination-aware counting model based on CSRNet [21]. Since there are three categories of illumination conditions in the dataset (cloudy, sunny and night), we first distinguish the day and night illumination. Then, we use different networks to handle different scenarios. We use geometry-adaptive kernels to tackle the different congested scenes. We use Gaussian Kernel to blur each head annotation. We crop patches from each image at different location for data augmentation.

A.5 Crowd Counting with Attentional Mechanism and Feature Pyramids (DevaNetv2)

Ye Tian, Chenzhen Duan, Xiaoqing Xu, Zhiwei Wei
{19s151092, 18s151541, 19s051052, 19s051024}@stu.hit.edu.cn

In the field of aerial image crowd counting, people usually use grid density maps as the label with "unprejudiced" neural network block, but there are a series of problems. On one hand, the scale and shape of people will change too much, according to the different shooting angles and flight heights of the UAV. On the other hand, the luminance of the image will change, because of the various shooting time. In order to solve the above problems, we use Gaussian density maps and useful data enhancement methods to improve the robustness of our model. We use attentional mechanism and feature pyramids to make model adjust the predicting results based on the global information. We also design an auxiliary loss to reduce the difficulty of model optimization. The network is implemented by the C^3 framework[1].

A.6 CSRNet on Drone-Based Scenarios (CSRNet+)

Florian Krüger, Thomas Golda
{florian.krueger, thomas.golda}@iosb.fraunhofer.de

CSRNet+ is modified from CSRNet [21] to perform crowd counting on drone-based scenarios. We pretrain our model using only the training set from DLR-Aerial Crowd Dataset (DLRACD) [2]. Then we finetune our model with the VisDrone2020-CC training dataset. For training on VisDrone2020-CC training dataset, we annotate the ground sampling distances (GSD) for the training set by hand and then generated density maps from those as in [2]. Furthermore, we upsample the training sequences to equalize the illumination distribution. For night sequences, we only double them since there are only 3 in the training set.

A.7 Mutil-scale Aware based SFANet (M-SFANet)

Zhipeng Luo, Bin Dong, Yuehan Yao, Zhenyu Xu
{luozp, dongb, yaoyh, xuzy}@blueai.com

M-SFANet modifies neural network architectures based on SFANet [44] for accurate and efficient crowd counting. Inspired by SFANet, the M-SFANet model is attached with two novel multi-scale-aware modules, called ASSP and CAN. The encoder of M-SFANet is enhanced with ASSP containing parallel atrous convolutional layers with different sampling rates and hence able to extract multi-scale features of the target object and incorporate larger context. To further deal with scale variation throughout an input image, we leverage contextual module called CAN which adaptively encodes the scales of the contextual information. The combination yields an effective model for counting in both dense and sparse crowd scenes. Based on SFANet decoder structure, M-SFANet's decoder has dual paths, for density map generation and attention map generation. During training stage, we use MSE Loss for density map and CrossEntropy Loss for attention map; The whole dataset is split into 8 folds, which means that 8 independent models were trained. During inference stage, we use separated min clip for nightly scene videos.

A.8 Scaled Cascade Network for Crowd Counting on Drone data (SCNet)

Omar Elharrouss, Noor Almaadeed, Khalid Abualsaud, Amr Mohamed, Tamer Khattab, Ali Al-Ali, Somaya Al-Maadeed
elharrouss.omar@gmail.com

[1] https://github.com/gjy3035/C-3-Framework.

We propose a crowd counting method based on deep convolutional neural networks (CNN) by extracting high-level features to generate density maps that represent an estimation of the crowd count with respect to the scale variations in the scene. Specifically, SCNet is a CNN-based model by adding a cascade network after the frontend block inspired by those used in SPN [6], CSRNet [21], and AGRD [27]. Cascade block is inspired by the Bi-directional cascade network in [13] used for edge detection. To handle the scale variations a Scale Enhancement Module (SEM) is introduced. The architecture employs sequential dilated convolution blocks with different kernels.

A.9 CSRNet on Segmented Scenes with Optical Flow (CSR-SSOF)

Siyang Pan, Shidong Liu, Binyu Zhang, Yanyun Zhao
{pansiyang, lsd215, jl-lagrange, zyy}@bupt.edu.cn

CSR-SSOF is derived from CSRNet [21] with several additional modules offering improvements. First, optical flow is computed based on the Gunnar Farneback algorithm [11] to extract temporal information. RGB images concatenated with corresponding optical flow are fed into the modified CSRNet which takes 5-channel data as input. Furthermore, we replace the last layer with the density map estimator (DME) in SANet [4] in order to generate high-resolution density maps. The patch-based test scheme in SANet is also applied in our method. Considering the interference of irrelevant regions in which crowds rarely appear, an improved semantic segmentation model HRNetV2 [30] is implemented to block out building areas. We extract the contours of segmentation maps following [31] and fill the small holes. The corrosion expansion morphology method is used to smooth and narrow down the boundaries of building areas. Besides, V channel of the HSV model is chosen for the division of day and night sequences and we take a series of measures to cope with the night ones. Night images are enhanced by RetinexNet [34] at the beginning to make the buried details visible for counting. In addition, we assume buildings in night sequences exist in pixels with below-average V values, so we directly binarize the original night images accordingly to perform segmentation. All final counting results are obtained by the integral of density maps within non-building areas. With partial parameters pre-trained on ImageNet, the modified CSRNet is finetuned on ShanghaiTech Part B [43] dataset and VisDrone2020 train set successively. For segmentation, HRNetV2 is firstly trained on UDD5 [7] dataset making use of Cityscapes [8] pre-trained weight. Then we manually annotate some images on VisDrone2020 train set distinguishing building areas and use the annotation to finetune our model. For the night image enhancement, we use RetinexNet model pre-trained on LOL [34] dataset and RAISE [9] dataset.

A.10 Soft Dilated Convolutional Neural Networks for Understanding the Highly Congested Scenes (Soft-CSRNET)

Bakour Imene, Bouchali Hadia Nesma
{imene.bakour, hadianesma.bouchali}@etu.usthb.dz

Soft-CSRNET is modified on the network for Congested Scene Recognition called CSRNet [21] to provide a data-driven and deep learning method that can understand highly congested scenes and perform accurate count estimation as well as present high-quality density maps. We try several possible configurations on the network to be able to optimize it and render it in real time, for that we focus on the accuracy of the counting, removing the minimum possible resolution of the density maps estimated The proposed network is composed of two major components: a convolutional neural network (CNN) as the front-end for 2D feature extraction and a dilated CNN for the back-end, which uses dilated kernels to deliver larger reception fields and to replace pooling operations. Our Soft-CSRNET contains fewer parameters and convolutional layers in the backend.

A.11 Context-Aware Crowd Counting (CANet)

Laihui Ding
xmyyzy123@qq.com

CANet [23] is an end-to-end network that leverages multiple receptive field scales to learn and combine feature at each image location. Thus the contextual information is encoded for predicting accurate crowd density.

A.12 MultI-view fuLly convoLutional nEural Network for crowd couNting In drone-captUred iMages (MILLENNIUM)

Giovanna Castellano, Ciro Castiello, Marco Cianciotta, Corrado Mencar, Gennaro Vessio
{giovanna.castellano, ciro.castiello, marco.cianciotta, corrado.mencar, gennaro.vessio}@uniba.it

We couple multi-view data and multi-functional deep learning for efficient crowd counting in aerial images. Specifically, we exploit the real-world RGB image and the corresponding crowd heatmap as multiple views of the same scene containing a crowd to create a powerful regression model for crowd counting. Two deep neural networks, one for each view, are jointly trained so that their weights are updated at the same time. After training, only the network that processes the real-world images is retained as final model for crowd counting. This provides an accurate light-weight model that is suitable to meet the limited computational resources of a UAV. The final model is able to provide the crowd count with an average processing speed of about 125 frames per second.

A.13 Scale Aggregation Network (SANet)

Shuang Qiu, Zhijian Zhao
{qiushuang, zhaozhijian}@supremind.com

SANet is the encoder-decoder based Scale Aggregation Network [4]. Specifically, the encoder is used to exploit multi-scale features by scale aggregation while the decoder is used to generate high-resolution density maps by transposed convolutions. To consider local correlation in density maps, both Euclidean loss and local pattern consistency loss are used to train the network for better performance. We finetune the network on additional datasets such as Shanghaitech B [43], and Crowd surveillance [39].

A.14 Residual Network Feature Pyramid Network_101 (ResNet-FPN101)

Muhammad Saqib, Sultan Daud Khan
muhammad.saqib@uts.edu.au, sultandaud@gmail.com

ResNet-FPN101 is the baseline crowd counting method. Specifically, we convert the dot-level annotation to bounding box-level annotation. We train and evaluate the network with ResNet-FPN-101 as a backbone architecture.

References

1. Arteta, C., Lempitsky, V., Zisserman, A.: Counting in the wild. In: Leibe, B., Matas, J., Sebe, N., Welling, M. (eds.) ECCV 2016. LNCS, vol. 9911, pp. 483–498. Springer, Cham (2016). https://doi.org/10.1007/978-3-319-46478-7_30
2. Bahmanyar, R., Vig, E., Reinartz, P.: MRCNet: crowd counting and density map estimation in aerial and ground imagery. CoRR abs/1909.12743 (2019)
3. Cai, Y., Wen, L., Zhang, L., Du, D., Wang, W., Zhu, P.: Rethinking object detection in retail stores. CoRR abs/2003.08230 (2020)
4. Cao, X., Wang, Z., Zhao, Y., Su, F.: Scale aggregation network for accurate and efficient crowd counting. In: Ferrari, V., Hebert, M., Sminchisescu, C., Weiss, Y. (eds.) ECCV 2018. LNCS, vol. 11209, pp. 757–773. Springer, Cham (2018). https://doi.org/10.1007/978-3-030-01228-1_45
5. Chan, A.B., Liang, Z.J., Vasconcelos, N.: Privacy preserving crowd monitoring: counting people without people models or tracking. In: CVPR (2008)
6. Chen, X., Bin, Y., Sang, N., Gao, C.: Scale pyramid network for crowd counting. In: WACV, pp. 1941–1950 (2019)
7. Chen, Yu., Wang, Y., Lu, P., Chen, Y., Wang, G.: Large-scale structure from motion with semantic constraints of aerial images. In: Lai, J.-H., et al. (eds.) PRCV 2018. LNCS, vol. 11256, pp. 347–359. Springer, Cham (2018). https://doi.org/10.1007/978-3-030-03398-9_30
8. Cordts, M., et al.: The cityscapes dataset for semantic urban scene understanding. In: CVPR, pp. 3213–3223 (2016)

9. Dang-Nguyen, D., Pasquini, C., Conotter, V., Boato, G.: RAISE: a raw images dataset for digital image forensics. In: Ooi, W.T., Feng, W., Liu, F. (eds.) MMSys, pp. 219–224 (2015)
10. Fang, Y., Zhan, B., Cai, W., Gao, S., Hu, B.: Locality-constrained spatial transformer network for video crowd counting. In: ICME, pp. 814–819 (2019)
11. Farnebäck, G.: Two-frame motion estimation based on polynomial expansion. In: Bigun, J., Gustavsson, T. (eds.) SCIA 2003. LNCS, vol. 2749, pp. 363–370. Springer, Heidelberg (2003). https://doi.org/10.1007/3-540-45103-X_50
12. Gao, G., Gao, J., Liu, Q., Wang, Q., Wang, Y.: CNN-based density estimation and crowd counting: a survey. CoRR abs/2003.12783 (2020)
13. He, J., Zhang, S., Yang, M., Shan, Y., Huang, T.: Bi-directional cascade network for perceptual edge detection. In: CVPR, pp. 3828–3837 (2019)
14. Hsieh, M., Lin, Y., Hsu, W.H.: Drone-based object counting by spatially regularized regional proposal network. In: ICCV (2017)
15. Hu, Y., Chang, H., Nian, F., Wang, Y., Li, T.: Dense crowd counting from still images with convolutional neural networks. J. Vis. Commun. Image Represent. **38**, 530–539 (2016)
16. Idrees, H., Saleemi, I., Seibert, C., Shah, M.: Multi-source multi-scale counting in extremely dense crowd images. In: CVPR, pp. 2547–2554 (2013)
17. Idrees, H., et al.: Composition loss for counting, density map estimation and localization in dense crowds. In: Ferrari, V., Hebert, M., Sminchisescu, C., Weiss, Y. (eds.) ECCV 2018. LNCS, vol. 11206, pp. 544–559. Springer, Cham (2018). https://doi.org/10.1007/978-3-030-01216-8_33
18. Jiang, X., et al.: Crowd counting and density estimation by trellis encoder-decoder networks. In: CVPR, pp. 6133–6142 (2019)
19. Leibe, B., Seemann, E., Schiele, B.: Pedestrian detection in crowded scenes. In: CVPR, pp. 878–885 (2005)
20. Lempitsky, V.S., Zisserman, A.: Learning to count objects in images. In: Lafferty, J.D., Williams, C.K.I., Shawe-Taylor, J., Zemel, R.S., Culotta, A. (eds.) NeurIPS, pp. 1324–1332 (2010)
21. Li, Y., Zhang, X., Chen, D.: CSRNet: dilated convolutional neural networks for understanding the highly congested scenes. In: CVPR, pp. 1091–1100 (2018)
22. Liu, N., Long, Y., Zou, C., Niu, Q., Pan, L., Wu, H.: ADCrowdNet: an attention-injective deformable convolutional network for crowd understanding. In: CVPR, pp. 3225–3234 (2019)
23. Liu, W., Salzmann, M., Fua, P.: Context-aware crowd counting. In: CVPR, pp. 5099–5108 (2019)
24. Liu, X., van de Weijer, J., Bagdanov, A.D.: Exploiting unlabeled data in CNNs by self-supervised learning to rank. TPAMI **41**(8), 1862–1878 (2019)
25. Loy, C.C., Gong, S., Xiang, T.: From semi-supervised to transfer counting of crowds. In: ICCV, pp. 2256–2263 (2013)
26. Ma, Z., Yu, L., Chan, A.B.: Small instance detection by integer programming on object density maps. In: CVPR, pp. 3689–3697 (2015)
27. Pan, X., Mo, H., Zhou, Z., Wu, W.: Attention guided region division for crowd counting. In: ICASSP, pp. 2568–2572 (2020)
28. Sam, D.B., Sajjan, N.N., Maurya, H., Babu, R.V.: Almost unsupervised learning for dense crowd counting. In: AAAI, pp. 8868–8875 (2019)
29. Shi, Z., Mettes, P., Snoek, C.: Counting with focus for free. In: ICCV, pp. 4199–4208 (2019)
30. Sun, K., et al.: High-resolution representations for labeling pixels and regions. CoRR abs/1904.04514 (2019)

31. Suzuki, S., Abe, K.: Topological structural analysis of digitized binary images by border following. CVGIP **30**(1), 32–46 (1985)
32. Wang, M., Wang, X.: Automatic adaptation of a generic pedestrian detector to a specific traffic scene. In: CVPR, pp. 3401–3408 (2011)
33. Wang, Q., Gao, J., Lin, W., Yuan, Y.: Learning from synthetic data for crowd counting in the wild. In: CVPR, pp. 8198–8207 (2019)
34. Wei, C., Wang, W., Yang, W., Liu, J.: Deep retinex decomposition for low-light enhancement. In: BMVC, p. 155 (2018)
35. Wen, L., et al.: Drone-based joint density map estimation, localization and tracking with space-time multi-scale attention network. CoRR abs/1912.01811 (2019)
36. Wu, B., Nevatia, R.: Detection of multiple, partially occluded humans in a single image by Bayesian combination of edgelet part detectors. In: ICCV, pp. 90–97 (2005)
37. Xu, C., et al.: Autoscale: learning to scale for crowd counting. CoRR abs/1912.09632 (2019)
38. Xu, C., Qiu, K., Fu, J., Bai, S., Xu, Y., Bai, X.: Learn to scale: generating multipolar normalized density maps for crowd counting. In: ICCV, pp. 8381–8389 (2019)
39. Yan, Z., et al.: Perspective-guided convolution networks for crowd counting. In: ICCV, pp. 952–961 (2019)
40. Zhang, A., et al.: Relational attention network for crowd counting. In: ICCV, pp. 6787–6796 (2019)
41. Zhang, C., Li, H., Wang, X., Yang, X.: Cross-scene crowd counting via deep convolutional neural networks. In: CVPR, pp. 833–841 (2015)
42. Zhang, L., Shi, M., Chen, Q.: Crowd counting via scale-adaptive convolutional neural network. In: WACV, pp. 1113–1121 (2018)
43. Zhang, Y., Zhou, D., Chen, S., Gao, S., Ma, Y.: Single-image crowd counting via multi-column convolutional neural network. In: CVPR, pp. 589–597 (2016)
44. Zhu, L., Zhao, Z., Lu, C., Lin, Y., Peng, Y., Yao, T.: Dual path multi-scale fusion networks with attention for crowd counting. CoRR abs/1902.01115 (2019)

VisDrone-DET2020: The Vision Meets Drone Object Detection in Image Challenge Results

Dawei Du[1], Longyin Wen[2], Pengfei Zhu[3](✉), Heng Fan[4], Qinghua Hu[3], Haibin Ling[4], Mubarak Shah[5], Junwen Pan[3], Apostolos Axenopoulos[34], Arne Schumann[31], Athanasios Psaltis[34], Ayush Jain[28], Bin Dong[36], Changlin Li[14], Chen Chen[14], Chengzhen Duan[7], Chongyang Zhang[33], Daniel Stadler[30], Dheeraj Reddy Pailla[22], Dong Yin[16], Faizan Khan[22], Fanman Meng[6], Guangyu Gao[15], Guosheng Zhang[17], Hansheng Chen[16], Hao Zhou[33], Haonian Xie[25], Heqian Qiu[6], Hongliang Li[6], Ioannis Athanasiadis[34], Jincai Cui[35], Jingkai Zhou[11], Jonghwan Ko[9], Joochan Lee[9], Jun Yu[25], Jungyeop Yoo[9], Lars Wilko Sommer[31], Lu Xiong[16], Michael Schleiss[21], Ming-Hsuan Yang[12], Mingyu Liu[6], Minjian Zhang[6], Murari Mandal[29], Petros Daras[34], Pratik Narang[28], Qiong Liu[11], Qiu Shi[20], Qizhang Lin[15], Rohit Ramaprasad[28], Sai Wang[36], Sarvesh Mehta[22], Shuai Li[13], Shuqin Huang[11], Sungtae Moon[8], Taijin Zhao[6], Ting Sun[24], Wei Guo[35], Wei Tian[16], Weida Qin[11], Weiping Yu[14], Wenxiang Lin[15], Xi Zhao[23], Xiaogang Jia[27], Xin He[32], Xingjie Zhao[24], Xuanxin Liu[18], Yan Ding[15], Yan Luo[33], Yang Xiao[13], Yi Wang[23], Yingjie Liu[26], Yongwoo Kim[10], Yu Sun[19], Yuehan Yao[36], Yuyao Huang[16], Zehui Gong[17], Zhenyu Xu[36], Zhipeng Luo[36], Zhiguo Cao[13], Zhiwei Wei[7], Zhongjie Fan[11], Zichen Song[6], and Ziming Liu[15]

[1] Kitware, Inc., Clifton Park, NY, USA
[2] JD Finance America Corporation, Mountain View, CA, USA
[3] Tianjin University, Tianjin, China
zhupengfei@tju.edu.cn
[4] Stony Brook University, New York, NY, USA
[5] University of Central Florida, Orlando, FL, USA
[6] University of Electronic Science and Technology of China, Chengdu, China
[7] Harbin Institute of Technology (Shenzhen), Shenzhen, China
[8] Korea Aerospace Research Institute, Daejeon, South Korea
[9] Sungkyunkwan University, Suwon, South Korea
[10] Sangmyung University, Cheonan, South Korea
[11] South China University of Technology, Guangzhou, China
[12] University of California at Merced, Merced, CA, USA
[13] Huazhong University of Science and Technology, Wuhan, China
[14] University of North Carolina at Charlotte, Charlotte, NC, USA
[15] Beijing Institute of Technology, Beijing, China
[16] Tongji University, Shanghai, China
[17] Guangdong University of Technology, Guangzhou, China
[18] Beijing Forestry University, Beijing, China
[19] Beihang University, Beijing, China

A. Bartoli and A. Fusiello (Eds.): ECCV 2020 Workshops, LNCS 12538, pp. 692–712, 2020.
https://doi.org/10.1007/978-3-030-66823-5_42

[20] Beijing Vion Technology, Inc., Beijing, China
[21] Fraunhofer FKIE, Wachtberg, Germany
[22] International Institute of Information Technology, Hyderabad, Hyderabad, India
[23] Xidian University, Xi'an, China
[24] Xi'an Jiaotong University, Xi'an, China
[25] Science and Technology of China, Hefei, China
[26] North University of China, Taiyuan, China
[27] Harbin Institute of Technology, Harbin, China
[28] Birla Institute of Technology and Science, Pilani, Pilani, India
[29] Indian Institute of Information Technology Kota, Jaipur, India
[30] Karlsruhe Institute of Technology, Karlsruhe, Germany
[31] Fraunhofer IOSB, Karlsruhe, Germany
[32] Wuhan University of Technology, Wuhan, China
[33] Shanghai Jiao Tong University, Shanghai, China
[34] Centre for Research and Technology Hellas, Thessaloniki, Greece
[35] Chongqing University, Chongqing, China
[36] DeepBlue Technology (Shanghai) Co., Ltd., Shanghai, China

Abstract. The Vision Meets Drone Object Detection in Image Challenge (VisDrone-DET 2020) is the third annual object detector benchmarking activity. Compared with the previous VisDrone-DET 2018 and VisDrone-DET 2019 challenges, many submitted object detectors exceed the recent state-of-the-art detectors. Based on the selected 29 robust detection methods, we discuss the experimental results comprehensively, which shows the effectiveness of ensemble learning and data augmentation in drone captured object detection. The full challenge results are publicly available at the website http://aiskyeye.com/leaderboard/.

Keywords: Drone · Object detection · Evaluation

1 Introduction

Object detection is a hot topic in computer vision community, which propels various industrial detection-based applications such as autonomous driving, anomaly detection, face detection and activity recognition. Although much progress has been made using deep learning based methods, it is still a difficult problem on real-world scenarios.

Despite of detection in general scenarios, our goal is to advance state-of-the-art object detection approaches on drone-captured scenes, which involves some unique challenging factors (*e.g.*, view point change, scale variation, occlusion and background clutter) in object detection. The studies are seriously limited by the lack of public drone based large-scale benchmarks. Following VisDrone-DET2018 Challenge [47] and VisDrone-DET2019 Challenge [8], we held the 3rd Vision Meets Drone Object Detection in Images Challenge (VisDrone-DET2020) on August 28, 2020, in conjunction with the 16-th European Conference on Computer Vision (ECCV 2020).

In this paper, we summarize 29 object detection algorithms submitted to this challenge, and provide a comprehensive performance evaluation for them. Theses algorithms are improved based state-of-the-art detectors that are recently published in top computer vision conferences or journals, *e.g.*, Cascade R-CNN [3], CenterNet [9,45], ATSS [42], YOLOv3 [28] and RetinaNet [21]. Specifically, there are 10 out of 29 detection methods that outperform the previous winners in VisDrone-DET2018 and VisDrone-DET2019. The complete experimental results can be found at our website http://www.aiskyeye.com/, which is useful to further promote the research on object detection on drone-captured scenes.

2 Related Work

With fast development of various effective detection framework, researchers focus on ensemble of complex models to improve the performance. Besides, it is crucial to apply data augmentation strategies to train the deep model if lack of training data. In the following, we briefly review the current ensemble learning and data augmentation strategies in object detection field.

2.1 Ensemble Learning

Ensemble learning contains several feature extractors from different backbones in parallel, which requires additional time cost to improve the accuracy. In [16], several detection models are ensembled to achieve the state-of-the-art performance on the 2016 COCO object detection challenge. Xu *et al.* [40] employ the Single Shot MultiBox Detector [23] as the backbone and combine ensemble learning with context modeling and multi-scale feature representation. Besides, Gao *et al.* [11] incorporate an ensemble of classification heads for both box predictor and region proposal predictor to reduce false positives of the mined bounding boxes. To reduce the computation cost, Chen and Shrivastava [6] develop the Group Ensemble Network that incorporates an ensemble of ConvNets in a single ConvNet by a shared-base and multi-head structure.

2.2 Data Augmentation

Inspired by image classification field, random cropping and multi-scale training are the most widely used among the data augmentation strategies for object detection. Some methods randomly erase or add objects to the image for improved accuracy [10,43]. Despite of hand-tuned ranges, Zoph *et al.* [49] investigate the application of a learned data augmentation policy on object detection performance. In [1], a comprehensive experiment is conducted and shows that CutMix [41] and Mosaic data augmentation is effective in YOLOv4 detector. Notably, Mosaic is a new data augmentation method that mixes 4 training images with different contexts, which allows detection of objects outside their normal context.

Fig. 1. Annotation exemplars in the VisDrone-DET2020 challenge. The dashed bounding box indicates occlusion of the object, and different colors indicate different categories of objects. Only some attributes are displayed for clarity. (Color figure online)

3 The VisDrone-DET2020 Challenge

3.1 Dataset

As shown in Fig. 1, we use the same dataset as that in the previous two challenges [8,47] for a fair comparison. Specifically, the challenge contains 6,471 images for training and 548 images for validation, and 3,190 images for testing. Among the testing set, we have 1,580 images in the **test-challenge** subset for workshop competition, and 1,610 images in the **test-dev** subset for public evaluation. Ten object categories are pre-defined, *i.e.*, *pedestrian*, *person*, *car*, *van*, *bus*, *truck*, *motor*, *bicycle*, *awning-tricycle*, and *tricycle*. Some rarely occurring special vehicles (*e.g.*, *machineshop truck*, *forklift truck*, and *tanker*) are ignored in evaluation.

The participators are required to submit the detection results of specific algorithm with detailed description to the evaluation server no more than 10 times. The best submission among ten times are used as the final result. We encourage the participants to use the provided training data, while also allow them to use additional training data. The use of external data must be indicated during submission. For a fair comparison, we rank the algorithms trained on external VisDrone test-dev set in the leaderboard individually. Notably, it is strictly forbidden to submit the same algorithm by different accounts. The teams that provide better performance than CornerNet [17] are offered co-authorship of this results paper.

For evaluation, we follow the MS COCO evaluation protocol [22] to rank the detection algorithms, *i.e.*, AP, AP50, AP75, AR1, AR10, AR100 and AR50 metrics. Specifically, AP is the primary metric and calculated by averaging over all Intersection over Union (IoU) thresholds in the range [0.50, 0.95] with the uniform step size 0.05 of all 10 object categories. AP50 and AP75 are the average precision at the IoU threshold of 0.50 and 0.75 respectively. Besides, we compute average recalls given 1, 10, 100 and 500 detection per image over all object categories and IoU thresholds.

3.2 Submission

We received 85 submissions from all over the world in the VisDrone-DET2020 Challenge, where 29 teams from 31 different research institutes developed robust detectors better than the state-of-the-art detection method CornerNet [17]. The VisDrone committee also reports the results of another 3 detectors including Light-RCNN [19], FPN [20] and Cascade R-CNN [3].

Among all the submissions, several top methods use ensemble model to improve the accuracy, *i.e.*, DPNetV3 (A.1), SMPNet (A.2), and ECascade R-CNN (A.8). Eleven algorithms are based on Cascade R-CNN [3] with various effective modules, including DBNet (A.3), DroneEye2020 (A.4), CDNet (A.6), CascadeAdapt (A.7), HR-Cascade++ (A.9), Cascade R-CNN++ (A.21), DMNet (A.16), CFPN (A.23), HRC (A.26), SSODD (A.28) and GabA-Cascade (A.29). Six methods are derived from anchor-free CenterNet [9,45], *i.e.*, FPAFS-CenterNet (A.10), MSC-CenterNet (A.11), CenterNet+ (A.12), CN-FaDhSa (A.14), HRNet (A.15), and Center-ClusterNet (A.24). Three detectors combine ATSS [42] in their networks, namely TAUN (A.5), ASNet (A.13) and HR-ATSS (A.22). Besides, HRD-Net (A.17) is based on the High-Resolution Detection Network [25] which takes multiple resolution inputs using multi-depth backbones. PG-YOLO (A.18) is a variant of YOLOv3 [28] with a polymorphic module to learn the multi-scale and multi-shape object features and a group attention module to refine the combined features. EFPN (A.19) is based on the feature pyramid network to exploit the semantic information of small objects by multi-branched dilated bottleneck and attention and augmented bottom-up pathway [13]. CRENet (A.20) denotes the Cluster Region Estimation Network to search cluster regions containing dense objects, which makes the detector focus on these regions to reduce background interference. DOHR-RetinaNet (A.25) is modified from RetinaNet [21]. IterDet (A.27) proposes an alternative iterative scheme, where a new subset of objects is detected at each iteration.

3.3 Overall Evaluation

The overall results of the submissions are presented in Table 1 and Table 2. Compared with the winner detectors HAL-Retina-Net in the VisDrone-DET2018 Challenge [47] and DPNet-ensemble in the VisDrone-DET2019 Challenge [8], there are top 10 methods in the VisDrone-DET2020 Challenge achieving better mAP score more than 32.0. By using test-dev dataset in the training phase, the top performer DPNetV3 (A.1) in Table 1 performs slightly better than the top performer DroneEye2020 (A.4), *i.e.*, 37.37 vs. 34.57.

As discussed above, ensemble of several networks is effective to improve the accuracy of object detection. In Table 1, DPNetV3 (A.1) ensembles a few powerful backbones such as HRNet-W40 [33], Res2Net [12], Balanced Feature Pyramid Network [26], and Cascade R-CNN paradigm [3]. SMPNet (A.2) ranks the second place with the mAP score of 35.98, which also uses different combinations of multiple models (*i.e.*, Cascade-RCNN [3], HRNet [33], and ATSS [42]) to fuse the detection results.

Table 1. Object detection results in the VisDrone-DET2020 Challenge (model trained with the `test-dev` subset).

Method	AP[%]	AP50[%]	AP75[%]	AR1[%]	AR10[%]	AR100[%]	AR500[%]
DPNetV3 (A.1)	37.37	62.05	39.10	**0.85**	7.96	42.03	53.78
SMPNet (A.2)	35.98	**59.53**	37.41	0.29	2.01	8.46	54.33
DBNet (A.3)	**35.73**	[illegible]	**36.92**	0.37	2.78	12.70	**52.57**
ECascade R-CNN (A.8)	34.09	56.77	35.30	**1.06**	**7.73**	**35.31**	49.57
FPAFS-CenterNet (A.10)	32.34	56.46	32.39	1.20	9.45	[illegible]	51.61
DOHR-RetinaNet (A.25)	21.68	44.59	18.73	0.55	5.64	28.89	39.48
SSODD (A.28)	19.65	34.75	19.50	0.44	3.91	27.63	27.63

On the other hand, the use of the Cascade-RCNN [3] framework has become wide-spread recently (*e.g.*, from (A.4) to (A.9)) due to its high performance and easy extensibility. Compared with the baseline Cascade-RCNN [3] method with the mAP score of 16.09, the submitted variants largely improve the performance by combining several effective modules. In Table 1, DBNet (A.3) improves Cascade R-CNN [3] by adding global context block [4], DCN [7] and double heads [39]. In Table 2, DroneEye2020 (A.4) is mainly based on Cascade R-CNN [3] with Recursive Feature Pyramid and Switchable Atrous Convolution [27], achieving comparable performance with 34.57 mAP. TAUN (A.5) uses mean teacher [36] to train the cascade DetectoRS model [3,27], which performs similarly as DroneEye2020 (A.4). CDNet (A.6) and CascadeAdapt (A.7) combine Cascade-RCNN [3] with deformable convolutions, and then improve the detection accuracy using several data augmentation strategies such as sub-image splitting and mosaic [1].

3.4 Category Based Evaluation

For comprehensive evaluation, we also report the detection results of each object category in Table 3 and Table 4. Compared with the results in the VisDrone-DET2019 Challenge [8], the top performers achieve much better accuracy in several categories, *e.g.*, *bus* and *awning-tricycle*. This is attributed to two reasons. First, ensemble learning takes full advantage of various backbones to deal with different poses and scales of objects especially in *bus*. Second, augmentation strategies can help training some categories (*e.g.*, *awning-tricycle*) lacking of training data.

In Table 3, It can be observed that the top 2 performers obtain the best results on almost all 10 categories. That is, using test-dev set in the training phase

Table 2. Object detection results in the VisDrone-DET2020 Challenge (model trained without the `test-dev` subset). $*$ indicates that the detection algorithm is submitted by the committee.

Method	AP[%]	AP50[%]	AP75[%]	AR1[%]	AR10[%]	AR100[%]	AR500[%]
DroneEye2020 (A.4)	34.57	**58.21**	35.74	0.28	1.92	6.93	[illegible]
TAUN (A.5)	[illegible]	59.42	**34.97**	0.14	0.72	12.81	49.80
CDNet (A.6)	**34.19**	57.52	[illegible]	0.80	8.12	[illegible]	52.62
CascadeAdapt (A.7)	34.16	[illegible]	34.50	0.84	**8.17**	39.96	47.86
HR-Cascade++ (A.9)	32.47	55.06	33.34	0.94	7.81	**37.93**	50.65
MSC-CenterNet (A.11)	31.13	54.13	31.41	0.27	1.85	6.12	50.48
CenterNet+ (A.12)	30.94	52.82	31.13	0.27	1.84	5.67	**50.93**
ASNet (A.13)	29.57	52.25	29.37	0.25	1.69	6.46	46.01
CN-FaDhSa (A.14)	28.52	49.50	28.86	0.26	1.76	6.32	48.06
HRNet (A.15)	27.39	49.90	26.71	0.80	7.67	33.67	46.16
DMNet (A.16)	27.33	48.44	27.31	0.65	7.15	32.91	37.06
HRD-Net (A.17)	26.93	45.45	27.77	0.27	2.58	35.38	35.38
PG-YOLO (A.18)	26.05	49.63	24.15	1.45	[illegible]	33.65	42.63
EFPN (A.19)	25.27	48.18	23.37	1.45	9.21	32.91	40.65
CRENet (A.20)	25.16	44.38	24.57	0.27	2.44	21.21	36.44
Cascade R-CNN++ (A.21)	24.66	43.53	24.71	0.25	1.70	7.97	40.42
HR-ATSS (A.22)	24.23	41.84	24.43	0.27	2.15	34.97	34.97
CFPN (A.23)	22.85	42.33	21.88	0.81	7.08	29.65	39.55
Center-ClusterNet (A.24)	22.72	41.45	22.13	1.01	7.75	28.56	33.85
HRC (A.26)	21.23	43.56	18.39	0.18	1.16	4.88	37.25
IterDet (A.27)	20.42	36.73	20.25	0.21	1.34	8.86	33.04
GabA-Cascade (A.29)	18.85	33.60	18.66	**1.09**	7.68	26.25	33.03
CornerNet* [17]	17.41	34.12	15.78	0.39	3.32	24.37	26.11
Light-RCNN* [19]	16.53	32.78	15.13	0.35	3.16	23.09	25.07
FPN* [20]	16.51	32.20	14.91	0.33	3.03	20.72	24.93
Cascade R-CNN* [3]	16.09	31.91	15.01	0.28	2.79	21.37	28.43

Table 3. Object detection results of each object category (model trained with the `test-dev` subset).

Method	ped.	person	bicycle	car	van	truck	tricycle	awn.	bus	motor
DPNetV3 (A.1)	38.03	22.10	18.68	[illegible]	[illegible]	[illegible]	38.21	28.06	54.10	32.39
SMPNet (A.2)	[illegible]	**20.08**	[illegible]	57.35	44.98	40.61	**34.67**	**26.70**	**53.33**	[illegible]
DBNet (A.3)	**35.73**	[illegible]	**15.75**	**55.24**	**43.92**	**39.87**	[illegible]	[illegible]	[illegible]	28.43
ECascade R-CNN (A.8)	34.66	18.94	12.64	55.07	42.82	38.14	32.74	24.74	52.06	**29.11**
FPAFS-CenterNet (A.10)	31.55	13.77	14.68	55.04	42.30	37.55	29.48	24.23	48.24	26.57
DOHR-RetinaNet (A.25)	23.31	10.66	6.35	44.74	29.10	26.64	17.99	13.00	27.70	17.38
SSODD (A.28)	19.39	6.97	2.77	42.77	24.70	20.79	15.93	12.38	35.57	15.18

generally improves the performance. Notably, CascadeAdapt (A.7) achieve the best mAP score on *awning-tricycle*, which adopts several augmentation methods such as mosaic in YOLOv4 [1]. Besides, Cascade R-CNN variants take top three places in term of each category.

Table 4. Object detection results of each object category (model trained without the `test-dev` subset). * indicates the detection algorithms submitted by the VisDrone Team.

Method	ped.	person	bicycle	car	van	truck	tricycle	awn.	bus	motor
DroneEye2020 (A.4)	35.70	**18.27**	11.02	56.51	42.01	37.61	35.41	**25.91**	50.37	**28.95**
TAUN (A.5)	**34.98**	16.05	17.23	54.62	41.71	38.67	35.10	26.53	48.49	29.06
CDNet (A.6)	35.61	19.15	**13.84**	55.77	42.12	**38.22**	32.97	25.42	**49.49**	29.28
CascadeAdapt (A.7)	31.61	15.63	12.83	53.59	43.17	38.29	**34.80**	32.09	51.49	28.13
HR-Cascade++ (A.9)	32.58	17.31	11.05	54.71	**42.37**	35.27	32.68	24.09	46.48	28.20
MSC-CenterNet (A.11)	33.70	15.23	12.07	55.19	40.47	34.08	29.24	21.63	42.23	27.45
CenterNet+ (A.12)	32.56	16.15	12.14	**55.35**	38.79	33.71	30.35	22.59	41.12	26.69
ASNet (A.13)	28.34	12.32	10.18	51.38	38.99	33.03	28.94	22.51	47.49	22.56
CN-FaDhSa (A.14)	30.52	12.89	9.85	52.52	38.14	32.91	25.85	22.00	39.89	20.61
HRNet (A.15)	27.61	13.35	11.79	50.78	36.46	29.89	24.42	21.03	35.16	23.36
DMNet (A.16)	29.06	13.30	9.59	50.27	30.54	32.04	27.15	17.60	39.68	24.03
HRD-Net (A.17)	26.83	12.13	8.05	48.97	32.73	31.38	26.16	18.39	42.10	22.57
PG-YOLO (A.18)	26.82	13.83	9.90	46.61	32.92	26.49	22.87	19.33	41.45	20.32
EFPN (A.19)	25.57	12.45	8.72	46.06	32.29	25.72	22.16	18.99	41.58	19.15
CRENet (A.20)	27.57	11.50	6.61	46.11	32.73	28.66	23.84	17.77	37.19	19.57
Cascade R-CNN++ (A.21)	24.81	8.83	4.79	48.96	35.45	29.01	24.36	19.95	33.46	17.02
HR-ATSS (A.22)	27.23	11.12	5.75	48.76	31.84	21.27	22.70	18.20	35.95	19.48
CFPN (A.23)	23.22	9.70	4.78	45.37	30.83	23.85	19.37	16.51	38.35	16.54
Center-ClusterNet (A.24)	24.02	9.14	5.70	48.43	26.43	27.14	19.36	14.90	33.30	18.76
HRC (A.26)	17.26	8.50	4.85	40.53	29.82	24.22	18.70	16.29	37.22	14.95
IterDet (A.27)	17.33	6.76	3.61	41.21	30.27	23.76	19.04	15.05	35.68	11.53
GabA-Cascade (A.29)	17.43	6.07	3.21	39.21	27.22	20.26	15.72	13.20	33.84	12.31
CornerNet* [17]	20.43	6.55	4.56	40.94	20.23	20.54	14.03	9.25	24.39	12.10
Light-RCNN* [19]	17.02	4.83	5.73	32.29	22.12	18.39	16.63	11.91	29.02	11.93
FPN* [20]	15.69	5.02	4.93	38.47	20.82	18.82	15.03	10.84	26.72	12.83
Cascade R-CNN* [3]	16.28	6.16	4.18	37.29	20.38	17.11	14.48	12.37	24.31	14.85

4 Conclusions

In this paper, we present the results of the VisDrone-DET2020 Challenge. It is the third annual object detector benchmarking activity, following the very successful VisDrone-DET2018 and VisDrone-DET2019 challenges. Evaluated on the same dataset, many submitted object detection methods set a new state-of-the-art. Specifically, the top performer is DPNetV3 (A.1) using test-dev set as training data, with the overall mAP score of 37.37. Without test-dev set, the top three performers are DroneEye2020 (A.4), TAUN (A.5), and CDNet (A.6), with the overall mAP score of more than 35.00. The experimental results indicate that ensemble learning of a few powerful detectors can largely boost the detection performance. Besides, Cascade R-CNN and ATSS are other popular detection frameworks. It is worth mentioning that the best detector DPNetV3 (A.1) improves the mAP score by over 6% than before, which shows the development of object detection in the past year. However, the best mAP score is still less than 40% and far from satisfactory in real applications. Meanwhile, the computational complexity of the submitted algorithms is another issue on the drone platform with limited resource. We hope we can provide a platform to advance state-of-the-art object detection methods on the drone-captured scenarios [46].

Acknowledgements. This work was supported in part by the National Natural Science Foundation of China under Grant 61876127 and Grant 61732011, in part by Natural Science Foundation of Tianjin under Grant 17JCZDJC30800.

A Submitted Detectors

In this appendix, we provide a short summary of all algorithms that were considered in the VisDrone-DET2020 Challenge.

A.1 Drone Pyramid Network V3 (DPNetV3)

Heqian Qiu, Zichen Song, Minjian Zhang, Mingyu Liu, Taijin Zhao, Fanman Meng, Hongliang Li
hqqiu@std.uestc.edu.cn, szc.uestc@gmail.com, jamiezhang722@outlook.com, myl8562@163.com, zhtjww@std.uestc.edu.cn, fmmeng@uestc.edu.cn, hlli@uestc.edu.cn

DPNetV3 is an ensemble model for object detection, see Fig. 2. First, it adopts HRNet-W40 [33] pre-trained on ImageNet dataset as our backbone network, which starts from a high-resolution subnetwork as the first stage, gradually adds high-to-low resolution subnetworks one by one to form more stages, and connects the mutli-resolution subnetworks in parallel. In addition, we also use Res2Net [12] as our backbone networks. To make features more robust for complex scenes, we introduce Balanced Feature Pyramid Network [26] with CARAFE (Content-Aware ReAssembly of FEatures) [37] and Deformable Convolution [7] into these backbone networks. Furthermore, we use Cascade R-CNN paradigm [3] to progressively refine detection boxes for accurate object localization. We ensemble them using weighted box fusion method.

A.2 Using Split, Mosaic and Paster Modules for Detecting Aerial Images (SMPNet)

Chengzhen Duan, Zhiwei Wei
{18S151541, 19S051024}@stu.hit.edu.cn

In order to improve the accuracy of aerial image detection, we propose adaptive split method, mosaic data enhancement method and resampling enhancement method. The adaptive split method adjusts the split absolute area according to the average target size in the split, so that the detector can focus on the narrower target scale range that is conducive to detection. After that, we calculate the scaling factor required by the target in split, and then scale the split proportionally. Then we cut the four splits and splice them into mosaics [1].

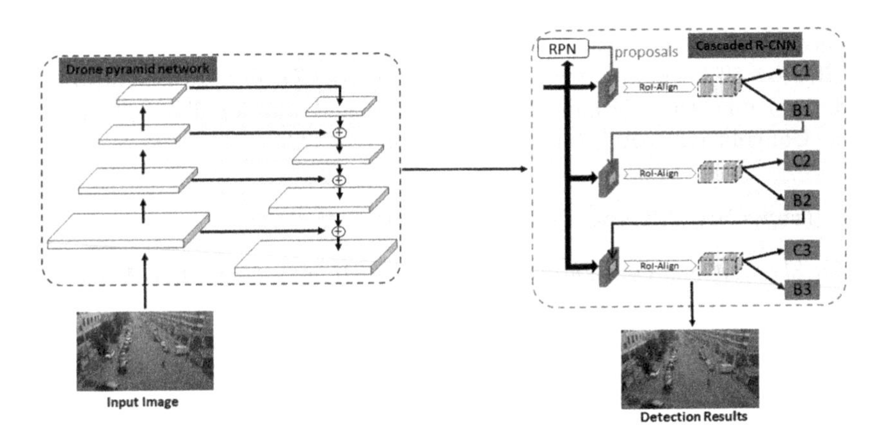

Fig. 2. The framework of DPNetv3.

In order to alleviate the problem of class imbalance, we use panoramic segmentation to build the target pool, and then paste the appropriate target from the target pool to the training sample. Different from the previous method of pasting the whole GT box [5], we only paste the target. We use multi-model to infer and fuse the detection results, including Cascade-RCNN [3] + HRNet [33], ATSS [42] + HRNet [33], and ATSS [42] + Res2Net [12] + general focal loss [18].

A.3 DeepBlueNet (DBNet)

Zhipeng Luo, Sai Wang, Zhenyu Xu, Yuehan Yao, Bin Dong
{luozp, wangs, xuzy, yaoyh, dongb}@deepblueai.com

DBNet adopts Cascade_x101_64-4d [3] as the pipeline and add a global context block [4] to improve the ability of the extractor which could get more information from global feature. Meanwhile, we use DCN [7] to reduce the effect of feature misalignment, and adaptive part localization for objects with different shapes. As for R-CNN part, we use Double-Head RCNN [39]. Thus object classification is enhanced by adding classification task in conv-head, as it is complementary to the classification in fc-head. That is, bounding box regression provides auxiliary supervision for fc-head. We ensemble multi-scale testing results as our final result.

A.4 Cascade R-CNN on Drone-Captured Scenarios (DroneEye2020)

Sungtae Moon, Joochan Lee, Jungyeop Yoo, Jonghwan Ko, Yongwoo Kim
stmoon@kari.re.kr, {maincold2, soso030, jhko}@skku.edu, yongwoo.kim @smu.ac.kr

DroneEye2020 is improved on Cascade R-CNN [3]. We divide original training images by 2×2 and horizontally flip all patches. If a divided patch has no objects, we exclude the patch when training. We use Cascade R-CNN [3] with ResNet-50 backbone, which is pretrained by COCO dataset. Notably, we use Recursive Feature Pyramid (RFP) for neck, and additionally use Switchable Atrous Convolution (SAC) for better performance [27].

A.5 Tricks are All yoU Need (TAUN)

Jingkai Zhou, Weida Qin, Zhongjie Fan, Shuqin Huang, Qiong Liu, Ming-Hsuan Yang
fs.jingkaizhou@gmail.com

TAUN is based on the cascade DetectoRS [3,27]. The backbone is HRNet-40 [38], and the neck is the original HRFPN neck. ATSS [42] is used as the assigner in the RPN [29]. We use multi-scale image crop (not SAIC [44] for saving time and memory) to augment training and testing data, use mean teacher [36] to train the model. When model testing, we use ratio, outside, and scale filters to post filter outline bounding boxes. The threshold of those filters are counted based on the training dataset.

A.6 Cascade RCNN with DCN (CDNet)

Shuai Li, Yang Xiao, Zhiguo Cao
{shuai_li1997, yang_xiao, zgcao}@hust.edu.cn

CDNet is based on Cascade RCNN [3] with ResNeXt101. Moreover, we add deformable convolutional network [7] for better performance. To reduce GPU memory and consider small objects, we split the train images into sub-image with size 416×416 and train the network between 416 and 416×2 size. In testing phase, we use Soft-NMS and multi-scale testing to achieve better accuracy.

A.7 Cascade Network with Test Time Self-supervised Adaptation (CascadeAdapt)

Weiping Yu, Chen Chen
wyu4@unc.edu, chen.chen@uncc.edu

CascadeAdapt is improved from Cascade R-CNN [3] with ResNet backbone and Deformable Convolutions. We adopt several augmentation methods such as mosaic in YOLOv4 [1]. We use double heads instead of traditional box head to output the detection results. Although We use weighted box fusion to ensemble several models and obtain prediction on the test-challenge set. After that, we obtain pseudo labels of the test-challenge set by setting the threshold, and then finetune the model by the pseudo labels for 2 epoch to remove a large amount of false detections. The performance can be further improved by using other tricks such as GN, test time augmentation, label smooth and GIOU loss.

A.8 Enhanced Cascade R-CNN for Drone (ECascade R-CNN)

Wenxiang Lin, Yan Ding, Qizhang Lin
{eutenacity, dingyan, 3120190071}@bit.edu.cn

ECascade R-CNN is an ensembling method based on the work in [32]. We additionally train a detector with the backbone of HRNetv2 [34], ResNet50 [14] and ResNet101 [14]. For the convenience, we call the additional detectors as HRDet, Res50Det and Res101Det and call the first detector as RXDet. Finally, we train two detectors (*i.e.*, RXDet and HRDet) with the trick for class variance and four more detectors (RXDet, HRDet, Res50Det and Res101Det) without that. The final result is the output of the ensemble of all six detectors.

A.9 Cascade R-CNN with High-Resolution Network and Enhanced Feature Pyramid (HR-Cascade++)

Hansheng Chen, Lu Xiong, Dong Yin, Yuyao Huang, Wei Tian
{1552083, xiong_lu, tjyd, huangyuyao}@tongji.edu.cn, tian-w@hotmail.com

HR-Cascade++ is based on the multi-stage detection architecture Cascade R-CNN [3], and is tuned specifically for dense small object detection. As the drone images include many small objects, we seek to obtain high resolution feature and improve the bounding box spatial accuracy. We adopt HRNet-W40 [38] as the backbone, which maintains high-resolution representations (1/4 of the original size) through the whole process. The feature pyramid network is enhanced with an additionally upsampled high resolution level (1/2 of the original image). This enables smaller and denser anchor generation, without resizing the original image. We adopt Quality Focal Loss [18] for R-CNN classification. Multi-scale and flip augmentations are applied in both training and testing. Photometric distortion is also used for training image augmentation. To reduce GPU memory consumption, the training images are cropped after resizing. If a ground truth bounding box is truncated during cropping, it is marked as ignore region when truncation ratio is greater than 50%. Soft-NMS is used for post-processing.

A.10 CenterNet with Feature Pyramid and Adaptive Feature Selection (FPAFS-CenterNet)

Zehui Gong
zehuigong@foxmail.com

FPAFS-CenterNet is based on CenterNet [45], because of its simplicity and high efficiency for object detection. CenterNet presents a new representation for detecting objects, in terms of their center locations. Other object properties, *e.g.*, object size, are regressed directly using the image features from the center locations. To achieve better performance, we have applied some useful modifications to CenterNet, with regard to the data augmentation, backbone network, feature fusion neck, and also, the detection loss. In order to extract more powerful features from input image, we employ CBNet [24] as our backbone network. We use BiFPN [35] as a feature fusion neck to enhance the information flow across the highly semantic and spatially finer features. We use GIOU loss [30], which is irrelevant to the object size.

A.11 CenterNet with Multi-Scale Cropping (MSC-CenterNet)

Xuanxin Liu, Yu Sun
liuxuanxin@bjfu.edu.cn, sunyv@buaa.edu.cn

MSC-CenterNet employs CenterNet [45] with Hourglass-104 as the base network and does not use the pre-trained models. Considering the larger scale range for the intra-class and inter-class objects, the model is trained with multi-scale cropping. The input resolution is 1024×1024 and the input region is cropped from the image with the scale randomly choose from $(0.9, 1.1, 1.3, 1.5, 1.8)$.

A.12 CenterNet for Small Object Detection (CenterNet+)

Qiu Shi
qiushi_0425@foxmail.com

CenterNet+ is CenterNet [45] with the hourglass feature extractor where there are three hourglass blocks. In order to improve the performance of our detector for small samples, we change the stride from 7 to 2 in each hourglass block. Besides, we adopt multi-scale training and multi-scale test to improve the detector performance.

CenterNet+ employs CenterNet [45] with the hourglass feature extractor where there are three hourglass blocks. In order to improve the performance of our detector for small samples, we change the stride from 5 to 2 in each hourglass block. Besides, we adopt multi-scale training and multi-scale test to improve the detector performance.

A.13 Aerial Surveillance Network (ASNet)

Michael Schleiss
michael.schleiss@fkie.fraunhofer.de

ASNet adopts the ATSS detector [42] based on the implementation from mmdetection v2.2. Standard settings are used if not stated otherwise. Our backbone is Res2Net [12] with 152 layers pretrained on ImageNet. In the neck we replace FPN with Carafe [37]. The neck has 384 output channels instead of 256 in the original implementation. Focal loss is replaced by generalized focal loss [30]. We use multi-scale training with sizes $[600, 800, 1000, 1200]$ for the shorter side. We train on a single gpu with batch size 4 for 12 epochs with a step wise learning schedule. Learning rate starts with 0.01 and is divided by 10 after epoch 8 and 11 respectively. We use no TTA and apply a scale of 1200 for the shorter side during testing.

A.14 CenterNet-Hourglass-104 (CN-FaDhSa)

Faizan Khan, Dheeraj Reddy Pailla, Sarvesh Mehta
{faizan.farooq, dheerajreddy.p}@students.iiit.ac.in, sarvesh.mehta@research.iiit.ac.in

CN-FaDhSa is modified from CenterNet [45]. Instead of using the default resolution of 512×512 during training, we train the model at the resolution of 1024×1024 and test at various scales of 2048×2048.

A.15 High-Resolution Net (HRNet)

Guosheng Zhang, Zehui Gong
249200734@qq.com
HRNet is similar to CenterNet [9]. However, we detect the object just as a single center point instead of triplets, and regress to object size $s = (h, w)$ for each object. In addition, we use the High-Resolution Net (HRNet) followed by FPN as the backbone, which is able to maintain high-resolution representations through the whole process.

A.16 Density Map Guided Object Detection (DMNet)

Changlin Li
cli33@uncc.edu

We use density crop+uniform crop to train baseline model and conduct fusion detection to obtain final detection. The baseline is Cascade R-CNN [3].

A.17 High-Resolution Detection Network (HRD-Net)

Ziming Liu, Guangyu Gao
liuziming.email@gmail.com, guangyugao@bit.edu.cn

To keep the benefits of high-resolution images without bringing up new problems, we propose the High-Resolution Detection Network (HRDNet) [25]. HRDNet takes multiple resolution inputs using multi-depth backbones. To fully take advantage of multiple features, we propose Multi-Depth Image Pyramid Network (MD-IPN) and Multi-Scale Feature Pyramid Network (MS-FPN) in HRDNet. MD-IPN maintains multiple position information using multiple depth backbones. Specifically, high-resolution input will be fed into a shallow network to reserve more positional information and reducing the computational cost while low-resolution input will be fed into a deep network to extract more semantics. By extracting various features from high to low resolutions, the MD-IPN is able to improve the performance of small object detection as well as maintaining the performance of middle and large objects. MS-FPN is proposed to align and fuse multi-scale feature groups generated by MD-IPN to reduce the information imbalance between these multi-scale multi-level features.

A.18 A Slimmer Network with Polymorphic and Group Attention Modules for More Efficient Object Detection in Aerial Images (PG-YOLO)

Wei Guo, Jincai Cui
{gwfemma, jinkaicui}@cqu.edu.cn

PG-YOLO is a YOLOv3 [28] based slimmer network for more efficient object detection in aerial images. Firstly, a polymorphic module (PM) is designed for simultaneously learning the multi-scale and multi-shape object features, so as to better detect the hugely different objects in aerial images. Then, a group attention module (GAM) is designed for better utilizing the diversiform concatenation features in the network. By designing multiple detection headers with adaptive anchors and above-mentioned two modules, the final one-stage network called PG-YOLO is obtained for realizing the higher detection accuracy.

A.19 Extended Feature Pyramid Network with Adaptive Scale Training Strategy and Anchors for Object Detection in Aerial Images (EFPN)

Wei Guo, Jincai Cui
{gwfemma, jinkaicui}@cqu.edu.cn

EFPN comes from the work in [13]. To enhance the semantic information of small objects in deep layers of the network, the extended feature pyramid network is proposed. Specifically, we use the multi-branched dilated bottleneck module in the lateral connections and an attention pathway to improve the detection accuracy for small objects. For better locating the objects. Besides, an adaptive scale training strategy is developed to enable the network to deal with multi-scale object detection, where adaptive anchors are achieved by a novel clustering method.

A.20 Cluster Region Estimation Network (CRENet)

Yi Wang, Xi Zhao
{wangyi0102, xizhao_1}@stu.xidian.edu.cn

Aerial images are increasingly used for critical tasks, such as traffic monitoring, pedestrian tracking, and infrastructure inspection. However, aerial images have the following main challenges: 1) small objects with non-uniform distribution; 2) the large difference in object size. In this paper, we propose a new network architecture, Cluster Region Estimation Network (CRENet), to solve these challenges. CRENet uses a clustering algorithm to search cluster regions containing dense objects, which makes the detector focus on these regions to reduce background interference and improve detection efficiency. However, not every cluster region can bring precision gain, so each cluster region is calculated a difficulty score, mining the difficult cluster region and eliminating the simple cluster region to speed up the detection. Finally, a Gaussian scaling function is used to scale the difficult cluster region to reduce the difference of object size.

A.21 Cascade R-CNN for Drone-Captured Scenes (Cascade R-CNN++)

Ting Sun, Xingjie Zhao
sunting9999@stu.xjtu.edu.cn, 1243273854@qq.com

We use Cascade R-CNN [3] as the baseline with four improvements: 1) We use Group normalization instead of Batch normalization; 2) We use online hard example mining to select positive and negative samples; 3) We use multi-scale testing; 4) We use two stronger backbones to train models and integrate them. Besides, we use ResNeXt as backbone to training, and Soft-Non maximum suppression instead of Non maximum suppression. At the same time, we use online hard example mining to select positive and negative samples in Region proposal networks.

A.22 HRNet Based ATSS for Object Detection (HR-ATSS)

Jun Yu, Haonian Xie
harryjun@ustc.edu.cn, xie233@mail.ustc.edu.cn

HR-ATSS is based on Adaptive Training Sample Selection (ATSS) [42], which can automatically select positive and negative samples according to statistical characteristics of object. We employ HRNet [33] as backbone to improve small object performance, where HRFPN [33] is adopted as the feature pyramid. Specifically, we adopt HRNet-W32 as backbone, HRFPN as the feature pyramid, and ATSS detection head to regress and classify objects. We adopt Synchronized BN instead of BN.

A.23 Concat Feature Pyramid Networks (CFPN)

Yingjie Liu
1497510582@qq.com

CFPN is improved from FPN [20] and Cascade R-CNN [3], which uses concatenation for lateral connections rather than the addition in FPN. Meanwhile, in the fast R-CNN stage, cascade architecture named Cascade R-CNN is utilized to refine the bounding box regression. ResNet-152 is used as the pre-trained backbone. In the training stage, we apply a manual adjustment on the learning rate to optimize detection performance. In the testing stage, we use Soft-NMS for better recall on the dense objects.

A.24 CenterNet+HRNet (Center-ClusterNet)

Xiaogang Jia
18846827115@163.com

The Center-ClusterNet detector is based on CenterNet [45] and HRNet [33]. We use MobileNetV3 [15] as the backbone to predict all centers of the objects. Then K-Means is used as a post-processing method to generate clusters. Both original images and cropped images are processed by the detector. Then the predicted bounding boxes are merged by standard NMS.

A.25 Deep Optimized High Resolution RetinaNet (DOHR-RetinaNet)

Ayush Jain, Rohit Ramaprasad, Murari Mandal, Pratik Narang
{f20170093, f20180224}@pilani.bits-pilani.ac.in, 2015rcp9525@mnit.ac.in, pratik.narang@pilani.bits-pilani.ac.in

DOHR-RetinaNet is based on RetinaNet [21], using ResNet101 backbone for feature extraction. FPN is utilized for creating semantically strong features and focal loss to mitigate class imbalance. The ResNet101 backbone is pretrained on the ImageNet dataset. Additionally, we use a combination of anchors consisting of 5 ratios and 5 scales which are optimized using the algorithm in [48] to improve detection of smaller objects. All our input images are resized such that the minimum side is 1728px and the maximum side is less than 3072px. The training images are augmented through random rotation, random flipping and brightness variation to help detect objects captured in dark backgrounds.

A.26 High Resolution Cascade R-CNN (HRC)

Daniel Stadler, Arne Schumann, Lars Wilko Sommer
daniel.stadler@kit.edu, {arne.schumann, lars.sommer}@iosb.fraunhofer.de

HRC is based on Cascade R-CNN [3] with FPN [20] and HRNetV2p-W32 [33] as backbone. We train four detectors with different anchor scales to account for varying object scales on randomly sampled image crops (608×608 pixels) of the VisDrone DET train and val set and use the SSD [23] data augmentation pipeline to enhance feature representation learning. For each class, the detector with best anchor scale is utilized. During test time, we follow a multi-scale strategy and, additionally, apply horizontal flipping. The resulting detections are combined via Soft-NMS [2]. To account for the large number of objects per frame, we increase the number of proposals and the maximum number of detections per image.

A.27 Iterative Scheme for Object Detection in Crowded Environments (IterDet)

Xin He
2962575697@whut.edu.cn

IterDet is an alternative iterative scheme, where a new subset of objects is detected at each iteration. Detected boxes from the previous iterations are passed to the network at the following iterations to ensure that the same object would not be detected twice. This iterative scheme can be applied to both one-stage and two-stage object detectors with just minor modifications of the training and inference procedures.

A.28 Small-Scale Object Detection for Drone Data (SSODD)

Yan Luo, Chongyang Zhang, Hao Zhou
{luoyan_bb, sunny_zhang, zhouhao_0039}@sjtu.edu.cn

SSODD is based on Cascade R-CNN [3] using ResNeXt-101 64x4d as backbone and FPN [20] as feature extractor. We also use deformable convolution to enhance our feature extractor. Our pretrained model is based on COCO dataset. Other techniques like multi-scale training and soft-nms are involved in our method. To detect small-scale objects, we crop each image into four parts, which are evenly distributed on the row image. The cropped four regions are fed into the framework with the multi-scale training technique, in which the scale is $(960, 720)$ and $(960, 640)$.

A.29 Cascade R-CNN Enhanced by Gabor-Based Anchoring (GabA-Cascade)

Ioannis Athanasiadis, Athanasios Psaltis, Apostolos Axenopoulos, Petros Daras
{athaioan, at.psaltis, axenop, daras}@iti.gr

GabA-Cascade is build upon Cascade R-CNN [3] which is enhanced by considering an additional set of anchors targeted explicitly at small objects. Inspired by [31], a set of simplified Gabor wavelets (SGWs) is applied on the input image resulting in an edge-enhanced version of the latter. Thereafter, the maximally stable extremal regions (MSERs) algorithm is applied on the edge-enhanced image extracting regions possible of containing an object, called edge anchors. As a next step, we aim at integrating the edge anchors into the Region Proposal Network (RPN). Due to edge anchors being of varying scale and having continuous center coordinates, some modifications are required so as to be compatible with the RPN training procedure. In order for the Feature Pyramid Network (FPN) feature maps to remain scale specific and have the bounding box regressor referring to identical shaped anchors, the edge anchors are refined to match the closest available shape and size hyper parameters configuration. The issue of edge anchor centers not being aligned with the pixel grid is addressed through rounding their centers. Furthermore, additional feature maps, dedicated to the edge anchors, are introduced with the purpose of minimizing the previously mentioned refinement. These feature maps correspond to different scales relevant to small objects and are identical to the feature map of the first FPN pyramid level. After the modifications described above the RPN is able to evaluate regions given both edge and regular anchors as input. Finally the rest of the object detection pipeline follows the cascade architecture as described in [3], deploying classifiers of increasing quality.

References

1. Bochkovskiy, A., Wang, C., Liao, H.M.: Yolov4: optimal speed and accuracy of object detection. CoRR abs/2004.10934 (2020)
2. Bodla, N., Singh, B., Chellappa, R., Davis, L.S.: Soft-NMS - improving object detection with one line of code. In: ICCV, pp. 5562–5570 (2017)
3. Cai, Z., Vasconcelos, N.: Cascade R-CNN: delving into high quality object detection. In: CVPR, pp. 6154–6162 (2018)
4. Cao, Y., Xu, J., Lin, S., Wei, F., Hu, H.: GCNet: non-local networks meet squeeze-excitation networks and beyond. In: ICCVW, pp. 1971–1980 (2019)
5. Chen, C., et al.: RRNet: a hybrid detector for object detection in drone-captured images. In: ICCV, pp. 100–108 (2019)
6. Chen, H., Shrivastava, A.: Group ensemble: learning an ensemble of convnets in a single convnet. CoRR abs/2007.00649 (2020)
7. Dai, J., et al.: Deformable convolutional networks. In: ICCV, pp. 764–773 (2017)
8. Du, D., et al.: VisDrone-DET2019: the vision meets drone object detection in image challenge results. In: ICCVW, pp. 213–226 (2019)
9. Duan, K., Bai, S., Xie, L., Qi, H., Huang, Q., Tian, Q.: CenterNet: keypoint triplets for object detection. In: ICCV, pp. 6568–6577 (2019)
10. Dwibedi, D., Misra, I., Hebert, M.: Cut, paste and learn: surprisingly easy synthesis for instance detection. In: ICCV, pp. 1310–1319 (2017)
11. Gao, J., Wang, J., Dai, S., Li, L., Nevatia, R.: NOTE-RCNN: noise tolerant ensemble RCNN for semi-supervised object detection. In: ICCV, pp. 9507–9516 (2019)
12. Gao, S., Cheng, M., Zhao, K., Zhang, X., Yang, M., Torr, P.H.S.: Res2Net: a new multi-scale backbone architecture. TPAMI (2019)
13. Guo, W., Li, W., Gong, W., Cui, J.: Extended feature pyramid network with adaptive scale training strategy and anchors for object detection in aerial images. Remote. Sens. **12**(5), 784 (2020)
14. He, K., Zhang, X., Ren, S., Sun, J.: Deep residual learning for image recognition. In: CVPR, pp. 770–778 (2016)
15. Howard, A., et al.: Searching for MobileNetV3. In: ICCV, pp. 1314–1324 (2019)
16. Huang, J., et al.: Speed/accuracy trade-offs for modern convolutional object detectors. In: CVPR, pp. 3296–3297 (2017)

17. Law, H., Deng, J.: CornerNet: detecting objects as paired keypoints. In: Ferrari, V., Hebert, M., Sminchisescu, C., Weiss, Y. (eds.) Computer Vision – ECCV 2018. LNCS, vol. 11218, pp. 765–781. Springer, Cham (2018). https://doi.org/10.1007/978-3-030-01264-9_45
18. Li, X., et al.: Generalized focal loss: learning qualified and distributed bounding boxes for dense object detection. CoRR abs/2006.04388 (2020)
19. Li, Z., Peng, C., Yu, G., Zhang, X., Deng, Y., Sun, J.: Light-head R-CNN: in defense of two-stage object detector. CoRR abs/1711.07264 (2017)
20. Lin, T., Dollár, P., Girshick, R.B., He, K., Hariharan, B., Belongie, S.J.: Feature pyramid networks for object detection. In: CVPR, pp. 936–944 (2017)
21. Lin, T., Goyal, P., Girshick, R.B., He, K., Dollár, P.: Focal loss for dense object detection. In: ICCV, pp. 2999–3007 (2017)
22. Lin, T.-Y., et al.: Microsoft COCO: common objects in context. In: Fleet, D., Pajdla, T., Schiele, B., Tuytelaars, T. (eds.) ECCV 2014. LNCS, vol. 8693, pp. 740–755. Springer, Cham (2014). https://doi.org/10.1007/978-3-319-10602-1_48
23. Liu, W., et al.: SSD: single shot multibox detector. In: Leibe, B., Matas, J., Sebe, N., Welling, M. (eds.) ECCV 2016. LNCS, vol. 9905, pp. 21–37. Springer, Cham (2016). https://doi.org/10.1007/978-3-319-46448-0_2
24. Liu, Y., et al.: CBNet: a novel composite backbone network architecture for object detection. In: AAAI, pp. 11653–11660 (2020)
25. Liu, Z., Gao, G., Sun, L., Fang, Z.: HRDNet: high-resolution detection network for small objects. CoRR abs/2006.07607 (2020)
26. Pang, J., Chen, K., Shi, J., Feng, H., Ouyang, W., Lin, D.: Libra R-CNN: towards balanced learning for object detection. In: CVPR, pp. 821–830 (2019)
27. Qiao, S., Chen, L., Yuille, A.L.: Detectors: detecting objects with recursive feature pyramid and switchable atrous convolution. CoRR abs/2006.02334 (2020)
28. Redmon, J., Farhadi, A.: Yolov3: an incremental improvement. CoRR abs/1804.02767 (2018)
29. Ren, S., He, K., Girshick, R.B., Sun, J.: Faster R-CNN: towards real-time object detection with region proposal networks. In: Cortes, C., Lawrence, N.D., Lee, D.D., Sugiyama, M., Garnett, R. (eds.) NeurIPS, pp. 91–99 (2015)
30. Rezatofighi, H., Tsoi, N., Gwak, J., Sadeghian, A., Reid, I.D., Savarese, S.: Generalized intersection over union: a metric and a loss for bounding box regression. In: CVPR, pp. 658–666 (2019)
31. Shao, F., Wang, X., Meng, F., Zhu, J., Wang, D., Dai, J.: Improved faster R-CNN traffic sign detection based on a second region of interest and highly possible regions proposal network. Sensors **19**(10), 2288 (2019)
32. Solovyev, R., Wang, W.: Weighted boxes fusion: ensembling boxes for object detection models. CoRR abs/1910.13302 (2019)
33. Sun, K., Xiao, B., Liu, D., Wang, J.: Deep high-resolution representation learning for human pose estimation. In: CVPR, pp. 5693–5703 (2019)
34. Sun, K., et al.: High-resolution representations for labeling pixels and regions. CoRR abs/1904.04514 (2019)
35. Tan, M., Pang, R., Le, Q.V.: EfficientDet: scalable and efficient object detection. In: CVPR, pp. 10778–10787 (2020)
36. Tarvainen, A., Valpola, H.: Mean teachers are better role models: weight-averaged consistency targets improve semi-supervised deep learning results. In: NeurIPS, pp. 1195–1204 (2017)
37. Wang, J., Chen, K., Xu, R., Liu, Z., Loy, C.C., Lin, D.: CARAFE: content-aware reassembly of features. In: ICCV, pp. 3007–3016 (2019)

38. Wang, J., et al.: Deep high-resolution representation learning for visual recognition. TPAMI (2020)
39. Wu, Y., et al.: Rethinking classification and localization in R-CNN. CoRR abs/1904.06493 (2019)
40. Xu, J., Wang, W., Wang, H., Guo, J.: Multi-model ensemble with rich spatial information for object detection. PR **99**, 107098 (2020)
41. Yun, S., Han, D., Chun, S., Oh, S.J., Yoo, Y., Choe, J.: CutMix: regularization strategy to train strong classifiers with localizable features. In: ICCV, pp. 6022–6031 (2019)
42. Zhang, S., Chi, C., Yao, Y., Lei, Z., Li, S.Z.: Bridging the gap between anchor-based and anchor-free detection via adaptive training sample selection. In: CVPR, pp. 9756–9765 (2020)
43. Zhong, Z., Zheng, L., Kang, G., Li, S., Yang, Y.: Random erasing data augmentation. In: AAAI, pp. 13001–13008 (2020)
44. Zhou, J., Vong, C., Liu, Q., Wang, Z.: Scale adaptive image cropping for UAV object detection. Neurocomputing **366**, 305–313 (2019)
45. Zhou, X., Wang, D., Krähenbühl, P.: Objects as points. CoRR abs/1904.07850 (2019)
46. Zhu, P., Wen, L., Du, D., Bian, X., Hu, Q., Ling, H.: Vision meets drones: past, present and future. CoRR abs/2001.06303 (2020)
47. Zhu, P., et al.: VisDrone-DET2018: the vision meets drone object detection in image challenge results. In: Leal-Taixé, L., Roth, S. (eds.) ECCV 2018. LNCS, vol. 11133, pp. 437–468. Springer, Cham (2019). https://doi.org/10.1007/978-3-030-11021-5_27
48. Zlocha, M., Dou, Q., Glocker, B.: Improving RetinaNet for CT lesion detection with dense masks from weak RECIST labels. In: Shen, D., et al. (eds.) MICCAI 2019. LNCS, vol. 11769, pp. 402–410. Springer, Cham (2019). https://doi.org/10.1007/978-3-030-32226-7_45
49. Zoph, B., Cubuk, E.D., Ghiasi, G., Lin, T., Shlens, J., Le, Q.V.: Learning data augmentation strategies for object detection. CoRR abs/1906.11172 (2019)

VisDrone-MOT2020: The Vision Meets Drone Multiple Object Tracking Challenge Results

Heng Fan[1], Dawei Du[2], Longyin Wen[3], Pengfei Zhu[4(✉)], Qinghua Hu[4], Haibin Ling[1], Mubarak Shah[5], Junwen Pan[4], Arne Schumann[10], Bin Dong[7], Daniel Stadler[8], Duo Xu[12], Filiz Bunyak[17], Guna Seetharaman[18], Guizhong Liu[6], V. Haritha[15], P. S. Hrishikesh[15], Jie Han[6], Kannappan Palaniappan[17], Kaojin Zhu[14], Lars Wilko Sommer[9], Libo Zhang[19], Linu Shine[15], Min Yao[19], Noor M. Al-Shakarji[16,17], Shengwen Li[13], Ting Sun[6], Wang Sai[7], Wentao Yu[6], Xi Wu[12], Xiaopeng Hong[6], Xing Wei[6], Xingjie Zhao[6], Yanyun Zhao[13], Yihong Gong[6], Yuehan Yao[7], Yuhang He[6], Zhaoze Zhao[11], Zhen Xie[12], Zheng Yang[14], Zhenyu Xu[7], Zhipeng Luo[7], and Zhizhao Duan[12]

[1] Stony Brook University, New York, NY, USA
[2] Kitware, Inc., Clifton Park, NY, USA
[3] JD Finance America Corporation, Mountain View, CA, USA
[4] Tianjin University, Tianjin, China
zhupengfei@tju.edu.cn
[5] University of Central Florida, Orlando, FL, USA
[6] Xi'an Jiaotong University, Xi'an, China
[7] DeepBlue Technology (Shanghai), Shanghai, China
[8] Karlsruhe Institute of Technology, Karlsruhe, Germany
[9] Fraunhofer IOSB, Karlsruhe, Germany
[10] Fraunhofer Center for Machine Learning, Karlsruhe, Germany
[11] Southwestern University of Finance and Economics, Chengdu, China
[12] Zhejiang University, Hangzhou, China
[13] Beijing University of Posts and Telecommunications, Beijing, China
[14] Xidian University, Xi'an, China
[15] College of Engineering Trivandrum, Thiruvananthapuram, India
[16] University of Technology, Baghdad, Iraq
[17] University of Missouri-Columbia, Columbia, MO, USA
[18] U.S. Naval Research Laboratory, Washington, DC, USA
[19] Institute of Software, Chinese Academy of Sciences, Beijing, China

Abstract. The Vision Meets Drone (VisDrone2020) Multiple Object Tracking (MOT) is the third annual UAV MOT tracking evaluation activity organized by the VisDrone team, in conjunction with European Conference on Computer Vision (ECCV 2020). The VisDrone-MOT2020 consists of 79 challenging video sequences, including 56 videos (∼24K frames) for training, 7 videos (∼3K frames) for validation and 17 videos (∼6K frames) for evaluation. All frames in these sequences are manually annotated with high-quality bounding boxes. Results of 12 participating MOT algorithms are presented and analyzed in detail. The challenging

A. Bartoli and A. Fusiello (Eds.): ECCV 2020 Workshops, LNCS 12538, pp. 713–727, 2020.
https://doi.org/10.1007/978-3-030-66823-5_43

results, video sequences as well as the evaluation toolkit are made available at http://aiskyeye.com/. By holding VisDrone-MOT2020 challenge, we hope to facilitate future research and applications of MOT algorithms on drone videos.

Keywords: Drone-based multiple object tracking · Drone · Performance evaluation

1 Introduction

The goal of multiple object tracking (MOT) is to simultaneously determine the identities of multiple moving target objects and estimate their trajectories in a video sequence. MOT is one of the most important components for video understanding in computer vision and has a long list of applications such as video surveillance, human-machine interaction, and robotics.

In order to boost the development of MOT, benchmarks have played a crucial role in developing and evaluating MOT algorithms. To this end, many MOT benchmarks have been proposed in recent years, such as UA-DETRAC [42], MOT challenge [13,32], TAO [12], and KITTI [17]. Nevertheless, these MOT benchmarks focus on general scenes for person or vehicle tracking. Recently, drone based algorithms and algorithms have drawn extensive attention owing to the flexibility of drone platform. Therefore, there are several attempts to constructing drone tracking datasets [14,20,33,38]. However, these drone benchmarks are often limited in size and do not aim at multiple object tracking. Based on these analysis, a large-scale MOT benchmark based on drone platform is still desired.

As mentioned by [42], a typical MOT algorithm consists of two components including an object detection component and a multi-object tracking component. In this challenge, unlike traditional MOT benchmark [32], we argue that it is more reasonable to assess the whole MOT system without using common prior detection results as inputs. For this purpose, we organize the challenge workshop "Vision Meets Drone Multiple Object Tracking (VisDrone-MOT2020)", in conjunction with European Conference on Computer Vision (ECCV 2020), followed by previous successful editions VisDrone-MOT2019 [45] (in conjunction with ICCV 2019) and VisDrone-VDT2018 [59] (in conjunction with ECCV 2018). We present 12 MOT algorithms with in-depth analysis and discussion. It is worth noting that many of these algorithms are modified based on existing state-of-the-art detection methods or multi-object tracking approaches. The results and videos of the VisDrone-MOT2020 challenge are available at http://aiskyeye.com/.

2 Related Work

2.1 Multi-Object Tracking

One of the most common strategy in MOT is tracking-by-detection. The approach of [44] introduces a new data association technique based on hierarchical relation hypergraph. The main idea is to formulate the MOT problem as a dense neighborhoods searching task on the dynamic affinity graph. The method in [22] proposes to leverage long-short term memory (LSTM) to incorporate temporal information of different targets into appearance modeling, which effectively improves the performance. In order to hand the noisy in detection results, the method of [57] utilizes a single object tracker and apply the tracking result for data association. The approach of [39] proposes to employ both spatial and temporal information for appearance modeling for improvement. The method of [21] takes motion information into consideration and proposes to combine low- and high-level cues for multi-object tracking. The approach in [43] proposes to learn a non-unified hypergraph for multi-object tracking. To fully explore powerful deep feature representation, the algorithm of [11] proposes an end-to-end architecture for feature learning, affinity estimation and multi-dimensional assignment. The method of [10] introduces an instance-aware tracker to combine single object tracking methods for MOT by encoding awareness both within and between target models, significantly boosting the performance. The approach of [5] learns a neural solver based on message passing networks for multi-object tracking in an end-to-end fashion. The approach of [3] exploits the bounding box regression in detection and apply it to improve the performance of multi-object tracking.

2.2 Similarity Learning in Person Re-Identification

The goal of person re-identification (Re-ID) is to recognize the person of interest from a set of gallery images. It is often adopted to deal with the data associate problem in MOT. The approach of [50] proposes an unsupervised strategy to learn multi-level descriptors from pixel-level, patch and image levels for person re-identification. The method in [27] introduce a filter pairing neural network to handle various misalignment problem in person re-identification. The work of [53] leverages salient cues from human body for recognition. The method in [55] proposes a graph learning method to deal with misalignment in person re-identification using local structures. The approach of [52] proposes to utilize deep learning method to learn an aligned representation for person re-identification. Further, this method is extend by [7] through incorporating bilinear coding to improve the robustness of feature representation. The method of [8] proposes to exploit high-order attention for person re-identification.

In addition to image based person re-identification, video based person re-identification has also drawn increasing interest. The work of [24] explores both global and local temporal information for video person re-identification. The approach of [25] utilizes 3D convolution network for video person re-identification. The work of [26] combines both motion and appearance information to improve video person re-identification.

2.3 Tracking Benchmark

Multi-object tracking is one of the most important problems in computer vision. In order to advance the research of MOT, many benchmarks have been proposed in recent years to evaluation different MOT approaches. Different from single object tracking benchmarks [16,47], constructing multi-object tracking benchmarks is more challenging as the number of targets in each frame are much larger than one. The KITTI benchmark [17] focuses on multi-object tracking in traffic scenes. MOT challenge 2016 [32] is one of the most popular MOT benchmarks for pedestrian tracking. Later, this challenge is extended to MOT challenge 2020 [13] by introducing more video sequences. The UA-DETRAC [42] proposes a MOT benchmark in various traffic sceneries and new protocols for evaluation. Considering the requirement of large-scale dataset in deep learning era, the recently proposed TAO [12] contributes a large set of videos for multi-object tracking. Especially, this dataset provides both training and evaluation videos for MOT. Despite the above MOT benchmarks, there is a lack of drone based MOT benchmark, which motivates the proposal of this challenge.

3 The VisDrone-MOT2020 Challenge

3.1 The VidDrone-MOT2020 Dataset

Similar to VisDrone-MOT2019 [45], VisDrone-MOT2020 uses the same video sequences for a fair comparison. In specific, VisDrone-MOT2020 consists of 79 video clips with around 70,000 frames in total. We divide VisDrone-MOT2020 into three subsets, including training set (56 video clips with around 24,000 frames), validation set (7 video clips with around 3,000 frames) and testing set (16 video clips with around 6,000 frames). Each frame in these videos are manually labeled in high quality. Similar to VisDrone-MOT2019 [45], we focus on five selected target classes in this challenge, including *pedestrian*, *car*, *van*, *bus* and *truck*.

As discussed early, VisDrone-MOT2020 does not provide the common detection results as inputs to the trackers, we encourage participants to use their own detectors to offer detection results. In this way, we are able to evaluate the complete multi-object tracking system more reliably. Following VisDrone-MOT2019 [45], we utilize the same evaluation protocol in [35] to evaluate the performance of submitted trackers. Specifically, each submitted tracker needs to generate a list of (axis-aligned) bounding box with confidence scores and the corresponding identities. Then the tracklets, which are formed by the bounding box detection results with the same identity, are sorted based on the average confidence scores over the detection results. Each tracklet is measured by its intersection over union (IoU) overlap with the groundtruth tracklet. If the IoU is larger than a pre-defined threshold (*i.e.*, 0.25, 0.50 and 0.75), the tracklet is correct. Finally, all participating algorithms are ranked by averaging the mean average precision (mAP) per object class over different thresholds. For more details, please refer to [35].

Table 1. The summary of the submitted MOT algorithms in the VisDrone-MOT2020 Challenge. GPUs for training, implementations (Python or Matlab), framework, pre-trained datasets ('V' indicates VisDrone-MOT2020 and 'C' indicates COCO [29]).

Method	GPU	Implementation	Framework	Pre-trained
COFE	TITAN Xp	Python	Cascade R-CNN [6]+OSNet [54]	V,C
SOMOT	Tesla V100	Python	Cascade R-CNN [6]	V,C
PAS tracker	Tesla V100	Python	Cascade R-CNN [6]+FPN [28]	V,C
Deepsort	n/a	Python	Fast R-CNN [18]+SORT [46]	C
YOLO-TRAC	Tesla P100	Python	YOLO-V5+V-IOU [4]	V,C
VDCT	GTX 1080Ti	Python	CenterTrack [56]+OSNet [54]	V,C
Cascade RCNN+IOU	n/a	Python	Cascade R-CNN [6]+IOU [4]	V,C
HTC+IOU	n/a	Python	HTC [9]+IOU [4]	V,C
HR-GNN	GTX 1080Ti	Python	HRNet [41]	V
TNT	Colab	Python	TNT [39]	V
anchor-free_mot	RTX 2080	Python	FairMOT [51]	V
SCTrack	GTX 960M	Matlab	Faster R-CNN [37]+YOLOv3 [36]	C

3.2 Submitted Trackers

In VisDrone-MOT2020 challenge, we receive 12 different multi-object tracking algorithms with detailed description, as described in Appendix A. Many of them are modified based on existing state-of-the-art detection models such as Cascade R-CNN [6], FPN [28] and HTC [9] and MOT methods such as SORT [46] and FairMOT [51]. Moreover, Re-ID techniques [31, 40, 54] are introduced to improve the accuracy of detection association.

All of these participating trackers are based on the tracking-by-detection framework. In specific, the submissions COFE (A.1), SOMOT (A.2), PAS tracker (A.3) and Cascade RCNN+IOU (A.7) apply the Cascade R-CNN detector [6] to obtain detection inputs. COFE (A.1) proposes a coarse-to-fine strategy to refine the tracking results in various kinds of vehicles (*e.g.*, van, bus, and car). SOMOT (A.2) relies on the embedding model by Multiple Granularity Network [40] to deal with detection association. PAS tracker (A.3) builds the similarity measure that integrates position, appearance and size information of objects, where the appearance of an object is represented by a feature vector computed with a re-identification model from [31]. The submissions Deepsort (A.4) and VDCT (A.6) are built based on the state-of-the-art MOT methods, *i.e.*, SORT [46] and CenterTrack [56]. The submission YOLO-TRAC (A.5) leverages the latest YOLO-v5 detector for multi-object tracking. The submission HTC+IOU (A.8) uses instance segmentation technique for MOT. The submission HR-GNN (A.9) employs stronger backbone to extract features and apply graph neural network for MOT. The submission TNT (A.10) combines temporal and appearance information together to develop a unified MOT framework. The submission anchor-free_mot (A.11) introduces anchor-free detector for multi-object tracking [51]. The submission SCTrack (A.12) develops a cascade architecture for multi-object tracking by exploiting various information. The summary of these trackers is shown in Table 1.

Table 2. Multi-object tracking results on the VisDrone-MOT2020 test-challenge set. The best three performers are highlighted by the red, green and **blue** fonts, respectively.

Method	AP	AP@0.25	AP@0.50	AP@0.75	AP_{car}	AP_{bus}	AP_{trk}	AP_{ped}	AP_{van}
COFE (A.1)	61.88	[illegible]	62.00	58.65	79.09	65.26	50.91	56.87	57.26
SOMOT (A.2)	57.65	70.06	60.13	42.75	68.52	62.10	47.08	[illegible]	[illegible]
PAS tracker (A.3)	**50.80**	**62.24**	**50.74**	**39.43**	**62.59**	**50.59**	**42.18**	**44.34**	**54.30**
Deepsort (A.4)	42.11	58.82	42.64	24.86	55.06	43.18	41.30	29.10	41.88
YOLO-TRAC (A.5)	42.10	52.94	41.86	31.49	52.81	48.98	39.17	28.92	40.59
VDCT (A.6)	35.76	45.86	35.46	25.96	56.94	24.62	28.16	34.00	35.06
Cascade RCNN+IOU (A.7)	27.23	36.14	28.25	17.31	49.56	16.27	30.18	10.78	29.36
HTC+IOU (A.8)	26.46	34.39	27.43	17.57	51.18	19.05	21.55	10.77	29.76
HR-GNN (A.9)	19.54	26.52	19.67	12.42	37.72	15.48	9.98	18.87	15.65
TNT (A.10)	6.55	10.93	7.00	1.70	1.88	19.51	2.07	1.96	7.32
anchor-free_mot (A.11)	4.88	9.73	3.38	1.53	10.69	2.04	1.51	5.89	4.26
SCTrack (A.12)	3.01	5.01	2.77	1.24	8.99	1.21	2.16	1.68	0.98

4 Results and Analysis

4.1 Overall Performance

The results of the submitted 12 multi-object tracking algorithms are presented in Table 2. COEF (A.1), SOMOT (A.2) and PAS tracker (A.3) achieve the top 3 AP score among all submissions, respectively. In specific, COFE obtains the AP score of 61.88, SOMOT AP score of 57.65 and PAS Tracker AP score of 50.80. In addition, under two out of three thresholds, COFE achieves the best performance on AP@0.50 and AP@0.75 with 62.00 and 58.65 scores. Moreover, it shows the best results for all five target classes. Further, we have observed that all these three solutions adopt the state-of-the-art detector Cascade R-CNN [6] to obtain inputs. We argue that the reason is that the targets in VisDrone-MOT2020 have various scales. The cascade strategy in Cascade R-CNN is essential for detecting these targets on drone-captured scenes. Moreover, the top 3 performers introduce the re-identification model in the MOT framework, resulting in considerable improvement. Following the top 3 MOT methods, Deepsort (A.4) and YOLO-TRAC (A.5) are typical tracking-by-detection frameworks without re-identification, achieving the AP score of 42.

It is worth noting that, compared with the solutions in VisDrone-MOT2019 [45], the submitted multi-object trackers in VisDrone-MOT2020 performs much better. In specific, the top three trackers in VisDrone-MOT2019 [45] achieve the AP scores of 43.94, 39.19 and 30.87, respectively. In comparison, the top three trackers COEF (A.1), SOMOT (A.2) and PAS tracker (A.3) respectively obtain the AP scores of 61.88, 57.65 and 50.80, significantly improving the performance. A potential reason accounting for this is that the submitted trackers in this challenge employ more powerful detection models and multi-object tracking components, leading to better performance.

4.2 Performance Analysis by Categories

In order to further analyze different tracking algorithms, we report the AP score for each target classes in VisDrone-MOT2020, as shown in Table 2. From Table 2, we observe that COFE (A.1) consistently achieves the best performance under all target categories. In specific, COFE achieves AP scores of 79.09, 65.26, 50.91, 56.87 and 57.26 for categories *car*, *bus*, *truck*, *pedestrian* and *van*. SOMOT (A.2) achieves the second best performance of AP scores 68.52, 62.10, 47.98, 54.94 and 54.69 for these five classes. PAS tracker (A.3) obtains the third best results of 62.59, 50.59, 42.18, 44.34 and 54.30 for five categories.

5 Discussion

It is challenging to develop a robust MOT approach on drone videos. Based on the results above, there is still large room to improvement for future research. Here we summarize some effective strategies for boosting MOT performance on drone-captured scenarios.

- **Robust detection.** Different from existing MOT benchmarks, we aim to evaluate the complete MOT system. According the top performers in this challenge, a robust detection model (*e.g.*, Cascade R-CNN [6]) with a simple detection association method (*e.g.*, Kalman Filter, Hungarian method [23] and SORT [46]) can achieve state-of-the-art performance compared with joint detection and tracking methods (*e.g.*, CenterTrack [56] and FairMOT [51]). Therefore, it is necessary to choose a robust detector to pursue better performance of multi-object tracking.
- **Motion information.** Although the current state-of-the-art MOT methods focus on modeling appearance information of the targets, motion information is also crucial for multi-object tracking, especially when targets have similar appearance or are occluded by others. In this challenge, the submission VDCT (A.6) incorporate motion information of targets and show promising results.

6 Conclusion

This paper concludes the VisDrone-MOT2020 challenge, which is the third annual UAV MOT tracking evaluation activity, in conjunction with ECCV 2020. In this challenge, we present a more challenging dataset consisting of 79 video clips with around 33K frames. 12 MOT algorithms based on existing state-of-the-art detectors or multi-object trackers are submitted and analyzed. The top three solutions are COFE (A.1), SOMOT (A.2) and PAS tracker (A.3) with the AP scores of 61.88, 57.65 and 50.80, respectively. The experiment shows the effectiveness of Cascade R-CNN detector and re-identification techniques in the tracking-by-detection MOT framework. We hope that this challenge can advance future research and applications of drone based MOT algorithms [58].

Acknowledgements. This work was supported in part by the National Natural Science Foundation of China under Grant 61876127 and Grant 61732011, in part by Natural Science Foundation of Tianjin under Grant 17JCZDJC30800.

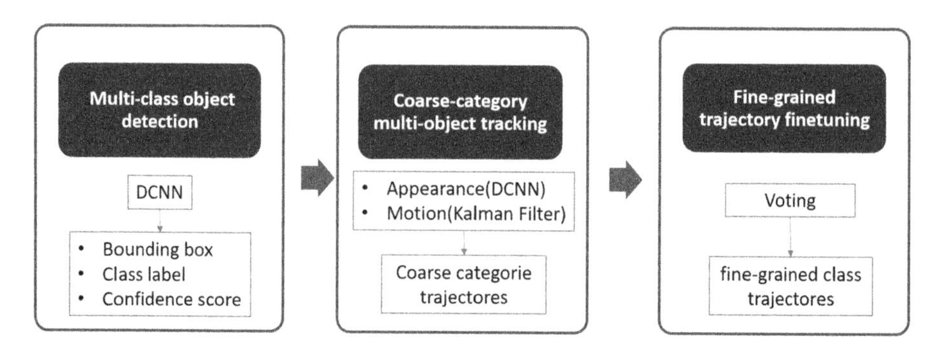

Fig. 1. The framework of COFE.

A Descriptions of Submitted Trackers

In the appendix, we summarize 12 trackers submitted in the VisDrone-MOT2020 Challenge, which are ordered according to the submissions of their final results.

A.1 Coarse-to-Fine Multi-Class Multi-Object Tracking (COFE)

Yuhang He, Wentao Yu, Jie Han, Xiaopeng Hong, Xing Wei and Yihong Gong
{hyh1379478,yu1034397129,hanjie1997}@stu.xjtu.edu.cn,
{hongxiaopeng,weixing,ygong}@mail.xjtu.edu.cn

COFE is proposed to track multiple targets in different categories under different scenarios. As shown in Fig. 1, the proposed method contains three major modules: 1) Multi-class object detection, 2) Coarse-category multi-object tracking, and 3) Fine-grained trajectory finetuning. Firstly, we use a Deep Convolutional Neural Network (DCNN) based object detector [6] to detect interested targets in the image plane, where each detection is denoted by a bounding box with a class label and a confidence score. Secondly, we track multiple targets in coarse categories, where fine-grained classes (such as van, bus, car) are summarized into coarse categories (*e.g.*, vehicle). For each coarse category, we perform multi-object tracking by exploiting the appearance and motion information of targets, where the appearance feature is extracted using a DCNN feature extractor [54] and the motion pattern of each target is modeled by a Kalman Filter. Finally, for each obtained trajectory, we finetune its fine-grained class label by a simple voting and refine the tracking results by post processing (*i.e.*, bounding box smoothing).

A.2 Simple Online Multi-Object Tracker (SOMOT)

Zhipeng Luo, Yuehan Yao, Zhenyu Xu, Bin Dong and Wang Sai
{luozp,yaoyh,xuzy,dongb,wangs}@deepblueai.com

Following separate detection and embedding model, we build a strong detector based on Cascade R-CNN [6] and a embedding model based on Multiple Granularity Network (MGN). For association step, we build simple online multi-object tracker, which is inspired by DeepSORT [46] and FairMOT [51]. For detector, Cascade R-CNN [6] pretrained on COCO [29] is applied. For embedding model, bag of tricks are used to improve the performance of MGN [40]. For association step, we initialize a number of tracklets based on the estimated boxes in the first frame. In the subsequent frames, we associate the boxes to the existing tracklets (all activated tracklets) according to their distances measured by embedding features. We update the appearance features of the trackers in each time step to handle appearance variations. Then, unmatched activated tracklets and estimated boxes are associated by their distance of Intersection over Union (IoU). Also, inactivated tracklets and estimated boxes are associated by their distance of IoU.

A.3 Position-, Appearance- and Size-aware Tracker (PAS tracker)

Daniel Stadler, Lars Wilko Sommer and Arne Schumann
daniel.stadler@kit.edu,{lars.sommer,arne.schumann}@iosb.fraunhofer.de

The PAS algorithm follows the tracking-by-detection paradigm. As detectors, we train two Cascade R-CNN [6] with FPN [28] on the VisDrone2020 MOT train and val set applying as backbone ResNeXt-101 [49] and HRNetV2p-W32 [41], respectively. Training is performed on randomly sampled image crops (608×608 pixels) and the SSD [30] data augmentation pipeline is used. To improve the quality of the detections, we utilize test-time strategies like horizontal flipping and multi-scale testing. Additionally, we generate category-specific expert models using weights from different epochs and from the two detectors with different backbones. For associating detections, we build a similarity measure that integrates position, appearance and size information of objects. A constant velocity model is assumed for the motion prediction of objects and a camera motion compensation model based on the Enhanced Correlation Coefficient Maximization [15] is also applied. The appearance of an object is represented by a feature vector computed with a re-identification model from [31] based on a ResNet-50 [19]. The association of tracks and new detections is solved by the Hungarian method [23]. Additionally, to remove false positive detections in crowded scenarios, a simple filtering approach considering the overlap of existing tracks and new detections is proposed. Finally, we remove short tracks with less than 10 frames and small tracks with a mean size of less than 100 pixels as most of them are false positives.

A.4 Simple Online and Realtime Tracking with a Deep Association (Deepsort)

Zhaoze Zhao
hanjie@smail.swufe.edu.cn

Simple Online and Realtime Tracking (SORT) [46] is a pragmatic approach to multiple object tracking with a focus on simple, effective algorithms. In this paper, we integrate appearance information to improve the performance of SORT. Due to this extension we are able to track objects through longer periods of occlusions, effectively reducing the number of identity switches. In spirit of the original framework we place much of the computational complexity into an offline pre-training stage where we learn a deep association metric on a large-scale person re-identification dataset. During online application, we establish measurement-to-track associations using nearest neighbour queries in visual appearance space.

A.5 YOLOv5 based V-IOU tracker (YOLO-TRAC)

Zhizhao Duan, Xi Wu, Duo Xu and Zhen Xie
{Duanai,21725018}@zju.edu.cn,wuxi9410@gmail.com,zjutxz@hotmail.com

Trac is a track by detection framework. We use YOLO-V5[1] as our detection network, and V-IOU Tracker [4] is used for tracking.

A.6 An improved multi-object tracking method for the VisDrone videos based on CenterTrack (VDCT)

Shengwen Li and Yanyun Zhao
{2019140337,zyy}@bupt.edu.cn

VDCT is improved from CenterTrack, which is a point-based framework that combines detection and tracking [56]. Its inputs include the current frame, the previous frame, and the tracked objects in the previous frame; and it outputs the displacements of tracked objects. Our improvements include: (1) The tracked objects which do not match within 20 frames are allowed to associate with objects detected in current frame by properly extending the survival time of the tracked objects. (2) The motion direction of adjacent frame objects usually does not change abruptly due to the continuity of object motion, so we calculate the dot product of the displacements of adjacent frame objects and decide whether to associate the objects. (3) We use the NIOU method [34] to perform non-maximum suppression on vehicle objects. (4) We adopt the hierarchical matching strategy in DeepSORT [46] to solve the long occlusion problem. (5) OSNet [54] is used to extract each trajectory's appearance feature, measure

[1] https://github.com/ultralytics/yolov5.

their distance from others and we simply merge two trajectories if their distance is close enough. The experimental results show the effectiveness of our improved method.

A.7 Cascade RCNN based IOU tracker (Cascade RCNN+IOU)

Ting Sun and Xingjie Zhao
sunting9999@stu.xjtu.edu.cn,1243273854@qq.com

We use Cascade R-CNN [6] as the detector with three improvements: (1) We use Group normalization [48] instead of Batch normalization; (2) We use online hard example mining to select positive and negative samples; (3) We use multiple scales to test our data; (4) We use two stronger backbones to train models and integrate them. Then, we perform detection association using the IOU tracker [4].

A.8 Hybrid task cascade based IOU tracker (HTC+IOU)

Ting Sun, Xingjie Zhao and Guizhong Liu
sunting9999@stu.xjtu.edu.cn,1243273854@qq.com

We use hybrid task cascade for instance segmentation [9] as the detector with three improvements: (1) We use Group normalization [48] instead of Batch normalization; (2) We use online hard example mining to select positive and negative samples; (3) We use multiple scales to test our data; (4) We use two stronger backbones to train models and integrate them. Then, we perform detection association using the IOU tracker [4].

A.9 Multi-object Tracking based on HRNet (HR-GNN)

Zheng Yang and Kaojin Zhu
151776257@qq.com,1320531351@qq.com

HR-GNN is built based on the detector using HRNet [41] as backbone. Then the tracking results are generated by using GNN to analyze the detection results.

A.10 Multi-object tracking with TrackletNet (TNT)

Haritha V, Melvin Kuriakose, Hrishikesh PS and Linu Shine
vakkatharitha@gmail.com

TNT is based on the work of [39] by merging temporal and appearance information together as a unified framework. We learn appearance similarity among tracklets by a graph model, where we use CNN features and intersection-over-union (IOU) with epipolar constraints to compensate camera movement between adjacent frames. Finally, the tracklets can be clustered into groups, resulting in trajectories with individual object IDs.

A.11 A simple baseline for one-shot multi-object tracking (anchor-free_mot)

Min Yao and Libo Zhang
libo@iscas.ac.cn

The anchor-free_mot method is based on FairMOT [51]. Specifically, we use the encoder-decoder network to extract feature maps. Then, two simple parallel heads are used to predict the bounding box and re-ID features of the targets, respectively. Notably, the targets are represented by points from the anchor-free object detection method.

A.12 Semantic Color Correlation Tracker (SCTrack)

Noor M. Al-Shakarji, Filiz Bunyak, Guna Seetharaman and Kannappan Palaniappan
{nmahyd,bunyak,palaniappank}@mail.missouri.edu, gunasekaran.seetharaman@rl.af.mil

SCTrack is a time-efficient detection-based multi-object tracking method. Specifically, we use a three-step cascaded data association scheme to combine a fast spatial distance only short-term data association, a robust tracklet linking step using discriminative object appearance models, and an explicit occlusion handling unit relying not only on tracked objects' motion patterns but also on environmental constraints such as presence of potential occluders in the scene. The details can be referred to [1,2].

References

1. Al-Shakarji, N.M., Bunyak, F., Seetharaman, G., Palaniappan, K.: Multi-object tracking cascade with multi-step data association and occlusion handling. In: AVSS (2018)
2. Al-Shakarji, N.M., Seetharaman, G., Bunyak, F., Palaniappan, K.: Robust multi-object tracking with semantic color correlation. In: AVSS (2017)
3. Bergmann, P., Meinhardt, T., Leal-Taixe, L.: Tracking without bells and whistles. In: ICCV (2019)
4. Bochinski, E., Eiselein, V., Sikora, T.: High-speed tracking-by-detection without using image information. In: AVSS (2017)
5. Brasó, G., Leal-Taixé, L.: Learning a neural solver for multiple object tracking. In: CVPR (2020)
6. Cai, Z., Vasconcelos, N.: Cascade R-CNN: delving into high quality object detection. In: CVPR (2018)
7. Chang, Z., et al.: Weighted bilinear coding over salient body parts for person re-identification. Neurocomputing **407**, 454–464 (2020)
8. Chen, B., Deng, W., Hu, J.: Mixed high-order attention network for person re-identification. In: ICCV (2019)
9. Chen, K., et al.: Hybrid task cascade for instance segmentation. In: CVPR (2019)

10. Chu, P., Fan, H., Tan, C.C., Ling, H.: Online multi-object tracking with instance-aware tracker and dynamic model refreshment. In: WACV (2019)
11. Chu, P., Ling, H.: FAMNet: joint learning of feature, affinity and multi-dimensional assignment for online multiple object tracking. In: ICCV (2019)
12. Dave, A., Khurana, T., Tokmakov, P., Schmid, C., Ramanan, D.: TAO: a large-scale benchmark for tracking any object. arXiv (2020)
13. Dendorfer, P., et al.: MOT20: a benchmark for multi object tracking in crowded scenes. arXiv (2020)
14. Du, D., et al.: The unmanned aerial vehicle benchmark: object detection and tracking. In: Ferrari, V., Hebert, M., Sminchisescu, C., Weiss, Y. (eds.) ECCV 2018. LNCS, vol. 11214, pp. 375–391. Springer, Cham (2018). https://doi.org/10.1007/978-3-030-01249-6_23
15. Evangelidis, G.D., Psarakis, E.Z.: Parametric image alignment using enhanced correlation coefficient maximization. PAMI **30**(10), 1858–1865 (2008)
16. Fan, H., et al.: LaSOT: a high-quality benchmark for large-scale single object tracking. In: CVPR (2019)
17. Geiger, A., Lenz, P., Stiller, C., Urtasun, R.: Vision meets robotics: the KITTI dataset. Int. J. Robot. Res. **32**(11), 1231–1237 (2013)
18. Girshick, R.: Fast R-CNN. In: ICCV (2015)
19. He, K., Zhang, X., Ren, S., Sun, J.: Deep residual learning for image recognition. In: CVPR (2016)
20. Hsieh, M.R., Lin, Y.L., Hsu, W.H.: Drone-based object counting by spatially regularized regional proposal network. In: ICCV (2017)
21. Keuper, M., Tang, S., Andres, B., Brox, T., Schiele, B.: Motion segmentation & multiple object tracking by correlation co-clustering. PAMI **42**(1), 140–153 (2018)
22. Kim, C., Li, F., Rehg, J.M.: Multi-object tracking with neural gating using bilinear LSTM. In: Ferrari, V., Hebert, M., Sminchisescu, C., Weiss, Y. (eds.) ECCV 2018. LNCS, vol. 11212, pp. 208–224. Springer, Cham (2018). https://doi.org/10.1007/978-3-030-01237-3_13
23. Kuhn, H.W.: The Hungarian method for the assignment problem. Naval Res. Logist. Q. **2**(1–2), 83–97 (1955)
24. Li, J., Wang, J., Tian, Q., Gao, W., Zhang, S.: Global-local temporal representations for video person re-identification. In: ICCV (2019)
25. Li, J., Zhang, S., Huang, T.: Multi-scale 3D convolution network for video based person re-identification. In: AAAI (2019)
26. Li, S., Yu, H., Hu, H.: Appearance and motion enhancement for video-based person re-identification. In: AAAI (2020)
27. Li, W., Zhao, R., Xiao, T., Wang, X.: DeepReID: deep filter pairing neural network for person re-identification. In: CVPR (2014)
28. Lin, T.Y., Dollár, P., Girshick, R., He, K., Hariharan, B., Belongie, S.: Feature pyramid networks for object detection. In: CVPR (2017)
29. Lin, T.Y., et al.: Microsoft COCO: common objects in context. In: Fleet, D., Pajdla, T., Schiele, B., Tuytelaars, T. (eds.) ECCV 2014. LNCS, vol. 8693, pp. 740–755. Springer, Cham (2014). https://doi.org/10.1007/978-3-319-10602-1_48
30. Liu, W., et al.: SSD: single shot multibox detector. In: Leibe, B., Matas, J., Sebe, N., Welling, M. (eds.) ECCV 2016. LNCS, vol. 9905, pp. 21–37. Springer, Cham (2016). https://doi.org/10.1007/978-3-319-46448-0_2
31. Luo, H., Gu, Y., Liao, X., Lai, S., Jiang, W.: Bag of tricks and a strong baseline for deep person re-identification. In: CVPRW (2019)
32. Milan, A., Leal-Taixé, L., Reid, I., Roth, S., Schindler, K.: MOT16: a benchmark for multi-object tracking. arXiv (2016)

33. Mueller, M., Smith, N., Ghanem, B.: A benchmark and simulator for UAV tracking. In: Leibe, B., Matas, J., Sebe, N., Welling, M. (eds.) ECCV 2016. LNCS, vol. 9905, pp. 445–461. Springer, Cham (2016). https://doi.org/10.1007/978-3-319-46448-0_27
34. Pan, S., Tong, Z., Zhao, Y., Zhao, Z., Su, F., Zhuang, B.: Multi-object tracking hierarchically in visual data taken from drones. In: ICCVW (2019)
35. Park, E., Liu, W., Russakovsky, O., Deng, J., Li, F.F., Berg, A.: Large Scale Visual Recognition Challenge 2017. http://image-net.org/challenges/LSVRC/2017
36. Redmon, J., Farhadi, A.: YOLOv3: an incremental improvement. arXiv (2018)
37. Ren, S., He, K., Girshick, R., Sun, J.: Faster R-CNN: towards real-time object detection with region proposal networks. In: NIPS (2015)
38. Robicquet, A., Sadeghian, A., Alahi, A., Savarese, S.: Learning social etiquette: human trajectory understanding in crowded scenes. In: Leibe, B., Matas, J., Sebe, N., Welling, M. (eds.) ECCV 2016. LNCS, vol. 9912, pp. 549–565. Springer, Cham (2016). https://doi.org/10.1007/978-3-319-46484-8_33
39. Wang, G., Wang, Y., Zhang, H., Gu, R., Hwang, J.: Exploit the connectivity: multi-object tracking with trackletnet. In: ACM MM, pp. 482–490 (2019)
40. Wang, G., Yuan, Y., Chen, X., Li, J., Zhou, X.: Learning discriminative features with multiple granularities for person re-identification. In: ACM MM (2018)
41. Wang, J., et al.: Deep high-resolution representation learning for visual recognition. PAMI (2020)
42. Wen, L., et al.: UA-DETRAC: a new benchmark and protocol for multi-object detection and tracking. Comput. Vis. Image Underst. **193**, 102907 (2020)
43. Wen, L., Du, D., Li, S., Bian, X., Lyu, S.: Learning non-uniform hypergraph for multi-object tracking. In: AAAI, pp. 8981–8988 (2019)
44. Wen, L., Li, W., Yan, J., Lei, Z., Yi, D., Li, S.Z.: Multiple target tracking based on undirected hierarchical relation hypergraph. In: CVPR (2014)
45. Wen, L., Zhang, Y., Bo, L., Shi, H., Zhu, R., et al.: VisDrone-MOT2019: the vision meets drone multiple object tracking challenge results. In: ICCVW, pp. 189–198 (2019)
46. Wojke, N., Bewley, A., Paulus, D.: Simple online and realtime tracking with a deep association metric. In: ICIP (2017)
47. Wu, Y., Lim, J., Yang, M.H.: Online object tracking: a benchmark. In: CVPR (2013)
48. Wu, Y., He, K.: Group normalization. In: Ferrari, V., Hebert, M., Sminchisescu, C., Weiss, Y. (eds.) ECCV 2018. LNCS, vol. 11217, pp. 3–19. Springer, Cham (2018). https://doi.org/10.1007/978-3-030-01261-8_1
49. Xie, S., Girshick, R., Dollár, P., Tu, Z., He, K.: Aggregated residual transformations for deep neural networks. In: CVPR (2017)
50. Yang, Y., Wen, L., Lyu, S., Li, S.Z.: Unsupervised learning of multi-level descriptors for person re-identification. In: AAAI (2017)
51. Zhan, Y., Wang, C., Wang, X., Zeng, W., Liu, W.: A simple baseline for multi-object tracking. arXiv (2020)
52. Zhao, L., Li, X., Zhuang, Y., Wang, J.: Deeply-learned part-aligned representations for person re-identification. In: ICCV (2017)
53. Zhao, R., Ouyang, W., Wang, X.: Unsupervised salience learning for person re-identification. In: CVPR (2013)
54. Zhou, K., Yang, Y., Cavallaro, A., Xiang, T.: Omni-scale feature learning for person re-identification. In: ICCV (2019)
55. Zhou, Q., et al.: Graph correspondence transfer for person re-identification. In: AAAI (2018)

56. Zhou, X., Koltun, V., Krähenbühl, P.: Tracking objects as points. arXiv (2020)
57. Zhu, J., Yang, H., Liu, N., Kim, M., Zhang, W., Yang, M.-H.: Online multi-object tracking with dual matching attention networks. In: Ferrari, V., Hebert, M., Sminchisescu, C., Weiss, Y. (eds.) ECCV 2018. LNCS, vol. 11209, pp. 379–396. Springer, Cham (2018). https://doi.org/10.1007/978-3-030-01228-1_23
58. Zhu, P., Wen, L., Du, D., Bian, X., Hu, Q., Ling, H.: Vision meets drones: past, present and future. CoRR abs/2001.06303 (2020)
59. Zhu, P., et al.: VisDrone-VDT2018: the vision meets drone video detection and tracking challenge results. In: Leal-Taixé, L., Roth, S. (eds.) ECCV 2018. LNCS, vol. 11133, pp. 496–518. Springer, Cham (2019). https://doi.org/10.1007/978-3-030-11021-5_29

VisDrone-SOT2020: The Vision Meets Drone Single Object Tracking Challenge Results

Heng Fan[1], Longyin Wen[2], Dawei Du[3], Pengfei Zhu[4](✉), Qinghua Hu[4], Haibin Ling[1], Mubarak Shah[5], Biao Wang[7], Bin Dong[15], Di Yuan[20], Dong Wang[8], Dongjie Zhou[23], Haoyang Sun[13], Hossein Ghanei-Yakhdan[11], Huchuan Lu[8], Javad Khaghani[10], Jinghao Zhou[13], Keyang Wang[24], Lei Pang[19], Lei Zhang[24], Li Cheng[10], Liting Lin[18], Lu Ding[22], Nana Fan[20], Peng Wang[13], Penghao Zhang[15], Ruiyan Ma[6], Seyed Mojtaba Marvasti-Zadeh[10], Shohreh Kasaei[12], Shuhao Chen[8], Simiao Lai[8], Tianyang Xu[17], Wentao He[23], Xiaojun Wu[16], Xin Hou[7], Xuefeng Zhu[16], Yanjie Gao[6], Yanyun Zhao[9], Yong Wang[21], Yong Xu[18], Yubo Sun[15], Yuting Yang[6], Yuxuan Li[6], Zezhou Wang[8], Zhenwei He[24], Zhenyu He[20], Zhipeng Luo[14], Zhongjian Huang[6], Zhongzhou Zhang[24], Zikai Zhang[13], and Zitong Yi[9]

[1] Stony Brook University, New York, NY, USA
[2] JD Finance America Corporation, Mountain View, CA, USA
[3] Kitware, Inc., Clifton Park, NY, USA
[4] Tianjin University, Tianjin, China
zhupengfei@tju.edu.cn
[5] University of Central Florida, Orlando, FL, USA
[6] Xidian University, Xi'an, China
[7] WeBank, Shenzhen, China
[8] Dalian University of Technology, Dalian, China
[9] Beijing University of Posts and Telecommunications, Beijing, China
[10] University of Alberta, Edmonton, Canada
[11] Yazd University, Yazd, Iran
[12] Sharif University of Technology, Tehran, Iran
[13] Northwestern Polytechnical University, Xi'an, China
[14] DeepBlue Technology (Shanghai), Shanghai, China
[15] DeepBlue Technology, Beijing, China
[16] Jiangnan University, Wuxi, China
[17] University of Surrey, Guildford, UK
[18] South China University of Technology, Guangzhou, China
[19] Chinese Academy of Sciences, Beijing, China
[20] Harbin Institute of Technology, Shenzhen, Shenzhen, China
[21] Sun Yat-Sen University, Guangzhou, China
[22] Shanghai Jiao Tong University, Shanghai, China
[23] Beijing Institute of Remote Sensing Equipment, Beijing, China
[24] Chongqing University, Chongqing, China

Abstract. The Vision Meets Drone (VisDrone2020) Single Object Tracking is the third annual UAV tracking evaluation activity organized by the VisDrone team, in conjunction with European Confer-

A. Bartoli and A. Fusiello (Eds.): ECCV 2020 Workshops, LNCS 12538, pp. 728–749, 2020.
https://doi.org/10.1007/978-3-030-66823-5_44

ence on Computer Vision (ECCV 2020). The VisDrone-SOT2020 Challenge presents and discusses the results of 13 participating algorithms in detail. By using ensemble of different trackers trained on several large-scale datasets, the top performer in VisDrone-SOT2020 achieves better results than the counterparts in VisDrone-SOT2018 and VisDrone-SOT2019. The challenging results, collected videos as well as the valuation toolkit are made available at http://aiskyeye.com/. By holding VisDrone-SOT2020 challenge, we hope to provide the community a dedicated platform for developing and evaluating drone-based tracking approaches.

Keywords: Drone-based single object tracking · Drone · Performance evaluation

1 Introduction

Visual object tracking has been one of the most fundamental topics in computer vision with a long list of applications such as video surveillance, intelligent vehicles, and robotics. In the past decades, tracking has been extensively studied and significant progress including stage-of-the-art algorithms [3,5,8,20,31,46,55] and various tracking benchmarks [2,17,26,36,45,59] has been made. However, existing tracking research mainly focuses on video captured by traditional cameras.

Recently, drones equipped with cameras have drawn increasing interest from the tracking community owing to its flexibility. In comparison to videos captured by traditional cameras, the drone videos are more difficult because of abrupt camera motion, extremely small target, frequent view point change in addition to challenges in normal videos. In order to facilitate research of drone-based tracking and its application, it is highly desired to build a dedicated platform with a large amount of drone video sequences in the community. To this end, several attempts have been made by introducing drone tracking benchmarks [14,34,43]. Nevertheless, these datasets only provide performance evaluation and are limited in scale for offering training videos, which motivates the proposal of a large-scale drone tracking benchmark for both training and evaluation of drone-based tracking algorithms.

For this purpose, the VisDrone team has compiled a dedicated large-scale drone tracking benchmark and organized challenges for single-object tracking, *i.e.*, VisDrone-SOT2018 [57] (in conjunction with ECCV 2020) and VisDrone-SOT2019 [16] (in conjunction with ICCV 2019). The VisDrone-SOT2018 consists of 132 videos and discusses 17 submitted trackers. Compared with VisDrone-SOT2018, VisDrone-SOT2019 introduces 35 new sequences and analyzes 22 participating algorithms. To further increase the diversity of drone videos and asses the performance of trackers in the wild, VisDrone-SOT2020 conducts extensive evaluation and discussion of 13 submitted tracking algorithms using the same dataset in VisDrone-SOT2019. We hope that the VisDrone-SOT2020 challenge offers an opportunity for researchers to discuss and share their ideas on drone-based tracking. All the challenge results, our benchmark and evaluation toolkit can be found at http://aiskyeye.com/.

2 Related Work

Visual object tracking has been extensively studied in recent decades. In this section, we will discuss the related tracking benchmarks and algorithms. More details refer to [67].

2.1 Visual Tracking Benchmark

Benchmarks have played an important role in advancing the research of visual tracking. OTB-2013 [58] is the first benchmark proposed for tracking. It consists of 51 manually annotated sequences with each labeled into 11 attributes. Later, the authors extend OTB-2013 to OTB-2015 [59] by introducing more videos. TC-128 [36] contributes 128 videos, aiming to analyze the impact of color information for tracking. The VOT [29] introduces a series of challenges from 2013 to 2019 with a small-scale of videos for single object tracking. ALOV [50] comprises 314 video sequences with the goal of tracking performance evaluation in various challenging situations. CDBT [39] provides 80 RGB-D videos by focusing the combination of RGB and depth information for tracking. NfS [21] consists of 100 videos with a high frame rate of 240 fps.

In the deep learning era, researchers in the tracking community have resorted to deep features for developing tracking algorithms and achieved state-of-the-art performance. However, the aforementioned datasets limited by their small scales. As a consequence, researchers have been forced to either utilize pre-trained model from ImageNet [13] or adopt video object detection dataset [47,49] for training, which may lead to sub-optimal performance due to domain gap. In order to alleviate this issue, a few large-scale tracking benchmarks have been proposed. LaSOT [17] introduces 1,400 sequences with more than 3.5M frames for both training and evaluation of tracking algorithms. Each frame is carefully and manually annotated with axis-aligned bounding box. GOT-10k [26] collects around 10,000 videos for constructing a large-scale benchmark with a focus on short-term tracking. TrackingNet provides 30K videos with each annotated using tracking algorithms. OxUvA [53] collects 366 sequences for long-term tracking and each video is sparsely labeled every 30 frames.

In recent years, drone-based object tracking has drawn extension attention in the community. To boost research on tracking in drone videos, there are several benchmarks. UAV123 [43] collects 123 drone videos with 113K frames for tracking. UAVDT [14] proposes a benchmark with 80K frames for object detection and tracking in drone videos. DTB70 [34] compiles a dataset of 70 drone sequences with 15K frames. Despite facilitating research on drone-based tracking, these benchmarks only provide performance evaluation for drone-based tracking algorithms, which is limited in offering enough videos for training of deep trackers. Tackling this problem, VisDrone-SOT2020 provides a large set of videos for both training and assessment of different trackers.

2.2 Visual Tracking Algorithm

Numerous tracking approaches have been proposed in the past decades. Among these methods, correlation filter based trackers have drawn extension attention owing to its high efficiency. The seminal work MOSSE [6] runs at an impressive speed of several hundred frames per second. KCF [25] extends this method using improved feature representation, achieving remarkable performance. Many later extensions based [6,25] have been done. To handle the problem of scale variation, the approaches of [10,35] leverage an additional scale filter to perform pyramid scale search. The methods in [11,33] propose to enhance correlation filter trackers with complex regularization techniques. Background contextual information is exploited in [22,44] to improve the robustness. Part-based strategy is used in [38] to handle the problem of occlusion for correlation filter tracking. The method in [15] incorporates structural information in local target parts variations using the global constraint in correlation filter tracking.

Inspired by the success of in various vision tasks [23,24,30], deep feature has been adopted for object tracking and demonstrated state-of-the-art results. The approaches in [12,18,41] replaced hand-crafted features with deep features in correlation filter tracking and achieved significant performance gains. To further explore the power of deep features in tracking, end-to-end deep tracking algorithms have been proposed. MDNet [46] introduces an end-to-end classification for tracking. This approach is extended by SANet [19] by exploring structure of target object. VITAL [51] improve the tracking performance by exploiting richer representation for classification. Despite excellent performance, these trackers are inefficient due to extensive deep feature extraction or network fine-tuning. To solve this problem, deep Siamese network has been proposed for tracking [3,52] and achieved a good balance between accuracy and speed. Many extensions based on Siamese tracking have been proposed. Among them, Siamese tracking with regional proposal network (RPN) [20,32,55,66,68] exhibits impressive results and fast speed owing to the effectiveness of RPN in dealing with scale changes. In addition to these methods, another trend is to combine discriminative localization model with a separate scale estimation model for tracking. ATOM [8] utilizes correlation filter method for target localization and an IoUNet [27] for scale estimation. Based on ATOM [8], DiMP [5] incorporates contextual background information for improvement.

2.3 Evaluation Metric

Performance evaluation is crucial for fair comparison of different trackers. There are many evaluation metrics proposed in recent years. The two most common evaluation metrics, precision and success plots, are introduced in [58] by measuring the distance and overlap of tracking result with groundtruth, respectively. These two metrics have been adopted by many later tracking benchmarks [17,36,45,59]. To address the scale sensitivity of precision plot to image scale, the work of [45] suggests an additional normalization precision plot. Different from these methods, the VOT challenge [29] adopts accuracy and robust-

Table 1. Comparison of VisDrone-SOT2020 with existing drone tracking benchmarks. The "Eva." and "Tra." in last column indicate evaluation and training, respectively.

Benchmark	Videos	Mean frames	Total frames	Attributes	Aim
UAV123 [43]	123	915	113K	12	Eva
UAV20L [43]	20	2,934	59K	12	Eva
DTB70 [34]	70	214	15K	12	Eva
UAVDT [14]	100	800	80K	n/a	Tra./Eva
VisDrone-SOT2018 [57]	132	803	106K	12	Tra./Eva
VisDrone-SOT2019 [16]	167	1,131	189K	12	Tra./Eva.
VisDrone-SOT2020	167	1,131	189K	12	Tra./Eva

Fig. 1. Example sequences and annotations of VisDrone-SOT2020. Best viewed in color and zooming-in. (Color figure online)

ness ranking mechanism for tracking performance evaluation. Average overlap is employeed in [26] for comparing different tracking approaches.

3 The VisDrone-SOT2020 Challenge

As discussed above, a large-scale dedicated platform for drone-based tracking is desired. Addressing this issue, VisDrone-SOT2020 introduce a drone tracking benchmark, aiming to provide both training and evaluation for drone-based tracking algorithms.

3.1 Dataset

In the VisDrone-SOT2020 Challenge, we collect 167 drone videos with more than 189K frames. Following VisDrone-SOT2019 [16], the VisDrone-SOT2020 dataset is divided into three subsets, including *training* set containing 86 videos with 70K frames, *validation* set containing 11 videos with 7K frames and *testing* set containing 95 videos with 145K frames. Compared to VisDrone-SOT2018 dataset with 132 videos and VisDrone-SOT2019 dataset with 167 videos, VisDrone-SOT2020 dataset introduce more challenging drone sequences. To reliably reflect the performance of tracking algorithms *in the wild*, the video sequences of these three subsets are captured from different scenarios with various drone platforms under varying conditions (*e.g.*, the weather and lighting conditions are different in videos of three subsets). In order to ensure the high quality of VisDrone-SOT2020 dataset, each frame is manually annotated with a bounding box. Moreover, in order to allow further performance analysis, we label each video sequence with 12 attributes, including *aspect ratio change* (ARC), *background clutter* (BC), *camera motion* (CM), *fast motion* (FM), *full occlusion* (FOC), *partial occlusion* (POC), *illumination variation* (IV), *low resolution* (LR), *out-of-view* (OV), *similar object* (SO), *scale variation* (SV) and *viewpoint change* (VC). Table 1 summarizes our VisDrone-SOT2020 dataset and comparisons with other drone tracking benchmarks. Figure 1 shows several example videos and annotations of VisDrone-SOT2020.

3.2 Evaluation Metric

Following [58], we adopt precision (PRE) and success (SUC) scores in one-pass evaluation (OPE) to assess different tracking approaches. The PRE score demonstrates the percentage of frames whose estimated average center location errors are within a given threshold distance (*e.g.*, 20 pixels) to the groundtruth. The SUC score is defined as the percentage of frames whose overlaps with groundtruth are larger than a given threshold (*i.e.*, 0.5). We refer readers to [58] for more details.

3.3 Submitted Trackers

In VisDrone-SOT2020, we summarize 13 tracking algorithms from 18 institutes, as described in the Appendix A. Many of them are improved based on recent state-of-the-art trackers from major computer vision conferences such as CVPR, ICCV and ECCV. Similar to VisDrone-SOT2018 [57] and VisDrone-SOT2019 [16], each participating algorithm can be submitted for at most 10 times. The performance is determined by the best results of ten submissions. As shown in Fig. 2, for training of tracking algorithms, one can use any existing tracking datasets (*e.g.*, LaSOT [17], TrackingNet [45], UAVDT [14], UAV123 [43], COCO [37], ImageNet DET [49], ImageNet VID, YouTube-BB [47], and GOT-10k [26]) in addition to the sequences from training set of VisDrone-SOT2020. Besides, other video object segmentation datasets are also allowed in the training phase, *e.g.*, YouTube-VOS[1] and DAVIS[2].

[1] https://youtube-vos.org/dataset/vos/.
[2] https://davischallenge.org/.

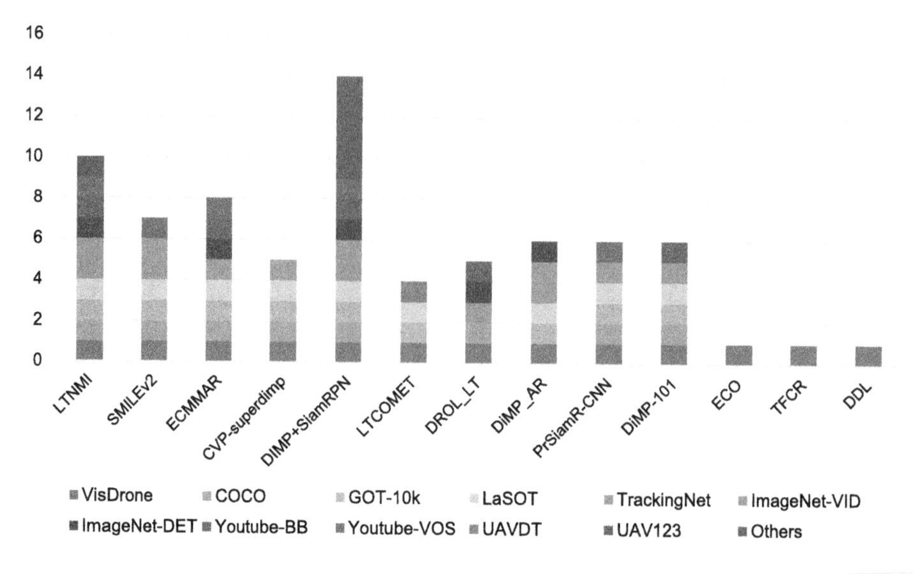

Fig. 2. The training datasets used in the VisDrone-SOT2020 Challenge.

Among the submitted trackers, 7 trackers are inspired by DiMP [5], including SMILEv2 (A.1), LTNMI (A.2), ECMMAR (A.3), CVP-superdimp (A.4), DIMP-SiamRPN (A.7), DiMP_AR (A.8) and DiMP-101 (A.10). One tracker PrSiamR-CNN (A.9) adopts a multi-stage re-detection framework for target tracking. One tracker DROL_LT (A.6) is the variation of Siamese tracking algorithm by introducing an online update strategy to adapt to appearance changes. One tracker LTCOMET (A.5) leverages contextual information for tracking. One tracker ECO (A.11) is based on discriminative correlation filter with deep features. One tracker TFCR (A.12) formulate tracking as regression problem. One tracker DDL (A.13) employs both deep feature and hand-crafted feature for tracking.

4 Results and Analysis

In this section, we evaluate all submitted tracking algorithms on the testing set of VisDrone-SOT2020 dataset and discuss several representative trackers in terms of short-term tracking, long-term tracking and different visual attributes, respectively. Finally, several potential research directions are concluded.

4.1 Overall Performance

We report the tracking results of each submitted tracking algorithm using precision and success scores, as shown in Fig. 4. As shown in Fig. 4, in term of precision score, SMILEv2 achieves the best results of 0.911. The approach of SMILEv2 is an ensemble of different tracking algorithms including three recent state-of-the-art DiMP, SiamMask and SORT. Besides, motion information obtained by

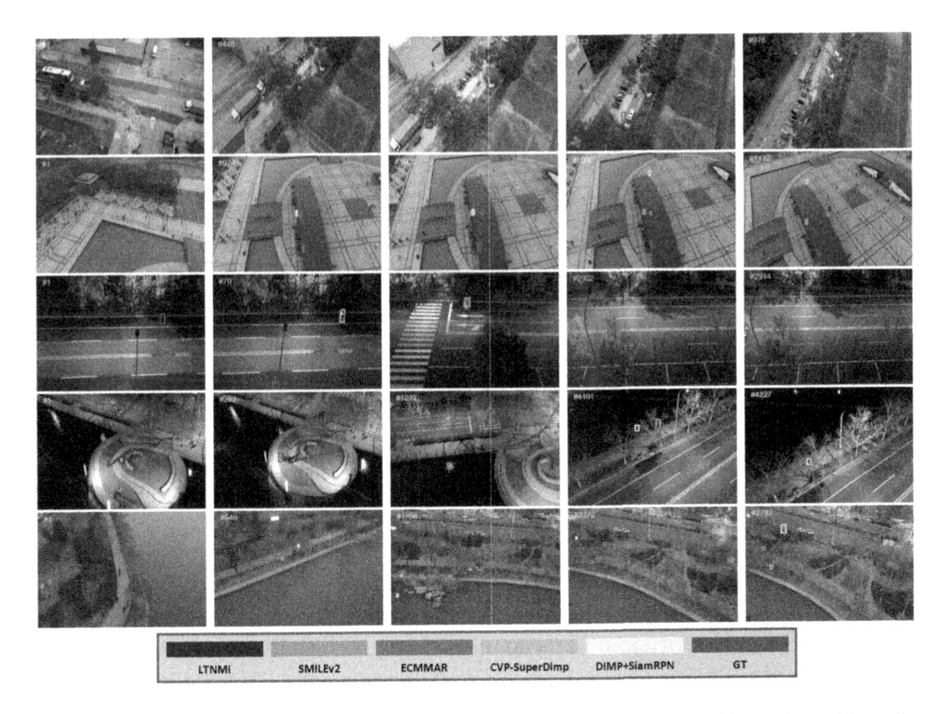

Fig. 3. Tracking results of top five trackers on VisDrone-SOT2020. Best viewed in color and zooming-in. (Color figure online)

optical flow is leveraged to improve the robustness of SMILEv2. LTNMI obtains the second best performance of 0.889 precision score. Similar to SMILEv2, the method of LTNMI utilizes several state-of-the-art tracking approaches DiMP and SiamRPN++ for tracking. In addition, it utilizes an addition detection Siam R-CNN module to re-locate the target object when it is lost. Moreover, an extra image enhancement module is used to improve the image quality for tracking. The method of ECMMAR achieves the third best result of 0.838 precision score. This method utilizes two tracking algorithms to locate the target object. In addition, to get the accurate bounding box of object, it employs a segmentation approach to refine the final tracking result. In term of success score, SMILEv2 and LTNMI both achieve the best results of 0.660. ECMMAR obtains the third best result of 0.630 success score. To better demonstrate these tracking algorithms, we show some qualitative results of the top five trackers on VisDrone-SOT2020 in Fig. 3.

It is worth noting that, compared to VisDrone-SOT2019, the performance of submitted trackers in VisDrone-SOT2020 are significantly improved. In specific, the success scores of the best three trackers in VisDrone-SOT2019 are 0.635, 0.617 and 0.594, respectively. In comparison with these trackers, the top three tracking algorithms in VisDrone-SOT2020 respectively achieve the success scores of 0.660, 0.660 and 0.630, demonstrating better performance.

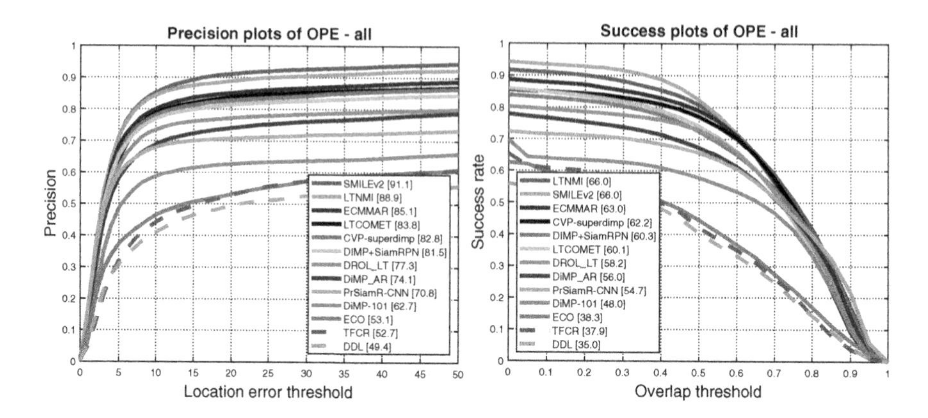

Fig. 4. The success and precision plots of the submitted trackers on all testing videos. The success and precision scores for each tracker are presented in the legend. Best viewed in color and zooming-in. (Color figure online)

The main reason accounting for this is the better baseline trackers used for VisDrone-SOT2020. For example, the most common baseline of the top trackers in VisDrone-SOT2019 is ATOM [8], while it is replaced with a stronger tracker DiMP [5] in VisDrone-SOT2020. In addition, deeper feature representation also helps improve the performance.

4.2 Short-Term Tracking

Following VisDrone-SOT2019, we further perform analysis of short-term tracking performance of different trackers. The results are demonstrated in Fig. 5. The top three trackers are LTNMI, SMILEv2 and LTCOMET. In specific, LTNMI achieves the 0.923 precision score and 0.765 success score. SMILEv2 obtains the 0.906 precision score and 0.734 success score. LTCOMET achieves the 0.889 precision score and 0.711 success score. An interesting observation is that, although SMILEv2 achieves the best overall performance, LTNMI performs more robust on short-term videos. We argue that the use of image enhancement method may help to improve tracking performance in short videos in LTNMI. In addition, LTCOMET outperforms ECMMAR on short-term videos, which shows the importance of contextual information in improving tracking performance.

4.3 Long-Term Tracking

Likewise, we conduct analysis of long-term tracking performance of each tracking algorithm. The results are shown in Fig. 6. The top three trackers are SMILEv2, ECMMAR and LTNMI, respectively. Specifically, SMILEv2 achieves 0.919 precision score and 0.555 success score. ECMMAR obtains 0.841 precision score and 0.544 success score. LTNMI achieves 0.841 precision score and 0.513 success score. We note that SMILEv2 significantly outperforms ECMMAR and LTNMI,

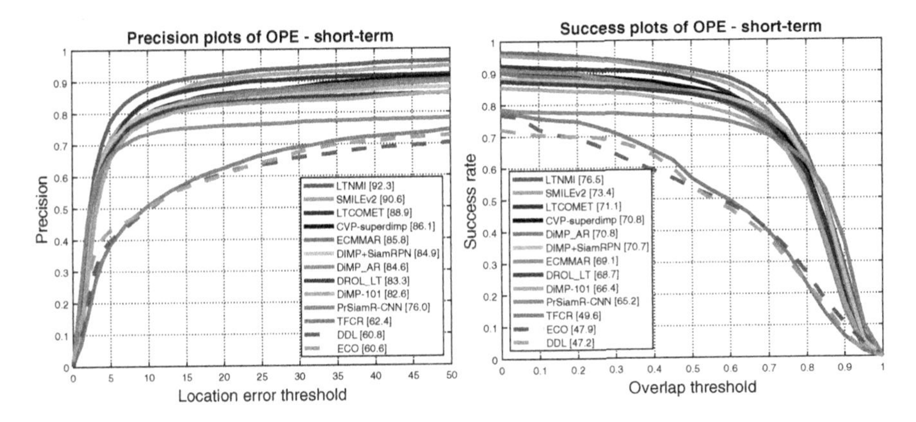

Fig. 5. The success and precision plots of the submitted trackers in term of short-term tracking. The success and precision scores for each tracker are presented in the legend. Best viewed in color and zooming-in. (Color figure online)

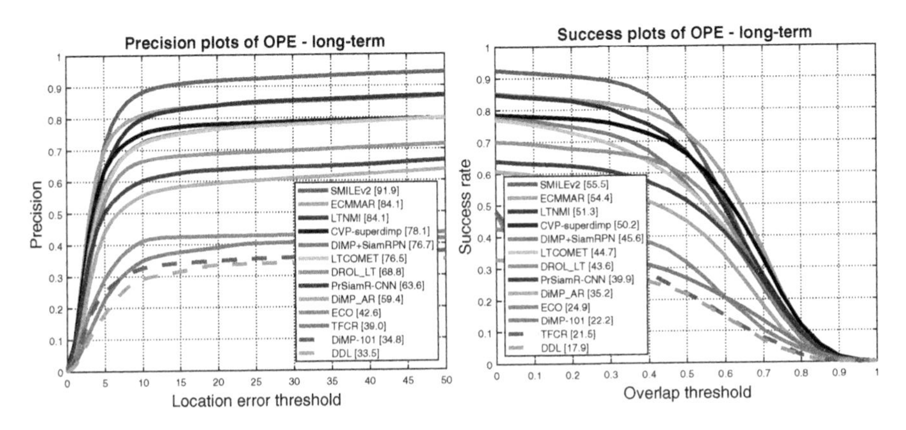

Fig. 6. The success and precision plots of the submitted trackers in term of long-term tracking. The success and precision scores for each tracker are presented in the legend. Best viewed in color and zooming-in. (Color figure online)

which demonstrates the effectiveness of incorporating motion information for long-term tracking.

4.4 Attribute-Based Analysis

In order to further analyze the performance of different tracking algorithms, we show the results of each tracker under all 12 attributes including ARC, BC, CM, FM, FOC, POC, IV, LR, OV, SO, SV and VC. The results are shown in Fig. 7 and 8. In term of precision score, SMILEv2 achieves the best performance on 8 out 12 attributes, including CM, FOC, IV, LR, OV, POC, SO and VC. LTNMI obtains the best result on 4 out of 12 attributes including ARC, BC,

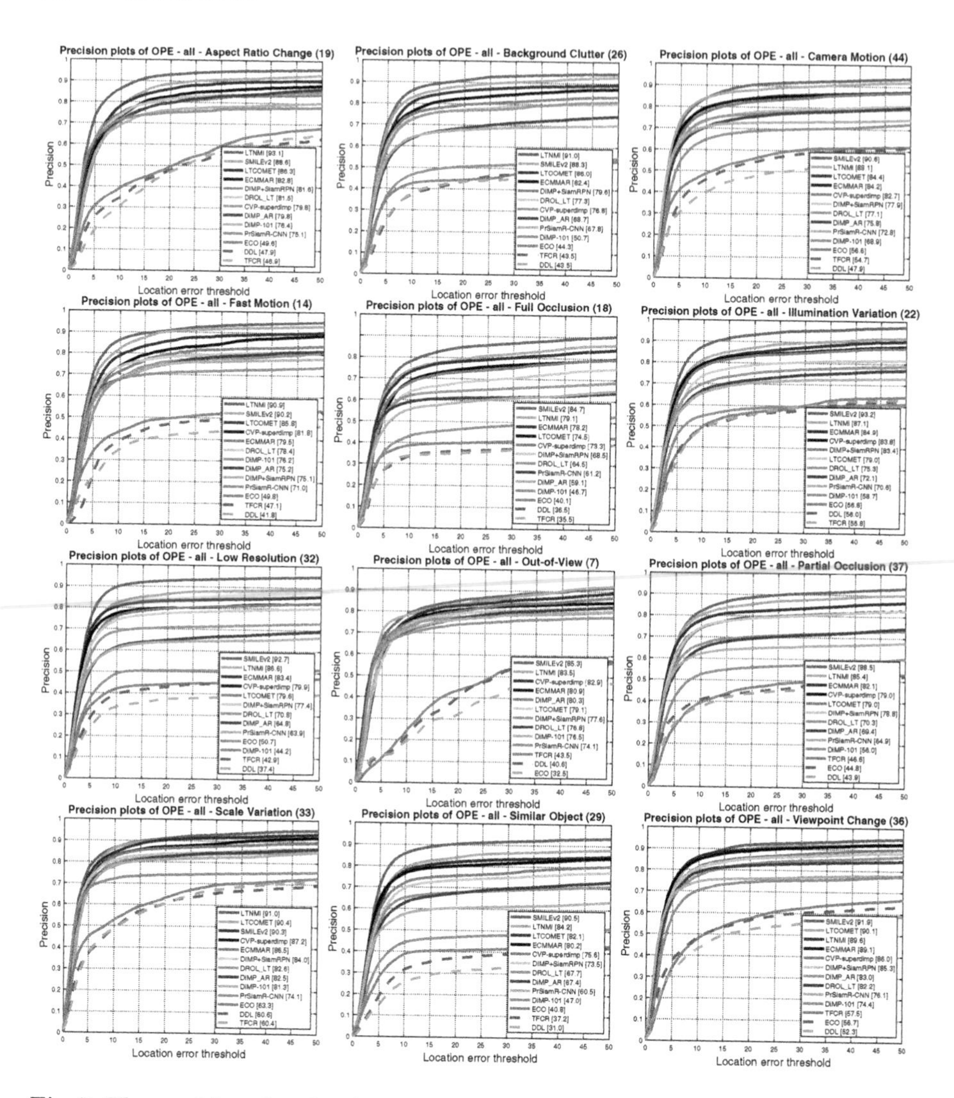

Fig. 7. The precision plots for the submitted trackers in different attributes. The number presented in the title indicates the number of sequences with that attribute. Best viewed in color and zooming-in. (Color figure online)

FM and SV, and it performs competitively on all other attributes. These finds show the effectiveness of applying multiple complementary trackers in tracking performance improvement. In term of success score, on the other hand, LTNMI achieves the best performance on 8 out of 12 attributes including ARC, BC, CM, FM, LTNMI, POC, SV and VC. The method os SMILEv2 obtains the best results on 4 out of 12 attributes including FOC, IV, LR and SO.

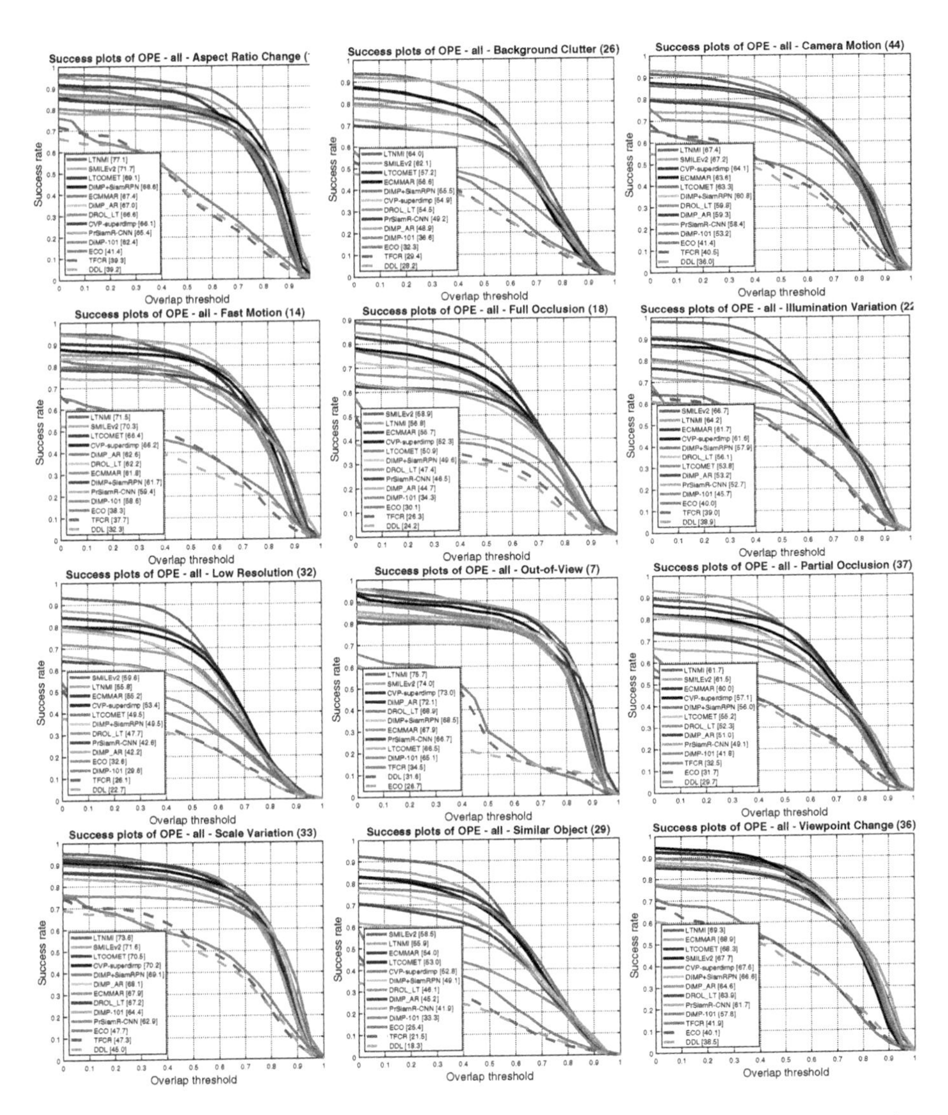

Fig. 8. The success plots for the submitted trackers in different attributes. The number presented in the title indicates the number of sequences with that attribute. Best viewed in color and zooming-in. (Color figure online)

4.5 Discussion

In order to provide guidance for designing robust tracker, we summarize several directions to work on in future research based on the analysis of submitted trackers.

Ensemble of Different Trackers. According to the analysis of submitted trackers, we observe that most of them utilize multiple tracking algorithms to locate the target, aiming to make full use of the advantage of each tracker for robustness. Especially, the top three trackers SMILEv2, LTNMI and ECMMAR utilize two or three state-of-the-art trackers proposed recently. We believe that this strategy will be widely adopted in the future. However, an interesting direction is how to combine different trackers in an adaptive way.

Integration of Tracking and Detection. In long-term tracking, the target object may frequently disappear and then re-enter the view. In such case, it is necessary to re-detect the target object when it appears again in the view. To this end, a possible solution is to integrate an additional detection model for tracking, as in LTNMI and DROL_LT. Since currently these method often utilize the detector explicitly, it is worth exploring how to build an end-to-end network that integrates both tracking and detection.

Incorporation of Motion Information. Motion information has been crucial for video understanding. In visual tracking, motion information can be used to filter out background distractors, especially for long-term tracking. The method of SMILEv2 effectively leverages such information and achieves the best overall performance.

5 Conclusion

In this paper, we introduce the VisDrone-SOT2020 challenge for drone-based tracking research, in conjunction with ECCV 2020. In specific, the VisDrone-SOT2020 dataset consists of 167 videos with more than 189K frames, aiming to provide a dedicated platform for both training and evaluation of drone tracking algorithm. In addition, we extensively evaluate and discuss 13 submitted tracking algorithms in depth, which serves as baselines on VisDrone-SOT2020 dataset. Based on the analysis of submitted trackers, we share some directions that are worth being explored in the future. By holding the VisDrone-SOT2020 challenge, we hope that it can facilitate the research on drone-based tracking algorithms. For the future, we plan to further expand the dataset and conduct more evaluation of different approaches.

Acknowledgements. This work was supported in part by the National Natural Science Foundation of China under Grant 61876127 and Grant 61732011, in part by Natural Science Foundation of Tianjin under Grant 17JCZDJC30800.

A Descriptions of Submitted Trackers

In the appendix, we summarize 13 trackers submitted in the VisDrone-SOT2020 Challenge, which are ordered according to the submissions of their final results.

A.1 Strategy and Motion Integrated Long-Term Experts-Version 2 (SMILEv2)

Yuxuan Li, Zhongjian Huang and Biao Wang
liyuxuan_xidian@126.com, huangzj@stu.xidian.edu.cn, biaowang@webank.com

SMILEv2 is combined with three kind of basic trackers and integrated in our IPIU-tracking framework. In this new framework, we are able to select different trackers in different situations by a semi-automatic way. As shown in Fig. 9, the framework has three parts, which are prediction module, tracking module and fix module. For prediction module, we introduce the Kalman filter and the optical flow method of VCN [61] as the object motion information and camera motion information respectively. For tracking module, we use three trackers including Dimp [5], SiamMask [56], SORT-MOT [4]. For fix module, we first obtain the output of prediction and tracking modules, and then judge the final result.

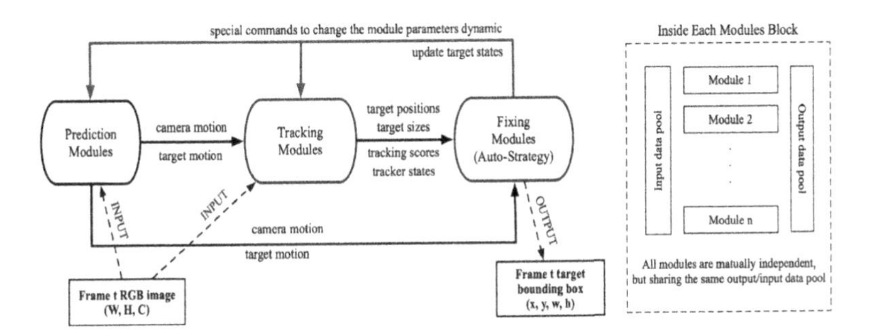

Fig. 9. The framework of SMILEv2.

A.2 Long-Term Tracking with Night-Enhancement and Motion Integrated (LTNMI)

Yuting Yang, Yanjie Gao, Ruiyan Ma and Xin Hou
{ytyang_1,yjgao}@stu.xidian.edu.cn, 3028408083@qq.com, xinhou@webank.com

LTNMI is a combination of ATOM [8], SiamRPN++ [31], Siam-RCNN [54] and Dimp [5]. We combined the ATOM and SiamRPN++ to get a better result, and then our method can give the reliability low limits of the above two systems on the condition of different confidence levels, which makes the systems more reliable respectively as different features play different role in the process of tracking based on their reliability. In addition, we improve the prediction of blurred scenes by using SIFT algorithm to match features. By estimating motion, the regression boxes can continue tracking the target in case of occlusion. When encountering dark or low-resolution scenes, we use threshold judgment and image brightness enhancement processing. We use MBLLEN [40] algorithm to process weak light enhancement. And then, we use Dimp to get the result of the sequences with weak light enhancement. At last, we use Siam-RCNN to find some lost frames. As a result, when the overlap are of fused result and the result generated by Siam-RCNN is nearly 95%, we conclude that the result generated by Siam-RCNN is better because of accurate detecting bounding box.

A.3 Ensemble of Classification and Matching Models with Alpha-Refine for UAV Tracking (ECMMAR)

Shuhao Chen, Zezhou Wang, Simiao Lai, Dong Wang and Huchuan Lu
{shuhaochn,zzwang}@mail.dlut.edu.cn, laisimiao1@gmail.com,
{wdice,lhchuan}@dlut.edu.cn

ECMMAR tracker is improved from Dimp [5] and SiamRPN++ [31] with online update module [65]. Dimp performs well in distinguishing distractors, while SiamRPN++ with the re-detection module performs well in detecting target when target disappeared by full occlusion or fast perspective conversion. The main modification are: 1) Develop an interactive mechanism to handle with long-term tracking and improve the robustness. 2) Muti-scale search regions are set to help to re-detect target when full occlusion or fast perspective conversion happened. 3) Use a refinement module [60] to refine the localized bounding box. 4) Employ a low-light image enhancement [62] method to deal with low-light scenes. 5) Fine-tune the superdimp pre-trained model and alpha-refine pre-trained model with visdrone2020 dataset. 6) Motion compensation is used when the camera viewing angle changes greatly. 7) Inertial motion is added when both tracker results are unreliable.

A.4 UAV Tracking with Extra Proposals Based on Corrected Velocity Prediction (CVP-superdimp)

Zitong Yi and Yanyun Zhao
{zitong.yi,zyy}@mail.dlut.edu.cn

CVP-superdimp is a robust tracking strategy under the circumstance of UAV tracking, especially for the nerve-wracking problem of fierce camera moving and long-term full occlusion. The base tracker follows [5,9], which contains two modules: object classification module based on DIMP and bounding box regression module based on prDIMP. Our proposed tracking strategy adds a new module of velocity prediction for both short-term and long-term, which can provide additional high-quality proposals for tracker searching in the next frame.

A.5 LTCOMET: Context-Aware IoU-Guided Network for Small Object Tracking (LTCOMET)

Seyed Mojtaba Marvasti-Zadeh, Javad Khaghani, Li Cheng, Hossein Ghanei-Yakhdan and Shohreh Kasaei
{mojtaba.marvasti,khaghani,lcheng5}@ualberta.ca,hghaneiy@yazd.ac.ir, kasaei@sharif.edu

To bridge the gap between aerial views tracking methods and modern trackers, the modified context-aware IoU-guided tracker (LTCOMET) is proposed that exploits the offline reference proposal generation strategy (same as COMET tracker [42]), multitask two-stream network [42], kindling the darkness (KinD) [64], and photo-realistic cascading residual network (PCARN) [1]. The network architecture is the same as [42] without using channel reduction after the multi-scale aggregation and fusion modules (MSAFs). The KinD employs a network for light adjustment and degradation removal, which is employed as a preprocessing of LTCOMET on target patches. Also, the LTCOMET employs the generator network of PCARN to recover high-resolution patches of target and its context from low-resolution ones. Furthermore, the proposed method uses a widowing search strategy when it loses the target. The proposed LTCOMET has been trained on a broad range of tracking datasets while it exploits various photometric and geometric distortions (*i.e.*, data augmentations) to improve the variability of target regions.

A.6 Discriminative and Robust Online Learning for Long Term Siamese Visual Tracking (DROL_LT)

Jinghao Zhou, Peng Wang, Haoyang Sun and Zikai Zhang
{jensen.zhoujh,zzkdemail}@gmail.com,{peng.wang,sunhaoyang}@gmail.com

DROL_LT is based on DROL [65]. DROL proposes an online module with an attention mechanism for offline Siamese networks to extract target-specific features under L2 error. DROL also proposes a filter update strategy adaptive to treacherous background noises for discriminative learning, and a template update strategy to handle large target deformations for robust learning. DROL_LT adds two modules to improve DROL results in long term tracking tasks. (1) A detector is added to help DROL recover the targets, which disappear and appear many times. ROI Align is used to extract the features from mixed offline feature maps with the bounding boxes information from detector. (2) A mechanism is designed to help tracker to decide when to update online classifiers and when to use detectors, which depends on a set of thresholds given from experience.

A.7 Discriminative and Robust Online Learning for Long Term Siamese Visual Tracking (DIMP-SiamRPN)

Zhipeng Luo, Penghao Zhang, Yubo Sun and Bin Dong
{luozp,zhangph,sunyb,Dongbin}@deepblueai.com

DIMP-SiamRPN is improved based on PrDIMP [9] and SiamRPN++ [31]. First, we use the frame numbers to divide videos in the challenge set into long-term videos and short-term videos. The short videos are tested using PrDIMP's hyper-parameter adjustment model to obtain the results. Daytime scenes in the long videos are tested by the SiamRPN++ model. In the SiamRPN++ model, we enlarge the instance size 15 pixels every frame, and the upper limit of the search threshold is 1000. In addition, when the target seems to be lost, we reset the center of search scope to the center of the image. Furthermore, we define a make-up strategy to deal with occlusion. As scenes of night in the long videos, we divide them into strong light scenes and dark scenes according to the light intensity, in which different inference parameters are used.

A.8 Discriminative Model Prediction and Accurate Re-detection for Drone Tracking (DiMP_AR)

Xuefeng Zhu, Xiaojun Wu and Tianyang Xu
{xuefeng_zhu95,xiaojun_wu_jnu,tianyang_xu}@163.com

DiMP_AR is based on the DiMP [5] by adding a re-detection module. We use the DiMP tracker as a local tracker to predict target state normally and the RT-MDNet [28] is used as a verifier to verify the prediction of DiMP. If the verification is above a predefined threshold, the normal local tracking is conducted in next frame. Otherwise, the re-detection module will be activated. Firstly, the faster R-CNN detector [48] is used to detect some highly possible target candidates in the whole image of next frame. Then the SiamRPN++ [31] tracker is employed to detect the search regions regarding the possible target candidates. When the target is regained, we switch to local tracking with the tracker DiMP.

A.9 Precise Visual Tracking by Re-detection (PrSiamR-CNN)

Zhongzhou Zhang, Lei Zhang, Keyang Wang and Zhenwei He
{zz.zhang,leizhang,wangkeyang,hzw}@cqu.edu.cn

PrSiamR-CNN is modified from recently proposed state-of-the-art single object tracker Siam R-CNN [54] by using extra training data from VisDrone-SOT2020.

A.10 Discriminative Model Prediction with Deeper ResNet-101 (DiMP-101)

Liting Lin and Yong Xu
l.lt@mail.scut.edu.cn

DiMP-101 is based on the DiMP [5] model, adopting deeper ResNet-101 as the backbone. With higher learning capacity of the feature extraction network, the performance of the tracking algorithm has been significantly improved to a new level.

A.11 ECO: Efficient Convolution Operators for Tracking (ECO)

Lei Pang
panglei2015@ia.ac.cn

ECO [7] is a discriminative correlation filter based tracker using deep features. This method introduces a factorized convolution operator and a compact generative model of the training sample distribution to reduce model parameters. In addition, it proposes a conservative model update strategy with improved robustness and reduced complexity. More details can be referred to [7].

A.12 Target-Focusing Convolutional Regression Tracking (TFCR)

Di Yuan, Nana Fan and Zhenyu He
dyuanhit@gmail.com

TFCR [63] is a target-focusing convolutional regression (CR) model for visual object tracking tasks. This model uses a target-focusing loss function to alleviate the influence of background noise on the response map of the current tracking image frame, which effectively improves the tracking accuracy. In particular, it can effectively balance the disequilibrium of positive and negative samples by reducing some effects of the negative samples that act on the object appearance model.

A.13 DDL-Tracker (DDL)

Yong Wang, Lu Ding, Dongjie Zhou and Wentao He
wangyong5@mail.sysu.edu.cn,dinglu@sjtu.edu.cn,13520071811@163.com, weishiinsky@126.com

DDL-tracker employs deep layers to extract features. Meanwhile, one HOG detector is trained online. If the tracking result is below a threshold, we use the results by the detector.

References

1. Ahn, N., Kang, B., Sohn, K.A.: Efficient deep neural network for photo-realistic image super-resolution. arXiv (2019)
2. Kristan, M., et al.: The sixth visual object tracking VOT2018 challenge results. In: Leal-Taixé, L., Roth, S. (eds.) ECCV 2018. LNCS, vol. 11129, pp. 3–53. Springer, Cham (2019). https://doi.org/10.1007/978-3-030-11009-3_1
3. Bertinetto, L., Valmadre, J., Henriques, J.F., Vedaldi, A., Torr, P.H.S.: Fully-convolutional Siamese networks for object tracking. In: Hua, G., Jégou, H. (eds.) ECCV 2016. LNCS, vol. 9914, pp. 850–865. Springer, Cham (2016). https://doi.org/10.1007/978-3-319-48881-3_56
4. Bewley, A., Ge, Z., Ott, L., Ramos, F., Upcroft, B.: Simple online and realtime tracking. In: ICIP (2016)
5. Bhat, G., Danelljan, M., Gool, L.V., Timofte, R.: Learning discriminative model prediction for tracking. In: ICCV (2019)
6. Bolme, D.S., Beveridge, J.R., Draper, B.A., Lui, Y.M.: Visual object tracking using adaptive correlation filters. In: CVPR (2010)
7. Danelljan, M., Bhat, G., Khan, F.S., Felsberg, M.: Eco: Efficient convolution operators for tracking. In: CVPR (2017)
8. Danelljan, M., Bhat, G., Khan, F.S., Felsberg, M.: ATOM: accurate tracking by overlap maximization. In: CVPR (2019)
9. Danelljan, M., Gool, L.V., Timofte, R.: Probabilistic regression for visual tracking. In: CVPR (2020)
10. Danelljan, M., Häger, G., Khan, F., Felsberg, M.: Accurate scale estimation for robust visual tracking. In: BMVC (2014)
11. Danelljan, M., Hager, G., Shahbaz Khan, F., Felsberg, M.: Learning spatially regularized correlation filters for visual tracking. In: ICCV (2015)
12. Danelljan, M., Robinson, A., Shahbaz Khan, F., Felsberg, M.: Beyond correlation filters: learning continuous convolution operators for visual tracking. In: Leibe, B., Matas, J., Sebe, N., Welling, M. (eds.) ECCV 2016. LNCS, vol. 9909, pp. 472–488. Springer, Cham (2016). https://doi.org/10.1007/978-3-319-46454-1_29
13. Deng, J., Dong, W., Socher, R., Li, L.J., Li, K., Fei-Fei, L.: ImageNet: a large-scale hierarchical image database. In: CVPR (2009)
14. Du, D., et al.: The unmanned aerial vehicle benchmark: object detection and tracking. In: Ferrari, V., Hebert, M., Sminchisescu, C., Weiss, Y. (eds.) ECCV 2018. LNCS, vol. 11214, pp. 375–391. Springer, Cham (2018). https://doi.org/10.1007/978-3-030-01249-6_23
15. Du, D., Wen, L., Qi, H., Huang, Q., Tian, Q., Lyu, S.: Iterative graph seeking for object tracking. TIP **27**(4), 1809–1821 (2018)
16. Du, D., et al.: VisDrone-SOT2019: the vision meets drone single object tracking challenge results. In: ICCVW (2019)
17. Fan, H., et al.: LaSOT: a high-quality benchmark for large-scale single object tracking. In: CVPR (2019)
18. Fan, H., Ling, H.: Parallel tracking and verifying: a framework for real-time and high accuracy visual tracking. In: ICCV (2017)
19. Fan, H., Ling, H.: SANet: structure-aware network for visual tracking. In: CVPRW (2017)
20. Fan, H., Ling, H.: Siamese cascaded region proposal networks for real-time visual tracking. In: CVPR (2019)

21. Galoogahi, H.K., Fagg, A., Huang, C., Ramanan, D., Lucey, S.: Need for speed: a benchmark for higher frame rate object tracking. In: ICCV (2017)
22. Galoogahi, H.K., Fagg, A., Lucey, S.: Learning background-aware correlation filters for visual tracking. In: ICCV (2017)
23. Girshick, R., Donahue, J., Darrell, T., Malik, J.: Rich feature hierarchies for accurate object detection and semantic segmentation. In: CVPR (2014)
24. He, K., Zhang, X., Ren, S., Sun, J.: Deep residual learning for image recognition. In: CVPR (2016)
25. Henriques, J.F., Caseiro, R., Martins, P., Batista, J.: High-speed tracking with kernelized correlation filters. TPAMI **37**(3), 583–596 (2015)
26. Huang, L., Zhao, X., Huang, K.: GOT-10k: a large high-diversity benchmark for generic object tracking in the wild. TPAMI (2019)
27. Jiang, B., Luo, R., Mao, J., Xiao, T., Jiang, Y.: Acquisition of localization confidence for accurate object detection. In: Ferrari, V., Hebert, M., Sminchisescu, C., Weiss, Y. (eds.) Computer Vision – ECCV 2018. LNCS, vol. 11218, pp. 816–832. Springer, Cham (2018). https://doi.org/10.1007/978-3-030-01264-9_48
28. Jung, I., Son, J., Baek, M., Han, B.: Real-time MDNet. In: Ferrari, V., Hebert, M., Sminchisescu, C., Weiss, Y. (eds.) ECCV 2018. LNCS, vol. 11208, pp. 89–104. Springer, Cham (2018). https://doi.org/10.1007/978-3-030-01225-0_6
29. Kristan, M., et al.: A novel performance evaluation methodology for single-target trackers. TPAMI **38**(11), 2137–2155 (2016)
30. Krizhevsky, A., Sutskever, I., Hinton, G.E.: ImageNet classification with deep convolutional neural networks. In: NIPS (2012)
31. Li, B., Wu, W., Wang, Q., Zhang, F., Xing, J., Yan, J.: SiamRPN++: evolution of Siamese visual tracking with very deep networks. In: CVPR (2019)
32. Li, B., Yan, J., Wu, W., Zhu, Z., Hu, X.: High performance visual tracking with Siamese region proposal network. In: CVPR (2018)
33. Li, F., Tian, C., Zuo, W., Zhang, L., Yang, M.H.: Learning spatial-temporal regularized correlation filters for visual tracking. In: CVPR (2018)
34. Li, S., Yeung, D.Y.: Visual object tracking for unmanned aerial vehicles: a benchmark and new motion models. In: AAAI (2017)
35. Li, Y., Zhu, J.: A scale adaptive kernel correlation filter tracker with feature integration. In: Agapito, L., Bronstein, M.M., Rother, C. (eds.) ECCV 2014. LNCS, vol. 8926, pp. 254–265. Springer, Cham (2015). https://doi.org/10.1007/978-3-319-16181-5_18
36. Liang, P., Blasch, E., Ling, H.: Encoding color information for visual tracking: algorithms and benchmark. TIP **24**(12), 5630–5644 (2015)
37. Lin, T.-Y., et al.: Microsoft COCO: common objects in context. In: Fleet, D., Pajdla, T., Schiele, B., Tuytelaars, T. (eds.) ECCV 2014. LNCS, vol. 8693, pp. 740–755. Springer, Cham (2014). https://doi.org/10.1007/978-3-319-10602-1_48
38. Liu, T., Wang, G., Yang, Q.: Real-time part-based visual tracking via adaptive correlation filters. In: CVPR (2015)
39. Lukezic, A., et al.: CDTB: a color and depth visual object tracking dataset and benchmark. In: ICCV (2019)
40. Lv, F., Lu, F., Wu, J., Lim, C.: MBLLEN: low-light image/video enhancement using CNNs. In: BMVC (2018)
41. Ma, C., Huang, J.B., Yang, X., Yang, M.H.: Hierarchical convolutional features for visual tracking. In: ICCV (2015)
42. Marvasti-Zadeh, S.M., Khaghani, J., Ghanei-Yakhdan, H., Kasaei, S., Cheng, L.: COMET: context-aware IoU-guided network for small object tracking. arXiv (2020)

43. Mueller, M., Smith, N., Ghanem, B.: A benchmark and simulator for UAV tracking. In: Leibe, B., Matas, J., Sebe, N., Welling, M. (eds.) ECCV 2016. LNCS, vol. 9905, pp. 445–461. Springer, Cham (2016). https://doi.org/10.1007/978-3-319-46448-0_27
44. Mueller, M., Smith, N., Ghanem, B.: Context-aware correlation filter tracking. In: CVPR (2017)
45. Müller, M., Bibi, A., Giancola, S., Alsubaihi, S., Ghanem, B.: TrackingNet: a large-scale dataset and benchmark for object tracking in the wild. In: Ferrari, V., Hebert, M., Sminchisescu, C., Weiss, Y. (eds.) ECCV 2018. LNCS, vol. 11205, pp. 310–327. Springer, Cham (2018). https://doi.org/10.1007/978-3-030-01246-5_19
46. Nam, H., Han, B.: Learning multi-domain convolutional neural networks for visual tracking. In: CVPR (2016)
47. Real, E., Shlens, J., Mazzocchi, S., Pan, X., Vanhoucke, V.: YouTube-BoundingBoxes: a large high-precision human-annotated data set for object detection in video. In: CVPR, pp. 7464–7473 (2017)
48. Ren, S., He, K., Girshick, R., Sun, J.: Faster R-CNN: towards real-time object detection with region proposal networks. In: NIPS (2015)
49. Russakovsky, O., et al.: ImageNet large scale visual recognition challenge. IJCV **115**(3), 211–252 (2015)
50. Smeulders, A.W., Chu, D.M., Cucchiara, R., Calderara, S., Dehghan, A., Shah, M.: Visual tracking: an experimental survey. TPAMI **36**(7), 1442–1468 (2014)
51. Song, Y., et al.: VITAL: visual tracking via adversarial learning. In: CVPR (2018)
52. Tao, R., Gavves, E., Smeulders, A.W.: Siamese instance search for tracking. In: CVPR (2016)
53. Valmadre, J., et al.: Long-term tracking in the wild: a benchmark. In: Ferrari, V., Hebert, M., Sminchisescu, C., Weiss, Y. (eds.) ECCV 2018. LNCS, vol. 11207, pp. 692–707. Springer, Cham (2018). https://doi.org/10.1007/978-3-030-01219-9_41
54. Voigtlaender, P., Luiten, J., Torr, P.H., Leibe, B.: Siam R-CNN: visual tracking by re-detection. In: CVPR (2020)
55. Wang, G., Luo, C., Xiong, Z., Zeng, W.: SPM-Tracker: series-parallel matching for real-time visual object tracking. In: CVPR (2019)
56. Wang, Q., Zhang, L., Bertinetto, L., Hu, W., Torr, P.H.: Fast online object tracking and segmentation: a unifying approach. In: CVPR (2019)
57. Wen, L., et al.: VisDrone-SOT2018: the vision meets drone single-object tracking challenge results. In: Leal-Taixé, L., Roth, S. (eds.) ECCV 2018. LNCS, vol. 11133, pp. 469–495. Springer, Cham (2019). https://doi.org/10.1007/978-3-030-11021-5_28
58. Wu, Y., Lim, J., Yang, M.H.: Online object tracking: a benchmark. In: CVPR (2013)
59. Wu, Y., Lim, J., Yang, M.H.: Object tracking benchmark. TPAMI **37**(9), 1834–1848 (2015)
60. Yan, B., Wang, D., Lu, H., Yang, X.: Alpha-Refine: boosting tracking performance by precise bounding box estimation. arXiv (2020)
61. Yang, G., Ramanan, D.: Volumetric correspondence networks for optical flow. In: NeurIPS (2019)
62. Ying, Z., Li, G., Ren, Y., Wang, R., Wang, W.: A new low-light image enhancement algorithm using camera response model. In: ICCVW (2017)
63. Yuan, D., Fan, N., He, Z.: Learning target-focusing convolutional regression model for visual object tracking. Knowl.-Based Syst. (2020)
64. Zhang, Y., Zhang, J., Guo, X.: Kindling the darkness: a practical low-light image enhancer. In: ACM MM (2019)

65. Zhou, J., Wang, P., Sun, H.: Discriminative and robust online learning for Siamese visual tracking. In: AAAI (2020)
66. Zhou, W., Wen, L., Zhang, L., Du, D., Luo, T., Wu, Y.: SiamMan: Siamese motion-aware network for visual tracking. CoRR abs/1912.05515 (2019)
67. Zhu, P., Wen, L., Du, D., Bian, X., Hu, Q., Ling, H.: Vision meets drones: past, present and future. CoRR abs/2001.06303 (2020)
68. Zhu, Z., Wang, Q., Li, B., Wu, W., Yan, J., Hu, W.: Distractor-aware Siamese networks for visual object tracking. In: Ferrari, V., Hebert, M., Sminchisescu, C., Weiss, Y. (eds.) ECCV 2018. LNCS, vol. 11213, pp. 103–119. Springer, Cham (2018). https://doi.org/10.1007/978-3-030-01240-3_7

Author Index

ited States